In addition to making the illustrations more colorful and visually pleasing, the use of four-color artwork makes the illustrations easier to understand. For example, forces and moments can be readily distinguished from all other information on the illustration because they are always designated by a big red arrow. Similarly, position vectors are always designated by a medium blue arrow; velocity and acceleration vectors, by a standard green arrow; and unit vectors by a small black arrow. Coordinate axes and dimension lines (linear and angular) are drawn with thin black lines. This still leaves many colors for depicting the various parts of objects.

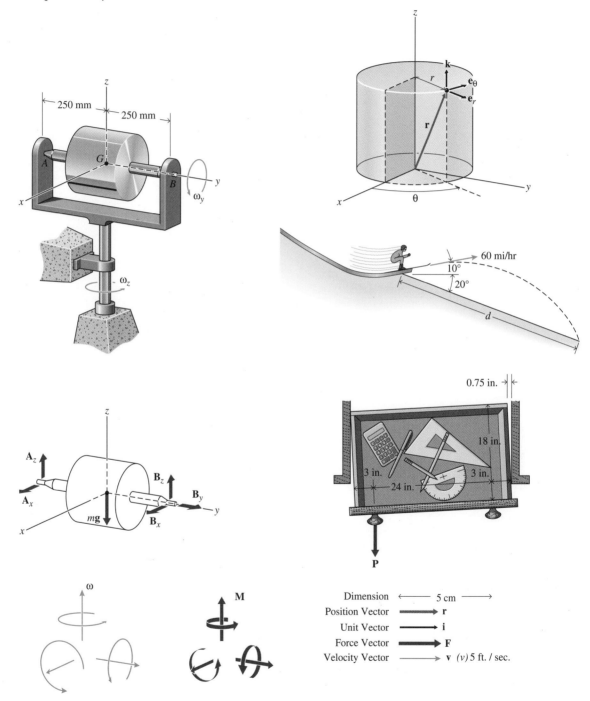

ENGINEERING MECHANICS
DYNAMICS

ENGINEERING MECHANICS
DYNAMICS

WILLIAM F. RILEY
Professor Emeritus
Iowa State University

LEROY D. STURGES
Iowa State University

JOHN WILEY & SONS, INC.
New York · Chichester · Brisbane · Toronto · Singapore

COVER: Designed by Laura Ierardi
Photograph by Oli Tennant-Tony Stone/Worldwide

ACQUISITIONS EDITOR	Charity Robey
DEVELOPMENTAL EDITOR	Christine Peckaitis
MARKETING MANAGER	Debra Riegert
PRODUCTION SUPERVISOR	Charlotte Hyland
DESIGN SUPERVISOR	Ann Marie Renzi
MANUFACTURING MANAGER	Andrea Price
COPY EDITING SUPERVISOR	Marjorie Shustak
PHOTO RESEARCHER	Hilary Newman
ILLUSTRATION SUPERVISOR	Sigmund Malinowski
ILLUSTRATION DEVELOPMENT	Boris Starosta
ELECTRONIC ILLUSTRATIONS	Precision Graphics

This book was set in Palatino by York Graphic Services and printed and bound by Von Hoffmann Press. The cover was printed by Phoenix Color Corp.

Library of Congress Cataloging in Publication Data:

Riley, William F. (William Franklin), 1925–
    Engineering mechanics : dynamics/William F. Riley, Leroy D. Sturges.
      p. cm.
    Includes index.
    ISBN 0-471-51242-7 (alk. paper)
    1. Dynamics. I. Sturges, Leroy D. II. Title.
  TA352.R55 1993
  620.1'04—dc20

92-30901
CIP

Printed in the United States of America

10 9 8 7 6 5 4 3 2 1

PREFACE

Our purpose in writing this dynamics book, together with the companion statics book, was to present a fresh look at the subject and to provide a more logical order of presentation of the subject material. We believe our order of presentation will give students a greater understanding of the material and will better prepare students for future courses and later professional life.

INTRODUCTION

This text has been designed for use in undergraduate engineering programs. Students are given a clear, practical, comprehensible, and thorough coverage of the theory normally presented in introductory mechanics courses. Application of the principles of dynamics to the solution of practical engineering problems is demonstrated. This text can also be used as a reference book by practicing aerospace, automotive, civil, mechanical, mining, and petroleum engineers.

Extensive use is made in this text of prerequisite course materials in mathematics and physics. Students entering a dynamics course that uses this book should have a working knowledge of introductory differential and integral calculus, should have taken an introductory course in vector algebra, and should have taken or be enrolled in an introductory course in differential equations.

Vector methods do not always simplify the solution of two-dimensional problems in dynamics. For three-dimensional problems, however, vector algebra provides a systematic procedure that often eliminates errors that might occur with a less systematic approach. In this book, vector algebra is used wherever it provides an efficient solution to a problem. If vector algebra offers no advantages, a scalar approach is used. Likewise, students are encouraged to develop the ability to select the mathematical tools most appropriate for the particular problem they are attempting to solve.

ORGANIZATION

This text breaks away from the traditional organization of first covering all aspects of particle dynamics and then covering all aspects of rigid-body dynamics. The description of particle kinematics (Chapter 13) is followed immediately by a description of rigid-body kinematics (Chapter 14). For a course covering both particle and rigid-body kinematics, we believe that it is more logical and efficient to cover all kinematics at one time and in one place.

Similarly, the use of the Force–Mass–Acceleration (Newton's Second Law) method to solve particle kinetics problems (Chapter 15) is followed immediately by the use of the Force–Mass–Acceleration method to solve rigid-body kinetics problems (Chapter 16). With this approach, the student first learns one method of solving kinetics problems and applies it consistently to a wide variety of problems. These four chapters form a complete, albeit brief, introduction to the dynamics of particles and rigid bodies.

The following four chapters present two alternative methods of solving certain types of problems. First, the Work–Energy method is used to solve particle kinetics problems (Chapter 17) and rigid-body kinetics problems (Chapter 18). Next, the Impulse–Momentum method is used to solve particle kinetics problems (Chapter 19) and rigid-body kinetics problems (Chapter 20). Neither method can be used to solve all types of problems. However, when these special methods apply, they generally give the solution with less effort than the Force–Mass–Acceleration method does.

The final chapter (Chapter 21) presents an introduction to Mechanical Vibrations. In this chapter, the principles of kinetics are used to solve a special class of problems that involve vibratory or oscillatory motion. This chapter is included for those instructors who feel that no introductory course is complete without at least an introduction to vibrations.

Since the discussion of particle kinetics in Chapter 15 does not depend on the discussion of rigid-body kinematics in Chapter 14, and similarly, the discussion of the work–energy method for particles in Chapter 17 does not depend on the discussion of rigid bodies in either Chapter 14 or Chapter 16, a course can be taught in the traditional way using this book by covering chapters in the order 13, 15, 17, 19, 14, 16, 18, 20, 21.

FEATURES

Engineering Emphasis

Throughout this book, a strong emphasis has been placed on the engineering significance of the subject area in addition to the mathematical methods of analysis. Many illustrative example problems have been integrated into the main body of the text at points where the presentation of a method can be best reinforced by the immediate illustration of the method. Students are usually more enthusiastic about a subject if they can see and appreciate its value as they proceed into the subject.

We believe that students can progress in a mechanics course only by understanding the physical and mathematical principles jointly, not by mere memorization of formulas and substitution of data to obtain answers to simple problems. Furthermore, we think that it is better to teach a few fundamental principles for solving problems than to teach a large number of special cases and trick procedures. Therefore the text aims to develop in the student the ability to analyze a given problem in a simple and logical manner and to apply a few fundamental, well-understood principles to its solution.

A conscientious effort has been made to present the material in a simple and direct manner, with the student's point of view constantly in mind.

Coverage of Kinematics Before Kinetics

The natural learning process begins with simple situations and progresses to more complicated ones. This book is organized in such a way that each principle is applied first to a particle, then to a system of particles, then to a rigid body subjected to a coplanar system of forces, and finally to the general case of a rigid body subjected to a three-dimensional system of forces.

Kinematics (the study of motion without reference to the causes of the motion) of particles and rigid bodies is carefully developed since mastery of this material is essential for a successful study of kinetics (the study of the relationship between the motion and the forces that cause the motion) of particles and rigid bodies.

Next, the three common methods for solving kinetics problems—namely, (1) force, mass, and acceleration; (2) work and energy; (3) impulse and momentum—are developed for use in a wide variety of problem situations. Each method is developed in turn and applied first to particles, then to a system of particles, then to rigid bodies under plane motion, and finally to general three-dimensional situations. Mastery of the force, mass, and acceleration method provides a means for solving all dynamics problems. The other two methods simply provide a more efficient means for solving some problems.

We believe that this approach represents both the logical and the desirable organization of material for an introductory course in dynamics. Kinematics is covered completely before kinetics is discussed. Kinetics, using Newton's Laws, is developed completely before the student is introduced to work–energy methods and impulse–momentum methods. This way, the student is not introduced to each of the methods in a disjointed fashion.

In addition, in each of the major sections of the book, the progression of material is from the slightly familiar (from a previous physics course) concepts of particle mechanics to the unfamiliar concepts of two-dimensional rigid-body mechanics to the more complex concepts of three-dimensional rigid-body motion. Therefore, the more challenging topics are spread more uniformly over the semester.

Other books organize the material into four categories: (1) dynamics of particles (kinematics of particles followed by an integrated presentation of the three methods for solution of particle kinetics problems); (2) dynamics of systems of particles (kinematics of systems of particles followed by an integrated presentation of the three methods for solution of kinetics problems); (3) dynamics of rigid bodies (kinematics of rigid bodies followed by an integrated presentation of the three methods for solution of rigid-body kinetics problems) in plane motion; and (4) three-dimensional kinematics and kinetics of particles and rigid bodies.

This method of presentation suffers from three difficulties. First, students find it hard to master simultaneously three methods of solution for a problem. Second, all the easy and familiar concepts (dealing with particle kinematics and kinetics) are covered in the first few weeks of the course and all the more difficult and less familiar concepts (dealing with rigid-body kinematics and kinetics) are dealt with in the last few weeks of the course. Third, if only the particle dynamics portion of the course is mastered, the student does not have the ability to perform in future courses requiring competency in rigid-body dynamics.

Free-body Diagrams

Most engineers consider the free-body diagram to be the single most important tool for the solution of mechanics problems. The free-body diagram is just as important in dynamics as it is in statics. It is our approach that, whenever an equation of equilibrium or an equation of motion is written, it must be accompanied by a complete, proper free-body diagram.

Problem-solving Procedures

Success in engineering mechanics courses depends, to a surprisingly large degree, on a well-disciplined method of problem solving and on the solution of a large number of problems. The student is urged to develop the ability to reduce problems to a series of simpler component problems that can be easily analyzed and combined to give the solution of the initial problem. Along with an effective methodology for problem decomposition and solution, the ability to present results in a clear, logical, and neat manner is emphasized throughout the text. A first course in mechanics is an excellent place to begin development of this disciplined approach which is so necessary in most engineering work.

Worked-out Examples

Worked-out examples are invaluable to students. Example problems were carefully chosen to illustrate the concepts being discussed. When a concept is presented, a worked-out example follows to illustrate the concept to the student. We have included approximately 120 worked-out examples in this book.

Homework Problems

This book contains a large selection of problems that illustrate the wide application of the principles of dynamics to the various fields of engineering. The problems in each set represent a considerable range of difficulty. We believe that a student gains mastery of a subject through application of basic theory to the solution of problems that appear somewhat difficult. Mastery, in general, is not achieved by solving a large number of simple but similar problems. The problems in this text require an understanding of the principles of dynamics without demanding excessive time for computational work.

Significant Figures

Results should always be reported as accurately as possible. However, results should not be reported to 10 significant figures merely because the calculator displays that many digits. One of the tasks in all engineering work is to determine the accuracy of the given data and the expected accuracy of the final answer. Results should reflect the accuracy of the given data.

In a textbook, however, it is not possible for students to examine or question the accuracy of the given data. It is also impractical, in an introductory course, to give error bounds on every number. Therefore, since an accuracy greater than about 0.2 percent is seldom possible for practical engineering problems, all given data in Example Problems and Homework Problems, regardless of the number of figures shown, will be assumed sufficiently accurate to justify rounding off the final answer to approximately this degree of accuracy (three to four significant figures).

Computer Problems

Many students come to school with computers as well as programmable calculators. In recognition of this fact, we include problems at the ends of most chapters that can be best solved using these tools. These problems are more than just an exercise in crunching numbers; each has been chosen to illustrate how the solution to the problem depends on the initial or final conditions or on some parameter of the problem. Computer problems appear at the end of most chapters, and are marked with a C before the problem number.

Review Problems

A set of review problems is provided at the end of each chapter. These problems are designed to test students on all the concepts covered in the chapter. Since the problems are not directly associated with any particular section, they often integrate topics covered in the chapter and thus can deal with more realistic applications than can a problem designed to illustrate a single concept.

SI vs. U.S. Customary Units

Most large engineering companies deal in an international marketplace. In addition, the use of the International System of Units (SI) is gaining acceptance in the United States. As a result, most engineers must be proficient in both the SI system and the U.S. Customary System (USCS) of units. In response to this need, both U.S. Customary units and SI units are used in approximately equal proportions in the text for both illustrative examples and homework problems. As an aid to the instructor in problem selection, all odd-numbered problems are given in USCS units and even-numbered problems in SI units.

Chapter Summaries

As an aid to students we have written a summary that appears at the end of each chapter. These sections provide a synopsis of the major concepts that are explained in the chapter and can be used by students as a review or study aid.

Answers Provided

Answers to about half of the problems are included in the back of the book. We believe that the first assignment on a given topic should include some problems for which the answers are given. Since the simpler problems are usually reserved for this first assignment, answers are provided for the first few problems of each article and thereafter are given for approximately half of the remaining problems. The problems whose answers are provided are indicated by an asterisk after the problem number.

DESIGN

Use of Color

One of the first things you'll notice when you open this book is that we have used a variety of colors. We believe that color will help students learn mechanics more effectively for two reasons: First, today's visually oriented students are more motivated by texts that depict the real world more accurately. Second, the careful color coding makes it easier for students to understand the figures and text.

Following are samples of figures found in the book. As you can see, force vectors are depicted as red arrows; velocity vectors are depicted as green arrows. Position vectors appear in blue; dimensions as a thin black line; and unit vectors in bold black. This pedagogical use of color is consistent throughout the book.

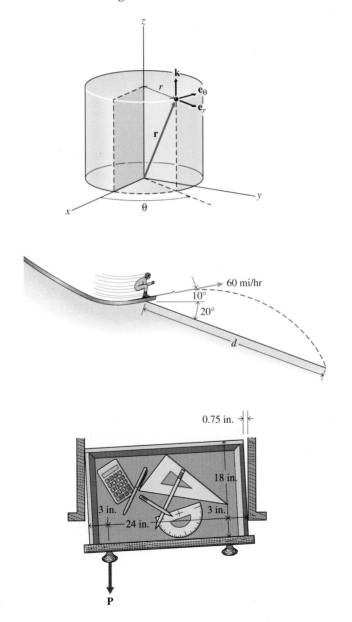

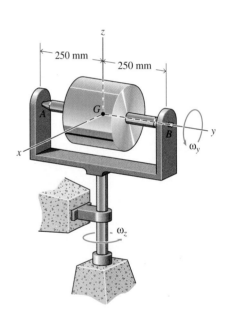

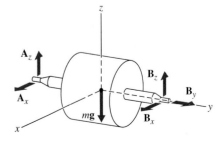

We have also used color to help students identify the most important study elements. For instance, example problems are always outlined in red and important equations appear in a green box.

Illustrations

One of the most difficult things for students to do is to visualize engineering problems. Over the years, students have struggled with the lack of realism in mechanics books. We think that mechanics illustrations should be as colorful and three-dimensional as life is. To hold students' attention, we developed the text illustrations with this point in mind.

We started with a basic sketch. Then a specialist in technical illustration added detail. Then the art studio created the figures using *Adobe Illustrator*©. All of these steps enabled us to provide you with the most realistic and accurate illustrations on the market.

Accuracy

After many years of teaching, we appreciate the importance of an accurate text. We have made an extraordinary effort to provide an error-free book. Every problem in the text has been worked out twice independently. Many of the problems have been worked out a third time independently.

Development Process

This book is the most extensively developed text ever published for the engineering market. The development process involved several steps.

1. **Market Research** A Wiley marketing specialist team of six senior sales representatives was formed to gather information to help focus and develop the text. An extensive market research survey was also sent to over 3,000 professors teaching Statics and Dynamics to home in on key market issues. Two focus groups consisting of professors teaching Statics and Dynamics were conducted to gain a clearer understanding of classroom needs as the texts took shape.

2. **Reviews** Professors from the United States and Canada carefully reviewed each draft of this manuscript. Their suggestions were carefully considered and incorporated whenever possible. Six additional reviewers were commissioned to evaluate one of the key components of the text—the problem sets.

3. **Manuscript and Illustration Development** A developmental editor worked with the authors to hone both the manuscript and the art sketches to their highest potential. A special art developer worked with the authors and the art studio to enhance the illustrations.

TECHNICAL PACKAGE FOR THE INSTRUCTOR

Solution Manual

After years of teaching, we realize the importance of an accurate solution manual that matches the quality of the text. For that reason, we have prepared the manual ourselves. The manual includes a complete solution for every problem in the book, and especially challenging problems are marked with an asterisk. Each solution appears with the original problem statement and, where appropriate, the problem figure. We do this for the convenience of the instructor, who no longer will have to refer to both book and solution manual in preparing for class. The manual also contains transparency masters for use in preparing overhead transparencies.

FOR THE STUDENT

Software

Our reviewers told us that they are generally dissatisfied with publisher-provided software. They also told us that students need software that is easy to use, provides reinforcement of basic concepts, and is highly interactive. With this in mind, we have worked with Intellipro, an engineering software developer, to produce a package that satisfies all these demands. The software consists of 30 problems, 10 from *Statics* and 20 from *Dynamics*. The software reinforces the importance of free-body diagrams by giving students practice in drawing them. The dynamics problems are animated to aid student visualization.

Study Guides

Mechanics can be a tough course, and sometimes students need extra help. Our study guide is written as a tool for developing student understanding and problem-solving skills. This study guide provides reinforcement of the major concepts in the text.

ACKNOWLEDGMENTS

Many people participated directly and indirectly in the preparation of this book. In particular, we wish to thank Rebecca Sidler for her careful review of the manuscript and for solving many problems in the two books. In addition to the authors, many present and former colleagues and students contributed ideas concerning methods of presentation, example problems, and homework problems. Final judgments concerning organization of material and emphasis of topics, however, were made by the authors. We will be pleased to receive comments from readers and will attempt to acknowledge all such communications.

We'd like to thank the following people for their suggestions and encouragement throughout the reviewing process.

H. J. Sneck	Rensselaer
Thomas Lardiner	University of Massachusetts
K. L. DeVries	University of Utah
John Easley	University of Kansas
Brian Harper	Ohio State University
Kenneth Oster	University of Missouri–Rolla
D. W. Yannitell	Louisiana State University

James Andrews	University of Iowa
D. A. DaDeppo	University of Arizona
Ed Hornsey	University of Missouri–Rolla
William Bingham	North Carolina State University
Robert Rankin	Arizona State University
David Taggart	University of Rhode Island
Allan Malvick	University of Arizona
Gaby Neunzert	Colorado School of Mines
Tim Hogue	Oklahoma State University
Bill Farrow	Marquette University
Matthew Ciesla	New Jersey Institute of Technology
William Lee	US Naval Academy
J. K. Al-Abdulla	University of Wisconsin
Erik G. Thompson	Colorado State University
Dr. Kumar	University of Pennsylvania
William Walston	University of Maryland
John Dunn	Northeastern University
Ron Anderson	Queen's University (Canada)
Duane Storti	University of Washington
Jerry Fine	Rose-Hulman Institute of Technology
Ravinder Chona	Texas A & M
Bahram Ravani	University of California–Davis
Paul C. Chan	New Jersey Institute of Technology
Wally Venable	West Virginia University
Eugene B. Loverich	North Arizona University
Kurt Keydel	Montgomery College
Francis Thomas	University of Kansas
Colonel Tezak	U.S. Military Academy

William F. Riley
Leroy D. Sturges

CONTENTS

LIST OF SYMBOLS

Unit Vectors

$\mathbf{i}, \mathbf{j}, \mathbf{k}$	Unit vectors in the x, y, z directions (rectangular coordinates)
$\mathbf{e}_n, \mathbf{e}_t$	Unit vectors in the n, t directions (normal and tangential coordinates)
$\mathbf{e}_r, \mathbf{e}_\theta$	Unit vectors in the r, θ directions (polar coordinates)

Kinematic Quantities

$\mathbf{r}$	Position vector
x, y, z	Rectangular components of the position vector
$\mathbf{v}$	Velocity vector
v_x, v_y, v_z	Rectangular components of the velocity vector
$\mathbf{v}_{Brel}$	Velocity of point B relative to the origin of a set of coordinate axes fixed in and rotating with a rigid body
$\mathbf{a}$	Acceleration vector
a_x, a_y, a_z	Rectangular components of the acceleration vector
$\mathbf{a}_{Brel}$	Acceleration of point B relative to the origin of a set of coordinate axes fixed in and rotating with a rigid body
$\boldsymbol{\omega}$	Angular velocity vector
$\omega_x, \omega_y, \omega_z$	Rectangular components of the angular velocity vector
$\boldsymbol{\alpha}$	Angular acceleration vector
$\alpha_x, \alpha_y, \alpha_z$	Rectangular components of the angular acceleration vector

Work and Energy Quantities

T_i, T_f	Kinetic energy (initial, final)
$U_{1 \to 2}$	Work done by a force and/or a moment as the particle or body moves from position 1 to position 2
$U_{1 \to 2}^{(c)}$	Work done by a conservative force
$U_{1 \to 2}^{(o)}$	Work done by a nonconservative force (or a force whose potential is not known)
V_i, V_f	Potential energy of a force (initial, final)

Impulse and Momentum Quantities

$\mathbf{L}$	Linear momentum vector
$\mathbf{H}_A$	Angular momentum vector (relative to point A)
e	Coefficient of restitution

Vibrational Quantities

ω_n	Undamped natural circular frequency
ω_d	Damped natural circular frequency
f_n	Undamped natural frequency
f_d	Damped natural frequency
τ_n	Undamped natural period
τ_d	Damped natural period
ζ	Damping ratio
Ω	Frequency of forced oscillation
δ	Logarithmic decrement

Miscellaneous Physical Constants

m	Mass of a particle or rigid body
W	Weight of a particle or rigid body
k	Spring constant
c	Damping coefficient
μ_s	Coefficient of static friction
μ_k	Coefficient of dynamic friction
$I_x, I_y, I_{xy}, \ldots$	Moments and products of inertia
k	Radius of gyration
G	Universal gravitational constant
M_e	Mass of the Earth
R_e	Radius of the Earth

12

GENERAL PRINCIPLES

The double Ferris wheel illustrates types of motion encountered in dynamics problems. The motions are curvilinear translation for the seats, fixed axis rotation for the arms, and general plane motion for the wheels.

12-1   INTRODUCTION TO DYNAMICS

Mechanics has been defined as the branch of the physical sciences that deals with the response of bodies to the action of forces. For convenience, the study of mechanics is divided into three parts: namely, the mechanics of rigid bodies, the mechanics of deformable bodies, and the mechanics of fluids. Rigid-body mechanics can be further subdivided into statics (equilibrium of a rigid body) and dynamics (motion of a rigid body).

The portion of mechanics known as statics developed early in recorded history because the principles of statics are needed in building construction. The builders of the pyramids of Egypt understood and used such devices as the lever, the pulley, and the inclined plane.

The portion of mechanics known as dynamics developed much later since the quantities involved (velocity and acceleration) depend on an accurate measurement of time. The experiments of Galileo Galilei (1564–1642) with falling bodies, pendulums, and rolling cylinders on an inclined plane started development of the field of dynamics. However, Galileo was handicapped in his work by a lack of adequate clocks for measuring the small time intervals involved in the experiments. Huygens (1629–1695) continued the work of Galileo and invented the pendulum clock. He also determined the acceleration of gravity and introduced theorems involving centrifugal force. Sir Isaac Newton (1642–1727) completed the formulation of the basic principles of mechanics with his discovery of the law of universal gravitation and his statement of the laws of motion. Newton's work on particles, based on geometry, was extended to rigid-body systems by Euler (1707–1793), who was also the first to use the term *moment of inertia*. D'Alembert (1717–1783) introduced the concept of an inertia force. The sum of the forces on a body in motion is zero when an inertia force is included. Previous work in mechanics, which had been based largely on astronomical observations and geometrical concepts, was formalized by Lagrange (1736–1813), who derived the generalized equations of motion, analytically, by using energy concepts. The derivation of these equations, which are known as Lagrange's equations, represented a significant advancement in the development of classical mechanics. Another major advancement was made by Coriolis (1792–1843), who showed how the introduction of additional terms could validate Newton's laws when the reference frame is rotating.

The next major advances in mechanics were made by Max Planck (1858–1947), who formulated quantum mechanics, and Albert Einstein (1879–1955), who formulated the theory of relativity (1905). These new theories did not repudiate Newtonian mechanics; they were simply more general. Newtonian mechanics was and is applicable to the prediction of the motion of bodies where the speeds are small compared to the speed of light.

12-2   NEWTON'S LAWS

The foundations for studies in engineering mechanics are the laws formulated and published by Sir Isaac Newton in 1687. In a treatise called "The Principia," Newton stated the basic laws governing the motion of a particle as:[1]

Newton's Laws of Motion

Law 1 Every body perseveres in its state of rest or of uniform motion in a straight line, except in so far as it is compelled to change that state by impressed forces.

Law 2 Change of motion is proportional to the moving force impressed, and takes place in the direction of the straight line in which such force is impressed.

Law 3 Reaction is always equal and opposite to action; that is to say, the actions of two bodies upon each other are always equal and directly opposite.

These laws, which have come to be known as "Newton's Laws of Motion," are commonly expressed today as:

Law 1 In the absence of external forces, a particle originally at rest or moving with a constant velocity will remain at rest or continue to move with a constant velocity along a straight line.

Law 2 If an external force acts on a particle, the particle will be accelerated in the direction of the force and the magnitude of the acceleration will be directly proportional to the force and inversely proportional to the mass of the particle.

Law 3 For every action there is an equal and opposite reaction. The forces of action and reaction between contacting bodies are equal in magnitude, opposite in direction, and collinear.

Newton's three laws were developed from a study of planetary motion (the motion of particles). During the eighteenth century, Leonhard Euler (1707–1783) extended Newton's work on particles to rigid-body systems.

The first law covers the case where a body is in equilibrium. Thus, the first law provides the foundation for the study of statics. The second law deals with accelerated motion of a body and provides the foundation for the study of dynamics.

The law that governs the mutual attraction between two isolated bodies was also formulated by Newton and is known as the **"Law of Gravitation."** The law of gravitation is very important in all studies involving the motion of planets, space craft, or artificial satellites.

12-3 FUNDAMENTAL QUANTITIES OF MECHANICS

The fundamental quantities of mechanics are space, time, mass, and force. Three of these quantities—space, time, and mass—are absolute quantities. This means that they are independent of each other and

---

[1] Dr. Ernst Mach, "Die Mechanik in ihrer Entwickelung historisch-kritisch dargestellt," Professor an der Universitat zu Wien. Mit 257 Abbildungen. Leipzig, 1893. First translated from the German by Thomas J. McCormack in 1902. *The Science of Mechanics*, 9th ed. The Open Court Publishing Company, LaSalle, Ill., 1942.

cannot be expressed in terms of the other quantities or in simpler terms. A force is not independent of the other three quantities but is related to the mass of the body and to the manner in which the velocity of the body varies with time. A brief description of these four quantities together with some other concepts of importance in dynamics follows:

Space is the geometric region commonly referred to as "the universe." The region extends without limit in all directions.

Time is the interval between two events. Measurement of this interval is made by making comparisons with some reproducible event such as the time required for the earth to rotate on its axis.

Matter is any substance that occupies space.

A **body** is matter bounded by a closed surface.

Inertia is the property of matter that causes resistance to a change in motion.

Mass is a quantitative measure of a body's resistance to a change in its motion.

A **force** is the action of one body on another body. Forces always occur in pairs, and the two forces have equal magnitude and opposite sense. The external effect of a force on a body is either development of resisting forces (reactions) on the body (statics problems) or accelerated motion of the body (dynamics problems).

A **particle** is a body without size or shape that can be assumed to occupy a single point in space. Mechanics problems are greatly simplified when the body can be treated as a particle.

A **rigid body** is a collection of particles that remain at fixed distances from each other at all times and under all conditions of loading. The rigid-body concept represents an idealization of the true situation since all real bodies will change shape to a certain extent when they are subjected to a system of forces. When it is assumed that the body is rigid (free of deformation), the material properties of the body are not required for the analysis of forces and their effects on the body. The bodies dealt with in this book, with the exception of deformable springs, will be considered to be rigid bodies.

The **position** of a point in space is specified by using linear and angular measurements with respect to a coordinate system whose origin is located at some reference point. The basic reference system used as an aid in solving mechanics problems is a primary inertial system, which is an imaginary set of rectangular axes that do not translate or rotate in space. Measurements made relative to this system are called absolute. The laws of Newtonian mechanics are valid for this reference system as long as any velocities involved are negligible with respect to the speed of light which is 300,000 km/s (186,000 mi/s). A reference frame fixed to the surface of the earth moves with respect to the primary inertial system; however, corrections to account for the absolute motion of the earth are insignificant and may be neglected for most engineering problems involving machines and structures that remain on the surface of the earth.

12-4 UNITS OF MEASUREMENT

The physical quantities used to express the laws of mechanics are mass, length, force, time, velocity, acceleration, and so on. These quantities can be divided into fundamental quantities and derived quantities.

Fundamental quantities cannot be defined in terms of other physical quantities. The number of quantities regarded as fundamental is the minimum number needed to give a consistent and complete description of all of the physical quantities ordinarily encountered in the subject area. Examples of quantities viewed as fundamental in mechanics are length and time.

Derived quantities are those whose defining operations are based on measurements of other physical quantities. Examples of derived quantities in mechanics are area, volume, velocity, and acceleration.

Mass and force are examples of quantities that may be viewed as either fundamental or derived. In the SI system of units, mass is regarded as a fundamental quantity and force is a derived quantity. In the U.S. Customary System of units, force is regarded as a fundamental quantity and mass is a derived quantity.

Units of Length   The magnitude of each of the fundamental quantities is defined by an arbitrarily chosen unit or "standard." The familiar yard, foot, and inch, for example, come from the old practice of using the human arm, foot, and thumb as length standards. The first truly international standard of length was a bar of platinum–iridium alloy, called the standard meter,[2] which is kept at the International Bureau of Weights and Measures in Sèvres, France. The distance between two fine lines engraved on gold plugs near the ends of the bar is defined to be one meter. Historically, the meter was intended to be one ten-millionth of the distance from the pole to the equator along the meridian line through Paris. Accurate measurements made after the standard meter bar was constructed show that it differs from its intended value by approximately 0.023 percent.

In 1961 an atomic standard of length was adopted by international agreement. The wavelength in vacuum of the orange-red line from the spectrum of isotope krypton 86 was chosen. One meter (m) is now defined to be 1,650,763.73 wavelengths of this light. The choice of an atomic standard offers advantages other than increased precision in length measurements. Krypton 86 is readily available everywhere; the material is relatively inexpensive, and all atoms of the material are identical and emit light of the same wavelength. The particular wavelength chosen is uniquely characteristic of krypton 86 and is very sharply defined.

The definition of the yard, by international agreement, is 1 yard = 0.9144 m, exactly.[3] Thus, 1 inch = 25.4 mm, exactly; and 1 foot = 0.3048 m, exactly.

---

[2]The United States has accepted the meter as a standard of length since 1893.

[3]*Guide for the Use of the International System of Units,* National Institute of Standards and Technology (NIST) Special Publication 811, September 1991.

Units of Time   Similarly, time can be measured in a number of ways. Since early times, the length of the day has been an accepted standard of time measurement. The internationally accepted standard unit of time, the second (s), was defined in the past as 1/86,400 of a mean solar day or 1/31,557,700 of a mean solar year. Time defined in terms of the rotation of the earth was based on astronomical observations. Since these observations require at least several weeks, a good secondary terrestrial measure, calibrated by astronomical observations, was required for practical use. Quartz crystal clocks, based on the electrically sustained natural periodic vibrations of a quartz wafer, are used as secondary time standards for scientific work. The best of these quartz clocks keep time for a year with a maximum error of 0.02 s. To meet the need for an even better time standard, an atomic clock has been developed that uses the periodic atomic vibrations of isotope cesium 133. The second based on this cesium clock was adopted as the time standard by the Thirteenth General Conference on Weights and Measures in 1967. The second is defined as the duration of 9,192,631,770 cycles of vibration of isotope cesium 133. The cesium clock provides a significant improvement over the accuracy associated with other methods based on astronomical observations. Two cesium clocks will differ by no more than one second after running 3000 years.

Units of Mass and Weight   The standard unit of mass, the kilogram (kg), is defined by a bar of platinum–iridium alloy that is kept at the International Bureau of Weights and Measures in Sèvres, France.

In the SI System of Units, the unit of force is derived and is called a newton (N). One newton is the force required to give one kilogram of mass an acceleration of one meter per second squared. Thus, $1 \, \text{N} = 1 \, \text{kg} \cdot \text{m/s}^2$.

In the U.S. Customary System of Units the unit of force is called a pound (lb) and is defined as the weight at sea level and at a latitude of 45 degrees of a platinum standard, which is kept at the National Institute of Standards and Technology (NIST) in Washington, D.C. This platinum standard has a mass of 0.453,592,43 kg. The unit of mass in this system is derived and is called a slug. One slug is the mass that is accelerated one foot per second squared by a force of one pound, or 1 slug equals $1 \, \text{lb} \cdot \text{s}^2/\text{ft}$. Since the weight of the platinum standard depends on the gravitational attraction of the earth, the U.S. Customary System is a gravitational rather than an absolute system of units.

As an aid to interpreting the physical significance of answers in SI units for those more accustomed to the U.S. Customary System, some conversion factors for the quantities normally encountered in mechanics are provided in Table 12-1.

12-5   DIMENSIONAL CONSIDERATIONS

All the physical quantities encountered in engineering mechanics can be expressed dimensionally in terms of three fundamental or base quantities: mass, length, and time, denoted, respectively by M, L, and T. The dimensions of other physical quantities follow from their definitions or from physical laws. For example, the dimension of velocity follows from the definition of velocity, quotient of length and time

TABLE 12-1 CONVERSION FACTORS BETWEEN THE SI AND U.S. CUSTOMARY SYSTEMS

Quantity	U.S. Customary to SI	SI to U.S. Customary
Length	1 in. = 25.40 mm	1 m = 39.37 in.
	1 ft = 0.3048 m	1 m = 3.281 ft
	1 mi = 1.609 km	1 km = 0.6214 mi
Area	1 in.^{2} = 645.2 mm^2	1 m^2 = 1550 in.^{2}
	1 ft^2 = 0.0929 m^2	1 m^2 = 10.76 ft^2
Volume	1 in.^{3} = 16.39(10^3) mm^3	1 mm^3 = 61.02(10^{-6}) in.^{3}
	1 ft^3 = 0.02832 m^3	1 m^3 = 35.31 ft^3
	1 gal = 3.785 L^a	1 L = 0.2642 gal
Velocity	1 in./s = 0.0254 m/s	1 m/s = 39.37 in./s
	1 ft/s = 0.3048 m/s	1 m/s = 3.281 ft/s
	1 mi/h = 1.609 km/h	1 km/h = 0.6214 mi/h
Acceleration	1 in./s^2 = 0.0254 m/s^2	1 m/s^2 = 39.37 in./s^2
	1 ft/s^2 = 0.3048 m/s^2	1 m/s^2 = 3.281 ft/s^2
Mass	1 slug = 14.59 kg	1 kg = 0.06854 slug
Second moment of area	1 in.^{4} = 0.4162(10^6) mm^4	1 mm^4 = 2.402(10^{-6}) in.^{4}
Force	1 lb = 4.448 N	1 N = 0.2248 lb
Distributed load	1 lb/ft = 14.59 N/m	1 kN/m = 68.54 lb/ft
Pressure or stress	1 psi = 6.895 kPa	1 kPa = 0.1450 psi
	1 ksi = 6.895 MPa	1 MPa = 145.0 psi
Bending moment or torque	1 ft·lb = 1.356 N·m	1 N·m = 0.7376 ft·lb
Work or energy	1 ft · lb = 1.356 J	1 J = 0.7376 ft·lb
Power	1 ft·lb/s = 1.356 W	1 W = 0.7376 ft·lb/s
	1 hp = 745.7 W	1 kW = 1.341 hp

^{a}Both L and l are accepted symbols for liter. Because "l" can easily be confused with the numeral "1," the symbol "L" is recommended for United States use by the National Institute of Standards and Technology (see NIST special publication 811, September 1991).

(L/T). From Newton's second law, force is defined as the product of mass and acceleration; therefore, force has the dimension ML/T^2. The dimensions of other physical quantities commonly encountered in engineering mechanics are given in Table 12-2.

12-5-1 Dimensional Homogeneity

When an equation is used to describe a physical process, the equation is said to be dimensionally homogeneous if the form of the equation does not depend on the units of measurement. For example, the equation describing the distance h a body travels when released at rest in the earth's gravitational field is $h = gt^2/2$, where h is the distance traveled, t is the time since release, and g is the gravitational acceleration. This equation is valid whether length is measured in feet, meters, or inches and whether time is measured in hours, years, or seconds, provided g is measured in the same units of length and time as h and t. Dimensionally homogeneous equations are usually preferred because of the potential confusion connected with the units of constants appearing in dimensionally inhomogeneous equations.

12-6 METHOD OF PROBLEM SOLVING

The principles of mechanics are few and relatively simple; however, the applications are infinite in their number, variety, and complexity. Success depends to a large degree on a well-disciplined method of

TABLE 12-2 DIMENSIONS OF THE PHYSICAL QUANTITIES OF MECHANICS

Physical Quantity	Dimension	Common Units	
		SI System	US Customary System
Length	L	m, mm	in., ft
Area	L^2	m^2, mm^2	$in.^2$, ft^2
Volume	L^3	m^3, mm^3	$in.^3$, ft^3
Angle	$1 (L/L)$	rad, degree	rad, degree
Time	T	s	s
Linear velocity	L/T	m/s	ft/s
Linear acceleration	L/T^2	m/s^2	ft/s^2
Angular velocity	$1/T$	rad/s	rad/s
Angular acceleration	$1/T^2$	rad/s^2	rad/s^2
Mass	M	kg	slug
Force	ML/T^2	N	lb
Moment of a force	ML^2/T^2	$N \cdot m$	$ft \cdot lb$
Pressure	M/LT^2	Pa, kPa	psi, ksi
Stress	M/LT^2	Pa, MPa	psi, ksi
Energy	ML^2/T^2	J	$ft \cdot lb$
Work	ML^2/T^2	J	$ft \cdot lb$
Power	ML^2/T^3	W	hp
Linear impulse	ML/T	$N \cdot s$	$lb \cdot s$
Momentum	ML/T	$N \cdot s$	$lb \cdot s$
Specific weight	M/L^2T^2	N/m^3	lb/ft^3
Density	M/L^3	kg/m^3	$slug/ft^3$
Second moment of area	L^4	m^4, mm^4	$in.^4$, ft^4
Moment of inertia	ML^2	$kg \cdot m^2$	$slug \cdot ft^2$

problem solving. Problem solving typically consists of the following phases.

Three Phases of Professional Problem Solving

1. Problem definition and identification
2. Model development and simplification
3. Mathematical solution and result interpretation.

The problem-solving method outlined in this section will prove useful for this course, for the engineering courses that follow, and for most situations encountered later in engineering practice.

Problem Definition and Identification Problems in engineering mechanics (statics, dynamics, and mechanics of deformable bodies) are concerned primarily with the external effects of a system of forces on a physical body. The approach usually followed in solving an engineering mechanics problem requires identification of all external forces acting on the body of interest. This can be accomplished by preparing a free-body diagram that shows the body of interest isolated from all other bodies and with all external forces applied.

Most engineers consider a free-body diagram to be the single most important tool for the solution of mechanics problems.

This simple tool provides a powerful means for distinguishing between cause (the external forces) and effect (motion or deformation of the body of interest) and helps focus attention on the principles and information required for solution of the problem.

Model Development and Simplification Since relationships between cause and effect are stated in mathematical form, the true physical situation must be represented by a mathematical model in order to obtain the required solution. Often it is necessary to make simplifying assumptions or approximations in setting up this model in order to solve the problem. The most common approximation is to treat most of the bodies in statics and dynamics problems as rigid bodies. No real body is absolutely rigid; however, the changes in shape of a real body usually have a negligible effect on the acceleration produced by a force system or on the reactions required to maintain equilibrium of the body. Considerations of the changes in shape under these circumstances would be an unnecessary complication of the problem.

Mathematical Solution and Result Interpretation Usually, an actual physical problem cannot be solved exactly or completely. However, even in complicated problems, a simplified model can provide good qualitative results. Appropriate interpretation of such results can lead to approximate predictions of physical behavior or be used to verify the "reasonableness" of more sophisticated analytical or numerical results. The engineer must constantly be aware of the actual physical problem being considered and of any limitations associated with the mathematical model being used. Assumptions must be continually evaluated to ensure that the mathematical problem being solved provides an adequate representation of the physical process or device of interest.

The most effective way to learn the material contained in engineering mechanics courses is to solve a variety of problems. In order to become an effective engineer, the student must develop the ability to reduce complicated problems to simple parts that can be easily analyzed and to present results of the work in a clear, logical, and neat manner. This can be accomplished by using the following sequence of steps.

Steps for Analyzing and Solving Problems

1. Read the problem carefully.
2. Identify the result requested.
3. Identify the principles to be used to obtain the result.
4. Tabulate the information provided.
5. Draw the appropriate free-body diagrams.
6. Apply the appropriate principles and equations.
7. Report the answer with the appropriate number of significant figures and the appropriate units.
8. Study the answer and determine if it is reasonable.

Accepted engineering practice requires that all work be neat and orderly since neatness is usually associated with clear and orderly thinking. Sloppy solutions that cannot be easily read and understood by others, because they contain superfluous or confusing details, have little or no value. The development of an ability to apply an orderly approach to problem solving constitutes a significant part of an engineering education. Also note that the problem identification, model simplification, and result interpretation phases of engineering problem solving are often more important than the mathematical solution phase.

12-7   SIGNIFICANCE OF NUMERICAL RESULTS

The accuracy of solutions to real engineering problems depends on three factors:

Factors Influencing Accuracy

1. The accuracy of the known physical data
2. The accuracy of the physical model
3. The accuracy of the computations performed

The accuracy obtained when solving practical engineering problems is usually determined by the accuracy of the physical data, which are seldom known with an accuracy greater than 0.2 percent. A practical rule for "rounding off" the final numbers obtained in the computations involved in engineering analysis, which provides answers to approximately this degree of accuracy, is to retain four significant figures for numbers beginning with the figure "1" and to retain three significant figures for numbers beginning with any figure from "2" through "9."

The speed and accuracy of pocket electronic calculators facilitates the numerical computations involved in the solution of engineering problems. However, the number of significant figures that can easily be obtained should not be taken as an indication of the accuracy of the solution. As noted previously, engineering data are seldom known to an accuracy greater than 0.2 percent; therefore, calculated results should always be "rounded off" to the number of significant figures that will yield the same degree of accuracy as the data on which they are based. Most of the example problems in this book are solved with the assumption that the data provided are accurate to three significant figures. Consequently, intermediate calculations are worked out to four significant figures and the answers are reported to four significant figures when the first digit is 1 and to 3 significant figures when the first digit is any other number. In problems with data given to only one or two significant figures, the answers are based on the assumption that all such data are exact.

For closed-form analytical predictions, the accuracy of the data and the adequacy of the model determine the accuracy of the results.

For numerical predictions, the computational accuracy of the algorithms used also influences the accuracy of the results.

An error can be defined as the difference between two quantities. The difference, for example, might be between an experimentally measured value and a computed theoretical value. An error may also result from the rounding off of numbers during a calculation. One method for describing an error is to state a percent difference (%D). Thus, for two numbers A and B, if it is desired to compare number A with number B, the percent difference between the two numbers is defined as

$$\%D = \frac{A - B}{B}(100)$$

In this equation, B is the reference value with which A is to be compared.

SUMMARY

Dynamics problems in which the motions are known are similar to statics problems, in that application of Newton's laws leads to equations that can be easily solved for the unknown forces. Dynamics problems in which some aspects of the motion are not known are more difficult. In these problems, application of Newton's second law normally results in a set of differential equations that may be relatively easy or very difficult to solve.

Physical quantities used to express the laws of mechanics can be divided into fundamental quantities and derived quantities. The magnitude of each fundamental quantity is defined by an arbitrarily chosen unit or "standard." The units used in the SI system are the meter (m) for length, the kilogram (kg) for mass, and the second (s) for time. The unit of force is a derived unit called a newton (N). In the U.S. Customary System of Units the units used are the foot (ft) for length, the pound (lb) for force, and the second (s) for time. The unit of mass is a derived unit called a slug. The U.S. Customary System is a gravitational rather than an absolute system of units.

The terms of an equation used to describe a physical process should not depend on the units of measurement (should be dimensionally homogeneous). If an equation is dimensionally homogeneous, the equation is valid for use with any system of units provided all quantities in the equation are measured in the same system. Use of dimensionally homogeneous equations eliminates the need for unit conversion factors.

Success in engineering depends to a large degree on a well-disciplined method of problem solving. Professional problem solving consists of three phases:

1. Problem definition and identification
2. Model development and simplification
3. Mathematical solution and result interpretation

Identification of all external forces acting on a body can be accomplished by preparing a free-body diagram that shows the body of interest isolated from all other bodies and with all external forces applied. In order to obtain a solution to most problems, the true physical situation must be represented by a mathematical model. A common approximation made in setting up this model is to treat the body as a rigid body. Even though no real body is absolutely rigid, changes in shape usually have a negligible effect on accelerations produced by a force system or on reactions required to maintain equilibrium of the body; therefore, considerations of changes in shape usually result in an unnecessary complication of the problem. Anytime a mathematical model is used in solving a problem, care must be exercised to ensure that the model and the associated mathematical problem being solved provides an adequate representation of the physical process or device that it represents.

The accuracy of solutions to real engineering problems depends on three factors:

1. The accuracy of the known physical data
2. The accuracy of the physical model
3. The accuracy of the computations performed

An accuracy greater than 0.2 percent is seldom possible. Calculated results should always be "rounded off" to the number of significant figures that will yield the same degree of accuracy as the data on which they are based.

KINEMATICS OF PARTICLES

A skier's flight trajectory is determined by takeoff velocity, gravity, and aerodynamic forces.

13-1 INTRODUCTION

The study of dynamics consists of two related topics: **kinematics,** which is the study of how objects move, and **kinetics,** which is the study of the relationship between the motion and the forces that cause the motion. Kinematics describes how the velocity and acceleration of a body change with time and with changes in the position of the body. Not only is a sound understanding of kinematics a necessary background for the later study of kinetics, but kinematics is an important field of study in its own right. The design of many machine parts to create specific motions is based almost exclusively on kinematics. The study of the motion of projectiles, rocket ships, and space satellites is also based heavily on kinematics. Kinematics is the subject of this chapter and of Chapter 14.

A **particle** is an object whose size can be ignored when studying its motion. Only the position of the mass center of the particle need be considered. The orientation of the object or rotation of the object plays no role in describing the motion of a particle. Particles can be very small or very large. Small does not always guarantee that an object can be modeled as a particle; large does not always prevent modeling an object as a particle. Whether a body is large or small relates to the length of path followed, the distance between bodies, or both. The kinematics of particles is covered in this chapter. The kinematics of rigid bodies (bodies for which the orientation and rotation are important) is covered in Chapter 14.

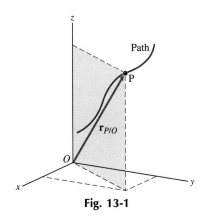

Fig. 13-1

13-2 POSITION, VELOCITY, AND ACCELERATION

Suppose a particle travels along a path as shown in Figure 13-1. At some instant of time the particle is at point P. In terms of the fixed, rectangular, Cartesian coordinate system indicated, the position of the particle is given by the position vector of P relative to the origin O, which can be written

$$\mathbf{r}_{P/O} = x\mathbf{i} + y\mathbf{j} + z\mathbf{k} \tag{13-1}$$

where the basis vectors, $\mathbf{i}$, $\mathbf{j}$, and $\mathbf{k}$, are unit vectors in the x-, y-, and z-coordinate directions, respectively. The $/O$ on the subscript indicates that the components (x,y,z) depend not only on the orientation of the coordinate system but also on the location of the origin O. In terms of a second fixed, rectangular, Cartesian, coordinate system oriented the same way as the first but having an origin at $\hat{O}$ (Fig. 13-2), the position of the particle is written

$$\mathbf{r}_{P/\hat{O}} = \hat{x}\mathbf{i} + \hat{y}\mathbf{j} + \hat{z}\mathbf{k} \tag{13-2}$$

By the triangle law of addition for vectors, these position vectors are related by

$$\mathbf{r}_{P/O} = \mathbf{r}_{\hat{O}/O} + \mathbf{r}_{P/\hat{O}} \tag{13-3}$$

where $\mathbf{r}_{\hat{O}/O}$ is a constant vector.

The difference in position of the particle at two instants of time is called the **displacement** of the particle. If the particle that is at P at time t is at Q at time $t + \Delta t$ (Fig. 13-3), then the displacement $\delta\mathbf{r}$ is given by

$$\delta\mathbf{r} = \mathbf{r}_{Q/O} - \mathbf{r}_{P/O} \tag{13-4}$$

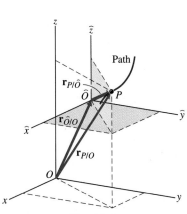

Fig. 13-2

Note that the displacement is independent of the location of the origin of the coordinate system since

$$\mathbf{r}_{Q/\hat{O}} - \mathbf{r}_{P/\hat{O}} = (\mathbf{r}_{Q/O} - \mathbf{r}_{\hat{O}/O}) - (\mathbf{r}_{P/O} - \mathbf{r}_{\hat{O}/O})$$
$$= \mathbf{r}_{Q/O} - \mathbf{r}_{P/O}$$

The **velocity** of a particle is defined to be the time rate of change of its position:

$$\mathbf{v}_P = \frac{d\mathbf{r}_{P/O}}{dt} = \dot{\mathbf{r}}_{P/O} = \lim_{\delta t \to 0} \frac{\delta \mathbf{r}}{\delta t} \tag{13-5}$$

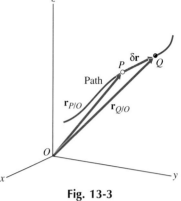

Fig. 13-3

Since the displacement $\delta\mathbf{r}$ is independent of the location of the origin of the coordinate system, it is obvious that the velocity $\mathbf{v}_P$ is also independent of the location of the origin of the coordinate system. Furthermore, it is obvious that the direction of the velocity $\mathbf{v}_P$ is in the direction of the displacement $\delta\mathbf{r}$ or tangent to the path of the particle.

Since the basis vectors are also constant, the velocity can be written in terms of components as

$$\mathbf{v}_P = v_x\mathbf{i} + v_y\mathbf{j} + v_z\mathbf{k} = \dot{x}\mathbf{i} + \dot{y}\mathbf{j} + \dot{z}\mathbf{k} \tag{13-6}$$

The **acceleration** of a particle is defined to be the time rate of change of its velocity

$$\mathbf{a}_P = \frac{d\mathbf{v}_P}{dt} = \dot{\mathbf{v}}_P = \frac{d^2\mathbf{r}_{P/O}}{dt^2} = \ddot{\mathbf{r}}_{P/O} \tag{13-7}$$

The acceleration is also independent of the location of the origin of the coordinate system. The acceleration can be written in terms of components as

$$\mathbf{a}_P = a_x\mathbf{i} + a_y\mathbf{j} + a_z\mathbf{k} = \dot{v}_x\mathbf{i} + \dot{v}_y\mathbf{j} + \dot{v}_z\mathbf{k}$$
$$= \ddot{x}\mathbf{i} + \ddot{y}\mathbf{j} + \ddot{z}\mathbf{k} \tag{13-8}$$

If a coordinate system can be found such that the y- and z-components of the position, velocity, and acceleration are all zero for all time, the motion is called **rectilinear motion.** In this case the particle moves in a straight line (along the x-axis) with varying speed and acceleration. Rectilinear motion is covered in Sections 13-3 and 13-4.

If the particle is not moving in rectilinear motion but a coordinate system can be found such that the z-components of the position, velocity, and acceleration are all zero for all time, the motion is called **plane curvilinear motion.** Kinematics of plane curvilinear motion is covered in Sections 13-5 and 13-6.

Motions for which no coordinate system can be found that makes at least one component of the position, velocity, and acceleration zero for all time are called **general curvilinear motion** or **space curvilinear motion.** Kinematics of space curvilinear motion is covered in Section 13-7.

13-3   RECTILINEAR MOTION

Motion in this case is along a straight line. The coordinate system will be oriented such that the x-axis coincides with the line of motion. For convenience this line will be assumed horizontal with positive to the right and negative to the left.

Since the position, velocity, and acceleration are completely prescribed by giving only their x-components, the vector notation will be dropped. The equations for position, displacement, velocity, and acceleration will be written simply as

$$r_{P/O} = x \qquad\qquad \delta x = x_Q - x_P \qquad\qquad \text{(13-9a,b)}$$
$$v_P = v = \dot{x} \qquad\qquad a_P = a = \dot{v} = \ddot{x} \qquad\qquad \text{(13-9c,d)}$$

where the sign of the number indicates whether the vector is in the direction of the positive x-axis or the negative x-axis. Thus, if x is positive, the particle is to the right of the origin; x negative, it's to the left of the origin. When δx is positive, the particle's final position Q is to the right of its initial position, P; δx negative, Q is to the left of P. If v is positive, the particle is moving to the right; v negative, it's moving to the left. When the velocity and acceleration have the same sign, the velocity is increasing and the particle is said to be **accelerating.** When the velocity and acceleration have opposite signs, the velocity is decreasing and the particle is said to be **decelerating.**

It should be noted that the displacement of a particle is not the same thing as the distance traveled by a particle. The displacement is the (vector) difference in position between the beginning and end of a particle's path. If a particle begins and ends at the same point, it's displacement is **0**. For example, the displacement of a train that travels from Baltimore to Washington, D.C., and back to Baltimore is **0**. The distance traveled, on the other hand, keeps track of what happens between the beginning and end of the path. It measures the total length of the path without regard to direction. The distance traveled is a scalar quantity—just a number. For the train that travels from Baltimore to Washington, D.C. (50 mi), and back (also 50 mi), the distance traveled is 100 mi. The distance traveled is always a positive number.

Equations 13-9 relate the four main variables of interest: position, velocity, acceleration, and time. In typical applications a relationship between two of the variables is given and it is desired to find the other two variables. Some of the more common combinations will be considered in the following sections.

13-3-1   Given $x(t)$

If the position is given as a function of time, the velocity and acceleration are found simply by differentiating,

$$v(t) = \frac{dx}{dt} \qquad \text{and} \qquad a(t) = \frac{dv}{dt} \qquad\qquad \text{(13-10)}$$

13-3-2   Given $v(t)$

When the velocity is given as a function of time, the acceleration can be found by differentiation as above (Eq. 13-10b). The position is found by integrating Eq. 13-10a

$$v(t) = \frac{dx}{dt}$$

which gives

$$\int_{x_0}^{x} ds = \int_{t_0}^{t} v(\tau)\, d\tau \qquad\qquad \text{(13-11a)}$$

or

$$x - x_0 = \int_{t_0}^{t} v(\tau)\, d\tau \qquad\qquad \text{(13-11b)}$$

13-3-3 Given $a(t)$

When the acceleration is given as a function of time, the velocity is found by integrating Eq. 13-10b

$$\frac{dv}{dt} = a(t)$$

which gives

$$\int_{v_0}^{v} dv = \int_{t_0}^{t} a(\tau)\, d\tau \qquad (13\text{-}12a)$$

or

$$v - v_0 = \int_{t_0}^{t} a(\tau)\, d\tau \qquad (13\text{-}12b)$$

The position is then found by integrating the velocity as before.

13-3-4 Given $a(x)$

When the acceleration is given as a function of position, the definition of the acceleration (Eq. 13-10b) must first be rewritten using the chain rule of differentiation:

$$\frac{dv}{dt} = \frac{dv}{dx}\frac{dx}{dt} = v\frac{dv}{dx} = a(x) \qquad (13\text{-}13)$$

Then integration gives

$$\int_{v_0}^{v} v\, dv = \int_{x_0}^{x} a(s)\, ds \qquad (13\text{-}14a)$$

or

$$\frac{v^2 - v_0^2}{2} = \int_{x_0}^{x} a(s)\, ds \qquad (13\text{-}14b)$$

Now that the velocity is known as a function of position, the position can be found as a function of time by integrating Eq. 13-10a

$$\frac{dx}{dt} = v(x)$$

which gives

$$\int_{x_0}^{x} \frac{ds}{v(s)} = \int_{t_0}^{t} d\tau = t - t_0 \qquad (13\text{-}15)$$

13-3-5 Given $a(v)$

When the acceleration is given as a function of velocity, the velocity can be found as a function of time by integrating Eq. 13-10b

$$\frac{dv}{dt} = a(v)$$

which gives

$$\int_{v_0}^{v} \frac{dv}{a(v)} = \int_{t_0}^{t} d\tau = t - t_0 \qquad (13\text{-}16)$$

Once the velocity is known as a function of time, the velocity can then be integrated as in Section 13-3-2 to get the position as a function of time.

Alternatively, the velocity can be found as a function of position by integrating Eq. 13-13

$$\frac{dv}{dt} = \frac{dv}{dx}\frac{dx}{dt} = v\frac{dv}{dx} = a(v)$$

to get

$$\int_{v_0}^{v} v\frac{dv}{a(v)} = \int_{x_0}^{x} ds = x - x_0 \tag{13-17}$$

13-3-6   Given a = constant

The case of constant acceleration (called **uniformly accelerated motion**) is included in the foregoing analysis since if the acceleration is constant, it can be treated as a function of time or of position or of velocity, whichever is convenient in the problem. However, the special case of uniformly accelerated motion (and the related special case of **uniform motion** in which $a = 0$) arises so often in mechanics that it is worth treating it as a separate case.

If the acceleration is constant, then the integrations of Section 13-3-3 can be performed immediately to get

$$v - v_0 = \int_{t_0}^{t} a\, d\tau = a(t - t_0) \tag{13-18}$$

and

$$x - x_0 = \int_{t_0}^{t} [v_0 + a(\tau - t_0)]d\tau$$

$$= v_0(t - t_0) + \frac{1}{2}a(t - t_0)^2 \tag{13-19}$$

Similarly, the integration of Section 13-3-4 gives

$$v^2 - v_0^2 = 2a(x - x_0) \tag{13-20}$$

It cannot be emphasized enough that Eqs. 13-18, 13-19, and 13-20 are valid only when the acceleration is a constant. A very common error is to use these equations when the acceleration is not constant.

13-3-7   Graphical Analysis

Typical graphs of the position, velocity, and acceleration versus time are drawn in Fig. 13-4. In these graphs the acceleration is the slope of the velocity graph since (by Eq. 13-10b)

$$a(t) = \frac{dv}{dt} = \text{slope of the velocity graph at } t$$

If the value of the acceleration is positive, then the velocity is increasing; if the value of the acceleration is negative, the velocity is decreas-

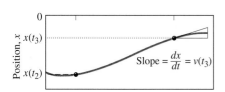

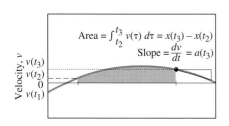

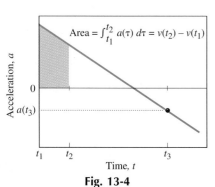

Fig. 13-4

ing. The more positive or negative the acceleration is, the faster the velocity is increasing or decreasing. Furthermore, the change in the velocity from time t_0 to t_1 is equal to the (signed) area under the a-t graph between those times since (by Eq. 13-12b)

$$v_1 - v_0 = \int_{t_0}^{t_1} a(\tau)\, d\tau = \left\{ \begin{array}{l} \text{area under the } a\text{-}t \text{ graph} \\ \text{between} \quad t_0 \quad \text{and} \quad t_1 \end{array} \right.$$

Similarly, the velocity is the slope of the position graph since (by Eq. 13-10a)

$$v(t) = \frac{dx}{dt} = \text{slope of the position graph at } t$$

If the value of the velocity is positive, then the position is increasing; if the value of the velocity is negative, the position is decreasing. Also, the change in position from time t_0 to t_1 is equal to the (signed) area under the v-t graph between those times since (by Eq. 13-11b)

$$x_1 - x_0 = \int_{t_0}^{t_1} v(\tau)\, d\tau = \left\{ \begin{array}{l} \text{area under the } v\text{-}t \text{ graph} \\ \text{between} \quad t_0 \quad \text{and} \quad t_1 \end{array} \right.$$

If the area is positive, the particle moves a distance $x_1 - x_0$ to the right (in the positive x-direction). If the area is negative, the particle moves a distance $|x_1 - x_0|$ to the left (in the negative x-direction). The displacement of the particle between time t_0 and t_1 is then the sum of all these positive and negative areas. The displacement may be positive or negative depending on the relative amounts of positive and negative areas. The distance traveled, on the other hand, is the sum of the positive areas and the absolute value of the negative areas. Mathematically, the distance traveled can be expressed as

$$\text{distance traveled} = \int_{t_0}^{t_1} |v(\tau)|\, d\tau$$

Although these area–slope relations are not usually useful for computing the position or velocity, they can be used with sketches of the position, velocity, and acceleration as a quick check of the results.

A particle moves along the y-axis with an acceleration given by $a(t) = 5 \sin \omega t$ ft/s² where $\omega = 0.7$ rad/s. Initially (at $t = 0$) the particle is 2 ft above the origin and is moving downward with a speed of 5 ft/s.

a. Determine the velocity and position of the particle as functions of time.
b. Show the position, velocity, and acceleration on a graph.
c. Determine the displacement of the particle δ between $t = 0$ s and $t = 4$ s.
d. Determine the total distance traveled by the particle s between $t = 0$ s and $t = 4$ s.

SOLUTION

a. Since the acceleration is given as a function of time, the velocity and position can be obtained simply by integrating the definitions. First,

$$\frac{dv}{dt} = a(t) = 5 \sin (0.7\, t)$$

which can be integrated immediately to get

$$v(t) = -5 - \frac{5}{0.7}[\cos (0.7\, t) - 1] \qquad \text{Ans.}$$

where the constant of integration has been chosen to satisfy the initial condition that $v = -5$ ft/s when $t = 0$ s. Then integrating

$$\frac{dy}{dt} = v(t) = -5 - \frac{5}{0.7}[\cos (0.7\, t) - 1]$$

gives

$$y(t) = 2 - 5t - \frac{5}{0.7}\left[\frac{\sin (0.7\, t)}{0.7} - t\right] \qquad \text{Ans.}$$

b. The position, velocity, and acceleration of the particle are sketched in Fig. 13-5. Notice that the value of the acceleration is positive for the entire first four seconds and hence the slope of the velocity graph is also positive for the entire first four seconds. Similarly, the value of the velocity is negative for about the first 1.8 s, during which time the slope of the position graph is also negative. After $t \cong 1.8$ s the value of the velocity is positive and the slope of the position graph is also positive. At $t \cong 1.8$ s the value of the velocity is zero; $v(t) = dy/dt = 0$ and the position takes on its minimum value there.

c. The displacement of the particle between $t = 0$ s and $t = 4$ s is just the difference in position at those two times

$$\delta y = y(4) - y(0) = 7.153 - 2 = 5.15 \text{ ft} \qquad \text{Ans.}$$

d. The distance traveled between $t = 0$ s and $t = 4$ s is greater than the displacement since the particle first moved below the origin and then came back up above the origin. The point where the particle turned around is found by determining when $dy/dt = 0$ (or equivalently, where $v(t) = 0$)

$$v(t) = -5 - \frac{5}{0.7}[\cos (0.7\, t) - 1] = 0$$

which gives $t = 1.809$ s. Then

$$s = |y(1.809) - y(0)| + |y(4) - y(1.809)|$$
$$= 5.858 + 11.011 = 16.87 \text{ ft} \qquad \text{Ans.}$$

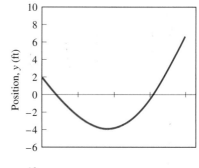

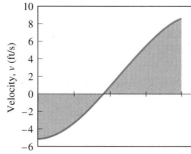

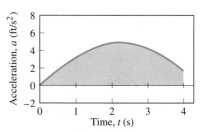

Fig. 13-5

EXAMPLE PROBLEM 13-2

A particle hanging from a spring moves with an acceleration that is proportional to its position and has the opposite sign. Suppose that $a(x) = -4x$ m/s^2 and that the velocity of the particle is 2 m/s upward when it passes through the origin.

a. Determine the velocity of the particle as a function of position.
b. If the particle is at the origin at $t = 1$ s, determine its position, velocity, and acceleration as a function of time.

SOLUTION

a. Since the acceleration is given as a function of position, the basic definition of acceleration needs to be rewritten using the chain rule

$$a(x) = \frac{dv}{dt} = \frac{dv}{dx}\frac{dx}{dt} = \frac{dv}{dx}v$$

Then the velocity is obtained by integrating this relationship

$$\int v \, dv = \int a(x) \, dx = \int (-4x) \, dx$$

which gives

$$\frac{v^2 - v_0^2}{2} = -2(x^2 - x_0^2)$$

Using the given conditions that $v = v_0 = 2$ m/s when $x = x_0 = 0$ and rearranging gives

$$v(x) = 2\sqrt{1 - x^2} \qquad \text{Ans.}$$

b. This last expression can now be integrated to get the position as a function of time. The definition gives

$$\frac{dx}{dt} = v(x) = 2\sqrt{1 - x^2} \qquad (a)$$

which can be rewritten

$$\frac{dx}{\sqrt{1 - x^2}} = 2 \, dt$$

Integration of this equation gives

$$\sin^{-1} x = 2t + \text{const} \qquad \text{or} \qquad x(t) = \sin(2t - 2) \qquad \text{Ans.}$$

where the constant of integration has been chosen to make $x = 0$ when $t = 1$ s. Substituting this expression into the given formula for the acceleration gives

$$a(t) = -4x = -4\sin(2t - 2) \qquad \text{Ans.}$$

The equation for the velocity as a function of time can be obtained either by substitution into Eq. a

$$v(x) = 2\sqrt{1 - x^2} = 2\sqrt{1 - \sin^2(2t - 2)} = 2\cos(2t - 2)$$

or by direct differentiation of the position

$$v(t) = \frac{dx}{dt} = 2\cos(2t - 2) \qquad \text{Ans.}$$

These results are graphed in Fig. 13-6.

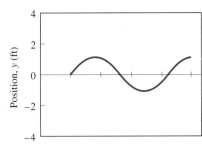
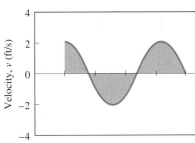
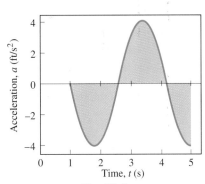

Fig. 13-6

EXAMPLE PROBLEM 13-3

The acceleration of a ball falling in air is given by

$$\frac{W}{g} a = -C_D \frac{1}{2} \rho v^2 A + W$$

where g is the acceleration of gravity ($= 32.2$ ft/s^2), W is the weight of the ball, C_D is its drag coefficient, A is its cross-sectional area ($= \pi r^2$), v is its velocity, and ρ is the density of air. Given that $W = 1$ lb, $r = 2.5$ in., $C_D = 1.0$, $\rho = 0.0023$ slugs/ft^3, and that the ball starts from rest, determine the velocity of the ball as a function of its height.

SOLUTION

Since the acceleration is given as a function of velocity,

$$a(v) = 32.2 - 0.005049v^2$$

the basic definition of acceleration needs to be rewritten using the chain rule

$$a(v) = \frac{dv}{dt} = \frac{dv}{dx}\frac{dx}{dt} = \frac{dv}{dx} v$$

This relationship can be rearranged and integrated

$$\int_0^v \frac{v \, dv}{32.2 - 0.005049v^2} = \int_0^y d\eta$$

to get

$$-\frac{1}{0.01010} \left[\ln (32.2 - 0.005049v^2) - \ln (32.2)\right] = y - 0$$

or

$$v = \left[\frac{32.2 \, (1 - e^{-0.0101y})}{0.005049}\right]^{1/2} \qquad \text{Ans.}$$

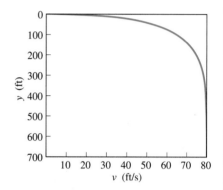

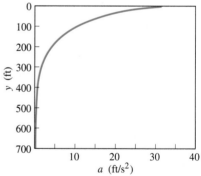

Fig. 13-7

where y and v are both measured positive downwards. This result is drawn in Fig. 13-7.

(Note that as the ball falls, the exponential term gets very small and the velocity approaches a constant value 79.9 ft/s. This is called the terminal velocity of the body.)

PROBLEMS

13-1–13-6 The position of a particle moving along the x-axis is given as a function of time. In each problem:

a. Compute the velocity of the particle as a function of time.
b. Compute the acceleration of the particle as a function of time.
c. Evaluate the position, velocity, and acceleration of the particle at $t = 5$ s.
d. Determine the total distance traveled by the particle between $t = 0$ and $t = 5$ s.
e. Sketch $x(t)$, $v(t)$, and $a(t)$; $0 \le t \le 10$ s.

13-1* $x(t) = 5t^2 - 8t + 6$ ft

13-2* $x(t) = 15 - 4t$ m

13-3 $x(t) = 3\,e^{-t/3}$ ft

13-4* $x(t) = 4 \sin t$ m

13-5 $x(t) = 12\,e^{-t/4} \sin 2t$ ft

13-6 $x(t) = 6t \sin 3t + 5 \cos 2t$ m

13-7–13-12 The velocity of a particle moving along the x-axis is given as a function of time; the position of the particle is given at some instant. In each problem:

a. Compute the position of the particle as a function of time.
b. Compute the acceleration of the particle as a function of time.
c. Evaluate the position, velocity, and acceleration of the particle at $t = 8$ s.
d. Determine the total distance traveled by the particle between $t = 5$ and $t = 8$ s.
e. Sketch $x(t)$, $v(t)$, and $a(t)$, $0 \le t \le 10$ s.

13-7* $v(t) = 10 - 16t$ ft/s
 $x(0) = 10$ ft

13-8* $v(t) = 8t^2 - 20$ m/s
 $x(20) = 60$ m

13-9 $v(t) = 3\,e^{-t/3}$ ft/s
 $x(3) = 20$ ft

13-10 $v(t) = 40 \cos 8t$ m/s
 $x(12) = 3$ m

13-11* $v(t) = 6t \sin 3t$ ft/s
 $x(6) = 10$ ft

13-12 $v(t) = 12\,e^{-t/4} \sin 3t$ m/s
 $x(0) = 0$ m

13-13–13-18 The acceleration of a particle moving along the x-axis is given as a function of time; the position and velocity of the particle are given at some instant. In each problem:

a. Compute the position of the particle as a function of time.
b. Compute the velocity of the particle as a function of time.
c. Evaluate the position, velocity, and acceleration of the particle at $t = 3$ s.
d. Determine the total distance traveled by the particle between $t = 3$ and $t = 8$ s.
e. Sketch $x(t)$, $v(t)$, and $a(t)$, $0 \le t \le 10$ s.

13-13* $a(t) = 5 - 3t$ ft/s^2
 $x(0) = 5$ ft $v(0) = 0$ ft/s

13-14* $a(t) = -9.81$ m/s^2
 $x(2) = 6$ m $v(2) = 12$ m/s

13-15 $a(t) = 12\,e^{-t/6}$ ft/s^2
 $x(8) = 60$ ft $v(8) = -5$ ft/s

13-16* $a(t) = 20 \sin 2t$ m/s^2
 $x(10) = 0$ m $v(10) = 5$ m/s

13-17 $a(t) = 6t \sin 3t$ ft/s^2
 $x(0) = 20$ ft $v(0) = 10$ ft/s

13-18 $a(t) = 24\,e^{-t/6} \sin 2t$ m/s^2
 $x(3) = 0$ m $v(3) = 0$ m/s

13-19* A ball hanging from the end of an elastic cord has an acceleration proportional to its position but of opposite sign

$$a(y) = -3y \text{ ft/s}^2$$

Determine the velocity of the ball when $y = 1$ ft if the ball is released from rest when $y = -2$ ft.

Fig. P13-19

13-20 A cart attached to a spring moves with an acceleration proportional to its position but of opposite sign

$$a(x) = -2x \text{ m/s}^2$$

Determine the velocity of the cart when $x = 3$ m if the cart has a velocity $v = 5$ m/s when $x = 0$ m.

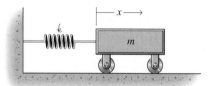

Fig. P13-20

13-21 The ball in Problem 13-19 passes through the point $y = 1$ ft with a positive velocity when $t = 5$ s. Determine the position, velocity, and acceleration of the ball as functions of time.

13-22* The cart in Problem 13-20 passes through the point $x = 3$ m with a positive velocity when $t = 3$ s. Determine the position, velocity, and acceleration of the cart as functions of time.

13-23 A ball is suspended between two elastic bands that are both stretched near their elastic limit. The acceleration in this case is not linear but is given by

$$a(x) = -3x - 5x^3 \text{ ft/s}^2$$

Determine the maximum velocity of the ball if it has a velocity $v = -4$ ft/s when $x = 1$ ft.

13-24* A cart is attached between two springs whose coils are very close together. The acceleration in this case is given by

$$a(x) = -x - 3x^2 \text{ m/s}^2$$

Determine the maximum position of the cart if it has a velocity $v = 2$ m/s when $x = -1$ m.

13-25* The acceleration of a rocket ship launched straight upward is given by (after the engines have stopped)

$$a = -g_0\frac{R^2}{(R + h)^2}$$

where g_0 is the acceleration of gravity at the surface of the earth (32.2 ft/s²), R is the radius of the earth (3960 mi), and h is the height of the rocket above the earth. Determine the maximum height attained by the rocket if the engines shut off at $h = 20$ mi and the velocity of the rocket at that point is 12,000 mi/h.

13-26 The acceleration of a rocket ship launched straight upward is given by (after the engines have stopped)

$$a = -g_0\frac{R^2}{(R + h)^2}$$

where g_0 is the acceleration of gravity at the surface of the earth (9.81 m/s²), R is the radius of the earth (6370 km), and h is the height of the rocket above the earth. Determine the escape velocity (the velocity necessary at engine burnout, $h = 30$ km, such that the maximum height of the rocket tends to infinity).

13-27* A spherical ball falling through air has an acceleration

$$a(v) = 32.2 - 0.001\ v^2 \text{ ft/s}^2$$

where the velocity is in feet per second and the positive direction is downward. Determine the velocity of the ball as a function of height if the ball has a downward velocity of 10 ft/s when $y = 0$. Also determine the terminal velocity of the ball.

Fig. P13-27

13-28 A spherical ball thrown upward through air has an acceleration

$$a(v) = -9.81 - 0.003\ v^2 \text{ m/s}^2$$

where the velocity is in meters per second and the positive direction is upward. Determine the velocity of the ball as a function of height if the ball is thrown upward with an initial velocity of 30 m/s. Also determine the maximum height reached by the ball.

13-29 Because the drag on objects moving through air increases as the square of the velocity, the acceleration of a bicyclist coasting down a slight hill is

$$a(v) = 0.4 - 0.0002v^2 \text{ ft/s}^2$$

where the velocity is in feet per second. Determine the velocity of the bicyclist as a function of distance if the velocity is zero when $x = 0$. Also determine the maximum velocity that the cyclist attains.

Fig. P13-29

13-30* A hockey puck sliding on a thin film of water on a horizontal surface has a deceleration directly proportional to its speed

$$a(v) = -0.50v \text{ m/s}^2 \qquad v > 0$$

where the velocity is in meters per second. If the puck has a velocity of 15 m/s when $x = 0$, determine its velocity as a function of distance and calculate the velocity of the puck when $x = 20$ m.

Fig. P13-30

13-31 For the bicyclist of Problem 13-29, determine the velocity as a function of time and calculate how long it takes the bicyclist to reach a velocity of 20 mi/h. (Assume that $t = 0$ when $x = 0$).

13-32* For the hockey puck of Problem 13-30, determine the velocity and position of the puck as functions of time. Also, calculate how long it takes the puck to slow to 0.1 m/s and its position at this time. (Let $t = 0$ when $x = 0$.)

13-33* The on-ramp to a freeway is 1200 ft long. A car starts up the ramp from zero speed. Determine the minimum acceleration the car must have to merge smoothly with traffic moving at 60 mi/h at the end of the ramp.

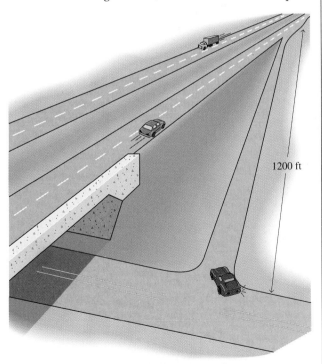

1200 ft

Fig. P13-33

13-34 A car traveling 100 km/h exits from a freeway into a rest area. Determine the minimum deceleration the car must have in order to slow to 15 km/h by the end of the exit ramp, which is only 300 m long.

13-35* A car has a maximum constant acceleration of 10 ft/s² and a maximum constant deceleration of 15 ft/s². Determine the minimum amount of time it would take to drive one mile assuming the car starts and ends at rest and never exceeds the speed limit (55 mi/h).

13-36 A small, electric car has a maximum constant acceleration of 1 m/s², a maximum constant deceleration of 2 m/s², and a top speed of 80 km/h. Determine the amount of time it would take to drive this car one kilometer—starting from and finishing at rest.

Fig. P13-36

13-37 Neglecting air resistance, a ball thrown upward in air has a downward acceleration of 32.2 ft/s². Determine the maximum initial velocity for which the height of the ball does not exceed 60 ft.

13-38* Neglecting air resistance, a sandbag dropped from a hot air balloon has a downward acceleration of 9.81 m/s². Determine the maximum height from which the sandbag can be dropped such that its velocity just before hitting the ground does not exceed 35 km/h.

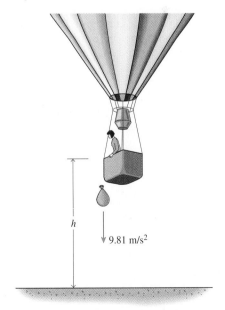

h

$\downarrow$ 9.81 m/s²

Fig. P13-38

13-39–13-40 Given the graphs of position versus time, construct the corresponding graphs of velocity versus time and acceleration versus time.

13-39

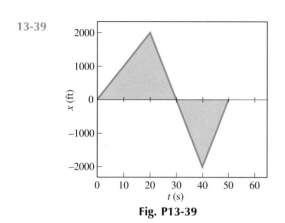

Fig. P13-39

13-40

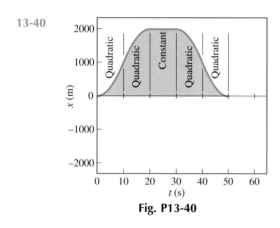

Fig. P13-40

13-41–13-44 Given the graphs of velocity versus time and the initial positions, construct the corresponding graphs of position versus time and acceleration versus time.

13-41 $x(0) = 0$ ft

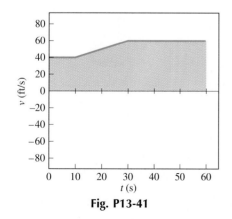

Fig. P13-41

13-42 $x(0) = 0$ m

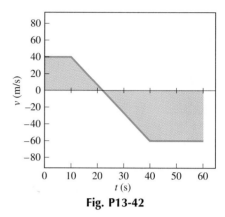

Fig. P13-42

13-43 $x(0) = 0$ ft

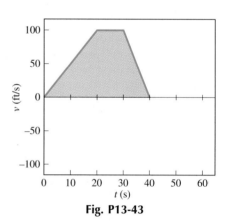

Fig. P13-43

13-44* $x(0) = 0$ m

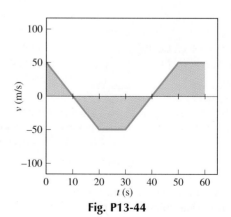

Fig. P13-44

13-45–13-48 Given the graphs of acceleration versus time, the initial positions, and the initial velocities, construct the corresponding graphs of position versus time, and velocity versus time.

13-45* $x(0) = 0$ ft; $v(0) = 0$ ft/s

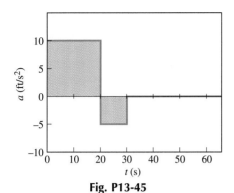

Fig. P13-45

13-46 $x(0) = 0$ m; $v(0) = 0$ m/s

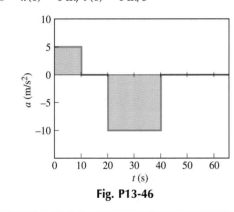

Fig. P13-46

13-47 $x(0) = 0$ ft; $v(0) = 25$ ft/s

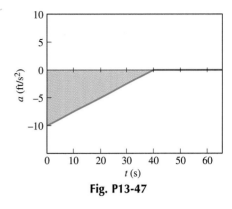

Fig. P13-47

13-48 $x(0) = 0$ m; $v(20) = 20$ m/s

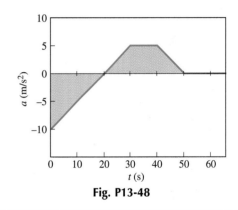

Fig. P13-48

13-4 RELATIVE MOTION ALONG A LINE

When two or more particles move in rectilinear motion, separate equations may be written to describe their motion. The particles may be moving along the same line or along separate lines. If the n particles are described by their various n coordinates but only m of the coordinates may be changed independently, then the system has m **degrees of freedom** (DOF). If $m = n$, then each particle can move independently of the others and the particles are said to be in **independent relative motion.** If $m < n$, then the motion of one or more of the particles is completely determined by the motion of the other particles and the particles are said to be in **dependent relative motion.**

Whether the particles are in independent or dependent relative motion, the motion of any particular particle can be written relative to the motion of one or more of the other particles. The need for relative description of motion arises often in engineering. For example, in structural applications, it is the relative position of two particles rather

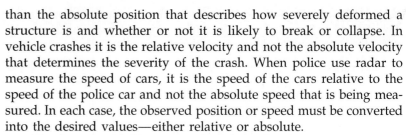

than the absolute position that describes how severely deformed a structure is and whether or not it is likely to break or collapse. In vehicle crashes it is the relative velocity and not the absolute velocity that determines the severity of the crash. When police use radar to measure the speed of cars, it is the speed of the cars relative to the speed of the police car and not the absolute speed that is being measured. In each case, the observed position or speed must be converted into the desired values—either relative or absolute.

The first part of this section describes relative motion in general and independent relative motion in particular. The second part applies the principles of relative motion to the case of dependent relative motion.

13-4-1 Independent Relative Motion

Let A and B be two particles moving along the same straight line as shown in Fig. 13-8. The positions x_A and x_B are both measured relative to the fixed origin O and are called the **absolute positions** of the particles. The position of particle B as measured from the moving particle A is denoted $x_{B/A}$ and is called the **relative position** of B measured with respect to A or more simply just the position of B relative to A. These positions are related simply by

$$x_B = x_A + x_{B/A} \tag{13-21}$$

(see Fig. 13-8). Then differentiating Eq. 13-21 with respect to time gives

$$v_B = v_A + v_{B/A} \tag{13-22}$$
$$a_B = a_A + a_{B/A} \tag{13-23}$$

That is, the velocity of particle B measured relative to particle A is the difference in the absolute velocities (velocities measured relative to a fixed coordinate system) of particles A and B. Similarly, the acceleration of particle B measured relative to particle A is the difference in the absolute accelerations of particles A and B.

Fig. 13-8

13-4-2 Dependent Relative Motion

In many engineering situations, two particles are not able to move independently. The motion of one particle will depend somehow on the motion of the other. A common dependency, or constraint, is that the particles are connected by a cord of fixed length (Fig. 13-9). In this case, an equation representing the constraint replaces Eq. 13-21.

Although both particles are moving in rectilinear motion, they need not be moving along the same straight line. Both particles must be measured relative to a fixed origin, but it is often convenient to use a different origin for each particle. Even when the two particles are moving along the same straight line and are measured relative to the same fixed origin, however, it is often convenient to set the positive direction for each particle separately.

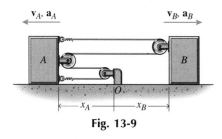

Fig. 13-9

A constraint equation is then written in terms of the coordinates of the individual particles. This constraint equation is differentiated to get the relationship between the absolute velocities and accelerations of the particles. Care must be taken to interpret the directions of positive velocity and positive acceleration in accordance with the assumed positive coordinate directions.

EXAMPLE PROBLEM 13-4

Two race cars start from rest at the same position. The acceleration of car A is

$$a_A = 15e^{-t/10} \text{ m/s}^2$$

while the acceleration of car B is

$$a_B = 10e^{-t/20} \text{ m/s}^2$$

Determine the distance at which car B overtakes car A and their relative velocity at that point.

SOLUTION

Integrating the given accelerations gives the velocities of the two cars

$$v_A = 150 \, (1 - e^{-t/10}) \qquad\qquad (a)$$
$$v_B = 200 \, (1 - e^{-t/20}) \qquad\qquad (b)$$

Integrating a second time gives the positions

$$x_A = 150 \, [t + 10 \, (e^{-t/10} - 1)] \qquad\qquad (c)$$
$$x_B = 200 \, [t + 20 \, (e^{-t/20} - 1)] \qquad\qquad (d)$$

The time at which car B overtakes car A is determined by setting Eqs. c and d equal

$$150 \, [t + 10 \, (e^{-t/10} - 1)] = 200 \, [t + 20 \, (e^{-t/20} - 1)]$$

or

$$50t + 4000e^{-t/20} - 1500e^{-t/10} - 2500 = 0$$

Solving this equation (using the Newton-Raphson method; see Appendix C) gives $t = 39.45$ s.

Now substituting this time in the position equations—Eqs. c and d—gives the position at which car B overtakes car A as

$$x_A = x_B = 4447 \text{ m} \cong 4450 \text{ m} \qquad\qquad \text{Ans.}$$

The velocity of the cars at the time car B overtakes car A is found by substituting the time into the velocity equations—Eqs. a and b—which gives

$$v_A = 147.1 \text{ m/s} \qquad v_B = 172.2 \text{ m/s}$$

The relative velocity is then

$$v_{B/A} = v_B - v_A = 25.1 \text{ m/s} \qquad\qquad \text{Ans.}$$

EXAMPLE PROBLEM 13-5

If particle A in Fig. 13-9 moves to the left with a speed of 20 ft/s, determine the speed of particle B. Also, if the speed of particle A decreases at a rate of 3 ft/s^2, determine the acceleration of particle B.

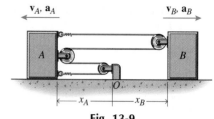

Fig. 13-9

SOLUTION

The positions of the two particles are related by the length of the rope, which is a constant

$$s = 4x_A + 2x_B + C \qquad (e)$$

where C is a constant to account for the length of rope that is wound around the pulleys (which is constant) and the distance between the centers of the pulleys and the particles (which is also constant). Differentiating Eq. e with respect to time gives

$$0 = 4v_A + 2v_B$$

or

$$v_B = -2v_A$$

where v_B is positive to the right (the same direction as x_B was measured) and v_A is positive to the left (the same direction as x_A was measured). Therefore

$$v_B = -40 \text{ ft/s} = 40 \text{ ft/s} \leftarrow \qquad \text{Ans.}$$

Differentiating Eq. e one more time gives

$$a_B = -2a_A = -2(-3) = 6 \text{ ft/s}^2 \rightarrow \qquad \text{Ans.}$$

EXAMPLE PROBLEM 13-6

If particle A in Fig. 13-10 moves to the left with a speed of 4 m/s, determine the motion of particle B.

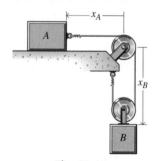

Fig. 13-10

SOLUTION

Measuring the positions of the particles from the center of the upper pulley as shown, the length of the rope is

$$s = x_A + 2x_B + C \qquad (f)$$

Differentiating Eq. f with respect to time gives

$$0 = v_A + 2v_B$$

or

$$v_B = -\frac{1}{2}v_A$$

Therefore

$$v_B = -2 \text{ m/s} = 2 \text{ m/s}\uparrow \qquad \text{Ans.}$$

(Note that the two particles do not have to move along the same line so long as they each move in rectilinear motion along some line and the constraint can be expressed in terms of their positions along the lines.)

PROBLEMS

13-49* Train A is traveling eastward at 80 mi/h while train B is traveling westward at 60 mi/h. Determine:

a. The velocity of train A relative to train B.
b. The velocity of train B relative to train A.

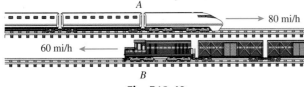

Fig. P13-49

13-50* Boat A travels down a straight river at 20 m/s while boat B travels up the river at 15 m/s. Determine:

a. The velocity of boat A relative to boat B.
b. The velocity of boat B relative to boat A.

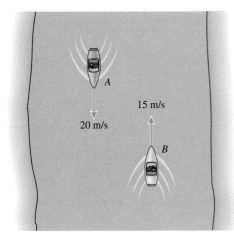

Fig. P13-50

13-51 A straight river flows at a speed of 5 mi/h. Boat A travels down river at 30 mi/h relative to the current while boat B travels up the river at 30 mi/h relative to the current. Determine:

a. The velocity of boat A relative to a stationary observer on the bank.
b. The velocity of boat B relative to boat A.

13-52* The jet stream flows from west to east at 50 m/s. Airplane A is flying from west to east at an indicated air speed (relative to the jet stream) of 150 m/s. Airplane B is flying from east to west at an indicated air speed of 150 m/s. Determine:

a. The actual speed (relative to the ground) of plane B.
b. The velocity of plane B relative to plane A.

13-53 The boats of Problem 13-51 are traveling between two towns 30 miles apart. Determine how long it will take:

a. For boat A to make the trip.
b. For boat B to make the trip.

13-54 The airplanes of Problem 13-52 are traveling between two cities 800 km apart. Determine how long it will take:

a. For plane A to make the trip.
b. For plane B to make the trip.

13-55* The boats of Problem 13-53 both start from their respective towns at 12:00 noon. Determine when and where the boats will meet.

13-56 The airplanes of Problem 13-54 both start from their respective cities at 8:00 A.M. Determine when and where the airplanes will meet.

13-57* A barge breaks away from its mooring and floats down river on the current at 10 ft/s. A tugboat goes after the runaway barge at a speed of 15 ft/s relative to the current. If the tugboat starts at a distance of 1500 ft behind the barge, determine the time it will take the tugboat to catch the barge and the total distance traveled by the tugboat in that time.

13-58* Spheres falling through still water fall at a steady speed, which is inversely proportional to their diameters. Sphere A is falling at a speed of 5 m/s. Sphere B is half as big and is falling at 10 m/s. If at some instant of time sphere A is ahead of sphere B by 20 m, determine the amount of time it will take sphere B to overtake sphere A and the total distance traveled by sphere B in that time.

13-59 Two cars travel between towns 50 miles apart. Both cars start at the same time but the first car travels at 50 mi/h while the second car travels at 30 mi/h. If the first car stops in the second town for 5 min and then returns (still at 50 mi/h), determine where the two cars will meet.

13-60* Two bicyclists start riding toward each other at 1:00 P.M. from towns 20 km apart. The first cyclist is riding with the wind and maintains a speed of 7 m/s. The second cyclist is riding against the wind at a speed of 5 m/s and stops to rest for 5 min every 4 km. When and where will the cyclists meet.

13-61 Two cars are separated by 60 ft and traveling in the same direction at 50 mi/h when the front car suddenly begins to brake at 12 ft/s^2. One second later the driver of the back car begins to brake at 15 ft/s^2. Determine the separation distance between the cars when they are both stopped.

13-62 Two cars are traveling in the same direction at 80 km/h when car A (the front car) suddenly begins to brake at 4 m/s². If the reaction time of the driver of car B (the back car) is 1 s and car B also brakes at 4 m/s², determine the safe following distance (the distance between the two cars such that car B will stop before hitting car A).

13-63* A motorcycle is stopped by the side of the road when a car passes at 50 mi/h. Twenty seconds later the motorcycle starts chasing the car. Assume that the motorcycle accelerates at 8 ft/s² until it reaches 60 mi/h and then travels at a constant speed. Find the amount of time it will take the motorcycle to overtake the car and the total distance traveled by the motorcycle in that time.

13-64 Two fighter airplanes are flying in the same direction at 1100 km/h and are separated by 3 km when the back airplane fires a missile at the front airplane. Determine:

a. The constant acceleration that the missile must have to catch the front airplane in 5 s.
b. The relative velocity of the missile to the front airplane at the time of impact.

13-65* A railroad car is loose on a siding and rolling at a constant speed of 8 mi/h. A switch engine dispatched to catch the runaway car has maximum acceleration of 3 ft/s², maximum deceleration of 5 ft/s², and maximum speed of 45 mi/h. Determine the minimum distance required to catch the runaway car. (Assume that the switch engine starts from rest when the runaway car is 500 ft down the track and that the relative velocity when the engine catches the car must be less than 3 mi/h.)

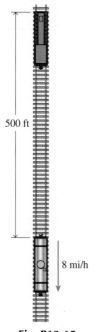

Fig. P13-65

13-66* In Fig. P13-66 block A is moving to the left with a speed of 1 m/s, its speed is decreasing at the rate of 0.5 m/s², and block C is stationary. Determine the velocity and acceleration of block B, the velocity of B relative to A, and the acceleration of B relative to A.

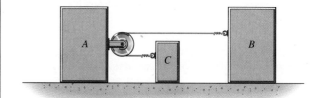

Fig. P13-66

13-67 In Fig. P13-67 the elevator E is moving downward at a speed of 3 ft/s and its speed is increasing at the rate of 0.3 ft/s². Determine the velocity and acceleration of the counterweight C, the velocity of C relative to E, and the acceleration of C relative to E.

Fig. P13-67

13-68* In Fig. P13-68 the elevator E is moving upward at a speed of 2 m/s and its speed is decreasing at the rate of 0.2 m/s². Determine the velocity and acceleration of the

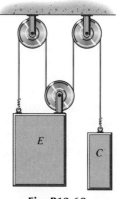

Fig. P13-68

counterweight C, the velocity of C relative to E, and the acceleration of C relative to E.

13-69 In Fig. P13-69 block B moves to the right with a speed of 10 ft/s, its speed is decreasing at the rate of 1 ft/s², and block C is stationary. Determine the velocity and acceleration of block A, the velocity of A relative to B, and the acceleration of A relative to B.

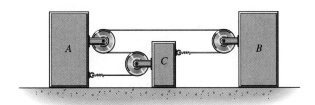

Fig. P13-69

13-70 In Fig. P13-70 block B moves to the right with a speed of 2 m/s, its speed is increasing at the rate of 0.3 m/s², and block C is stationary. Determine the speed and acceleration of block A, the velocity of B relative to A, and the acceleration of B relative to A.

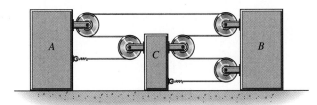

Fig. P13-70

13-71* In Fig. P13-71 the winch W reels in cable at the constant rate of 5 ft/s. If the block that the winch is mounted on is stationary, determine the velocity of block A.

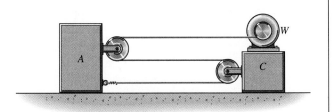

Fig. P13-71

13-72 In Fig. P13-72 the winch W is drawing in cable at a constant rate of 2 m/s. Determine the velocity of the counterweight C relative to the elevator E.

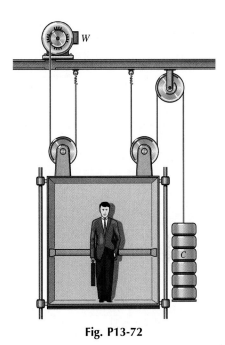

Fig. P13-72

13-73* In Fig. P13-73 block A is moving to the left with a speed of 3 ft/s and its speed is increasing at the rate of 0.6 ft/s². Determine the velocity and acceleration of block B.

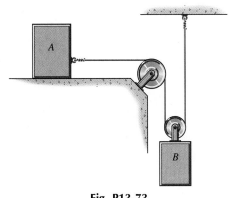

Fig. P13-73

13-74* In Fig. P13-74 block A is moving to the right with a speed of 5 m/s and its speed is decreasing at the rate of 0.2 m/s². Determine the velocity and acceleration of block B.

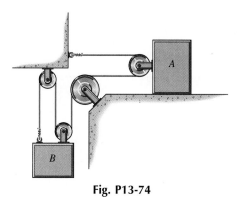

Fig. P13-74

13-75 In Fig. P13-75 block B is moving downward with a speed of 5 ft/s and its speed is decreasing at the rate of 0.2 ft/s². Determine the velocity and acceleration of block A.

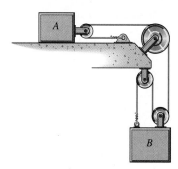

Fig. P13-75

13-76* Repeat Problem 13-66 for the case where block C is moving to the right with a speed of 2 m/s and the speed of C is decreasing at the rate of 0.2 m/s². Also determine the velocity of B relative to C and the acceleration of B relative to C.

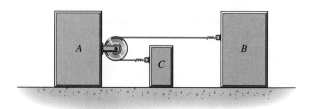

Fig. P13-66

13-77 Repeat Problem 13-69 for the case where block C is moving to the right with a speed of 2 ft/s and the speed of C is increasing at the rate of 0.5 ft/s². Also determine the velocity of A relative to C and the acceleration of A relative to C.

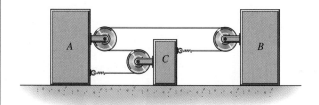

Fig. P13-69

13-78 Repeat Problem 13-70 for the case where block C is moving to the left with a speed of 1 m/s and the speed of C is increasing at the rate of 0.5 m/s². Also determine the velocity of B relative to C and the acceleration of B relative to C.

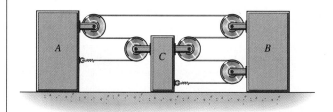

Fig. P13-70

13-79* Repeat Problem 13-71 for the case where block C is moving to the left with a speed of 1 ft/s and the speed of C is increasing at the rate of 0.5 ft/s². Also determine the velocity of A relative to C and the acceleration of A relative to C.

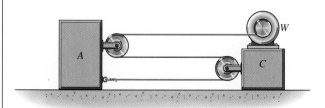

Fig. P13-71

13-5 PLANE CURVILINEAR MOTION

When motion occurs in a single plane, two coordinates are required to describe the motion. The choice of which coordinates to use in a particular problem will depend on the geometry of the problem, on the way in which data are given for the problem, and on the type of solution that is desired. Three of the more commonly used coordinate systems used to represent the motion are rectangular coordinates, polar coordinates, and normal/tangential coordinates. These will be discussed in turn in the next three sections.

13-5-1 Rectangular Coordinates

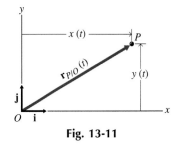

Fig. 13-11

In a rectangular coordinate system (in a plane), the position of a particle is described by giving its distance from two, fixed, orthogonal lines (Fig. 13-11). These two lines are called the x- and y-axes and the coordinates are called the x- and y-components of the position. Unit vectors along the x- and y-axes will be denoted by $\mathbf{i}$ and $\mathbf{j}$, respectively. Although the x- and y-coordinate directions ($\mathbf{i}$ and $\mathbf{j}$) need not be horizontal and vertical, once they are chosen, they must remained fixed.

The position of a particle P with respect to the origin O of the fixed coordinate system is given by (Fig. 13-11)

$$\mathbf{r}_{P/O}(t) = x(t)\mathbf{i} + y(t)\mathbf{j} \tag{13-24}$$

where $x(t)$ is the (time dependent) x-component of the position and $y(t)$ is the (time dependent) y-component of the position. The displacement of the particle between times t_1 and $t_2 > t_1$ is (Fig. 13-12)

$$\begin{aligned} \delta\mathbf{r} &= \mathbf{r}_{P/O}(t_2) - \mathbf{r}_{P/O}(t_1) \\ &= [x(t_2) - x(t_1)]\mathbf{i} + [y(t_2) - y(t_1)]\mathbf{j} \end{aligned}$$

Since the directions as well as the magnitudes of the unit vectors $\mathbf{i}$ and $\mathbf{j}$ are fixed, their derivatives are zero. Then the velocity and acceleration of the particle are

$$\begin{aligned} \mathbf{v}_P(t) &= v_x(t)\mathbf{i} + v_y(t)\mathbf{j} \\ &= \dot{\mathbf{r}}_{P/O}(t) = \dot{x}(t)\mathbf{i} + \dot{y}(t)\mathbf{j} \end{aligned} \tag{13-25}$$

and

$$\begin{aligned} \mathbf{a}_P(t) &= a_x(t)\mathbf{i} + a_y(t)\mathbf{j} \\ &= \dot{\mathbf{v}}(t) = \dot{v}_x(t)\mathbf{i} + \dot{v}_y(t)\mathbf{j} \\ &= \ddot{\mathbf{x}}(t) = \ddot{x}(t)\mathbf{i} + \ddot{y}(t)\mathbf{j} \end{aligned} \tag{13-26}$$

respectively.

The rectangular coordinate system is usually the most convenient one to use when the x- and y-components of the motion are specified

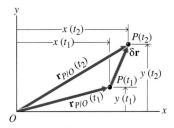

Fig. 13-12

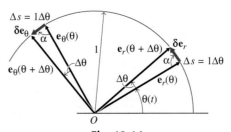

Fig. 13-13

separately from each other, do not depend on each other, or both. Typical examples are motion plotted on map grids (where x may be the longitude and y may be the latitude of the particle) and trajectory motion (where x may be the distance measured along the ground and y may be the height of the particle above the ground).

13-5-2   Polar Coordinates (Radial/Transverse Coordinates)

In a polar coordinate system, the position of a particle is described by giving the distance from a fixed point and the angular displacement relative to a fixed line (Fig. 13-13). The coordinate directions ($\mathbf{e}_r$ and $\mathbf{e}_\theta$) are taken to be radially outward from the fixed point and perpendicular to the radial line in the direction of increasing θ.

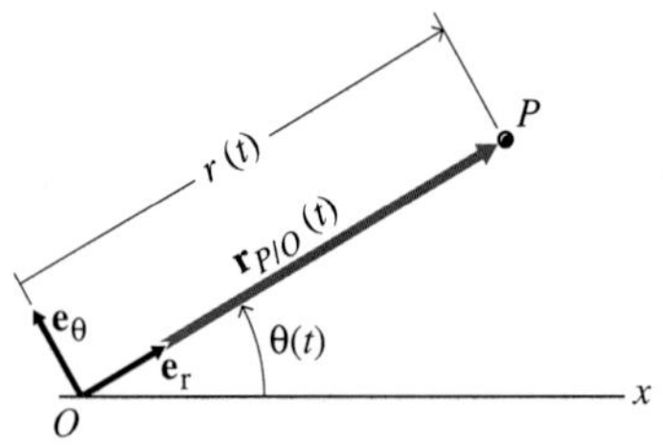

In polar coordinates, the position of particle P with respect to the origin O is given by (Fig. 13-13)

$$\mathbf{r}_{P/O}(t) = r(t)\mathbf{e}_r \tag{13-27}$$

where $r(t)$ is the (time dependent) r-component of the position. The dependence of the position vector on the angle $\theta(t)$ is hidden in the unit vector $\mathbf{e}_r$, which depends on θ (which may depend on time).

Since the directions of the unit vectors $\mathbf{e}_r$ and $\mathbf{e}_\theta$ are not necessarily fixed, their changes must be considered when the position vector, Eq. 13-27, is differentiated. The derivative of $\mathbf{e}_r$ with respect to time is calculated using the chain rule of differentiation

$$\dot{\mathbf{e}}_r = \frac{d\mathbf{e}_r}{dt} = \frac{d\mathbf{e}_r}{d\theta}\frac{d\theta}{dt} \tag{13-28a}$$

where

$$\frac{d\mathbf{e}_r}{d\theta} = \lim_{\Delta\theta\to 0}\frac{\mathbf{e}_r(\theta+\Delta\theta)-\mathbf{e}_r(\theta)}{\Delta\theta} \tag{13-28b}$$

But in the limit as $\Delta\theta\to 0$, the distance $|\mathbf{e}_r(\theta+\Delta\theta)-\mathbf{e}_r(\theta)|$ tends to the arc length along a unit circle $\Delta s = 1\,\Delta\theta$ and the angle α tends to 90° (see Fig. 13-14). Therefore the vector $\mathbf{e}_r(\theta+\Delta\theta)-\mathbf{e}_r(\theta)$ has magnitude $\Delta\theta$ and points in the $\mathbf{e}_\theta$-direction and

$$\dot{\mathbf{e}}_r = \dot{\theta}\lim_{\Delta\theta\to 0}\frac{\Delta\theta\,\mathbf{e}_\theta}{\Delta\theta} = \dot{\theta}\mathbf{e}_\theta \tag{13-28c}$$

where $\dot{\theta} = d\theta/dt$.

Fig. 13-14

Similarly the derivative of $\mathbf{e}_\theta$ with respect to time can be calculated

$$\dot{\mathbf{e}}_\theta = \frac{d\mathbf{e}_\theta}{dt} = \frac{d\mathbf{e}_\theta}{d\theta}\frac{d\theta}{dt} \tag{13-29a}$$

where

$$\frac{d\mathbf{e}_\theta}{d\theta} = \lim_{\Delta\theta \to 0} \frac{\mathbf{e}_\theta(\theta + \Delta\theta) - \mathbf{e}_\theta(\theta)}{\Delta\theta} \tag{13-29b}$$

But in the limit as $\Delta\theta \to 0$, the distance $|\mathbf{e}_\theta(\theta + \Delta\theta) - \mathbf{e}_\theta(\theta)|$ again tends to the arc length along a unit circle $\Delta s = 1\,\Delta\theta$ and the angle α again tends to 90° (see Fig. 13-14). Therefore the vector $\mathbf{e}_\theta(\theta + \Delta\theta) - \mathbf{e}_\theta(\theta)$ has magnitude $\Delta\theta$ and points in the negative $\mathbf{e}_r$-direction and

$$\dot{\mathbf{e}}_\theta = \dot{\theta} \lim_{\Delta\theta \to 0} \frac{-\Delta\theta\,\mathbf{e}_r}{\Delta\theta} = -\dot{\theta}\mathbf{e}_r \tag{13-29c}$$

An alternative way to evaluate the derivatives of $\mathbf{e}_r$ and $\mathbf{e}_\theta$ with respect to θ, which may be easier for the student to understand and remember, is to write $\mathbf{e}_r$ and $\mathbf{e}_\theta$ in terms of their rectangular components and then take the derivatives. With reference to Fig. 13-15,

$$\mathbf{e}_r = \cos\theta\,\mathbf{i} + \sin\theta\,\mathbf{j} \tag{13-30a}$$
$$\mathbf{e}_\theta = -\sin\theta\,\mathbf{i} + \cos\theta\,\mathbf{j} \tag{13-30b}$$

The derivatives are then

$$d\mathbf{e}_r/d\theta = -\sin\theta\,\mathbf{i} + \cos\theta\,\mathbf{j} = \mathbf{e}_\theta \tag{13-30c}$$
$$d\mathbf{e}_\theta/d\theta = -\cos\theta\,\mathbf{i} - \sin\theta\,\mathbf{j} = -\mathbf{e}_r \tag{13-30d}$$

which is the same as above.

The velocity of the particle can now be computed

$$\begin{aligned}\mathbf{v}_P(t) = v_r\mathbf{e}_r + v_\theta\mathbf{e}_\theta &= \dot{\mathbf{r}}_{P/O}(t) \\ &= \dot{r}\mathbf{e}_r + r\dot{\mathbf{e}}_r = \dot{r}\mathbf{e}_r + r\dot{\theta}\mathbf{e}_\theta \end{aligned} \tag{13-31}$$

Finally, the acceleration is computed

$$\begin{aligned}\mathbf{a}_P(t) = a_r\mathbf{e}_r + a_\theta\mathbf{e}_\theta &= \dot{\mathbf{v}}_P(t) \\ &= \ddot{r}\mathbf{e}_r + \dot{r}\dot{\mathbf{e}}_r + (\dot{r}\dot{\theta} + r\ddot{\theta})\mathbf{e}_\theta + r\dot{\theta}\dot{\mathbf{e}}_\theta \\ &= \ddot{r}\mathbf{e}_r + \dot{r}(\dot{\theta}\mathbf{e}_\theta) + (\dot{r}\dot{\theta} + r\ddot{\theta})\mathbf{e}_\theta - r\dot{\theta}(\dot{\theta}\mathbf{e}_r) \\ &= (\ddot{r} - r\dot{\theta}^2)\mathbf{e}_r + (r\ddot{\theta} + 2\dot{r}\dot{\theta})\mathbf{e}_\theta \end{aligned} \tag{13-32}$$

For the special case of a particle in circular motion, r = constant, Eqs. 13-31 and 13-32 reduce to

$$\mathbf{v}_P(t) = r\dot{\theta}\mathbf{e}_\theta \tag{13-33}$$
$$\mathbf{a}_P(t) = -r\dot{\theta}^2\mathbf{e}_r + r\ddot{\theta}\mathbf{e}_\theta \tag{13-34}$$

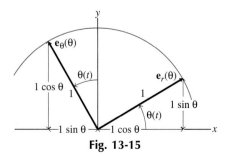

Fig. 13-15

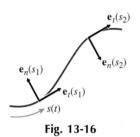

Fig. 13-16

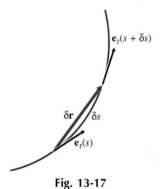

Fig. 13-17

The polar coordinate system is usually the most convenient one to use when the position of a particle is measured relative to a fixed point (such as a radar tracking an airplane in a plane) or when a particle is fixed on or moves along a rotating arm.

13-5-3   Normal and Tangential Coordinates

In some problems the motion is specified by giving the path that the particle is moving along and the speed of the particle at each point along the path. Coordinates are chosen at each point along the path with unit vectors $\mathbf{e}_t$ tangential to the path and pointing in the direction of motion and $\mathbf{e}_n$ normal to the path and pointing toward the center of curvature (Fig. 13-16).

The velocity of the particle has direction $\mathbf{e}_t$ and a magnitude equal to the rate at which the particle moves along the path. To see that this is so, draw the position of the particle at two instants of time (Fig. 13-17). For δt small, the magnitude of the displacement is nearly the same as the distance along the curve δs and the direction of the displacement tends to the direction of the unit tangent vector $\mathbf{e}_t$. The velocity is then

$$\mathbf{v}(t) = \lim_{\delta t \to 0} \frac{\delta \mathbf{r}(t)}{\delta t} = \lim_{\delta t \to 0} \frac{\delta s(t)}{\delta t}\, \mathbf{e}_t$$
$$= \dot{s}\mathbf{e}_t = v\mathbf{e}_t \tag{13-35}$$

where $v = \dot{s}$ is the magnitude of the velocity and the direction of the unit tangent vector $\mathbf{e}_t$ varies with position (which varies with time).

Since the direction of the unit vector $\mathbf{e}_t$ is not fixed, its change must also be considered when differentiating the velocity to find the acceleration. Using the chain rule of differentiation, the derivative of $\mathbf{e}_t$ with respect to time is calculated by

$$\frac{d\mathbf{e}_t}{dt} = \frac{d\mathbf{e}_t}{ds}\frac{ds}{dt} \tag{13-36a}$$

To evaluate the derivative of the unit tangent vector $\mathbf{e}_t$ with respect to s, let s be the position of the particle at time t and $s + \Delta s$ be its position at time $t + \Delta t$ (Fig. 13-18). Draw a circle having its center at the intersection of $\mathbf{e}_n(s)$ and $\mathbf{e}_n(s + \Delta s)$ that passes through the points s and

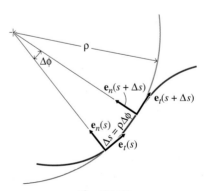

Fig. 13-18

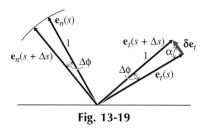

Fig. 13-19

$s + \Delta s$. The relationship between the unit tangent and unit normal vectors at s and $s + \Delta s$ is shown in Figure 13-19. Then

$$\frac{d\mathbf{e}_t}{ds} = \lim_{\Delta s \to 0} \frac{\mathbf{e}_t(s + \Delta s) - \mathbf{e}_t(s)}{\Delta s} \qquad (13\text{-}36b)$$

But in the limit as $\Delta s \to 0$, the distance $|\mathbf{e}_t(s + \Delta s) - \mathbf{e}_t(s)|$ tends to the arc length along a unit circle $1\,\Delta\phi$ and the angle α tends to $90°$. Therefore the vector $\mathbf{e}_t(s + \Delta s) - \mathbf{e}_t(s)$ has magnitude $\Delta\phi$ and points in the $\mathbf{e}_n$-direction and

$$\frac{d\mathbf{e}_t}{ds} = \lim_{\Delta s \to 0} \frac{\Delta\phi}{\Delta s}\,\mathbf{e}_n(s) \qquad (13\text{-}36c)$$

But from Fig. 13-18, $\Delta s = \rho\,\Delta\phi$ so finally

$$\dot{\mathbf{e}}_t = \dot{s}\lim_{\Delta s \to 0} \frac{\Delta s}{\rho\,\Delta s}\,\mathbf{e}_n(s) = \frac{\dot{s}}{\rho}\,\mathbf{e}_n(s) \qquad (13\text{-}36d)$$

where

$$\dot{s} = \lim_{\Delta t \to 0} \frac{\Delta s}{\Delta t} = \lim_{\Delta t \to 0} \frac{\rho\,\Delta\phi}{\Delta t} = \rho\dot{\phi} \qquad (13\text{-}36e)$$

The acceleration of a particle in normal and tangential coordinates then is given by

$$\mathbf{a}(t) = a_t\mathbf{e}_t + a_n\mathbf{e}_n = \dot{\mathbf{v}}(t) = \ddot{s}\mathbf{e}_t + \dot{s}\dot{\mathbf{e}}_t$$
$$= \ddot{s}\mathbf{e}_t + \frac{\dot{s}^2}{\rho}\,\mathbf{e}_n \qquad (13\text{-}37)$$

For the special case of a particle in circular motion, $\rho = r = \text{const}$, $\mathbf{e}_r = -\mathbf{e}_n$ (since $\mathbf{e}_r$ points outward from the center of the circle and $\mathbf{e}_n$ points toward the center of curvature), and $\mathbf{e}_t = \mathbf{e}_\theta$. Then

$$\mathbf{v}(t) = v_t\mathbf{e}_t = r\dot{\theta}\mathbf{e}_\theta \qquad (13\text{-}38)$$

and

$$\mathbf{a}(t) = \ddot{s}\mathbf{e}_\theta + \frac{\dot{s}^2}{\rho}\,(-\mathbf{e}_r) = r\ddot{\theta}\mathbf{e}_\theta - \frac{(r\dot{\theta})^2}{r}\,\mathbf{e}_r$$
$$= -r\dot{\theta}^2\mathbf{e}_r + r\ddot{\theta}\mathbf{e}_\theta \qquad (13\text{-}39)$$

But Eqs. 13-38 and 13-39 are the same as Eqs. 13-33 and 13-34, which were derived for the velocity and acceleration in polar coordinates for the case where r is a constant.

Normal and tangential coordinates are the most convenient to use when particles move along a surface of known shape. In such cases, the normal accelerations are required to determine the contact force between the particle and the surface. When the contact force becomes negative as in the design of roller coasters, special tracks must be used to keep the particle following the curve. Also, the normal contact force is often required to compute the tangential (friction) force and thereby determine the tangential acceleration and velocity.

13-5-4 Summary of Equations

For convenience, the major equations of plane curvilinear motion are summarized as follows.

Rectangular Coordinates

$$\mathbf{r}_{P/O}(t) = x(t)\mathbf{i} + y(t)\mathbf{j}$$
$$\mathbf{v}_P(t) = \dot{x}(t)\mathbf{i} + \dot{y}(t)\mathbf{j}$$
$$= v_x(t)\mathbf{i} + v_y(t)\mathbf{j}$$
$$\mathbf{a}_P(t) = \ddot{x}(t)\mathbf{i} + \ddot{y}(t)\mathbf{j}$$
$$= \dot{v}_x(t)\mathbf{i} + \dot{v}_y(t)\mathbf{j}$$
$$= a_x(t)\mathbf{i} + a_y(t)\mathbf{j}$$

Polar Coordinates

$$\mathbf{r}_{P/O}(t) = r(t)\mathbf{e}_r$$
$$\mathbf{v}_P(t) = \dot{r}\mathbf{e}_r + r\dot{\theta}\mathbf{e}_\theta$$
$$= v_r\mathbf{e}_r + v_\theta\mathbf{e}_\theta$$
$$\mathbf{a}_P(t) = (\ddot{r} - r\dot{\theta}^2)\mathbf{e}_r + (r\ddot{\theta} + 2\dot{r}\dot{\theta})\mathbf{e}_\theta$$
$$= a_r\mathbf{e}_r + a_\theta\mathbf{e}_\theta$$

Normal and Tangential Coordinates

$$\mathbf{v}_P(t) = \dot{s}(t)\mathbf{e}_t = v_t\mathbf{e}_t$$
$$\mathbf{a}_P(t) = \ddot{s}\mathbf{e}_t + \frac{\dot{s}^2}{\rho}\mathbf{e}_n$$
$$= a_t\mathbf{e}_t + a_n\mathbf{e}_n$$

When air resistance is neglected, a bullet fired through the air has a downward acceleration of 32.2 ft/s^2. If the bullet has an initial velocity of 750 ft/s with an angle of 30° above the horizontal, determine:

a. The maximum height reached by the bullet.
b. The range of the bullet (that is, where it hits the ground).

SOLUTION

a. Choose rectangular coordinates with the x-axis horizontal (positive in the direction of motion) and the y-axis vertical (positive upward). Then the acceleration of the bullet can be written

$$\mathbf{a}(t) = -32.2\mathbf{j} \text{ ft/s}^2 = \dot{v}_x(t)\mathbf{i} + \dot{v}_y(t)\mathbf{j}$$

The x- and y-components of this equation are independent of each other and can be integrated separately to get

$$v_x(t) = C_1$$
$$v_y(t) = -32.2t + C_2$$

Using the initial condition that at $t = 0$, the velocity is

$$\mathbf{v}_0 = 750 \cos 30°\mathbf{i} + 750 \sin 30°\mathbf{j} \text{ ft/s}$$
$$= 649.5\mathbf{i} + 375.0\mathbf{j} \text{ ft/s}$$

gives

$$C_1 = 649.5 \text{ ft/s} \quad \text{and} \quad C_2 = 375.0 \text{ ft/s}$$

Then, integrating the velocity

$$\mathbf{v}(t) = 649.5\mathbf{i} + (375.0 - 32.2t)\mathbf{j} \text{ ft/s}$$
$$= \dot{x}(t)\mathbf{i} + \dot{y}(t)\mathbf{j}$$

gives

$$x(t) = 649.5t \text{ ft}$$
$$y(t) = (375.0t - 16.1t^2) \text{ ft}$$

where the constants of integration are both zero since initially both x and y are zero.

Now, the maximum height occurs when the y-component of the velocity changes from positive (upward) to negative (downward). This occurs when

$$v_y(t) = 375.0 - 32.2t = 0$$

or at $t = 11.65$ s. The height at this time is

$$y(11.65) = y_{max} = 2184 \text{ ft} \qquad \text{Ans.}$$

b. To find the time when the bullet again reaches the ground, the height is set to zero

$$y(t) = (375.0t - 16.1t^2) = 0$$

which gives either $t = 0$ s or $t = 23.29$ s. The $t = 0$ s solution corresponds to the initial position while the $t = 23.29$ s solution corresponds to the bullet falling back to earth. The x-position of the particle at this time is

$$x(23.29) = x_{range} = 15130 \text{ ft} \qquad \text{Ans.}$$

When air resistance is neglected, a cannonball fired through the air has a downward acceleration of g. If the cannonball has an initial velocity of v_0 with an angle of θ above the horizontal, determine the angle θ_r that will give the maximum range.

SOLUTION

Choose rectangular coordinates with the x-axis horizontal (positive in the direction of motion) and the y-axis vertical (positive upward). Integrating the acceleration

$$\mathbf{a}(t) = -g\mathbf{j} = \dot{v}_x(t)\mathbf{i} + \dot{v}_y(t)\mathbf{j}$$

and using the given initial velocity

$$\mathbf{v}_0 = v_0 \cos\theta\,\mathbf{i} + v_0 \sin\theta\,\mathbf{j}$$

gives

$$\begin{aligned}\mathbf{v}(t) &= v_0 \cos\theta\,\mathbf{i} + (v_0 \sin\theta - gt)\mathbf{j} \\ &= \dot{x}(t)\mathbf{i} + \dot{y}(t)\mathbf{j}\end{aligned}$$

Integrating again gives

$$x(t) = v_0 t \cos\theta \qquad\qquad (a)$$

$$y(t) = \left(v_0 t \sin\theta - \frac{1}{2}gt^2\right) \qquad\qquad (b)$$

To find the time $\hat{t}_r$ when the cannonball again reaches the ground, the height is set to zero

$$y(t) = \left(v_0 \hat{t}_r \sin\theta - \frac{1}{2}g\hat{t}_r^2\right) = 0$$

which gives

$$\hat{t}_r = \frac{2v_0 \sin\theta}{g}$$

The x-position of the particle at this time is

$$\begin{aligned}x(\hat{t}_r) = x_{\text{range}} &= \frac{2v_0^2 \sin\theta \cos\theta}{g} \\ &= \frac{v_0^2 \sin 2\theta}{g} \qquad\qquad (c)\end{aligned}$$

The angle that gives the maximum x_{range} is then found by differentiating Eq. c with respect to θ and setting the derivative equal to zero

$$\frac{dx_{\text{range}}}{d\theta} = \frac{2v_0^2 \cos 2\theta_r}{g} = 0$$

which gives

$$\theta_r = 45° \qquad\qquad \text{Ans.}$$

EXAMPLE PROBLEM 13-9

A radar tracking an airplane gives the coordinates of the plane as $r(t)$ and $\theta(t)$ (Fig. 13-20). At some instant of time, $\theta = 40°$ and $r = 6400$ ft. From successive measurements of r and θ, the derivatives at this instant are estimated as $\dot{r} = 312$ ft/s, $\dot{\theta} = -0.039$ rad/s, $\ddot{r} = 9.751$ ft/s², and $\ddot{\theta} = 0.003807$ rad/s². Calculate the velocity and acceleration of the airplane at this instant.

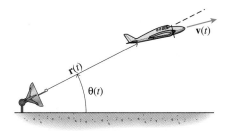

Fig. 13-20

SOLUTION

Choosing polar coordinates centered at the radar as indicated (Fig. 13-20), the radial component of velocity is

$$v_r = \dot{r} = 312 \text{ ft/s}$$

and the θ-component is

$$v_\theta = r\dot{\theta} = (6400)(-0.039) = -250 \text{ ft/s}$$

The resultant then has magnitude

$$v = \sqrt{312^2 + 250^2} = 400 \text{ ft/s} \qquad \text{Ans.}$$

and acts at an angle

$$\phi_v = \tan^{-1}\frac{250}{312} = 38.7° \qquad \text{Ans.}$$

measured clockwise from the radial direction (Fig. 13-21a).

The radial component of acceleration is

$$a_r = \ddot{r} - r\dot{\theta}^2 = (9.751) - (6400)(-0.039)^2 = 0.017 \text{ ft/s}^2$$

and the θ-component is

$$a_\theta = r\ddot{\theta} + 2\dot{r}\dot{\theta} = (6400)(0.003807) + 2(312)(-0.039)$$
$$= 0.029 \text{ ft/s}^2$$

The resultant then has magnitude

$$a = \sqrt{0.017^2 + 0.029^2} = 0.034 \text{ ft/s}^2 \qquad \text{Ans.}$$

and acts at an angle

$$\phi_a = \tan^{-1}\frac{0.029}{0.017} = 59.6° \qquad \text{Ans.}$$

measured counterclockwise from the radial direction (Fig. 13-21b).

It should be noted at this point that the numbers obtained for the acceleration terms are probably not very accurate. They resulted from subtracting two numbers that were accurate to the second decimal place at best. Therefore the answer probably has no more than one significant figure.

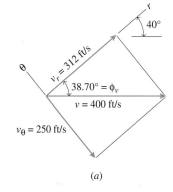

(a)

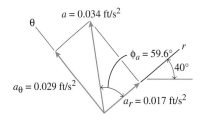

(b)

Fig. 13-21

A cam lobe has a shape given by $r = 20 + 15 \cos \theta$ mm (Fig. 13-22). Pin P slides in a slot along arm AB and is held against the cam by a spring. The arm AB rotates counterclockwise about A at a rate of 30 rev/min. Given that $\theta = 0$ at $t = 0$:

a. Determine the velocity and acceleration of the pin.
b. Evaluate the expressions of part a for the velocity and acceleration at $t = 0.75$ s.
c. Show the velocity and acceleration of part b on a suitable sketch.

   (Assume that the pin is very small so that the center of the pin follows the contour of the cam lobe.)

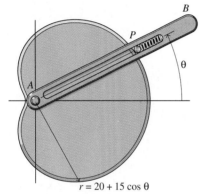

$r = 20 + 15 \cos \theta$

Fig. 13-22

SOLUTION

a. First, integrating the given angular velocity

$$\dot{\theta} = \frac{(30 \text{ rev/min})}{(60 \text{ sec/min})} (2\pi \text{ rad/rev}) = \pi \text{ rad/s}$$

gives

$$\theta = \pi t \text{ rad}$$

where the constant of integration is zero since $\theta = 0$ at $t = 0$. Also, since the rotation rate is a constant,

$$\ddot{\theta} = 0$$

   Next, differentiating the radial function

$$r = 20 + 15 \cos \theta = 20 + 15 \cos \pi t$$

gives

$$\dot{r} = -15\pi \sin \pi t$$

and

$$\ddot{r} = -15\pi^2 \cos \pi t$$

Now, the velocity components can be computed

$$v_r = \dot{r} = -15\pi \sin \pi t \qquad \text{Ans.}$$
$$v_\theta = r\dot{\theta} = (20 + 15 \cos \pi t)(\pi)$$
$$= 20\pi + 15\pi \cos \pi t \qquad \text{Ans.}$$

Similarly, the acceleration components are

$$a_r = \ddot{r} - r\dot{\theta}^2 = (-15\pi^2 \cos \pi t) - (20 + 15 \cos \pi t)(\pi)^2$$
$$= -\pi^2(20 + 30 \cos \pi t) \qquad \text{Ans.}$$
$$a_\theta = r\ddot{\theta} + 2\dot{r}\dot{\theta} = (20 + 15 \cos \pi t)(0) + 2(-15\pi \sin \pi t)(\pi)$$
$$= -30\,\pi^2 \sin \pi t \qquad \text{Ans.}$$

b. At $t = 0.75$ s, $\theta = 3\pi/4$ rad $= 135°$. The velocity and acceleration components are

$$v_r = -15\pi \sin 3\pi/4 = -33.3 \text{ mm/s} \qquad \text{Ans.}$$
$$v_\theta = 20\pi + 15\pi \cos 3\pi/4 = 29.5 \text{ mm/s} \qquad \text{Ans.}$$
$$a_r = -\pi^2(20 + 30 \cos 3\pi/4)$$
$$= 11.97 \text{ mm/s}^2 \qquad \text{Ans.}$$
$$a_\theta = -30\,\pi^2 \sin 3\pi/4 = -209.4 \text{ mm/s}^2 \qquad \text{Ans.}$$

c. The magnitude of the velocity is

$$v = \sqrt{33.3^2 + 29.5^2} = 44.5 \text{ mm/s}$$

and the direction of the velocity is

$$\phi_v = \tan^{-1}\frac{29.5}{33.3} = 41.5°$$

measured clockwise from the negative r-direction. The magnitude of the acceleration is

$$a = \sqrt{11.97^2 + 209.4^2} = 209.7 \text{ mm/s}^2$$

and the direction of the acceleration is

$$\phi_a = \tan^{-1}\frac{209.4}{11.97} = 86.7°$$

measured clockwise from the positive r-direction. These values are shown in Fig. 13-23.

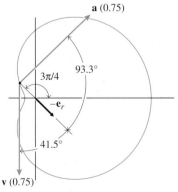

Fig. 13-23

45

EXAMPLE PROBLEM 13-11

A turn in a country highway has a radius of curvature that varies from infinity at the beginning and end of the turn to ρ_{min} at the middle of the turn. If the tires of a car going around the turn begin to slide when the normal acceleration reaches 12 ft/s², determine:

a. The maximum constant speed at which the car can go around the turn for $\rho_{min} = 500$ ft.

b. The smallest ρ_{min} for which the car can go around the turn at 60 mi/h.

SOLUTION

a. The normal component of acceleration is given by

$$a_n = v^2/\rho$$

Using $\rho = \rho_{min}$ and solving for v gives the maximum speed the car can have as

$$v = \sqrt{(\rho_{min}\, a_n)} = \sqrt{(500)(12)}$$
$$= 77.5 \text{ ft/s} \qquad\qquad \text{Ans.}$$

b. Solving for ρ and using $v = 60$ mi/h $= 88$ ft/s gives the smallest ρ_{min} that the turn can have as

$$\rho_{min} = \frac{v^2}{a_n} = \frac{88^2}{12} = 645 \text{ ft} \qquad\qquad \text{Ans.}$$

A box slides down a chute, which is bent in the shape of a hyperbola (Fig. 13-24). When the box reaches the point $x = 5$ m, it has a speed of 5 m/s and the speed is decreasing at the rate of 0.5 m/s^2. Determine the normal and tangential components of the acceleration of the box.

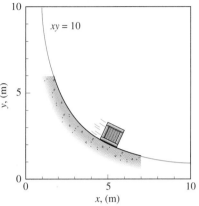

Fig. 13-24

SOLUTION

The tangential direction is found by computing the slope of the curve

$$\frac{dy}{dx} = \frac{d}{dx}\left(\frac{10}{x}\right) = -\frac{10}{x^2} = \tan\phi$$

So at $x = 5$ m

$$\phi = \tan^{-1}\left(-\frac{10}{5^2}\right) = -21.80°$$

(below the horizontal). The tangential component of the acceleration is then

$$\mathbf{a}_t = \dot{v}\,\mathbf{e}_t = -0.5 \text{ m/s}^2 \searrow 21.80° \qquad\qquad \text{Ans.}$$

The normal component of the acceleration is

$$a_n = v^2/\rho$$

where the radius of curvature is given by (see any elementary calculus book)

$$\frac{1}{\rho} = \frac{\left|\dfrac{d^2y}{dx^2}\right|}{\left[1 + \left(\dfrac{dy}{dx}\right)^2\right]^{3/2}}$$

and the absolute value is to guarantee that ρ is positive. Calculating the second derivative gives

$$\frac{d^2y}{dx^2} = \frac{20}{x^3}$$

so that at $x = 5$ m

$$\frac{1}{\rho} = \frac{\left|\dfrac{20}{5^3}\right|}{\left[1 + \left(-\dfrac{10}{5^2}\right)^2\right]^{3/2}} = 0.1281 \text{ m}^{-1}$$

Finally, the normal component of the acceleration is

$$\mathbf{a}_n = (5)^2(0.1281)\,\mathbf{e}_n$$
$$= 3.20 \text{ m/s}^2 \measuredangle 68.2° \qquad\qquad \text{Ans.}$$

PROBLEMS

13-80* An airplane flying horizontally at 300 km/h drops a bomb from an altitude of 2 km. The acceleration of the bomb is 9.81 m/s², downward. Determine the horizontal distance traveled by the bomb before it hits the ground.

13-81* A cannon firing at a hilltop target has a muzzle velocity of 600 ft/s. If the acceleration of the cannonball is 32.2 ft/s², downward, and the horizontal and vertical distances to the target are 0.5 mi and 0.25 mi, respectively, determine the angle at which the cannon should be fired.

13-82 Pin P of Fig. P13-82 slides in the horizontal and vertical grooves attached to collars A and B. Collar A slides in a horizontal plane with a position given by $x(t) = 10 \cos 3t$ mm while collar B slides in a vertical plane with a position given by $y(t) = 10 \sin 4t$ mm.

a. Compute the velocity of the pin, $\mathbf{v}_p(t)$.
b. Compute the acceleration of the pin, $\mathbf{a}_p(t)$.
c. Sketch the position of the pin for $0 < t < 2$ s.
d. Evaluate the velocity $\mathbf{v}_p(t)$ and the acceleration $\mathbf{a}_p(t)$ at $t = 0.5$ s and show them on the sketch of part c.

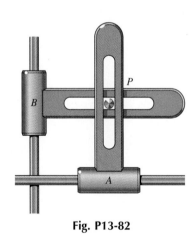

Fig. P13-82

13-83 Repeat Problem 13-82 for the case where the position of collar A is given by $x(t) = e^{-t}/5$ ft and collar B by $y(t) = \sin 2t$ ft.

13-84* A 10-mm diameter marble rolls off a horizontal step 4 m (=h) high (Fig. P13-84). Determine the minimum and maximum speed v_0 the marble can have if it is to pass through a 200-mm (=D) wide hole 2 m (=d) away from

the bottom of the step. (The acceleration of the marble is 9.81 m/s², downward.)

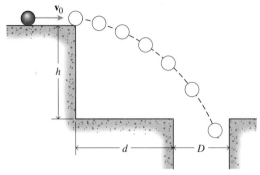

Fig. P13-84

13-85 A $\frac{1}{2}$-in. diameter marble rolls off a horizontal step 8 ft (=h) high with an initial speed of $v_0 = 5 \pm 0.5$ ft/s (Fig. P13-84). Determine the minimum size hole D that this marble will be certain to fall through. (The acceleration of the marble is 32.2 ft/s², downward.)

13-86 Using Fig. P13-84, repeat Problem 13-84 for a step height of $h = 1$ m.

13-87* Using Fig. P13-85, repeat Problem 13-85 for a step height of $h = 2$ ft.

13-88 A boy standing 5 m (=d) away from the bottom of a building is trying to throw a small ball through a 1 m (=H) high window 7 m (=h) up (Fig. P13-88). If the ball is to pass through the window at the peak of its trajectory, determine:

a. The initial velocity (magnitude and direction) required to just clear the bottom of the window.
b. The initial velocity (magnitude and direction) required to just clear the top of the window.
(The acceleration of the ball is 9.81 m/s², downward.)

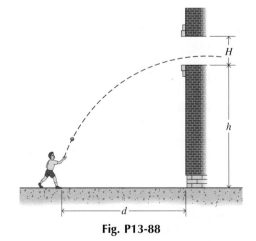

Fig. P13-88

13-89* A boy standing 20 ft (=*d*) away from the bottom of a building is trying to throw a small ball through a 3 ft (=*H*) high window 20 ft (=*h*) up (Fig. P13-88). If the initial speed of the ball is $v_0 = 50$ ft/s, determine the range of initial angles θ_0 with which the ball can be thrown. (The acceleration of the ball is 32.2 ft/s², downward.)

13-90* In a game of baseball, the ball leaves the hitter's bat at 1 m above the ground, at an angle of 25° to the horizontal, and with an initial speed of v_0 (Fig. P13-90). If the ball just clears the center-field fence 120 m away:

a. Determine the initial speed of the ball.
b. Determine the maximum height attained by the ball.
c. Determine the length of time it will take the ball to travel from the bat to the fence.

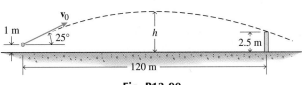

Fig. P13-90

13-91 A radar is tracking a rocket (Fig. P13-91). At some instant of time, the distance *r* and angle θ are measured as 10 mi and 30°, respectively. From successive measurements, the derivatives $\dot{r}$, $\ddot{r}$, $\dot{\theta}$, and $\ddot{\theta}$ are estimated to be 650 ft/s, 165 ft/s², 0.031 rad/s, and 0.005 rad/s², respectively. Determine the velocity and acceleration of the rocket.

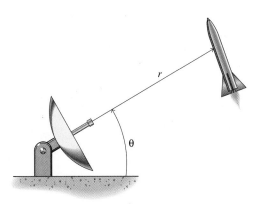

Fig. P13-91

13-92* A particle is following a spiral path given by $r(t) = 5\theta/3$ where $\theta(t)$ is in radians and *r* is in mm. Given that $\dot{\theta} = 10/t$ rad/s and that $\theta = 0$ when $t = 1$ s.

a. Compute the velocity of the particle $\mathbf{v}(t)$.
b. Compute the acceleration of the particle $\mathbf{a}(t)$.
c. Sketch the position of the particle for 1 s $< t < 2$ s.
d. Evaluate the velocity $\mathbf{v}(t)$ and the acceleration $\mathbf{a}(t)$ when $\theta = 2\pi$ rad and show them on the sketch of part c.

13-93 A particle is following a path given by $r(t) = 5 \sin\theta \cos^2\theta$ where $\theta(t)$ is in radians and *r* is in inches. Given that $\dot{\theta} = 2$ rad/s (constant) and that $\theta = 0$ when $t = 0$.

a. Compute the velocity of the particle $\mathbf{v}(t)$.
b. Compute the acceleration of the particle $\mathbf{a}(t)$.
c. Sketch the position of the particle $0 < t < 2$ s.
d. Evaluate $\mathbf{v}(t)$ and $\mathbf{a}(t)$ when $\theta = \pi$ rad and show them on the sketch of part c.

13-94 A particle is following a path given by $r(t) = 50 \cos 3\theta$ where $\theta(t)$ is in radians and *r* is in millimeters. Given that $\dot{\theta} = 2.5$ rad/s (constant) and that $\theta = 0$ when $t = 0$:

a. Compute the velocity of the particle $\mathbf{v}(t)$.
b. Compute the acceleration of the particle $\mathbf{a}(t)$.
c. Sketch the position of the particle for $0 < t < 2$ s.
d. Evaluate the velocity $\mathbf{v}(t)$ and the acceleration $\mathbf{a}(t)$ when $\theta = 2\pi$ rad and show them on the sketch of part c.

13-95* A collar that slides along a horizontal rod has a pin that is constrained to move in the slot of arm AB (Fig. P13-95). The arm oscillates with angular position given by $\theta(t) = 90 - 30 \cos t$ where $\theta(t)$ is in degrees and *t* is in seconds. For $t = 5$ s:

a. Determine the radial distance $r(t)$.
b. Determine the velocity components $v_r(t)$ and $v_\theta(t)$.
c. Determine the acceleration components $a_r(t)$ and $a_\theta(t)$.
d. Verify that the velocity vector and acceleration vector are both directed along the horizontal rod.

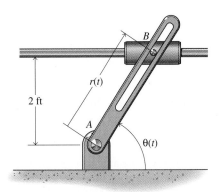

Fig. P13-95

49

13-96 A collar that slides around a circular wire has a pin that is constrained to move in the slot of arm *AB* (Fig. P13-96). The arm rotates counterclockwise at a constant angular speed of 2 rad/s. When the arm is 30° above the horizontal:

a. Determine the radial distance $r(t)$.
b. Determine the velocity components $v_r(t)$ and $v_\theta(t)$.
c. Determine the acceleration components $a_r(t)$ and $a_\theta(t)$.
d. Verify that the velocity vector is directed along the wire.

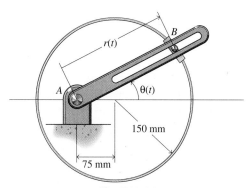

Fig. P13-96

13-97* If the speed of the particle of Problem 13-93 is not to exceed 10 ft/s, determine the maximum, constant value that $\dot{\theta}$ can be.

13-98* If the speed of the particle of Problem 13-94 is not to exceed 5 m/s, determine the maximum, constant value that $\dot{\theta}$ can be.

13-99 If the acceleration of the particle of Problem 13-93 is not to exceed 30 ft/s², determine the maximum, constant value that $\dot{\theta}$ can be.

13-100* If the acceleration of the particle of Problem 13-94 is not to exceed 15 m/s², determine the maximum, constant value that $\dot{\theta}$ can be.

13-101 A car is traveling around a curve as shown in Fig. P13-101. At some instant of time, the car has a speed of 45 mi/h in a direction 30° north of east, its speed is increasing at a rate of 5 ft/s², and the radius of curvature is 450 ft.

Determine the acceleration (magnitude and direction) of the car.

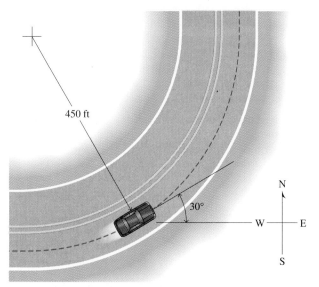

Fig. P13-101

13-102 The car of Fig. P13-102 has a speed of 100 km/h increasing at the rate of 5 m/s² at the instant shown. If the radius of curvature at the bottom of the hill is 80 m, determine the acceleration (magnitude and direction) of the car.

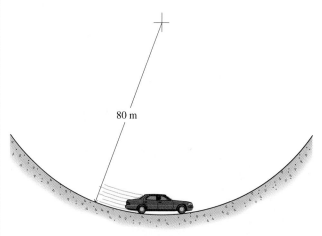

Fig. P13-102

13-103* A car drives over the top of a hill that has a radius of curvature of 110 ft (Fig. P13-103). If the normal component of acceleration necessary to keep the car on the road becomes greater than that provided by gravity, the car will become airborne. Determine the maximum constant speed at which the car can go over the hill.

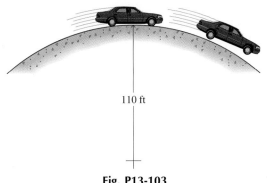

110 ft

Fig. P13-103

13-104 When the total acceleration of a car going around a curve exceeds one-third of the gravitational acceleration, the tires of the car will begin to slide. For a car increasing in speed at 2 m/s² around a corner having a radius of curvature of 60 m, determine the speed at which the tires will begin to slide.

13-105* Repeat Problem 13-104 for a car increasing in speed at 5 ft/s² and a radius of curvature of 200 ft.

13-106* If the car of Problem 13-104 has a speed of 100 km/h, determine the minimum radius of curvature for which the tires will not slide.

13-107 If the car of Problem 13-105 has a speed of 80 mi/h, determine the minimum radius of curvature for which the tires will not slide.

13-108* The chute of Fig. P13-108 is parabolic in shape; that is,

$$f(x) = x^2 - 6x + 9 \text{ m}$$

A marble rolling down the chute passes point A ($x_0 = 5$ m) with a speed of 3 m/s, which is increasing at the rate of 5 m/s².

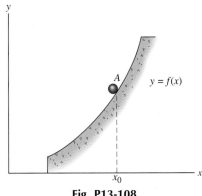

Fig. P13-108

a. Determine the normal and tangential components a_n and a_t of the acceleration of the marble as it passes point A.
b. Determine the angle between the velocity vector and the acceleration vector at point A.

13-109 The chute of Fig. P13-108 is elliptic in shape; that is,

$$f(x) = 1 - 0.5\sqrt{4 - x^2} \text{ ft}$$

A marble rolling down the chute passes point A ($x_0 = 1.5$ ft) with a speed of 12 ft/s, which is increasing at the rate of 8 ft/s².

a. Determine the normal and tangential components a_n and a_t of the acceleration of the marble as it passes point A.
b. Determine the angle between the velocity vector and the acceleration vector at point A.

13-110 The chute of Fig. P13-108 is hyperbolic in shape; that is,

$$f(x) = \frac{6}{(5 - x)} \text{ m}$$

A marble rolling down the chute passes point A ($x_0 = 3$ m) with a speed of 2 m/s, which is increasing at the rate of 3 m/s².

a. Determine the normal and tangential components a_n and a_t of the acceleration of the marble as it passes point A.
b. Determine the angle between the velocity vector and the acceleration vector at point A.

13-111* A ski slope has a shape given by

$$y = 0.003\,(x - 150)^2$$

where x and y are both in feet. When a skier is at $x = 100$ ft, he has a speed of 30 ft/s, which is increasing at the rate of 4 ft/s².

a. Determine the normal and tangential components a_n and a_t of the acceleration of the skier at this point.
b. Determine the angle between the velocity vector and the acceleration vector of the skier at this point.

13-112 A ski slope has a shape given by

$$y = \frac{400}{x + 15}$$

where x and y are both in meters. When a skier is at $x = 20$ m, she has a speed of 15 m/s, which is increasing at the rate of 2 m/s².

a. Determine the normal and tangential components a_n and a_t of the acceleration of the skier at this point.
b. Determine the angle between the velocity vector and the acceleration vector of the skier at this point.

13-6   RELATIVE MOTION IN A PLANE

The motion of two separate particles moving in plane curvilinear motion can be related just as was the motion of two particles moving in rectilinear motion in Section 13-4. The difference, of course, is that now the relative motion, like the individual motions, must be described with vectors.

The relationship between the positions of the individual particles and the relative position is obtained from the vector law of addition (Fig. 13-25)

$$\mathbf{r}_{Q/O} = \mathbf{r}_{P/O} + \mathbf{r}_{Q/P} \tag{13-40}$$

where $\mathbf{r}_{Q/P}$ is the position of particle Q relative to the position of particle P. Differentiating Eq. 13-40 with respect to time gives

$$\mathbf{v}_Q = \mathbf{v}_P + \mathbf{v}_{Q/P} \tag{13-41}$$

$$\mathbf{a}_Q = \mathbf{a}_P + \mathbf{a}_{Q/P} \tag{13-42}$$

That is, the velocity of particle Q measured relative to particle P is the difference in the absolute velocities (velocities measured relative to a fixed coordinate system) of particles Q and P. Similarly, the acceleration of particle Q measured relative to particle P is the difference in the absolute accelerations of particles Q and P.

The individual terms of these equations can be written in any convenient coordinate system: rectangular, polar, normal/tangential. However, all components must be converted to a common coordinate system (usually the rectangular coordinate system) before they can be added.

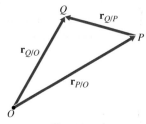

Fig. 13-25

EXAMPLE PROBLEM 13-13

An airplane is trying to fly straight north (Fig. 13-26). However, a 20 m/s wind blowing due east carries the airplane off course unless the airplane flies at an angle to the desired direction. If the speed of the airplane is 250 km/h, determine:

a. The direction in which the airplane must fly so that it travels due north.
b. The time required for the airplane to fly 250 km in the northerly direction.

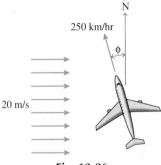

Fig. 13-26

SOLUTION

a. In a coordinate system in which east corresponds to the x-direction and north corresponds to the y-direction, the wind velocity is

$$\mathbf{v}_w = 20\mathbf{i} \text{ m/s}$$

and the desired velocity of the airplane is

$$\mathbf{v}_a = V\mathbf{j} \text{ m/s}$$

The velocity of the airplane relative to the wind is

$$\mathbf{v}_{a/w} = -250 \sin \phi \mathbf{i} + 250 \cos \phi \mathbf{j} \text{ km/h}$$
$$= -69.44 \sin \phi \mathbf{i} + 69.44 \cos \phi \mathbf{j} \text{ m/s}$$

where ϕ is the heading (west of north) at which the airplane must fly so that its absolute velocity will be due north.
  Putting these all together gives

$$\mathbf{v}_a = \mathbf{v}_w + \mathbf{v}_{a/w}$$

or

$$V\mathbf{j} = 20\mathbf{i} + (-69.44 \sin \phi \mathbf{i} + 69.44 \cos \phi \mathbf{j})$$

Then the x-component of this equation gives the heading

$$\phi = \sin^{-1} \frac{20}{69.44}$$
$$= 16.74° \text{ (west of north)} \qquad \text{Ans.}$$

and the y-component gives the apparent speed of the airplane in the northerly direction

$$V = 69.44 \cos 16.74° = 66.50 \text{ m/s} = 239.4 \text{ km/h}$$

b. The time required to fly 250 km to the north is then

$$t = \frac{250}{239.4} = 1.044 \text{ h} \qquad \text{Ans.}$$

Two bicyclists are riding around a circular track (Fig. 13-27). Cyclist 1 rides around the inside of the track where the radius is 200 ft, while cyclist 2 rides around the outside of the track where the radius is 210 ft. Both start at $\theta = 0$ and $v = 0$ at $t = 0$. Both accelerate at a constant rate of 2 ft/s² until they reach a speed of 20 ft/s after which they maintain a constant speed. When the first cyclist reaches B determine:

a. The angular position of cyclist 2, θ_2.
b. The relative position $\mathbf{r}_{2/1}$.
c. The relative velocity $\mathbf{v}_{2/1}$.
d. The relative acceleration $\mathbf{a}_{2/1}$.

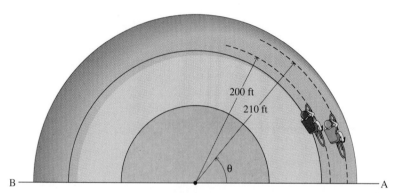

Fig. 13-27

SOLUTION

a. For cyclist 1, $r_1 = 200$ ft = constant, $\dot{r}_1 = 0$, $\ddot{r}_1 = 0$. Therefore the velocity and acceleration are given by

$$\mathbf{v}_1 = \dot{r}_1 \, \mathbf{e}_r + r_1 \dot{\theta}_1 \, \mathbf{e}_\theta = r_1 \dot{\theta}_1 \, \mathbf{e}_\theta$$
$$\mathbf{a}_1 = (\ddot{r}_1 - r_1 \dot{\theta}_1^2) \, \mathbf{e}_r + (r_1 \ddot{\theta}_1 + 2\dot{r}_1 \dot{\theta}_1) \, \mathbf{e}_\theta$$
$$= -r_1 \dot{\theta}_1^2 \, \mathbf{e}_r + r_1 \ddot{\theta}_1 \, \mathbf{e}_\theta$$

Initially, the acceleration along the track (θ-component) is 2 ft/s² = constant, so

$$\ddot{\theta}_1 = \frac{2}{200} = 0.010 \text{ rad/s}^2 \qquad (a)$$

Integrating Eq. a gives

$$\dot{\theta}_1 = 0.010t \text{ rad/s}$$

and

$$\theta_1 = 0.005t^2 \text{ rad}$$

Cyclist 1 accelerates until his speed reaches 20 ft/s or until

$$20 = (200)(0.010t)$$

which gives $t = 10$ s. The angular speed and position at this time are

$$\dot{\theta}_1 = 0.10 \text{ rad/s}$$
$$\theta_1 = 0.50 \text{ rad} \ (= 28.6°)$$

After $t = 10$ s, the speed (and therefore the angular speed) of cyclist 1 remains constant

$$\dot{\theta}_1 = 0.10 \text{ rad/s} = \text{constant}$$

Integrating to get the angular position as a function of time gives

$$\theta_1 = 0.10t - 0.50 \text{ rad}$$

where the constant of integration has been chosen so that $\theta_1 = 0.50$ rad when $t = 10$ s. Finally, the time at which cyclist 1 is at B ($\theta_1 = 180° = \pi$ rad) is found

$$t = \frac{\pi + 0.50}{0.10} = 36.42 \text{ s}$$

at which time his position, velocity, and acceleration are

$$\mathbf{r}_1 = 200 \, \mathbf{e}_r \text{ ft} = -200 \, \mathbf{i} \text{ ft}$$
$$\mathbf{v}_1 = (200)(0.10) \, \mathbf{e}_\theta = 20 \, \mathbf{e}_\theta \text{ ft/s} = -20 \, \mathbf{j} \text{ ft/s}$$
$$\mathbf{a}_1 = -(200)(0.10)^2 \, \mathbf{e}_r = -2 \, \mathbf{e}_r \text{ ft/s}^2 = 2 \, \mathbf{i} \text{ ft/s}^2$$

Similarly for cyclist 2, $r_2 = 210$ ft $=$ constant, $\dot{r}_2 = 0$, and $\ddot{r}_2 = 0$. Initially,

$$\ddot{\theta} = (2)/(210) = 0.00952 \text{ rad/s}^2$$
$$\dot{\theta}_2 = 0.00952t \text{ rad/s}$$
$$\theta_2 = 0.00476t^2 \text{ rad}$$

Cyclist 2 also reaches a speed of 20 ft/s at

$$t = (20)/(210)(0.00952) = 10 \text{ s}$$

at which time his angular speed and angular position are

$$\dot{\theta}_2 = 0.0952 \text{ rad/s}$$
$$\theta_2 = 0.476 \text{ rad} \ (= 27.3°)$$

After $t = 10$ s, the speed (and therefore the angular speed) of cyclist 2 remains constant so that

$$\dot{\theta}_2 = 0.0952 \text{ rad/s}$$

and integrating with respect to time gives

$$\theta_2 = 0.0952t - 0.476 \text{ rad}$$

where the constant of integration has been chosen such that $\theta_2 = 0.476$ rad when $t = 10$ s. Then the position, velocity, and acceleration of cyclist 2 at $t = 36.42$ s (the time when cyclist 1 is at B) is

$$\theta_2 = 2.991 \text{ rad} \ (= 171.4°) \qquad\qquad \text{Ans.}$$
$$\mathbf{r}_2 = 210\mathbf{e}_r \text{ ft} = -207.6\mathbf{i} + 31.55\mathbf{j} \text{ ft}$$
$$\mathbf{v}_2 = 20\mathbf{e}_\theta \text{ ft/s} = -3.005\mathbf{i} - 19.77\mathbf{j} \text{ ft/s}$$
$$\mathbf{a}_2 = -1.903\mathbf{e}_r \text{ ft/s}^2 = 1.881\mathbf{i} - 0.286\mathbf{j} \text{ ft/s}^2$$

b. Now the relative position can be computed

$$\mathbf{r}_{2/1} = \mathbf{r}_2 - \mathbf{r}_1 = -7.6\mathbf{i} + 31.55\mathbf{j} \text{ ft} \qquad\qquad \text{Ans.}$$

c. Similarly, the relative velocity is

$$\mathbf{v}_{2/1} = \mathbf{v}_2 - \mathbf{v}_1 = 17.00\mathbf{i} - 19.77\mathbf{j} \text{ ft/s} \qquad\qquad \text{Ans.}$$

d. Finally, the relative acceleration is

$$\mathbf{a}_{2/1} = \mathbf{a}_2 - \mathbf{a}_1$$
$$= -0.119\mathbf{i} - 0.286\mathbf{j} \text{ ft/s}^2 \qquad\qquad \text{Ans.}$$

PROBLEMS

13-113* A boat is trying to travel straight across a river as shown in Fig. P13-113. The river is 2000 ft wide and has a current of 5 mi/h. If the boat is traveling at 15 mi/h, determine

a. The time T required to travel straight from A to B.
b. The angle ϕ at which the boat must head to travel straight from A to B.

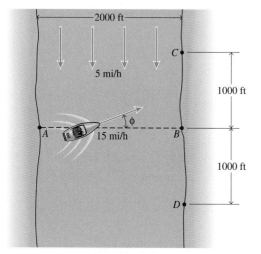

Fig. P13-113

13-114* A man distributing newspapers by car tosses a bundle of papers from the car as shown in Fig. P13-114. If the car is traveling at 15 km/h and the papers are tossed with a velocity of 5 m/s relative to the car and perpendicular to the motion of the car, determine

a. The velocity $\mathbf{v}_p$ of the papers relative to the sidewalk.
b. The angle ϕ between the velocities $\mathbf{v}_p$ and $\mathbf{v}_c$.

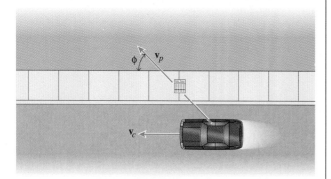

Fig. P13-114

13-115 For the boat of Problem 13-113, determine the time T and angle ϕ required to travel directly from

a. A to C.
b. A to D.

13-116* Rain is falling with a speed of 30 m/s and an angle of 20° to the vertical (Fig. P13-116). For a car traveling into the rain, determine

a. The angle ϕ at which the rain appears to strike the windshield if the car is traveling at 60 km/h.
b. The speed of the car for which $\phi = 90°$.

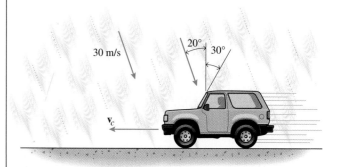

Fig. P13-116

13-117 Rain falls (vertical component of velocity) at 90 ft/s and is blown sideways (horizontal component of velocity) by the wind at 15 ft/s (Fig. P13-117). For a man walking at 6 ft/s, determine the angle ϕ at which the man should hold the umbrella (the angle of the relative velocity) if the man is walking

a. With the wind.
b. Into the wind.

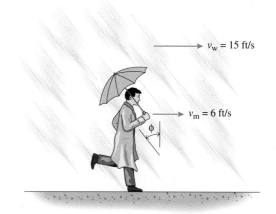

Fig. P13-117

13-118 Using Fig. P13-116, repeat Problem 13-116 for a car traveling with the rain.

13-119* Two boats leave a dock at the same time ($t = 0$) as shown in Fig. P13-119. Boat A travels at a steady speed of 15 mi/h while boat B travels at a steady speed of 20 mi/h. For $t = 30$ s, determine

a. The distance d between the boats.
b. The rate $\dot{d}$ at which the boats are separating.

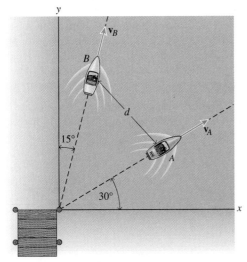

Fig. P13-119

13-120 Two airplanes are flying straight and level at the same altitude, as shown in Fig. P13-120. At $t = 0$ s the distances AC and BC are 20 km and 30 km, respectively. The planes maintain constant speeds; $v_A = 300$ km/h and $v_B = 400$ km/h. Determine

a. The relative position of the planes $\mathbf{r}_{B/A}$ at $t = 3$ min.
b. The relative velocity of the planes $\mathbf{v}_{B/A}$ at $t = 3$ min.
c. The distance d separating the planes at $t = 3$ min.
d. The time T at which the separation distance is a minimum.

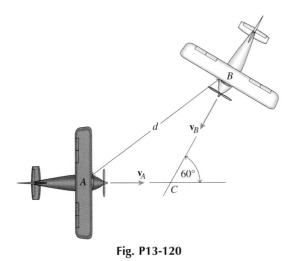

Fig. P13-120

13-121* Rollers A and B are attached to the ends of a rigid rod 5 ft long (Fig. P13-121). Roller B travels in a horizontal slot with a constant speed of 1 ft/s to the right while roller A travels in a vertical slot.

a. Determine the position $\mathbf{r}_A$, velocity $\mathbf{v}_A$, and acceleration $\mathbf{a}_A$ of roller A as functions of s; $0 \le s \le 5$ ft.
b. For $s = 3$ ft, determine the relative position $\mathbf{r}_{A/B}$, relative velocity $\mathbf{v}_{A/B}$, and relative acceleration $\mathbf{a}_{A/B}$.
c. Show that the relative position and relative velocity of part b are perpendicular.

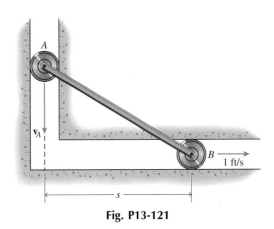

Fig. P13-121

13-122* A boy throws a ball from a window 10 m above the ground, as shown in Fig. P13-122. The initial speed of the ball is 10 m/s and the ball has a constant downward acceleration of 9.81 m/s². A second boy runs along the ground at a speed of 5 m/s and catches the ball on the run. Determine

a. The distance x at which the boy catches the ball.
b. The relative velocity $\mathbf{v}_{B/A}$ of the ball with respect to the boy when he catches the ball.

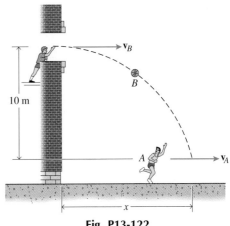

Fig. P13-122

57

13-123 A football player throws a pass to a receiver, as shown in Fig. P13-123. The initial velocity of the football is 35 ft/s at $\theta = 30°$ and the ball has a constant downward acceleration of 32.2 ft/s². If the receiver runs at a constant speed of 15 ft/s, determine

a. The distance x at which the receiver catches the football.
b. The relative velocity $\mathbf{v}_{B/A}$ of the ball with respect to the receiver when he catches the ball.

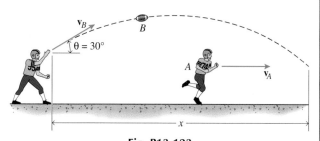

Fig. P13-123

13-124* In Problem 13-122 the second boy is at $x = 2$ m when the ball is thrown. Determine

a. The initial speed of the ball $\mathbf{v}_B$ such that the boy can catch it on the run.
b. The distance x at which the boy will catch the ball.
c. The relative velocity $\mathbf{v}_{B/A}$ of the ball with respect to the boy when he catches the ball.

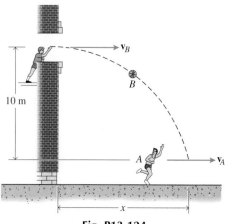

Fig. P13-124

13-125 In Problem 13-123 the receiver is at $x = 30$ ft when the ball is thrown. Determine

a. The initial speed of the football v_B such that the receiver can catch it on the run.
b. The distance x at which the receiver will catch the football.
c. The relative velocity $\mathbf{v}_{B/A}$ of the ball with respect to the receiver when he catches the ball.

13-126 Two boys are playing catch on a hill as shown in Fig. P13-126. The first boy throws the ball with an initial speed of 10 m/s in a horizontal direction and the ball has a constant downward acceleration of 9.81 m/s². The second boy runs at a constant speed of 5 m/s. Determine

a. The distance s at which the second boy catches the ball.
b. The relative velocity $\mathbf{v}_{B/A}$ of the ball with respect to the boy when he catches the ball.

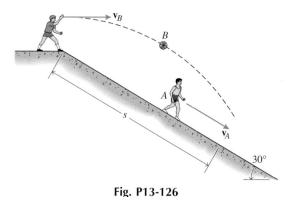

Fig. P13-126

13-127* Two boys are playing catch on a hill as shown in Fig. P13-126. The first boy throws the ball with an initial speed of v_B in a horizontal direction and the ball has a constant downward acceleration of 32.2 ft/s². The second boy starts at $s = 20$ ft and runs at a constant speed of 15 ft/s. Determine

a. The initial speed of the ball v_B such that the boy can catch it on the run.
b. The distance s at which the second boy will catch the ball.
c. The relative velocity $\mathbf{v}_{B/A}$ of the ball with respect to the boy when he catches the ball.

13-128 An airplane towing a glider (Fig. P13-128) is flying straight and level at a constant speed of 70 m/s. The tow rope is 50 m long and makes an angle θ of 10° with the horizontal. If $\dot{\theta} = 0.40$ rad/s, and $\ddot{\theta} = -0.25$ rad/s², determine

a. The rate of climb of the glider v_{By}.
b. The acceleration of the glider $\mathbf{a}_B$.

13-129* An airplane towing a glider (Fig. P13-128) is flying straight and level at a constant speed of 150 mi/h. The tow rope is 200 ft long and makes an angle θ of 10° with the horizontal. If the glider is climbing at an angle α of 15°, determine

a. The rate of change of the tow rope angle $\dot{\theta}$.
b. The velocity of the glider $\mathbf{v}_B$.

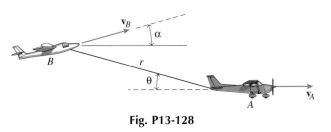

Fig. P13-128

13-7 SPACE CURVILINEAR MOTION

Three coordinates are required to describe motion along a curve in three-dimensional space. The most commonly used coordinate systems are the rectangular coordinate system and the cylindrical coordinate system, which are described in detail. A less commonly used coordinate system—the spherical coordinate system—is also described briefly. Although a modified version of the normal and tangential coordinate system can be made for general three-dimensional motion, it is not of general interest since the plane of motion (called the osculating plane and defined by the tangent and the principal normal directions) varies from point to point along the curve and from instant to instant in time.

13-7-1 Rectangular Coordinates

The three-dimensional rectangular coordinate system starts with the rectangular coordinates x and y (Section 13-5-1) and then adds a z-coordinate—the distance from the x-y plane (Fig. 13-28). The unit vector in the z-direction is called $\mathbf{k}$ and the position of a particle is

$$\mathbf{r}(t) = x(t)\mathbf{i} + y(t)\mathbf{j} + z(t)\mathbf{k} \qquad (13\text{-}43)$$

Like $\mathbf{i}$ and $\mathbf{j}$ the unit vector $\mathbf{k}$ is constant in both magnitude and direction so that the derivatives of the position are

$$\mathbf{v}(t) = \dot{\mathbf{r}}(t) = \dot{x}(t)\mathbf{i} + \dot{y}(t)\mathbf{j} + \dot{z}(t)\mathbf{k} \qquad (13\text{-}44)$$
$$\mathbf{a}(t) = \dot{\mathbf{v}}(t) = \ddot{\mathbf{r}}(t) = \ddot{x}(t)\mathbf{i} + \ddot{y}(t)\mathbf{j} + \ddot{z}(t)\mathbf{k} \qquad (13\text{-}45)$$

Again, the (x,y)-coordinate directions need not lie in a horizontal or vertical plane. The coordinate directions can be chosen arbitrarily so long as they are orthogonal. Once chosen, however, they must remain fixed.

13-7-2 Cylindrical Coordinates

The cylindrical coordinate system starts with the two-dimensional polar coordinate system of Section 13-5-2 and adds a z-coordinate— the distance from the r-θ plane (Fig. 13-29). The unit vector in the

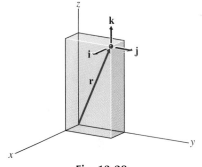

Fig. 13-28

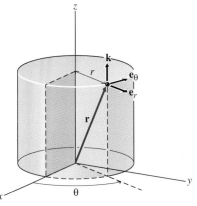

Fig. 13-29

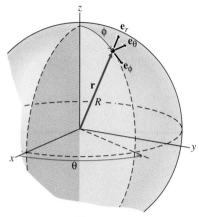

Fig. 13-30

z-direction is again called **k** and is again constant in both magnitude and direction. Since **k** is fixed and $\mathbf{e}_r$ and $\mathbf{e}_\theta$ are independent of the *z*-coordinate, the *r*- and θ-components of the position, velocity, and acceleration are the same as for polar coordinates, and the *z*-components are the same as for rectangular coordinates:

$$\mathbf{r}(t) = r(t)\mathbf{e}_r + z(t)\mathbf{k} \tag{13-46}$$

$$\mathbf{v}(t) = \dot{\mathbf{r}}(t) = \dot{r}\mathbf{e}_r + r\dot{\theta}\mathbf{e}_\theta + \dot{z}\mathbf{k} \tag{13-47}$$

$$\mathbf{a}(t) = \dot{\mathbf{v}}(t) = \ddot{\mathbf{r}}(t) = (\ddot{r} - r\dot{\theta}^2)\mathbf{e}_r + (r\ddot{\theta} + 2\dot{r}\dot{\theta})\mathbf{e}_\theta + \ddot{z}\mathbf{k} \tag{13-48}$$

As with rectangular coordinates, the (r,θ)-coordinate directions need not lie in a horizontal or vertical plane. The cylindrical coordinate system is usually used for a body rotating about an axis. The *z*-axis is usually chosen along the axis of rotation.

13-7-3   Spherical Coordinates

The spherical coordinate system describes the position of a particle in terms of a radial distance and two angles, as in Fig. 13-30. The θ-coordinate is measured in a plane as in polar and cylindrical coordinates. However, the distance from the θ-plane is given by ϕ, the angle between the position vector and the normal to the θ-plane. The three unit vectors $\mathbf{e}_R$, $\mathbf{e}_\theta$, and $\mathbf{e}_\phi$ are perpendicular to each other and point in the sense that their coordinates increase. Clearly, the directions of all three unit vectors depend on both angles θ and ϕ, which in turn depend on time, and any time derivatives must consider variations in all three unit vectors with respect to both coordinates.

Therefore, although the expression for position in spherical coordinates is very simple

$$\mathbf{r}(t) = R(t)\mathbf{e}_R \tag{13-49}$$

derivation of the expressions for the velocity and the acceleration in spherical coordinates are not. Since the derivation is not central to understanding the kinematics, it will not be given. However, the results are included here for completeness:

$$\mathbf{v}(t) = \dot{\mathbf{r}}(t) = \dot{R}\mathbf{e}_R + R\dot{\theta}\sin\phi\,\mathbf{e}_\theta + R\dot{\phi}\,\mathbf{e}_\phi \tag{13-50}$$

$$
\begin{aligned}
\mathbf{a}(t) = \dot{\mathbf{v}}(t) &= \ddot{\mathbf{r}}(t) \\
&= (\ddot{R} - R\dot{\phi}^2 - R\dot{\theta}^2\sin^2\phi)\mathbf{e}_R \\
&\quad + (R\ddot{\theta}\sin\phi + 2\dot{R}\dot{\theta}\sin\phi + 2R\dot{\theta}\dot{\phi}\cos\phi)\mathbf{e}_\theta \\
&\quad + (R\ddot{\phi} + 2\dot{R}\dot{\phi} - R\dot{\theta}^2\sin\phi\cos\phi)\mathbf{e}_\phi
\end{aligned} \tag{13-51}
$$

The spherical coordinate system is most often used in radar observations of aircraft or spacecraft positions and in describing the position and motion of robotic arms.

EXAMPLE PROBLEM 13-15

A particle following a space curve has a velocity given by

$$\mathbf{v}(t) = 12\,t^2\mathbf{i} + 16t^3\mathbf{j} + \sin \pi t\mathbf{k} \text{ ft/s}$$

If at $t = 0$ the particle has the position $\mathbf{r}_0 = 4\mathbf{j} + 3\mathbf{k}$ ft, find

a. The acceleration of the particle $\mathbf{a}(t)$.
b. The position of the particle $\mathbf{r}(t)$.

SOLUTION

a. The acceleration is obtained by simply differentiating the velocity to get

$$\mathbf{a}(t) = 24t\mathbf{i} + 48t^2\mathbf{j} + \pi \cos \pi t\mathbf{k} \text{ ft/s}^2 \qquad \text{Ans.}$$

b. The position of the particle is obtained by integrating the velocity, giving

$$\mathbf{r}(t) = 4t^3\mathbf{i} + 4t^4\mathbf{j} - (1/\pi) \cos \pi t\mathbf{k} + \mathbf{C}$$

where $\mathbf{C}$ is a constant of integration to be determined using the initial condition. At $t = 0$

$$\mathbf{r}(0) = -(1/\pi)\mathbf{k} + \mathbf{C} = \mathbf{r}_0 = 4\mathbf{j} + 3\mathbf{k} \text{ ft}$$

Therefore

$$\mathbf{C} = 4\mathbf{j} + (3 + 1/\pi)\mathbf{k} \text{ ft}$$

and

$$\begin{aligned}\mathbf{r}(t) = 4t^3\mathbf{i} + 4(1 + t^4)\mathbf{j} \\ + [3 + 1/\pi - (1/\pi) \cos \pi t]\mathbf{k} \text{ ft}\end{aligned} \qquad \text{Ans.}$$

The exit ramp of a parking garage is in the shape of a helix

$$\mathbf{r}(\theta) = 15 + 3 \sin \theta \, \text{m}$$

which drops 6 m for each complete revolution. For a car traveling down the ramp such that $\dot{\theta} = 0.3$ rad/s = constant:

a. Determine the velocity and acceleration when $\theta = 0°$.
b. Determine the velocity and acceleration when $\theta = 90°$.
c. Show that the velocity and acceleration are perpendicular when $\theta = 90°$.

SOLUTION

a. The position vector in cylindrical coordinates is

$$\mathbf{r}(t) = r(t)\mathbf{e}_r + z(t)\mathbf{k}$$

where

$r(t) = 15 + 3 \sin \theta \, \text{m}$
$z(t) = A - (6\theta/2\pi) \, \text{m}$
$\theta(t) = B + 0.3t \, \text{rad}$

and A and B are constants. The velocity and acceleration are given by

$$\mathbf{v}(t) = \dot{r}\mathbf{e}_r + r\dot{\theta}\mathbf{e}_\theta + \dot{z}\mathbf{k}$$
$$\mathbf{a}(t) = (\ddot{r} - r\dot{\theta}^2)\mathbf{e}_r + (r\ddot{\theta} + 2\dot{r}\dot{\theta})\mathbf{e}_\theta + \ddot{z}\mathbf{k}$$

where

$\dot{r} = 3\dot{\theta}\cos\theta \, \text{m/s}$ $\ddot{r} = 3\ddot{\theta}\cos\theta - 3\dot{\theta}^2 \sin\theta \, \text{m/s}^2$
$\dot{\theta} = 0.3 \, \text{rad/s}$ $\ddot{\theta} = 0 \, \text{rad/s}^2$
$\dot{z} = -\dfrac{6\dot{\theta}}{2\pi} \, \text{m/s} = -0.286 \, \text{m/s}$ $\ddot{z} = 0 \, \text{m/s}^2$

Then when $\theta = 0°$

$$r = 15 \, \text{m} \qquad \dot{r} = 0.9 \, \text{m/s} \qquad \ddot{r} = 0 \, \text{m/s}^2$$

and

$$\mathbf{v} = 0.900\mathbf{e}_r + 4.500\mathbf{e}_\theta - 0.286\mathbf{k} \, \text{m/s} \qquad \text{Ans.}$$
$$\mathbf{a} = -1.350\mathbf{e}_r + 0.540\mathbf{e}_\theta \, \text{m/s}^2 \qquad \text{Ans.}$$

b. When $\theta = 90°$

$$r = 18 \, \text{m} \qquad \dot{r} = 0 \, \text{m/s} \qquad \ddot{r} = -0.270 \, \text{m/s}^2$$

and

$$\mathbf{v} = 5.400\mathbf{e}_\theta - 0.286\mathbf{k} \, \text{m/s} \qquad \text{Ans.}$$
$$\mathbf{a} = -1.890\mathbf{e}_r \, \text{m/s}^2$$

c. Checking the scalar product of the velocity and acceleration vectors of part b gives

$$\mathbf{v} \cdot \mathbf{a} = 0$$

which shows that the velocity and acceleration vectors are indeed perpendicular. Ans.

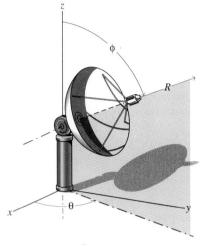

The radar of Fig. 13-31 is tracking an airplane. At the instant shown, the position of the airplane is given by $R = 65,000$ ft, $\theta = 110°$, and $\phi = 60°$. Comparison with previous positions allows the estimation of the derivatives $\dot{R} = -285$ ft/s, $\ddot{R} = 15$ ft/s^2, $\dot{\theta} = 9.0(10^{-3})$ rad/s, $\ddot{\theta} = 20(10^{-6})$ rad/s^2, $\dot{\phi} = 2.5(10^{-3})$ rad/s, and $\ddot{\phi} = 80(10^{-6})$ rad/s^2. For this instant, determine:

a. The velocity and acceleration of the airplane in spherical coordinates (R, ϕ, θ).
b. The velocity and acceleration of the airplane in rectangular coordinates in which the z-axis corresponds to the axis $\phi = 0°$ and the x-axis corresponds to the axis $\phi = 90°$ and $\theta = 0°$.
c. The magnitudes of the velocity and acceleration of the airplane.

Fig. 13-31

SOLUTION

a. The velocity of the airplane is given in spherical coordinates by Eq. 13-50

$$\mathbf{v}(t) = \dot{R}\mathbf{e}_R + R\dot{\theta}\sin\phi\,\mathbf{e}_\theta + R\dot{\phi}\mathbf{e}_\phi$$
$$= -285\mathbf{e}_R + 506.6\mathbf{e}_\theta + 162.5\mathbf{e}_\phi \text{ ft/s} \qquad \text{Ans.}$$

and the acceleration is given in spherical coordinates by Eq. 13-51

$$\mathbf{a}(t) = (\ddot{R} - R\dot{\phi}^2 - R\dot{\theta}^2\sin^2\phi)\mathbf{e}_R$$
$$+ (R\ddot{\theta}\sin\phi + 2\dot{R}\dot{\theta}\sin\phi + 2R\dot{\theta}\dot{\phi}\cos\phi)\mathbf{e}_\theta$$
$$+ (R\ddot{\phi} + 2\dot{R}\dot{\phi} - R\dot{\theta}^2\sin\phi\cos\phi)\mathbf{e}_\phi$$
$$= 10.65\mathbf{e}_R - 1.854\mathbf{e}_\theta + 1.495\mathbf{e}_\phi \text{ ft/s}^2 \qquad \text{Ans.}$$

b. With reference to Fig. 13-30, the spherical coordinate unit vectors $\mathbf{e}_R$, $\mathbf{e}_\theta$, and $\mathbf{e}_\phi$ can be related to the rectangular unit vectors $\mathbf{i}$, $\mathbf{j}$, and $\mathbf{k}$ by

$$\mathbf{e}_R = \sin\phi\cos\theta\,\mathbf{i} + \sin\phi\sin\theta\,\mathbf{j} + \cos\phi\,\mathbf{k}$$
$$\mathbf{e}_\theta = -\sin\theta\,\mathbf{i} + \cos\theta\,\mathbf{j}$$

and (since R-ϕ-θ form a right-handed orthogonal coordinate system)

$$\mathbf{e}_\phi = \mathbf{e}_\theta \times \mathbf{e}_R$$

Then when $\theta = 110°$ and $\phi = 60°$

$$\mathbf{e}_R = -0.2962\mathbf{i} + 0.8138\mathbf{j} + 0.5000\mathbf{k}$$
$$\mathbf{e}_\theta = -0.9397\mathbf{i} - 0.3420\mathbf{j}$$
$$\mathbf{e}_\phi = -0.1710\mathbf{i} + 0.4699\mathbf{j} - 0.8660\mathbf{k}$$

and

$$\mathbf{v}(t) = -285\,(-0.2962\mathbf{i} + 0.8138\mathbf{j} + 0.5000\mathbf{k})$$
$$+506.6\,(-0.9397\mathbf{i} - 0.3420\mathbf{j})$$
$$+162.5\,(-0.1710\mathbf{i} + 0.4699\mathbf{j} - 0.8660\mathbf{k})$$
$$= -419.4\mathbf{i} - 328.9\mathbf{j} - 283.2\mathbf{k} \text{ ft/s} \qquad \text{Ans.}$$
$$\mathbf{a}(t) = 10.65\,(-0.2962\mathbf{i} + 0.8138\mathbf{j} + 0.5000\mathbf{k}$$
$$-1.854\,(-0.9397\mathbf{i} - 0.3420\mathbf{j})$$
$$+1.495\,(-0.1710\mathbf{i} + 0.4699\mathbf{j} - 0.8660\mathbf{k})$$
$$= -1.666\mathbf{i} + 10.00\mathbf{j} + 4.03\mathbf{k} \text{ ft/s}^2 \qquad \text{Ans.}$$

c. The magnitudes of the velocity and acceleration can be computed from the components in either the spherical or rectangular coordinate system. Using the spherical coordinate components

$$v = \sqrt{285^2 + 506.6^2 + 162.5^2} = 604 \text{ ft/s} \qquad \text{Ans.}$$
$$a = \sqrt{10.65^2 + 1.854^2 + 1.495^2} = 10.91 \text{ ft/s}^2 \qquad \text{Ans.}$$

PROBLEMS

13-130* The three-dimensional motion of a particle is described by the relations

$$x = 6 \sin 6t \text{ m} \qquad y = 3\sqrt{3} \cos 6t \text{ m}$$
$$z = 3 \cos 6t \text{ m}$$

Compute the acceleration of the particle and show that it has constant magnitude.

13-131* The three-dimensional motion of a particle is described by the relation

$$r = 5t^2\mathbf{i} + 3t\mathbf{j} + 15t^3\mathbf{k} \text{ ft}$$

Compute the velocity and acceleration of the particle.

13-132 An eagle riding an updraft travels along an elliptical helical path described by the relations

$$x = 15 \cos 0.2t \text{ m} \qquad y = 10 \sin 0.2t \text{ m}$$
$$z = 0.8t \text{ m}$$

Compute the velocity and acceleration of the eagle at $t = 80$ s.

13-133* The three-dimensional motion of a particle is described by the relations

$$x = 2 \sin 3t \text{ ft} \qquad y = 1.5t \text{ ft}$$
$$z = 2 \cos 3t \text{ ft}$$

a. Compute the velocity and acceleration of the particle at $t = 25$ s.
b. Show that the velocity and acceleration are perpendicular for any value of t.

13-134 The three-dimensional motion of a particle is described by the relations

$$r = 5(1 - e^{-t}) \text{ m} \qquad \theta = 2\pi t \text{ rad}$$
$$z = 3 \sin 3\theta \text{ m}$$

Compute the velocity and acceleration of the particle for

a. $t = 0$ s.
b. $t = 3$ s.
c. $t = 100$ s.

13-135 The three-dimensional motion of a particle on the surface of a right circular cylinder is described by the relations

$$r = 2 \text{ ft} \qquad \theta = \pi t \text{ rad} \qquad z = \sin 6\theta \text{ ft}$$

Compute the velocity and acceleration of the particle at $t = 3$ s.

13-136* The three-dimensional motion of a particle on the surface of a right circular cylinder is described by the relations

$$r = 2 \text{ m} \qquad \theta = \pi t \text{ rad} \qquad z = \sin^2 4\theta \text{ m}.$$

Compute the velocity and acceleration of the particle at $t = 5$ s.

13-137 The three-dimensional motion of a particle is described by the relations

$$r = 5 \sin 3\theta \text{ ft} \qquad \theta = 2\pi t \text{ rad} \qquad z = \theta/4 \text{ ft}$$

Compute the velocity and acceleration of the particle at $t = 2$ s.

13-138* The three-dimensional motion of a particle on the surface of a right circular cone is described by the relations

$$r = z \tan \beta \text{ m} \qquad \theta = 2\pi t \text{ rad} \qquad z = \frac{h\theta}{2\pi} \text{ m}$$

where $\beta = 30°$ is the apex angle of the cone and $h = 0.25$ m is the distance the particle rises in one trip around the cone. Calculate the velocity and acceleration of the particle for:

a. $t = 0$ s.
b. $z = 1$ m.

13-139* The three-dimensional motion of a particle on the surface of a right circular cone 3 ft tall is described by the relations

$$r = z \tan \beta \text{ ft} \qquad \theta = 2\pi t \text{ rad} \qquad z = \frac{h\theta}{2\pi} \text{ ft}$$

where $\beta = 20°$ is the apex angle of the cone and $h = 0.5$ ft is the distance the particle rises in one trip around the cone. Calculate the velocity and acceleration of the particle at:

a. The apex of the cone.
b. The top of the cone.

13-140 An airplane is descending in a circular pattern that has a constant radius of 250 m. If the airplane has a horizontal speed of 75 m/s (constant) and a downward speed of 5 m/s (which is increasing at a rate of 2 m/s²), determine the acceleration of the airplane.

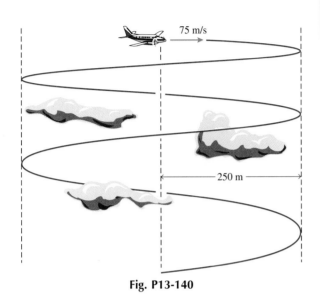

75 m/s

250 m

Fig. P13-140

13-141* A car is traveling down the exit ramp of a parking garage at a constant speed of 10 mi/h. The ramp is a circle of diameter 120 ft and drops 20 ft for every full revolution ($\theta = 2\pi$ rad). Determine the magnitude of the car's acceleration as it moves down the ramp.

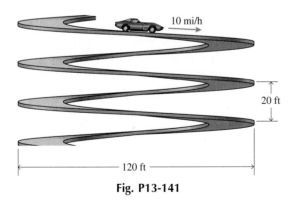

10 mi/h

20 ft

120 ft

Fig. P13-141

13-142 The crane of Fig. P13-142 rotates about the axis CD at a constant rate of 3 rad/min. At the same time the 20 m long boom AB is being lowered at a constant rate of 5 rad/min. Compute the velocity and acceleration of point B when $\phi = 30°$.

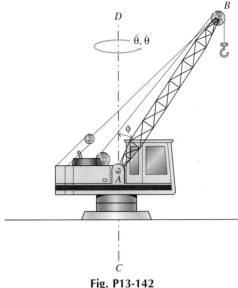

D

$\ddot{\theta}, \dot{\theta}$

B

ϕ

A

C

Fig. P13-142

13-143 The boom AB of the crane shown in Fig. P13-142 is 75 ft long. When $\phi = 30°$ the crane is rotating about the axis CD with $\dot{\theta} = 3$ rad/min, $\ddot{\theta} = -1$ rad/min², $\dot{\phi} = -5$ rad/min, and $\ddot{\phi} = 2$ rad/min². Calculate the acceleration of point B.

13-144* The crane of Fig. P13-142 rotates about the axis CD at a constant rate ω. At the same time the 20 m long boom AB is being lowered at a constant rate of 3 rad/min. Determine the maximum rotation rate ω for which the acceleration of point B does not exceed 0.25 m/s² when $\phi = 30°$.

13-145 An airplane is flying due west at a constant speed of 300 mi/h and a constant altitude of 5000 ft. The ground track of the airplane passes 3000 ft to the north of a radar tracking site. Determine the rates $\dot{\theta}$, $\ddot{\theta}$, $\dot{\phi}$, and $\ddot{\phi}$ at which the radar dish must be turned to track the airplane when it is 6000 ft east of the radar site.

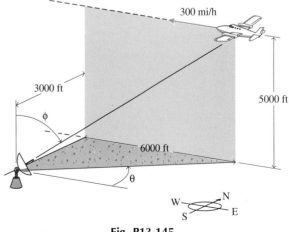

300 mi/h

3000 ft

ϕ

5000 ft

6000 ft

θ

W N
S E

Fig. P13-145

13-146* An airplane is flying due west at a constant speed of 100 m/s and a constant altitude of 1500 m. The ground track of the airplane passes 2 km to the north of a radar tracking site. Determine the rates $\dot{\theta}$, $\ddot{\theta}$, $\dot{\phi}$, and $\ddot{\phi}$ at which the radar dish must be turned to track the airplane when it is directly north of the radar site.

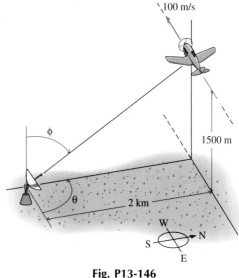

Fig. P13-146

SUMMARY

Kinematics is the study of how particles move. It describes how the velocity and acceleration of a body change with time and with changes in the position of the body. A sound understanding of kinematics is a necessary background for the study of kinetics, which relates the motion to the forces that cause it.

A particle is an object whose size can be ignored when studying its motion. Only the position of the mass center of a particle need be considered. The orientation of the object or rotation of the object plays no role in describing the motion of a particle. Particles can be very small or very large. Small does not always guarantee that an object can be modeled as a particle; large does not always prevent modeling an object as a particle. Whether a body is large or small relates to the length of path followed, the distance between bodies, or both.

The kinetic quantities used to describe the motion of a particle are time, position (including displacement and total distance traveled), velocity, and acceleration. The various kinematic quantities are related by differential equations. The problems of kinematics consist of determining one or more of the preceding quantities from the given data.

The velocity of a particle is the time rate of change of its position

$$\mathbf{v}_P = \frac{d\mathbf{r}_{P/O}}{dt} = \dot{\mathbf{r}}_{P/O} \tag{13-5}$$

Unlike the position vector $\mathbf{r}_{P/O}$, the velocity is independent of the location of the origin of the coordinate system. The direction of the velocity $\mathbf{v}_P$ is tangent to the path of the particle. In terms of fixed Cartesian coordinates, the velocity is

$$\mathbf{v}_P = v_x\mathbf{i} + v_y\mathbf{j} + v_z\mathbf{k} = \dot{x}\mathbf{i} + \dot{y}\mathbf{j} + \dot{z}\mathbf{k} \tag{13-6}$$

The acceleration of the particle P is the time rate of change of its velocity

$$\mathbf{a}_P = \frac{d\mathbf{v}_P}{dt} = \dot{\mathbf{v}}_P = \frac{d^2\mathbf{r}_{P/O}}{dt^2} = \ddot{\mathbf{r}}_{P/O} \tag{13-7}$$

The acceleration is also independent of the location of the origin of the coordinate system. In terms of fixed Cartesian coordinates, the acceleration of the particle P is

$$\begin{aligned} \mathbf{a}_P &= a_x\mathbf{i} + a_y\mathbf{j} + a_z\mathbf{k} = \dot{v}_x\mathbf{i} + \dot{v}_y\mathbf{j} + \dot{v}_z\mathbf{k} \\ &= \ddot{x}\mathbf{i} + \ddot{y}\mathbf{j} + \ddot{z}\mathbf{k} \end{aligned} \tag{13-8}$$

Rectilinear motion is motion along a straight line. If the coordinate system is oriented such that the x-axis coincides with the line of motion, the position, velocity, and acceleration are completely prescribed by giving only their x-components. That is, the position vector, velocity vector, and acceleration vector may be specified by giving their "signed magnitudes" x, $v = \dot{x}$, and $a = \ddot{x}$, respectively. Positive values for x, v, and a indicates that the *vectors* are in the positive coordinate direction, whereas negative values indicates that the *vectors* are in the negative coordinate direction.

When two or more particles move in rectilinear motion, separate equations may be written to describe their motion. The particles may be moving along the same line or along separate lines. If the position of one particle depends on the position of another or several other particles, a constraint equation can be written relating the positions of the two particles. The constraint equation is then differentiated to get the relative velocity and relative acceleration equations.

In most problems, the acceleration of a particle is derived from the forces that act on the particle. Depending on the nature of the forces, the acceleration may be a function of time, a function of velocity, or a function of the position of the particle. The velocity and position of the particle are obtained by integrating the definitions of acceleration and velocity.

The choice of which coordinates to use in a particular problem will depend on the geometry of the problem, on the way in which data are given for the problem, and on the type of solution desired. Three of the more commonly used coordinate systems used to represent the motion are: rectangular coordinates, most convenient when the x- and y-components of the motion are specified independently of each other and do not depend on each other; polar coordinates, most convenient when the position of a particle is measured relative to a fixed point or is moving along a rotating arm; and normal/tangential coordinates, most convenient when particles move along a surface of known shape.

REVIEW PROBLEMS

13-147* Find the average acceleration for

a. A jet launched from an aircraft carrier goes from 0 to 120 mi/h in 4 s.
b. A drag racer uses a parachute to slow from 100 mi/h to 20 mi/h in 5 s.

13-148* A tow truck is pulling a car up a 25° incline using pulleys as shown in Fig. P13-148. If the tow truck accelerates at a rate of 0.8 m/s², determine the speed and acceleration of the car 5 s after the tow truck starts from rest.

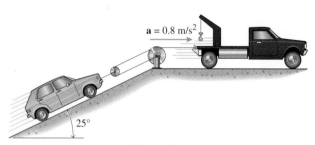

Fig. P13-148

13-149 In baseball the distance between the pitcher and the plate is 60 ft (Fig. P13-149). If the pitcher throws a fastball with an initial speed of 96 mi/h, determine:

a. The distance a that the ball will drop if the ball is pitched horizontally ($\theta_0 = 0$).
b. The initial angle θ_0 for which the baseball will reach the catcher at its initial level ($a = 0$).
c. The maximum height that the ball will attain if thrown at the angle θ_0 of part b.

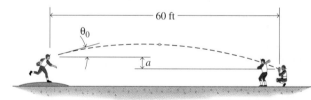

Fig. P13-149

13-150* A police car is parked in a school zone when a car speeds by. The police car starts from rest just as the car passes it, accelerates at 2 m/s² until it reaches a speed of 80 km/h, and then maintains a constant speed. If the speed of the first car is constant at 50 km/h, determine how far the police car will have to chase the speeder before catching him.

13-151 A radar tracking an airplane gives the coordinates of the plane as $r(t)$ and $\theta(t)$ (Fig. P13-151). At some instant of time, $\theta = 60°$ and $r = 10,000$ ft. From successive measurements of r and θ, the derivatives at this instant are estimated as $\dot{r} = 250$ ft/s, $\dot{\theta} = -0.0433$ rad/s, $\ddot{r} = 10.75$ ft/s², and $\ddot{\theta} = -7.8(10^{-4})$ rad/s². For this instant determine:

a. The velocity and acceleration of the airplane.
b. The radius of curvature of the airplane's path.

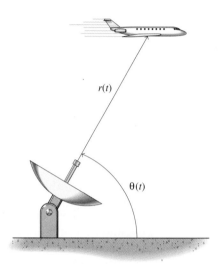

Fig. P13-151

13-152 Two cars are approaching each other on a narrow straight road. Car A has an initial speed of 60 km/h and car B has an initial speed of 30 km/h. If both drivers apply their brakes when they are 45 m apart and both cars slow down at a rate of 3 m/s², determine:

a. If the two cars will collide.
b. The relative speed of the two cars when they collide (if they do in fact collide).

13-153* A shotputter tosses a shot at 40° to the horizontal from a height of 6 ft (Fig. P13-153). If the shot lands 50 ft away, determine:

a. The initial speed v_0 of the shot.
b. The maximum height h attained by the shot.
c. The distance d at which the maximum height occurs.

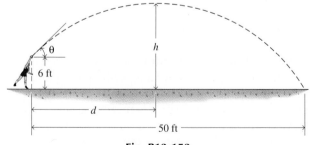

Fig. P13-153

13-154 A cam lobe has a shape given by $r = 20 + 15 \cos \theta$ mm (Fig. P13-154). Pin P slides in a slot along arm AB and is held against the cam by a spring. The arm AB rotates counterclockwise about A at a constant rate of 30 rev/min. Given that $\theta = 0$ at $t = 0$, determine:

a. The velocity $\mathbf{v}$ and acceleration $\mathbf{a}$ of the pin at $t = 0.6$ s.
b. The radius of curvature of the path at $t = 0.6$ s.

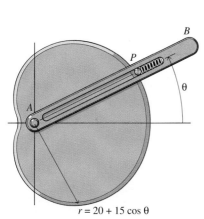

Fig. P13-154

13-155* The car and truck shown in Fig. P13-155a are both traveling at 50 mi/h when the car decides to pass the truck. If the car accelerates at 4 ft/s², passes the truck, and returns to the right lane when it is 35 ft ahead of the truck (Fig. P13-155b), determine:

a. The distance the car travels while passing the truck.
b. The speed of the car when it pulls back into the right lane.

(a)

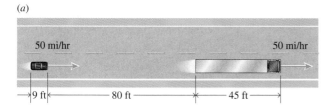

(b)

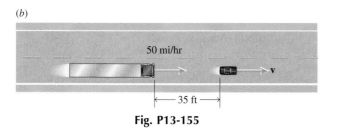

Fig. P13-155

13-156* In Fig. P13-156 block A is moving to the right with a speed of 4 m/s; the speed is decreasing at the rate of 0.15 m/s². At the instant shown, $d_A = 8$ m and $d_B = 6$ m. Determine the relative velocity $\mathbf{v}_{B/A}$ and relative acceleration $\mathbf{a}_{B/A}$.

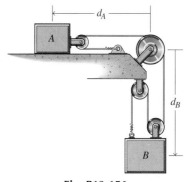

Fig. P13-156

13-157 A skier leaves the end of a ski jump at 60 mi/h and 10° above the horizontal (Fig. P13-157). Determine:

a. The maximum height the skier will rise from the end of the ski jump.
b. The time of flight of the jump.
c. The distance of the jump (the distance d measured along the slope).

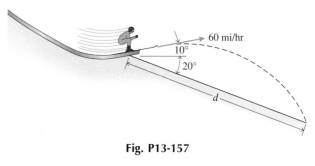

Fig. P13-157

13-158* A car is traveling at 90 km/h on a road that is parallel to a railroad track when it comes upon a train. If the train is 800 m long and moving at 65 km/h, determine the length of time it will take the car to pass the train if they are traveling:

a. In the same direction.
b. In opposite directions.

13-159 In Fig. P13-159 block A is moving to the left with a speed of 3 ft/s; the speed is increasing at the rate of 0.8 ft/s^2. At the instant shown $d_A = 6$ ft and $d_B = 8$ ft. Determine the relative velocity $\mathbf{v}_{B/A}$ and relative acceleration $\mathbf{a}_{B/A}$.

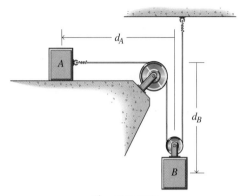

Fig. P13-159

13-160 A particle is following a path given by $r(t) = 50 \cos 3\theta$ where $\theta(t)$ is in radians and r is in millimeters. If $\dot\theta = 3$ rad/s (constant) and $\theta = 0$ when $t = 0$, determine:

a. The velocity $\mathbf{v}$ and acceleration $\mathbf{a}$ of the particle when $t = 0.8$ s.
b. The radius of curvature of the path when $t = 0.8$ s.

13-161* A basketball player shoots the ball at the basket which is 25 ft away, as shown in Fig. P13-161. Determine:

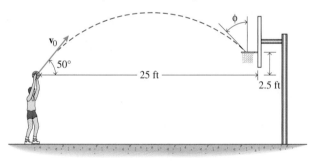

Fig. P13-161

a. The required initial speed v_0.
b. The angle ϕ that the trajectory of the ball will make with the vertical when the ball goes through the basket.

13-162* A high-speed train travels from Washington, D.C., to Philadelphia (a distance of 80 km) in just 35 min. The top speed of the train is 225 km/h. If the train decelerates twice as quickly as it accelerates, determine:

a. The deceleration of the train.
b. The distance the train travels at its maximum speed.

13-163 In Fig. P13-163 block B is moving downward with a speed of 5 ft/s; the speed is decreasing at the rate of 0.25 ft/s^2. At the instant shown, $d_A = 12$ ft and $d_B = 9$ ft. Determine the relative velocity $\mathbf{v}_{B/A}$ and relative acceleration $\mathbf{a}_{B/A}$.

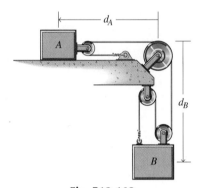

Fig. P13-163

13-164 A particle is following a path given by $r(t) = 125 \sin \theta \cos^2\theta$ where $\theta(t)$ is in radians and r is in millimeters. If $\dot\theta = 2$ rad/s (constant) and $\theta = 0$ when $t = 0$, determine:

a. The velocity $\mathbf{v}$ and acceleration $\mathbf{a}$ of the particle when $t = 0.6$ s.
b. The radius of curvature of the path when $t = 0.6$ s.

Computer Problems

C13-165 Rain is falling with a speed of 90 ft/s and an angle of 20° to the vertical (Fig. P13-165). Plot the angle ϕ at which the rain appears to strike the car's windshield as a function of the car's speed v_c $(0 \leq v_C \leq 80$ mi/h),

a. For a car traveling into the rain.
b. For a car traveling with the rain.

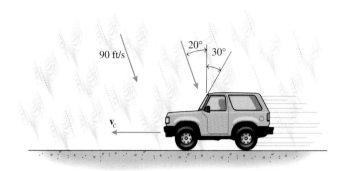

Fig. P13-165

C13-166 Arm AC of the cam follower mechanism shown in Fig. P13-166 is rotating at a constant angular speed of $\omega = 150$ rev/min. A spring holds the pin B against the cam lobes. If the equation that describes the shape of the cam lobes is

$$R = 125 + 50 \cos 3\theta$$

where R is in millimeters, calculate and plot the magnitude of the velocity v_B and acceleration a_B of the pin B as a function of θ $(0 \leq \theta \leq 180°)$. Is the shape of the curves the same if the angular speed ω is doubled?

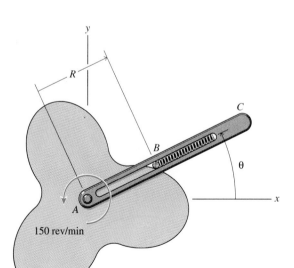

Fig. P13-166

C13-167 An airplane A is flying due north at a constant speed of 480 mi/h when a jet fighter J is sent to intercept it as shown in Fig. P13-167. (Both planes are at the same altitude). If the fighter flies at a constant speed of 600 mi/h,

a. Determine the minimum time required for the fighter to intercept the target airplane and the position of the resulting intercept.
b. A simpler intercept strategy, which does not require complicated calculations, is for the fighter to pursue a flight path such that it is always flying directly toward the target. If the fighter pilot uses this strategy, compute the position of the jet as a function of time until it intercepts the target plane. (Use the Euler method for solving differential equations described in Appendix C to compute the position of the jet.) Plot the position of both airplanes at 1-min intervals until the jet intercepts the target. Determine the time and position of the intercept. Compare this result with the minimum time solution of part a.

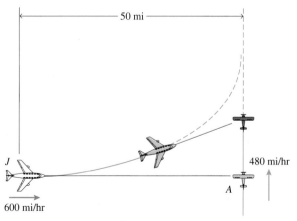

Fig. P13-167

C13-168 A basketball player shoots the ball at the basket with an initial speed v_0 and initial angle θ_0 as shown in Fig. P13-168. For the ball to go through the hoop, it must be coming down at an angle ϕ less than 70°. A further constraint is that the maximum arch of the ball should not be greater than $h = 3$ m.

a. Calculate and plot the initial velocity v_0 versus initial angle θ_0 ($30° \le \theta_0 \le 70°$) for successful shots. (Don't count shots that bounce off the backboard and into the hoop!)

b. Since the hoop is a little larger than the ball, a successful shot need not hit the center of the basket. However, shots that enter the basket at a low angle ϕ must be more precise than shots that enter the basket at a high angle. Assume that the ball can be short or long by

10 mm	for	$65° \le \phi \le 70°$
25 mm	for	$60° \le \phi \le 65°$
50 mm	for	$50° \le \phi \le 60°$
100 mm	for	$40° \le \phi \le 50°$
125 mm	for	$30° \le \phi \le 40°$
150 mm	for	$15° \le \phi \le 30°$
175 mm	for	$0° \le \phi \le 15°$

and plot the range of initial velocities that are acceptable for each angle θ_0 ($30° \le \theta_0 \le 70°$).

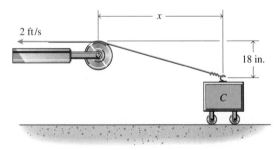

Fig. P13-168

C13-169 The cart C shown in Fig. P13-169 is being pulled to the left by a winch. The hook on top of the cart is curved such that the cable will release when $x = 2$ in. If the winch is reeling in cable at a constant rate of 2 ft/s:

a. Calculate and plot the speed v_C and acceleration a_C of the cart as a function of its position x ($-2 \le x \le 6$ ft).
b. Plot the position x, speed v_C, and acceleration a_C of the cart as a function of time t ($0 \le t \le 3$ s).

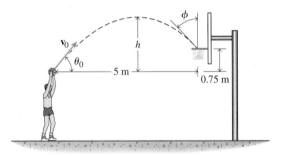

Fig. P13-169

C13-170 A fox starts chasing a rabbit as shown in Fig. P13-170a. The rabbit runs at a constant speed of 6 m/s around a circular path of radius 12 m and the fox runs at a constant speed of 7 m/s. The fox follows a path that is always directly toward the rabbit's current position.

a. Use the Euler method for solving differential equations (see Appendix C) to compute the position of the fox as a function of time until it catches the rabbit (assume that the fox catches the rabbit when the distance between them is less than 0.1 m). Plot the position of both animals at 0.5 s intervals until the fox catches the rabbit. Instead of always running toward the rabbit, where should the fox run to catch the rabbit in the shortest amount of time?

b. Repeat part a for the case where the rabbit runs around a circle away from the fox (Fig. P13-170b) and for the case where the rabbit runs a zigzag path away from the fox (Fig. P13-170c).

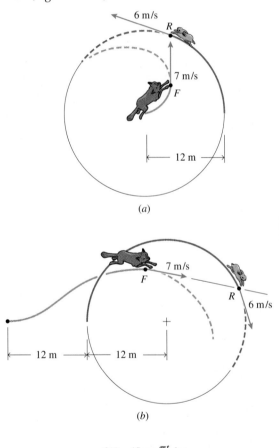

(a)

(b)

$\theta(t) = 45 \cos \frac{\pi t}{5}$ deg

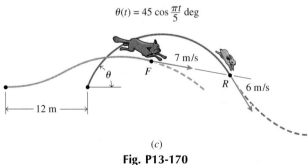

(c)

Fig. P13-170

KINEMATICS OF RIGID BODIES

An end loader exhibits several types of rigid body motion. As the loader moves, the body is in translation and the wheels and the bucket exhibit general plane motion.

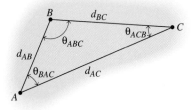

Fig. 14-1

14-1   INTRODUCTION

The kinematics of particles was analyzed in the preceding chapter. In that analysis, the location of a particle at all instants of time was all that was necessary to fully describe the motion of the particle. For solid bodies, however, a full description of the motion requires giving both the location and the orientation of the body. Kinematics of solid bodies involves both linear and angular quantities.

All solid bodies considered in this chapter and in the rest of the book will be considered to be rigid. In a rigid body, the distance between all pairs of particles is fixed and independent of time (Fig. 14-1). Clearly, if the distances between all pairs of particles are fixed then so are the angles between triples of points (Fig. 14-1).

No real body is absolutely rigid, of course. However, for most engineering applications, the deformations due to applied forces are relatively small, and the changes in the shape of the body due to applied forces have a negligible effect on the acceleration produced by a force system or on the forces required to produce a given motion. After the kinetic analysis is completed, the deformations should be computed. If the deformations are large, the kinematic and kinetic analyses may have to be repeated taking into account the deformation.

There are five general types of rigid body motion to be considered:

1. **Translation**   In the translation of a rigid body, the orientation of every straight line in the body is fixed. That is, horizontal lines remain horizontal, vertical lines remain vertical, and so on. A motion for which one line is always aligned with the velocity, such as horizontal lines on the body of a car driving on a straight level road (Fig. 14-2a), is called **rectilinear translation.** In rectilinear translation, every particle in the body follows a straight-line path in the direction of the motion. In **curvilinear translation** the orientation of every straight line is still fixed but individual particles do not follow straight-line paths (Fig. 14-2b). In **coplanar translation,** the path of each particle—whether straight or curved—remains in a single plane.

2. **Rotation about a Fixed Axis**   In rotation about a fixed axis, one straight line in the body, the axis of rotation, is fixed. Particles that are not on the axis travel in circular paths centered on the axis (Fig. 14-2c). If the axis of rotation does not intersect the body, the body may be imagined to be extended to include the axis of rotation (Fig. 14-2d). That is, for the purpose of kinematics, the motion of the body is the same as it would be if it were part of a larger rigid body which includes the axis of rotation. Since each circular path is contained in a single plane, the rotation of a body about a fixed axis is a plane motion.

3. **General Plane Motion**   In a plane motion, each particle in the body remains in a single plane. Coplanar translation and rotation about a fixed axis are specific types of plane motion in which lines in the body are fixed in special ways. Any other plane motion is called general plane motion (Fig. 14-2e).

4. **Rotation about a Fixed Point**   In rotation about a fixed point, one point in the body is fixed (Fig. 14-2f). Particles move along paths on the surface of a sphere centered at the point.

5. **General Motion**   All other motions are called general motion.

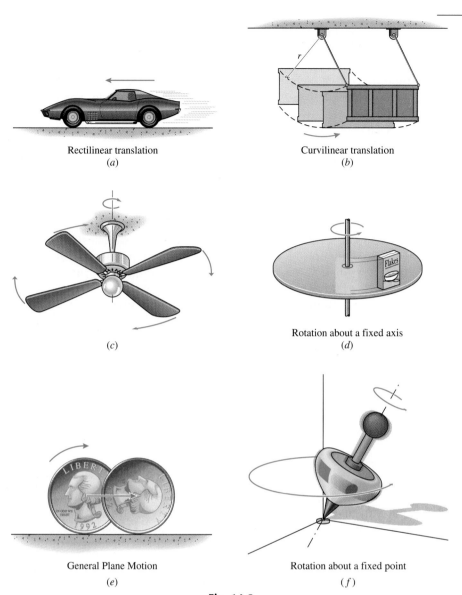

Rectilinear translation
(a)

Curvilinear translation
(b)

(c)

Rotation about a fixed axis
(d)

General Plane Motion
(e)

Rotation about a fixed point
(f)

Fig. 14-2

14-2 TRANSLATION

In the translation of a rigid body, the orientation of every straight line in the body is fixed. That is, horizontal lines remain horizontal, vertical lines remain vertical, and so on. If A and B are two arbitrary points in the body, the position of the points are related by the triangle law of addition for vectors:

$$\mathbf{r}_B = \mathbf{r}_A + \mathbf{r}_{B/A} \qquad (14\text{-}1)$$

where $\mathbf{r}_A$ and $\mathbf{r}_B$ are the absolute positions of the particles A and B, respectively, and $\mathbf{r}_{B/A}$ is the position of B relative to A. Since the relative position $\mathbf{r}_{B/A}$ is constant in both magnitude (since the body is rigid) and direction (since the body is translating), its derivative is zero and the time derivative of Eq. 14-1 gives simply

$$\mathbf{v}_B = \mathbf{v}_A \qquad (14\text{-}2)$$

where $\mathbf{v}_A$ and $\mathbf{v}_B$ are the absolute velocities of the particles A and B, respectively. That is, the velocity of every particle in a translating body is the same as the velocity of every other particle in the body.

Equation 14-2 can be differentiated with respect to time again to get

$$\mathbf{a}_B = \mathbf{a}_A \qquad (14\text{-}3)$$

where $\mathbf{a}_A$ and $\mathbf{a}_B$ are the absolute accelerations of the particles A and B, respectively. Equation 14-3 says that the acceleration of every particle in a translating body is also the same as the acceleration of every other particle in the body.

Since the motion of every point is the same as every other point, no distinction need be made between the motion of particle A and the motion of particle B. It is called simply the motion of the body. Since the size, shape, and orientation of the body are not important in describing the motion of the body, the kinematics of the particles that make up a rigid body undergoing translation is identical to the kinematics of particle motion discussed in Chapter 13. All the results of Chapter 13 apply, and no further discussion of a rigid body in translation will be made here.

14-3   PLANAR MOTION

In a plane motion, each particle in the body remains in a single plane. Since all points along lines perpendicular to a plane have the same motion, only the motion in a single plane need be considered. In the discussion that follows, the plane that contains the mass center, called the plane of motion, will be used.

Since particles cannot move out of the plane of motion, the position of a rigid body in plane motion is completely determined by giving the location of one point and the orientation of one line in the plane of motion (Fig. 14-3). The orientation of the line may be given either by giving the angle it makes with a fixed direction (Fig. 14-3a) or by giving the location of any two points on the line (Fig. 14-3b). The motion of the entire body can be determined from the motion of this one point and the motion of the line.

It is important to note that the angular motion of lines in the plane of motion is the same for every straight line in a rigid body. For example, consider the body of Fig. 14-4 on which have been drawn two lines separated by a fixed angle β. Both lines are in the plane of motion, and the angles between these lines and a fixed reference direction are θ_{AB} and θ_{CD}, respectively, as shown. From Fig. 14-4 these angles are related by

$$\theta_{CD} = \theta_{AB} + \beta \qquad (a)$$

As the body moves, the angles θ_{AB} and θ_{CD} will change. Since the body

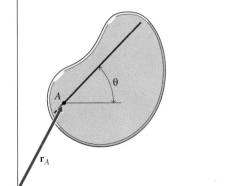

(a)

(b)

Fig. 14-3

is rigid, however, the angle β is fixed and differentiation of Eq. a with respect to time gives

$$\omega_{CD} = \dot{\theta}_{CD} = \dot{\theta}_{AB} = \omega_{AB} \qquad (b)$$

where ω, the time rate of change of angular position, is called the **angular velocity.** Equation b gives that the angular velocity of every line in the body is the same. Therefore, $\omega_{AB} = \omega_{CD}$ will be called simply the *angular velocity of the body.*

Equation b can be differentiated with respect to time again to get

$$\alpha_{CD} = \dot{\omega}_{CD} = \ddot{\theta}_{CD} = \ddot{\theta}_{AB} = \dot{\omega}_{AB} = \alpha_{AB} \qquad (c)$$

where α, the time rate of change of angular velocity, is called the **angular acceleration.** Like the angular velocity, the angular acceleration of every line in the body is the same, and $\alpha_{AB} = \alpha_{CD}$ will be called simply the *angular acceleration of the body.*

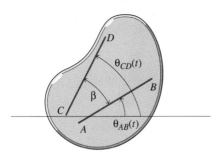

Fig. 14-4

14-4   ROTATION ABOUT A FIXED AXIS

It has been noted that the position of a rigid body in plane motion is completely determined by giving the location of one point and the orientation of one line in the plane of motion. The motion of the entire body can be determined from the motion of this one point and the motion of the line. For rotation about a fixed axis, however, the point on the axis remains on the axis—always. Therefore, the motion of the entire body can be determined from the motion of a line.

14-4-1   Motion of a Line in Fixed Axis Rotation

In fixed axis rotation the position of the body is completely determined by giving the angular position θ of any line in the plane of motion. The time derivatives of the angular position give the angular velocity $\omega(t)$

$$\frac{d\theta}{dt} = \omega(t) \qquad (14\text{-}4)$$

and angular acceleration $\alpha(t)$

$$\frac{d^2\theta}{dt^2} = \frac{d\omega}{dt} = \alpha(t) \qquad (14\text{-}5)$$

of the rigid body.

Eqs. 14-4 and 14-5 relating the angular position, angular velocity, and angular acceleration of a rigid body are analogous to the equations relating the position, velocity, and acceleration of a particle in rectilinear motion developed in Section 13-3. Just as the equations of Section 13-3 were integrated to get general relationships between the position, velocity, and acceleration of a particle in rectilinear motion, these equations can be integrated to get general relationships between the angular position, angular velocity, and angular acceleration of a rigid body.

In particular, if the angular acceleration is known as a function of time, then it can be integrated to get the angular velocity

$$\omega(t) - \omega_0 = \int_0^t \alpha(\tau)\, d\tau \qquad (14\text{-}6)$$

and the angular position

$$\theta(t) - \theta_0 = \int_0^t \omega(\tau)\, d\tau \tag{14-7}$$

as functions of time. For the special case in which the angular acceleration is constant, these integrals may be evaluated immediately to get

$$\omega(t) = \omega_0 + \alpha t \tag{14-8}$$

and

$$\theta(t) = \theta_0 + \omega_0 t + \frac{1}{2}\alpha t^2 \tag{14-9}$$

If the angular acceleration is known as a function of angular position instead of time, then using the chain rule of differentiation gives

$$\alpha(\theta) = \frac{d\omega}{dt} = \frac{d\omega}{d\theta}\frac{d\theta}{dt} = \omega\frac{d\omega}{d\theta}$$

which can be integrated to get the angular velocity as a function of angular position

$$\frac{\omega_2^2}{2} - \frac{\omega_1^2}{2} = \int_{\theta_1}^{\theta_2} \alpha(\phi)\, d\phi \tag{14-10}$$

Equations 14-6 through 14-10 are directly analogous to the formulas developed in Section 13-3 for a particle in rectilinear motion. All the results derived there also apply to the rotation of a rigid body about a fixed axis by simply replacing x with θ, v with ω, and a with α.

14-4-2   Motion of a Point in Fixed Axis Rotation

In fixed axis rotation, particles that are not on the axis travel in circular paths centered on the axis. If $\mathbf{r}_P$ is the position vector of point P measured relative to the axis of rotation (Fig. 14-5), then the velocity of point P can be written in n-t coordinates[1] (Eq. 13-38)

$$\mathbf{v}_P = r_P\omega\mathbf{e}_t \tag{14-11a}$$

where $\mathbf{e}_t = \mathbf{e}_\theta$ is a unit vector tangent to the circular path at P (perpendicular to $\mathbf{r}_P$). The x- and y-components of this velocity are easily found to be

$$\mathbf{v}_P = -r_P\omega \sin\theta\, \mathbf{i} + r_P\omega \cos\theta\, \mathbf{j} \tag{14-11b}$$

where θ is the angle between the position vector $\mathbf{r}_P$ and the x-axis.

The velocity of P can also be written in terms of a vector angular velocity $\boldsymbol{\omega}$ defined by

$$\boldsymbol{\omega} = \omega\mathbf{k}$$

(Fig. 14-6). The direction of this vector represents the axis about which

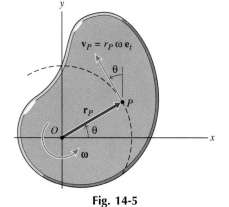

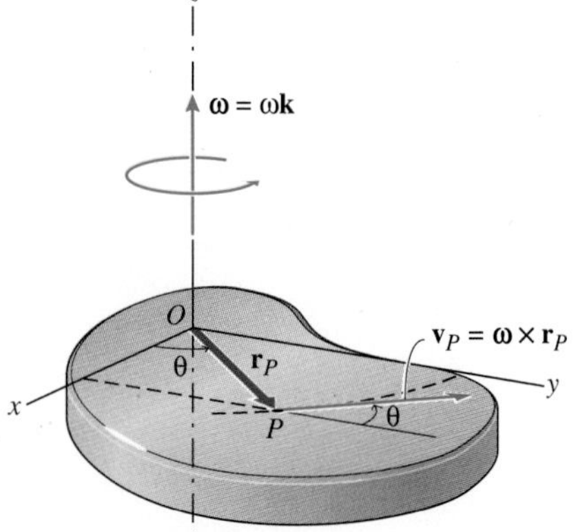

Fig. 14-5

Fig. 14-6

[1] Equations 14-11a and 14-12a could just as easily have been written in terms of r-θ coordinates. As shown in Section 13-5, when a particle travels around a circular path of constant radius, the expressions for the velocity in the n-t coordinate system (Eq. 13-38) and in the r-θ coordinate system (Eq. 13-33) are the same. Similarly, when a particle travels around a circular path of constant radius, the expressions for the acceleration in the n-t coordinate system (Eq. 13-38) and in the r-θ coordinate system (Eq. 13-33) are the same.

the body is rotating. The sense of rotation is according to the *right-hand rule*. That is, if you hold your right hand with your thumb pointing in the direction of the vector, your fingers curl in the sense of the rotation (in this case, counterclockwise when the plane of motion is viewed downward along the z-axis). Then the vector product

$$\boldsymbol{\omega} \times \mathbf{r}_P$$

gives a vector **b**, which is perpendicular to $\boldsymbol{\omega}$ and perpendicular to $\mathbf{r}_P$. Hence the vector **b** lies in the plane of motion and points in the direction of the velocity $\mathbf{v}_P$. Furthermore, since $\boldsymbol{\omega}$ and $\mathbf{r}_P$ are perpendicular, the vector **b** has magnitude $r_P \omega \sin 90° = r_P \omega$. Therefore, the vector **b** is the velocity of point P,

$$\mathbf{v}_P = \boldsymbol{\omega} \times \mathbf{r}_P = r_P \omega \mathbf{e}_t \qquad (14\text{-}11c)$$

Evaluating the vector product of Eq. 14-11c in terms of x-y coordinates gives

$$\begin{aligned} \mathbf{v}_P &= (\omega \mathbf{k}) \times (r_P \cos \theta\, \mathbf{i} + r_P \sin \theta\, \mathbf{j}) \\ &= -r_P \omega \sin \theta\, \mathbf{i} + r_P \omega \cos \theta\, \mathbf{j} \end{aligned} \qquad (14\text{-}11d)$$

which is the same as Eq. 14-11b above.

The acceleration of point P moving along the circular path about the axis of rotation will have both a tangential component and a normal component (Fig. 14-7)

$$\mathbf{a}_P = (\mathbf{a}_P)_t + (\mathbf{a}_P)_n = r_P \alpha \mathbf{e}_t + r_P \omega^2 \mathbf{e}_n \qquad (14\text{-}12a)$$

The x- and y-components of the acceleration are

$$\begin{aligned} \mathbf{a}_P = &-r_P \alpha \sin \theta\, \mathbf{i} + r_P \alpha \cos \theta\, \mathbf{j} \\ &-r_P \omega^2 \cos \theta\, \mathbf{i} - r_P \omega^2 \sin \theta\, \mathbf{j} \end{aligned} \qquad (14\text{-}12b)$$

By analogy with the velocity of P, the tangential component of the acceleration can be written

$$(\mathbf{a}_P)_t = \boldsymbol{\alpha} \times \mathbf{r}_P \qquad (14\text{-}12c)$$

where $\boldsymbol{\alpha}$ is the vector angular acceleration defined by

$$\boldsymbol{\alpha} = \alpha \mathbf{k}$$

(Fig. 14-8). The direction of the vector $\boldsymbol{\alpha}$ is in accordance with the right-hand rule the same as $\boldsymbol{\omega}$. Also, since $\mathbf{k}$, $\mathbf{e}_t$, and $\mathbf{e}_n$ are perpendicular unit vectors,

$$\boldsymbol{\omega} \times \mathbf{v}_P = (\omega \mathbf{k}) \times (r_P \omega \mathbf{e}_t) = r_P \omega^2 \mathbf{e}_n = (\mathbf{a}_P)_n \qquad (14\text{-}12d)$$

and therefore the acceleration of P is

$$\begin{aligned} \mathbf{a}_P &= (\mathbf{a}_P)_t + (\mathbf{a}_P)_n = \boldsymbol{\alpha} \times \mathbf{r}_P + \boldsymbol{\omega} \times \mathbf{v}_P \\ &= \boldsymbol{\alpha} \times \mathbf{r}_P + \boldsymbol{\omega} \times (\boldsymbol{\omega} \times \mathbf{r}_P) \end{aligned} \qquad (14\text{-}12e)$$

(Note that the parentheses are needed in the last term since

$$\boldsymbol{\omega} \times (\boldsymbol{\omega} \times \mathbf{r}_P) \neq (\boldsymbol{\omega} \times \boldsymbol{\omega}) \times \mathbf{r}_P = \mathbf{0}$$

and the order in which the cross-products are performed is important.)

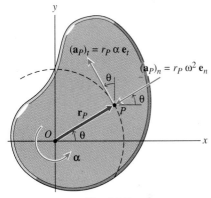

Fig. 14-7

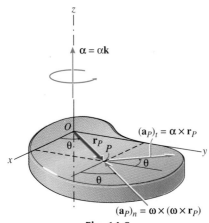

Fig. 14-8

EXAMPLE PROBLEM 14-1

The turntable of a record player attains its operating speed of $33\frac{1}{3}$ rpm in 5 revolutions after being turned on. Determine the initial angular acceleration α_0 of the turntable if:

a. The angular acceleration is constant, $\alpha = \alpha_0 = $ constant.
b. The angular acceleration decreases linearly with angular velocity from α_0 when $\omega = 0$ to $\alpha_0/4$ when $\omega = 33\frac{1}{3}$ rpm.

SOLUTION

a. First, convert the angular velocity to rad/s

$$\frac{(33\frac{1}{3} \text{ rpm})(2\pi \text{ rad/rev})}{60 \text{ s/min}} = 3.491 \text{ rad/s}$$

and the angular displacement to radians

$$(5 \text{ rev})(2\pi \text{ rad/rev}) = 10\pi \text{ rad}$$

Then since a relationship between the angular displacement and angular velocity is desired, integrate

$$\alpha = \omega\frac{d\omega}{d\theta} = \alpha_0$$

to get

$$\frac{3.491^2}{2} = \alpha_0 (10\pi)$$

or

$$\alpha_0 = 0.1939 \text{ rad/s}^2 \qquad\qquad \text{Ans.}$$

b. This time the angular acceleration is to decrease linearly with angular velocity so

$$\alpha(\omega) = \alpha_0 (1 - 0.2148\omega)$$

and

$$\int_0^{3.491} \frac{\omega d\omega}{1 - 0.2148\omega} = \alpha_0 \int_0^{10\pi} d\theta$$

which gives

$$\alpha_0 = 0.4390 \text{ rad/s}^2 \qquad\qquad \text{Ans.}$$

EXAMPLE PROBLEM 14-2

An 80-mm diameter gear rotates about an axle through its center O (Fig. 14-9). At some instant of time, the angular velocity of the gear is 2 rad/s counterclockwise and is increasing at the rate of 1 rad/s^2. Determine the acceleration (magnitude and direction) of tooth A at this instant.

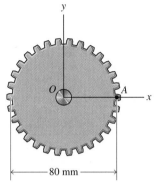

Fig. 14-9

SOLUTION

The acceleration of a point on a rigid body in fixed axis rotation is

$$\mathbf{a}_A = r\alpha\mathbf{e}_t + r\omega^2\mathbf{e}_n = (40)(1)\mathbf{j} + (40)(2)^2(-\mathbf{i})$$
$$= -160\mathbf{i} + 40\mathbf{j} \text{ mm/s}^2$$
$$= 164.9 \text{ mm/s}^2 \, \text{\textbackslash}\, 14.04° \qquad\qquad \text{Ans.}$$

Alternatively, using the vector product form for the acceleration

$$\mathbf{a}_A = \boldsymbol{\alpha} \times \mathbf{r} + \boldsymbol{\omega} \times (\boldsymbol{\omega} \times \mathbf{r})$$
$$= (1\mathbf{k}) \times (40\mathbf{i}) + (2\mathbf{k}) \times [(2\mathbf{k}) \times (40\mathbf{i})]$$
$$= 40\mathbf{j} + (2\mathbf{k}) \times (80\mathbf{j}) = 40\mathbf{j} + 160(-\mathbf{i})$$
$$= -160\mathbf{i} + 40\mathbf{j} \text{ mm/s}^2$$
$$= 164.9 \text{ mm/s}^2 \, \text{\textbackslash}\, 14.04° \qquad\qquad \text{Ans.}$$

as before.

PROBLEMS

14-1* The angular position of a gear is given by

$$\theta = 5e^{-t} \sin 2t \text{ rad}$$

where t is in seconds. Determine the angular velocity and the angular acceleration of the gear when $t = 2$ s.

14-2* The angular velocity of a gear is given by

$$4\omega^2 + 9\theta^2 = 25$$

where $\omega = \dot{\theta}$ and $\theta = \theta(t)$. Determine

a. The angular acceleration of the gear as a function of the angular position.
b. The angular position of the gear as a function of time.

14-3 The electric motor shown in Fig. P14-3 gives the grinding wheel a constant angular acceleration when it is turned on. If the motor attains its operating speed of 3600 rpm within 3 s after it is turned on, determine the angular acceleration of the grinding wheel.

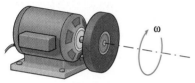

Fig. P14-3

14-4* The electric motor of Problem 14-3 gives the grinding wheel a constant angular acceleration of 150 rad/s^2 when it is turned on. Determine

a. The time required for the motor to achieve its operating speed of 3600 rpm after it is turned on.
b. The number of revolutions the grinding wheel turns before the motor attains its operating speed.

14-5 When the electric motor of Problem 14-3 is turned off, friction in the bearings causes the grinding wheel to coast to rest in 2 min with a constant angular deceleration. Determine the angular deceleration of the wheel caused by the friction.

14-6 Friction in the bearings of the electric motor of Problem 14-3 causes a constant angular deceleration of 3 rad/s^2 when the motor is shut off. Determine

a. The time required for the grinding wheel to coast to rest from the operating speed of 3600 rpm.
b. The number of revolutions that the grinding wheel turns before it comes to rest.

14-7* A variable-torque drive motor gives a circular disk an angular acceleration that varies linearly with angular position as shown in Fig. P14-7. If the angular velocity of the disk is 10 rad/s when $\theta = 0$, determine the angular velocity of the disk after 50 revolutions.

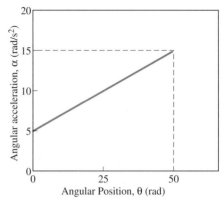

Fig. P14-7

14-8 A variable-torque drive motor gives a circular disk an angular acceleration inversely proportional to its angular velocity

$$\alpha = \frac{k}{\omega} \text{ rad/s}^2$$

where ω is in rad/s, k is a constant, and $\omega = 0$ when $\theta = 0$. If the angular velocity of the disk is 40 rad/s after 25 revolutions, determine the angular velocity of the disk after 50 revolutions.

14-9* A variable-torque drive motor gives a circular disk an angular acceleration of

$$\alpha = \left(\frac{\omega}{16} - 8\right)^2 \text{ rad/s}^2$$

where ω is in rad/s and $\omega = \theta = 0$ at $t = 0$. Determine

a. The time it takes the motor to rotate the disk 50 revolutions.
b. The angular velocity of the disk after 50 revolutions.

14-10* A variable-torque drive motor gives a circular disk an angular acceleration of

$$\alpha = 8 - 0.5\omega$$

where ω is in rad/s and $\omega = \theta = 0$ at $t = 0$. Determine

a. The time it takes the motor to rotate the disk 50 revolutions.
b. The angular velocity of the disk after 50 revolutions.

14-11 A small block B rotates with the horizontal turntable A of Fig. P14-11. The distance between the block and the axis of rotation is $r = 3$ in.; the angular acceleration of the turntable is $\alpha = 5$ rad/s^2 = constant; and the initial angular velocity is zero. Determine

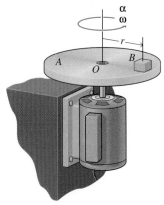

Fig. P14-11

a. The time t_1 when the normal and tangential components of the acceleration of the block are equal.
b. The number of revolutions N of the turntable between $t = 0$ and $t = t_1$.
c. The angular velocity of the turntable at $t = t_1$.

14-12* A small block B rotates with the horizontal turntable A of Fig. P14-11. The distance between the block and the axis of rotation is $r = 50$ mm; the angular acceleration of the turntable is $\alpha = 5$ rad/s^2 = constant; and the initial angular velocity is zero. Determine

a. The time t_1 when the magnitude of the acceleration of the block is 4 m/s^2.
b. The number of revolutions N of the turntable between $t = 0$ and $t = t_1$.
c. The angular velocity of the turntable at $t = t_1$.

14-13 A small block B rotates with the horizontal turntable A of Fig. P14-11. The distance between the block and the axis of rotation is $r = 5$ in.; the angular acceleration of the turntable is $\alpha = -2$ rad/s^2 = constant; and the initial angular velocity is $\omega = 15$ rad/s. Determine the angle ϕ between the acceleration of the block $\mathbf{a}_B$ and the radial line OB at $t = 5$ s.

14-14 A small block B rotates with the horizontal turntable A of Fig. P14-11. The distance between the block and the axis of rotation is $r = 80$ mm; the angular acceleration of the turntable is $\alpha = -3$ rad/s^2 = constant; and the initial angular velocity is $\omega = 15$ rad/s. Determine the time t_1 when the angle ϕ between the acceleration of the block $\mathbf{a}_B$ and the radial line OB is 30°.

14-15* A small block B rotates with the horizontal turntable A of Fig. P14-11. The distance between the block and the axis of rotation is $r = 4$ in.; the angular acceleration of the turntable is $\alpha = 2$ rad/s^2 = constant; and the initial angular velocity is zero. Compute and sketch

a. The magnitude of the acceleration of the block $\mathbf{a}_B$ as a function of the angle of rotation θ for the first two revolutions of the turntable.

b. The angle ϕ between the acceleration of the block $\mathbf{a}_B$ and the radial line OB as a function of the angle of rotation θ for the first two revolutions of the turntable.

14-16 The bicycle sprocket of Fig. P14-16 has a diameter of 200 mm. At some instant of time, a link of the chain has a velocity $v_A = 0.4$ m/s and an acceleration $a_A = 0.1$ m/s^2. For this instant, determine

a. The angular velocity ω of the sprocket.
b. The angular acceleration α of the sprocket.
c. The acceleration $\mathbf{a}_B$ of tooth B of the sprocket.

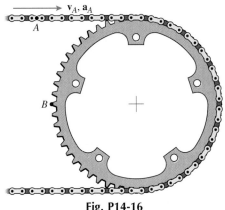

Fig. P14-16

14-17* The 6-in. diameter bicycle sprocket of Fig. P14-16 is driven by an electric motor that gives it a constant angular acceleration α. If link A on the chain has a velocity of 20 ft/s just 5 s after the motor is started, determine

a. The angular acceleration α of the sprocket.
b. The angular velocity ω of the sprocket at the instant when $v_A = 20$ ft/s.
c. The acceleration $\mathbf{a}_B$ of tooth B on the sprocket at the instant when $v_A = 20$ ft/s.

14-18* Weights A and B are supported by cords wrapped around a stepped drum as shown in Fig. P14-18. At the instant shown, weight A has a downward velocity of 2 m/s and its speed is decreasing at a rate of 1.5 m/s^2. For this instant, determine

a. The acceleration of weight B.
b. The acceleration of point D on the rim of the drum.

Fig. P14-18

14-19 The rock tumbler C shown in Fig. P14-19 is rotating on two small wheels A and B. When the motor is started, it rotates wheel A with a constant angular acceleration α_A. If the tumbler reaches its operating speed of 20 rpm in 3 s after being started, determine

a. The angular acceleration of wheel A.
b. The number of revolutions the tumbler rotates in the first 3 s.

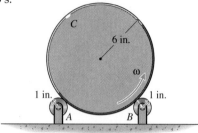

Fig. P14-19

14-20* Initially, disk B of Fig. P14-20 is at rest and disk A is rotating at 600 rpm. When the disks are brought together, they slip for a period of 10 s, during which the angular acceleration of each disk is constant. At the end of the 10 s, the disks roll without slipping on one another and disk A has reached a final angular velocity of 250 rpm. Determine the angular acceleration of each disk and the final angular velocity of disk B.

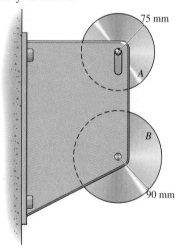

Fig. P14-20

14-21 Initially, disk B of Fig. P14-21 is rotating clockwise at 200 rpm and disk A is rotating counterclockwise at 500 rpm. When the disks are brought together, they slip for a period of 5 s, during which the angular acceleration of each disk is constant. At the end of the 5 s, the disks roll without slipping on one another and disk B has reached a final angular velocity of 250 rpm clockwise. Determine the angular acceleration of each disk and the final angular velocity of disk A.

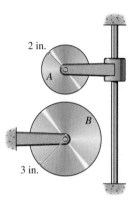

Fig. P14-21

14-22 It is desired to bring the two disks of Fig. P14-20 together without slipping. Initially, disk B is at rest and disk A is rotating at 600 rpm. If disk A is given a constant angular deceleration of 3 rad/s² and disk B is given a constant angular acceleration of 5 rad/s², determine the time at which the two disks can be brought together without slipping and the angular velocity of each disk at that time.

14-23* It is desired to bring the two disks of Fig. P14-21 together without slipping. Initially, disk B is at rest and disk A is rotating at 750 rpm. If disk A is given a constant angular deceleration of 5 rad/s² and disk B is given a constant angular acceleration of 8 rad/s², determine the time at which the two disks can be brought together without slipping and the angular velocity of each disk at that time.

14-5 GENERAL PLANE MOTION

A general plane motion is any plane motion for which lines in the body rotate but no point in the body is fixed. It will be seen that general plane motions are a superposition of translation and fixed axis rotation.

There are two general approaches to solving general plane motion problems: absolute motion analysis and relative motion analysis. In the first approach, geometric relationships are written that describe the constraints that act on the body and its interaction with other bodies. These relationships are then used to describe the location and motion of other points on the body. The second approach uses the concept of the relative motion of particles developed in Section 13-6. Since the distance between two points in a rigid body is fixed, the expressions for the relative velocity and relative acceleration take particularly simple forms that depend only on the angular velocity and angular acceleration of the body.

Either approach may be used to solve any particular problem. Some problems are easily described geometrically and are easily handled by the absolute motion approach. Problems that are not easily described geometrically are usually solved using the relative motion approach. For many problems the choice is a matter of personal preference.

14-5-1 Absolute Motion Analysis

Equations relating the angular motion of the rigid body and the motion of some point of the rigid body can be obtained from careful geometric analysis of the relationship between points and lines on a rigid body. The location of some point on the body is first obtained as a function of the angular orientation of the body. Then the time derivatives of this relationship give the velocity and acceleration of the point as functions of the angular orientation, the angular velocity, and the angular acceleration of the body.

Since the absolute motion approach relies totally on the geometric description of the body or bodies in a problem, no general formulas can be derived. Specific formulas must be derived for each specific problem, as illustrated in Example 14-3.

EXAMPLE PROBLEM 14-3

Derive an expression relating the position of a point on the rim of a wheel and the rotation of the wheel as it rolls without slipping on a stationary horizontal surface. Use the expression to:

a. Give the velocity of the point as a function of θ and ω.
b. Show that the velocity of the point of contact between the wheel and the surface is instantaneously zero.
c. Give the acceleration of the point as a function of θ, ω, and α.
d. Show that the acceleration of the point of contact with the surface is normal to the surface and is not zero.

SOLUTION

a. Let A, B, C, and D be points on the rim of the wheel as shown in Fig. 14-10. When the wheel rotates through an angle θ, the center of the wheel will move from O to O' and point C will rotate down to contact the surface at C'. The location of point A can be written in terms of the angle θ

$$\mathbf{r}_A = (x + r \sin \theta) \mathbf{i} + (r + r \cos \theta) \mathbf{j}$$

where $x = \overline{OO'} = \overline{BC'}$ is the distance traveled by the center of the wheel as it rotates. Since the wheel does not slip as it rolls, the distance $\overline{BC'}$ must equal the arc length $\overline{BC} = r\theta$ and therefore

$$\mathbf{r}_A = (r\theta + r \sin \theta) \mathbf{i} + (r + r \cos \theta) \mathbf{j} \qquad \text{Ans.}$$

(The path described by this equation is called a cycloid and is drawn in Fig. 14-11a).

The velocity of point A is given by the time derivative of the position

$$\mathbf{v}_A = \dot{\mathbf{r}}_A = \frac{d\mathbf{r}_A}{d\theta} \frac{d\theta}{dt}$$
$$= r\omega[(1 + \cos \theta) \mathbf{i} - \sin \theta \, \mathbf{j}] \qquad \text{Ans.}$$

The x- and y-components of the velocity are drawn in Fig. 14-11b.

b. Evaluating the velocity of A when $\theta = 180°$ (and A is in contact with the surface) gives

$$\mathbf{v}_A(180°) = r\omega[(1 - 1) \mathbf{i} - 0 \, \mathbf{j}] = \mathbf{0} \qquad \text{Ans.}$$

Therefore the contact point is instantaneously at rest: it is an instantaneous center of zero velocity.

c. The acceleration of A is given by the time derivative of the velocity

$$\mathbf{a}_A = \dot{\mathbf{v}}_A = r[\alpha(1 + \cos \theta) - \omega^2 \sin \theta] \mathbf{i}$$
$$- r[\alpha \sin \theta + \omega^2 \cos \theta] \mathbf{j} \qquad \text{Ans.}$$

d. Evaluating the acceleration of A when $\theta = 180°$ (and A is in contact with the surface) gives

$$\mathbf{a}_A = r\omega^2 \, \mathbf{j} \qquad \text{Ans.}$$

Therefore the contact point is not an instantaneous center of zero acceleration. The acceleration of the contact point is toward the center of the wheel and is perpendicular to the surface on which the wheel rolls.

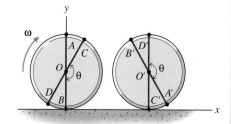

Fig. 14-10

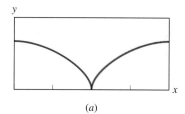

(a)

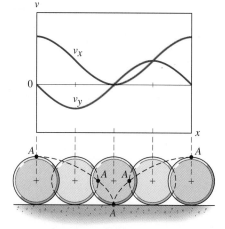

(b)

Fig. 14-11

PROBLEMS

14-24* Rod *CD* of Fig. P14-24 moves in the horizontal direction causing lever *AB* to rotate. The circular disk has a radius $r = 50$ mm, and at the instant shown $\theta = 30°$ and the velocity of the plunger is 7 m/s to the right. Determine the angular velocity ω of the lever *AB*.

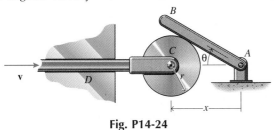

Fig. P14-24

14-25* Rod *CD* of Fig. P14-24 moves in the horizontal direction causing lever *AB* to rotate. The circular disk has a radius $r = 2$ in., and at the instant shown $x = 5$ in. and the velocity of the plunger is 15 ft/s to the right. Determine the angular velocity ω of the lever *AB*.

14-26 For the plunger/rod mechanism of Problem 14-24, express the angular velocity ω of the lever *AB* in terms of v, the velocity of the rod *CD*; r, the radius of the circular disk; and θ, the angle lever *AB* makes with the horizontal.

14-27* For the plunger/rod mechanism of Problem 14-24 express the angular velocity ω of the lever *AB* in terms of v, the velocity of the rod *CD*; r, the radius of the circular disk; and the distance x.

14-28 Rotation of the circular cam of Fig. P14-28 causes the plunger to move up and down. The radius of the cam is

$r = 50$ mm, and it is mounted on a shaft $b = 35$ mm from its center. At the instant shown, the cam is rotating at a constant angular speed of $\omega = 15$ rad/s and $\theta = 60°$. Determine the velocity and acceleration of the plunger at this instant.

14-29 For the cam/rod mechanism of Problem 14-28, express the velocity v and acceleration a of the plunger in terms of ω and α, the angular velocity and angular acceleration of the cam; r, the radius of the circular cam; and the angle θ.

14-30* The mechanism shown in Fig. P14-30 is used to convert the rotary motion of the arm *AB* into translational motion of the plunger *CD*. At the instant shown, the $b = 0.2$ m long arm *AB* is rotating counterclockwise at a constant angular speed of $\omega = 12$ rad/s, and $\theta = 60°$. Determine the velocity and acceleration of the plunger at this instant.

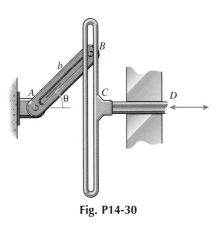

Fig. P14-30

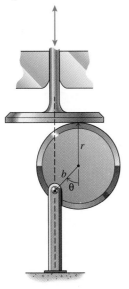

Fig. P14-28

14-31 For the mechanism of Problem 14-30 express the velocity and acceleration of the plunger in terms of θ, ω, α, and b, the angular position, angular velocity, angular acceleration, and length of the rod *AB*.

14-32* Sliders A and B of Fig. P14-32 are constrained to move in vertical and horizontal slots, respectively, and are connected by a $d = 800$ mm long rigid rod. At the instant shown $\theta = 75°$ and slider B is moving to the right with a constant speed of 25 mm/s. Determine the angular velocity and angular acceleration of bar AB at this instant.

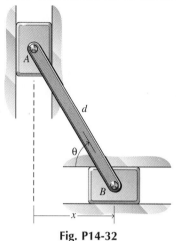

Fig. P14-32

14-33* Sliders A and B of Fig. P14-32 are constrained to move in vertical and horizontal slots, respectively, and are connected by a $d = 3$ ft long rigid rod. At the instant shown $x = 1$ ft and the slider B is moving to the right with a constant speed of 0.5 ft/s. Determine the velocity and acceleration of the slider A at this instant.

14-34 For the slider/rod system of Problem 14-32, express the angular velocity and angular acceleration of rod AB in terms of v_B and a_B, the velocity and acceleration of slider B; d, the length of the rod AB; and the angle θ.

14-35* For the slider/rod system of Problem 14-32, express the velocity and acceleration of the slider A in terms of v_B and a_B, the velocity and acceleration of slider B; d, the length of the rod AB; and the distance x.

14-36 The 2-m long rod AB of Fig. P14-36 slides on a

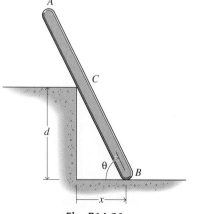

Fig. P14-36

$d = 1$ m high step. If end B of the rod is made to move to the right at a constant speed of 0.25 m/s, determine the angular velocity of the rod at the instant when the angle $\theta = 50°$.

14-37 The 7-ft long rod AB of Fig. P14-36 slides on a $d = 4$ ft high step. If end B of the rod is made to move to the right at a constant speed of 0.5 ft/s, determine the angular velocity of the rod at the instant when $x = 2$ ft.

14-38* For the rod of Problem 14-36, determine the velocity of point C, the point on the rod in contact with the corner of the step.

14-39 For the rod of Problem 14-36, express the angular velocity of the rod AB in terms of the position x and speed $\dot{x}$ of B, the height of the step d, and the angle θ.

14-40* As the circular cam of Fig. P14-40 rotates, the follower rod moves up and down. The radius of the cam is $r_1 = 40$ mm, and it is mounted on a shaft $b = 25$ mm from its center. The radius of the smaller disk is $r_2 = 30$ mm. At the instant shown, the cam is rotating at a constant angular speed of $\omega = 10$ rad/s and $\theta = 30°$. Determine the velocity and acceleration of the follower rod at this instant.

Fig. P14-40

14-41* For the cam/rod mechanism of Problem 14-40 express the velocity and acceleration of the follower rod in terms of ω and α, the angular velocity and angular acceleration of the cam; r_1 and r_2, the radii of the circular disks; and the angle θ.

14-5-2   Relative Velocity

If A and B are any two particles, their positions are related by the triangle law of addition for vectors

$$\mathbf{r}_B = \mathbf{r}_A + \mathbf{r}_{B/A} \tag{14-13a}$$

where $\mathbf{r}_A$ and $\mathbf{r}_B$ are the absolute positions of the particles A and B, respectively, and $\mathbf{r}_{B/A}$ is the position of B relative to A. The time derivative of Eq. 14-1 gives

$$\mathbf{v}_B = \mathbf{v}_A + \mathbf{v}_{B/A} \tag{14-13b}$$

where $\mathbf{v}_A$ is the absolute velocity (measured relative to a fixed coordinate system) of particle A, $\mathbf{v}_B$ is the absolute velocity of particle B, and $\mathbf{v}_{B/A}$ is the relative velocity of particle B (measured relative to particle A). Equations 14-13a and 14-13b apply to any two particles—whether they are part of a rigid body or not.

If particles A and B are two particles in a rigid body, however, then the distance between them is constant and particle B appears to travel a circular path around particle A. Therefore, the relative velocity $\mathbf{v}_{B/A}$ is given by (Eq. 14-11)

$$\mathbf{v}_{B/A} = r_{B/A}\omega\mathbf{e}_t = \omega\mathbf{k} \times \mathbf{r}_{B/A} \tag{14-13c}$$

where $\mathbf{r}_{B/A} = \mathbf{r}_B - \mathbf{r}_A$, $\mathbf{e}_t$ is a unit vector tangent to the relative motion (tangent to the circle centered at A), and ω is the angular velocity of the body. Then

$$\begin{aligned}
\mathbf{v}_B &= \mathbf{v}_A + \mathbf{v}_{B/A} = \mathbf{v}_A + r_{B/A}\omega\mathbf{e}_t \\
&= \mathbf{v}_A + \omega\mathbf{k} \times \mathbf{r}_{B/A}
\end{aligned} \tag{14-13d}$$

Therefore, the velocity of particle B consists of the sum of two parts: $\mathbf{v}_A$, which represents a translation of the entire body with particle A; and $r_{B/A}\omega\mathbf{e}_t$, which represents fixed axis rotation of the body about A (Fig. 14-12). In words, Eq. 14-13d says: *The velocity of any point B of a rigid body consists of a translation of the rigid body with point A plus a rotation of the rigid body about point A.*

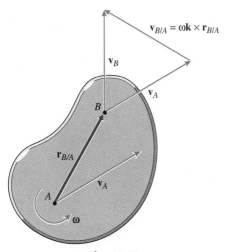

Fig. 14-12

The relative velocity equation (Eq. 14-13d) is a vector equation that, for planar motion, has two independent scalar components. Therefore the relative velocity equation can be used to find the two components of the velocity of some point $\mathbf{v}_B$ when the angular velocity of the body ω and the velocity of some other point in the body $\mathbf{v}_A$ are known. This equation can also be solved when the directions of the velocities $\mathbf{v}_A$ and $\mathbf{v}_B$ are known (for example, if they slide along fixed guides) and one of the three magnitudes v_A, v_B, or ω is given.

When two or more rigid bodies are pinned together as in Fig. 14-13, relative velocity equations can be written for each of the bodies separately. One of the points used in each of the equations should be the common point (point B of Fig. 14-13) connecting the two bodies; its velocity will be the same for each body. The other point in each equation should be some other point (A or C) whose velocity is known or is to be found. Then the velocities and angular velocities of the bodies can be related by equating the two expressions for the velocity of the common point

$$\mathbf{v}_B = \mathbf{v}_A + \mathbf{v}_{B/A} = \mathbf{v}_C + \mathbf{v}_{B/C}$$

or

$$\mathbf{v}_A + \omega_{AB}\mathbf{k} \times \mathbf{r}_{B/A} = \mathbf{v}_C + \omega_{BC}\mathbf{k} \times \mathbf{r}_{B/C}$$

This equation can be solved for any two unknowns if the other quantities are given.

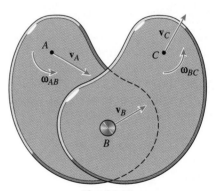

Fig. 14-13

EXAMPLE PROBLEM 14-4

The 3-m long ladder AB slides along a corner as shown in Fig. 14-14. When the angle $\theta = 30°$, the bottom end of the ladder is moving to the right with a constant speed of 2.0 m/s. Determine the velocity of the top end of the ladder and the angular velocity of the ladder at this instant.

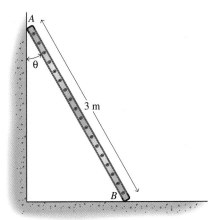

Fig. 14-14

SOLUTION

A coordinate system is chosen with x to the right and y up. Then the velocity of B is given as $\mathbf{v}_B = 2\,\mathbf{i}$ m/s, the velocity of A is $\mathbf{v}_A = v_A\mathbf{j}$, and the relative velocity equation gives

$$v_A\mathbf{j} = 2\mathbf{i} + \mathbf{v}_{A/B} \qquad (a)$$

But, from Fig. 14-15, the relative velocity is given by

$$\mathbf{v}_{A/B} = 3\omega\mathbf{e}_t = 3\omega(-\cos 30°\,\mathbf{i} - \sin 30°\,\mathbf{j})$$

Then the x-component of Eq. a gives

$$0 = 2 - 3\omega \cos 30°$$

or

$$\omega = 0.770 \text{ rad/s (CCW)} \qquad \text{Ans.}$$

and the y-component gives

$$v_A = -(3)(0.770) \sin 30° = -1.155 \text{ m/s}$$

or

$$\mathbf{v}_A = 1.155 \text{ m/s} \downarrow \qquad \text{Ans.}$$

Alternatively, the relative velocity term can be computed using the vector cross-product

$$\mathbf{v}_{A/B} = \omega\mathbf{k} \times (-3 \sin 30°\,\mathbf{i} + 3 \cos 30°\,\mathbf{j})$$
$$= -3\omega \sin 30°\,\mathbf{j} - 3\omega \cos 30°\,\mathbf{i}$$

which gives the same result as before.

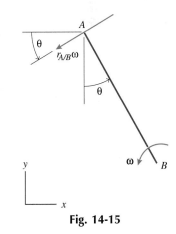

Fig. 14-15

EXAMPLE PROBLEM 14-5

The wheel of the slider-crank mechanism shown in Fig. 14-16 is rotating coun-terclockwise at a constant rate of 10 rad/s. Determine the velocity of the slider $\mathbf{v}_B$ and the angular velocity of the crank arm ω_{AB} when $\theta = 60°$.

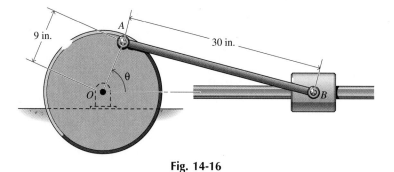

Fig. 14-16

SOLUTION

The wheel and the crank arm AB are separate rigid bodies connected at point A. The velocity of O, the axle of the wheel, is zero and the velocity of B, the slider, has only a horizontal component. The velocity of the common point is written relative to the velocities of these points and the angular velocities of each of the bodies

$$\mathbf{v}_A = \mathbf{v}_B + \mathbf{v}_{A/B} = \mathbf{v}_O + \mathbf{v}_{A/O} \qquad (b)$$

Using a coordinate system with x to the right and y up and referring to Fig. 14-17, the relative velocity terms are

$$\mathbf{v}_{A/B} = 30\omega_{AB}(-\sin\phi\,\mathbf{i} - \cos\phi\,\mathbf{j})\text{ in./s}$$
$$\mathbf{v}_{A/O} = 9\omega_{OA}(-\sin 60°\,\mathbf{i} + \cos 60°\,\mathbf{j})\text{ in./s}$$

where the angle ϕ is determined using the Law of Sines (Fig. 14-17c)

$$\frac{\sin\phi}{9} = \frac{\sin 60°}{30}$$

or

$$\phi = 15.06°$$

Then the x- and y-components of Eq. b

$$v_B - 30\omega_{AB}\sin 15.06° = -(9)(10)\sin 60°$$
$$-30\omega_{AB}\cos 15.06° = (9)(10)\cos 60°$$

give

$$\omega_{AB} = -1.553\text{ rad/s}$$
$$= 1.553\text{ rad/s (CW)} \qquad \text{Ans.}$$

and

$$v_B = -90.0\text{ in./s}$$

or

$$\mathbf{v}_B = 90.0\text{ in./s} \leftarrow \qquad \text{Ans.}$$

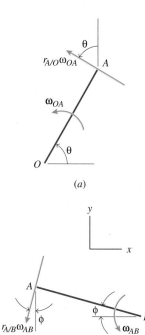

(a)

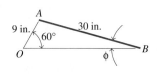

(b)

(c)

Fig. 14-17

PROBLEMS

14-42* Sliders A and B of Fig. P14-42 are constrained to move in vertical and horizontal slots, respectively, and are connected by a $d = 800$ mm long rigid rod. At the instant shown $\theta = 75°$ and slider B is moving to the right with a constant speed of 25 mm/s. Determine the velocity of slider A and the angular velocity of bar AB at this instant.

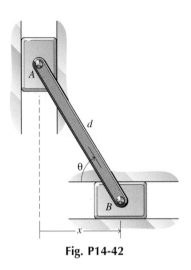

Fig. P14-42

14-43* The 7-ft long rod AB of Fig. P14-43 rests on a 4-ft high step. If end B of the rod is made to move to the right at a constant speed of 0.5 ft/s, determine the angular velocity of the rod at the instant when $x = 2$ ft.

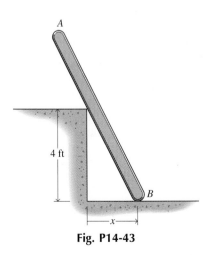

Fig. P14-43

14-44 The piston of Fig. P14-44 is connected to the crankshaft by a 650-mm long connecting rod. At the instant shown, the angular velocity of the crankshaft is 360 rpm clockwise. Determine the velocity of the piston at this instant.

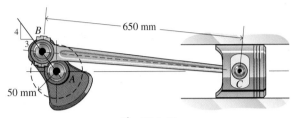

Fig. P14-44

14-45* Movement of the solenoid plunger shown in Fig. P14-45 causes a gear to rotate. If at the instant shown the angular velocity of the gear is $\omega_0 = 4$ rad/s counterclockwise, determine the angular velocity ω_{AB} of rod AB and the velocity $\mathbf{v}_A$ of the plunger.

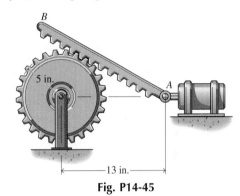

Fig. P14-45

14-46 Arm BC of Fig. P14-46 is attached to the sliding block and is made to rotate clockwise at a constant rate of 2 rpm. Determine the velocity of the block v_C and the angular velocity of the link AB at the instant shown when $\theta = 75°$.

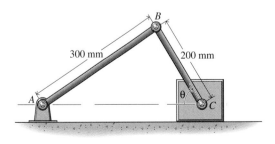

Fig. P14-46

14-47 A control rod *BC* is attached to a rotating crank *AB* as shown in Fig. P14-47. At the instant shown, the control rod *BC* is sliding through the pivoted guide *D* at a rate of 15 in./s. Determine the angular velocity ω_{AB} of the crank at this instant.

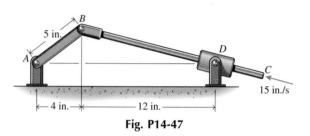

Fig. P14-47

14-48* The wheel shown in Fig. P14-48 is rotating at a constant angular velocity of 120 rpm counterclockwise. For the instant shown, determine the angular velocity ω_{AB} of bar *AB* and the velocity $\mathbf{v}_A$ of the slider.

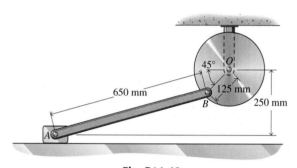

Fig. P14-48

14-49 The wheel of Fig. P14-49 has a velocity of 2 ft/s up the inclined plane and a clockwise angular velocity of 5 rad/s. Determine the rate of slip at the contact point *A* (the relative velocity between the point on the wheel and the stationary surface).

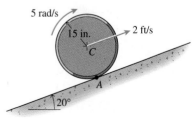

Fig. P14-49

14-50* At the instant shown roller *C* of Fig. P14-50 has a velocity of 250 mm/s up the channel. Determine the angular velocities of both bars and the velocity of pin *B* at this instant.

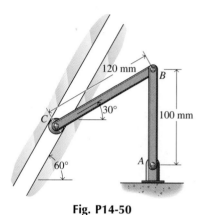

Fig. P14-50

14-51* The crank *AB* of Fig. P14-51 has a constant counterclockwise angular velocity of 60 rpm. If at the instant shown $\theta = 40°$, determine the angular velocity of member *BC* and the rate of slip at the contact point *D* (the relative velocity between the point on the arm *BC* and the stationary pivot point).

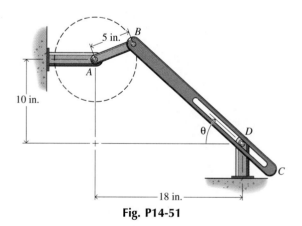

Fig. P14-51

14-52 The gears of an automatic transmission consist of a ring gear *R*, a sun gear *S*, three (or more) equal-sized planet gears *P*, and a planet carrier *C*, which is pin-connected to the center of the planet gears (Fig. P14-52). Various gear ratios (ratio of the output angular velocity ω_{out} to the input angular velocity ω_{in}) are achieved by holding one of the parts fixed and driving the others. Determine the gear ratio of the transmission shown if the ring gear is held fixed, the sun gear is the input ($\omega_{in} = \omega_S$), and the planet carrier is the output ($\omega_{out} = \omega_C$).

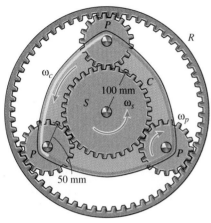

Fig. P14-52

14-53* The triangular plate of Fig. P14-53 oscillates as link AB rotates. Determine the angular velocity of the plate and the velocity of point D at the instant shown where link AB is rotating counterclockwise at 60 rpm.

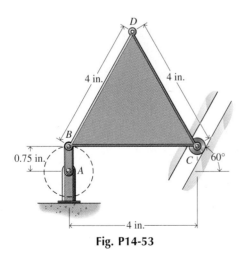

Fig. P14-53

14-54* Motion of the scissors mechanism shown in Fig. P14-54 is controlled by the horizontal action of the hydraulic cylinders attached at A and C. At the instant shown, A has a velocity to the right of 75 mm/s and C has a velocity

to the left of 50 mm/s. Determine the angular velocities of both bars and the velocity of point B at this instant.

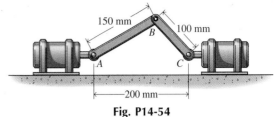

Fig. P14-54

14-55 The handle of the mechanism shown in Fig. P14-55 has a velocity of 5 in./s downward for some short interval of its motion. Determine the angular velocity of bar AB and the velocity of point A at the instant shown.

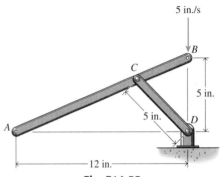

Fig. P14-55

14-56* In the four-bar linkage shown in Fig. P14-56 control link AB has a counterclockwise angular velocity of 100 rpm during a short interval of its motion. Bar BC is horizontal and bar CD is vertical when $\theta = 90°$. Determine the angular velocity of bar BC and the velocity of point E when $\theta = 90°$.

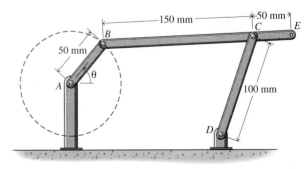

Fig. P14-56

14-57* Repeat Problem 14-55 for the case where the vertical velocity of point B is replaced by a velocity of 5 in./s horizontally to the right.

14-58 Repeat Problem 14-56 for $\theta = 0°$.

14-59* In the four-bar linkage shown in Fig. P14-59 control link *CD* has a constant clockwise angular velocity of 120 rpm. Determine the angular velocity of bars *AB* and *BC* at the instant shown in which *B* is directly above *D*.

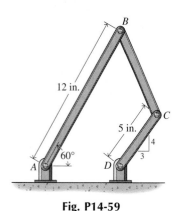

Fig. P14-59

14-60 For Problem 14-50 determine the maximum velocity of roller *C* for which the velocity of point *B* is less than 1 m/s and the angular velocity of bar *AB* is less than 2 rad/s at the instant shown.

14-61 For Problem 14-53 determine the maximum angular velocity of link *AB* for which the velocity of point *D* is less than 6 in./s and the angular velocity of the plate is less than 1 rad/s at the instant shown.

14-62* For Problem 14-56 determine the maximum angular velocity of link *AB* for which the velocity of point *E* is less than 1 m/s and the angular velocity of bar *CD* is less than 1.5 rad/s when $\theta = 0°$.

14-63 For Problem 14-57 determine the maximum horizontal velocity of point *B* for which the velocity of point *A* is less than 10 in./s at the instant shown.

14-64* For Problem 14-58 determine the maximum angular velocity of link *AB* for which the velocity of point *E* is less than 1 m/s and the angular velocity of bar *CD* is less than 1.5 rad/s when $\theta = 0°$.

14-65* For Problem 14-59 determine the maximum angular velocity of link *CD* for which the angular velocity of

bars *AB* and *BC* are both less than 3.0 rad/s at the instant shown.

14-66 Arm *D* of the mechanism shown in Fig. P14-66 has a sleeve that slides freely on the rod *BC*. The control link *AB* has an angular velocity of 60 rpm counterclockwise when in the position shown. For this position determine

a. The angular velocity of the sleeve body *D*.
b. The velocity of point *C*.
c. The velocity of rod *BC* relative to the sleeve body *D*.

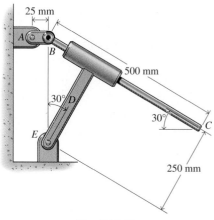

Fig. P14-66

14-67* The Geneva mechanism of Fig. P14-67 is used to produce intermittent motion. Both wheels rotate on fixed axles through their centers. If wheel *A* has a constant clockwise angular velocity of 30 rpm, determine the angular velocity of wheel *B* when $\theta = 30°$.

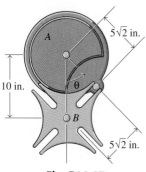

Fig. P14-67

14-5-3   Instantaneous Center of Zero Velocity

In the general plane motion of a rigid body, no point is fixed for all time. However, at any instant of time, it is always possible to find a point on the body (or the body extended) that has zero velocity.[2] This point is called the **instantaneous center of zero velocity** or simply the **instantaneous center.**

It is important to recognize that the instantaneous center of zero velocity for a rigid body in general plane motion is not fixed. The acceleration of the point that is the instantaneous center is usually not zero. Therefore, different points of the rigid body will be instantaneous centers at different instants of time and the location of the instantaneous center of zero velocity will move with respect to time.

In order to locate the instantaneous center, suppose that A and B are any two points in the rigid body whose velocity is known and that point C is an instantaneous center of zero velocity. Point C may lie in the body or in the body extended. Then, since $\mathbf{v}_C = \mathbf{0}$, the relative velocity equation (Eq. 14-13) gives

$$\mathbf{v}_A = \omega\mathbf{k} \times \mathbf{r}_{A/C}$$

and point C must lie on the line through A perpendicular to $\mathbf{v}_A$. Similarly,

$$\mathbf{v}_B = \omega\mathbf{k} \times \mathbf{r}_{B/C}$$

and point C must lie on the line through B perpendicular to $\mathbf{v}_B$. Assuming that $\mathbf{v}_A$ and $\mathbf{v}_B$ are not parallel, these two lines will intersect, and the point of intersection is point C (Fig. 14-18a).

If the velocities of points A and B are parallel, then the instantaneous center must lie on the line joining them. Since the magnitude of the relative velocity is ωr, the instantaneous center will be a distance $r_{A/C} = v_A/\omega$ from point A and a distance $r_{B/C} = v_B/\omega$ from point B, and its location can be found by similar triangles as shown in Figs. 14-18b and 14-18c.

If the velocities of points A and B are the same at any instant, then the body is instantaneously in translation and $\omega = 0$. This case can be included in the foregoing analysis if the instantaneous center is considered to be at infinity.

Once the instantaneous center is located, the velocity of any other point in the body is found using the relative velocity equation (Eq. 14-13)

$$\mathbf{v}_D = \mathbf{v}_C + \mathbf{v}_{D/C} = \omega\mathbf{k} \times \mathbf{r}_{D/C}$$

When two or more bodies are pinned together, an instantaneous center of zero velocity can be found for each body. In general, the location of these separate instantaneous centers will not coincide. The location of each of the instantaneous centers can be found as before. Since the velocity of the point joining two bodies is the same for each body, the instantaneous centers of both bodies must lie on a single line through the common point of the two bodies (Fig. 14-19).

Use of the instantaneous center is not required to solve any particular problem. It is just another way of expressing the relative velocity equation.

[2] The case of ω being instantaneously zero requires that the point be at infinity.

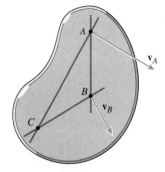

(a)

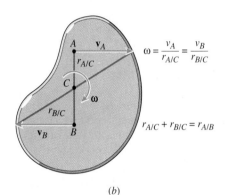

$$\omega = \frac{v_A}{r_{A/C}} = \frac{v_B}{r_{B/C}}$$

$$r_{A/C} + r_{B/C} = r_{A/B}$$

(b)

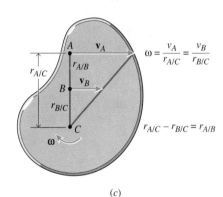

$$\omega = \frac{v_A}{r_{A/C}} = \frac{v_B}{r_{B/C}}$$

$$r_{A/C} - r_{B/C} = r_{A/B}$$

(c)

Fig. 14-18

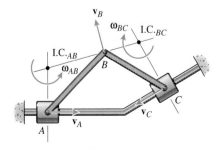

Fig. 14-19

At the instant shown in Fig. 14-20, slider A is moving to the right with a speed of 3 m/s. Find the location of the instantaneous center of zero velocity and use it to find the angular velocity of the arm ω_{AB} and the velocity of the slider $\mathbf{v}_B$.

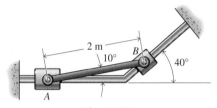

Fig. 14-20

SOLUTION

Since slider A moves in the horizontal direction, the instantaneous center of zero velocity must lie on the vertical line through A (Fig. 14-21). Similarly, the instantaneous center must lie on the line through B perpendicular to the right guide bar. The intersection of these two lines, point C, is the desired point. By the Law of Sines,

$$\frac{r_{A/C}}{\sin 60°} = \frac{2}{\sin 40°}$$
$$r_{A/C} = 2.69 \text{ m}$$

Therefore, the instantaneous center of zero velocity is located 2.69 m vertically above A.   Ans.

Since C is an instantaneous center of zero velocity,

$$\mathbf{v}_A = \mathbf{v}_C + \mathbf{v}_{A/C} = \mathbf{0} + \mathbf{v}_{A/C}$$

and the velocity of point A is given by

$$v_A = r_{A/C}\omega$$

so

$$\omega = 1.113 \text{ rad/s (CCW)} \qquad \text{Ans.}$$

Also since C is an instantaneous center of zero velocity,

$$\mathbf{v}_B = \mathbf{v}_C + \mathbf{v}_{B/C} = \mathbf{0} + \mathbf{v}_{B/C}$$

and the velocity of point B is given by

$$v_B = r_{B/C}\omega$$

Again using the Law of Sines to find the distance $r_{B/C}$ gives

$$\frac{r_{B/C}}{\sin 80°} = \frac{2}{\sin 40°}$$
$$r_{B/C} = 3.06 \text{ m}$$

Therefore, the velocity of slider B is

$$v_B = r_{B/C}\omega = 3.41 \text{ m/s} \measuredangle 40° \qquad \text{Ans.}$$

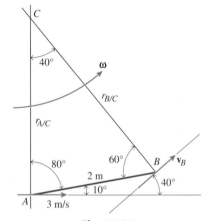

Fig. 14-21

PROBLEMS

14-68* Repeat Problem 14-42 (p. 93) using instantaneous center principles.

14-69* Repeat Problem 14-43 (p. 93) using instantaneous center principles.

14-70 Repeat Problem 14-44 (p. 93) using instantaneous center principles.

14-71* Repeat Problem 14-53 (p. 95) using instantaneous center principles.

14-72 Repeat Problem 14-56 (p. 95) using instantaneous center principles.

14-73 Repeat Problem 14-57 (p. 95) using instantaneous center principles.

14-74* Repeat Problem 14-58 (p. 95) using instantaneous center principles.

14-75 Repeat Problem 14-59 (p. 96) using instantaneous center principles.

14-76* The square plate of Fig. P14-76 is attached to a short link at A and a roller at C. If at the instant shown, link OA has a counterclockwise angular velocity of 4 rad/s, determine the angular velocity of the plate and the velocity of the roller C using instantaneous center principles.

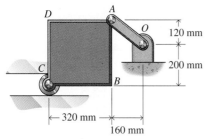

Fig. P14-76

14-77* For some short period of its motion, slider D of Fig. P14-77 has a velocity of 3 ft/s up the channel. For the

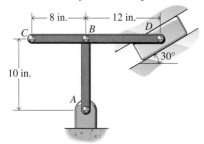

Fig. P14-77

instant shown, determine the angular velocities of both bars and the velocity of point B using instantaneous center principles.

14-78 The crank wheel OA of Fig. P14-78 is rotating counterclockwise at a steady angular velocity of 180 rpm. For the position shown determine the velocity of the slider B and the angular velocity of bar AB using instantaneous center principles.

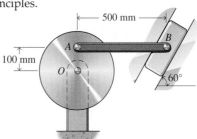

Fig. P14-78

14-79* The 1-ft diameter wheels shown in Fig. P14-79 roll along a horizontal plane without slipping and are connected by a 3-ft long rod AB. The pins A and B are 4 in. from the centers of the wheels. At the instant shown, the velocity of point P is 2 ft/s to the right. Determine the velocity of point Q and the angular velocity of bar AB at this instant using instantaneous center principles.

Fig. P14-79

14-80 A toy jeep is being driven on a treadmill as shown in Fig. P14-80. The treadmill is moving at 2 m/s and the absolute velocity of the jeep is 0.3 m/s. If the diameter of the wheels is 50 mm, determine the angular velocity of the wheels and the velocity $\mathbf{v}_P$ of the point on the front of a wheel.

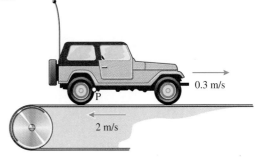

Fig. P14-80

14-81 Repeat Problem 14-79 when the wheels and bar are arranged as shown in Fig. P14-81.

Fig. P14-81

14-82* A stepped wheel rolls without slipping on its hub as shown in Fig. P14-82. If the velocity of the center C is 4.5 m/s to the right at the instant shown, determine

a. The angular velocity of the wheel ω.
b. The velocities of points A, D, and E.

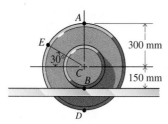

Fig. P14-82

14-83 The centers of the three gears shown in Fig. P14-83 are connected by smooth pins to the arm ABC, which rotates with an angular velocity of 5 rad/s counterclockwise. If the larger gear is fixed and does not rotate, determine the angular velocity of the smaller gears and the velocity of tooth D using instantaneous center principles.

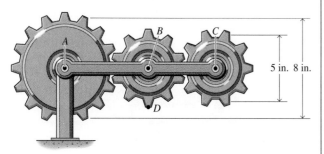

Fig. P14-83

14-84* The 0.5-m diameter wheel of Fig. P14-84 rolls without slipping inside a 2.0-m diameter fixed drum. If the center of the wheel has a velocity of 1.5 m/s to the right when it passes the bottom of the drum, determine the angular velocity of the wheel at that instant using instantaneous center principles.

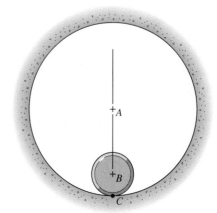

Fig. P14-84

14-85* The centers of the two gears shown in Fig. P14-85 are connected by smooth pins to the arm ABC, which rotates at a constant angular velocity of 5 rad/s clockwise about the fixed point A. The larger gear rolls on the inside of a fixed toothed drum. Determine the angular velocity of each of the gears and the velocity of tooth D for the position shown using instantaneous center principles.

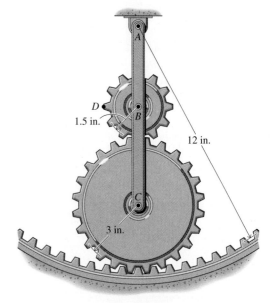

Fig. P14-85

14-5-4 Relative Acceleration

Taking two time derivatives of the relative position equation (Eq. 14-13a) gives the relationship between the accelerations of particles A and B

$$\mathbf{a}_B = \mathbf{a}_A + \mathbf{a}_{B/A} \qquad (14\text{-}14a)$$

where $\mathbf{a}_A$ is the absolute acceleration of particle A, $\mathbf{a}_B$ is the absolute acceleration of particle B, and $\mathbf{a}_{B/A}$ is the relative acceleration of particle B (measured relative to particle A). Equation 14-14a applies to any two particles—whether they are part of a rigid body or not.

If particles A and B are two particles in a rigid body, however, then the distance between them is constant and particle B appears to travel a circular path around particle A. Therefore, the relative acceleration $\mathbf{a}_{B/A}$ is given by (Eq. 14-12)

$$\begin{aligned}\mathbf{a}_{B/A} &= (\mathbf{a}_{B/A})_t + (\mathbf{a}_{B/A})_n = \alpha\mathbf{k} \times \mathbf{r}_{B/A} + \omega\mathbf{k} \times (\omega\mathbf{k} \times \mathbf{r}_{B/A}) \\ &= r_{B/A}\alpha\mathbf{e}_t + r_{B/A}\omega^2\mathbf{e}_n\end{aligned} \qquad (14\text{-}14b)$$

and

$$\begin{aligned}\mathbf{a}_B &= \mathbf{a}_A + \mathbf{a}_{B/A} = \mathbf{a}_A + \alpha\mathbf{k} \times \mathbf{r}_{B/A} + \omega\mathbf{k} \times (\omega\mathbf{k} \times \mathbf{r}_{B/A}) \\ &= \mathbf{a}_A + r_{B/A}\alpha\mathbf{e}_t + r_{B/A}\omega^2\mathbf{e}_n\end{aligned} \qquad (14\text{-}14c)$$

Therefore, the acceleration of particle B consists of two parts: $\mathbf{a}_A$, which represents a translation of the entire body with particle A; and $r_{B/A}\alpha\,\mathbf{e}_t + r_{B/A}\omega^2\,\mathbf{e}_n$, which represents fixed axis rotation of the body about A (Fig. 14-22).

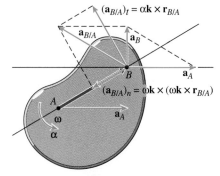

Fig. 14-22

The use of the relative acceleration equation (Eq. 14-14) in solving problems involving a single body or multiple connected bodies is analogous to the use of the relative velocity equation. Since the normal component of the relative acceleration equation involves the angular velocity ω, the relative velocity problem usually must be solved before the relative acceleration problem can be solved. For multiple connected bodies, the acceleration of the common point of the two rigid bodies is written in terms of the acceleration of some other point (whose acceleration is known or desired) on each of the rigid bodies.

The instantaneous center of zero velocity for a rigid body in general plane motion is not fixed. The acceleration of the point that is the instantaneous center is usually not zero. Therefore, this point must not be used to compute accelerations. An analysis similar to that used to locate the instantaneous center of zero velocity can also be used to locate an instantaneous center of zero acceleration. However, the instantaneous center of zero acceleration is not usually useful in the solution of simple problems.

EXAMPLE PROBLEM 14-7

For the conditions and instant specified in Example 14-4, determine the angular acceleration of the ladder and the acceleration of the top end of the ladder.

SOLUTION

The acceleration of end B of the ladder is given as $\mathbf{a}_B = \mathbf{0}$ and the acceleration of end A of the ladder must be along the wall $\mathbf{a}_A = a_A\mathbf{j}$. Then the relative acceleration equation gives

$$a_A\mathbf{j} = \mathbf{a}_B + \mathbf{a}_{A/B} = 0 + (\mathbf{a}_{A/B})_t + (\mathbf{a}_{A/B})_n \tag{c}$$

But from Fig. 14-23 the relative acceleration terms are given by

$$(\mathbf{a}_{A/B})_t = 3\alpha\mathbf{e}_t = 3\alpha(-\cos 30° \, \mathbf{i} - \sin 30° \, \mathbf{j})$$

and

$$(\mathbf{a}_{A/B})_n = 3\omega^2\mathbf{e}_n = 3\omega^2 (\sin 30° \, \mathbf{i} - \cos 30° \, \mathbf{j})$$

where from Example Problem 14-4 $\omega = 0.770$ rad/s. Then the x-component of Eq. c gives

$$0 = -3\alpha \cos 30° + 3(0.770)^2 \sin 30°$$
$$\alpha = 0.342 \text{ rad/s}^2 \text{ (CCW)} \qquad \text{Ans.}$$

and the y-component gives

$$a_A = -(3)(0.342) \sin 30° - (3)(0.770)^2 \cos 30°$$
$$= -2.053 \text{ m/s}^2$$

or

$$\mathbf{a}_A = 2.053 \text{ m/s}^2 \downarrow \qquad \text{Ans.}$$

Alternatively, the relative acceleration terms can be computed using the vector cross-product

$$(\mathbf{a}_{A/B})_t = \alpha\mathbf{k} \times (-3 \sin 30° \, \mathbf{i} + 3 \cos 30° \, \mathbf{j})$$
$$= -3 \, \alpha \sin 30° \, \mathbf{j} - 3\alpha \cos 30° \, \mathbf{i}$$
$$(\mathbf{a}_{A/B})_n = \omega\mathbf{k} \times [\omega\mathbf{k} \times (-3 \sin 30° \, \mathbf{i} + 3 \cos 30° \, \mathbf{j})]$$
$$= \omega\mathbf{k} \times [-3\omega \sin 30° \, \mathbf{j} - 3\omega \cos 30° \, \mathbf{i}]$$
$$= 3\omega^2 \sin 30° \, \mathbf{i} - 3\omega^2 \cos 30° \, \mathbf{j}$$

which again gives the same result as before.

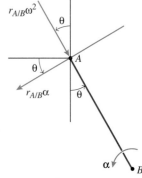

Fig. 14-23

SOLUTION

The acceleration of O, the axle of the wheel, is zero, and the acceleration of B, the slider, has only a horizontal component. Also, from Example Problem 14-5, $\omega_{OA} = 10$ rad/s (CCW), $\alpha_{OA} = 0$ rad/s^2, $\omega_{AB} = 1.533$ rad/s (CW), and the angle between bar AB and the horizontal is $\phi = 15.06°$. The acceleration of the common point A is written relative to the accelerations of points O and B

$$\mathbf{a}_A = \mathbf{a}_B + \mathbf{a}_{A/B} = \mathbf{a}_O + \mathbf{a}_{A/O}$$

or

$$a_B\mathbf{i} + (\mathbf{a}_{A/B})_t + (\mathbf{a}_{A/B})_n = \mathbf{0} + (\mathbf{a}_{A/O})_t + (\mathbf{a}_{A/O})_n \qquad (d)$$

But from Fig. 14-24 the relative acceleration terms are given by

$$(\mathbf{a}_{A/B})_t = 30\alpha_{AB}(-\sin\phi\,\mathbf{i} - \cos\phi\,\mathbf{j})$$
$$= -7.795\alpha_{AB}\mathbf{i} - 28.97\alpha_{AB}\mathbf{j} \text{ in./s}^2$$
$$(\mathbf{a}_{A/B})_n = 30\omega_{AB}^2(\cos\phi\,\mathbf{i} - \sin\phi\,\mathbf{j})$$
$$= 68.08\mathbf{i} - 18.32\mathbf{j} \text{ in./s}^2$$
$$(\mathbf{a}_{A/O})_t = 9\alpha_{OA}(-\sin 60°\,\mathbf{i} + \cos 60°\,\mathbf{j}) = 0 \text{ in./s}^2$$

and

$$(\mathbf{a}_{A/O})_n = 9\omega_{OA}^2(-\cos 60°\,\mathbf{i} - \sin 60°\,\mathbf{j})$$
$$= -450.0\mathbf{i} - 779.4\mathbf{j} \text{ in./s}^2$$

Then the y-component of Eq. d gives

$$-28.97\alpha_{AB} - 18.32 = -779.4$$

or

$$\alpha_{AB} = 26.3 \text{ rad/s}^2 \text{ (CCW)} \qquad \text{Ans.}$$

and the x-component gives

$$a_B - 7.795\alpha_{AB} + 68.08 = -450.0$$
$$a_B = -313 \text{ in./s}^2$$
$$\mathbf{a}_B = 313 \text{ in./s}^2 \leftarrow \qquad \text{Ans.}$$

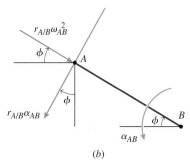

Fig. 14-24

PROBLEMS

14-86* For Problem 14-42 determine the acceleration of slider A and the angular acceleration of bar AB at the instant when $\theta = 75°$.

14-87* For Problem 14-43 determine the acceleration of A and the angular acceleration of the rod AB at the instant when $x = 2$ ft.

14-88 The crankshaft of Problem 14-44 has an angular acceleration of 5 rad/s² clockwise at the instant shown. Determine the acceleration of the piston and the angular acceleration of the connecting rod BC at this instant.

14-89* In Problem 14-45 the velocity of the plunger is decreasing at the rate of 0.25 ft/s². Determine the angular acceleration α_0 of the gear and the angular acceleration α_{AB} of AB at the instant shown.

14-90 For Problem 14-46 determine the acceleration of the block a_C and the angular acceleration of the link α_{AB} at the instant when $\theta = 75°$.

14-91 In Problem 14-53 the angular acceleration of link AB is 10 rad/s² counterclockwise. Determine the angular acceleration of the plate and the acceleration of point D at the instant shown.

14-92* In Problem 14-56 the angular acceleration of link AB is 12 rad/s² counterclockwise. Determine the angular acceleration of bar BC and the acceleration of point E when $\theta = 90°$.

14-93 For Problem 14-59 determine the angular acceleration of bars AB and BC at the instant shown.

14-94* In Problem 14-76 the velocity of point C is slowing down at the rate of 50 mm/s². Determine the acceleration of points A and B at this instant.

14-95* The wheel shown in Fig. P14-95 rolls without slipping on the horizontal surface. At the instant shown the

wheel has an angular velocity of 6 rad/s and an angular acceleration of 2 rad/s², both clockwise. Determine the angular acceleration of the links AB and BC at this instant.

14-96 The 400-mm diameter wheel shown in Fig. P14-96 rolls without slipping on the horizontal surface. Bar AB is 750 mm long and is attached to the wheel by a smooth pin 150 mm from the center. At the instant shown the center of the wheel has a velocity of 1.5 m/s to the left and an acceleration of 0.80 m/s² to the right. Determine the angular acceleration of the wheel α_0 and the angular acceleration of the bar α_{AB} at this instant.

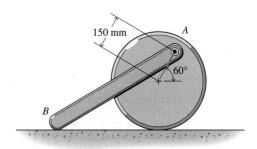

Fig. P14-96

14-97* The axles of two wheels are rigidly attached to a vehicle as shown in Fig. P14-97. The velocity and acceleration of the vehicle are 30 ft/s to the right and 3 ft/s² to the left, respectively. The wheels are further connected by a horizontal bar AB. If the smaller wheel rolls without slipping, determine the angular acceleration of both wheels and the relative acceleration between the larger wheel and the surface at point C.

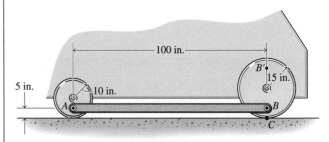

Fig. P14-97

14-98 The two gears shown in Fig. P14-98 rotate about their respective shafts. Bars AB and BC are each 125 mm long. At the instant shown the larger gear is rotating counterclockwise at a constant angular velocity of $\omega_A = 12$ rad/s. Determine the angular acceleration of the two

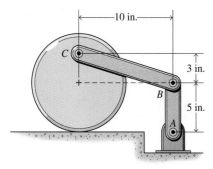

Fig. P14-95

bars α_{AB} and α_{BC} and the acceleration of pin A relative to pin C.

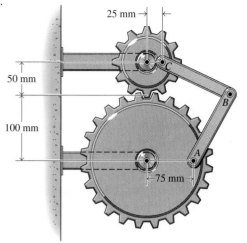

Fig. P14-98

14-99 Repeat Problem 14-97 if the right end of the 100-in. long bar AB is connected at point B' instead of at B.

14-100* In Problem 14-98 the large gear is rotating counterclockwise at $\omega_A = 5$ rad/s. Determine the angular acceleration of the large gear α_A for which the angular accelerations of the two bars are equal $\alpha_{AB} = \alpha_{BC}$.

14-101 In Problem 14-85 the arm ABC has a counterclockwise angular acceleration of 2 rad/s². Determine the angular acceleration of each of the gears and the acceleration of tooth D for the position shown.

14-102* In Problem 14-84 the speed of the center of the wheel is decreasing at a rate of 0.5 m/s² when the wheel passes the bottom of the drum. Determine the angular acceleration of the wheel and the acceleration of point C (the point of the wheel in contact with the drum) at that instant. (*Hint:* Pretend that the center of the drum A and the center of the wheel B are connected by a rigid rod. Relate the angular motions of the wheel and the imaginary rod through the given motion of the center of the wheel.)

14-103* For the Geneva mechanism of Problem 14-67 determine the angular acceleration of wheel B when $\theta = 30°$.

14-6   MOTION RELATIVE TO ROTATING AXES

Thus far in the chapter, the position, velocity, and acceleration of the individual particles have all been described using a single fixed coordinate system. The relative position, relative velocity, and relative acceleration have also all been described using the same fixed coordinate system. For the types of problems considered thus far, this approach has been adequate, straightforward, and relatively simple to use.

There are several other types of problems, however, for which it is convenient to describe the position or motion of one of the particles relative to a rotating coordinate system. Problems of this type include the following:

1. The motion is observed from a coordinate system that is rotating. For example, the Earth rotates, and coordinate systems fixed to the earth are rotating coordinate systems. The effect of the Earth's rotation in describing the motion of swings, baseballs, bicycles, airplanes, and so on is so small that it is not even considered. However, the effect is not small in describing the motion of rockets and spacecraft as observed from a rotating Earth.
2. The motions of two points are somehow related but are not equal and are not on the same rigid body. For example, some mechanisms are connected by pins that slide in grooves or slots. Relative motion is conveniently specified by giving the translational and rotational motion of the member containing the slot, the shape of the slot, and the rate of travel of the pin along the slot.

3.   The solution of kinetic problems involving the rotation of irregu-
larly shaped rigid bodies. Moments and products of inertia depend on the coordinate system used to describe them. If the axes
are fixed but the body rotates, then the moments of inertia will
change unless the body has certain symmetries. If the coordinate
axes are allowed to rotate with the body, however, the moments
and products of inertia will be constant.

Of course, the rotation of the coordinate system needs to be taken into
account in the derivatives for velocity and acceleration.

14-6-1   Position

In order to see how the rotation of the coordinate system affects the
description of the motion, let A and B be any two points undergoing
plane motion. In terms of a fixed X-Y coordinate system, the locations
of A and B are given by the position vectors

$$\mathbf{r}_A = X_A\mathbf{i} + Y_A\mathbf{j} \qquad \text{and} \qquad \mathbf{r}_B = X_B\mathbf{i} + Y_B\mathbf{j}$$

where $\mathbf{i}$ and $\mathbf{j}$ are unit vectors along the X- and Y-axes, respectively.
Then the triangle law of addition of two vectors gives $\mathbf{r}_B = \mathbf{r}_A + \mathbf{r}_{B/A}$. If
$\mathbf{r}_{B/A}$ is measured in the fixed X-Y coordinate system, this is exactly the
result used in the first part of this chapter.

However, suppose that point A represents a particle on a rigid
body that is rotating with angular velocity $\boldsymbol{\omega} = \dot{\theta}\,\mathbf{k}$ and angular acceleration $\boldsymbol{\alpha} = \ddot{\theta}\,\mathbf{k}$ (Fig. 14-25). Further suppose that the motion (position,
velocity, and acceleration) of A is easy to describe in terms of the fixed
coordinate system. Suppose that point B, on the other hand, represents
a particle that moves in a prescribed manner relative to the rotating
rigid body—perhaps a pin that slides in a slot. Although it may be
easy to describe the motion (position, velocity, and acceleration) of
point B relative to the rotating body, it may not be very easy to describe its motion relative to the fixed X-Y coordinate system.

Instead, let x-y be a rotating coordinate system that is attached to
and rotates with the rigid body. Point A will be chosen for the origin of
the rotating coordinate system. Then, in terms of this rotating coordinate system, the relative position vector is

$$\mathbf{r}_{B/A} = x\mathbf{e}_x + y\mathbf{e}_y$$

where the unit vectors along the rotating x- and y-axes have been de-
noted $\mathbf{e}_x$ and $\mathbf{e}_y$ to distinguish them from the fixed unit vectors $\mathbf{i}$ and $\mathbf{j}$
and to emphasize that they are functions of time. Therefore, the posi-
tion of B is given by

$$r_B = \mathbf{r}_A + \mathbf{r}_{B/A} = \mathbf{r}_A + (x\mathbf{e}_x + y\mathbf{e}_y) \tag{14-15}$$

in which x, y, $\mathbf{r}_A = X_A\,\mathbf{i} + Y_A\,\mathbf{j}$, $\mathbf{e}_x = \cos\theta\,\mathbf{i} + \sin\theta\,\mathbf{j}$, and $\mathbf{e}_y = -\sin\theta\,\mathbf{i} + \cos\theta\,\mathbf{j}$ are all presumed known as functions of time.

14-6-2   Velocity

The relationship between the absolute and relative velocities is obtained by differentiating the relative position equation (Eq. 14-15) with
respect to time to get

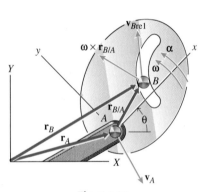

Fig. 14-25

$$\mathbf{v}_B = \mathbf{v}_A + \frac{d\mathbf{r}_{B/A}}{dt} = \mathbf{v}_A + \frac{d(x\mathbf{e}_x + y\mathbf{e}_y)}{dt}$$

$$= \mathbf{v}_A + \frac{dx}{dt}\mathbf{e}_x + x\frac{d\mathbf{e}_x}{dt} + \frac{dy}{dt}\mathbf{e}_y + y\frac{d\mathbf{e}_y}{dt}$$

$$= \mathbf{v}_A + \mathbf{v}_{\text{Brel}} + x\frac{d\mathbf{e}_x}{dt} + y\frac{d\mathbf{e}_y}{dt} \qquad (14\text{-}16a)$$

in which $\mathbf{v}_{\text{Brel}} = \dot{x}\mathbf{e}_x + \dot{y}\mathbf{e}_y$ is the velocity of B relative to (as measured in) the rotating x-y coordinate system. The last two terms arise because the directions of the unit vectors $\mathbf{e}_x$ and $\mathbf{e}_y$ vary with time due to the rotation of the x-y axes.

The derivatives of the unit vectors $\mathbf{e}_x$ and $\mathbf{e}_y$ are evaluated using the chain rule of differentiation, which gives

$$\frac{d\mathbf{e}_x}{dt} = \frac{d\mathbf{e}_x}{d\theta}\frac{d\theta}{dt}$$

where

$$\frac{d\mathbf{e}_x}{d\theta} = \lim_{\Delta\theta \to 0} \frac{\mathbf{e}_x(\theta + \Delta\theta) - \mathbf{e}_x(\theta)}{\Delta\theta}$$

But in the limit as $\Delta\theta \to 0$, the distance $|\mathbf{e}_x(\theta + \Delta\theta) - \mathbf{e}_x(\theta)|$ tends to the arc length along a unit circle $\Delta s = 1\,\Delta\theta$ and the angle β tends to 90° (Fig. 14-26). Therefore the vector $\mathbf{e}_x(\theta + \Delta\theta) - \mathbf{e}_x(\theta)$ has magnitude $\Delta\theta$ and points in the $\mathbf{e}_y$-direction and

$$\frac{d\mathbf{e}_x}{dt} = \dot{\theta}\lim_{\Delta\theta \to 0}\frac{\Delta\theta\mathbf{e}_y}{\Delta\theta} = \dot{\theta}\mathbf{e}_y = \boldsymbol{\omega} \times \mathbf{e}_x \qquad (14\text{-}16b)$$

where $\boldsymbol{\omega} = \omega\mathbf{k}$ and $\omega = \dot{\theta} = d\theta/dt$.

Similarly the derivative of $\mathbf{e}_y$ with respect to time can be calculated using the chain rule

$$\frac{d\mathbf{e}_y}{dt} = \frac{d\mathbf{e}_y}{d\theta}\frac{d\theta}{dt}$$

where

$$\frac{d\mathbf{e}_y}{d\theta} = \lim_{\Delta\theta \to 0}\frac{\mathbf{e}_y(\theta + \Delta\theta) - \mathbf{e}_y(\theta)}{\Delta\theta}$$

But in the limit as $\Delta\theta \to 0$, the distance $|\mathbf{e}_y(\theta + \Delta\theta) - \mathbf{e}_y(\theta)|$ again tends to the arc length along a unit circle $\Delta s = 1\,\Delta\theta$ and the angle β again tends to 90° (Fig. 14-26). Therefore the vector $\mathbf{e}_y(\theta + \Delta\theta) - \mathbf{e}_y(\theta)$ has magnitude $\Delta\theta$ and points in the negative $\mathbf{e}_x$-direction and

$$\frac{d\mathbf{e}_y}{dt} = \dot{\theta}\lim_{\Delta\theta \to 0}\frac{-\Delta\theta\mathbf{e}_x}{\Delta\theta} = -\dot{\theta}\mathbf{e}_x = \boldsymbol{\omega} \times \mathbf{e}_y \qquad (14\text{-}16c)$$

Substituting these results back into the relative velocity equation (Eq. 14-16a) gives

$$\mathbf{v}_B = \mathbf{v}_A + \mathbf{v}_{\text{Brel}} + (x\boldsymbol{\omega} \times \mathbf{e}_x + y\boldsymbol{\omega} \times \mathbf{e}_y)$$
$$= \mathbf{v}_A + \boldsymbol{\omega} \times \mathbf{r}_{B/A} + \mathbf{v}_{\text{Brel}} \qquad (14\text{-}16d)$$

where $\mathbf{v}_A$, $\mathbf{v}_B$, and $\boldsymbol{\omega}$ are all measured relative to the fixed X-Y coordinate system; $\mathbf{r}_{B/A}$ and $\mathbf{v}_{\text{Brel}}$ are measured relative to the rotating x-y

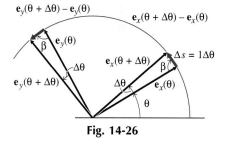

Fig. 14-26

coordinate system. Of course, all of the vectors in Eq. 14-16d must be expressed in a common coordinate system before the vector sums and product can be performed. Either $\mathbf{r}_{B/A}$ and $\mathbf{v}_{Brel}$ must be expressed in the fixed X-Y coordinate system (using $\mathbf{e}_x = \cos\theta\mathbf{i} + \sin\theta\mathbf{j}$ and $\mathbf{e}_y = -\sin\theta\mathbf{i} + \cos\theta\mathbf{j}$) or $\mathbf{v}_A$ and $\mathbf{v}_B$ must be expressed in the rotating x-y coordinate system (using $\mathbf{i} = \cos\theta\mathbf{e}_x - \sin\theta\mathbf{e}_y$ and $\mathbf{j} = \sin\theta\mathbf{e}_x + \cos\theta\mathbf{e}_y$). The choice is based solely on the form in which the data are given and the form in which the results are desired.

If A and B are two points fixed on the same rigid body, then $\mathbf{v}_{Brel} = \mathbf{0}$, $\boldsymbol{\omega}$ is the angular velocity of the body, and Eq. 14-16d reduces to Eq. 14-13d. The added complexity of the rotating coordinate system is not needed, nor is it useful for this type of problem.

If A is a point fixed in a rotating rigid body and B is a pin sliding in a slot in the body (Fig. 14-25), then $\mathbf{v}_A + \boldsymbol{\omega} \times \mathbf{r}_{B/A}$ is the velocity point B would have if it were attached to the rigid body rather than moving relative to it. The last term $\mathbf{v}_{Brel}$ is the additional velocity point B has because of its motion along the slot. The direction of $\mathbf{v}_{Brel}$ is tangent to the slot as shown.

14-6-3   Acceleration

The relationship between the absolute and relative accelerations is obtained by differentiating the relative velocity equation (Eq. 14-16d) with respect to time to get

$$\mathbf{a}_B = \mathbf{a}_A + \frac{d\boldsymbol{\omega}}{dt} \times \mathbf{r}_{B/A} + \boldsymbol{\omega} \times \frac{d\mathbf{r}_{B/A}}{dt} + \frac{d\mathbf{v}_{Brel}}{dt} \qquad (14\text{-}17a)$$

From the calculation for relative velocity,

$$\frac{d\mathbf{r}_{B/A}}{dt} = \mathbf{v}_{Brel} + \boldsymbol{\omega} \times \mathbf{r}_{B/A} \qquad (14\text{-}17b)$$

A similar calculation for the derivative of $\mathbf{v}_{Brel}$ gives

$$\begin{aligned}
\frac{d\mathbf{v}_{Brel}}{dt} &= \frac{d\,(\dot{x}\mathbf{e}_x + \dot{y}\mathbf{e}_y)}{dt} \\
&= (\ddot{x}\mathbf{e}_x + \ddot{y}\mathbf{e}_y) + \left(\dot{x}\frac{d\mathbf{e}_x}{dt} + \dot{y}\frac{d\mathbf{e}_y}{dt}\right) \\
&= \mathbf{a}_{Brel} + (\dot{x}\boldsymbol{\omega} \times \mathbf{e}_x + \dot{y}\boldsymbol{\omega} \times \mathbf{e}_y) \\
&= \mathbf{a}_{Brel} + \boldsymbol{\omega} \times \mathbf{v}_{Brel} \qquad (14\text{-}17c)
\end{aligned}$$

where $\mathbf{a}_{Brel} = \ddot{x}\mathbf{e}_x + \ddot{y}\mathbf{e}_y$ is the acceleration of B relative to (as measured in) the rotating x-y coordinate system. Substituting Eqs. 14-17b and 14-17c into Eq. 14-17a and rearranging terms yields

$$\begin{aligned}
\mathbf{a}_B = \mathbf{a}_A + \boldsymbol{\alpha} \times \mathbf{r}_{B/A} + \boldsymbol{\omega} \times (\boldsymbol{\omega} \times \mathbf{r}_{B/A}) \\
+ \mathbf{a}_{Brel} + 2\,\boldsymbol{\omega} \times \mathbf{v}_{Brel} \qquad (14\text{-}17d)
\end{aligned}$$

where $\mathbf{a}_A$, $\mathbf{a}_B$, $\boldsymbol{\omega}$, and $\boldsymbol{\alpha}$ are all measured relative to the fixed X-Y coordinate system; $\mathbf{r}_{B/A}$, $\mathbf{v}_{Brel}$, and $\mathbf{a}_{Brel}$ are measured relative to the rotating x-y coordinate system. Again, the vectors in Eq. 14-17d must be expressed in a common coordinate system before the vector sums and products can be performed. Either the fixed X-Y coordinate system or the rotating x-y coordinate system can be used. The choice is based solely on the form in which the data are given and the form in which the results are desired.

If A and B are two points fixed on the same rigid body, then $\mathbf{v}_{Brel} = \mathbf{a}_{Brel} = \mathbf{0}$, $\boldsymbol{\omega}$ and $\boldsymbol{\alpha}$ are the angular velocity and angular acceleration of the body, and Eq. 14-17d reduces to Eq. 14-14c. The added complexity of the rotating coordinate system is not needed, nor is it useful for this type of problem.

If A is a point fixed in a rotating rigid body and B is a pin sliding in a slot in the body (Fig. 14-25), then $\mathbf{a}_A + \boldsymbol{\alpha} \times \mathbf{r}_{B/A} + \boldsymbol{\omega} \times (\boldsymbol{\omega} \times \mathbf{r}_{B/A})$ is the acceleration point B would have if it were attached to the rigid body rather than moving relative to it. The term $\mathbf{a}_{Brel}$ is the additional acceleration point B has because of its motion along the slot. The remaining term $2\,\boldsymbol{\omega} \times \mathbf{v}_{Brel}$, called the **Coriolis acceleration,** has no simple interpretation. As indicated by the vector cross-product, the Coriolis acceleration will always be perpendicular to both $\boldsymbol{\omega}$ (it will be in the plane of motion) and $\mathbf{v}_{Brel}$ (it will be perpendicular to the slot along which the pin travels).

The orientation, location of the origin, angular velocity, and angular acceleration of the rotating coordinate system should be selected to simplify the calculation of the various terms in the relative velocity and relative acceleration equations. For example, the origin A should be a point whose absolute velocity and absolute acceleration are easily obtained. The angular velocity and angular acceleration of the rotating frame should be chosen so that the velocity and acceleration of point B are easy to calculate relative to the rotating coordinate system. The orientation of the rotating coordinate system relative to the fixed coordinate system should be chosen such that the components of the various vectors are easy to describe.

Finally, the Eqs. 14-16 and 14-17 are equally valid for describing the relative motion of individual particles and of points on rigid bodies. Although they were derived for plane motion ($\boldsymbol{\omega} = \omega \mathbf{k}$, $\mathbf{r}_{B/A} = x\mathbf{e}_x + y\mathbf{e}_y$, etc.), it will be seen in the next section that the vector forms of these equations are valid for general three-dimensional motion as well.

Car B is traveling along a straight road at a constant speed of 60 mi/h while car A is traveling around a circular curve of radius 500 ft at a constant speed of 45 mi/h (Fig. 14-27). Determine the velocity and acceleration that car B appears to have to an observer riding in and turning with car A at the instant shown.

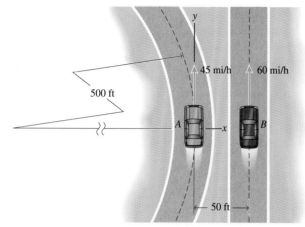

Fig. 14-27

SOLUTION

In terms of fixed r-θ coordinates with origin at the center of the turn,

$$\mathbf{v}_A = r\dot{\theta}\mathbf{e}_\theta \qquad 66\mathbf{e}_\theta = 500\ \dot{\theta}\mathbf{e}_\theta$$

Therefore $\dot{\theta} = 0.1320$ rad/s. Also, since the speed of car A is constant, $\ddot{\theta} = 0$ and the acceleration of car A is

$$\mathbf{a}_A = (\ddot{r} - r\dot{\theta}^2)\mathbf{e}_r + (r\ddot{\theta} + 2\dot{r}\dot{\theta})\mathbf{e}_\theta = -(500)(0.1320)^2\mathbf{e}_r = -8.712\mathbf{e}_r\ \text{ft/s}^2$$

The moving x-y coordinate system is fixed in and moving with car A. Since these coordinates are always aligned with the r-θ coordinates above, the rotation rates of the x-y coordinate system are $\boldsymbol{\omega} = 0.1320\mathbf{k}$ rad/s and $\boldsymbol{\alpha} = \mathbf{0}$. Then the relative velocity equation (Eq. 14-16d) is

$$\mathbf{v}_B = \mathbf{v}_A + \boldsymbol{\omega} \times \mathbf{r}_{B/A} + \mathbf{v}_{Brel}$$

where

$$\mathbf{v}_A = 45\mathbf{e}_y\ \text{mi/h} = 66\mathbf{e}_y\ \text{ft/s} \qquad \mathbf{v}_B = 60\mathbf{e}_y\ \text{mi/h} = 88\mathbf{e}_y\ \text{ft/s}$$
$$\boldsymbol{\omega} \times \mathbf{r}_{B/A} = (0.1320\mathbf{k}) \times (50\mathbf{e}_x) = 6.600\mathbf{e}_y\ \text{ft/s}$$

Solving for $\mathbf{v}_{Brel}$ gives

$$\mathbf{v}_{Brel} = 15.40\mathbf{e}_y\ \text{ft/s} = 10.50\mathbf{e}_y\ \text{mi/h} \qquad\qquad \text{Ans.}$$

The relative acceleration equation (Eq. 14-17d) is

$$\mathbf{a}_B = \mathbf{a}_A + \boldsymbol{\alpha} \times \mathbf{r}_{B/A} + \boldsymbol{\omega} \times (\boldsymbol{\omega} \times \mathbf{r}_{B/A}) + \mathbf{a}_{Brel} + 2\boldsymbol{\omega} \times \mathbf{v}_{Brel}$$

where

$$\mathbf{a}_B = \mathbf{0} \qquad \mathbf{a}_A = -8.712\mathbf{e}_x\ \text{ft/s}^2 \qquad \boldsymbol{\alpha} \times \mathbf{r}_{B/A} = \mathbf{0}$$
$$\boldsymbol{\omega} \times (\boldsymbol{\omega} \times \mathbf{r}_{B/A}) = (0.1320\mathbf{k}) \times (6.600\mathbf{e}_y) = -0.871\mathbf{e}_x\ \text{ft/s}^2$$
$$2\boldsymbol{\omega} \times \mathbf{v}_{Brel} = 2(0.1320\mathbf{k}) \times (15.40\mathbf{e}_y) = -4.066\mathbf{e}_x\ \text{ft/s}^2$$

Solving for $\mathbf{a}_{Brel}$ gives

$$\mathbf{a}_{Brel} = 13.65\mathbf{e}_x\ \text{ft/s}^2 \qquad\qquad \text{Ans.}$$

EXAMPLE PROBLEM 14-10

As the 400-mm long arm BC of the mechanism shown in Fig. 14-28 oscillates, the collar C slides up and down the arm AD. Given $\phi = 1.5 \sin \pi t$ rad where t is in seconds, determine the rotation rate ω of arm AD and the speed v of the slider along arm AD when $t = \frac{1}{3}$ s.

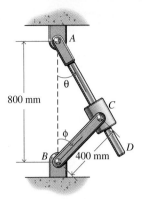

Fig. 14-28

SOLUTION

The x-y coordinate system will be chosen to rotate with arm AD and to have its origin at A as shown in Fig. 14-29. When $t = \frac{1}{3}$ s

$$\phi = 1.299 \text{ rad} = 74.43° \qquad \dot{\phi} = 2.356 \text{ rad/s}$$
$$\ddot{\phi} = -12.821 \text{ rad/s}^2$$

and

$$\mathbf{v}_C = (400)(2.356)\mathbf{e}_\phi = 942.48 \text{ mm/s} \; \searrow 74.43°$$

Also, by the Law of Cosines

$$\overline{AC}^2 = 800^2 + 400^2 - 2(800)(400) \cos 74.43°$$

or

$$\overline{AC} = 792.60 \text{ mm}$$

and then by the Law of Sines

$$\frac{\sin \theta}{400} = \frac{\sin 74.43°}{792.60} \qquad \theta = 29.09°$$

Now the relative velocity equation (Eq. 14-16d) is

$$\mathbf{v}_C = \mathbf{v}_A + \boldsymbol{\omega} \times \mathbf{r}_{C/A} + \mathbf{v}_{Crel}$$

where in terms of the rotating coordinate system (see Fig. 14-29)

$$\mathbf{v}_A = \mathbf{0} \qquad \mathbf{v}_{Crel} = v\mathbf{e}_x$$
$$\boldsymbol{\omega} \times \mathbf{r}_{C/A} = \omega \mathbf{k} \times 792.60\mathbf{e}_x = 792.60 \, \omega \mathbf{e}_y \text{ mm/s}$$
$$\mathbf{v}_C = 942.48(\cos 13.52° \, \mathbf{e}_x - \sin 13.52° \, \mathbf{e}_y) \text{ mm/s}$$

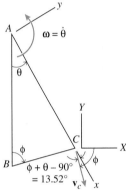

Fig. 14-29

Then the $\mathbf{e}_y$ component of the relative velocity equation gives

$$\omega = \dot{\theta} = -0.278 \text{ rad/s} \qquad \boldsymbol{\omega} = 0.278 \text{ rad/s} \downarrow \qquad \text{Ans.}$$

and the $\mathbf{e}_x$ component gives

$$v = 916.4 \text{ mm/s (outward)} \qquad \text{Ans.}$$

Alternatively, the components of the relative velocity equation can be written in terms of the fixed X-Y coordinate system (see Fig. 14-29)

$$\mathbf{v}_A = \mathbf{0}$$
$$\mathbf{v}_{Crel} = v \sin 29.09° \, \mathbf{i} - v \cos 29.09° \, \mathbf{j}$$
$$\boldsymbol{\omega} \times \mathbf{r}_{C/A} = \omega \mathbf{k} \times 792.60(\sin 29.09 \, \mathbf{i} - \cos 29.09 \, \mathbf{j})$$

and

$$\mathbf{v}_C = 942.48(\cos 74.43 \, \mathbf{i} - \sin 74.43 \, \mathbf{j})$$

Then the $\mathbf{i}$ and $\mathbf{j}$ components of the relative velocity equation are

$$692.6 \, \omega + 0.486 \, v = 253.0$$
$$385.3 \, \omega - 0.874 \, v = -908.0$$

Solving these simultaneous equations gives the same results as before.

In the mechanism of Fig. 14-30, arm AB rotates clockwise at a constant rate of 6 rpm while the pin P moves outward along a radial slot in the rotating disk at a constant rate of 1.0 in./s. At the instant shown, $r = 3$ in., $\omega = 12$ rpm, $\alpha = 0.1$ rad/s², both clockwise. Determine the absolute velocity and acceleration of the pin P at this instant.

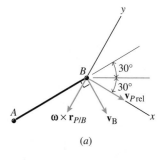

Fig. 14-30

SOLUTION

Choose the rotating x-y coordinate system with origin at B and x-axis aligned with the slot as shown in Fig. 14-31a. Then the angular velocity of the x-y coordinate system is

$$\omega = (-12\mathbf{k}\ \text{rev/min})\left(\frac{2\pi\ \text{rad/rev}}{60\ \text{sec/min}}\right) = -1.2566\mathbf{k}\ \text{rad/s.}$$

The relative velocity equation (Eq. 14-16d) is

$$\mathbf{v}_P = \mathbf{v}_B + \boldsymbol{\omega} \times \mathbf{r}_{P/B} + \mathbf{v}_{\text{Prel}}$$

where in terms of the rotating coordinate system (Fig. 14-31a)

$$\mathbf{v}_B = (18)\frac{(6)(2\pi)}{60} \diagdown 60°$$
$$= 9.795\mathbf{e}_x - 5.655\mathbf{e}_y\ \text{in./s}$$
$$\mathbf{v}_{\text{Prel}} = 1.0\mathbf{e}_x\ \text{in./s}$$
$$\boldsymbol{\omega} \times \mathbf{r}_{P/B} = (-1.2566\mathbf{k}) \times (3\mathbf{e}_x) = -3.770\mathbf{e}_y\ \text{in./s}$$

Therefore

$$\mathbf{v}_P = 10.79\mathbf{e}_x - 9.42\mathbf{e}_y\ \text{in./s} \qquad \text{Ans.}$$

or (see Fig. 14-31b)

$$\mathbf{v}_P = 14.32\ \text{in./s} \diagdown 71.1° \qquad \text{Ans.}$$

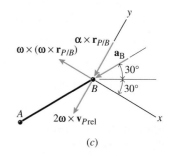

The relative acceleration equation (Eq. 14-17d) is

$$\mathbf{a}_P = \mathbf{a}_B + \boldsymbol{\alpha} \times \mathbf{r}_{P/B} + \boldsymbol{\omega} \times (\boldsymbol{\omega} \times \mathbf{r}_{P/B})$$
$$+ \mathbf{a}_{\text{Prel}} + 2\,\boldsymbol{\omega} \times \mathbf{v}_{\text{Prel}}$$

where in terms of the rotating coordinate system (Fig. 14-31c)

$$\mathbf{a}_B = (18)\left[\frac{(6)(2\pi)}{60}\right]^2 \diagup 30°$$
$$= -3.553\mathbf{e}_x - 6.154\mathbf{e}_y\ \text{in./s}^2$$
$$\boldsymbol{\alpha} \times \mathbf{r}_{P/B} = (-0.1\mathbf{k}) \times (3\mathbf{e}_x) = -0.3\mathbf{e}_y\ \text{in./s}^2$$
$$\boldsymbol{\omega} \times (\boldsymbol{\omega} \times \mathbf{r}_{P/B}) = (-1.2566\mathbf{k}) \times (-3.770\mathbf{e}_y)$$
$$= -4.737\mathbf{e}_x\ \text{in./s}^2$$
$$2\,\boldsymbol{\omega} \times \mathbf{v}_{\text{Prel}} = 2(-1.2566\mathbf{k}) \times (1.0\mathbf{e}_x)$$
$$= -2.513\mathbf{e}_y\ \text{in./s}^2$$
$$\mathbf{a}_{\text{Prel}} = \mathbf{0}$$

Therefore

$$\mathbf{a}_P = -8.29\mathbf{e}_x - 8.97\mathbf{e}_y\ \text{in./s}^2 \qquad \text{Ans.}$$

or (see Fig. 14-31d)

$$\mathbf{a}_P = 12.21\ \text{in./s}^2 \diagup 17.3° \qquad \text{Ans.}$$

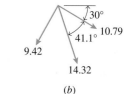

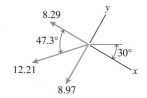

Fig. 14-31

PROBLEMS

14-104* Car *A* is traveling along a straight road at a constant speed of 90 km/h while car *B* is traveling around a circular curve of radius 150 m at a constant speed of 70 km/h (Fig. P14-104). Determine the velocity and acceleration that car *A* appears to have to an observer riding in and turning with car *B* at the instant shown.

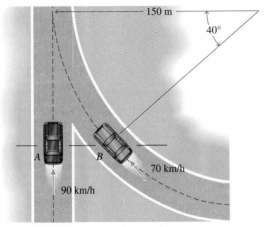

Fig. P14-104

14-105* Car *A* is traveling along a straight road at a constant speed of 65 mi/h while car *B* is traveling around a circular curve of radius 600 ft at a speed of 45 mi/h (Fig. P14-105). If the speed of car *B* is decreasing at the rate of 10 ft/s², determine the velocity and acceleration that car *A* appears to have to an observer riding in and turning with car *B* at the instant shown.

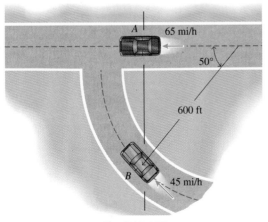

Fig. P14-105

14-106* Car *A* is traveling along a straight road at a constant speed of 80 km/h while car *B* is traveling around a circular curve of radius 125 m at a speed of 50 km/h (Fig. P14-106). If the speed of car *B* is increasing at the rate of 5 m/s², determine the velocity and acceleration that car *A* appears to have to an observer riding in and turning with car *B* at the instant shown.

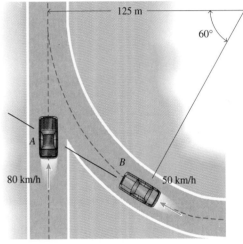

Fig. P14-106

14-107 Car *A* is traveling around a circular curve of radius 500 ft at a constant speed of 45 mi/h while car *B* is traveling around a circular curve of radius 750 ft at a constant speed of 60 mi/h (Fig. P14-107). Determine the velocity and acceleration that car *A* appears to have to an observer riding in and turning with car *B* at the instant shown.

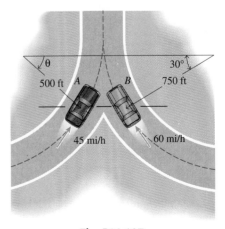

Fig. P14-107

14-108 Car A is traveling along a straight road at a constant speed of 60 km/h while car B is traveling around a circular curve of radius 100 m at a speed of 35 km/h (Fig. P14-108). If the speed of car B is decreasing at the rate of 1.5 m/s², determine the velocity and acceleration that car A appears to have to an observer riding in and turning with car B at the instant shown.

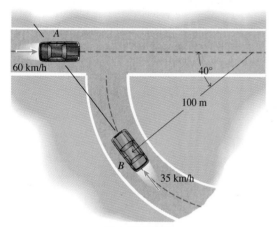

Fig. P14-108

14-109* For the conditions of Problem 14-107, determine the velocity and acceleration that car B appears to have to an observer riding in and turning with car A at the instant shown. (Are these values the negatives of the velocity and acceleration determined in Problem 14-107?)

14-110 For the mechanism of Example 14-10, determine the angular acceleration α_{AD} of arm AD and the acceleration a of the slider along arm AD when $t = \frac{1}{3}$ s.

14-111* In the mechanism of Fig. P14-111, arm AB is rotating counterclockwise with an angular velocity of 2 rad/s at the instant shown. Determine the angular velocity ω_{CD} of arm CD and the speed v of the slider along arm CD at this instant.

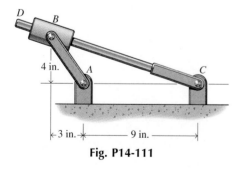

Fig. P14-111

14-112* In the mechanism of Fig. P14-112, arm AB is rotating counterclockwise with an angular velocity of 2 rad/s at the instant shown. Determine the angular velocity ω_{CD}

of arm CD and the speed v of the slider along arm AB at this instant.

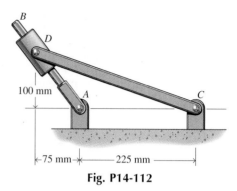

Fig. P14-112

14-113 Determine the angular acceleration α_{CD} of arm CD and the acceleration a of the slider along arm CD for the mechanism of Problem 14-111 if the angular velocity ω_{AB} is constant.

14-114* Determine the angular acceleration α_{CD} of arm CD and the acceleration a of the slider along arm AB for the mechanism of Problem 14-112 if the angular velocity ω_{AB} is constant.

14-115 Determine the angular acceleration α_{CD} of arm CD and the acceleration a of the slider along arm CD for the mechanism of Problem 14-111 if the angular velocity ω_{AB} is decreasing at the rate of 0.5 rad/s².

14-116 Determine the angular acceleration α_{CD} of arm CD and the acceleration a of the slider along arm AB for the mechanism of Problem 14-112 if the angular velocity ω_{AB} is decreasing at the rate of 0.5 rad/s².

14-117* As the 6-in. long arm AB of the mechanism shown in Fig. P14-117 oscillates, the pin B slides up and down in the slot of arm CD. Given that $\theta = \cos \pi t$ rad where t is in seconds, determine the angular velocity ω_{CD} and the angular acceleration α_{CD} of the arm CD when $t = \frac{1}{2}$ s.

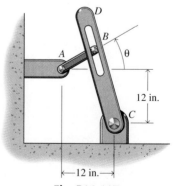

Fig. P14-117

14-118 As the 75-mm long arm AB of the mechanism shown in Fig. P14-118 oscillates, the pin B slides up and down in the slot of arm CD. Given that $\theta = 3 \sin 2\pi t$ rad where t is in seconds, determine the angular velocity ω_{CD} and the angular acceleration α_{CD} of the arm CD when $t = 0.1$ s.

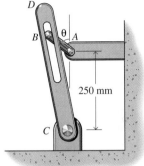

Fig. P14-118

14-119* The wheel of Fig. P14-119 is rotating clockwise at a constant rate of 120 rpm. Pin D is fastened to the wheel at a point 5 in. from its center and slides in the slot of arm AB. Determine the angular velocity ω_{AB} and angular acceleration α_{AB} of arm AB at the instant shown.

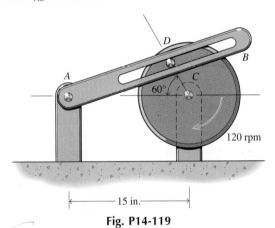

Fig. P14-119

14-120* As the arm of Fig. P14-120 rotates, the string attached to the slider gets wrapped around the fixed shaft at A and pulls the slider B inward at a rate of $r\omega$ relative to the arm. If the arm is rotating at a constant rate of 60 rpm, determine the absolute velocity $\mathbf{v}_B$ and absolute acceleration $\mathbf{a}_B$ of the slider when it is 400 mm from the shaft and $r = 10$ mm.

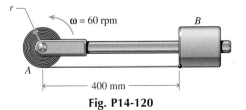

Fig. P14-120

14-121 As the arm of Fig. P14-121 rotates, a peg A in the underside of the slider follows the spiral groove in a fixed plate and pulls the slider outward. The spiral is given by $r = 0.035\ \theta^2$ where θ is in radians and r is in inches, and the arm is rotating at a constant rate of $\dot\theta = 1.5$ rad/s. Determine the absolute velocity $\mathbf{v}_A$ and absolute acceleration $\mathbf{a}_A$ of the slider when $r = 6$ in.

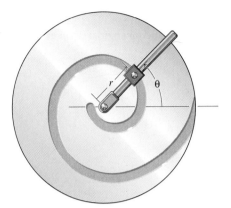

Fig. P14-121

14-122* The slotted plate of Fig. P14-122 has a constant clockwise angular velocity of 15 rad/s. The slider A has a constant speed relative to the slot of $u = 100$ mm/s. If $a = 200$ mm and $b = 0$ mm, determine the absolute velocity $\mathbf{v}_A$ and the absolute acceleration $\mathbf{a}_A$ of the slider. Repeat for a counterclockwise angular velocity of 15 rad/s.

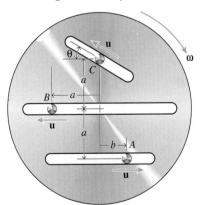

Fig. P14-122

14-123 The slotted plate of Fig. P14-122 has a constant clockwise angular velocity of 15 rad/s. The slider B has a constant speed relative to the slot of $u = 10$ in/s. If $a = 0$ in, determine the absolute velocity $\mathbf{v}_B$ and the absolute acceleration $\mathbf{a}_B$ of the slider. Repeat for a counterclockwise angular velocity of 15 rad/s.

14-124 Repeat Problem 14-122 for $b = 50$ mm.

14-125* Repeat Problem 14-123 for $a = 8$ in.

115

14-126 The slotted plate of Fig. P14-122 has an angular velocity of 15 rad/s and an angular acceleration of 5 rad/s², both clockwise. The slider C has a constant speed relative to the slot of $u = 100$ mm/s. If $a = 200$ mm and $\theta = 20°$, determine the absolute velocity $\mathbf{v}_C$ and the absolute acceleration $\mathbf{a}_C$ of the slider. Repeat for a counterclockwise angular velocity and angular acceleration.

14-127* The slotted plate of Fig. P14-127 has an angular velocity of 15 rad/s and an angular acceleration of 5 rad/s²,

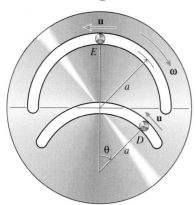

Fig. P14-127

both clockwise. The slider D has a constant speed relative to the slot of $u = 10$ in./s. If $a = 8$ in. and $\theta = 0°$, determine the absolute velocity $\mathbf{v}_D$ and the absolute acceleration $\mathbf{a}_D$ of the slider. Repeat for a counterclockwise angular velocity and angular acceleration.

14-128* Repeat Problem 14-126 for $\theta = 40°$.

14-129 Repeat Problem 14-127 for $\theta = 60°$.

14-130* The slotted plate of Fig. P14-127 has an angular velocity of 15 rad/s and an angular acceleration of 5 rad/s², both clockwise. The slider E has a constant speed relative to the slot of $u = 100$ mm/s. If $a = 200$ mm, determine the absolute velocity $\mathbf{v}_E$ and the absolute acceleration $\mathbf{a}_E$ of the slider. Repeat for a counterclockwise angular velocity and angular acceleration.

14-7 THREE-DIMENSIONAL MOTION OF A RIGID BODY

The three-dimensional motion of rigid bodies studied in this section is considerably more complex than the two-dimensional motion treated in earlier sections of this chapter. Not only do particles of the body move in three-dimensional space, but also the directions of the angular velocity and the angular acceleration vectors vary with time. Rather than being simply useful for describing the motion, vector analysis will be absolutely required to describe the motion of bodies in three dimensions.

Before going on to the general three-dimensional motion of a rigid body or the special case of rotation about a fixed point, it is first necessary to consider some of the complicating aspects of rotations of rigid bodies in three dimensions.

14-7-1 Euler's Theorem

Euler's theorem states that *When a rigid body rotates about a fixed point, any position of the body is obtainable from any other position of the body by a single rotation about some axis through the fixed point.*

In order to prove Euler's theorem, consider the motion of a rigid body which rotates about a fixed point A. Point B represents the position of an arbitrary particle at some instant of time, and point B' represents its position at some later time (Fig. 14-32). Since the body is rigid, the particle B must move on a spherical surface of radius R centered at

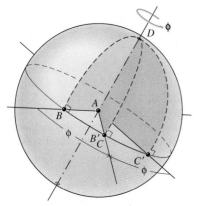

Fig. 14-32

A. The spherical shell shown represents the possible positions of particle *B* during the motion.

The particle that occupies point *B'* in the initial position of the body will be called *C* and will be moved to *C'* in the final position by this same motion. Since the final position of particle *B* is the same as the initial position of particle *C*, both particles are the same distance from *A* and both particles move on the same spherical surface.

The configuration of a rigid body is fixed by any three points in it. Therefore, the proof of the theorem requires showing that the motion of the body that takes particle *B* to *B'* can be obtained by a single rotation about some axis through *A* and that the same rotation takes the particle *C* to *C'*.

Since points B, $B' = C$, and C' are all on the same sphere, the great-circle arcs $\overline{BC}$ (distance between particles *B* and *C* in the initial position of the body) and $\overline{B'C'}$ (distance between particles *B* and *C* in the final position of the body) must be equal by rigidity of the body. On the sphere with center *A* and radius *R*, construct the great circles that bisect the arcs $\overline{BB'}$ and $\overline{CC'}$ orthogonally. These two circles intersect at two points, one of which is labeled *D* on Fig. 14-32. Finally, draw the great-circle arcs $\overline{BD}$, $\overline{B'D} = \overline{CD}$, and $\overline{C'D}$. By their construction, these arcs will be equal. Therefore the two spherical triangles *BB'D* and *CC'D* are congruent and the angle ϕ between the tangents to $\overline{BD}$ and $\overline{B'D}$ at *D* is equal to the angle between the tangents to $\overline{CD}$ and $\overline{C'D}$ at *D*. Then a rotation of magnitude ϕ in the proper sense about *AD* will bring *B* to *B'* and *C* to *C'*, determining the final position of the body from the initial position as stated in Euler's theorem.

14-7-2   Finite Rotations (Are Not Vectors)

It follows from Euler's theorem that the motion during a time interval Δt of a rigid body with a fixed point can be considered as a rotation of angle $\Delta\theta$ about a certain axis. This might be expressed much like a vector having its direction along the axis of rotation and magnitude equal to the amount of rotation. For example, the expression $\boldsymbol{\phi} = \phi\mathbf{e}_{AD}$ might be used to designate the rotation of Fig. 14-32. However, even though these expressions have a magnitude and direction, they do not obey the rules of vector addition and are not vectors unless the rotations are infinitesimally small.

It is easily shown by means of a simple example that finite rotations do not obey the rules for addition of vectors. Take a book and set up a coordinate system as shown in Fig. 14-33a. Let $\Delta\boldsymbol{\theta}_x = 90° \mathbf{i}$ and $\Delta\boldsymbol{\theta}_y = 90° \mathbf{j}$ represent counterclockwise rotations of 90° about the *x*- and *y*-axes, respectively. Then the rotation $\Delta\boldsymbol{\theta}_x + \Delta\boldsymbol{\theta}_y$ (that is, the rotation $\Delta\boldsymbol{\theta}_x$ followed by the rotation $\Delta\boldsymbol{\theta}_y$) results in the final position shown in Fig. 14-33b. However, the rotation $\Delta\boldsymbol{\theta}_y + \Delta\boldsymbol{\theta}_x$ (that is, $\Delta\boldsymbol{\theta}_y$ followed by $\Delta\boldsymbol{\theta}_x$) results in the final position shown in Fig. 14-33c. Obviously, these final positions are not the same, and the "sum" of the rotations depends on the order in which they are written

$$\Delta\boldsymbol{\theta}_x + \Delta\boldsymbol{\theta}_y \neq \Delta\boldsymbol{\theta}_y + \Delta\boldsymbol{\theta}_x$$

Therefore, finite rotations are not vectors.

Quantities such as finite rotations are known as "pseudo vectors." Although representable by directed line segments, they do not add

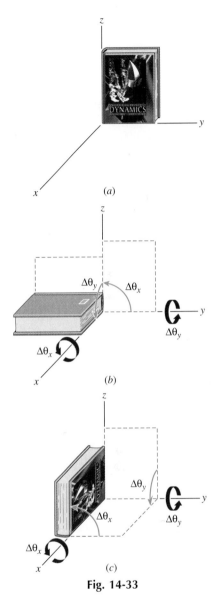

Fig. 14-33

properly as vectors and they are not vector quantities. For this and other reasons, finite rotations are difficult quantities to work with. Although they must be dealt with in advanced dynamics problems, the computation of finite rotations will not be required in this first course in dynamics.

14-7-3   Infinitesimal Rotations

Although finite rotations do not combine vectorially, rotations that are small enough in magnitude do combine properly and are vectors. As depicted in Fig. 14-34, the infinitesimal rotations $d\boldsymbol{\theta}_1$ and $d\boldsymbol{\theta}_2$ of a rigid body about the fixed point A would move the point P first to Q_1 and then to S if the rotation $d\boldsymbol{\theta}_1$ were applied first. If the rotation $d\boldsymbol{\theta}_2$ were applied first, then P would move first to Q_2 and then to S'. Although these motions take place on the surface of a sphere of radius R, for infinitesimal rotations the curvature of the sphere is of negligible effect, the sides of the displacement figure are essentially parallel, and $S = S'$. Then the total displacement of point P is given by

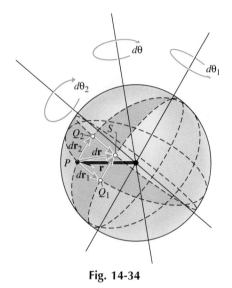

Fig. 14-34

$$
\begin{aligned}
d\mathbf{r} &= d\mathbf{r}_1 + d\mathbf{r}_2 = d\mathbf{r}_2 + d\mathbf{r}_1 \\
&= d\boldsymbol{\theta}_1 \times \mathbf{r} + d\boldsymbol{\theta}_2 \times \mathbf{r} = d\boldsymbol{\theta}_2 \times \mathbf{r} + d\boldsymbol{\theta}_1 \times \mathbf{r} \\
&= (d\boldsymbol{\theta}_1 + d\boldsymbol{\theta}_2) \times \mathbf{r} = (d\boldsymbol{\theta}_2 + d\boldsymbol{\theta}_1) \times \mathbf{r} \\
&= d\boldsymbol{\theta} \times \mathbf{r}
\end{aligned}
\tag{14-18}
$$

where

$$
d\boldsymbol{\theta} = d\boldsymbol{\theta}_1 + d\boldsymbol{\theta}_2
$$

is a single resultant rotation about the axis shown and the order of the vector addition is not important.

14-7-4   Rotation about a Fixed Point

Since $d\boldsymbol{\theta}$ is now known to be a vector, its derivative with respect to time is also a vector. This vector

$$
\boldsymbol{\omega} = \dot{\boldsymbol{\theta}} = \frac{d\boldsymbol{\theta}}{dt}
$$

is called the angular velocity vector. The direction of the angular velocity vector $\boldsymbol{\omega}$ represents the axis about which the body is rotating and its magnitude is the rate of rotation about that axis. For rotation about a fixed point, however, the direction of this axis is not fixed. Therefore, both the magnitude and the direction of $\boldsymbol{\omega}$ will be functions of time.

Angular acceleration is the derivative of the angular velocity with respect to time:

$$
\boldsymbol{\alpha} = \dot{\boldsymbol{\omega}}
$$

Since both the direction and magnitude of $\boldsymbol{\omega}$ are functions of time, the derivative must take into account variations in both quantities. In general, the direction of $\boldsymbol{\alpha}$ is not the same as the direction of $\boldsymbol{\omega}$.

The velocity of any point in the rigid body is given by the derivative of its position with respect to time. If the displacement of Eq. 14-18 takes place in time dt, then

$$
\frac{d\mathbf{r}}{dt} = \frac{d\boldsymbol{\theta}}{dt} \times \mathbf{r}
$$

and the velocity of the particle at point P is given by

$$
\boldsymbol{v}_P = \boldsymbol{\omega} \times \mathbf{r}_P
\tag{14-19}
$$

The acceleration of point P is given by the derivative of the velocity with respect to time:

$$\mathbf{a}_P = \frac{d\mathbf{v}_P}{dt} = \frac{d\boldsymbol{\omega}}{dt} \times \mathbf{r}_P + \boldsymbol{\omega} \times \frac{d\mathbf{r}_P}{dt}$$
$$= \boldsymbol{\alpha} \times \mathbf{r}_P + \boldsymbol{\omega} \times \mathbf{v}_P = \boldsymbol{\alpha} \times \mathbf{r}_P + \boldsymbol{\omega} \times (\boldsymbol{\omega} \times \mathbf{r}_P) \qquad (14\text{-}20)$$

Although these equations have the same form as the equations for plane motion, it is important to remember that both the magnitudes and directions of $\boldsymbol{\omega}$ and $\boldsymbol{\alpha}$ vary with time and that the direction of $\boldsymbol{\alpha}$ is not the same as the direction of $\boldsymbol{\omega}$.

14-7-5   General Motion of a Rigid Body

If A and B are any two particles, their positions are related by the triangle law of addition for vectors

$$\mathbf{r}_B = \mathbf{r}_A + \mathbf{r}_{B/A} \qquad (14\text{-}21)$$

where $\mathbf{r}_A$ and $\mathbf{r}_B$ are the absolute positions of the particles A and B, respectively, and $\mathbf{r}_{B/A}$ is the position of B relative to A. The time derivative of Eq. 14-21 gives the relative velocity equation

$$\mathbf{v}_B = \mathbf{v}_A + \mathbf{v}_{B/A}$$

where $\mathbf{v}_A$ and $\mathbf{v}_B$ are the absolute velocities of particles A and B, respectively, and $\mathbf{v}_{B/A}$ is the velocity of B relative to A. If particles A and B are two particles in a rigid body, however, then the distance between them is constant and particle B appears to travel on a spherical surface about A. Therefore, the relative velocity $\mathbf{v}_{B/A}$ is given by (Eq. 14-19)

$$\mathbf{v}_{B/A} = \boldsymbol{\omega} \times \mathbf{r}_{B/A}$$

Therefore

$$\boldsymbol{v}_B = \mathbf{v}_A + \mathbf{v}_{B/A} = \mathbf{v}_A + \boldsymbol{\omega} \times \mathbf{r}_{B/A} \qquad (14\text{-}22)$$

and the velocity of particle B consists of the sum of two parts: $\mathbf{v}_A$, which represents a translation of the entire body with particle A; and $\boldsymbol{\omega} \times \mathbf{r}_{B/A}$, which represents rotation about point A. Similarly, taking the time derivative of Eq. 14-22 gives

$$\mathbf{a}_B = \mathbf{a}_A + \mathbf{a}_{B/A} = \mathbf{a}_A + \boldsymbol{\alpha} \times \mathbf{r}_{B/A} + \boldsymbol{\omega} \times (\boldsymbol{\omega} \times \mathbf{r}_{B/A}) \qquad (14\text{-}23)$$

and the acceleration of particle B also consists of the sum of two parts: $\mathbf{a}_A$, which represents a translation of the entire body with particle A; and $\boldsymbol{\alpha} \times \mathbf{r}_{B/A} + \boldsymbol{\omega} \times (\boldsymbol{\omega} \times \mathbf{r}_{B/A})$, which represents rotation about point A.

Again it must be cautioned that the directions of $\boldsymbol{\omega}$ and $\boldsymbol{\alpha}$ are not fixed and the computation of the relative velocity and the relative acceleration must not be done in such a manner as to assume that they are fixed.

14-7-6   Three-Dimensional Motion Relative to Rotating Axes

The development of the relative velocity and relative acceleration equations for three-dimensional motion relative to rotating axes parallels that of planar motion in Section 14-6. The difference is that for three-dimensional motion, the directions of the angular velocity and angular acceleration vectors are not fixed as they were for two-dimensional motion. While the variation of these directions with time must be included in the differentiations, the vector form of Eqs. 14-16d and 14-17d properly accounts for them, and the equations are correct for both two- and three-dimensional motion.

In a manner analogous to the development of Section 14-6, consider the general three-dimensional motion of two points A and B whose locations are specified in the fixed XYZ-coordinate system by the position vectors

$$\mathbf{r}_A = X_A\mathbf{i} + Y_A\mathbf{j} + Z_A\mathbf{k}$$
$$\mathbf{r}_B = X_B\mathbf{i} + Y_B\mathbf{j} + Z_B\mathbf{k}$$

and $\mathbf{i}$, $\mathbf{j}$, and $\mathbf{k}$ are unit vectors along the X-, Y-, and Z-axes, respectively. The relative position, however, will be written relative to a coordinate system xyz, which has its origin attached to A and rotates with an angular velocity $\boldsymbol{\omega}$ and an angular acceleration $\boldsymbol{\alpha}$ relative to the fixed coordinate system XYZ. Therefore,

$$\mathbf{r}_{B/A} = x\mathbf{e}_x + y\mathbf{e}_y + z\mathbf{e}_z$$

where the unit vectors along the x-, y-, and z-axes have again been denoted $\mathbf{e}_x$, $\mathbf{e}_y$, and $\mathbf{e}_z$ to distinguish them from $\mathbf{i}$, $\mathbf{j}$, and $\mathbf{k}$ and to emphasize that they are functions of time. Then by vector addition the absolute position vectors $\mathbf{r}_A$ and $\mathbf{r}_B$ and the relative position vector $\mathbf{r}_{B/A}$ are related by

$$\mathbf{r}_B = \mathbf{r}_A + (x\mathbf{e}_x + y\mathbf{e}_y + z\mathbf{e}_z) \tag{14-24}$$

The relationship between the absolute and relative velocities is obtained by differentiating the relative position equation (Eq. 14-24) with respect to time to get

$$\mathbf{v}_B = \mathbf{v}_A + \frac{d\mathbf{r}_{B/A}}{dt} = \mathbf{v}_A + \frac{d(x\mathbf{e}_x + y\mathbf{e}_y + z\mathbf{e}_z)}{dt}$$
$$= \mathbf{v}_A + \frac{dx}{dt}\mathbf{e}_x + x\frac{d\mathbf{e}_x}{dt} + \frac{dy}{dt}\mathbf{e}_y + y\frac{d\mathbf{e}_y}{dt} + \frac{dz}{dt}\mathbf{e}_z + z\frac{d\mathbf{e}_z}{dt}$$
$$= \mathbf{v}_A + \mathbf{v}_{Brel} + \left(x\frac{d\mathbf{e}_x}{dt} + y\frac{d\mathbf{e}_y}{dt} + z\frac{d\mathbf{e}_z}{dt}\right) \tag{14-25a}$$

in which $\mathbf{v}_{Brel} = \dot{x}\mathbf{e}_x + \dot{y}\mathbf{e}_y + \dot{z}\mathbf{e}_z$ is the velocity of B relative to (as measured in) the rotating xyz-coordinate system.

The derivatives of the unit vectors in the parentheses of Eq. 14-25a could be evaluated in a manner analogous to that used in Section 14-6. Alternatively, the unit vector $\mathbf{e}_x$ can be thought of as the position of an imaginary particle that rotates about the point A with angular velocity $\boldsymbol{\omega}$. Then, the derivative of $\mathbf{e}_x$ with respect to time would be the velocity of the imaginary particle and is given by

$$\frac{d\mathbf{e}_x}{dt} = \boldsymbol{\omega} \times \mathbf{e}_x$$

Similarly,

$$\frac{d\mathbf{e}_y}{dt} = \boldsymbol{\omega} \times \mathbf{e}_y \quad \text{and} \quad \frac{d\mathbf{e}_z}{dt} = \boldsymbol{\omega} \times \mathbf{e}_z$$

Substituting these results back into Eq. 14-25a gives

$$\mathbf{v}_B = \mathbf{v}_A + \mathbf{v}_{Brel} + (x\boldsymbol{\omega} \times \mathbf{e}_x + y\boldsymbol{\omega} \times \mathbf{e}_y + z\boldsymbol{\omega} \times \mathbf{e}_z)$$
$$= \mathbf{v}_A + \boldsymbol{\omega} \times \mathbf{r}_{B/A} + \mathbf{v}_{Brel} \tag{14-25b}$$

where $\mathbf{v}_A$, $\mathbf{v}_B$, and $\boldsymbol{\omega}$ are all measured relative to the fixed XYZ-coordinate system; $\mathbf{r}_{B/A}$ and $\mathbf{v}_{Brel}$ are measured relative to the rotating xyz-coordinate system.

The relationship between the absolute and relative accelerations is obtained by differentiating the relative velocity equation (Eq. 14-25b) with respect to time to get

$$\boldsymbol{a}_B = \mathbf{a}_A + \frac{d\boldsymbol{\omega}}{dt} \times \mathbf{r}_{B/A} + \boldsymbol{\omega} \times \frac{d\mathbf{r}_{B/A}}{dt} + \frac{d\mathbf{v}_{Brel}}{dt} \qquad (14\text{-}26a)$$

From the calculation for relative velocity,

$$\frac{d\mathbf{r}_{B/A}}{dt} = \mathbf{v}_{Brel} + \boldsymbol{\omega} \times \mathbf{r}_{B/A} \qquad (14\text{-}26b)$$

A similar calculation for the derivative of $\mathbf{v}_{Brel}$ with respect to time gives

$$\begin{aligned}
\frac{d\mathbf{v}_{Brel}}{dt} &= \frac{d\,(\dot{x}\mathbf{e}_x + \dot{y}\mathbf{e}_y + \dot{z}\mathbf{e}_z)}{dt} \\
&= (\ddot{x}\mathbf{e}_x + \ddot{y}\mathbf{e}_y + \ddot{z}\mathbf{e}_z) + \left(\dot{x}\frac{d\mathbf{e}_x}{dt} + \dot{y}\frac{d\mathbf{e}_y}{dt} + \dot{z}\frac{d\mathbf{e}_z}{dt} \right) \\
&= \mathbf{a}_{Brel} + (\dot{x}\boldsymbol{\omega} \times \mathbf{e}_x + \dot{y}\boldsymbol{\omega} \times \mathbf{e}_y + \dot{z}\boldsymbol{\omega} \times \mathbf{e}_z) \\
&= \mathbf{a}_{Brel} + \boldsymbol{\omega} \times \mathbf{v}_{Brel} \qquad (14\text{-}26c)
\end{aligned}$$

where $\mathbf{a}_{Brel} = \ddot{x}\mathbf{e}_x + \ddot{y}\mathbf{e}_y + \ddot{z}\mathbf{e}_z$ is the acceleration of point B relative to (as measured in) the rotating xyz-coordinate system. Substituting Eqs. 14-26b and 14-26c into Eq. 14-26a and rearranging terms yields

$$\begin{aligned}
\mathbf{a}_B = \mathbf{a}_A &+ \boldsymbol{\alpha} \times \mathbf{r}_{B/A} + \boldsymbol{\omega} \times (\boldsymbol{\omega} \times \mathbf{r}_{B/A}) \\
&+ 2\boldsymbol{\omega} \times \mathbf{v}_{Brel} + \mathbf{a}_{Brel} \qquad (14\text{-}26d)
\end{aligned}$$

where $\mathbf{a}_A$, $\mathbf{a}_B$, $\boldsymbol{\omega}$, and $\boldsymbol{\alpha}$ are all measured relative to the fixed XYZ-coordinate system; $\mathbf{r}_{B/A}$, $\mathbf{v}_{Brel}$, and $\mathbf{a}_{Brel}$ are measured relative to the rotating xyz-coordinate system; and the term $2\boldsymbol{\omega} \times \mathbf{v}_{Brel}$ is called the Coriolis acceleration.

Equations 14-26 and 14-27 require the sum of several vectors. In order to add all these vectors together, their components need to be written in a common coordinate system. The components may be written in either the fixed XYZ-coordinate system or the rotating xyz-coordinate system. The choice is based solely on the form in which the data are given and the form in which the results are desired.

The orientation, location of the origin, angular velocity, and angular acceleration of the rotating coordinate system should be selected to simplify the calculation of the various terms in the relative velocity and relative acceleration equations. For example, the origin A should be a point whose absolute velocity and absolute acceleration are easily obtained. The angular velocity and angular acceleration of the rotating frame should be chosen so that the velocity and acceleration of point B are easy to calculate relative to the rotating coordinate system. The orientation of the rotating coordinate system relative to the fixed coordinate system should be chosen such that the components of the various vectors are easy to describe.

The 400-mm diameter disk of Fig. 14-35 is rigidly attached to a 600-mm long axle and rolls without slipping on a fixed surface in the x-y plane. The axle, which is perpendicular to the disk, is attached to a ball and socket joint at A and is free to pivot about A. As the disk and axle rotate about their own axis with angular velocity ω_1, the axle also rotates about a vertical axis with angular velocity ω_2. If $\omega_1 = 5$ rad/s and $\dot{\omega}_1 = 20$ rad/s^2 at the instant shown, determine

a. The total angular velocity $\boldsymbol{\omega}$ and total angular acceleration $\boldsymbol{\alpha}$ of the disk at this instant.

b. The velocity $\mathbf{v}_C$ and acceleration $\mathbf{a}_C$ of point C on the rim of the disk at this instant.

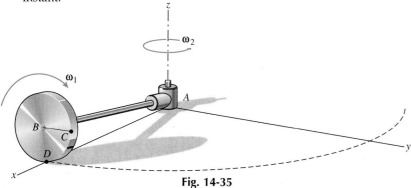

Fig. 14-35

SOLUTION

a. The rod AB makes an angle of $18.43°$ with the x-axis (Fig. 14-36a). In accordance with the right-hand rule, the angular velocity $\boldsymbol{\omega}_1$ points from B to A and has components

$$\boldsymbol{\omega}_1 = \omega_1 \mathbf{e}_{BA} = -5 \cos 18.43° \, \mathbf{i} - 5 \sin 18.43° \, \mathbf{k} \text{ rad/s}$$
$$= -4.744\mathbf{i} - 1.581\mathbf{k} \text{ rad/s} \qquad (a)$$

where $\mathbf{e}_{BA}$ is a unit vector pointing from B to A. Since the wheel rolls without slipping, the arc length $\overline{DD'}$ on the wheel is equal to the arc length $\overline{DD'}$ on the surface (Fig. 14-36b)

$$s = 200 \, \theta_1 = 632.46 \, \theta_2 \qquad (b)$$

Differentiating Eq. b with respect to time gives the relationship between the angular velocities ω_1 and ω_2

$$\omega_2 = 0.3162 \, \omega_1 \qquad (c)$$

Then
$$\boldsymbol{\omega}_2 = 1.581\mathbf{k} \text{ rad/s}$$
$$\boldsymbol{\omega} = \boldsymbol{\omega}_1 + \boldsymbol{\omega}_2 = -4.74\mathbf{i} \text{ rad/s} \qquad \text{Ans.}$$

The angular acceleration $\boldsymbol{\alpha}$ is the derivative of the angular velocity

$$\boldsymbol{\alpha} = \dot{\boldsymbol{\omega}} = \dot{\boldsymbol{\omega}}_1 + \dot{\boldsymbol{\omega}}_2$$

where the derivatives must account for changes in the directions as well as changes in the magnitudes of $\boldsymbol{\omega}_1$ and $\boldsymbol{\omega}_2$. Following the discussion of Section 14-7 for the derivative of the unit vector $\mathbf{e}_{BA}$, the derivative of $\boldsymbol{\omega}_1$ is

$$\dot{\boldsymbol{\omega}}_1 = \dot{\omega}_1 \mathbf{e}_{BA} + \omega_1 \dot{\mathbf{e}}_{BA} = \dot{\omega}_1 \mathbf{e}_{BA} + \omega_1 (\boldsymbol{\omega}_2 \times \mathbf{e}_{BA})$$
$$= (-20 \cos 18.43° \, \mathbf{i} - 20 \sin 18.43° \, \mathbf{k})$$
$$+ 5[(1.581\mathbf{k}) \times (-\cos 18.43° \, \mathbf{i} - \sin 18.43° \, \mathbf{k})]$$
$$= -18.97\mathbf{i} - 7.50\mathbf{j} - 6.323\mathbf{k} \text{ rad/s}^2$$

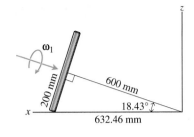

(a)

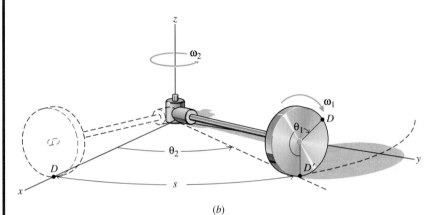

(b)

Fig. 14-36

Differentiating Eq. *c* with respect to time gives

$$\dot{\omega}_2 = 0.3162\dot{\omega}_1$$

and, since the direction of ω_2 is constant,

$$\dot{\boldsymbol{\omega}}_2 = 6.324\mathbf{k} \text{ rad/s}^2$$

Therefore

$$\boldsymbol{\alpha} = -18.97\mathbf{i} - 7.50\mathbf{j} \text{ rad/s}^2 \qquad \text{Ans.}$$

b. The position of point C relative to the point about which the disk is rotating is

$$\mathbf{r}_C = 600 \cos 18.43° \, \mathbf{i} + 200\mathbf{j} + 600 \sin 18.43° \, \mathbf{k}$$
$$= 569.2\mathbf{i} + 200\mathbf{j} + 189.69\mathbf{k} \text{ mm}$$

Then the velocity of point C is given by Eq. 14-19

$$\mathbf{v}_C = \boldsymbol{\omega} \times \mathbf{r}_C = -4.744\mathbf{i} \times (569.2\mathbf{i} + 200\mathbf{j} + 189.69\mathbf{k})$$
$$= 899.9\mathbf{j} - 948.8\mathbf{k} \text{ mm/s}$$
$$\cong 900\mathbf{j} - 949\mathbf{k} \text{ mm/s} \qquad \text{Ans.}$$

The acceleration of point C is given by Eq. 14-20

$$\mathbf{a}_C = \boldsymbol{\alpha} \times \mathbf{r}_C + \boldsymbol{\omega} \times \mathbf{v}_C$$
$$= (-18.97\mathbf{i} - 7.50\mathbf{j}) \times (569.2\mathbf{i} + 200\mathbf{j} + 189.69\mathbf{k})$$
$$\quad -4.744\mathbf{i} \times (899.9\mathbf{j} - 948.8\mathbf{k})$$
$$= -1423\mathbf{i} - 903\mathbf{j} - 3790\mathbf{k} \text{ mm/s}^2 \qquad \text{Ans.}$$

The slender rod of Fig. 14-37 is connected to the sliders A and B by ball and socket joints. If slider A is moving in the negative x-direction at a constant speed of 6 in./s, determine

a. The velocity $\mathbf{v}_B$ and acceleration $\mathbf{a}_B$ of slider B at the instant shown.

b. The angular velocity $\boldsymbol{\omega}$ and angular acceleration $\boldsymbol{\alpha}$ of the rod at the instant shown. (Assume that the rod is not rotating about its own axis.)

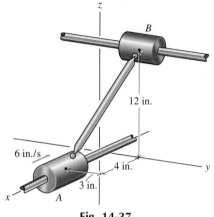

Fig. 14-37

SOLUTION

a. Letting x represent the position of slider A and y represent the position of slider B, the length of the rod AB at any instant of time is

$$\sqrt{(x^2 + y^2 + 12^2)} = 13 \qquad (a)$$

Squaring both sides of Eq. a and differentiating with respect to time gives

$$2x\dot{x} + 2y\dot{y} = 0 \qquad (b)$$

where $\dot{x} = v_A = -6$ in./s and $\dot{y} = v_B$. At the instant shown, $x = 4$ in. and $y = 3$ in. Therefore,

$$v_B = 8 \text{ in./s}$$

or
$$\mathbf{v}_B = 8\mathbf{j} \text{ in./s} \qquad \text{Ans.}$$

Differentiating Eq. b with respect to time gives

$$\dot{x}^2 + x\ddot{x} + \dot{y}^2 + y\ddot{y} = 0$$

where $\ddot{x} = a_A = 0$ and $\ddot{y} = a_B$. Therefore,

$$a_B = -100/3 \text{ in./s}^2$$

or
$$\mathbf{a}_B = -33.3\mathbf{j} \text{ in./s}^2 \qquad \text{Ans.}$$

b. The relative velocity equation is

$$\mathbf{v}_B = \mathbf{v}_A + \mathbf{v}_{B/A} = \mathbf{v}_A + \boldsymbol{\omega} \times \mathbf{r}_{B/A}$$

where $\mathbf{v}_A = -6\mathbf{i}$ in./s, $\mathbf{v}_B = 8\mathbf{j}$ in./s, and the position of B relative to A is

$$\mathbf{r}_{B/A} = -4\mathbf{i} + 3\mathbf{j} + 12\mathbf{k} \text{ in.}$$

Therefore

$$\begin{aligned}
8\mathbf{j} &= -6\mathbf{i} + (\omega_x\mathbf{i} + \omega_y\mathbf{j} + \omega_z\mathbf{k}) \times (-4\mathbf{i} + 3\mathbf{j} + 12\mathbf{k}) \\
&= (12\omega_y - 3\omega_z - 6)\mathbf{i} - (4\omega_z + 12\omega_x)\mathbf{j} + (3\omega_x + 4\omega_y)\mathbf{k} \qquad (c)
\end{aligned}$$

Although Eq. c is a vector equation, its three components

$$x: \quad 12\omega_y - 3\omega_z = 6 \qquad (d)$$
$$y: \quad 12\omega_x + 4\omega_z = -8 \qquad (e)$$
$$z: \quad 3\omega_x + 4\omega_y = 0 \qquad (f)$$

are not sufficient to find the three unknown components of the angular velocity. The relative velocity of the ends of bar AB are independent of the rotation of bar AB about its own axis and Eqs. d, e, and f allow infinitely many solutions for the angular velocity $\boldsymbol{\omega}$ differing by the rate of rotation about AB. This ambiguity is removed by the assumption that the bar is not rotating about its own axis.

The assumption that bar AB is not rotating about its own axis is equivalent to saying that the component of $\boldsymbol{\omega}$ in the direction of the bar is zero:

$$\boldsymbol{\omega} \cdot \mathbf{r}_{B/A} = (\omega_x \mathbf{i} + \omega_y \mathbf{j} + \omega_z \mathbf{k}) \cdot (-4\mathbf{i} + 3\mathbf{j} + 12\mathbf{k})$$
$$= -4\omega_x + 3\omega_y + 12\omega_z = 0 \qquad (g)$$

Solving Eqs. d, e, and g simultaneously gives

$$\omega_x = -0.5680 \text{ rad/s}; \qquad \omega_y = 0.4260 \text{ rad/s}$$
$$\omega_z = -0.2959 \text{ rad/s}$$

or

$$\boldsymbol{\omega} = -0.568\mathbf{i} + 0.426\mathbf{j} - 0.296\mathbf{k} \text{ rad/s} \qquad \text{Ans.}$$

Similarly, the relative acceleration equation

$$\mathbf{a}_B = \mathbf{a}_A + \mathbf{a}_{B/A} = \mathbf{a}_A + \boldsymbol{\alpha} \times \mathbf{r}_{B/A} + \boldsymbol{\omega} \times \mathbf{v}_{B/A} \qquad (h)$$

where $\mathbf{a}_A = \mathbf{0}$, $\mathbf{a}_B = -33.33\mathbf{j}$ in./s^2, and $\mathbf{v}_{B/A} = \mathbf{v}_B - \mathbf{v}_A = 6.000\mathbf{i} + 8.000\mathbf{j}$ in./s. Therefore, the relative acceleration equation, Eq. h,

$$-33.33\mathbf{j} = \mathbf{0} + (\alpha_x \mathbf{i} + \alpha_y \mathbf{j} + \alpha_z \mathbf{k}) \times (-4\mathbf{i} + 3\mathbf{j} + 12\mathbf{k})$$
$$+ (-0.5680\mathbf{i} + 0.4260\mathbf{j} - 0.2959\mathbf{k}) \times (6.00\mathbf{i} + 8.00\mathbf{j})$$
$$= [(12\alpha_y - 3\alpha_z)\mathbf{i} - (4\alpha_z + 12\alpha_x)\mathbf{j} + (3\alpha_x + 4\alpha_y)\mathbf{k}]$$
$$+ [2.367\mathbf{i} - 1.7754\mathbf{j} - 7.100\mathbf{k}] \qquad (i)$$

has the three components

$$x: \quad 12\alpha_y - 3\alpha_z = -2.367 \qquad (j)$$
$$y: \quad 12\alpha_x + 4\alpha_z = 31.555 \qquad (k)$$
$$z: \quad 3\alpha_x + 4\alpha_y = 7.100 \qquad (l)$$

Again using the assumption that the bar is not rotating about its own axis,

$$\boldsymbol{\alpha} \cdot \mathbf{r}_{B/A} = (\alpha_x \mathbf{i} + \alpha_y \mathbf{j} + \alpha_z \mathbf{k}) \cdot (-4\mathbf{i} + 3\mathbf{j} + 12\mathbf{k})$$
$$= -4\alpha_x + 3\alpha_y + 12\alpha_z = 0 \qquad (m)$$

Finally, solving Eqs. k, l, and m simultaneously gives

$$\alpha_x = 2.367 \text{ rad/s}^2 \qquad \alpha_y = 0.000 \text{ rad/s}^2$$
$$\alpha_z = 0.789 \text{ rad/s}^2$$

or

$$\boldsymbol{\alpha} = 2.367\mathbf{i} + 0.789\mathbf{k} \text{ rad/s}^2 \qquad \text{Ans.}$$

EXAMPLE PROBLEM 14-14

The fire truck ladder shown in Fig. 14-38 is being raised at a constant rate of $\dot{\theta}_2 = 0.5$ rad/s. Simultaneously, it is rotating about a vertical axis at a constant rate of $\dot{\theta}_1 = 0.8$ rad/s and is being extended at a constant rate of $\dot{s} = 1.5$ m/s. Determine the velocity $\mathbf{v}_B$ and acceleration $\mathbf{a}_B$ of the end of the ladder when $s = 10$ m and $\theta_2 = 30°$.

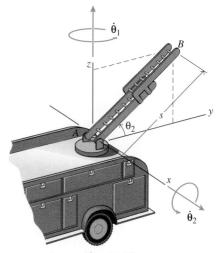

Fig. 14-38

SOLUTION

The rotating xyz-coordinate system is chosen with its origin at A as shown on Fig. 14-38. The rotation rate of the xyz system is chosen so that the ladder is always in the yz-plane. Then the relative velocity equation (Eq. 14-25b) is

$$\mathbf{v}_B = \mathbf{v}_A + \boldsymbol{\omega} \times \mathbf{r}_{B/A} + \mathbf{v}_{Brel}$$

where
$$\mathbf{v}_A = \mathbf{0}$$
$$\boldsymbol{\omega} = \dot{\theta}_1 \mathbf{e}_z$$
$$\boldsymbol{\omega} \times \mathbf{r}_{B/A} = \dot{\theta}_1 \mathbf{e}_z \times (s \cos \theta_2 \, \mathbf{e}_y + s \sin \theta_2 \, \mathbf{e}_z)$$
$$= -(0.8)(10) \cos 30° \, \mathbf{e}_x = -6.928 \, \mathbf{e}_x$$

and (Fig. 14-39a)

$$\mathbf{v}_{Brel} = [\dot{s} \cos \theta_2 \, \mathbf{e}_y + \dot{s} \sin \theta_2 \, \mathbf{e}_z$$
$$- s\dot{\theta}_2 \sin \theta_2 \, \mathbf{e}_y + s\dot{\theta}_2 \cos \theta_2 \, \mathbf{e}_z]$$
$$= [1.5 \cos 30° - (10)(0.5) \sin 30°] \, \mathbf{e}_y$$
$$+ [1.5 \sin 30° + (10)(0.5) \cos 30°] \, \mathbf{e}_z$$
$$= -1.201 \mathbf{e}_y + 5.080 \mathbf{e}_z$$

Therefore

$$\mathbf{v}_B = -6.928 \mathbf{e}_x - 1.201 \mathbf{e}_y + 5.080 \mathbf{e}_z \qquad \text{Ans.}$$

The relative acceleration equation (Eq. 14-26d) is

$$\mathbf{a}_B = \mathbf{a}_A + \boldsymbol{\alpha} \times \mathbf{r}_{B/A} + \boldsymbol{\omega} \times (\boldsymbol{\omega} \times \mathbf{r}_{B/A}) + \mathbf{a}_{Brel} + 2\boldsymbol{\omega} \times \mathbf{v}_{Brel}$$

where
$$\mathbf{a}_A = \mathbf{0}$$
$$\boldsymbol{\alpha} \times \mathbf{r}_{B/A} = \mathbf{0}$$
$$\boldsymbol{\omega} \times (\boldsymbol{\omega} \times \mathbf{r}_{B/A}) = \dot{\theta}_1 \mathbf{e}_z \times (-6.928 \mathbf{e}_x) = -5.543 \mathbf{e}_y$$
$$2\boldsymbol{\omega} \times \mathbf{v}_{Brel} = 2\dot{\theta}_1 \mathbf{e}_z \times (-1.201 \mathbf{e}_y + 5.080 \mathbf{e}_z) = 1.922 \mathbf{e}_x$$

and (Fig. 14-39b)

$$\mathbf{a}_{Brel} = (\ddot{s} - s\dot{\theta}_2^2)(\cos \theta_2 \, \mathbf{e}_y + \sin \theta_2 \, \mathbf{e}_z)$$
$$+ (s\ddot{\theta}_2 + 2\dot{s}\dot{\theta}_2)(-\sin \theta_2 \, \mathbf{e}_y + \cos \theta_2 \, \mathbf{e}_z)$$
$$= [0 - (10)(0.5)^2][\cos 30° \, \mathbf{e}_y + \sin 30° \, \mathbf{e}_z]$$
$$+ [0 + 2(1.5)(0.5)][-\sin 30° \, \mathbf{e}_y + \cos 30° \, \mathbf{e}_z]$$
$$= -2.915 \mathbf{e}_y + 0.049 \mathbf{e}_z$$

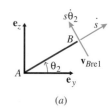

(a)

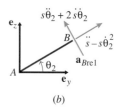

(b)

Fig. 14-39

Therefore

$$\mathbf{a}_B = 1.922 \mathbf{e}_x - 8.458 \mathbf{e}_y + 0.049 \mathbf{e}_z \text{ rad/s}^2 \qquad \text{Ans.}$$

PROBLEMS

14-131* Sketch the final position of the book of Fig. 14-33a after successive rotations of

$$\Delta\theta_x = 90° \qquad \Delta\theta_y = 90° \qquad \Delta\theta_z = 90°$$

Also determine the single rotation (axis and angle) that is equivalent to this combination of rotations.

14-132* Sketch the final position of the book of Fig. 14-33a after successive rotations of

$$\Delta\theta_z = 180° \qquad \Delta\theta_y = 90° \qquad \Delta\theta_x = 180°$$

Also determine the single rotation (axis and angle) that is equivalent to this combination of rotations.

14-133 Sketch the final position of the book of Fig. 14-33a after successive rotations of

$$\Delta\theta_z = 90° \qquad \Delta\theta_y = 90° \qquad \Delta\theta_z = 90° \qquad \Delta\theta_x = 90°$$

Also determine the single rotation (axis and angle) that is equivalent to this combination of rotations.

14-134* Sketch the final position of the book of Fig. 14-33a after successive rotations of

$$\Delta\theta_x = 90° \qquad \Delta\theta_y = 90° \qquad \Delta\theta_x = -90° \qquad \Delta\theta_z = 90°$$

14-135 Determine which of the following rotations of an object will result in the same final position of the object:

a. $\Delta\theta_x = 90°$, $\Delta\theta_y = 90°$

b. $\Delta\theta_x = 90°$, $\Delta\theta_z = 90°$

c. $\Delta\theta_x = 90°$, $\Delta\theta_z = -90°$

d. $\Delta\theta_y = -90°$, $\Delta\theta_x = 90°$

e. $\Delta\theta_x = 90°$, $\Delta\theta_y = 90°$, $\Delta\theta_z = 90°$

f. $\Delta\theta_y = 90°$, $\Delta\theta_z = 90°$, $\Delta\theta_x = -90°$

g. $\Delta\theta_z = -90°$, $\Delta\theta_x = 90°$, $\Delta\theta_y = 90°$

h. $\Delta\theta_y = -90°$, $\Delta\theta_x = 90°$, $\Delta\theta_y = 90°$, $\Delta\theta_x = 90°$

i. $\Delta\theta_x = 90°$, $\Delta\theta_y = 90°$, $\Delta\theta_x = -90°$, $\Delta\theta_y = -90°$

14-136 The 400-mm diameter disk of Fig. P14-136 is rigidly attached to a 750-mm long axle and rolls without slipping on a fixed surface in the xy-plane. The axle, which is perpendicular to the disk, is attached to a ball and socket joint at A and is free to pivot about A. As the disk rotates about its axle with angular velocity ω_1, the axle also rotates about a vertical axis with angular velocity ω_2. If $\omega_1 = 2$ rad/s and $\dot{\omega}_1 = -5$ rad/s² at the instant shown, determine:

a. The total angular velocity ω and total angular acceleration α of the disk at this instant.

b. The velocity $\mathbf{v}_C$ and acceleration $\mathbf{a}_C$ of point C on the rim of the disk at this instant.

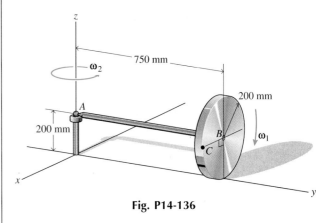

Fig. P14-136

14-137* The 15-in. long pipe AB of Fig. P14-137 is rotating about a vertical axis at a rate of ω_z. At the same time, the 12-in. long pipe BC is rotating about AB at a rate of ω_2 and the 10-in. diameter disk rotates about pipe BC at a rate of ω_1. For the instant shown (when BC is in the horizontal plane, $\omega_1 = 5$ rad/s = constant, $\omega_2 = \dot{\omega}_2 = 0$, $\omega_z = 3$ rad/s, and $\dot{\omega}_z = -10$ rad/s²), determine

a. The total angular velocity ω and total angular acceleration α of the disk.

b. The velocity $\mathbf{v}_D$ and acceleration $\mathbf{a}_D$ of point D on the rim of the disk.

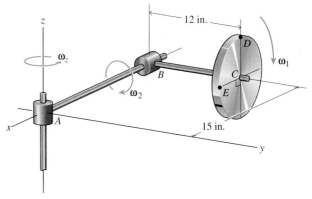

Fig. P14-137

14-138 The axle of the 250-mm diameter disk of Fig. P14-138 is mounted on and rotates with the 800-mm diameter turntable. If at the instant shown $\omega_1 = 10$ rad/s, $\dot\omega_1 = 40$ rad/s^2, $\omega_2 = 3$ rad/s, and $\dot\omega_2 = -25$ rad/s^2, determine

a. The total angular velocity $\boldsymbol{\omega}$ and total angular acceleration $\boldsymbol{\alpha}$ of the disk at this instant.
b. The velocity $\mathbf{v}_D$ and acceleration $\mathbf{a}_D$ of point D on the rim of the disk at this instant.

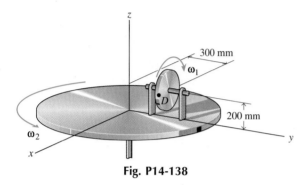

Fig. P14-138

14-139* The 15-in. long pipe AB of Fig. P14-137 is rotating about a vertical axis at a rate of ω_z. At the same time, the 12-in. long pipe BC is rotating about AB at a rate of ω_2 and the 10-in. diameter disk rotates about pipe BC at a rate of ω_1. For the instant shown (when BC is in the horizontal plane, $\omega_1 = 5$ rad/s = constant, $\omega_2 = 3$ rad/s, $\dot\omega_2 = -10$ rad/s^2, $\omega_z = \dot\omega_z = 0$), determine

a. The total angular velocity $\boldsymbol{\omega}$ and total angular acceleration $\boldsymbol{\alpha}$ of the disk.
b. The velocity $\mathbf{v}_E$ and acceleration $\mathbf{a}_E$ of point E on the rim of the disk.

14-140* An electric motor spins the fan blades of Fig. P14-140 at a constant rate of $\omega_1 = 600$ rpm. At the same time, the motor rotates about a vertical axis at a constant rate of $\omega_z = 5$ rpm. For the instant shown, determine the total angular velocity $\boldsymbol{\omega}$ and total angular acceleration $\boldsymbol{\alpha}$ of the fan blades.

14-141 The 48-in. long slender rod of Fig. P14-141 is connected to the sliders A and B by ball-and-socket joints. At the instant shown, $x = 20$ in. and slider A is moving in the negative x-direction at a constant speed of 18 in./s. Determine

a. The velocity $\mathbf{v}_B$ and acceleration $\mathbf{a}_B$ of slider B at this instant.
b. The angular velocity $\boldsymbol{\omega}$ and angular acceleration $\boldsymbol{\alpha}$ of the rod at this instant. (Assume that the rod is not rotating about its own axis.)

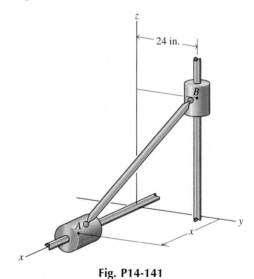

Fig. P14-141

14-142* The 1200-mm long slender rod of Fig. P14-142 is connected to the sliders A and B by ball-and-socket joints. At the instant shown, $y = 750$ mm and slider B is moving in the positive y-direction at a constant speed of 100 mm/s. Determine

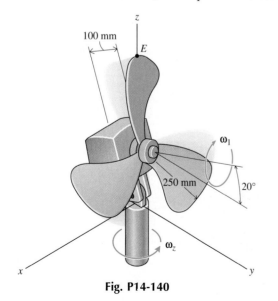

Fig. P14-140

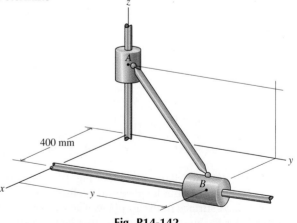

Fig. P14-142

a. The velocity $\mathbf{v}_A$ and acceleration $\mathbf{a}_A$ of slider A at this instant.
b. The angular velocity $\boldsymbol{\omega}$ and angular acceleration $\boldsymbol{\alpha}$ of the rod at this instant. (Assume that the rod is not rotating about its own axis.)

14-143 The 50-in. long slender rod of Fig. P14-143 is connected to the sliders A and B by ball and socket joints. When slider B crosses the x-axis ($x_B = 24$ in., $y_B = 0$ in., $z_B = 0$ in.), the velocity and acceleration of slider A are $\dot{y} = 18$ in./s and $\ddot{y} = -6$ in./s^2, respectively. For this instant determine:

a. The velocity $\mathbf{v}_B$ and acceleration $\mathbf{a}_B$ of slider B.
b. The angular velocity $\boldsymbol{\omega}$ and angular acceleration $\boldsymbol{\alpha}$. (Assume that the rod is not rotating about its own axis.)

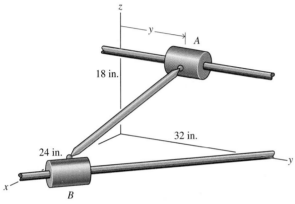

Fig. P14-143

14-144 Assume that the position of slider B of Problem 14-142 is given by $y(t) = 1000 \sin nt$ where t is in seconds, y is in millimeters, and $n = 1$ rad/s. Determine

a. The velocity $\mathbf{v}_A$ and acceleration $\mathbf{a}_A$ of slider A when $t = 0.8$ s.
b. The angular velocity $\boldsymbol{\omega}$ and angular acceleration $\boldsymbol{\alpha}$ of the rod when $t = 0.8$ s. (Assume that the rod is not rotating about its own axis.)

14-145* Assume that the position of slider A of Problem 14-141 is given by $x(t) = 24 \sin nt$ where t is in seconds, x is in inches, and $n = 1$ rad/s. Determine:

a. The velocity $\mathbf{v}_B$ and acceleration $\mathbf{a}_B$ of slider B when $t = 0.5$ s.
b. The angular velocity $\boldsymbol{\omega}$ and angular acceleration $\boldsymbol{\alpha}$ of the rod when $t = 0.5$ s. (Assume that the rod is not rotating about its own axis.)

14-146 The 600-mm diameter wheel of Fig. P14-146 is rotating at a constant rate of $\omega_0 = 5$ rad/s. The 1000-mm long slender rod AB is connected to the rim of the wheel at

A and to the slider B by ball-and-socket joints. For the instant shown when $\theta = 90°$, determine:

a. The velocity $\mathbf{v}_B$ and acceleration $\mathbf{a}_B$ of slider B at this instant.
b. The angular velocity $\boldsymbol{\omega}$ and angular acceleration $\boldsymbol{\alpha}$ of the rod at this instant. (Assume that the rod is not rotating about its own axis.)

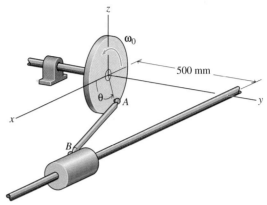

Fig. P14-146

14-147* The crank OA of Fig. P14-147 is rotating with $\dot{\theta} = 3$ rad/s and $\ddot{\theta} = 10$ rad/s^2. The 36-in. long slender rod is connected to the crank at A and to the slider B by ball-and-socket joints. For the instant shown when $\theta = 0°$, determine:

a. The velocity $\mathbf{v}_B$ and acceleration $\mathbf{a}_B$ of slider B at this instant.
b. The angular velocity $\boldsymbol{\omega}$ and angular acceleration $\boldsymbol{\alpha}$ of the rod at this instant. (Assume that the rod is not rotating about its own axis.)

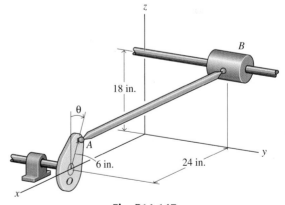

Fig. P14-147

14-148* The fire truck ladder of Example Problem 14-14 is rotating about a vertical axis at a constant rate of $\omega_1 = 0.8$ rad/s with $\dot{s} = 0$, $\ddot{s} = -2.5$ m/s^2, $\dot{\theta}_2 = 0$, and $\ddot{\theta}_2 = -1.5$ rad/s^2 when $s = 10$ m and $\theta_2 = 30°$. Determine the velocity $\mathbf{v}_B$ and acceleration $\mathbf{a}_B$ of the end of the ladder at this instant.

14-149   A bead B is sliding along a bent rod, which is rotating about the x-axis (Fig. P14-149). At the instant shown, the rod is in the xz-plane and $\omega = 5$ rad/s, $\dot{\omega} = 18$ rad/s^2, $s = 8$ in., $\dot{s} = 1$ in./s, and $\ddot{s} = -2.5$ in./s^2. Determine the velocity $\mathbf{v}_B$ and acceleration $\mathbf{a}_B$ of the bead at this instant.

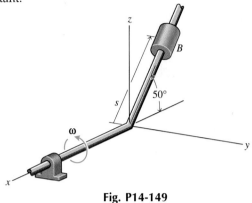

Fig. P14-149

14-150*   A bead B slides along a slot in a 500-mm diameter disk, which is rotating about a vertical axis (Fig. P14-150). At the instant shown, $\omega = 3$ rad/s, $\dot{\omega} = 8$ rad/s^2, $s = 200$ mm, $\dot{s} = 250$ mm/s, and $\ddot{s} = -50$ mm/s^2. Determine the velocity $\mathbf{v}_B$ and acceleration $\mathbf{a}_B$ of the bead at this instant.

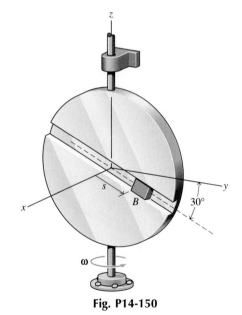

Fig. P14-150

14-151*   A bead B is sliding along a circular wire ring, which is rotating about the y-axis (Fig. P14-151). At the instant shown, the 20-in. diameter ring is in the y-z plane and $\omega = 8$ rad/s, $\dot{\omega} = 12$ rad/s^2, $\theta = 30°$, and $\dot{\theta} = 10$ rad/s = constant. Determine the velocity $\mathbf{v}_B$ and acceleration $\mathbf{a}_B$ of the bead at this instant.

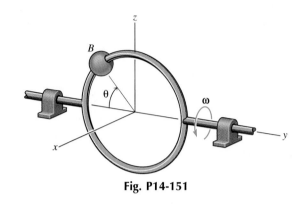

Fig. P14-151

14-152   Repeat Problem 14-150 for the case $s = 0$.

SUMMARY

Kinematics is the study of how objects move. For solid bodies, a full description of the motion requires giving both the location and the orientation of the body. Kinematics of solid bodies involves both linear and angular quantities. A sound understanding of kinematics is a necessary background for the study of kinetics, which relates the motion to the forces that cause it.

All solid bodies will be considered to be rigid. In a rigid body, the distance between all pairs of particles is fixed and independent of time. Also, the angles between all triples of points is fixed.

In the translation of a rigid body, the orientation of every straight line in the body is fixed: horizontal lines remain horizontal and vertical lines remain vertical. This means that the motion of every point in the rigid body is the same as every other point in the rigid body. The kinematics of the particles that make up a rigid body undergoing translation is identical to the kinematics of particle motion.

In a plane motion of a rigid body, each particle in the body remains in a single plane. Coplanar translation and rotation about a fixed axis are specific types of plane motion. A general plane motion is any plane motion for which lines in the body rotate but no point in the body is fixed. General plane motions of a rigid body consist of a translation of the body with some point plus a rotation of the body about that point.

There are two general approaches to solving general plane motion problems: absolute motion analysis and relative motion analysis. In the absolute motion approach, geometric relationships are written that describe the constraints that act on the body and its interaction with other bodies. These relationships are then used to describe the location and motion of other points on the body. The relative motion approach uses the rigidity of the body to relate the velocity and acceleration of two different points in the same rigid body. Since the distance between any two points in a rigid body is fixed, the expressions for the relative velocity and the relative acceleration take particularly simple forms, which depend only on the angular velocity and angular acceleration of the body.

Either approach may be used to solve a particular problem. Some problems are easily described geometrically and are easily handled by the absolute motion approach. Problems that are not easily described geometrically are usually solved using the relative motion approach. For many problems the choice is a matter of personal preference.

In the general plane motion of a rigid body, no point is fixed for all time. However, at any instant of time, it is always possible to find a point on the body (or the body extended) that has zero velocity. Once the instantaneous center is located, the velocity of any other point in the body is found using the relative velocity equation. Use of the instantaneous center is not required to solve any problem. It is just another way of expressing the relative velocity equation.

The instantaneous center of zero velocity for a rigid body in general plane motion is not fixed. Therefore, different points of the rigid body will be instantaneous centers at different instants of time and the location of the instantaneous center of zero velocity will move with respect to time. The instantaneous center of zero velocity must not be used to compute accelerations.

There are several types of problems for which it is convenient to describe the position or motion of one particle relative to a rotating coordinate system. In particular, some mechanisms are connected by pins that slide in grooves or slots. Relative motion is conveniently specified by giving the translational and rotational motion of the member containing the slot, the shape of the slot, and the rate of travel of the pin along the slot. Differentiation of the relative position equation to get the relative velocity and relative acceleration equations must take into account the rotation of the coordinate system. This gives three new terms: $\mathbf{v}_{Brel}$, $\mathbf{a}_{Brel}$, and $2\boldsymbol{\omega} \times \mathbf{v}_{Brel}$ in the relative velocity and relative acceleration equations.

REVIEW PROBLEMS

14-153*   The mechanism shown in Fig. P14-153 is a simplified sketch of a printing press. As the crank arm AB rotates ($\dot{\theta} = 5$ rpm = constant), the printing drum C moves back and forth across the paper. For the instant shown ($\theta = 50°$) determine

a.   The angular velocity $\boldsymbol{\omega}_C$ and angular acceleration $\boldsymbol{\alpha}_C$ of the print drum.
b.   The velocity $\mathbf{v}_D$ and acceleration $\mathbf{a}_D$ of point D on the surface of the drum.

a.   Of point A on the surface of the shaft.
b.   Of point C at the middle of the roller bearing.
c.   Of point B on the surface of the roller bearing.
d.   Of point D on the surface of the roller bearing.

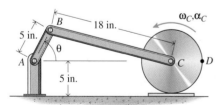

Fig. P14-153

14-154   The case of the roller bearing shown in Fig. P14-154 is fixed, whereas the inner shaft rotates at a constant rate of 5000 rpm. If the roller bearings roll without slipping on the 50-mm diameter shaft and on the 60-mm diameter bearing, determine the velocity $\mathbf{v}$ and acceleration $\mathbf{a}$

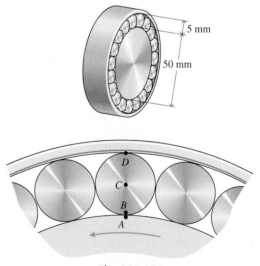

Fig. P14-154

14-155 The truck shown in Fig. P14-155 is initially at rest at a stop light. When the light turns green the truck accelerates at a constant rate of 0.8 ft/s² and the 3-ft diameter tank begins to roll backward with a constant angular acceleration of $\alpha = 0.025$ rad/s². Determine

a. How far the truck travels before the tank rolls off the back of the truck.
b. The velocity $\mathbf{v}_C$ of the center of the tank and the angular velocity ω of the tank when it falls off the back of the truck.
c. The slip velocity (relative velocity between the tank and the road) when the tank hits the road.

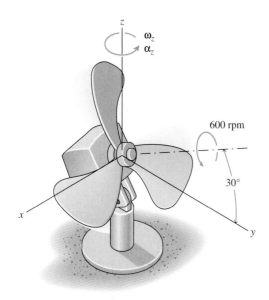

Fig. P14-155

14-156 An electric motor spins the fan blades of Fig. P14-156 at a constant rate of 600 rpm. At the same time, the motor rotates about a vertical axis with an angular velocity ω_z and angular acceleration α_z. Determine the total angular velocity $\boldsymbol{\omega}$ and angular acceleration $\boldsymbol{\alpha}$ of the fan blades for the instant when $\omega_z = 3$ rad/s and $\alpha_z = 12$ rad/s².

14-157 A 12-ft long beam AB is being raised by a winch as shown in Fig. P14-157. The winch accelerates at a constant rate of 0.05 rad/s² until its angular velocity is 10 rpm, after which its angular velocity is constant. If the system starts from rest with $\theta = 90°$, determine the velocity $\mathbf{v}_B$ and acceleration $\mathbf{a}_B$ when

a. $\theta = 60°$.   b. $\theta = 30°$.   c. $\theta = 0°$.

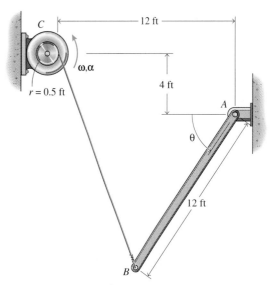

Fig. P14-157

14-158* The cutting blade of a rotary lawn mower is rotating at a constant rate of 900 rpm clockwise as viewed from above (Fig. P14-158). The cutting blade is 800 mm long and the lawn mower is being pushed forward at a constant speed of $v = 3$ m/s. For the instant shown in which the blade is perpendicular to the velocity $\mathbf{v}$

a. Determine the velocity $\mathbf{v}$ and acceleration $\mathbf{a}$ of the tips of the blades.
b. Find the instantaneous center of zero velocity of the blade.

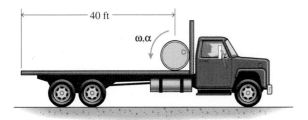

Fig. P14-156

Fig. P14-158

14-159 The tank shown in Fig. P14-159 is moving forward at a constant speed of 30 mi/h. At the instant shown, the turret is facing forward and rotating at a constant rate of $\omega_1 = 2$ rad/s while the gun barrel is being raised at a rate of $\omega_2 = 0.5$ rad/s and $\dot{\omega}_2 = 0.03$ rad/s^2. If the gun barrel is 10 ft long, determine the velocity $\mathbf{v}_A$ and acceleration $\mathbf{a}_A$ of the end of the gun barrel.

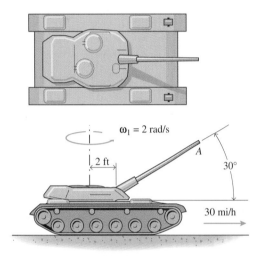

Fig. P14-159

14-160* A small block B rotates with the horizontal turntable A of Fig. P14-160. The distance between the block and the axis of rotation is 200 mm, and the turntable starts from rest. If the block will begin to slip when its acceleration exceeds 0.6$\mathbf{g}$, determine N, the number of revolutions at which slip begins; ω, the angular velocity of the turntable when slip begins; and θ, the angle between the acceleration of the block and the radial direction when slip begins for:

a. $\alpha = 1.0$ rad/s^2.
b. $\alpha = 10.0$ rad/s^2.
c. $\alpha = 20.0$ rad/s^2.

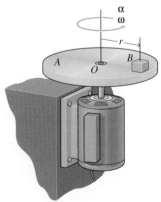

Fig. P14-160

14-161* The tank of Problem 14-159 fires a shell at the instant shown. If the shell leaves the gun barrel with a speed of 850 ft/s relative to the gun barrel

a. Determine the absolute velocity $\mathbf{v}_s$ of the shell when it leaves the gun barrel.
b. Determine where the shell will land (the shell has a constant downward acceleration of 32.2 ft/s^2 after leaving the gun barrel and the ground is flat and horizontal).
c. Compare the answer of part b with where the shell would have landed if the tank had been stationary.

14-162 A small block B rotates with the horizontal turntable A of Fig. P14-160. The block will begin to slip when its acceleration exceeds 0.6 g. The turntable starts from rest and accelerates to 30 rpm in 1 rev, after which the angular velocity is constant. Determine the maximum distance r for which the block will not slip if

a. $\alpha = $ constant
b. α decreases linearly with respect to θ from $\alpha = \alpha_0$ when $\omega = 0$ to $\alpha = 0$ after 1 rev.

14-163* At the instant shown, the truck of Fig. P14-163 has a velocity of 30 mi/h and is accelerating at a rate of 5 ft/s^2. If end A of the 10-ft long bar AB is sliding backward at a rate of 2 ft/s relative to the truck, determine

a. Determine the velocity $\mathbf{v}_G$ and acceleration $\mathbf{a}_G$ of the center of the bar AB.
b. Locate C, the instantaneous center of zero velocity for bar AB.

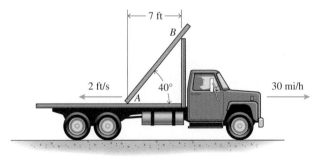

Fig. P14-163

14-164 The shuttle C of Fig. P14-164 is made to oscillate back and forth by the rotation of the 0.50-m diameter wheel D. If the wheel is rotating at a constant angular rate of 30 rpm, determine the velocity $\mathbf{v}_C$ and acceleration $\mathbf{a}_C$ of the shuttle for the instant shown in which arm AB is horizontal.

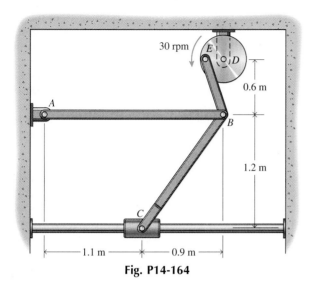

Fig. P14-164

14-165 The radar antenna of Fig. P14-165 is tracking an airplane. At the instant shown, the radar dish is rotating about a vertical axis at a constant rate of 0.4 rad/s, $\phi = 30°$,

$\dot\phi = -0.5$ rad/s, and $\ddot\phi = 0.02$ rad/s^2. For this instant determine:

a. The angular velocity $\boldsymbol{\omega}$ and angular acceleration $\boldsymbol{\alpha}$ of the antenna.

b. The velocity $\mathbf{v}_H$ and acceleration $\mathbf{a}_H$ of the signal horn H.

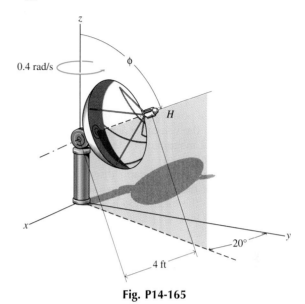

Fig. P14-165

Computer Problems

C14-166 A 600-mm diameter wheel rolls without slipping on the horizontal surface as shown in Fig. P14-166. The 1-m long rod AB is attached to the wheel at a point 250 mm from its center, and end A slides freely along the surface. If the center of the wheel has a constant speed of 1.2 m/s to the right and $\theta = 0$ when $t = 0$, calculate and plot

a. The velocity v_A of end A of the rod as a function of time $(0 \le t \le 3 \text{ s})$.

b. The angular acceleration α_{AB} of the rod as a function of time $(0 \le t \le 3 \text{ s})$.

c. The acceleration $\mathbf{a}_G$ of the mass center of the rod as a function of time $(0 \le t \le 3 \text{ s})$.

C14-167 The slider-crank mechanism of Fig. P14-167 is an idealization of an automobile crank shaft, connecting rod, and piston. If the crank shaft is rotating at a constant rate $\dot\theta$

a. Write expressions for a_C, the acceleration of the piston; $\dot\phi$, the angular velocity of the connecting rod; and $\ddot\phi$, the angular acceleration of the connecting rod in terms of θ, $\dot\theta$, ℓ_{AB}, and ℓ_{BC}.

b. Using $\ell_{AB} = 3$ in., $\ell_{BC} = 7$ in., and $\dot\theta = 4800$ rpm, plot a_C, $\dot\phi$, and $\ddot\phi$, as functions of θ $(0 \le \theta \le 360°)$.

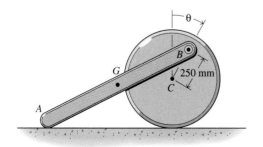

Fig. P14-166

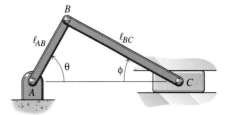

Fig. P14-167

C14-168 A stepped wheel rolls without slipping on a pair of rails as shown in Fig. P14-168. If the center of the wheel moves at a constant speed v_C and the x-coordinate of point A is zero when $\theta = 0°$:

a. Write expressions for the position (x_A, y_A), velocity (v_{Ax}, v_{Ay}), and acceleration (a_{Ax}, a_{Ay}) of point A in terms of the radii r_1 and r_2, the velocity v_C, and the angle θ.
b. For $r_1 = 75$ mm, $r_2 = 150$ mm, and $v_C = 0.5$ m/s, plot the position of point A (y_A versus x_A) for one-and-a-half revolutions of the wheel ($0 \le \theta \le 450°$). Draw the radial line from the center to point A for $\theta = 0°$, 30°, 60°, 90°, 120°, ...
c. On a copy of the position graph, draw a line from A in the direction of the velocity $\mathbf{v}_A$ and with a length proportional to the magnitude of the velocity for $\theta = 0°$, 30°, 60°, 90°, 120°, ...
d. On a copy of the position graph, draw a line from A in the direction of the acceleration $\mathbf{a}_A$ and with a length proportional to the magnitude of the acceleration for $\theta = 0°$, 30°, 60°, 90°, ...

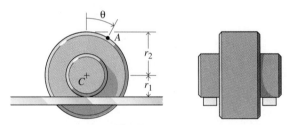

Fig. P14-168

C14-169 A Geneva wheel (Fig. P14-169) is a mechanism often used to create intermittent motion. The size and location of the input wheel A is such that the pin P enters and exits the slots smoothly. If the input wheel rotates with a constant angular speed of $\omega_A = 5$ rad/s:

a. Compute and plot the angular position θ_B of the Geneva wheel versus time for one complete revolution of the input wheel (that is, for $0° \le \theta_B \le 360°$). Let $\theta_B = 0°$ at $t = 0$.
b. Compute and plot the angular velocity ω_B of the Geneva wheel versus time for one complete revolution of the input wheel.
c. Compute and plot the angular acceleration α_B of the Geneva wheel versus time for one complete revolution of the input wheel.

C14-170 The mechanism shown in Fig. P14-170 is used to advance the film in a movie projector. As the input link AB rotates, the hook at P alternately engages the film and pulls it to the left, then lifts away from the film and moves back to the right. If link AB rotates at a constant angular speed of $\omega_{AB} = 900$ rpm.

a. Compute and plot the motion of the film pawl P (the position y_P versus x_P) for one complete revolution of the link AB.
b. Compute and plot the horizontal and vertical components of the velocity of the film pawl P as a function of time. (Let $t = 0$ when AB and CD are both vertical as shown.)
c. Compute and plot the horizontal and vertical components of the acceleration of the film pawl P as a function of time.

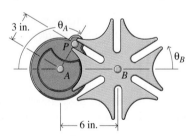

Fig. P14-169

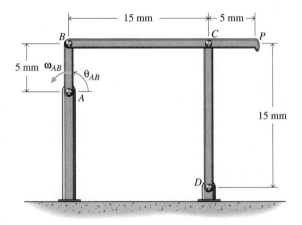

Fig. P14-170

KINETICS OF PARTICLES: NEWTON'S LAWS

Race cars rounding a curve experience an inward acceleration that must be generated by frictional forces if the roadway is level. The roadway can be banked to reduce the frictional forces required.

15-1   INTRODUCTION

Previously, in Chapters 13 (kinematics of particles) and 14 (kinematics of rigid bodies), the motions of particles and rigid bodies were studied without considering the forces required to produce the motion. Relationships were developed that describe how the velocity and the acceleration of a body change either with time or with a change in position. In an earlier course in statics, methods were developed for determining the resultant force **R** and the resultant couple **C** for any force system that can act on a body. When the resultant of the force system is zero, the body is in equilibrium (at rest or moving with a constant velocity). When the resultant of the force system is not zero, the body experiences accelerated motion. Unbalanced forces and the motions they produce (kinetics) will be the topic for the remaining chapters of this book.

The motion a body experiences when it is subjected to an unbalanced system of forces can be established by using three different methods: (1) force, mass, and acceleration method, (2) work and energy method, and (3) impulse and momentum method. The most useful method for a particular problem depends on the nature of the force system (constant or variable) and the information sought (reactions, velocities, accelerations, etc.).

15-2   EQUATIONS OF MOTION

Prior to the time of Galileo and Newton, it was generally assumed that a body at rest was in its natural state; therefore, some kind of force was required to keep it moving. The great contribution of Newton to the science of mechanics was his realization that a force is not needed to keep a body moving once it has been set in motion and that the effect of a force is to change the velocity, not maintain the velocity.

15-2-1   Newton's Second Law

Newton's three laws of motion, as they are commonly expressed today, were listed in Section 12-2. The first law pertaining to a particle at rest or moving with a constant velocity and the third law governing action and reaction between interacting bodies were used in developing the concepts of statics. Newton's second law of motion, which relates the accelerated motion of a particle to the forces producing the motion, forms the basis for studies in dynamics. It has previously been noted that Newton's first law, which covers the case of particle equilibrium, is a special case of his second law. When the resultant force on a particle is zero ($\mathbf{R} = \mathbf{0}$), the acceleration of the particle is zero ($\mathbf{a} = \mathbf{0}$); therefore, the particle is either at rest or moving with a constant velocity (in equilibrium). A modern statement of Newton's second law, as presented in Section 12-2, is:

Law 2.   If an external force acts on a particle, the particle will be accelerated in the direction of the force, and the magnitude of the acceleration will be directly proportional to the force and inversely proportional to the mass of the particle.

Mathematically, Newton's second law is expressed as:

$$\mathbf{a} = k\frac{\mathbf{F}}{m} \qquad (15\text{-}1)$$

where

a is the acceleration of the particle
F is the force acting on the particle
m is the mass of the particle
k is a proportionality constant, which depends on the units selected for the acceleration, force, and mass. A system in which $k = 1$ has consistent kinetic units.

With $k = 1$, Eq. 15-1 can be written in its familiar form

$$\mathbf{F} = m\mathbf{a} \qquad (15\text{-}2)$$

At the present time, two systems with consistent kinetic units are used by engineers in the United States: the International System of Units (SI units), and the U. S. customary system of units. In the SI system, the base quantities are length (m), mass (kg), and time (s). The unit of force, called a newton (N), is defined as the force required to give a mass of 1 kg an acceleration of 1 m/s². The SI system is an absolute system since the three base units are the same in any environment (on the earth, on the moon, anywhere in space, etc.). In the SI system, the weight **W** of a body (the force of gravity), like any other force, is expressed in newtons. Thus, from Newton's second law, the magnitude W of the weight of a body of mass m is

$$W = mg \qquad (15\text{-}3)$$

A system in which the base quantities are length (ft), force (lb), and time (s) continues to be widely used in engineering in the United States. The unit of time (the second) is the same as the corresponding SI unit. The unit of length (the foot) is defined as 0.3048 m. The unit of force (the pound) is defined as the weight at sea level and at a latitude of 45° of a platinum standard having a mass of 0.453 592 43 kg. Since the unit of force depends on the gravitational attraction of the earth, U. S. customary units do not form an absolute system. The unit of mass in the U. S. customary system is the slug. By definition, a slug of mass receives an acceleration of 1 ft/s² when a force of 1 lb is applied to it. For kinetics problems, where forces, masses, and accelerations are involved, if the weight W of a body is given in pounds, the mass m in slugs can be obtained from the expression

$$m = \frac{W}{g} \qquad (15\text{-}4)$$

where g is the acceleration of gravity.

Equation 15-2 expresses the fact that the magnitudes of **F** and **a** are proportional and that the vectors **F** and **a** have the same direction. Equation 15.2 is valid both for constant forces and for forces that vary with time (in either magnitude or direction).

When Eq. 15-2 is used to solve kinetics problems, the reference axes for the acceleration measurements must be fixed in space (have a constant orientation with respect to the stars). Such a system of axes is called a newtonian frame of reference or a primary inertial system.

When a system of reference axes is attached to the earth, the measured acceleration is not the absolute acceleration required for application of Eq. 15-2, owing to rotation of the earth on its axis and the acceleration of the earth with respect to the sun as the earth moves in its orbit around the sun. For most engineering problems on the surface of the earth, the corrections required to compensate for the acceleration of the earth with respect to the primary system are negligible, and the accelerations measured with respect to axes attached to the surface of the earth may be treated as absolute. Equation 15-2 is not valid, however, if **a** represents a relative acceleration measured with respect to a system of moving axes on earth. Also, the acceleration components of the earth's motion must be considered when dealing with such problems as flight paths for spacecraft and ballistic missile trajectories.

The internationally accepted values of g relative to the earth at sea level and at a latitude of $45°$ are 9.80665 m/s^2 and 32.1740 ft/s^2. These values of g are commonly rounded off for routine engineering work to 9.81 m/s^2 or 32.2 ft/s^2. The internationally accepted absolute values of g at sea level and at a latitude of $45°$, which should be used when a primary inertial system of axes is employed, are 9.8236 m/s^2 and 32.2295 ft/s^2.

15-2-2   Equations of Motion for a Single Particle

When a system of forces $\mathbf{F}_1, \mathbf{F}_2, \mathbf{F}_3, \cdots, \mathbf{F}_n$ acts on a particle, their resultant is a force $\mathbf{R}$ with a line of action through the mass center of the particle, since any system of forces acting on a particle must be a concurrent force system. The motion of the particle resulting from the action of the resultant $\mathbf{R}$ is given by Newton's second law of motion as

$$\mathbf{R} = \Sigma \mathbf{F} = m\mathbf{a} \tag{15-5}$$

If the resultant force $\mathbf{R}$ and the acceleration $\mathbf{a}$ are written in terms of rectangular cartesian coordinates, Eq. 15-5 becomes

$$\Sigma(F_x\mathbf{i} + F_y\mathbf{j} + F_z\mathbf{k}) = m(a_x\mathbf{i} + a_y\mathbf{j} + a_z\mathbf{k}) \tag{15-6}$$

The component form of this vector equation is

$$\begin{aligned} R_x &= \Sigma \mathbf{F}_x = m\mathbf{a}_x \\ R_y &= \Sigma \mathbf{F}_y = m\mathbf{a}_y \\ R_z &= \Sigma \mathbf{F}_z = m\mathbf{a}_z \end{aligned} \tag{15-7}$$

Similarly, Eq. 15-6 can be written in scalar form as

$$\begin{aligned} R_x &= \Sigma F_x = ma_x \\ R_y &= \Sigma F_y = ma_y \\ R_z &= \Sigma F_z = ma_z \end{aligned} \tag{15-8}$$

For many problems in particle kinetics, it is convenient to have the acceleration of the particle expressed as a function of the position (x,y,z) of the particle. For these cases, combining Eqs. 15-5 and 13-8 yields

$$\Sigma(F_x\mathbf{i} + F_y\mathbf{j} + F_z\mathbf{k}) = m(\ddot{x}\mathbf{i} + \ddot{y}\mathbf{j} + \ddot{z}\mathbf{k})$$

The scalar components of this vector equation are

$$\Sigma F_x = ma_x = m\frac{d^2x}{dt^2} = m\ddot{x}$$

$$\Sigma F_y = ma_y = m\frac{d^2y}{dt^2} = m\ddot{y} \qquad (15\text{-}9)$$

$$\Sigma F_z = ma_z = m\frac{d^2z}{dt^2} = m\ddot{z}$$

When any of these equations of motion for a single particle are used in the solution of a problem, a sign convention must be established. After a system of reference axes has been established, a convenient sign convention (see Fig. 15-1) shows force, velocity, and acceleration components with the same sign as the associated reference axis (a positive force, velocity, or acceleration component acts in the direction of the corresponding positive coordinate axis). Unknown force, velocity, and acceleration components are assumed to be positive and are shown as positive quantities on any motion (kinetic) diagram used in the solution of the problem. Students may find it useful to show the $m\mathbf{a}$ vectors on a separate diagram next to the free-body diagram used for the forces. The assumed direction for an unknown quantity is verified if the unknown is evaluated as a positive quantity.

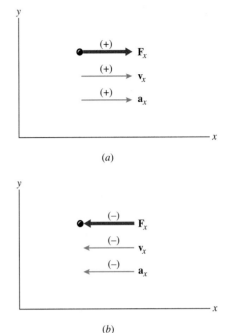

Fig. 15-1

15-2-3 Equations of Motion for a System of Particles

The equations of motion for a system of particles can be obtained by applying Newton's second law to each particle within the system. As an example, consider the collection of n particles shown in Fig. (15-2a). The ith particle has a mass m_i, and its location is specified with respect to an appropriate system of reference axes by using the position vector $\mathbf{r}_i$ from the origin of coordinates. Each particle in the system (see Fig. 15-2b) may be subjected to a system of external forces with a resultant $\mathbf{R}_i$ and a system of internal forces $\mathbf{f}_{i1}, \mathbf{f}_{i2}, \mathbf{f}_{i3}, \mathbf{f}_{ij}, \cdots, \mathbf{f}_{in}$. The internal forces result from elastic interactions between particles and from electric or magnetic effects. The internal force exerted by particle p_j on particle p_i is denoted as $\mathbf{f}_{ij}$. Applying Newton's second law to the particle yields

$$\mathbf{R}_i + \sum_{j=1}^{n} \mathbf{f}_{ij} = m_i\mathbf{a}_i \qquad (15\text{-}10)$$

In the summation of internal forces, $\mathbf{f}_{ii}$ is zero since particle p_i does not exert a force on itself.

If a force $\mathbf{f}_{ij}$ is exerted on particle p_i by particle p_j, Newton's third law requires that a force $\mathbf{f}_{ji}$, which is equal in magnitude, opposite in direction, and collinear with $\mathbf{f}_{ij}$, is exerted on particle p_j by particle p_i. Thus,

$$\mathbf{f}_{ij} + \mathbf{f}_{ji} = \mathbf{0} \qquad (15\text{-}11)$$

An equation of motion for the system is obtained by summing the equations of motion for the n individual particles. Thus

$$\sum_{i=1}^{n} \mathbf{R}_i + \sum_{i=1}^{n} \left(\sum_{j=1}^{n} \mathbf{f}_{ij} \right) = \sum_{i=1}^{n} m_i\mathbf{a}_i \qquad (15\text{-}12)$$

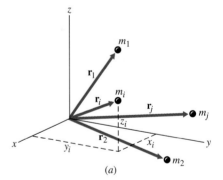

(a)

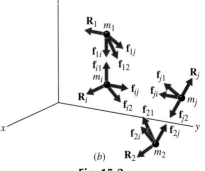

(b)

Fig. 15-2

Since all the internal forces of the system exist as equal, opposite, and collinear pairs, they sum to zero and Eq. 15-12 reduces to

$$\mathbf{R} = \sum_{i=1}^{n} \mathbf{R}_i = \sum_{i=1}^{n} m_i \mathbf{a}_i \qquad (15\text{-}13)$$

Equation 15-13 indicates that the resultant $\mathbf{R}$ of the external system of applied forces acting on a system of particles is equal to the resultant of the $m\mathbf{a}$ inertia vectors of the particles of the system. The quantity $m\mathbf{a}$ is frequently referred to as an inertia force; however, since it is neither a contact force nor a gravitational force (weight), many people avoid using the word force with the inertia vector $m\mathbf{a}$.

Equation 15-13 can be written in an alternate form if the mass center of the system of particles is considered. The mass center of the system is the point G defined by the position vector $\mathbf{r}_G$, which satisfies the relation

$$m\mathbf{r}_G = \sum_{i=1}^{n} m_i \mathbf{r}_i \qquad (15\text{-}14)$$

where

$$m = \sum_{i=1}^{n} m_i$$

is the total mass of the system of particles. Differentiating Eq. 15-14 with respect to time yields

$$m\dot{\mathbf{r}}_G = \sum_{i=1}^{n} m_i \dot{\mathbf{r}}_i$$

$$m\ddot{\mathbf{r}}_G = \sum_{i=1}^{n} m_i \ddot{\mathbf{r}}_i$$

which can be written

$$m\mathbf{a}_G = \sum_{i=1}^{n} m_i \mathbf{a}_i \qquad (15\text{-}15)$$

Combining Eqs. 15-13 and 15-15 yields

$$\mathbf{R} = m\mathbf{a}_G \qquad (15\text{-}16)$$

This vector equation can be written in component form as

$$\begin{aligned}
\Sigma F_x &= \mathbf{R}_x = m\mathbf{a}_{Gx} \\
\Sigma F_y &= \mathbf{R}_y = m\mathbf{a}_{Gy} \\
\Sigma F_z &= \mathbf{R}_z = m\mathbf{a}_{Gz}
\end{aligned} \qquad (15\text{-}17)$$

Equations 15-16 and 15-17 are mathematical expressions of the "principle of motion of the mass center" of a system of particles. Equations 15-17 for a system of particles are identical in form to Eqs. 15-7 for a single particle. This correspondence indicates that a system of particles can be treated as a single particle with the mass of the system concentrated at the mass center G if the resultant $\mathbf{R}$ of the external forces applied to the particles of the system has a line of action that passes through the mass center G of the system. In fact, any body can be considered a particle when applying Eqs. 15-17. In general, however, the line of action of the resultant force $\mathbf{R}$ will not pass through the mass center of the system, and the resultant will consist of a resultant force $\mathbf{R}$, which passes through the mass center G, and a resultant couple $\mathbf{C}$. Rotational motion resulting from the couple $\mathbf{C}$ is discussed in the next chapter on rigid-body kinetics.

15-3 RECTILINEAR MOTION

The kinematics of a particle subjected to rectilinear motion was described in Section 13-3. Motion in this case is along a straight line and, if the coordinate system is oriented such that the x-axis coincides with the line of motion, the position, velocity, and acceleration of the particle are completely described by their x-components. Thus,

$$\mathbf{r} = x\mathbf{i}$$
$$\mathbf{v} = \dot{\mathbf{r}} = \dot{x}\mathbf{i} \tag{15-18}$$
$$\mathbf{a} = \ddot{\mathbf{r}} = \ddot{x}\mathbf{i}$$

For rectilinear motion along the x-axis, Eqs. 15-7 for the particle reduce to

$$\Sigma\mathbf{F}_x = m\mathbf{a}_x$$
$$\Sigma\mathbf{F}_y = 0 \tag{15-19}$$
$$\Sigma\mathbf{F}_z = 0$$

For this type of motion, the vector notation can be dropped and the sign of a quantity can be used to indicate whether the sense of a vector quantity is in the direction of the positive or negative x-axis. Four types of problems are encountered that involve rectilinear motion.

Case 1. $F = $ constant.   For rectilinear motion problems involving a constant force, application of Newton's second law yields

$$\ddot{x} = \frac{F}{m} \tag{a}$$

Integrating twice with respect to time t yields

$$\dot{x} = \frac{F}{m}t + C_1$$

$$x = \frac{1}{2}\frac{F}{m}t^2 + C_1 t + C_2$$

The two constants C_1 and C_2 can be determined from the initial conditions for the specific problem being considered.

Case 2. $F = $ function of time.   For rectilinear motion problems involving a force that varies as a function of time, application of Newton's second law yields

$$\ddot{x} = \frac{F(t)}{m} \tag{b}$$

When the function $F(t)$ is known, Eq. b can be integrated twice with respect to time to obtain expressions for the velocity $\dot{x}$ and the position x. Two constants of integration C_1 and

C_2 will appear in the expressions, which can be evaluated from the initial conditions for the specific problem.

Case 3. F = function of position. For rectilinear motion problems involving a force that varies as a function of position, application of Newton's second law yields

$$\ddot{x} = \frac{F(x)}{m} \qquad (c)$$

Equation c can be transformed to a more useful form by noting that

$$\ddot{x} = \frac{d\dot{x}}{dt} = \frac{d\dot{x}}{dx}\frac{dx}{dt} = \dot{x}\frac{d\dot{x}}{dx} \qquad (d)$$

Thus, Eq. c can be expressed as

$$\dot{x}\,d\dot{x} = \frac{F(x)}{m}\,dx \qquad (e)$$

When the function $F(x)$ is known, Eq. e can be integrated to obtain $\dot{x}$ as a function of x. Also since $\dot{x} = dx/dt$, the expression obtained from the first integration can be integrated to obtain a relationship between position x and time t. Constants resulting from the integrations can be evaluated from the initial conditions for the specific problem.

Case 4. F = function of velocity. For rectilinear motion problems involving a force that varies as a function of velocity, application of Newton's second law yields either

$$\ddot{x} = \frac{d\dot{x}}{dt} = \frac{F(\dot{x})}{m} \qquad (f)$$

or

$$\ddot{x} = \dot{x}\frac{d\dot{x}}{dx} = \frac{F(\dot{x})}{m} \qquad (g)$$

When a relationship between velocity and time is sought, Eq. f yields

$$dt = \frac{m\,d\dot{x}}{F(\dot{x})}$$

When a relationship between velocity and position is sought, Eq. g yields

$$dx = \frac{m\dot{x}\,d\dot{x}}{F(\dot{x})}$$

In either case constants resulting from the integrations are evaluated by using the initial conditions for the specific problem.

The following examples illustrate the procedure for solving problems involving rectilinear motion of particles.

A 90-lb block rests on a horizontal surface as shown in Fig. 15-3a. Determine the displacement, velocity, and acceleration of the block 3 s after the 50-lb force **F** is applied if the horizontal surface is smooth.

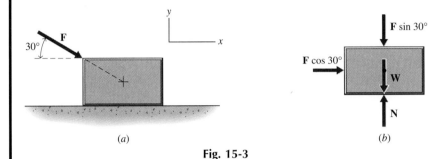

(a)   (b)

Fig. 15-3

SOLUTION

A free-body diagram of the block resting on a smooth surface is shown in Fig. 15-3b. The force **F** is represented by its rectangular components. The scalar equations of motion for the block (Eqs. 15-19) are

$$\Sigma F_x = ma_x = m\frac{d^2x}{dt^2} \qquad \Sigma F_y = 0$$

Thus,

$$\Sigma F_y = N - W - F \sin 30° = 0$$
$$N = W + F \sin 30° = 90 + 50 \sin 30° = 115 \text{ lb}$$
$$\Sigma F_x = F \cos 30° = 50 \cos 30° = 43.30 = ma_x$$

Since the sum of forces is constant, the acceleration a_x is constant. Thus,

$$a_x = \frac{43.30}{m} = \frac{43.30}{90/32.2} = 15.49 \text{ ft/s}^2 \qquad \text{Ans.}$$

Integrating the constant acceleration $a_x = dv_x/dt = 15.49$ ft/s² yields

$$v_x = \frac{dx}{dt} = 15.49t + C_1$$

Since $v_x = 0$ (no initial velocity) at $t = 0$, $C_1 = 0$. Thus, at $t = 3$ s,

$$v_x = 15.49t = 15.49(3) = 46.47 = 46.5 \text{ ft/s} \qquad \text{Ans.}$$

Integrating the velocity $v_x = dx/dt = 15.49t$ then yields

$$x = \frac{1}{2}15.49t^2 + C_2$$

Since $x = 0$ (no initial displacement) at $t = 0$, $C_2 = 0$. At $t = 3$ s,

$$x = \frac{1}{2}15.49t^2 = \frac{1}{2}15.49(3)^2 = 69.7 \text{ ft} \qquad \text{Ans.}$$

Two bodies A and B with masses $m_A = 50$ kg and $m_B = 60$ kg are connected by a rope that passes over a pulley as shown in Fig. 15-4a. Assume that the rope and pulley have negligible mass and that the length of the rope remains constant. The kinetic coefficient of friction μ_k between block A and the inclined surface is 0.25. Determine the tension in the cord and the acceleration of block A after the blocks are released from rest.

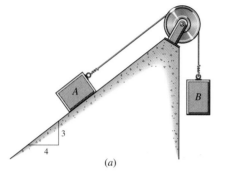
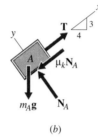

(a) (b) (c)

Fig. 15-4

SOLUTION

Free-body diagrams for blocks A and B are shown in Figs. 15-4b and 15-4c, respectively. Applying the scalar equations of motion (Eqs. 15-19) to block A yields

$$\Sigma F_y = 0 \qquad\qquad N_A - \frac{4}{5}m_A g = 0$$

$$N_A = \frac{4}{5}(50)(9.81)$$

$$= 392.4 \text{ N}$$

$$\Sigma F_x = m_A a_{Ax} \qquad T - \mu_k N_A - \frac{3}{5}m_A g = m_A a_{Ax}$$

$$T - (0.25)(392.4) - \frac{3}{5}(50)(9.81) = 50a_{Ax}$$

$$T - 392.4 = 50a_{Ax} \qquad (a)$$

Since the length of the rope is constant, $a_{By} = -a_{Ax}$. Applying Eqs. 15-19 to block B yields

$$\Sigma F_y = m_B a_{By} \qquad\qquad T - m_B g = -m_B a_{Ax}$$
$$= -m_B a_{Ax} \qquad\qquad T - 60(9.81) = -60a_{Ax}$$
$$T - 588.6 = -60a_{Ax} \qquad (b)$$

Solving Eq. a and b simultaneously yields

$$a_{Ax} = 1.784 \text{ m/s}^2 \qquad\qquad \text{Ans.}$$
$$T = 482 \text{ N}$$

The two blocks shown in Fig. 15-5a are at rest on a horizontal surface when a force **F** is applied to block B. Blocks A and B weigh 45 lb and 75 lb, respectively. The static coefficient of friction μ_s between the two blocks is 0.25, and the kinetic coefficient of friction μ_k between the horizontal surface and block B is 0.20. Determine the maximum force **F** that can be applied to block B before the blocks no longer move together.

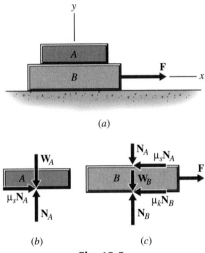

(a)

(b)   (c)

Fig. 15-5

SOLUTION

Free-body diagrams for blocks A and B are shown in Figs. 15-5b and 15-5c, respectively. Applying the scalar equations of motion (Eqs. 15-19) to block A yields

$$\Sigma F_y = 0 \qquad N_A - W_A = 0$$
$$N_A = W_A = 45 \text{ lb}$$
$$\Sigma F_x = ma_x \qquad \mu_s N_A = m_A a_x$$
$$0.25(45) = \frac{45}{32.2} a_x$$

Since $\mu_s N_A$ is the maximum frictional force that can be developed at the interface between blocks A and B,

$$a_x(\text{max}) = \frac{32.2}{45}(0.25)(45) = 8.05 \text{ ft/s}^2$$

Applying Eqs. 15-19 to block B yields

$$\Sigma F_y = 0 \qquad N_B - N_A - W_B = 0$$
$$N_B = N_A + W_B$$
$$= 45 + 75 = 120 \text{ lb}$$
$$\Sigma F_x = ma_x \qquad F - \mu_s N_A - \mu_k N_B = m_B a_x$$
$$F = m_B a_x + \mu_s N_A + \mu_k N_B$$
$$= \frac{75}{32.2}(8.05) + 0.25(45) + 0.20(120)$$
$$= 54.0 \text{ lb} \qquad\qquad \text{Ans.}$$

The sled shown in Fig. 15-6a is used to test small solid-propellant rockets. The combined mass of the sled and rocket is 1000 kg. From the burn characteristics of the propellant, it is known that the thrust provided by the rocket during motion of the sled can be expressed as

$$F = a + bt - ct^2$$

where F is in newtons and t is in seconds. If the sled is released at rest when the thrust of the rocket is 10 kN and the sled travels 700 m and attains a velocity of 150 m/s during a 10-s test run, determine

a. Values for the constants a, b, and c,
b. The maximum and minimum accelerations experienced by the sled during the test.

Neglect friction between the sled and the rails and the reduction in mass of the propellant during the test.

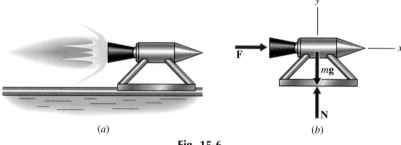

(a)   (b)

Fig. 15-6

SOLUTION

A free-body diagram of the rocket and sled and the reference system being used for the analysis is shown in Fig. 15-6b. Applying the scalar equation of motion (Eq. 15-19) in the x-direction yields

$$\Sigma F_x = ma_x = m\ddot{x} \qquad a + bt - ct^2 = m\ddot{x}$$

a. Since the force is expressed as a function of time, the velocity and displacement of the sled can be obtained simply by integrating the equation of motion in the x-direction. Thus,

$$m\dot{x} = at + \frac{1}{2}bt^2 - \frac{1}{3}ct^3 + C_1$$

$$mx = \frac{1}{2}at^2 + \frac{1}{6}bt^3 - \frac{1}{12}ct^4 + C_1 t + C_2$$

From the initial conditions:

$$\dot{x} = 0 \quad \text{at} \quad t = 0 \qquad\qquad C_1 = 0$$
$$x = 0 \quad \text{at} \quad t = 0 \qquad\qquad C_2 = 0$$
$$F = 10(10^3)\,\text{N} \quad \text{at} \quad t = 0 \qquad a = 10(10^3)\,\text{N} \qquad \text{Ans.}$$

From the end-of-test conditions:

$$\dot{x} = 150\,\text{m/s} \quad \text{at} \quad t = 10\,\text{s} \qquad 1000(150) = 10(10^3)(10) + \frac{1}{2}b(10)^2 - \frac{1}{3}c(10)^3$$

which simplifies to

$$15b - 100c = 15000 \qquad\qquad (a)$$

$$x = 700\,\text{m} \quad \text{at} \quad t = 10\,\text{s} \quad 1000(700) = \frac{1}{2}(10)(10^3)(10)^2 + \frac{1}{6}b(10)^3 - \frac{1}{12}c(10)^4$$

which simplifies to

$$b - 5c = 1200 \qquad\qquad (b)$$

Simultaneous solution of Eqs. a and b yields

$$b = 1800\,\text{N/s} \qquad\qquad \text{Ans.}$$
$$c = 120\,\text{N/s}^2 \qquad\qquad \text{Ans.}$$

Therefore

$$F = 10{,}000 + 1800t - 120t^2$$

b.

$$a_x = \frac{F}{m} = 10 + 1.8t - 0.12t^2$$

For maximum or minimum acceleration

$$\frac{da_x}{dt} = 1.8 - 0.24t = 0 \qquad t = 7.50\,\text{s}$$

At $t = 7.50$ s

$$a_x = 10 + 1.8(7.50) - 0.12(7.50)^2 = 16.75\,\text{m/s}^2$$

At $t = 0$ s

$$a_x = 10.0\,\text{m/s}^2$$

At $t = 10$ s

$$a_x = 16.0\,\text{m/s}^2$$

Therefore, at $t = 7.50$ s

$$a_x = a_{\text{max}} = 16.75\,\text{m/s}^2 \qquad\qquad \text{Ans.}$$

At $t = 0$ s

$$a_x = a_{min} = 10.00\,\text{m/s}^2 \qquad\qquad \text{Ans.}$$

Motion of block A ($W = 805$ lb) on the inclined surface of Fig. 15-7a is resisted by friction ($\mu = 0.10$) and a linear spring ($k = 25$ lb/ft). If the block is released at rest with the spring unstretched, determine, during the first phase of motion down the inclined surface,

a. The maximum displacement of the block from its rest position.
b. The velocity of the block when it is 15 ft from its rest position.
c. The time required for the block to move 15 ft from its rest position.
d. The acceleration of the block as it begins to move up the inclined surface.

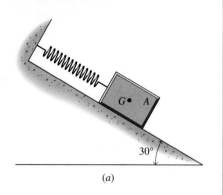

(a)

SOLUTION

A free-body diagram of the block for the first phase of the motion (down the inclined surface) is shown in Fig. 15-7b. The scalar equations of motion for the block (Eqs. 15-19) are

$+\nearrow \Sigma F_y = 0$ $\qquad$ $N - W \cos 30° = 0$

$\qquad\qquad\qquad\qquad\qquad N = 805 \cos 30° = 697.2$ lb

$+\searrow \Sigma F_x = m\ddot{x}$ $\qquad$ $W \sin 30° - \mu N - F = m\ddot{x}$

$\qquad\quad 805 \sin 30° - 0.10(697.2) - 25x = \dfrac{805}{32.2}\ddot{x}$

which can be simplified to

$$\ddot{x} = 13.31 - x$$

However,

$$\ddot{x} = \frac{d}{dt}\dot{x} = \frac{d}{dx}\dot{x}\frac{dx}{dt} = \dot{x}\frac{d\dot{x}}{dx}$$

Therefore

$$\dot{x}\, d\dot{x} = (13.31 - x)\, dx$$

Integrating both sides yields

$$\frac{1}{2}\dot{x}^2 = 13.31x - \frac{1}{2}x^2 + C_1$$

From the initial condition $\dot{x} = 0$ at $x = 0$, $C_1 = 0$. Thus

$$\dot{x} = \frac{dx}{dt} = \left[26.62x - x^2\right]^{1/2}$$

a. The maximum displacement occurs when $\dot{x} = 0$. Thus,

$$x(26.62 - x) = 0 \qquad x_{max} = 26.62 \text{ ft}\searrow \qquad\qquad \text{Ans.}$$

b. When the displacement of the block is 15 ft,

$$\dot{x} = [26.62(15) - (15)^2]^{1/2} = 13.20 \text{ ft/s}\searrow \qquad\qquad \text{Ans.}$$

c. The time is obtained by integrating the expression

$$dt = \frac{dx}{[26.62x - x^2]^{1/2}}$$

which has the solution

$$t = \sin^{-1}\!\left(\frac{x - 13.31}{13.31}\right) + C_2$$

From the initial condition $x = 0$ at $t = 0$, $C_2 = \pi/2$. Thus,

$$t = \sin^{-1}\!\left(\frac{x - 13.31}{13.31}\right) + \frac{\pi}{2}$$

$$= \sin^{-1}\!\left(\frac{15 - 13.31}{13.31}\right) + \frac{\pi}{2} = 1.698 \text{ s} \qquad\qquad \text{Ans.}$$

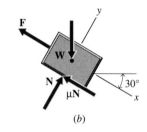

(b)

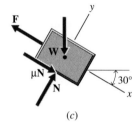

(c)

Fig. 15-7

d. A free-body diagram of the block for the second phase of the motion (up the inclined surface) is shown in Fig. 15-7c. The equation of motion at the beginning of this phase of the motion yields

$$\Sigma F_x = m\ddot{x} \qquad\qquad W\sin 30° + \mu N - F = m\ddot{x}$$

$$805\sin 30° + 0.10(697.2) - 25(26.62) = \frac{805}{32.2}\ddot{x}$$

From which

$$\ddot{x} = a_x = -7.73 = 7.73 \text{ ft/s}^2 \nwarrow \quad \text{Ans.}$$

EXAMPLE PROBLEM 15-6

A particle falls under the force of gravity in a medium that exerts a resisting force proportional to the velocity of the particle. Develop equations for the velocity and displacement of the particle. The velocity and displacement of the particle are zero at time $t = 0$.

SOLUTION

A free-body diagram of the particle and the reference system being used is shown in Fig. 15-8. Applying the scalar equation of motion (Eq. 15-19) in the x-direction to the particle yields

$$\Sigma F_x = ma_x$$

$$W - F_R = mg - kv = m\frac{dv}{dt}$$

Separating the variables (v and t) and integrating yields

$$\ln(mg - kv) = -\frac{k}{m}t + C_1$$

Since $v = 0$ when $t = 0$,

$$C_1 = \ln(mg)$$

$$\ln(mg - kv) = -\frac{k}{m}t + \ln(mg)$$

$$\frac{mg - kv}{mg} = e^{-kt/m} \qquad \text{or} \qquad mg - kv = mge^{-kt/m}$$

$$v = \frac{dx}{dt} = \frac{mg}{k}(1 - e^{-kt/m}) \qquad\qquad \text{Ans.}$$

Integrating yields

$$x = \frac{mg}{k}t + \frac{m^2g}{k^2}e^{-kt/m} + C_2$$

Since $x = 0$ when $t = 0$,

$$C_2 = -\frac{m^2g}{k^2}$$

Therefore

$$x = \frac{m^2g}{k^2}(e^{-kt/m} - 1) + \frac{mg}{k}t \qquad\qquad \text{Ans.}$$

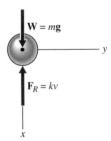

$\mathbf{W} = m\mathbf{g}$

y

$\mathbf{F}_R = kv$

x

Fig. 15-8

PROBLEMS

In the following problems, all ropes, cords, and cables are assumed to be flexible, inextensible, and of negligible mass. All pins and pulleys have negligible mass and are frictionless.

15-1* A box weighing 322 lb rests on the floor of an elevator (Fig. P15-1). Determine the force exerted by the box on the floor of the elevator if

a. The elevator starts up with an acceleration of 10 ft/s².
b. The elevator starts down with an acceleration of 8 ft/s².

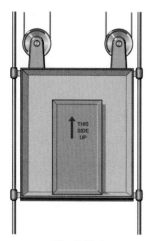

Fig. P15-1

15-2* Determine the constant force **F** required to accelerate an automobile (*m* = 1000 kg) on a level road from rest to 20 m/s in 10 s (see Fig. P15-2).

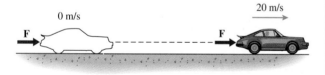

Fig. P15-2

15-3* A 200-lb block rests on a horizontal surface as shown in Fig. P15-3. Determine

a. The magnitude of the force **F** required to produce an acceleration of 5 ft/s² if the surface is smooth.

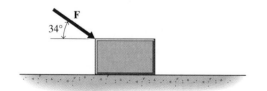

Fig. P15-3

b. The acceleration that a 100-lb force **F** would produce if the kinetic coefficient of friction μ_k between the block and the surface is 0.25.

15-4* A block of ice with a mass of 15 kg slides on a horizontal surface for 20 m before it stops (Fig. P15-4). If its initial speed was 15 m/s, determine

a. The force of friction between the block and the surface.
b. The kinetic coefficient of friction μ_k.

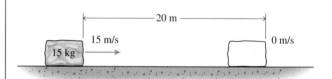

Fig. P15-4

15-5 When a television set is placed on a spring-operated platform scale on the floor of an elevator, the scale indicates a weight of 50 lb when the elevator is at rest. Determine

a. The acceleration of the elevator when the scale indicates a weight of 40 lb.
b. The weight indicated by the scale when the acceleration of the elevator is 10 ft/s² upward.

15-6 An automobile with a mass of 1500 kg is moving along a level road at a constant speed of 60 km/h (Fig. P15-6). The automobile accelerates at a constant rate and reaches a speed of 80 km/h in 5 s. Determine

a. The force required to produce this acceleration.
b. The distance traveled by the automobile during the 5-s interval that it is accelerating.

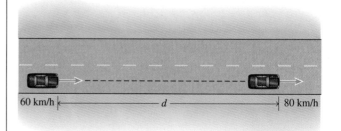

Fig. P15-6

15-7* A force of 20 lb is applied to a 25-lb block as shown in Fig. P15-7. Let $x = 0$ and $v = 0$ when $t = 0$ and determine the velocity and displacement of the block at $t = 5$ s if

a. The inclined plane supporting the block is smooth.
b. The kinetic coefficient of friction μ_k between the inclined plane and the block is 0.25.

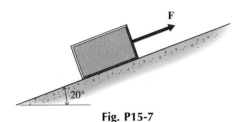

Fig. P15-7

15-8* A block with a mass of 20 kg is pushed up an inclined plane by a horizontal force **F** of 200 N as shown in Fig. P15-8. The kinetic coefficient of friction μ_k between the inclined plane and the block is 0.10. If $v = 0$ and $x = 0$ when $t = 0$, determine

a. The acceleration of the block
b. The time required for the block to travel 15 m.
c. The velocity of the block after it has traveled 10 m.

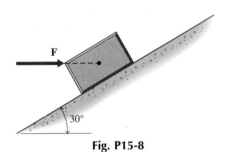

Fig. P15-8

15-9 Blocks A and B, which weigh 30 and 60 lb, respectively, are connected by a rope as shown in Fig. P15-9. The coefficients of kinetic friction are 0.20 for block A and 0.15 for block B. If the force **F** applied to the cable is 40 lb, determine

a. The acceleration of block B.
b. The velocity of block A after 5 s.

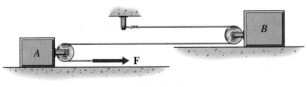

Fig. P15-9

15-10 A freight elevator contains three crates as shown in Fig. P15-10. The mass of the elevator cage is 750 kg and the masses of crates A, B, and C are 300 kg, 200 kg, and 100 kg, respectively. During a short interval of the lift, the elevator experiences an upward acceleration of 8 m/s². During this interval determine

a. The tension in the elevator cable.
b. The force exerted on crate A by the floor of the elevator.
c. The force exerted by crate B on crate C.

Fig. P15-10

15-11* Three boxes connected with cables rest on a horizontal surface as shown in Fig. P15-11. The weights of boxes A, B, and C are 200 lb, 150 lb, and 300 lb, respectively. The kinetic coefficient of friction μ_k between the surface and the boxes is 0.20. If the force **F** applied to the boxes is 175 lb, determine the acceleration of the boxes and the tensions in the cables between the boxes.

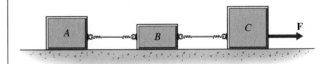

Fig. P15-11

15-12* A particle released from an elevated platform reaches a speed of 10 m/s before a constant resisting force is applied. During the next 10 m of fall, the speed is reduced to 5 m/s. If the particle has a mass of 5 kg, determine the force exerted on the particle.

15-13 A 2000-lb elevator cage, which is descending a mine shaft at a speed of 25 ft/s, is brought to rest with a uniform acceleration in a distance of 50 ft (Fig. P15-13). Determine the acceleration and the tension in the elevator cable while the cage is coming to rest.

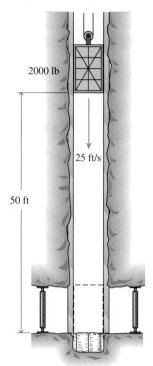

Fig. P15-13

15-14 Two bodies A (m_A = 50 kg) and B (m_B = 25 kg) are connected by a cable as shown in Fig. P15-14. Five seconds after the bodies are released from rest, body B has a velocity of 10 m/s downward. Determine

a. The acceleration of body A.
b. The tension in the cable.
c. The kinetic coefficient of friction μ_k for body A.

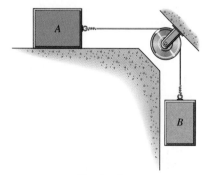

Fig. P15-14

15-15* The chute shown in Fig. P15-15 is 20 ft long and is used to transfer boxes from the street into the basement of a store. The kinetic coefficient of friction μ_k between the

box and the chute is 0.25. The kinetic coefficient of friction μ_k between the box and the basement floor is 0.40. If a 30-lb box is given an initial velocity of 10 ft/s when it is placed on the chute, determine

a. The velocity of the box as it leaves the end of the chute.
b. The distance the box slides on the basement floor after it leaves the end of the chute.

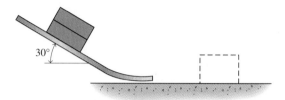

Fig. P15-15

15-16* The cart shown in Fig. P15-16 has a mass of 200 kg and is traveling to the right with a velocity of 5 m/s. Determine

a. The acceleration of the cart during its travel up the inclined plane.
b. The distance d that the cart will travel up the inclined plane before coming to rest.

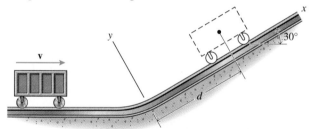

Fig. P15-16

15-17 Blocks A and B, which weigh 30 and 50 lb, respectively, are connected by a rope as shown in Fig. P15-17. The kinetic coefficients of friction μ_k are 0.35 for block A and 0.15 for block B. During motion of the blocks down the inclined plane, determine

a. The acceleration of block B.
b. The tension in the rope.

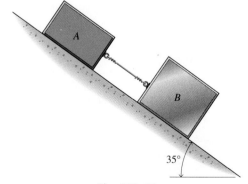

Fig. P15-17

15-18 Two bodies A and B, with masses of 25 kg and 30 kg, respectively, are shown in Fig. P15-18. During motion of the bodies, determine the acceleration of body A and the tension in the cable connecting the bodies if

a. The horizontal surface is smooth.
b. The kinetic coefficient of friction μ_k for body B is 0.20.

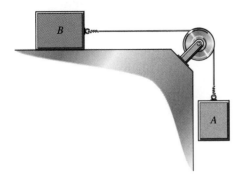

Fig. P15-18

15-19* Blocks A and B, which weigh 200 and 80 lb, respectively, are connected by a rope as shown in Fig. P15-19. Determine

a. The acceleration of block B.
b. The tension in the rope.

Fig. P15-19

15-20* Blocks A ($m_A = 25$ kg) and B ($m_B = 40$ kg) in Fig. P15-20 are connected by flexible cables to sheaves, which have diameters of 300 mm and 150 mm, respec-

Fig. P15-20

tively. The two sheaves are fastened together and are both weightless and frictionless. Find the tensions in the cables after the bodies are released from rest.

15-21 Two carts are connected by a cable that passes over a small pulley as shown in Fig. P15-21. The weights of carts A and B are 500 lb and 400 lb, respectively. After the carts are released from rest determine

a. The acceleration of the carts.
b. The tension in the cable.
c. The distance moved during the first 10 s of motion.

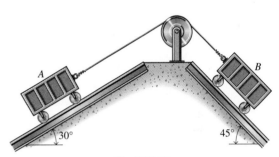

Fig. P15-21

15-22 The box shown in Fig. P15-22 has a mass of 100 kg and a velocity of 15 m/s as it begins to ascend the inclined plane. The static and kinetic coefficients of friction μ_S and μ_k between the plane and the box are 0.30 and 0.25, respectively. Determine

a. The acceleration of the box as it ascends the inclined plane.
b. The distance traveled by the box before it comes to rest.
c. The velocity of the box when it returns to the bottom of the inclined plane.

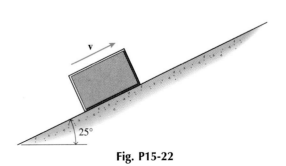

Fig. P15-22

15-23* Two blocks A and B, connected by a flexible cable, are released from rest in the positions shown in Fig. P15-23. The kinetic coefficient of friction μ_k between block A and the inclined surface is 0.15. Block B strikes the horizontal surface 3 s after being released. If the weight of body A is 50 lb, determine

a. The acceleration of body B.
b. The weight of body B.
c. The tension in the cable while the blocks are in motion.

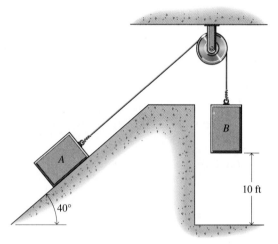

Fig. P15-23

15-24* Two bodies A and B, with masses of 25 kg and 30 kg, respectively, are shown in Fig. P15-24. The kinetic coefficient of friction for body A is 0.20 and the system is released from rest. During motion of the bodies, determine

a. The acceleration of body A.
b. The tension in the cable connecting the bodies.
c. The velocity of the bodies 5 s after motion begins.

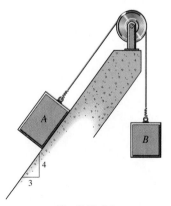

Fig. P15-24

15-25 Two bodies A and B, with weights of 50 lb and 45 lb, respectively, are shown in Fig. P15-25. The kinetic coefficient of friction μ_k for body B is 0.20 and the system is released from rest. During motion of the bodies, determine

a. The acceleration of body A.
b. The tension in the cable connecting the bodies.
c. The distance traveled by body B during the first 5 s of motion.

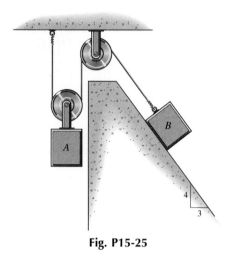

Fig. P15-25

15-26 Two bodies A and B, with masses of 40 kg and 30 kg, respectively, are shown in Fig. P15-26. The kinetic coefficient of friction μ_k for body A is 0.25 and the system is released from rest. During motion of the bodies, determine

a. The acceleration of body A.
b. The tension in the cable connecting the bodies.
c. The velocity of body B after 5 s of motion.

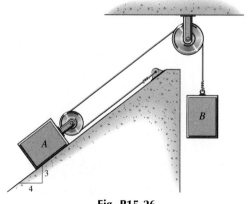

Fig. P15-26

15-27* Two bodies A and B, which weigh 30 lb and 20 lb, respectively, are shown in Fig. P15-27. The kinetic coefficient of friction between block A and the inclined surface is 0.30. The horizontal surface supporting block B is smooth. When the blocks are in the position shown, block B is moving to the right with a velocity of 5 ft/s. Determine

a. The tension in the cable connecting the bodies.
b. The time required for body B to come to rest.
c. The distance traveled by body B before it comes to rest.

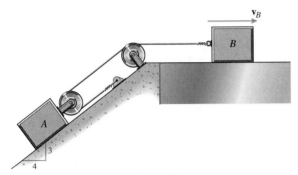

Fig. P15-27

15-28* The masses of bodies A and B in Fig. P15-28 are 15 kg and 10 kg, respectively. The kinetic coefficient of friction μ_k between body A and the horizontal surface is 0.20 and the bodies are released from rest. During motion of the bodies, determine

a. The tension in the cable connecting the bodies.
b. The velocity of body B after 5 s of motion.
c. The distance moved by body B during the first 5 s of motion.

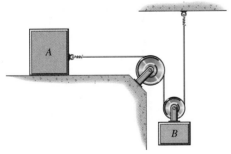

Fig. P15-28

15-29 The two bodies shown in Fig. P15-29 are connected by a flexible flat belt that passes over a cylindrical surface. The static and kinetic coefficients of friction μ_s and μ_k between the belt and the surface are 0.15 and 0.10, respectively. The weight of body A is 100 lb and the weight of body B is 200 lb. Determine

a. The acceleration of body A.
b. The tension in the segment of the belt between body A and the cylindrical surface.

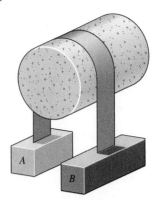

Fig. P15-29

15-30 Two bodies A and B, with masses of 25 kg and 30 kg, respectively, are shown in Fig. P15-30. The static and kinetic coefficients of friction μ_s and μ_k are 0.25 and 0.20, respectively. If the force $\mathbf{F}$ applied to body B is 100 N, determine

a. The acceleration of body A.
b. The tension in the cable connecting the bodies.
c. The distance traveled by body A during the first 5 s force $\mathbf{F}$ is applied.

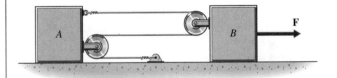

Fig. P15-30

15-31* The 644-lb block shown in Fig. P15-31 is moving to the left with a velocity of 20 ft/s when the force $\mathbf{F}$ is applied. The magnitude of the force is given by the expression $F = 40 + 12t$, where F is in pounds and t is in seconds. If the surface is smooth, determine

a. The time required to bring the block to rest.
b. The distance moved while coming to rest.
c. The position of the block 10 s after the force $\mathbf{F}$ is applied.

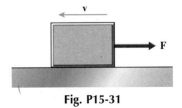

Fig. P15-31

15-32* The speed of a 12,500-kg navy fighter plane landing on a carrier is reduced from 216 km/h to rest by the plane's brakes and the cable arrester system. The arresting force provided by the plane's brakes is a constant 90 kN. The arresting force provided by the cable system can be expressed in equation form as $F = 850{,}000t - 425{,}000t^2$, where F is in newtons and t is in seconds. Determine

a. The maximum acceleration experienced by the pilot.
b. The time required for the arresting operation.
c. The distance traveled by the plane during the arrest.

15-33 The block shown in Fig. P15-33 weighs 50 lb. The coefficient of kinetic friction between the block and the inclined surface is 0.20. At the instant shown, the velocity of the block is 20 ft/s down the inclined surface. If a resisting force $F_R = 0.50v$, where F_R is in pounds and v is in feet per second, is applied at this instant, determine

a. The velocity of the block after 5 s.
b. The distance the block moves during the first 5 s that the resisting force is applied.
c. The terminal velocity of the block.

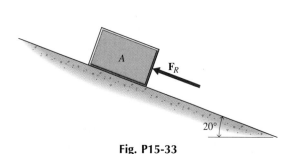

Fig. P15-33

15-34 A 5-kg projectile is fired vertically upward from the earth with a velocity of 300 m/s. Determine

a. The maximum height attained by the projectile if air resistance is neglected.
b. The maximum height attained by the projectile if the drag force exerted by the air is $F_D = 0.006v^2$, where F_D is in newtons and v is in meters per second.
c. The velocity of the projectile when it returns to earth if the drag force of the air is acting.

15-35* Blocks A and B of Figure P15-35 weigh 25 and 50 lb, respectively. The blocks are at rest and the spring ($k = 25$ lb/ft) is unstretched when the blocks are in the position shown. Determine the velocity and acceleration of block B when it is 1 ft below its initial position.

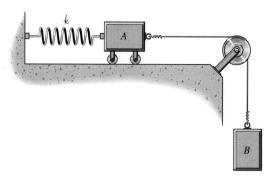

Fig. P15-35

15-36* A skydiver whose mass is 55 kg jumps from a stationary balloon at a height of 3000 m (Fig. P15-36). The drag force exerted on her body by the air when she is in a spread-eagle posture can be expressed as $F_D = 0.180v^2$, where F_D is in newtons and v is in meters per second. Determine her terminal velocity and the time required to reach 95 percent of her terminal velocity.

$F_D = 0.180v^2$

Fig. P15-36

15-37 The flexible chain shown in Fig. P15-37 weighs 0.50 lb/ft. The kinetic coefficient of friction between the chain and the horizontal surface is 0.20. If the chain is re-

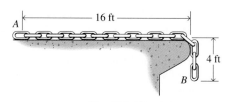

Fig. P15-37

leased at rest in the position shown, determine the velocity of the chain at the instant it becomes completely vertical and the time required for end A to leave the surface.

15-38 The mass of block A of Figure P15-38 is 10 kg. The block is at rest and the spring ($k = 25$ N/m) is unstretched when the block is in the position shown. One second after the block is released to move, determine

a. The velocity and acceleration of the block.
b. The tension T in the cable.

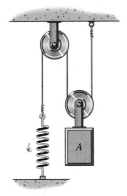

Fig. P15-38

15-39* The ball shown in Fig. P15-39 weighs 0.33 lb. The free length of the spring ($k = 5$ lb/in.) is 20 in. If the ball is released at rest in the position shown and friction between the ball and the tube is negligible, determine

a. The velocity of the ball as it exits the tube.
b. The time required for the ball to exit the tube.

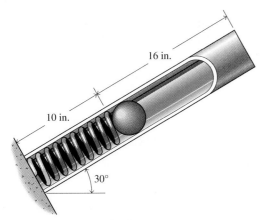

Fig. P15-39

15-40* A circular disk with a mass of 2 kg is supported by two identical springs ($k = 400$ N/m) as shown in Fig. P15-40. The free length of a spring is 300 mm. If the disk is released at rest with the springs horizontal, determine the velocity of the disk when it has fallen 100 mm below its initial position.

Fig. P15-40

15-41 The two carts A and B, which weigh 200 lb and 300 lb, respectively, are connected by a cable as shown in Fig. P15-41. Determine the acceleration of both carts and the tension in the cable if $F = 50$ lb and

a. $v_B = 0$ ft/s at the instant shown.
b. $v_B = 10$ ft/s at the instant shown.

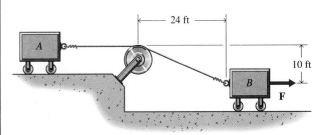

Fig. P15-41

15-42 The masses of blocks A and B of Figure P15-42 are 30 and 20 kg, respectively. The pulleys have negligible mass and are very small. Determine the acceleration of both blocks and the tension in the cable for the position shown if

a. $v_A = 0$ m/s.
b. $v_A = 5$ m/s downward.

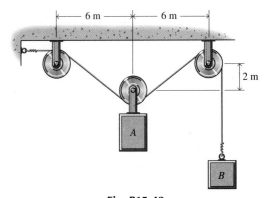

Fig. P15-42

15-4 CURVILINEAR MOTION

The kinematics of a particle subjected to curvilinear motion was described in Sections 13-5 and 13-7. Motion in this case is along a curved path. If a coordinate system can be found such that the z-components of the position, velocity, and acceleration are zero for all times, the motion is called plane curvilinear motion. Motions for which no coordinate system can be found that makes at least one component of the position, velocity, and acceleration zero for all times is called space curvilinear motion.

15-4-1 Plane Curvilinear Motion

When motion occurs in a plane, two coordinates are required to describe the motion. Three coordinate systems used to describe curvilinear motion in a plane are rectangular coordinates, polar coordinates, and normal/tangential coordinates.

Rectangular Coordinates In a rectangular coordinate system, the position of the particle is described by using its distance from two reference axes (say the x- and y-axes). The equations for position, velocity, and acceleration are

$$\mathbf{r} = x\mathbf{i} + y\mathbf{j}$$
$$\mathbf{v} = \dot{\mathbf{r}} = \dot{x}\mathbf{i} + \dot{y}\mathbf{j} \tag{15-20}$$
$$\mathbf{a} = \ddot{\mathbf{r}} = \ddot{x}\mathbf{i} + \ddot{y}\mathbf{j}$$

For plane curvilinear motion in rectangular coordinates, Eqs. 15-7 for the particle reduce to

$$\Sigma\mathbf{F}_x = m\mathbf{a}_x$$
$$\Sigma\mathbf{F}_y = m\mathbf{a}_y \tag{15-21}$$
$$\Sigma\mathbf{F}_z = \mathbf{0}$$

Combining Eqs. 15-20 and 15-21 yields the scalar equations

$$\Sigma F_x = ma_x = m\ddot{x}$$
$$\tag{15-22}$$
$$\Sigma F_y = ma_y = m\ddot{y}$$

Equations 15-22 indicate that plane curvilinear motion in rectangular coordinates is simply a superposition of rectilinear motions along the x- and y-axes.

Polar Coordinates In a polar coordinate system, the position of the particle is described by using a distance r from a fixed point and an angular displacement θ from a fixed line. Unit vectors $\mathbf{e}_r$ and $\mathbf{e}_\theta$ are directed radially outward from the fixed point and in the direction of increasing θ, respectively. Equations for position, velocity, and acceleration are

$$\mathbf{r} = r\,\mathbf{e}_r$$
$$\mathbf{v} = \dot{\mathbf{r}} = \dot{r}\,\mathbf{e}_r + r\dot{\theta}\,\mathbf{e}_\theta \tag{15-23}$$
$$\mathbf{a} = \ddot{\mathbf{r}} = (\ddot{r} - r\dot{\theta}^2)\,\mathbf{e}_r + (r\ddot{\theta} + 2\dot{r}\dot{\theta})\,\mathbf{e}_\theta$$

For plane curvilinear motion in polar coordinates, Eqs. 15-7 for the particle become

$$\Sigma \mathbf{F}_r = m\mathbf{a}_r$$
$$\Sigma \mathbf{F}_\theta = m\mathbf{a}_\theta \tag{15-24}$$
$$\Sigma \mathbf{F}_z = \mathbf{0}$$

Combining Eqs. 15-23 and 15-24 yields the scalar equations

$$\Sigma F_r = ma_r = m(\ddot{r} - r\dot{\theta}^2)$$
$$\Sigma F_\theta = ma_\theta = m(r\ddot{\theta} + 2\dot{r}\dot{\theta}) \tag{15-25}$$

Normal and Tangential Coordinates   In a normal and tangential coordinate system, unit vectors $\mathbf{e}_t$ and $\mathbf{e}_n$ are directed tangent to the path (in the direction of motion) and normal to the path (toward the center of curvature), respectively, at each point along the path of the particle. Equations for the velocity and acceleration at position s along the path are

$$\mathbf{v} = \dot{s}\,\mathbf{e}_t \tag{15-26}$$

$$\mathbf{a} = \ddot{s}\,\mathbf{e}_t + \frac{\dot{s}^2}{\rho}\,\mathbf{e}_n$$

For plane curvilinear motion in normal and tangential coordinates, Eqs. 15-7 for the particle reduce to

$$\Sigma \mathbf{F}_t = m\mathbf{a}_t$$
$$\Sigma \mathbf{F}_n = m\mathbf{a}_n \tag{15-27}$$
$$\Sigma \mathbf{F}_z = \mathbf{0}$$

Combining Eqs. 15-26 and 15-27 yields the scalar equations

$$\Sigma F_t = ma_t = m\ddot{s} \tag{15-28}$$

$$\Sigma F_n = ma_n = m\frac{\dot{s}^2}{\rho} = m\frac{v^2}{\rho}$$

15-4-2   Space Curvilinear Motion

When motion occurs along a curve in three-dimensional space, three coordinates are required to describe the motion. Three coordinate systems used to describe curvilinear motion in space are rectangular coordinates, cylindrical coordinates, and spherical coordinates.

Rectangular Coordinates   The rectangular coordinate system used for three-dimensional curvilinear motion of a particle is a direct extension of the rectangular system used for plane problems. The equations for position, velocity, and acceleration are

$$\mathbf{r} = x\mathbf{i} + y\mathbf{j} + z\mathbf{k}$$
$$\mathbf{v} = \dot{\mathbf{r}} = \dot{x}\mathbf{i} + \dot{y}\mathbf{j} + \dot{z}\mathbf{k} \tag{15-29}$$
$$\mathbf{a} = \ddot{\mathbf{r}} = \ddot{x}\mathbf{i} + \ddot{y}\mathbf{j} + \ddot{z}\mathbf{k}$$

For curvilinear motion in space, Newton's second law for the particle, Eqs. 15-7, is

$$\Sigma \mathbf{F}_x = m\mathbf{a}_x$$
$$\Sigma \mathbf{F}_y = m\mathbf{a}_y \tag{15-30}$$
$$\Sigma \mathbf{F}_z = m\mathbf{a}_z$$

Combining Eqs. 15-29 and 15-30 yields the scalar equations

$$\Sigma F_x = ma_x = m\ddot{x}$$
$$\Sigma F_y = ma_y = m\ddot{y} \tag{15-31}$$
$$\Sigma F_z = ma_z = m\ddot{z}$$

Cylindrical Coordinates   The cylindrical coordinate system used for three-dimensional curvilinear motion of a particle is a direct extension of the polar coordinate system used for plane problems. Equations for position, velocity, and acceleration are

$$\mathbf{r} = r\mathbf{e}_r + z\mathbf{k}$$
$$\mathbf{v} = \dot{\mathbf{r}} = \dot{r}\mathbf{e}_r + r\dot{\theta}\mathbf{e}_\theta + \dot{z}\mathbf{k} \tag{15-32}$$
$$\mathbf{a} = \ddot{\mathbf{r}} = (\ddot{r} - r\dot{\theta}^2)\,\mathbf{e}_r + (r\ddot{\theta} + 2\dot{r}\dot{\theta})\mathbf{e}_\theta + \ddot{z}\mathbf{k}$$

For curvilinear motion in space, Newton's second law for the particle, Eqs. 15-7, is

$$\Sigma \mathbf{F}_r = m\mathbf{a}_r$$
$$\Sigma \mathbf{F}_\theta = m\mathbf{a}_\theta \tag{15-33}$$
$$\Sigma \mathbf{F}_z = m\mathbf{a}_z$$

Combining Eqs. 15-32 and 15-33 yields the scalar equations

$$\Sigma F_r = ma_r = m(\ddot{r} - r\dot{\theta}^2)$$
$$\Sigma F_\theta = ma_\theta = m(r\ddot{\theta} + 2\dot{r}\dot{\theta}) \tag{15-34}$$
$$\Sigma F_z = ma_z = m\ddot{z}$$

Spherical Coordinates   In a spherical coordinate system, the position of the particle is described in terms of a radial distance R and two angles θ and ϕ. Equations for the position, velocity, and acceleration are

$$\mathbf{r} = R\mathbf{e}_R$$
$$\mathbf{v} = \dot{\mathbf{r}} = \dot{R}\mathbf{e}_R + R\dot{\theta}\sin\phi\,\mathbf{e}_\theta + R\dot{\phi}\,\mathbf{e}_\phi \tag{15-35}$$
$$\mathbf{a} = \ddot{\mathbf{r}} = (\ddot{R} - R\dot{\phi}^2 - R\dot{\theta}^2\sin^2\phi)\,\mathbf{e}_R$$
$$\qquad + (R\ddot{\theta}\sin\phi + 2\dot{R}\dot{\theta}\sin\phi + R\dot{\theta}\dot{\phi}\cos\phi)\,\mathbf{e}_\theta$$
$$\qquad + (R\ddot{\phi} + 2\dot{R}\dot{\phi} - R\dot{\theta}^2\sin\phi\cos\phi)\,\mathbf{e}_\phi$$

For curvilinear motion in space, Newton's second law for the particle, Eqs. 15-7, is

$$\Sigma \mathbf{F}_R = m\mathbf{a}_R$$
$$\Sigma \mathbf{F}_\theta = m\mathbf{a}_\theta \tag{15-36}$$
$$\Sigma \mathbf{F}_\phi = m\mathbf{a}_\phi$$

Combining Eqs. 15-35 and 15-36 yields the scalar equations

$$\Sigma F_R = ma_R = m(\ddot{R} - R\dot{\phi}^2 - R\dot{\theta}^2\sin^2\phi)$$
$$\Sigma F_\theta = ma_\theta = m(R\ddot{\theta}\sin\phi + 2\dot{R}\dot{\theta}\sin\phi + R\dot{\theta}\dot{\phi}\cos\phi) \tag{15-37}$$
$$\Sigma F_\phi = ma_\phi = m(R\ddot{\phi} + 2\dot{R}\dot{\phi} - R\dot{\theta}^2\sin\phi\cos\phi)$$

The following examples illustrate the procedure for solving problems involving curvilinear motion of a particle in a plane and in space.

EXAMPLE PROBLEM 15-7

A 30-lb projectile is fired horizontally with an initial velocity of 750 ft/s from the top of a hill, which is 500 ft above the surrounding area. Determine the range R of the projectile (horizontal distance traveled) and the elapsed time before it strikes the ground. Neglect air resistance.

SOLUTION

A free-body diagram for the projectile is shown in Fig. 15-9. The projectile moves in a vertical plane under the action of the earth's gravity. The equations of motion for the projectile in rectangular coordinates (Eqs. 15-22) are

$$\Sigma F_x = ma_x = m\ddot{x} \qquad (a)$$
$$\Sigma F_y = ma_y = m\ddot{y} \qquad (b)$$

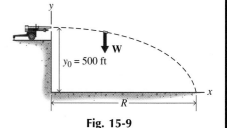

Fig. 15-9

From Eq. a

$$\Sigma F_x = m\ddot{x} = 0$$

Integrating yields

$$\dot{x} = C_1$$
$$x = C_1 t + C_2$$

From the initial conditions,

$$\dot{x} = v_0 \qquad \text{when } t = 0 \qquad C_1 = v_0$$
$$x = 0 \qquad \text{when } t = 0 \qquad C_2 = 0$$

From Eq. b

$$\Sigma F_y = m\ddot{y} = \frac{W}{g}\ddot{y} = -W \qquad \text{or} \qquad \ddot{y} = -g$$

Integrating yields

$$\dot{y} = -gt + C_3$$
$$y = -\frac{1}{2}gt^2 + C_3 t + C_4$$

From the initial conditions,

$$\dot{y} = 0 \qquad \text{when } t = 0 \qquad C_3 = 0$$
$$y = y_0 \qquad \text{when } t = 0 \qquad C_4 = y_0$$

Therefore,

$$\dot{x} = v_o \qquad (c)$$
$$x = v_0 t \qquad (d)$$
$$\dot{y} = -gt \qquad (e)$$
$$y = -\frac{1}{2}gt^2 + y_0 \qquad (f)$$

The projectile will strike the ground when $y = 0$. Therefore, from Eq. f

$$t = \sqrt{\frac{2y_0}{g}} = \sqrt{\frac{2(500)}{32.2}} = 5.573 = 5.57 \text{ s} \qquad \text{Ans.}$$

The range R is obtained from Eq. d. Thus,

$$R = v_0 t = 750(5.573) = 4180 \text{ ft} \qquad \text{Ans.}$$

A sphere of mass m is attached to the top end of a slender vertical rod of negligible mass as shown in Fig. 15-10. When the sphere is given a small displacement, rotation of the system about the pin at point O is initiated. Determine the linear velocity **v** of the sphere and the force **P** in the rod when the rod is in a horizontal position if $m = 5$ kg and $R = 2$ m.

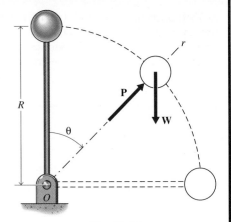

Fig. 15-10

SOLUTION

A free-body diagram of the sphere (Fig. 15-10) shows that two forces act on the sphere; the weight **W** and the reaction of the rod **P**. The equations of motion for the sphere in polar coordinates (Eqs. 15-25) are

$$\Sigma F_r = ma_r = m(\ddot{r} - r\dot{\theta}^2) \qquad (a)$$
$$\Sigma F_\theta = ma_\theta = m(r\ddot{\theta} + 2\dot{r}\dot{\theta}) \qquad (b)$$

Since the length of the slender rod is fixed, $r = R$, $\dot{r} = 0$, and $\ddot{r} = 0$. Thus, from Eq. b

$$mg \sin \theta = mR\ddot{\theta}$$
$$\ddot{\theta} = \frac{g}{R} \sin \theta$$

Multiplying both sides by $\dot{\theta}$ and integrating yields

$$\frac{1}{2}\dot{\theta}^2 = -\frac{g}{R} \cos \theta + C$$

Since $\dot{\theta} = 0$ when $\theta = 0$,

$$C = \frac{g}{R}$$

Therefore

$$\dot{\theta} = \sqrt{\frac{2g}{R}(1 - \cos \theta)}$$

When $\theta = \pi/2$,

$$\dot{\theta} = \sqrt{\frac{2g}{R}}$$

Therefore

$$v = R\dot{\theta} = R\sqrt{\frac{2g}{R}}$$
$$= \sqrt{2gR} = \sqrt{2(9.81)(2)} = 6.26 \text{ m/s}\downarrow \qquad \text{Ans.}$$

From Eq. a

$$P - mg \cos \theta = -mR\dot{\theta}^2$$

When $\theta = \pi/2$,

$$P = -mR\left(\sqrt{\frac{2g}{R}}\right)^2$$
$$= -2(5)(9.81) = -98.1 \text{ N} = 98.1 \text{ N} \leftarrow \qquad \text{Ans.}$$

A conical pendulum consisting of a 10-lb sphere supported by a 6-ft cord revolves about a vertical axis with a constant angular velocity $\dot{\theta}$ so that the string is inclined 30° to the vertical, as shown in Fig. 15-11. Determine the tension T in the string and the linear velocity v of the sphere.

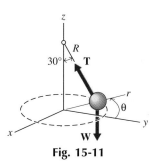

Fig. 15-11

SOLUTION

A free-body diagram for the sphere is shown in Fig. 15-11. The sphere moves on a circular path in a horizontal plane under the influence of the two forces **T** and **W**. The equations of motion for the sphere in cylindrical coordinates (Eqs. 15-34) are

$$\Sigma F_r = ma_r = m(\ddot{r} - r\dot{\theta}^2) \qquad (a)$$

$$\Sigma F_\theta = ma_\theta = m(r\ddot{\theta} + 2\dot{r}\dot{\theta}) \qquad (b)$$

$$\Sigma F_z = ma_z = m\ddot{z} \qquad (c)$$

Since the sphere moves with a constant angular velocity on a circular path in a horizontal plane, $r = R \sin 30° = 6 \sin 30° = 3$ ft, $\dot{r} = 0$, $\ddot{r} = 0$, $\ddot{\theta} = 0$, and $\ddot{z} = 0$. Thus, from Eq. c

$$\Sigma F_z = T \cos 30° - 10 = 0 \qquad T = 11.547 = 11.55 \text{ lb} \qquad \text{Ans.}$$

From Eq. a

$$\Sigma F_r = -T \sin 30° = -\frac{W}{g} r\dot{\theta}^2$$

$$\dot{\theta}^2 = \frac{Tg \sin 30°}{Wr} = \frac{11.547(32.2) \sin 30°}{10(3)} = 6.197$$

$$\dot{\theta} = 2.489 \text{ rad/s}$$

$$v = r\dot{\theta} = 3(2.489) = 7.467 = 7.47 \text{ ft/s} \qquad \text{Ans.}$$

EXAMPLE PROBLEM 15-10

A sphere with a mass of 3 kg slides along a rod (see Fig. 15-12a), which is bent in a vertical plane into a shape that can be described by the equation $y = 8 - \frac{1}{2}x^2$, where x and y are measured in meters. When $x = 2$ m, the collar is moving along the rod at a speed of 5 m/s and the speed is increasing at a rate of 3 m/s^2. Determine the normal $\mathbf{F}_n$ and tangential $\mathbf{F}_t$ components of the force being exerted on the sphere by the rod at this time.

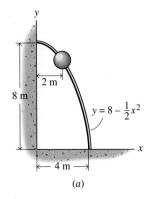

(a)

SOLUTION

A free-body diagram for the sphere is shown in Fig. 15-12b. The sphere moves on the curved path in the vertical plane under the influence of the forces $\mathbf{F}_n$, $\mathbf{F}_t$, and the weight $\mathbf{W}$. The equations of motion for the sphere in normal and tangential coordinates (Eqs. 15-28) are

$$\Sigma F_t = ma_t = m\ddot{s} \qquad (a)$$

$$\Sigma F_n = ma_n = m\frac{\dot{s}^2}{\rho} = m\frac{v^2}{\rho} \qquad (b)$$

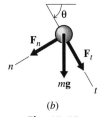

(b)

Fig. 15-12

For the curved path

$$y = 8 - \frac{1}{2}x^2 \qquad \frac{dy}{dx} = -x \qquad \frac{d^2y}{dx^2} = -1$$

$$\rho = \frac{[1 + (dy/dx)^2]^{3/2}}{d^2y/dx^2} = -(1 + x^2)^{3/2}$$

At the point (2,6)

$$\frac{dy}{dx} = -x = -2 = \text{slope}$$

$$\theta_x = \tan^{-1}(-2) = -63.43° \qquad \theta = 63.43°$$

$$|\rho| = (1 + x^2)^{3/2} = (1 + 2^2)^{3/2} = 11.180 \text{ m}$$

From Eq. a with $a_t = 3$ m/s^2

$$\Sigma F_t = ma_t$$
$$F_t + mg \sin 63.43° = ma_t$$
$$F_t = ma_t - mg \sin 63.43°$$
$$= 3(3) - 3(9.81) \sin 63.43° = -17.32 \text{ N}$$
$$\mathbf{F}_t = -17.32 \ \mathbf{e}_t \text{ N} \qquad \text{Ans.}$$

From Eq. b with $v_t = 5$ m/s

$$\Sigma F_n = ma_n = m\frac{v^2}{\rho}$$

$$F_n + mg \cos 63.43° = m\frac{v^2}{\rho}$$

$$F_n = m\frac{v^2}{\rho} - mg \cos 63.43°$$

$$= 3\frac{(5)^2}{11.180} - 3(9.81) \cos 63.43° = -6.455 = -6.46 \text{ N}$$

$$\mathbf{F}_n = -6.46 \ \mathbf{e}_n \text{ N} \qquad \text{Ans.}$$

PROBLEMS

15-43* An airplane in level flight drops a 1000-lb bomb from an altitude of 30,000 ft, as shown in Fig. P15-43. If the speed of the plane is 450 mi/h when the bomb is released, determine the horizontal distance from the point of release to the point of impact and the time of flight for the bomb. Neglect air resistance.

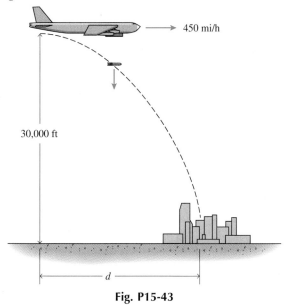

Fig. P15-43

15-44* A projectile with a mass of 15 kg is fired on a level surface with an initial velocity of 300 m/s and at an angle of 30° with the horizontal. Determine the maximum height h reached by the projectile, the range R (horizontal distance traveled) of the projectile, and the elapsed time before it strikes the ground. Neglect air resistance.

15-45 A water ski ramp is set at an angle of 25° as shown in Fig. P15-45. The speed of a 180-lb skier as he leaves the end of the ramp after releasing his tow rope is 20 mi/h. If air resistance is negligible, determine

a. The maximum height h attained by the skier.
b. The distance R from the end of the ramp to his point of landing.

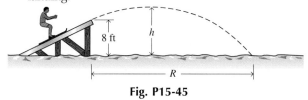

Fig. P15-45

15-46 When a mortar launches a 5-kg projectile, the angle of inclination θ (with respect to the horizontal) can be set, but the muzzle velocity v_0 is fixed (see Fig. P15-46). If air resistance is neglected, determine

a. The angle of inclination required for maximum range across a level field.
b. The maximum range and time of flight if $v_0 = 100$ m/s.

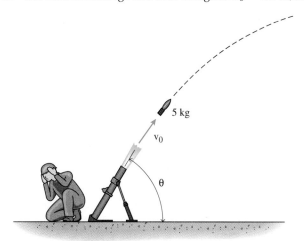

Fig. P15-46

15-47* A 100-lb projectile is fired with an initial velocity of 1500 ft/s and at an angle of 45° with the horizontal from the top of a hill, which is 750 ft above the surrounding area. Determine the range R (horizontal distance traveled) of the projectile and the elapsed time before it strikes the ground. Neglect air resistance.

15-48* An airplane is descending at an angle of 20° with respect to the horizontal when it drops a bomb (Fig. P15-48). If the altitude at the time of release is 5000 m and the speed of the plane 750 km/h, determine the range (horizontal distance traveled) of the bomb and the elapsed time before it strikes the ground. Neglect air resistance.

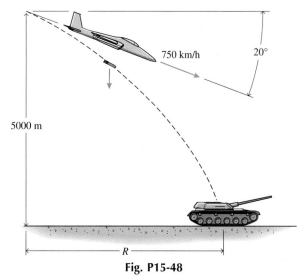

Fig. P15-48

15-49 A 2-oz tennis ball is thrown vertically upward with an initial speed of 35 ft/s. If a strong cross wind exerts a constant horizontal force of 0.01 lb on the ball, determine the horizontal distance traveled by the tennis ball before it returns to the ground.

15-50 A 100-g apple falls from a tree branch 3 m above the ground. As it falls, a strong cross wind exerts a constant horizontal force of 0.05 N on the apple. Determine the horizontal distance traveled by the apple as it falls to the ground.

15-51* The velocity of a particle at a point in its path can be expressed as $\mathbf{v}_1 = 25\mathbf{i} - 40\mathbf{j}$ ft/s. Thirty seconds later, the velocity has changed to $\mathbf{v}_2 = -75\mathbf{i} + 82\mathbf{j}$ ft/s. If the weight of the particle is 50 lb, determine the magnitude and direction of the constant force required to produce this change in motion.

15-52* The velocity of a particle at a point in its path can be expressed as $\mathbf{v}_1 = 96\mathbf{i} + 72\mathbf{j}$ m/s. Twenty-five seconds later, the velocity has changed to $\mathbf{v}_2 = 36\mathbf{i} - 12\mathbf{j}$ m/s. If the mass of the particle is 5 kg, determine the magnitude and direction of the constant force required to produce this change in motion.

15-53 The 2-lb paperweight, shown in Fig. P15-53, starts from rest at point A and slides on the surface of the smooth 6-ft diameter cylinder until it leaves the surface at point B. Determine the angle θ.

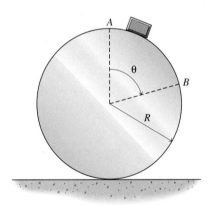

Fig. P15-53

15-54 An automobile with a mass of 1600 kg passes over the crest of a hill that has a radius of curvature of 70 m. Determine the maximum speed possible if the tires remain in contact with the surface of the road.

15-55* A 5-lb ball attached to a 6-ft cord swings through a full circle in a vertical plane as shown in Fig. P15-55. If the velocity of the ball is 15 ft/s at the top of the circle, determine the tension in the cord and the linear velocity of the ball

a. When angle θ is 45°.
b. When angle θ is 270°.

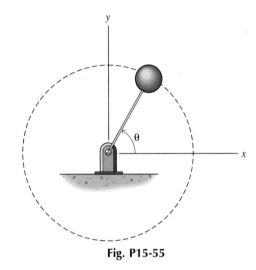

Fig. P15-55

15-56* An automobile with a mass of 1500 kg is traveling at a speed of 100 km/h. Determine the total normal force N between the tires and the road when the car passes over

a. The top of a hump in the road that has a radius of curvature of 100 m.
b. The bottom of a dip in the road that has a radius of curvature of 120 m.

15-57 Two bodies A ($W_A = 50$ lb) and B ($W_B = 75$ lb) and the frame on which they rest rotate about a vertical axis at a constant angular velocity of 50 rpm, as shown in Fig. P15-57. If friction between the bodies and the frame is negligible, determine

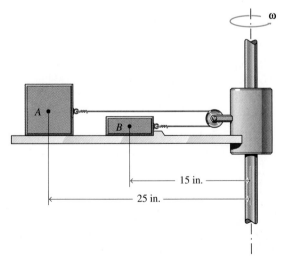

Fig. P15-57

a. The tension T in the cable connecting the bodies.
b. The force exerted on body B by the stop.

15-58 A sphere with a mass of 3 kg is supported by a rod with negligible mass and a cord as shown in Fig. P15-58. Determine the tension T in the rod

a. When the sphere is in the position shown in the figure.
b. Immediately after the cord is cut.
c. When the sphere is at the bottom of its swing.

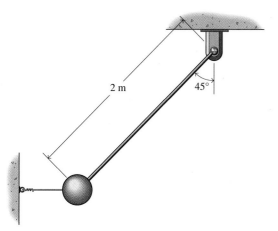

Fig. P15-58

15-59* A highway is designed for traffic moving at 65 mi/h. Along a certain portion of the highway, the radius of a curve is 900 ft. The curve is banked, as shown in Fig. P15-59, so that no friction is required to keep cars on the road. Determine

a. The required angle of banking (angle θ) of the road.
b. The minimum coefficient of friction between the tires and the road that would keep traffic from skidding at this speed if the curve is not banked.

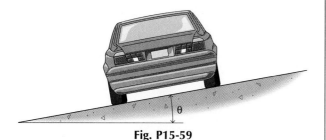

Fig. P15-59

15-60* A curve of 200-m radius on a level road is banked at the correct angle for a speed of 65 km/h. If an automobile rounds this curve at 100 km/h, determine the minimum coefficient of friction required between the tires and the road so that the automobile will not skid.

15-61 The circular disk shown in Fig. P15-61 rotates in a horizontal plane. A 3-lb block rests on the disk 8 in. from the axis of rotation. The static coefficient of friction between the block and the disk is 0.50. If the disk starts from rest with a constant angular acceleration of 0.5 rad/s^2, determine the length of time required for the block to begin to slip.

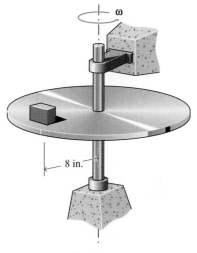

Fig. P15-61

15-62 A block ($m = 5$ kg) rests on a frame (friction is negligible) that can be rotated about a vertical axis as shown in Fig. P15-62. When the frame is not rotating, the tension in the spring is 80 N. Determine the force exerted on the block by the stop when the frame is rotating at a constant angular velocity of 30 rpm.

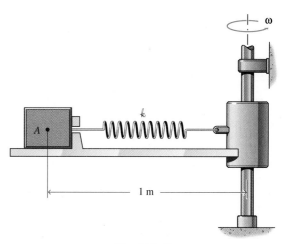

Fig. P15-62

15-63* A 10-lb sphere is attached to a vertical rod with two cords as shown in Fig. P15-63. When the system rotates about the axis of the rod, the cords extend as shown in the figure. Determine

a. The tensions in the two cords when the angular velocity ω of the system is 5 rad/s.
b. The angular velocity ω of the system when cord B is taut but carrying no load.

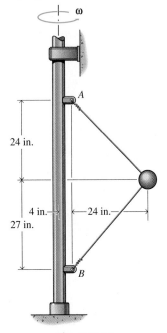

Fig. P15-63

15-64* A small sphere with a mass of 0.50 kg is mounted on the circular hoop shown in Fig. P15-64 and is free to slide (friction is negligible) when the hoop is rotated. Determine the angle θ and the force **F** exerted by the hoop on the sphere when the hoop is rotating about a vertical diameter at a constant angular velocity of 120 rpm.

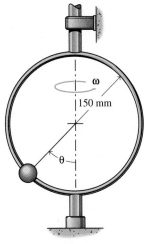

Fig. P15-64

15-65 A car on the track of a roller coaster ride at an amusement park is shown in Fig. P15-65. The car and its four occupants weigh 900 lb. The speed of the car is 35 mi/h when it passes point A on the track. If the car and its occupants are treated as a particle, determine

a. The speed of the car at the bottom of the loop.
b. The force exerted on the car by the track at point B.
c. The force exerted on the car by the track at point C.
d. The minimum speed required at the top of the loop to keep the car in contact with the track.

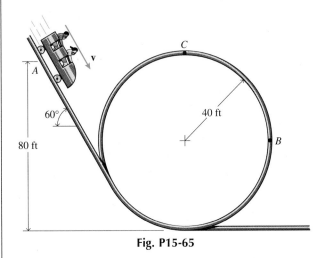

Fig. P15-65

15-66 A block with a mass of 5 kg rests on a smooth conical surface that is revolved about a vertical axis with a constant angular velocity ω. The block is attached to the rotating shaft with a cable as shown in Fig. P15-66. Determine

a. The tension in the cable when the system is rotating at 20 rpm.
b. The angular velocity in revolutions per minute when the force between the conical surface and the block is zero.

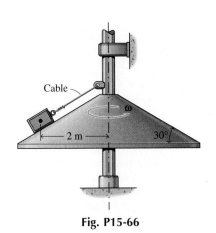

Fig. P15-66

15-67* A 3-lb sphere slides along a rod that is bent in a vertical plane into a shape that can be described by the equation $x^2 = 8y$, where x and y are measured in feet. When the sphere is at the point $(-8,8)$, as shown in Fig. P15-67, it is moving along the rod at a speed of 15 ft/s and is slowing down at a rate of 3 ft/s^2. Determine the normal and tangential components of the force being exerted on the sphere by the rod at this time.

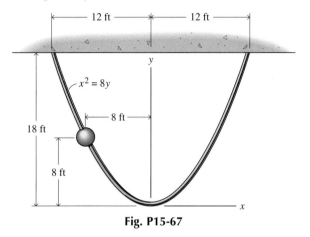

Fig. P15-67

15-68* A particle P of mass $m = 1.5$ kg is released from rest in the position shown in Fig. P15-68 and slides down the rod, which has been bent to form a circular arc of radius $R = 2$ m in a vertical plane. If the circular portion of the rod is smooth but the kinetic coefficient of friction between the particle and the rod is 0.10 on the straight portion of the rod, determine

a. The force exerted by the rod on the particle at a point 1 m below the point of release.
b. The distance d traveled by the particle along the straight portion of the rod before coming to rest.

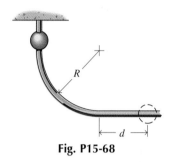

Fig. P15-68

15-5 CENTRAL FORCE MOTION

The motion of a particle moving under the influence of a force directed toward a fixed point is called central-force motion. Common examples of central-force motion include the motion of planets about the sun and the motion of the moon and artificial satellites about the Earth. From observations of the motions of planets about the sun, J. Kepler (1571–1630) deduced the following three laws, which govern central-force motion.[1]

Kepler's Laws of Planetary Motion

Law 1: The planets move about the sun in elliptical orbits with the sun at one focus.

Law 2: The radius vector joining each planet with the sun describes equal areas in equal times.

Law 3: The cubes of the mean distances of the planets from the sun are proportional to the squares of their times of revolution.

---

[1] Dr. Ernst Mach, "The Science of Mechanics," 9th ed., The Open Court Publishing Company, LaSalle, Illinois, 1942. Originally published in German in 1893 and translated from German to English by Thomas J. McCormack in 1902.

Newton's law of universal gravitation gives the magnitude of the force **F** between two masses separated by a distance r as

$$F = \frac{Gm_1m_2}{r^2} \tag{15-38}$$

where G is the universal gravitational constant. Consider the case where m_1 is a very large mass that can be considered fixed in space and m_2 is a small mass that moves in the xy-plane under the action of the force **F** exerted by mass m_1 on mass m_2. By using polar coordinates with the origin fixed at mass m_1, the motion of mass m_2 is given by Eqs. 15-25 as

$$\Sigma F_r = m_2 a_r = m_2(\ddot{r} - r\dot{\theta}^2) = -\frac{Gm_1m_2}{r^2} \tag{a}$$

$$\Sigma F_\theta = m_2 a_\theta = m_2(r\ddot{\theta} + 2\dot{r}\dot{\theta}) = 0 \tag{b}$$

Integrating Eq. b yields

$$r^2\dot{\theta} = \hbar \quad \text{(a constant)} \tag{c}$$

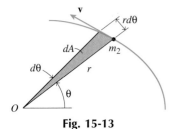

Fig. 15-13

The physical significance of Eq. c can be visualized (see Fig. 15-13) by considering the area generated by the radius vector r as it turns through an angle $d\theta$ in time dt. The shaded area dA, shown in Fig. 15-13, is a triangle; therefore,

$$dA = \frac{1}{2}(r)(r\,d\theta) = \frac{1}{2}r^2\,d\theta$$

The area dA and Eq. c are related by the expression

$$2\frac{dA}{dt} = r^2\frac{d\theta}{dt} = r^2\dot{\theta} = \hbar \tag{15-39}$$

The quantity $dA/dt = \hbar/2$ is called the areal speed and is a constant for any central-force system. Equation 15-39 is a mathematical statement of Kepler's second law of planetary motion.

The equation for the path of a particle subjected to a central force is obtained from Eqs. a and c. The derivatives are simplified by using the substitution $u = 1/r$. From Eq. 15-39,

$$\dot{\theta} = \frac{d\theta}{dt} = \frac{\hbar}{r^2} = \hbar u^2 \tag{d}$$

$$\dot{r} = \frac{dr}{dt} = \frac{dr}{d\theta}\frac{d\theta}{dt} = \frac{\hbar}{r^2}\frac{dr}{d\theta} = -\hbar\frac{du}{d\theta} \tag{e}$$

$$\ddot{r} = \frac{d\dot{r}}{dt} = \frac{d\dot{r}}{d\theta}\frac{d\theta}{dt} = \frac{\hbar}{r^2}\frac{d\dot{r}}{d\theta} = -\hbar^2u^2\frac{d^2u}{d\theta^2} \tag{f}$$

Substituting Eqs. d and f into Eq. a yields

$$\frac{d^2u}{d\theta^2} + u = \frac{Gm_1}{\hbar^2} \tag{g}$$

The solution of this differential equation (which can be verified by direct substitution) is

$$u = \frac{1}{r} = C\cos(\theta + \beta) + \frac{Gm_1}{\hbar^2}$$

where C and β are constants of integration to be determined from the initial conditions of the problem. Choosing the x-axis so that $\theta = 0$ when r is a minimum (u is a maximum, assuming C is positive) makes $\beta = 0$. Thus,

$$\frac{1}{r} = C \cos \theta + \frac{Gm_1}{\hbar^2} \qquad (15\text{-}40)$$

Solving Eq. 15.40 for r yields

$$r = \frac{\hbar^2/(Gm_1)}{1 + [C\hbar^2/(Gm_1)] \cos \theta} \qquad (15\text{-}41)$$

Equation 15-41 is the equation, in polar form, of a conic section (ellipse, parabola, or hyperbola). The origin of the coordinate system (the force center O) is a focus of the conic section, and the polar axis ($\theta = 0$) is an axis of symmetry.

The eccentricity e of a conic section is defined as

$$e = \frac{C}{Gm_1/\hbar^2} = \frac{C\hbar^2}{Gm_1} \qquad (15\text{-}42)$$

Equation 15-41 for r can be written in terms of the eccentricity e as

$$r = \frac{\hbar^2}{Gm_1} \frac{1}{1 + e \cos \theta} \qquad (15\text{-}43)$$

Equation 15-43 predicts three different types of paths for the particle, depending on the eccentricity e,

---

1. When $e > 1$, the radius $r \rightarrow \infty$ as $\cos \theta \rightarrow -1/e$. The path is a hyperbola. Many comets follow hyperbolic trajectories through the solar system.
2. When $e = 1$, the radius $r \rightarrow \infty$ as $\cos \theta \rightarrow -1$ ($\theta = \pm 180°$). The path is a parabola. Spacecraft leaving the earth for other points in the solar system may follow a parabolic path.
3. When $e < 1$, the radius r remains finite for all values of θ. The path is an ellipse. For the particular case when $e = 0$, the radius r is constant and the path is a circle. Spacecraft and other satellites in earth orbit follow elliptical or circular paths.

The different types of paths are shown in Fig. 15-14. A second branch of the hyperbola (not shown on Fig. 15-14) corresponds to a repulsive central force field rather than an attractive central force field. Equation 15-43 is a mathematical statement of Kepler's first law.

The velocity of a particle at any point on its path is obtained from differentiating Eq. 15-43 with respect to time. Thus,

$$r = \frac{\hbar^2}{Gm_1} \frac{1}{1 + e \cos \theta}$$

$$\dot{r} = \frac{\hbar^2}{Gm_1} \frac{e \sin \theta}{(1 + e \cos \theta)^2} \dot{\theta}$$

$$\dot{\theta} = \frac{\hbar}{r^2} = \frac{G^2 m_1^2}{\hbar^3}(1 + e \cos \theta)^2$$

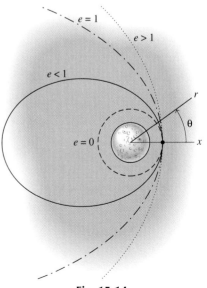

Fig. 15-14

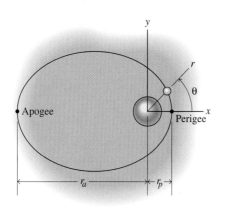

Fig. 15-15

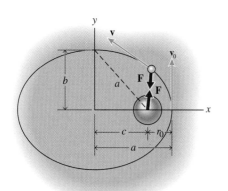

Fig. 15-16

which yields

$$\dot{r} = \frac{Gm_1}{\hbar} e \sin \theta$$

$$r\dot{\theta} = \frac{Gm_1}{\hbar}(1 + e \cos \theta)$$

$$v = \sqrt{\dot{r}^2 + (r\dot{\theta})^2}$$

$$= \frac{Gm_1}{\hbar}\sqrt{e^2 + 2e \cos \theta + 1} \qquad (15\text{-}44)$$

For planetary motion about the sun and for artificial earth satellites, ϵ is less than 1 and the orbits are ellipses (see Fig. 15-15). The minimum distance from the focus to the particle is called the perigee r_p of the orbit and occurs when $\theta = 0°$. The maximum distance is called the apogee r_a and occurs when $\theta = 180°$. Thus, from Eq. 15-43

$$r_p = r_{min} = \frac{\hbar^2}{Gm_1(1 + e)}$$

$$(15\text{-}45)$$

$$r_a = r_{max} = \frac{\hbar^2}{Gm_1(1 - e)}$$

The semimajor axis a, the semiminor axis b, and the area A of the ellipse (see Figs. 15-15 and 15-16) are

$$a = \frac{1}{2}(r_a + r_p)$$

$$= \frac{\hbar^2}{2Gm_1(1 - e)} + \frac{\hbar^2}{2Gm_1(1 + e)} = \frac{\hbar^2}{Gm_1(1 - e^2)} \qquad (15\text{-}46)$$

$$b = \sqrt{a^2 - c^2}$$

$$= \sqrt{a^2 - (a - r_p)^2} = a\sqrt{1 - e^2} \qquad (15\text{-}47)$$

$$A = \pi ab = \pi a^2\sqrt{1 - e^2} \qquad (15\text{-}48)$$

The period T (time for one revolution) can be obtained by using Eq. 15-39:

$$dA = \frac{\hbar}{2} dt$$

$$\int_0^A dA = \frac{\hbar}{2}\int_0^T dt$$

$$T = \frac{2A}{\hbar} = \frac{2\pi a^2}{\hbar}(1 - e^2)^{1/2} = \frac{2\pi a^2}{\hbar}\left(\frac{\hbar^2}{Gm_1 a}\right)^{1/2} \qquad (15\text{-}49)$$

From which

$$\frac{T^2}{a^3} = \frac{4\pi^2}{Gm_1} \qquad \text{or} \qquad T = \left[\frac{4\pi^2 a^3}{Gm_1}\right]^{1/2} \qquad (15\text{-}50)$$

Equation 15-50 is a mathematical statement of Kepler's third law.

For space probes and satellites launched from the earth, the very large mass m_1 in the previous equations is the mass of the earth m_e. The astronomical data required for solution of the following example problems is listed in Appendix B (Table B-8).

A rocket transports a satellite to a point 250 mi above the earth's surface, as shown in Fig. 15-17. Determine the velocity (parallel to the earth's surface) required to place the satellite in a circular orbit.

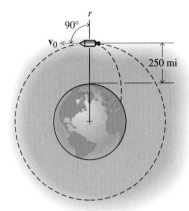

Fig. 15-17

SOLUTION

After the powered flight of the rocket ends and the satellite is given an initial velocity $\mathbf{v}_0$ parallel to the earth's surface, the satellite is in free flight and subjected only to the gravitational attraction of the earth. The motion of the satellite is described by Eq. 15-40, and the velocity at any point in its flight path is given by Eq. 15-44. For a circular orbit, $e = 0$ and Eq. 15-44 becomes

$$v = \frac{Gm_1}{h}\sqrt{e^2 + 2e\cos\theta + 1} = \frac{Gm_e}{h} = v_0 \qquad (a)$$

Equation a indicates that the velocity is constant and equal to the initial velocity v_0. The constant h can be determined by using Eq. 15-39 and the initial conditions for the launch; namely, when $r = r_0$, $v = v_0$. Thus,

$$h = r^2\dot{\theta} = r_0^2\frac{v_0}{r_0} = r_0 v_0 \qquad (b)$$

From Eqs. a and b

$$v_0 = \frac{Gm_e}{h} = \frac{Gm_e}{r_0 v_0} \qquad \text{or} \qquad v_0^2 = \frac{Gm_e}{r_0}$$

At an altitude $h = 250$ miles,

$$r_0 = r_e + h = 3960 + 250 = 4210 \text{ mi}$$

Therefore

$$v_0 = \left[\frac{Gm_e}{r_0}\right]^{1/2} = \left[\frac{3.439(10^{-8})(4.095)(10^{23})}{4210(5280)}\right]^{1/2} = 2.517(10^4) \text{ ft/s}$$

$$= 17{,}160 \text{ mi/h} \qquad \text{Ans.}$$

A rocket transports a satellite to a point 800 km above the earth's surface. Determine the velocity (parallel to the earth's surface) required to place the satellite

a. In an elliptical orbit with a maximum altitude of 8000 km.
b. On a parabolic flight path out of the earth's gravitational field.

SOLUTION

Once the satellite is in free flight and subjected only to the earth's gravitational field, the motion is described by Eq. 15-40, and the velocity at any point in its flight path is given by Eq. 15-44.

a. For an elliptical orbit with

$$r_p = r_e + h_p = 6.371(10^6) \text{ m} + 0.800(10^6) \text{ m} = 7.171(10^6) \text{ m}$$
$$r_a = r_e + h_a = 6.371(10^6) \text{ m} + 8.000(10^6) \text{ m} = 14.371(10^6) \text{ m}$$

equations 15-45 yield

$$r_a = \frac{\hbar^2}{Gm_1(1 - e)} \qquad r_p = \frac{\hbar^2}{Gm_1(1 + e)}$$

$$e = \frac{r_a - r_p}{r_a + r_p} = \frac{14.371(10^6) - 7.171(10^6)}{14.371(10^6) + 7.171(10^6)} = 0.3342$$

At $\theta = 0°$, Eq. 15-44 becomes

$$v = \frac{Gm_1}{\hbar}\sqrt{e^2 + 2e \cos \theta + 1} = \frac{Gm_e}{\hbar}(1 + e) = v_0 \qquad (a)$$

The constant $\hbar$ can be determined by using Eq. 15-39 and the initial conditions for the launch; namely, when $r = r_0$, $v = v_0$. Thus,

$$\hbar = r^2\dot{\theta} = r_0^2\frac{v_0}{r_0} = r_0v_0 \qquad (b)$$

From Eqs. a and b

$$v_0 = \frac{Gm_e}{\hbar}(1 + e) = \frac{Gm_e}{r_0v_0}(1 + e) \qquad \text{or} \qquad v_0^2 = \frac{Gm_e}{r_0}(1 + e)$$

With $r_0 = r_p = 7.171(10^6)$ m,

$$v_0 = \left[\frac{Gm_e}{r_0}(1 + e)\right]^{1/2} = \left[\frac{6.673(10^{-11})(5.976)(10^{24})}{7.171(10^6)}(1 + 0.3342)\right]^{1/2}$$
$$= 8.614(10^3) \text{ m/s} = 8.61 \text{ km/s} \qquad \text{Ans.}$$

b. For a parabolic flight trajectory, $e = 1$; therefore,

$$v_0 = \left[\frac{Gm_e}{r_0}(1 + e)\right]^{1/2} = \left[\frac{2Gm_e}{r_0}\right]^{1/2}$$
$$= \left[\frac{2(6.673)(10^{-11})(5.976)(10^{24})}{7.171(10^6)}\right]^{1/2}$$
$$= 10.546(10^3) \text{ m/s} = 10.55 \text{ km/s} \qquad \text{Ans.}$$

The velocity v_0 associated with a parabolic flight trajectory is the minimum velocity required for escape from the earth's gravitational field and is commonly referred to as the escape velocity v_{esc}.

The maximum speed of a satellite in an elliptical orbit ($e = 0.25$) is 16,000 mi/h. Determine

a. The maximum and minimum distances (in miles) from the surface of the earth to the satellite's trajectory.
b. The period of the elliptical orbit.

SOLUTION

a. For an elliptical orbit, the maximum velocity occurs at r_p, where it is parallel to the surface of the earth. Thus, from Eq. 15-39,

$$h = r^2\dot{\theta} = r_p^2\frac{v_p}{r_p}$$
$$= r_pv_p = r_pv_{max}$$

Then from Eqs. 15-45 with $v_{max} = 16,000$ mi/h $= 23,467$ ft/s:

$$r_p = \frac{Gm_e(1 + e)}{v_{max}^2}$$
$$= \frac{3.439(10^{-8})(4.095)(10^{23})(1 + 0.25)}{(23,467)^2}$$
$$= 3.197(10^7) \text{ ft} = 6055 \text{ mi}$$
$$r_a = \frac{1 + e}{1 - e}r_p$$
$$= \frac{1 + 0.25}{1 - 0.25}(6055) = 10,090 \text{ mi}$$
$$h_p = r_p - r_e \qquad\qquad\qquad \text{Ans.}$$
$$= 6055 - 3960 = 2095 \text{ mi}$$
$$h_a = r_a - r_e \qquad\qquad\qquad \text{Ans.}$$
$$= 10,090 - 3960 = 6130 \text{ mi}$$

b. For the ellipse

$$a = \frac{1}{2}(r_a + r_p)$$
$$= \frac{1}{2}[5.328(10^7) + 3.197(10^7)]$$
$$= 4.263(10^7) \text{ ft}$$

From Eq. 15-50

$$T = \left[\frac{4\pi^2a^3}{Gm_e}\right]^{1/2}$$
$$= \left\{\frac{4\pi^2[4.263(10^7)]^3}{3.439(10^{-8})(4.095)(10^{23})}\right\}^{1/2}$$
$$= 1.4737(10^4) \text{ s} = 4.09 \text{ h} \qquad\qquad \text{Ans.}$$

A space probe launched from the surface of the earth (see Fig. 15-18) is traveling parallel to the surface of the earth, at an altitude of 1200 km, when the powered portion of the flight ends with a velocity of 12.00 km/s. Determine

a. The eccentricity e of the trajectory.
b. The velocity of the space probe when the distance from the center of the earth to the space probe is 100,000 km.
c. The maximum angle θ for this flight trajectory.

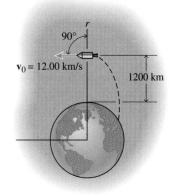

Fig. 15-18

SOLUTION

a. Since the space probe is traveling parallel to the surface of the earth when the powered portion of the flight ends, the initial conditions for the flight (when $\theta = 0°$) are

$$r_0 = r_e + h_0 = 6.371(10^6) + 1.200(10^6) = 7.571(10^6) \text{ m}$$
$$v_0 = 12.00 \text{ km/s} = 12.00(10^3) \text{ m/s}$$

From Eq. 15-39

$$h = r^2\dot{\theta} = r_0^2\frac{v_0}{r_0} = r_0v_0 = 7.571(10^6)12.00(10^3)$$
$$= 9.085(10^{10}) \text{ m}^2/\text{s}$$

From Eq. 15-43 with $r = r_0$ when $\theta = 0°$

$$e = \frac{h^2}{Gm_er_0} - 1$$
$$= \frac{[9.085(10^{10})]^2}{6.673(10^{-11})(5.976)(10^{24})(7.571)(10^6)} - 1 = 1.734$$

b. The velocity is determined by using Eqs. 15-43 and 15-44. From Eq. 15-43

$$\cos\theta = \frac{h^2 - Gm_er}{Gm_ere}$$
$$= \frac{[9.085(10^{10})]^2 - 6.673(10^{-11})(5.976)(10^{24})(100)(10^6)}{6.673(10^{-11})(5.976)(10^{24})(100)(10^6)(1.734)}$$
$$= -0.4573 \qquad \theta = 117.2°$$

From Eq. 15-44

$$v = \frac{Gm_e}{h}\sqrt{e^2 + 2e\cos\theta + 1}$$
$$= \frac{6.673(10^{-11})(5.976)(10^{24})}{9.085(10^{10})}\sqrt{1.734^2 + 2(1.734)\cos 117.2° + 1}$$
$$= 6.830(10^3) \text{ m/s} = 6.83 \text{ km/s} \qquad\qquad \text{Ans.}$$

c. The maximum angle θ_{max} occurs as $r \to \infty$. From Eq. 15-43, $r \to \infty$ as

$$1 + e\cos\theta \to 0$$

Therefore

$$\cos\theta_{max} = -\frac{1}{e} = -\frac{1}{1.734}$$
$$\theta_{max} = 125.2° \qquad\qquad \text{Ans.}$$

PROBLEMS

The astronomical data required for solution of the following problems is listed in Appendix B (Table B-8).

15-69* A rocket transports a satellite to a point 500 mi above the earth's surface. Determine the velocity (parallel to the earth's surface) required to place the satellite

a. In a circular orbit.
b. On a parabolic flight path.

15-70* A rocket transports a satellite to a point 900 km above the earth's surface (Fig. P15-70). Determine the velocity (parallel to the earth's surface) required to place the satellite

a. In a circular orbit.
b. On a hyperbolic flight path with an eccentricity of 1.25.

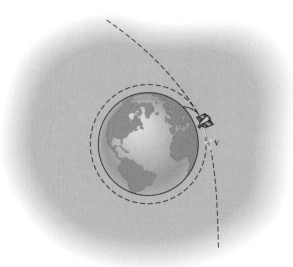

Fig. P15-70

15-71 It is desired to place a satellite in a circular polar orbit such that successive ground tracks at the equator are spaced 2000 mi apart. Determine the required altitude of the circular orbit.

15-72 A satellite is to be placed into an equatorial circular orbit so that it always remains over the same point on the earth's surface. Determine the radius of the orbit in kilometers and the orbital speed of the satellite in kilometers per hour.

15-73* A rocket is traveling in a circular orbit 600 mi above the surface of the earth. Determine the minimum change in speed needed to permit the rocket to escape the earth's gravitational field.

15-74* A rocket is in a circular orbit at an altitude of 500 km. During a very short interval of an orbit, the engines increase the velocity (tangent to the orbit) by 1000 m/s. Determine

a. The eccentricity e of the new orbit.
b. The altitude and orbital speed of the rocket at the highest point of its new orbit.

15-75 At an altitude of 500 mi above the surface of the earth, a satellite is inserted into orbit with a speed of 20,000 mi/h parallel to the surface of the earth. Determine

a. The eccentricity e of the orbit.
b. The maximum and minimum altitudes for the satellite's trajectory.

15-76 The altitude of a satellite in an elliptical orbit around the earth is 1600 km at apogee and 600 km at perigee (see Fig. P15-76). Determine

a. The eccentricity e of the orbit.
b. The orbital speeds at apogee and perigee.
c. The period of the orbit.

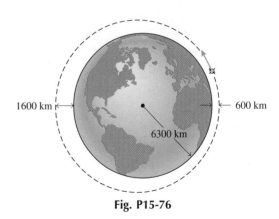

Fig. P15-76

15-77* A satellite is traveling in a circular orbit 500 mi above the surface of the earth. Determine

a. The change in velocity required to place the satellite in an elliptical orbit with an eccentricity of 0.30.
b. The altitude and orbital speed of the satellite at the highest point of its new elliptical orbit.

15-78* The maximum distance from the center of the earth to a satellite in an elliptical orbit ($e = 0.25$) is 15,000 km. Determine

a. The altitude of the satellite at perigee.
b. The velocity of the satellite at apogee and perigee.
c. The period of the elliptical orbit.

15-79 The altitude of a satellite in an elliptical orbit around the earth is 1500 mi at apogee and 900 mi at perigee. Determine

a. The eccentricity e of the orbit.
b. The orbital speeds at apogee and perigee.
c. The period of the orbit.

15-80 The Apollo lunar lander is in an orbit about the moon with a perigee altitude of 50 km and an apogee altitude of 500 km above the lunar surface (see Fig. P15-80). Determine

a. The eccentricity of the orbit.
b. The velocity of the lunar lander at its perigee.
c. The orbital period of the lunar lander.

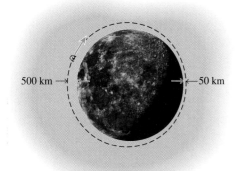

Fig. P15-80

15-81* A satellite is in a circular orbit 250 mi above the surface of the earth. Determine the new perigee and apogee altitudes of the satellite if an on-board engine:

a. Increases the speed of the satellite by 800 ft/s.
b. Gives the satellite a radial (outward) component of velocity of 800 ft/s.

15-82* A satellite is in an elliptic orbit with a perigee altitude of 1000 km and an apogee altitude of 9000 km. Determine the r- and θ-components of the satellite's velocity as it crosses the minor axis of the ellipse (point B of Fig. P15-82).

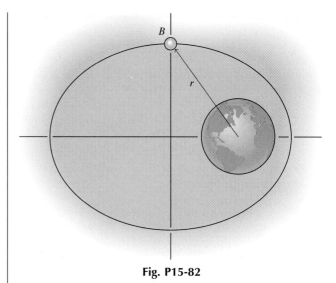

Fig. P15-82

15-83 The altitude of a satellite in an elliptical orbit around the earth is 21,000 mi at apogee and 2500 mi at perigee, as shown in Fig. P15-83. Determine

a. The eccentricity e of the orbit.
b. The orbital speeds at apogee and perigee.
c. The radial distance r from the center of the earth and the velocity v when $\theta = 150°$.

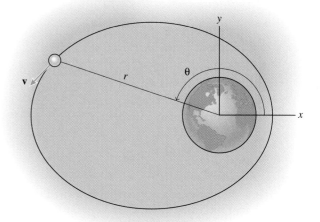

Fig. P15-83

15-84 A satellite is traveling in a circular orbit 5000 km above the surface of the earth, as shown in Fig. P15-84. At point A, the velocity is increased to put the satellite in an elliptical orbit with a maximum altitude of 10,000 km at point B. Determine

a. The eccentricity e of the elliptical orbit.
b. The change in velocity required at point A to place the satellite in the elliptical orbit.

c. The change in velocity required at point B to change the elliptical orbit to the higher circular orbit.
d. The period of the higher circular orbit.

a. The amount by which the speed of satellite A must be increased to put it on the elliptic orbit.
b. The amount by which the speed of satellite A must be increased at point C to complete the maneuver.

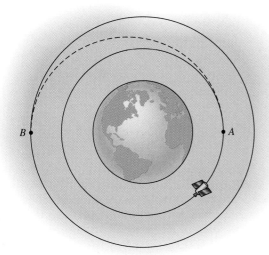

Fig. P15-84

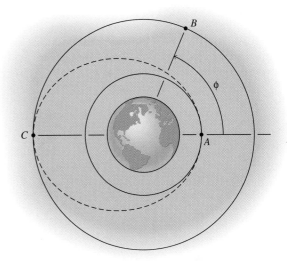

Fig. P15-86

15-85* A space vehicle is traveling in a circular orbit 300 mi above the surface of the earth (see Fig. P15-85). If air resistance is neglected, determine

a. The change in speed required to bring the vehicle back to earth at a point 180° from the point of retrorocket firing.
b. The time required for the descent.

15-87 The lunar lander is in a circular orbit 50 mi above the surface of the moon (Fig. P15-87). Determine

a. The amount by which the velocity must be decreased to land after one-quarter orbit (at B).
b. The speed with which the lander would strike the moon if it did not use its retrorockets to slow its descent.
c. The angle that the velocity of part b makes with the radial direction.

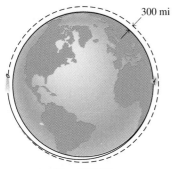

300 mi

Fig. P15-85

15-86* Satellites A and B are in circular orbits 200 km and 800 km, respectively, above the surface of the earth (see Fig. P15-86). If satellite A is to rendezvous with satellite B at point C by using the elliptic orbit shown, determine

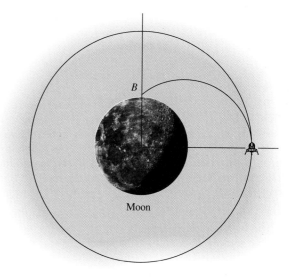

Moon

Fig. P15-87

15-88 The lunar lander, which is sitting on the surface of the moon, wants to return to the command module, which is in a circular orbit 80 km above the moon's surface (Fig. P15-88). Determine

a. The velocity (magnitude and direction) with which the lunar lander must leave the moon's surface to rendezvous with the command module as shown.

b. The amount by which the speed of the lunar lander must be increased at its apogee to complete its rendezvous with the command module.

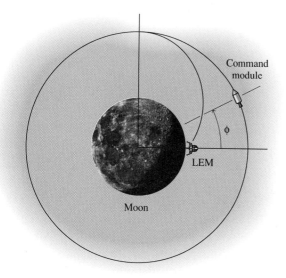

Fig. P15-88

15-89* A satellite is being placed into a low earth orbit by a launch vehicle. When the rocket motor of the launch vehicle shuts off, the satellite is at an altitude of 80 mi above the earth and has a velocity of 26,000 ft/s. As a result of guidance error (see Fig. P15-89), the satellite is injected into orbit at an angle of 85° with respect to the radius vector rather than parallel to the surface of the earth. Determine

a. The equation of the planned (parallel injection) orbit.
b. The equation of the actual orbit.
c. The velocity and altitude at apogee of the planned orbit.
d. The velocity and altitude at apogee of the actual orbit.
e. The altitude at perigee of the actual orbit.

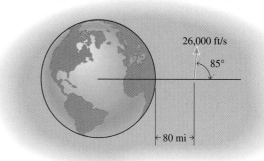

Fig. P15-89

15-90* A rocket launched from the surface of the earth has a speed of 8.85 km/s when powered flight ends at an altitude of 550 km. The flight path of the rocket at this time is inclined at an angle of 84° with respect to a radial line through the center of the earth. Determine

a. The eccentricity e of the trajectory.
b. The altitude of the rocket at perigee.
c. The velocity of the rocket at apogee and perigee.
d. The period of the orbit.

15-91 A meteoroid is first observed approaching the earth along a hyperbolic flight path when it is 250,000 mi from the center of the earth (Fig. P15-91). If the speed of the meteoroid at this time is 5000 mi/h at $\theta = 150°$, determine

a. The eccentricity e of the trajectory.
b. The closest approach of the meteoroid to the earth's surface.
c. The velocity of the meteoroid when it is closest to the earth's surface.

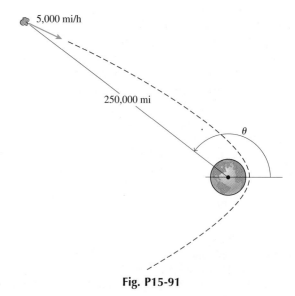

Fig. P15-91

15-92 Two satellites are in the same circular orbit 1000 km above the earth's surface with satellite A leading satellite B by 2500 km (Fig. P15-92). Satellite B proposes to "catch up" to satellite A by "slowing down" into the elliptic orbit shown. Determine the amount by which satellite B must slow down to catch satellite A after

a. One period in the elliptic orbit.
b. Two periods in the elliptic orbit.

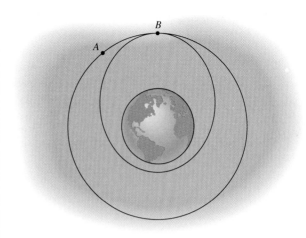

Fig. P15-92

SUMMARY

The basic law governing the motion of a particle is Newton's second law, which relates the accelerated motion of a particle to the forces producing the motion. Mathematically, Newton's second law is expressed as

$$\mathbf{F} = m\mathbf{a} \qquad (15\text{-}2)$$

Equation 15-2 expresses the fact that the magnitudes of $\mathbf{F}$ and $\mathbf{a}$ are proportional and that the vectors $\mathbf{F}$ and $\mathbf{a}$ have the same direction. Equation 15-2 is valid both for constant forces and for forces that vary with time. The system of axes used for the acceleration measurements must be a primary inertial system. Any *nonrotating system of axes,* which translates with a constant velocity with respect to the primary system, is equally satisfactory. For most engineering problems *on the surface of the earth,* the corrections required to compensate for the acceleration of the earth with respect to the primary system are negligible, and the accelerations measured with respect to axes attached to the surface of the earth may be treated as absolute.

The equation of motion for a system of particles can be obtained by using Newton's second law for each individual particle of the system and summing the results to obtain an equation for the motion of the mass center G of the system. The result is

$$\mathbf{R} = m\mathbf{a}_G \qquad (15\text{-}16)$$

Equation 15-16 is valid for any type of motion and shows that the equation of motion for a system of particles is the same as the equation of motion for a single particle located at the mass center of the system and having a mass equal to the total mass of the system. Any body can be considered to be a particle when applying this equation.

Motion of a particle along a straight line is known as rectilinear motion, and if the coordinate system is oriented such that the x-axis coincides with the line of motion, the position, velocity, and acceleration of the particle are completely described by their x-components:

$$\mathbf{r} = x\mathbf{i} \qquad \mathbf{v} = \dot{\mathbf{r}} = \dot{x}\mathbf{i} \qquad \mathbf{a} = \ddot{\mathbf{r}} = \ddot{x}\mathbf{i}$$

Equation 15-2 for the particle then reduces to

$$\Sigma\mathbf{F}_x = m\mathbf{a}_x \qquad \Sigma\mathbf{F}_y = \mathbf{0} \qquad \Sigma\mathbf{F}_z = \mathbf{0} \qquad (15\text{-}19)$$

Motion of a particle along a curved path is known as curvilinear motion. When the motion occurs in a plane, two coordinates are required to describe the motion. The three coordinate systems used to describe plane curvilinear motion are rectangular coordinates, polar coordinates, and normal/tangential coordinates.

With a rectangular coordinate system, the position of the particle is described by using its distance from two reference axes (say the x- and y-axes). The equations for position, velocity, and acceleration are

$$\mathbf{r} = x\mathbf{i} + y\mathbf{j} \qquad \mathbf{v} = \dot{\mathbf{r}} = \dot{x}\mathbf{i} + \dot{y}\mathbf{j} \qquad \mathbf{a} = \ddot{\mathbf{r}} = \ddot{x}\mathbf{i} + \ddot{y}\mathbf{j}$$

Equation 15-2 for the particle then reduces to

$$\Sigma\mathbf{F}_x = m\mathbf{a}_x \qquad \Sigma\mathbf{F}_y = m\mathbf{a}_y \qquad \Sigma\mathbf{F}_z = \mathbf{0} \qquad (15\text{-}21)$$

In a polar coordinate system, the position of the particle is described by using a distance r from a fixed point and an angular displacement θ from a fixed line. Unit vectors $\mathbf{e}_r$ and $\mathbf{e}_\theta$ are directed radially outward from the fixed point and in the direction of increasing θ, respectively. Equations for position, velocity, and acceleration are

$$\mathbf{r} = r\mathbf{e}_r \qquad \mathbf{v} = \dot{\mathbf{r}} = \dot{r}\mathbf{e}_r + r\dot{\theta}\,\mathbf{e}_\theta$$
$$\mathbf{a} = \ddot{\mathbf{r}} = (\ddot{r} - r\dot{\theta}^2)\,\mathbf{e}_r + (r\ddot{\theta} + 2\dot{r}\dot{\theta})\,\mathbf{e}_\theta \qquad (15\text{-}23)$$

Equation 15-2 for the particle then reduces to

$$\Sigma\mathbf{F}_r = m\mathbf{a}_r = m(\ddot{r} - r\dot{\theta}^2)\mathbf{e}_r$$
$$\Sigma\mathbf{F}_\theta = m\mathbf{a}_\theta = m(r\ddot{\theta} + 2\dot{r}\dot{\theta})\,\mathbf{e}_\theta \qquad (15\text{-}24)$$
$$\Sigma\mathbf{F}_z = \mathbf{0}$$

In a normal and tangential coordinate system, unit vectors $\mathbf{e}_t$ and $\mathbf{e}_n$ are directed tangent to the path (in the direction of motion) and normal to the path (toward the center of curvature), respectively, at each point along the path of the particle. Equations for the velocity and acceleration at position s along the path are

$$\mathbf{v} = \dot{s}\mathbf{e}_t \qquad \mathbf{a} = \ddot{s}\,\mathbf{e}_t + \frac{\dot{s}^2}{\rho}\mathbf{e}_n \qquad (15\text{-}26)$$

Equation 15-2 for the particle then reduces to

$$\Sigma\mathbf{F}_t = m\mathbf{a}_t = m\ddot{s}\,\mathbf{e}_t \qquad \Sigma\mathbf{F}_n = m\mathbf{a}_n = m\frac{\dot{s}^2}{\rho}\mathbf{e}_n \qquad \Sigma\mathbf{F}_z = \mathbf{0} \qquad (15\text{-}27)$$

The motion of a particle moving under the influence of a force directed toward a fixed point is called central-force motion. Common examples include the motion of planets about the sun and the motion of the moon and artificial satellites about the Earth. The force involved in these examples is given by Newton's law of universal gravitation, which states that the force **F** between two masses m_1 and m_2 separated by a distance r has the magnitude

$$F = \frac{Gm_1m_2}{r^2}$$
(15-38)

where G is the universal gravitational constant.

REVIEW PROBLEMS

15-93* A 25,000-lb navy fighter plane launched from a carrier increases in speed from 20 mi/h (the carrier speed) to 150 mi/h in 2.2 s. Determine

a. The constant force applied by the catapult.
b. The distance traveled by the plane during the launch.

15-94* A Saturn V rocket has a mass of 2.75(10⁶) kg and a thrust of 33(10⁶) N. Determine

a. The initial vertical acceleration of the rocket.
b. The rocket's velocity 10 s after liftoff.
c. The time required to reach an altitude of 10,000 m.

15-95 The velocity of a baseball ($W = 0.33$ lb) when it is hit by a bat changes from 80 mi/h in a given direction to 90 mi/h in the opposite direction. If the ball and the bat are in contact for 0.005 s, what constant force must be exerted on the ball by the bat?

15-96 A skydiver, whose mass is 80 kg, is falling 85 m/s when he opens his parachute. His speed is reduced to 5 m/s during the next 60 m of fall. Determine the average force exerted on his body by the parachute during this interval.

15-97* Blocks A and B of Figure P15-97 weigh 60 and 40 lb, respectively. If the blocks are released at rest in the position shown, determine

a. The velocity of block B after it has moved 10 ft.
b. The tension in the cable supporting block A.

15-98* A 10-kg block A is released at rest on a 20-kg wedge B as shown in Fig. P15-98. If all surfaces are smooth (frictionless), determine

a. The normal force between the block and the wedge.
b. The acceleration of the block.
c. The acceleration of the wedge.

Fig. P15-98

15-99 The weight of block A of Figure P15-99 is 50 lb. The block is at rest and the spring ($k = 10$ lb/ft) is unstretched when the block is released to move. Determine

a. The velocity of the block after it has moved 3 ft.
b. The maximum distance the block moves from its initial position.

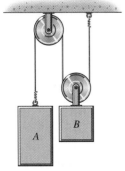

Fig. P15-97

Fig. P15-99

15-100 The rocket engine of a missile having a mass of 15,000 kg produces a thrust of 200 kN. If the missile is launched in a vertical direction, determine its velocity and vertical height after 1 min if

a. Air resistance is negligible.
b. Air resistance produces a drag force $F_D = 0.25v^2$, where F_D is in newtons and v is in meters per second.

15-101* A 50-lb projectile is fired, as shown in Fig. P15-101, with an initial velocity of 1500 ft/s. If air resistance can be neglected, determine the radius of curvature of the projectile's path when it is at its peak.

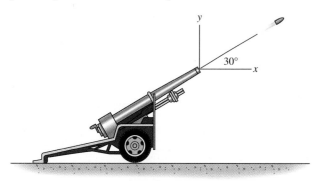

Fig. P15-101

15-102* The circular disk shown in Fig. P15-102 is rotating in a vertical plane. Motion of body A (m = 1 kg) in the smooth radial slot is resisted by a spring attached to the hub of the disk. When the disk is in the position shown, its angular velocity is 100 rad/s clockwise and its angular acceleration is 25 rad/s² counterclockwise. At this instant, determine

a. The force exerted on body A by the disk.
b. The spring constant k if the rest position of body A is 350 mm from the axis of the disk.

Fig. P15-102

15-103 A 1-lb particle P is driven along a circular slot in a vertical plane by a slotted bar, which is rotating about a fixed point A as shown in Fig. P15-103. The angular velocity of the bar is 25 rad/s clockwise and the angular acceleration is 20 rad/s² counterclockwise. If all surfaces are smooth, determine the forces exerted on the particle when (a) $\theta = 60°$ and (b) $\theta = 120°$.

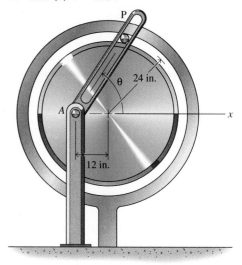

Fig. P15-103

15-104 A sphere S with a mass of 5 kg is attached to a 1-kg block B, which is free to slide in a smooth horizontal slot as shown in Fig. P15-104. The mass of the rod connecting the sphere to the block is negligible. If the system is released at rest in the position shown, determine

a. The tension in the rod as motion begins.
b. The acceleration of the block as motion begins.

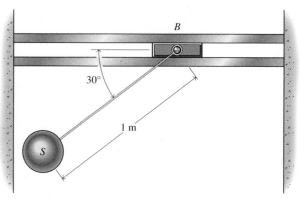

Fig. P15-104

15-105 A 2-lb sphere slides along a smooth rod, which is bent in a vertical plane into a shape that can be described by the equation $y = \sqrt{x}$, where x and y are measured in feet. When the sphere is in position A or B of Fig. P15-105, its speed is 10 ft/s to the left. If the unstretched length of the spring ($k = 3$ lb/ft) is 2 ft, determine the acceleration of the sphere and the force exerted on the sphere by the rod

a. When the sphere is in position A.
b. When the sphere is in position B.

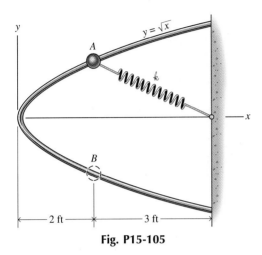

Fig. P15-105

15-106 A block B ($m = 0.50$ kg) moves along a smooth circular slot in a vertical plane, as shown in Fig. P15-106. When the block is in the position shown, its speed is 20 m/s up and to the left. If the unstretched length of the spring ($k = 25$ N/m) is 300 mm, determine the acceleration of the block and the force exerted on the block by the surface of the slot.

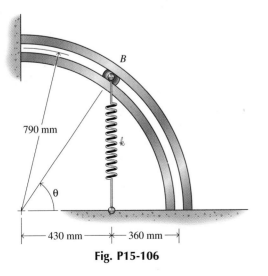

Fig. P15-106

15-107* Determine the orbital speed and period of a satellite in a circular orbit 1000 mi above the earth's surface.

15-108* A rocket transports a satellite to a point 1500 km above the earth's surface. Determine the velocity (parallel to the earth's surface) required to place the satellite

a. In a circular orbit.
b. On a parabolic flight path.
c. On a hyperbolic flight path with an eccentricity of 1.40.

15-109 At an altitude of 750 mi above the surface of the earth, a satellite is inserted into orbit with a speed of 18,000 mi/h parallel to the surface of the earth. Determine

a. The eccentricity e of the orbit.
b. The maximum and minimum altitudes for the satellite's trajectory.

15-110 The altitude of a satellite in an elliptical orbit around the earth is 3600 km at apogee and 900 km at perigee. Determine

a. The eccentricity e of the orbit.
b. The orbital speeds at apogee and perigee.
c. The period of the orbit.

C15-111 When air resistance is neglected, it is easily shown that the trajectory of objects moving near the surface of the earth are parabolas. However, all objects moving through a fluid (such as air) experience a drag force that is proportional to the square of their speed and that acts in a direction opposite to their velocity.

Suppose that a small ball is thrown upward with an initial speed v_0 and an initial angle of θ_0 to the horizontal. At some point in its trajectory the ball will be acted on by the forces of gravity W and wind drag

$$D = C_D \frac{1}{2} \rho v^2 A$$

where C_D is the drag coefficient (may be taken as approximately one-half for spheres at moderate speeds), ρ is the mass density of the air through which the ball is moving, and $A = \pi r^2$ is the cross-sectional area of the ball (Fig. P15-111).

a. If a tennis ball ($W = 2.0$ oz, $r = 1.25$ in.) is thrown with an initial speed $v_0 = 60$ ft/s through air ($\rho = 0.002377$ slug/ft^3), use the Euler method for solving differential equations (see Appendix C) to compute the position of the ball as a function of time until it returns to the ground for initial angles $\theta_0 = 15°$, $30°$, $45°$, and $60°$.
b. Plot the trajectory (y versus x) for each of the initial angles above on a single graph. Also plot on the same graph the trajectory of the ball neglecting wind resistance for each initial angle.
c. Compute and plot the range (the horizontal distance between the initial and final points of the trajectory) for various initial angles θ_0 ($30° \leq \theta_0 \leq 60°$). When wind resistance is neglected, the maximum range is attained when $\theta_0 = 45°$. What angle gives the maximum range when wind resistance is included?
d. Repeat the calculations for an initial velocity $v_0 = 30$ ft/s. Does the angle that gives the maximum range depend on the initial speed when wind resistance is included?

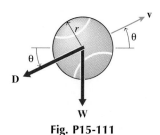

Fig. P15-111

C15-112 A small block slides down the inside of a circular bowl as shown in Fig. P15-112. If the mass of block $m = 2$ kg, the radius of the bowl $r = 1.5$ m, the coefficients of friction $\mu_s = \mu_k = 0.3$, and the block is released from rest with $\theta = \theta_0$,

a. Use the Euler method for solving differential equations (see Appendix C) to compute the position of the block as a function of time until it comes to rest for initial angles $\theta_0 = 80°$, $60°$, $45°$, and $30°$.
b. For each of the initial angles, plot the angular position of the block θ, the speed of the block v, and the friction force acting on the block F as a function of time t ($0 \leq t \leq 5$ s). (Be careful to ensure that the friction always opposes the motion.)
c. For what initial angle θ_0 will the block stop just as it gets to the bottom of the bowl? For what initial angle θ_0 will the block stop at the highest position on the opposite side of the bowl? For what initial angle θ_0 will the block stop at the highest position on the same side of the bowl (possibly after sliding past the bottom and coming back)? Do the results depend on the weight of the object?

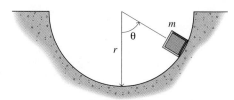

Fig. P15-112

C15-113 A small paper weight ($W = 2.5$ lb) slides on the outside of a 2-ft radius cylinder as shown in Fig. P15-113. If friction may be neglected and the weight starts from rest when $\theta = 0°$,

a. Plot the speed of the weight v as a function of the angular position θ ($0 \leq \theta \leq \beta$) where β is the angle at which the weight loses contact with the cylinder (the normal force becomes zero).
b. Plot the normal force N between the weight and the cylinder as a function of the angular position θ ($0 \leq \theta \leq \beta$).
c. Plot the angular position θ as a function of time t ($0 \leq t \leq t_\beta$). (You may need to use the Euler method for solving differential equations. See Appendix C.)
d. Repeat the problem for $\mu_s = \mu_k = 0.3$ and ($\phi_s \leq \theta \leq \beta$) where $\phi_s = \tan^{-1} \mu_s$ is the angle of static friction. (Assume that the weight starts from rest at $\theta = \phi_s$.)

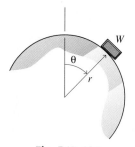

Fig. P15-113

C15-114 The 10-kg cart shown in Fig. P15-114 is being pulled to the left by a constant force $P = 10$ N applied to the end of the cord. If the cart starts from rest when $x = 8$ m, calculate and plot

a. The speed of the cart v as a function of its position x ($-3 \le x \le 8$ m).

b. The position of the cart x as a function of time t ($0 \le t \le 5$ s). (You may need to use the Euler method for solving differential equations. See Appendix C.)

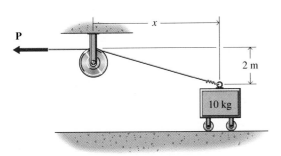

Fig. P15-114

C15-115 The pair of carts shown in Fig. P15-115 are being pulled to the left by a constant force $P = 2$ lb. Cart A weighs 10 lb, cart B weighs 20 lb, and the system starts from rest when $x = 24$ ft. Use the Euler method for solving differential equations (see Appendix C) to calculate the position and velocity of both carts as a function of time. Then plot:

a. The velocities of both carts v_A and v_B as a function of x ($-5 \le x \le 24$ ft).

b. The tension in the connecting cord T as a function of x ($-5 \le x \le 24$ ft).

c. The position of cart B as a function of time t ($0 \le t \le 5$ s).

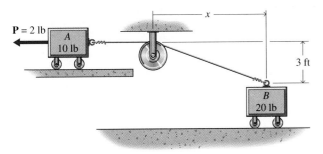

Fig. P15-115

C15-116 A 30-kg mass is suspended by a cord, which passes around three small pulleys as shown in Fig. P15-116. A constant force $P = 200$ N is applied to the other end of the cord. If the system starts from rest when $y = 2$ m, use the Euler method for solving differential equations (see Appendix C) to calculate the position and velocity of the mass as a function of time. Then plot

a. The position of the mass y as a function of time t ($0 \le t \le 10$ s).

b. The velocity of the mass v as a function of time t ($0 \le t \le 10$ s).

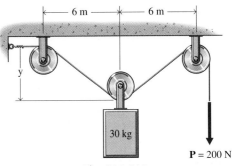

Fig. P15-116

C15-117 The pair of blocks shown in Fig. P15-117 are suspended by a cord, which passes around three small pulleys. Block A weighs 60 lb, block B weighs 40 lb, and the system starts from rest when $y = 6$ ft. Use the Euler method for solving differential equations (see Appendix C) to calculate the position and velocity of both blocks as a function of time. Then plot

a. The position y of block A as a function of time t ($0 \le t \le 15$ s).

b. The velocities of both blocks v_A and v_B as a function of time t ($0 \le t \le 15$ s).

c. The tension in the connecting cord T as a function of time t ($0 \le t \le 15$ s).

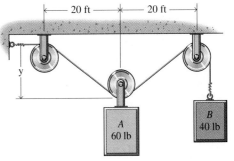

Fig. P15-117

KINETICS OF RIGID BODIES: NEWTON'S LAWS

Both forces and moments are required to orient the Hubble Space Telescope prior to its deployment by the Discovery crew.

16-1 INTRODUCTION

Rigid bodies can be viewed as a collection of particles; therefore, the relationships developed in Chapter 15 for the motion of a system of particles can be utilized for rigid bodies. In this chapter, extensive use is made of Eq. 15-16, which relates the resultant $\mathbf{R}$ of the applied external forces to the acceleration $\mathbf{a}_G$ of the mass center G for the special case where the line of action of the resultant $\mathbf{R}$ passes through the mass center G of the system. For the more general case where the resultant of the applied external forces consists of a resultant force $\mathbf{R}$ that passes through the mass center G plus a couple $\mathbf{C}$, the body experiences both rotation and translation, and additional equations are needed to relate the moments of the external forces to the angular motion of the body.

16-2 EQUATIONS FOR PLANE MOTION

Newton's laws apply only to the motion (translation) of a single particle; therefore, they are inadequate to describe the complete motion of a rigid body, which may include both translation and rotation. In this section, Newton's laws of motion are extended to cover plane motion of a rigid body. Later, in Section 16-5, Newton's laws will be further extended to cover the general case of three-dimensional motion of a rigid body. These laws (either planar or three-dimensional) provide differential equations that relate the linear and angular accelerated motion of the body to the forces and moments producing the motion. These equations can be used to determine

1. The instantaneous accelerations due to known forces and moments, or
2. The forces and moments required to produce a prescribed motion.

In Chapter 15 the "principle of motion of the mass center" of a system of particles was developed. Since a rigid body can be viewed as a collection of particles that remain at fixed distances with respect to each other, the motion of the mass center G of a rigid body is given by Eq. 15-16 as

$$\mathbf{R} = m\mathbf{a}_G \tag{15-16}$$

where

$\mathbf{R}$ is the resultant of the forces acting on the rigid body at a given instant of time.
m is the mass of the rigid body.
$\mathbf{a}_G$ is the instantaneous linear acceleration of the mass center of the rigid body in the direction of the resultant force $\mathbf{R}$.

This vector equation can be written in component scalar form as the following three equations:

$$\begin{aligned} \Sigma F_x = R_x = ma_{Gx} \\ \Sigma F_y = R_y = ma_{Gy} \\ \Sigma F_z = R_z = ma_{Gz} \end{aligned} \tag{15-17}$$

Since Eq. 15-16 was obtained by the simple process of summing forces, no information is provided regarding the location of the line of action of the resultant force $\mathbf{R}$. The mass center G of a rigid body moves (translates) as though the rigid body were a single particle of mass m subjected to the resultant force $\mathbf{R}$. The actual motion of most rigid bodies consists of the superposition of the translation produced by the resultant force $\mathbf{R}$ and the rotation produced by the moment of the resultant force $\mathbf{R}$ when its line of action does not pass through the mass center G of the body.

Consider the rigid body of arbitrary shape shown in Fig. 16-1a. The XYZ-coordinate system is fixed in space. The xyz-coordinate system is attached to the body at point A. The displacement of an element of mass dm with respect to point A is given by the vector $\boldsymbol{\rho}$ and with respect to the origin O of the XYZ-coordinate system by the vector $\mathbf{R}$. The displacement of point A with respect to the origin O of the XYZ system is given by the vector $\mathbf{r}$. The resultant external and internal forces acting on the element of mass dm are $\mathbf{F}$ and $\mathbf{f}$, respectively. The moment produced about point A by the forces $\mathbf{F}$ and $\mathbf{f}$ is

$$dM_A = \boldsymbol{\rho} \times (\mathbf{F} + \mathbf{f}) \qquad (a)$$

But from Newton's second law

$$\mathbf{F} + \mathbf{f} = dm\, \mathbf{a}_{dm} = dm\, \ddot{\mathbf{R}} \qquad (b)$$

Thus, from Eqs. a and b,

$$dM_A = \boldsymbol{\rho} \times (\mathbf{F} + \mathbf{f}) = (\boldsymbol{\rho} \times \mathbf{a}_{dm})\, dm \qquad (c)$$

The acceleration $\mathbf{a}_{dm}$ for a rigid body in either plane motion or general three-dimensional motion can be written

$$\mathbf{a}_{dm} = \mathbf{a}_A + (\dot{\boldsymbol{\omega}} \times \boldsymbol{\rho}) + [\boldsymbol{\omega} \times (\boldsymbol{\omega} \times \boldsymbol{\rho})] \qquad (14\text{-}29)$$

Substituting Eq. 14-29 into Eq. c and integrating yields

$$\mathbf{M}_A = \int_m (\boldsymbol{\rho} \times \mathbf{a}_A)\, dm + \int_m [\boldsymbol{\rho} \times (\dot{\boldsymbol{\omega}} \times \boldsymbol{\rho})]\, dm$$
$$+ \int_m \{\boldsymbol{\rho} \times [\boldsymbol{\omega} \times (\boldsymbol{\omega} \times \boldsymbol{\rho})]\}\, dm \qquad (d)$$

Plane motion of a rigid body is defined as motion in which all elements of the body move in parallel planes. The parallel plane that passes through the mass center G of the body is known as the "plane of motion." Thus, as shown in Fig. 16-1b, the angular velocity and angular acceleration vectors $\boldsymbol{\omega}$ and $\boldsymbol{\alpha}$, respectively, are parallel to each other and are perpendicular to the plane of motion. If the xyz-coordinate system is chosen so that the motion is parallel to the xy-plane, then $a_{Az} = \omega_x = \omega_y = 0$. The angular velocity of the body is $\omega_z = \omega$, and the angular acceleration is $\dot{\omega}_z = \alpha$. For motion in the xy-plane, the different terms appearing in Eq. d, when point A is located in the plane of motion, are evaluated as follows:

$$\boldsymbol{\rho} \times \mathbf{a}_A = \begin{vmatrix} \mathbf{i} & \mathbf{j} & \mathbf{k} \\ x & y & z \\ a_{Ax} & a_{Ay} & 0 \end{vmatrix} = -za_{Ay}\,\mathbf{i} + za_{Ax}\,\mathbf{j} + (xa_{Ay} - ya_{Ax})\,\mathbf{k} \qquad (e)$$

$$\dot{\boldsymbol{\omega}} \times \boldsymbol{\rho} = \begin{vmatrix} \mathbf{i} & \mathbf{j} & \mathbf{k} \\ 0 & 0 & \alpha \\ x & y & z \end{vmatrix} = -y\alpha\,\mathbf{i} + x\alpha\,\mathbf{j}$$

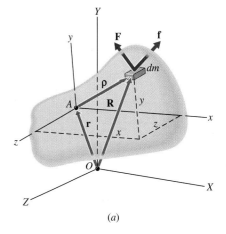

(a)

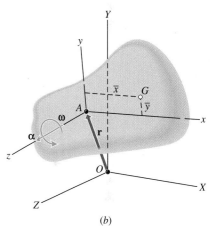

(b)

Fig. 16-1

Similarly

$$\boldsymbol{\rho} \times (\dot{\boldsymbol{\omega}} \times \boldsymbol{\rho}) = -xz\alpha\, \mathbf{i} - yz\alpha\, \mathbf{j} + (x^2 + y^2)\alpha\mathbf{k} \qquad (f)$$
$$\boldsymbol{\omega} \times \boldsymbol{\rho} = -y\omega\mathbf{i} + x\omega\mathbf{j}$$
$$\boldsymbol{\omega} \times (\boldsymbol{\omega} \times \boldsymbol{\rho}) = -x\omega^2\mathbf{i} - y\omega^2\mathbf{j}$$
$$\boldsymbol{\rho} \times [\boldsymbol{\omega} \times (\boldsymbol{\omega} \times \boldsymbol{\rho})] = yz\omega^2\mathbf{i} - zx\omega^2\mathbf{j} \qquad (g)$$

Consider now the rectangular components of the moment $\mathbf{M}_A$.

$$\mathbf{M}_A = M_{Ax}\mathbf{i} + M_{Ay}\mathbf{j} + M_{Az}\mathbf{k}$$
$$= \int_m (\boldsymbol{\rho} \times \mathbf{a}_A)\, dm + \int_m [\boldsymbol{\rho} \times (\dot{\boldsymbol{\omega}} \times \boldsymbol{\rho})]\, dm$$
$$+ \int_m \{\boldsymbol{\rho} \times [\boldsymbol{\omega} \times (\boldsymbol{\omega} \times \boldsymbol{\rho})]\}\, dm \qquad (h)$$

Substituting Eqs. e, f, and g into Eq. h yields the following general expressions for the three components of the moment at point A:

$$M_{Ax} = -a_{Ay} \int_m z\, dm - \alpha \int_m zx\, dm + \omega^2 \int_m yz\, dm$$

$$M_{Ay} = a_{Ax} \int_m z\, dm - \alpha \int_m yz\, dm - \omega^2 \int_m zx\, dm \qquad (16\text{-}1)$$

$$M_{Az} = a_{Ay} \int_m x\, dm - a_{Ax} \int_m y\, dm + \alpha \int_m (x^2 + y^2)\, dm$$

Integrals of the form $\int_m x\, dm$ are first moment expressions that are normally studied in detail in most statics courses. Integrals of the form $\int_m x^2\, dm$ and $\int_m xy\, dm$ are similar to the expressions previously encountered in statics for second moments of area and mixed second moments of area. The integrals in Eqs. 16-1 represent the inertia properties of the rigid body and are known as moments of inertia and products of inertia, respectively. A complete discussion of moments and products of inertia is presented in Appendix A of the book, together with solved examples and an extensive selection of homework problems. A brief review of moments and products of inertia is provided in the following section for those students who have covered the topic in some detail in a previous statics course. The following section of the book can be omitted by those students who have a firm understanding of the subject.

The first moments, moment of inertia, and products of inertia appearing in Eqs. 16-1 are

$$\int_m x\, dm = \bar{x}\, m \qquad\qquad \int_m zx\, dm = I_{Azx}$$

$$\int_m y\, dm = \bar{y}\, m \qquad\qquad \int_m yz\, dm = I_{Ayz}$$

$$\int_m z\, dm = \bar{z}\, m = 0 \qquad \int_m (x^2 + y^2)\, dm = I_{Az} \qquad (16\text{-}2)$$

Equations 16-1 written in terms of the first moments and the moments and products of inertia given in Eqs. 16-2 become

$$M_{Ax} = -\alpha\, I_{Azx} + \omega\, I_{Ayz}$$
$$M_{Ay} = -\alpha\, I_{Ayz} - \omega^2\, I_{Azx} \qquad (16\text{-}3)$$
$$M_{Az} = a_{Ay}\, \bar{x}m - a_{Ax}\, \bar{y}m + \alpha\, I_{Az}$$

This set of equations relates the moments of the external forces acting on the rigid body to the angular velocities and inertia properties of the body. The moments of the forces and the moments and products of inertia are with respect to x-, y-, and z-axes through point A that are fixed in the body. If the x-, y-, and z-axes are not fixed in the body, the moments and products of inertia will be functions of time. Equations 16-2 clearly show the dependence of the moment about a given axis on the angular velocity ω about the z-axis. Alternatively, the equations show that moments M_{Ax} and M_{Ay} may be required to maintain planar motion about the z-axis.

For most dynamics problems involving planar motion, a considerable simplification of Eqs. 16-3 is possible. If the body is symmetric about the plane of motion, the product of inertia terms vanish ($I_{Ayz} = I_{Azx} = 0$) and Eqs. 16-3 become

$$M_{Ax} = 0$$
$$M_{Ay} = 0 \qquad \qquad (16\text{-}4)$$
$$M_{Az} = a_{Ay}\,\bar{x}m - a_{Ax}\,\bar{y}m + \alpha\,I_{Az}$$

Finally, by selecting the origin of the xyz-coordinate system at the center of mass G of the body $\bar{x} = \bar{y} = 0$ and Eqs. 16-4 reduce to

$$M_{Gx} = 0$$
$$M_{Gy} = 0 \qquad \qquad (16\text{-}5)$$
$$M_{Gz} = I_{Gz}\,\alpha$$

Equations 16-3 through 16-5 together with Eqs. 15-17 provide the relationships required to solve a wide variety of plane motion problems. An extensive discussion of plane motion is presented in Section 16-4, following the discussion of moments and products of inertia in the next section.

16-3 MOMENTS AND PRODUCTS OF INERTIA

In the previous analysis of the motion of a rigid body, integrals were encountered that involve the product of the mass of a small element of the body and the square of its distance from a line of interest. This product is called the second moment of the mass of the element or, more frequently, the moment of inertia of the element.

16-3-1 Moment of Inertia

The moment of inertia dI of the element of mass dm shown in Fig. 16-2 about the axis OO is defined as

$$dI = r^2\,dm$$

The moment of inertia of the entire body about axis OO is defined as

$$I = \int_m r^2\,dm \qquad \qquad (16\text{-}6)$$

Since both the mass of the element and the distance squared from the axis to the element are always positive, the moment of inertia of a mass is always a positive quantity.

Moments of inertia have the dimensions of mass multiplied by length squared, ML^2. Common units for the measurement of moment

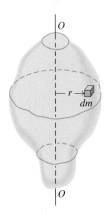

Fig. 16-2

of inertia in the SI system are $kg \cdot m^2$. In the U. S. customary system, force, length, and time are selected as the fundamental quantities, and mass has the dimensions FT^2L^{-1}. Therefore, moment of inertia has the units $lb \cdot s^2 \cdot ft$. If the mass of the body W/g is expressed in slugs ($lb \cdot s^2/ft$), the units of measurement of moment of inertia in the U. S. customary system are $slug \cdot ft^2$.

The moments of inertia of a body with respect to an xyz-coordinate system can be determined by considering an element of mass, as shown in Fig. 16-3. From the definition of moment of inertia,

$$dI_x = r_x^2 \, dm = (y^2 + z^2) \, dm$$

Similar expressions can be written for the y- and z-axes. Thus,

$$I_x = \int_m r_x^2 \, dm = \int_m (y^2 + z^2) \, dm$$

$$I_y = \int_m r_y^2 \, dm = \int_m (z^2 + x^2) \, dm \tag{16-7}$$

$$I_z = \int_m r_z^2 \, dm = \int_m (x^2 + y^2) \, dm$$

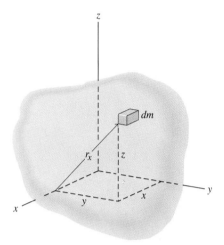

Fig. 16-3

When integration methods are used to determine the moment of inertia of a body with respect to an axis, the mass of the body can be divided into elements in various ways. Depending on the way the element is chosen, single, double, or triple integration may be required. The geometry of the body usually determines whether Cartesian or polar coordinates are used.

In some instances, a body can be regarded as a system of particles. The moment of inertia of a system of particles with respect to a line of interest is the sum of the moments of inertia of the particles with respect to the given line. Thus, if the masses of the particles of a system are denoted by $m_1, m_2, m_3, \dots, m_n$, and the distances of the particles from a given line are denoted by $r_1, r_2, r_3, \cdots, r_n$, the moment of inertia of the system can be expressed as

$$I = \Sigma m r^2 = m_1 r_1^2 + m_2 r_2^2 + m_3 r_3^2 + \cdots + m_n r_n^2$$

Moments of inertia for thin plates are relatively easy to determine. For example, consider the thin plate shown in Fig. 16-4. The plate has a uniform density ρ, a uniform thickness t, and a cross-sectional area A. The moments of inertia about x-, y-, and z-axes are by definition

$$I_{xm} = \int_m y^2 \, dm = \int_V y^2 \rho \, dV = \int_A y^2 \rho t \, dA = \rho t \int_A y^2 \, dA = \rho t \, I_{xA}$$

$$I_{ym} = \int_m x^2 \, dm = \int_V x^2 \rho \, dV = \int_A x^2 \rho t \, dA$$

$$= \rho t \int_A x^2 \, dA = \rho t \, I_{yA} \tag{16-8}$$

$$I_{zm} = \int_m (x^2 + y^2) \, dm = \rho t \, I_{yA} + \rho t \, I_{xA} = \rho t \, (I_{yA} + I_{xA})$$

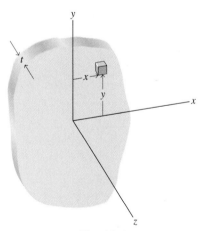

Fig. 16-4

where the subscripts m and A denote moments of inertia and second moments of area, respectively. Since the equations for the moments of inertia of thin plates contain the expressions for the second moments of area, the results listed in Appendix B (Table B-3) for second moments of areas can be used for moments of inertia by simply multiplying the results listed in the table by ρt.

Moments of inertia with respect to x-, y-, and z-axes for the general three-dimensional body can be determined by using Eqs. 16-7. If the density of the body is uniform, the element of mass dm can be expressed in terms of the element of volume dV of the body as $dm = \rho \, dV$. Equations 16-7 then become

$$I_x = \rho \int_V (y^2 + z^2) \, dV$$

$$I_y = \rho \int_V (z^2 + x^2) \, dV \qquad (16\text{-}9)$$

$$I_z = \rho \int_V (x^2 + y^2) \, dV$$

If the density of the body is not uniform, it must be expressed as a function of position and retained within the integral sign.

Frequently in engineering practice, a body of interest can be broken up into a number of simple shapes, such as cylinders, spheres, plates, and rods, for which the moments of inertia have been evaluated and tabulated. The moment of inertia of the composite body, with respect to any axis, is equal to the sum of the moments of inertia of the separate parts of the body with respect to the specified axis. For example,

$$I_x = \int_m (y^2 + z^2) \, dm$$

$$= \int_{m_1} (y^2 + z^2) \, dm_1 + \int_{m_2} (y^2 + z^2) \, dm_2 + \cdots + \int_{m_n} (y^2 + z^2) \, dm_n$$

$$= I_{x1} + I_{x2} + I_{x3} + \cdots + I_{xn}$$

When one of the component parts is a hole, its moment of inertia must be subtracted from the moment of inertia of the larger part to obtain the moment of inertia for the composite body. Appendix B (Table B-5) contains a listing of the moments of inertia for some frequently encountered shapes such as rods, plates, cylinders, spheres, and cones.

16-3-2 Radius of Gyration

The definition of moment of inertia (Eq. 16-6) indicates that the dimensions of moment of inertia are mass multiplied by a length squared. As a result, the moment of inertia of a body can be expressed as the product of the mass m of the body and a length k squared. This length k is defined as the radius of gyration of the body. Thus, the moment of inertia I of a body with respect to a given line can be expressed as

$$I = mk^2 \qquad \text{or} \qquad k = \sqrt{\frac{I}{m}} \qquad (16\text{-}10)$$

The radius of gyration of the mass of a body with respect to any axis can be viewed as the distance from the axis to the point where the total mass must be concentrated to produce the same moment of inertia with respect to the axis as does the actual (or distributed) mass.

The radius of gyration for masses is very similar to the radius of gyration for areas discussed in Section 10-2-3. The radius of gyration for masses is not the distance from the given axis to any fixed point in the body such as the mass center. The radius of gyration of the mass of a body with respect to any axis is always greater than the distance from the axis to the mass center of the body. There is no useful physical interpretation for a radius of gyration; it is merely a convenient means of expressing the moment of inertia of the mass of a body in terms of its mass and a length.

16-3-3   Parallel-Axis Theorem for Moments of Inertia

The parallel-axis theorem for moments of inertia is very similar to the parallel-axis theorem for second moments of area discussed in Section 10-2-1. Consider the body shown in Fig. 16-5, which has an xyz-coordinate system with its origin at the mass center G of the body and a parallel $x'y'z'$-coordinate system with its origin at point O'. Observe in the figure that

$$x' = \bar{x} + x$$
$$y' = \bar{y} + y$$
$$z' = \bar{z} + z$$

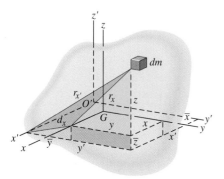

Fig. 16-5

The distance d_x between the x'- and x-axes is

$$d_x = \sqrt{\bar{y}^2 + \bar{z}^2}$$

The moment of inertia of the body about an x'-axis that is parallel to the x-axis through the mass center is by definition

$$I_{x'} = \int_m r_{x'}^2 \, dm = \int_m [(\bar{y} + y)^2 + (\bar{z} + z)^2] \, dm$$
$$= \int_m (y^2 + z^2) \, dm + \bar{y}^2 \int_m dm + 2\bar{y} \int_m y \, dm + \bar{z}^2 \int_m dm + 2\bar{z} \int_m z \, dm$$

However,

$$\int_m (y^2 + z^2) \, dm = I_{xG}$$

and, since the x- and y-axes pass through the mass center G of the body,

$$\int_m y \, dm = 0 \qquad \int_m z \, dm = 0$$

Therefore,

$$I_{x'} = I_{xG} + (\bar{y}^2 + \bar{z}^2)m = I_{xG} + d_x^2 m$$
$$I_{y'} = I_{yG} + (\bar{z}^2 + \bar{x}^2)m = I_{yG} + d_y^2 m \qquad (16\text{-}11)$$
$$I_{z'} = I_{zG} + (\bar{x}^2 + \bar{y}^2)m = I_{zG} + d_z^2 m$$

Equation 16-11 is the parallel-axis theorem for moments of inertia. The subscript G indicates that the x-axis passes through the mass center G of the body. Thus, if the moment of inertia of a body with respect to an axis passing through its mass center is known, the moment of inertia of the body with respect to any parallel axis can be found, without integrating, by use of Eqs. 16-11.

A similar relationship exists between the radii of gyration for the two axes. Thus, if the radii of gyration for the two parallel axes are denoted by k_x and $k_{x'}$, the above equation may be written

$$k_{x'}^2 m = k_{xG}^2 m + d_x^2 m$$

Hence

$$k_{x'}^2 = k_{xG}^2 + d_x^2$$
$$k_{y'}^2 = k_{yG}^2 + d_y^2 \qquad (16\text{-}12)$$
$$k_{z'}^2 = k_{zG}^2 + d_z^2$$

Note: Equations 16-11 and 16-12 are valid only for transfers to or from *xyz*-axes passing through the mass center of the body. **They are not valid for transfers between two arbitrary axes.**

16-3-4 Product of Inertia

In analyses of the motion of rigid bodies, expressions are sometimes encountered that involve the product of the mass of a small element and the coordinate distances from a pair of orthogonal coordinate planes. This product, which is similar to the mixed second moment of an area, is called the product of inertia of the element. For example, the product of inertia of the element shown in Fig. 16-6 with respect to the *xz*- and *yz*-planes is by definition

$$dI_{xy} = xy \, dm \qquad (16\text{-}13)$$

The sum of the products of inertia of all elements of mass of the body with respect to the same orthogonal planes is defined as the product of inertia of the body. The three products of inertia for the body shown in Fig. 16-6 are

$$I_{xy} = \int_m xy \, dm$$

$$I_{yz} = \int_m yz \, dm \qquad (16\text{-}14)$$

$$I_{zx} = \int_m zx \, dm$$

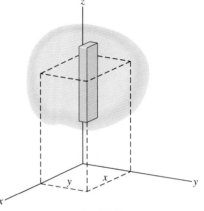

Fig. 16-6

Products of inertia, like moments of inertia, have the dimensions of mass multiplied by a length squared ML^2. Common units for the measurement of product of inertia in the SI system are $kg \cdot m^2$. In the U. S. customary system, common units are $slug \cdot ft^2$.

The product of inertia of a body can be positive, negative, or zero, since the two coordinate distances have independent signs. The product of inertia will be positive for coordinates with the same sign and negative for coordinates with opposite signs. The product of inertia will be zero if either of the planes is a plane of symmetry since pairs of elements on opposite sides of the plane of symmetry will have positive and negative products of inertia that will add to zero in the summation process.

The integration methods used to determine moments of inertia apply equally well to products of inertia. Depending on the way the element is chosen, single, double, or triple integration may be required. Moments of inertia for thin plates were related to second moments of area for the same plate. Likewise, products of inertia can be related to the mixed second moments for the plates. If the plate has a uniform density ρ, a uniform thickness t, and a cross-sectional area A, the products of inertia are by definition:

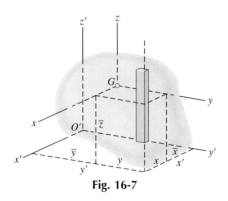

Fig. 16-7

$$I_{xym} = \int_m xy\, dm = \int_v xy\, \rho\, dV = \int_A xy\, \rho t\, dA = \rho t \int_A xy\, dA = \rho t\, I_{xyA}$$

$$I_{yzm} = \int_m yz\, dm = 0 \qquad (16\text{-}15)$$

$$I_{zxm} = \int_m zx\, dm = 0$$

where the subscripts m and A denote products of inertia of mass and mixed second moments of area, respectively. The products of inertia I_{yzm} and I_{zxm} for a thin plate are zero since the x- and y-axes are assumed to lie in the midplane of the plate (plane of symmetry).

A parallel-axis theorem for products of inertia can be developed that is very similar to the parallel-axis theorem for mixed second moments of area discussed in Section 10-2-5. Consider the body shown in Fig. 16-7, which has an xyz-coordinate system with its origin at the mass center G of the body and a parallel $x'y'z'$-coordinate system with its origin at point O' of the body. Observe in the figure that

$$x' = \bar{x} + x$$
$$y' = \bar{y} + y$$
$$z' = \bar{z} + z$$

The product of inertia $I_{x'y'}$ of the body with respect to the $x'z'$- and $y'z'$-planes is by definition

$$I_{x'y'} = \int_m x'y'\, dm = \int_m (\bar{x} + x)(\bar{y} + y)\, dm$$

$$= \int_m \bar{x}\bar{y}\, dm + \int_m \bar{x}y\, dm + \int_m \bar{y}x\, dm + \int_m xy\, dm$$

Since $\bar{x}$ and $\bar{y}$ are the same for every element of mass dm,

$$I_{x'y'} = \bar{x}\bar{y} \int_m dm + \bar{x} \int_m y\, dm + \bar{y} \int_m x\, dm + \int_m xy\, dm$$

However,

$$\int_m xy\, dm = I_{xy}$$

and, since the x- and y-axes pass through the mass center G of the body,

$$\int_m y\, dm = 0 \qquad \int_m x\, dm = 0$$

Therefore,

$$I_{x'y'} = I_{xyG} + \bar{x}\bar{y}\, m$$
$$I_{y'z'} = I_{yzG} + \bar{y}\bar{z}\, m \qquad (16\text{-}16)$$
$$I_{z'x'} = I_{zxG} + \bar{z}\bar{x}\, m$$

Equations 16-16 are the parallel-axis theorem for products of inertia. The subscript G indicates that the x- and y-axes pass through the mass center G of the body. Thus, if the product of inertia of a body with respect to a pair of orthogonal planes that pass through its mass center is known, the product of inertia of the body with respect to any other pair of parallel planes can be found, without integrating, by use of Eqs. 16-16.

16-3-5   Principal Moments of Inertia

In some instances, in the dynamic analysis of bodies, principal axes and maximum and minimum moments of inertia, which are similar to maximum and minimum second moments of an area, must be determined. Again, the problem is one of transforming known or easily calculated moments and products of inertia with respect to one coordinate system (such as an *xyz*-coordinate system along the edges of a rectangular prism) to a second *x'y'z'*-coordinate system through the same origin *O* but inclined with respect to the *xyz* system. A complete development of the equations required for determining principal moments of inertia is presented in Appendix A. Also included in Appendix A is a selection of solved example problems and a number of homework problems involving moment of inertia, radius of gyration, product of inertia, and principal moment of inertia determinations.

16-4   TRANSLATION, ROTATION, AND GENERAL PLANE MOTION OF RIGID BODIES

Plane motion problems can be classified in three distinct categories, which depend on the nature of the motion: (1) translation, (2) fixed-axis rotation, and (3) general plane motion. The equations for general plane motion were developed in Section 16-2. Translation and fixed-axis rotation are special cases of general plane motion.

For a body of arbitrary shape, the equations for general plane motion developed in Section 16-2 are given by Eqs. 15-17 and 16-3 as

$$\Sigma F_x = ma_{Gx} \qquad \Sigma M_{Ax} = -\alpha I_{Azx} + \omega^2 I_{Ayz}$$
$$\Sigma F_y = ma_{Gy} \qquad \Sigma M_{Ay} = -\alpha I_{Ayz} - \omega^2 I_{Azx} \qquad (16\text{-}17)$$
$$\Sigma F_z = 0 \qquad \Sigma M_{Az} = a_{Ay}\bar{x}m - a_{Ax}\bar{y}m + \alpha I_{Az}$$

16-4-1   Translation

The motion of a rigid body is defined as translation when every straight line in the body remains parallel to its initial position during the motion. During translation, there is no angular motion ($\omega = \alpha = 0$); therefore, all parts of the body experience the same linear acceleration **a**. Translation can occur only when the resultant of the external forces acting on the body is a force **R** whose line of action passes through the mass center *G* of the body. For translation, with the origin of the *xyz*-coordinate system at the mass center *G* of the body ($\bar{x} = \bar{y} = 0$), Eqs. 16-17 for general plane motion reduce to

$$\Sigma F_x = ma_{Gx} \qquad \Sigma F_y = ma_{Gy} \qquad \Sigma M_{Gz} = 0 \qquad (16\text{-}18)$$

When a body is experiencing translation of the type illustrated in Fig. 16-8, the *x*-axis can be selected parallel to the acceleration **a**$_G$, in which case the acceleration component a_{Gy} is zero. When the mass center of the body is traveling on a plane curve, as illustrated in Fig. 16-9, it is often convenient to select the *x*- and *y*-axes in the directions of the instantaneous normal and tangential components of the acceleration. If moments of the external forces are summed about a point other than the mass center (say point *A*), the moment equation must be modified to account for the effects of a_{Gx} and a_{Gy}. Thus

$$\Sigma M_{Az} = a_{Gy}\bar{x}m - a_{Gx}\bar{y}m \qquad (16\text{-}19)$$

The procedure for solving motion problems involving translation is illustrated in the following examples.

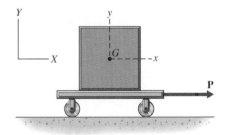

Fig. 16-8

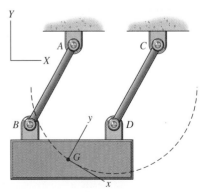

Fig. 16-9

A 16 × 20-ft hangar door, which weighs 800 lb, is supported by two rollers as shown in Fig. 16-10a. A force **F** of 300 lb is applied to open the door. Determine the acceleration of the door and the support forces exerted on the door by the two rollers. Neglect frictional forces and the mass of the rollers.

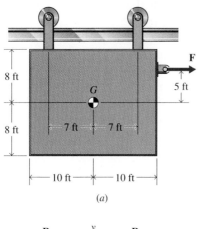

(a)

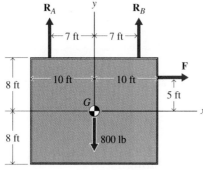

(b)

Fig. 16-10

SOLUTION

A free-body diagram for the door is shown in Fig. 16-10b, with the origin of an xyz-coordinate system placed at the mass center G. Since motion of the mass center of the door will be along a horizontal straight line, the type of motion is translation ($\omega = \alpha = \mathbf{a}_{Gy} = 0$) as long as the two rollers remain in contact with the rail. The equations of motion (Eqs. 16-18) are

$$+\rightarrow \Sigma F_x = ma_{Gx} \qquad 300 = \frac{800}{32.2}a_{Gx} \qquad a_{Gx} = 12.08 \text{ ft/s}^2 \qquad \text{Ans.}$$

$$+\uparrow \Sigma F_y = 0 \qquad R_A + R_B - 800 = 0 \qquad (a)$$

$$+\downarrow \Sigma M_{Gz} = 0 \qquad R_B(7) - R_A(7) - 300(5) = 0 \qquad (b)$$

Solving Eqs. a and b simultaneously yields

$$R_A = 293 \text{ lb} \qquad R_B = 507 \text{ lb} \qquad \text{Ans.}$$

The 1400 kg automobile shown in Fig. 16-11a has a wheel base of 3 m. Its center of mass is located 1.30 m behind the front axle and 0.5 m above the surface of the pavement. If the automobile has a rear-wheel drive and the coefficient of friction between the tires and the pavement is 0.80, determine the maximum acceleration that the vehicle can develop while traveling up the 15° incline.

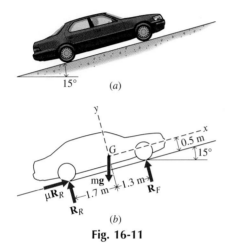

(a)

(b)

Fig. 16-11

SOLUTION

A free-body diagram of the automobile is shown in Fig. 16-11b, with the origin of an xyz-coordinate system placed at the mass center G. Since the vehicle has a rear-wheel drive, a frictional driving force is shown only for the rear wheels. Motion of the mass center of the automobile will be along a straight line inclined 15° with respect to the horizontal; therefore, the motion is translation ($\omega = \alpha = a_{Gy} = 0$) as long as the wheels remain in contact with the pavement. The equations of motion (Eqs. 16-18) are

$$+\nwarrow \Sigma F_y = 0 \qquad R_F + R_R - mg \cos 15° = 0$$
$$R_F + R_R = 1400(9.81) \cos 15° = 13{,}266 \text{ N} \qquad (a)$$
$$+\downarrow \Sigma M_{Gz} = 0 \qquad R_F(1.3) - R_R(1.7) + 0.80R_R(0.5) = 0$$
$$R_F - R_R = 0 \qquad (b)$$

Solving Eqs. a and b simultaneously yields

$$R_F = R_R = 6633 \text{ N}$$
$$+\nearrow \Sigma F_x = ma_{Gx} \qquad \mu R_R - mg \sin 15° = ma_{Gx}$$
$$0.80(6633) - 1400(9.81) \sin 15° = 1400a_{Gx}$$
$$a_{Gx} = 1.251 \text{ m/s}^2 \qquad \text{Ans.}$$

A 90-lb triangular plate is supported by two cables as shown in Fig. 16-12a. When the plate is in the position shown, the angular velocity of the cables is 4 rad/s counterclockwise. At this instant, determine

a. The acceleration of the mass center of the plate.
b. The tension in each of the cables.

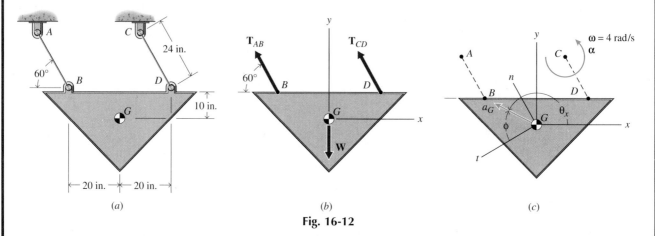

Fig. 16-12

SOLUTION

A free-body diagram of the plate is shown in Fig. 16-12b, with the origin of an xyz-coordinate system placed at the mass center G. Since points B and D of the plate move along parallel circular paths, the motion of the plate is curvilinear translation. The equations of motion (Eqs. 16-18) are

$$\Sigma F_x = ma_{Gx} \qquad \Sigma F_y = ma_{Gy} \qquad \Sigma M_{Gz} = 0$$

or in terms of normal and tangential coordinates (see Fig. 16-12c)

$$\Sigma F_n = ma_{Gn} \qquad \Sigma F_t = ma_{Gt} \qquad \Sigma M_{Gz} = 0$$

a. The acceleration of the mass center of the plate is obtained from the normal and tangential components ($\mathbf{v}_B = \mathbf{v}_D = \mathbf{v}_G$ and $\mathbf{a}_B = \mathbf{a}_D = \mathbf{a}_G$). Thus

$$a_{Gn} = r\omega^2 = \frac{24}{12}(4)^2 = 32.0 \text{ ft/s}^2$$

$$+\swarrow\Sigma F_t = ma_{Gt} \qquad W\sin 30° = \frac{W}{g}a_{Gt}$$

$$a_{Gt} = g\sin 30° = 32.2\sin 30° = 16.1 \text{ ft/s}^2$$

Therefore

$$a_G = \sqrt{(a_{Gn})^2 + (a_{Gt})^2} = \sqrt{(32.0)^2 + (16.1)^2} = 35.8 \text{ ft/s}^2 \qquad \text{Ans.}$$

$$\phi = \tan^{-1}\frac{a_{Gn}}{a_{Gt}} = \tan^{-1}\frac{32.0}{16.1} = 63.3° \qquad \text{Ans.}$$

b. The tensions in the cables are obtained from the equations $\Sigma F_n = ma_{Gn}$ and $\Sigma M_{Gz} = 0$. Thus,

$$+\nwarrow\Sigma F_n = ma_{Gn} \qquad T_{AB} + T_{CD} - W\cos 30° = \frac{W}{g}a_{Gn}$$

$$T_{AB} + T_{CD} = 90\cos 30° + \frac{90}{32.2}(32)$$

$$T_{AB} + T_{CD} = 167.38 \qquad (a)$$

$+\mathcal{L}\ \Sigma M_{Gz} = 0$ $T_{AB} \sin 30° (10) - T_{AB} \cos 30° (20)$
$+ T_{CD} \sin 30° (10) + T_{CD} \cos 30° (20) = 0$
$T_{AB} - 1.8117T_{CD} = 0$ (b)

From Eqs. a and b

$$T_{AB} = 107.9 \text{ lb} \qquad \text{Ans.}$$
$$T_{CD} = 59.5 \text{ lb} \qquad \text{Ans.}$$

Solution using rectangular coordinates and vector analysis

The resultant force **R** can be written in Cartesian vector form as

$$\mathbf{R} = (-\cos 60° \, T_{AB} - \cos 60° \, T_{CD})\mathbf{i} + (\sin 60° \, T_{AB} + \sin 60° \, T_{CD} - 90)\mathbf{j}$$
$$= (-0.5T_{AB} - 0.5T_{CD})\mathbf{i} + (0.866T_{AB} + 0.866T_{CD} - 90)\mathbf{j}$$

In a similar manner, the acceleration $\mathbf{a}_G$ can be written in Cartesian vector form (in terms of the angular velocity ω and the angular acceleration α) as

$$\mathbf{a}_G = (-\cos 60° \, r\omega^2 + \sin 60° \, r\alpha)\mathbf{i} + (\sin 60° \, r\omega^2 + \cos 60° \, r\alpha)\mathbf{j}$$
$$= (-16 + 1.732\alpha)\mathbf{i} + (27.71 + \alpha)\mathbf{j}$$

From the equation $\mathbf{R} = m\mathbf{a}_G$.
From the **i** terms

$$T_{AB} + T_{CD} = 89.44 - 9.682\alpha$$

From the **j** terms

$$T_{AB} + T_{CD} = 193.36 + 3.228\alpha$$

which requires that

$$\alpha = -8.049 \text{ rad/s}^2$$

Thus

$$\mathbf{a}_G = -29.94\mathbf{i} + 19.66\mathbf{j}$$

and

$$a_G = \sqrt{(a_{Gx})^2 + (a_{Gy})^2} = \sqrt{(29.94)^2 + (19.66)^2} = 35.8 \text{ ft/s}^2 \qquad \text{Ans.}$$
$$\theta_x = \tan^{-1} \frac{a_{Gy}}{a_{Gx}} = \tan^{-1} \frac{19.66}{-29.94} = 146.7° \qquad \text{Ans.}$$

Also

$$T_{AB} + T_{CD} = 167.37 \qquad (c)$$

From the equation $\mathbf{C}_G = \mathbf{0}$

$$\mathbf{C}_G = (-1.6667\mathbf{i} + 0.8333\mathbf{j}) \times (-0.5T_{AB}\mathbf{i} + 0.866T_{AB}\mathbf{j})$$
$$+ (1.6667\mathbf{i} + 0.8333\mathbf{j}) \times (-0.5T_{CD}\mathbf{i} + 0.866T_{CD}\mathbf{j}) = \mathbf{0}$$

which yields

$$(-1.0267T_{AB} + 1.8601T_{CD})\mathbf{k} = \mathbf{0}$$

or

$$T_{AB} - 1.8117T_{CD} = 0 \qquad (d)$$

From Eqs. c and d

$$T_{AB} = 107.9 \text{ lb} \qquad \text{Ans.}$$
$$T_{CD} = 59.5 \text{ lb} \qquad \text{Ans.}$$

PROBLEMS

In the following problems, all ropes, cords, and cables are assumed to be flexible, inextensible, and of negligible mass. All pins and pulleys have negligible mass and are frictionless unless specified otherwise.

16-1* The block shown in Fig. P16-1 weighs 900 lb. The kinetic coefficient of friction μ between the block and the horizontal surface is 0.20. Determine the acceleration of the block and the reactions at contact points A and B when a force **P** of 250 lb is applied to the block.

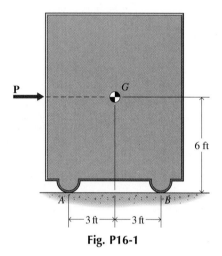

Fig. P16-1

16-2* The block shown in Fig. P16-2 has a mass of 350 kg. The kinetic coefficient of friction μ between the block and the horizontal surface is 0.15. Determine the acceleration of the block and the reactions at contact points A and B when a force **P** of 750 N is applied to the block.

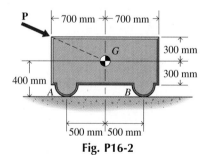

Fig. P16-2

16-3 A crate weighing 600 lb slides down an inclined surface as shown in Fig. P16-3. If the coefficient of kinetic friction between the crate and the inclined surface is 0.30, determine the normal and frictional forces exerted on the crate at contact surfaces A and B.

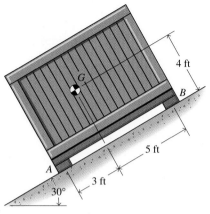

Fig. P16-3

16-4 A cabinet with a mass of 75 kg is moved across a horizontal surface as shown in Fig. P16-4. Determine the maximum force **P** that can be applied without tipping the cabinet.

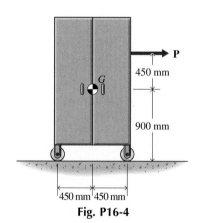

Fig. P16-4

16-5* The rear-wheel drive automobile shown in Fig. P16-5 has a weight of 3100 lb. The static coefficient of friction μ between the tires and the pavement is 0.70. Determine the minimum time required for the automobile to accelerate uniformly from rest to a speed of 60 mi/h.

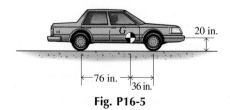

Fig. P16-5

16-6* The small rear-wheel drive truck shown in Fig. P16-6 has a mass of 1750 kg and is carrying a 400-kg load. The center of mass of the truck is 0.75 m behind the front axle; the center of mass of the crate is 0.5 m in front of the rear axle. If the static coefficient of friction between the pavement and the tires is 0.85 and the crate is securely tied down, determine the minimum time required for the truck:

a. To accelerate uniformly from rest to 90 km/h.
b. To decelerate uniformly from 90 km/h to rest.

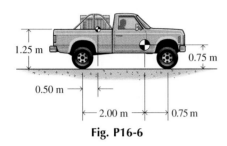

Fig. P16-6

16-7 Solve Problem 16-5 if the automobile has a front-wheel drive.

16-8 Solve Problem 16-6 if the truck has a front-wheel drive.

16-9* A crate rests on a small cart as shown in Fig. P16-9. The weights of the crate and cart are 850 lb and 100 lb, respectively. The static coefficient of friction between the crate and the cart is 0.25. Determine the support reactions at wheels A and B when a force $\mathbf{P}$ of 150 lb is applied to the cart.

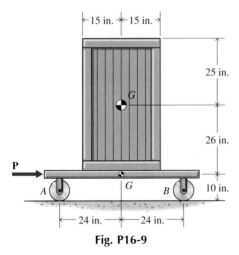

Fig. P16-9

16-10* A slab of material ($m = 1000$ kg) is supported on a rail by two shoes, as shown in Fig. P16-10. The coefficient of kinetic friction between the rail and the shoes is 0.25. Find the vertical reactions of the rail on the shoes when a force $\mathbf{F}$ of 2.50 kN is applied to the slab.

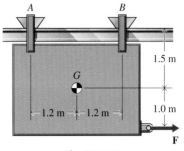

Fig. P16-10

16-11 When the two blocks A and B, shown in Fig. P16-11, are released from rest, block A slides up the inclined plane without tipping. Block A weighs 500 lb and the coefficient of friction μ between block A and the inclined plane is 0.2. Determine the maximum permissible weight for block B.

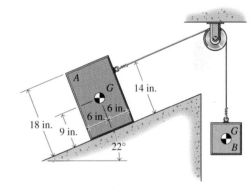

Fig. P16-11

16-12 A crate rests on a small cart as shown in Fig. P16-12. The masses of the crate and cart are 150 kg and 25 kg, respectively. The static coefficient of friction between the crate and the cart is 0.10. If the crate does not slip or tip, determine the maximum permissible mass for block B.

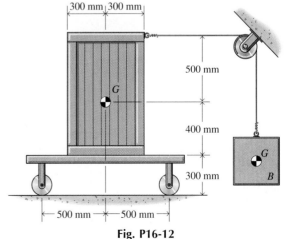

Fig. P16-12

16-13* A slab of material weighing 500 lb is supported on an inclined rail by shoes A and B as shown in Fig. P16-13. The kinetic coefficient of friction between the rail and shoes is 0.20. Determine the normal and frictional forces exerted by the rail on the shoes while the slab is moving along the rail.

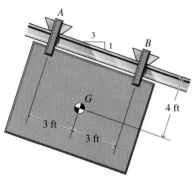

Fig. P16-13

16-14* A crate with a mass of 1000 kg rests on the back of a rear-wheel drive flatbed truck as shown in Fig. P16-14. The mass of the truck is 2500 kg. The center of mass of the truck is located 2 m behind the front axle and 0.85 m above the pavement. The static coefficient of friction between the crate and the bed is 0.25. If the crate does not slip or tip, determine the maximum permissible acceleration and the minimum static coefficient of friction between the tires and the pavement required to achieve this acceleration.

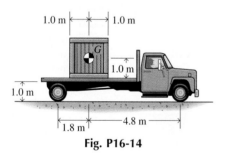

Fig. P16-14

16-15 A homogeneous plate, which weighs 100 lb, is suspended by two cables A and B of equal length as shown in Fig. P16-15. The plate swings in a vertical plane and is subjected to a horizontal force **F** of 20 lb. In the position shown, the cables are rotating counterclockwise with an angular velocity of 5 rad/s. At this instant, determine the angular acceleration of the cables and the tensions in the cables.

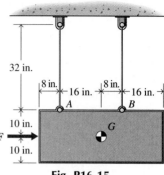

Fig. P16-15

16-16 The thin plate shown in Fig. P16-16 has a mass of 10 kg. The plate is supported in a vertical plane by two links A and B and a flexible cord C. Determine the acceleration of the mass center G of the plate and the force in each link immediately after the cord C is cut. Neglect the mass of the links.

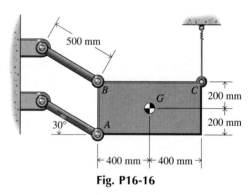

Fig. P16-16

16-17* Rod AB, shown in Fig. P16-17, has a uniform cross section and weighs 75 lb. The rod connects two wheels, which roll without slipping on a horizontal surface. The wheels rotate counterclockwise with a constant angular velocity of 25 rad/s. Determine the vertical component of the force exerted on the bar by the pin at B as a function of angular position θ.

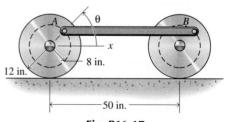

Fig. P16-17

16-18* A frame with a mass of 25 kg is supported in a vertical plane by two pairs of links and a cable as shown in Fig. P16-18. A block with a mass of 10 kg rests on the frame. If the cable breaks, determine the force carried by each pair of links and the force exerted on the block by the frame after the links have rotated 30° from their initial horizontal position. Neglect the masses of the links.

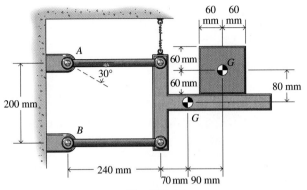

Fig. P16-18

16-19 Bar *AB* of Fig. P16-19 has a uniform cross section and weighs 60 lb. It is attached to the cart with a smooth pin at *A* and rests against a smooth surface at *B*. A force **P** is applied to the cart, which produces an acceleration of 15 ft/s² to the right. Determine

a. The forces exerted on the bar at supports *A* and *B*.
b. The magnitude of the force **P** if the cart weighs 50 lb.
c. The magnitude of the force **P** required to make the reaction zero at support *B* of the bar.

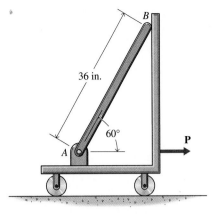

Fig. P16-19

16-20 The 100-mm diameter rollers of the conveyor system shown in Fig. P16-20 are driven at a constant angular velocity of 25 rad/s. The block being transported by the system has a mass of 250 kg and is moving to the right with a velocity of 1.0 m/s. The coefficient of friction between the block and the rollers is 0.25. At the instant shown, determine the acceleration of the mass center of the block and the vertical components of the forces exerted on the block by the rollers at *A* and *B*.

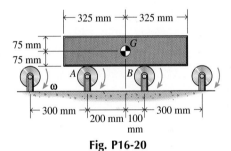

Fig. P16-20

16-4-2   Fixed-Axis Rotation

When all elements of a body travel in circular paths about a fixed axis, the motion is defined as fixed-axis rotation. A rigid body symmetric with respect to the plane of motion ($I_{Gzx} = I_{Gyz} = 0$) and rotating about a fixed axis through the mass center G of the body ($\bar{x} = \bar{y} = 0$) is illustrated in Fig. 16-13. For this case, $\mathbf{a}_G = \mathbf{0}$; therefore, the equations for general plane motion, Eqs. 16-17, reduce to

$$\Sigma F_x = ma_{Gx} = 0$$
$$\Sigma F_y = ma_{Gy} = 0 \qquad \Sigma M_{Gz} = I_{Gz}\alpha \qquad \text{(16-20)}$$

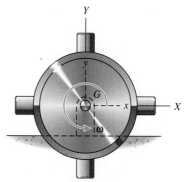

Fig. 16-13

Rotation about fixed axes other than those through the mass center G of the body are often encountered. An example of a body that is symmetric with respect to the plane of motion ($I_{Gzx} = I_{Gyz} = 0$) is illustrated in Fig. 16-14. For this type of rotation, $\mathbf{a}_A = \mathbf{0}$ and the equations for general plane motion, Eqs. 16-17, reduce to

$$\Sigma F_x = ma_{Gx} = -m\bar{x}\omega^2$$
$$\Sigma F_y = ma_{Gy} = m\bar{x}\alpha \qquad \Sigma M_{Az} = I_{Az}\alpha \qquad \text{(16-21)}$$

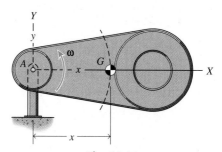

Fig. 16-14

The moment equation $\Sigma M_{Az} = I_{Az}\alpha$ of Eqs. 16-21 can be obtained from the moment equation $\Sigma M_{Gz} = I_{Gz}\alpha$ of Eqs. 16-20 by noting for fixed-axis rotation about an arbitrary point A that

$$\Sigma M_{Az} = \Sigma M_{Gz} + ma_{Gy}\bar{x} - ma_{Gx}\bar{y}$$

For the case illustrated in Fig. 16-14, $\bar{y} = 0$. Therefore,

$$\Sigma M_{Az} = \Sigma M_{Gz} + ma_{Gy}\bar{x} = I_{Gz}\alpha + m(\bar{x}\alpha)\bar{x} = (I_{Gz} + m\bar{x}^2)\alpha = I_{Az}\alpha$$

The procedure for solving motion problems involving fixed-axis rotation is illustrated in the following examples.

EXAMPLE PROBLEM 16-4

The rod AB shown in Fig. 16-15a has a constant cross section and a mass of 10 kg. As a result of the rotation of crank C, rod AB oscillates in a vertical plane. In the position shown, its angular velocity ω is 10 rad/s clockwise and its angular acceleration α is 40 rad/s² counterclockwise. Determine the force exerted by the link between the rod and the crank and the force exerted on rod AB by the pin at support A.

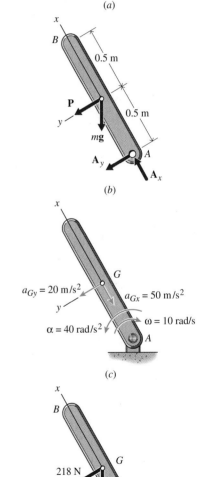

SOLUTION

A free-body diagram of rod AB is shown in Fig. 16-15b. Motion of the bar is rotation about a fixed axis not passing through the center of mass G of the rod. With the origin of an xyz-coordinate system placed at the mass center G of the rod, the equations of motion are

$$\Sigma F_x = ma_{Gx} \qquad \Sigma F_y = ma_{Gy} \qquad \Sigma M_{Gz} = I_{Gz}\alpha$$

The linear components of the acceleration of the mass center, as shown in Fig. 16-15c, are

$$a_{Gx} = -r\omega^2 = -0.5(-10)^2 = -50 \text{ m/s}^2$$
$$a_{Gy} = r\alpha = 0.5(40) = 20 \text{ m/s}^2$$

From Appendix B (Table B-5):

$$I_{Gz} = \frac{1}{12}mL^2 = \frac{1}{12}(10)(1)^2 = 0.8333 \text{ kg} \cdot \text{m}^2$$

$+\nwarrow\Sigma F_x = A_x - mg\sin 60° = ma_{Gx}$
$\qquad A_x - 10(9.81)\sin 60° = 10(-50)$ $\qquad\qquad A_x = -415.0$ N

$+\downarrow\Sigma M_{Gz} = I_{Gz}\alpha$
$\qquad -A_y(0.5) = 0.8333(40)$ $\qquad\qquad\qquad\qquad A_y = -66.66$ N

$+\swarrow\Sigma F_y = P + A_y + mg\cos 60° = ma_{Gy}$
$\qquad P - 66.66 + 10(9.81)\cos 60° = 10(20)$

$\qquad P = 217.6 = 218$ N $\qquad\qquad\qquad\qquad\qquad$ Ans.
$\qquad A = \sqrt{A_x^2 + A_y^2} = \sqrt{(-415)^2 + (-66.66)^2} = 420$ N $\qquad$ Ans.

$$\theta = \tan^{-1}\frac{A_y}{A_x} = \frac{-66.66}{-415.0} = 9.125 = 9.13°$$

These results are shown in Fig. 16-15d.

With the origin of a parallel xyz-coordinate system placed at the fixed axis of rotation, the equations of motion are

$$\Sigma F_x = -m\bar{x}\omega^2 \qquad \Sigma F_y = m\bar{x}\alpha \qquad \Sigma M_{Az} = I_{Az}\alpha$$
$$I_{Az} = I_{Gz} + m\bar{x}^2 = 0.8333 + 10(0.5)^2 = 3.333 \text{ kg} \cdot \text{m}^2$$

$+\downarrow\Sigma M_{Az} = I_{Az}\alpha$
$\qquad P(0.5) + 0.5(mg\cos 60°) = 3.333(40)$
$\qquad\qquad P = 2[3.333(40) - 0.5(10)(9.81)\cos 60°] = 218$ N Ans.

By summing moments about the fixed axis A, force $\mathbf{P}$ is obtained directly. Once $\mathbf{P}$ is known, the two remaining equations can be used to determine the reaction at support A.

Fig. 16-15

A container weighing 850 lb is moved by using a winch and cable as shown in Fig. 16-16a. The cylindrical drum of the winch weighs 100 lb and its radius of gyration with respect to the axis of rotation is 1.75 ft. The kinetic coefficient of friction between the container and the horizontal surface is 0.25. If the container slides on the horizontal surface but does not tip, determine

a. The maximum tension permitted in the cable.
b. The acceleration of the container when the maximum tension is applied.
c. The maximum couple **C** that can be applied to the winch.

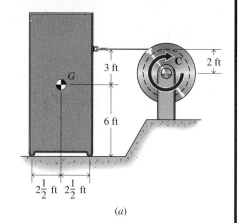

(a)

SOLUTION

Free-body diagrams for the container and winch are shown in Fig. 16-16b. When the container is on the verge of tipping, only the vertical support reactions R_{Ay} at the bottom right corners of the container will exist. The motion of the container is translation ($\omega = \alpha = \mathbf{a}_{Gy} = 0$); the motion of the winch drum is rotation ($\mathbf{a}_{Gx} = \mathbf{a}_{Gy} = 0$). The origin of an xyz-coordinate system has been placed at the mass center G of each body ($x = y = 0$).

a. The equations of motion for the container are

$$\Sigma F_x = ma_{Gx} \qquad \Sigma F_y = 0 \qquad \Sigma M_{Gz} = 0$$
$$+\uparrow \Sigma F_y = R_{Ay} - 850 = 0 \qquad R_{Ay} = 850 \text{ lb}$$
$$+\downarrow \Sigma M_{Gz} = R_{Ay}(2.5) - \mu R_{Ay}(6) - T(3)$$
$$= 850(2.5) - 0.25(850)(6) - T(3) = 0$$
$$T = 283.3 \text{ lb} \qquad\qquad \text{Ans.}$$

b.
$$+\rightarrow \Sigma F_x = T - \mu R_{Ay} = ma_{Gx}$$
$$= 283.3 - 0.25(850) = \frac{850}{32.2} a_{Gx} \qquad a_{Gx} = 2.682 \text{ ft/s}^2 \qquad \text{Ans.}$$

c. The equations of motion for the winch are

$$\Sigma F_x = 0 \qquad \Sigma F_y = 0 \qquad \Sigma M_{Gz} = I_{Gz}\alpha$$

The motions of the container and the winch are related by the kinematic equation

$$x_G = r\theta$$

From which

$$\ddot{x}_G = r\ddot{\theta} \qquad \text{or} \qquad a_G = r\alpha$$

Thus,

$$\alpha = \frac{a_G}{r} = \frac{2.682}{2} = 1.341 \text{ rad/s}^2$$
$$+\downarrow \Sigma M_{Gz} = I_{Gz}\alpha$$
$$C - T(r) = mk^2\alpha$$
$$C - 283.3(2) = \frac{100}{32.2}(1.75)^2(1.341)$$
$$C = 579 \text{ ft} \cdot \text{lb}\downarrow \qquad\qquad \text{Ans.}$$

Fig. 16-16

(b)

The mass of the unbalanced wheel A shown in Fig. 16-17a is 40 kg, and its radius of gyration with respect to the axis of rotation is 150 mm. A cable attached to the wheel supports a 25-kg block B. A constant 30 N·m couple $\mathbf{C}$ is applied to the wheel as shown. When the wheel is in the position shown, its angular velocity is 5 rad/s clockwise. At this instant, determine the tension T in the cable and the force $\mathbf{A}$ exerted on the wheel by the pin at support A.

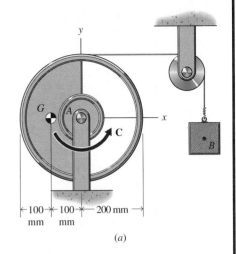

(a)

SOLUTION

Free-body diagrams for the wheel and block are shown in Fig. 16-17b. The motion of the block is translation ($a_{Gx} = 0$); the motion of the wheel is fixed-axis rotation ($a_{Gx} = r_G\omega^2$ and $a_{Gy} = -r_G\alpha$) about an axis that does not pass through the mass center G of the wheel. The equations of motion for the wheel are

$$\Sigma F_x = ma_{Gx} \qquad \Sigma F_y = ma_{Gy} \qquad \Sigma M_{Az} = I_{Az}\alpha$$

$$+\rightarrow \Sigma F_x = ma_{Gx} = mr_G\omega^2$$
$$A_x + T = 40(0.100)(-5)^2 = 100 \qquad (a)$$

$$+\uparrow \Sigma F_y = ma_{Gy} = -mr_G\alpha$$
$$A_y - 392.4 = -40(0.100)\alpha = -4.0\alpha \qquad (b)$$

$$+\downarrow \Sigma M_{Az} = I_{Az}\alpha = mk_A^2\alpha$$
$$30 + 392.4(0.100) - T(0.200) = 40(0.150)^2\alpha$$
$$T - 346.2 = -4.5\alpha \qquad (c)$$

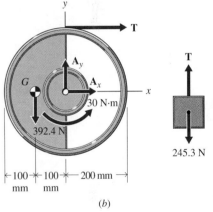

(b)

Fig. 16-17

Equations a, b, and c contain four unknowns; therefore, an additional equation is required for their solution. The additional equation is obtained from the equation of motion of the block. Thus

$$+\uparrow \Sigma F_y = ma_{By}$$
$$T - 245.3 = 25a_{By}$$

The acceleration a_{By} is related to the angular acceleration α of the wheel by the expression $a_{By} = 0.200\alpha$; therefore,

$$T - 245.3 = 25a_{By} = 25(0.200)\alpha = 5.0\alpha \qquad (d)$$

Solving Eqs. a, b, c, and d for A_x, A_y, T, and α yields

$$\alpha = 10.621 \text{ rad/s}^2$$
$$T = 298.4 = 298 \text{ N} \qquad \text{Ans.}$$
$$A_x = -198.4 = 198.4 \text{ N} \leftarrow$$
$$A_y = +349.9 \text{ N}$$
$$A = \sqrt{A_x^2 + A_y^2} = \sqrt{(-198.4)^2 + (349.9)^2} = 402 \text{ N} \qquad \text{Ans.}$$
$$\theta_x = \tan^{-1}\frac{A_y}{A_x} = \tan^{-1}\left(\frac{+349.9}{-198.4}\right) = 119.6° \qquad \text{Ans.}$$

PROBLEMS

In the following problems, all ropes, cords, and cables are assumed to be flexible, inextensible, and of negligible mass. All pins and pulleys have negligible mass and are frictionless unless specified otherwise.

16-21* The cable drum shown in Fig. P16-21 weighs 600 lb and has a radius of gyration of 2.5 ft. Block B weighs 500 lb. When a torque **T** of 1250 ft·lb is being applied to the cable drum, determine the acceleration of body B and the tension in the cable.

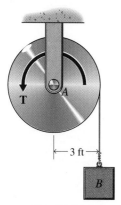

Fig. P16-21

16-22* A horizontal force **F** of 250 N is applied to a cable that is wrapped around the inner drum of the compound pulley being used to lift block B in Fig. P16-22. The pulley has a mass of 20 kg and its radius of gyration with respect to the axis of rotation is 160 mm. If block B has a mass of 10 kg, determine the angular acceleration of the pulley and the tension in the cable connected to block B.

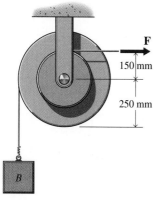

Fig. P16-22

16-23 Two blocks A and B are supported by cables wrapped around a compound cable drum as shown in Fig. P16-23. The drum weighs 60 lb and has a radius of gyration of 7.50 in. with respect to its axis of rotation. Blocks A and B weigh 50 and 90 lb, respectively. During motion of the system, determine the tensions in the two cables and the angular acceleration of the cable drum.

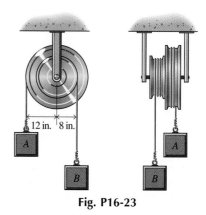

Fig. P16-23

16-24 A torque **T** of 300 N·m is applied to pulley A of the belt drive shown in Fig. P16-24. Pulley A is a solid circular disk with a mass of 15 kg. Pulley B and the cable drum have a combined mass of 75 kg and a radius of gyration with respect to the axis of rotation of 150 mm. The mass of block C is 150 kg. Determine the angular acceleration of pulley B and the tension in the cable.

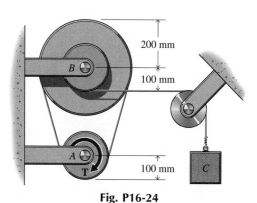

Fig. P16-24

16-25* The solid circular disk A of Fig. P16-25 weighs 50 lb and rotates about the smooth pin at O. Block B weighs 20 lb. During motion of the system, determine the angular acceleration of disk A, the tension in the cable, and the horizontal and vertical components of the force exerted on disk A by the pin at O.

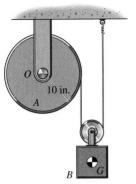

Fig. P16-25

16-26* Bar AB of Fig. P16-26 rotates in a horizontal plane with a constant angular velocity of 15 rad/s. A slender bar C with a uniform cross section and a mass of 2 kg supports a sphere D, which has a mass of 4 kg, at the end of bar AB. A cable keeps bar C vertical. Determine the tension in the cable and the force exerted by pin B on bar C.

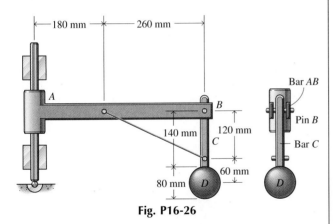

Fig. P16-26

16-27 A homogeneous 4-ft diameter cylinder that weighs 2000 lb rests on the bed of a flatbed truck as shown in Fig. P16-27a. The blocks shown in Fig. P16-27b are used to pre-

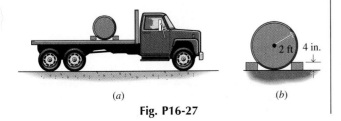

Fig. P16-27

vent the cylinder from rolling as the truck accelerates. Determine the acceleration of the truck required to start the cylinder rolling over the block.

16-28 The slender bar shown in Fig. P16-28 is rotating counterclockwise in a vertical plane about a smooth pin at support A. The mass of the bar is 15 kg. When the bar is in the position shown, its angular velocity is 10 rad/s. At this instant, determine the angular acceleration of the bar and the magnitude and direction of the force exerted on the bar by the pin at support A.

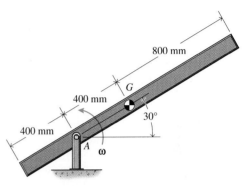

Fig. P16-28

16-29* The slender bar AB shown in Fig. P16-29 is rotating counterclockwise in a vertical plane about the smooth pin at support A. The bar has a uniform cross section and weighs 25 lb. When the bar is in the position shown, its angular velocity is 6 rad/s. At this instant, determine the angular acceleration of the bar and the magnitude and direction of the force exerted on the bar by the pin at support A.

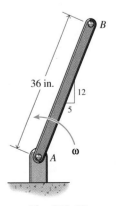

Fig. P16-29

16-30* Bar AB of Fig. P16-30 has a uniform cross section and a mass of 30 kg. The bar is held at rest in a vertical position and when released rotates in a vertical plane about the smooth pin at A. Determine the horizontal and vertical components of the pin reaction at support A when $\theta = 90°$.

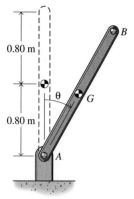

Fig. P16-30

16-31 A circular disk of uniform thickness that weighs 50 lb rotates in a vertical plane about a pin at point A as shown in Fig. P16-31. In the position shown (the diameter through pin A is horizontal), the angular velocity is 10 rad/s, counterclockwise. At this instant, determine the angular acceleration of the disk and the horizontal and vertical components of the force exerted on the disk by the pin at support A.

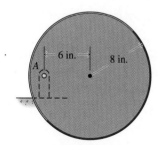

Fig. P16-31

16-32 The mass of the solid circular disk A, shown in Fig. P16-32, is 50 kg. A cable wrapped around a shallow groove

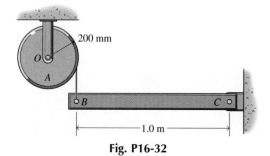

Fig. P16-32

in the disk is attached to bar BC, which has a uniform cross section and a mass of 25 kg. In the position shown, bar BC is horizontal and is rotating counterclockwise in a vertical plane with an angular velocity of 5 rad/s. Determine the angular acceleration of disk A, the tension in the cable, and the horizontal and vertical components of the reaction at support C.

16-33* The slender bar AB, shown in Fig. P16-33, has a uniform cross section and weighs 20 lb. Determine the acceleration of the mass center of the bar and the reaction at support A immediately after the cord at support B is cut.

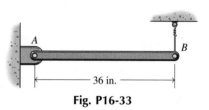

Fig. P16-33

16-34* A circular disk of uniform thickness with a mass of 25 kg rotates in a vertical plane about a pin at point A as shown in Fig. P16-34. In the position shown (the diameter through pin A is vertical), the angular velocity is 20 rad/s counterclockwise, and the magnitude of the couple $\mathbf{C}$ is 50 N · m. At this instant, determine the angular acceleration of the disk and the horizontal and vertical components of the force exerted on the disk by the pin at support A.

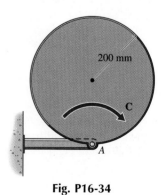

Fig. P16-34

16-35 A rectangular plate of uniform thickness is supported by a pin and a cable as shown in Fig. P16-35. The plate weighs 100 lb. If the cable at B breaks, determine the acceleration of the mass center of the plate and the reaction at support A at the instant motion begins.

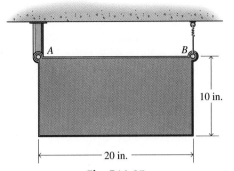

Fig. P16-35

16-36 A semicircular plate of uniform thickness is supported by a pin and a cable as shown in Fig. P16-36. The plate has a mass of 80 kg. If the cable at B breaks, determine the acceleration of the mass center of the plate and the reaction at support A at the instant motion begins.

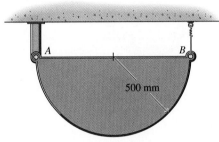

Fig. P16-36

16-37* The 24-in. diameter unbalanced disk shown in Fig. P16-37 weighs 125 lb, and its radius of gyration with respect to the fixed axis of rotation is 8.50 in. A cable wrapped around the circumference of the disk is attached

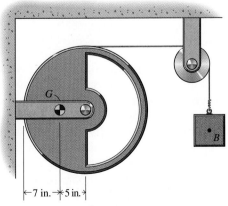

Fig. P16-37

to a block B, which weighs 50 lb. If the system is released from rest in the position shown, determine the tension in the cable and the acceleration of block B.

16-38* The unbalanced disk A shown in Fig. P16-38 has a mass of 20 kg. Its mass center is located 250 mm from the fixed axis of rotation O, and the radius of gyration with respect to the fixed axis of rotation is 350 mm. A cable wrapped around a shallow groove in the disk supports a 25 kg block. In the position shown, the angular velocity of the disk is 10 rad/s clockwise. Determine the tension in the cable and the horizontal and vertical components of the force exerted on the disk by the pin at O.

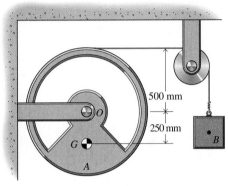

Fig. P16-38

16-39 A brake for regulating the descent of a body is shown in Fig. P16-39. The rotating parts of the brake (cable sheave and brake drum) weigh 250 lb and have a radius of gyration with respect to their axis of rotation of 4.25 in. The kinetic coefficient of friction between the brake pad and drum is 0.50. The weight of body C is 1000 lb. When a force P of 150 lb is being applied to the brake lever, determine the tension in the cable, the acceleration of body C, and the horizontal and vertical components of the reaction at support B of the brake lever.

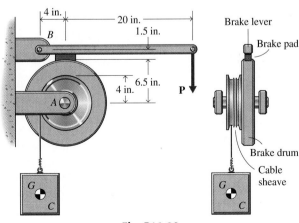

Fig. P16-39

16-40 The speed of a rotating system is controlled with a brake as shown in Fig P16-40. The rotating parts of the system have a mass of 300 kg and a radius of gyration with respect to the axis of rotation of 200 mm. The kinetic coefficient of friction between the brake pad and brake drum is 0.50. When a force **P** of 500 N is being applied to the brake lever, determine the horizontal and vertical components of the reaction at support B of the brake lever and the time t required to reduce the speed of the system from 1000 rpm to rest.

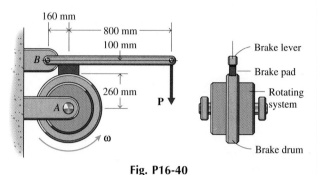

Fig. P16-40

16-41 The flywheel shown in Fig. P16-41 is rotating at a constant angular velocity of 50 rad/s in a counterclockwise

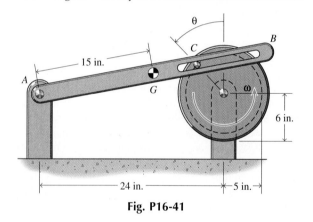

Fig. P16-41

direction. Bar AB weighs 20 lb and has a radius of gyration with respect to its mass center of 9.15 in. Determine the horizontal and vertical components of the force exerted by the pin at support A of bar AB when angle $\theta = 60°$. The slot in bar AB is smooth.

16-42 The flywheel shown in Fig. P16-42 is rotating with a constant angular velocity of 30 rad/s. Bar AB has a mass of 15 kg and a radius of gyration with respect to its mass center of 325 mm. The coefficient of friction between the pin at C and the slot in the bar is 0.10. When the bar is in the position shown, determine the horizontal and vertical components of the force exerted on the bar by the pin at support A.

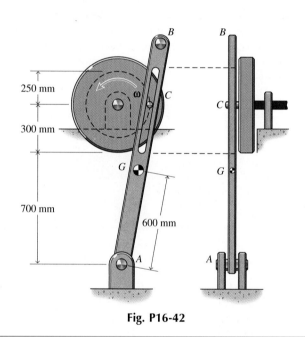

Fig. P16-42

16-4-3 General Plane Motion

Three forms of plane motion are illustrated in Fig. 16-18 where a piston is connected to a flywheel with a connecting rod AB. Clearly, the motion of the flywheel is rotation about a fixed axis, and the motion of the piston is rectilinear translation. The motion of the connecting rod AB is an example of general plane motion. As the flywheel rotates through an angle θ (see Fig. 16-18b), pin A moves a distance $s_A = R\theta$ along a circular path. The movement of pin B can be visualized as a superposition of the displacements resulting from a curvilinear translation of the rod, as shown in Fig. 16-18b, and a rotation of the rod about pin A as shown in Fig. 16-18c. As a result of these two displacements, pin B

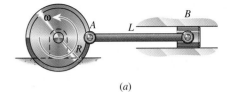

(a)

moves a distance s_B along a horizontal path. Thus, the general plane motion of rod AB is a superposition of translation and rotation about a fixed axis. The reduced form of Eqs. 16-17, which describe the motion of rod AB if the origin of the coordinate system is placed at pin A and the x- and y-axes are oriented along and normal to the axis of the rod ($\bar{y} = 0$), respectively, is

$$\Sigma F_x = ma_{Gx} \qquad \Sigma F_y = ma_{Gy}$$
$$\Sigma M_{Az} = a_{Ay}\bar{x}m + \alpha I_{Az} \qquad (16\text{-}22)$$

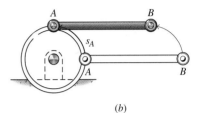

(b)

If the origin of the coordinate system is located at the mass center G of the rod, Eqs. 16-22 reduce to

$$\Sigma F_x = ma_{Gx} \qquad \Sigma F_y = ma_{Gy}$$
$$\Sigma M_{Gz} = \alpha I_{Gz} \qquad (16\text{-}23)$$

For those cases where the body is not symmetric with respect to the plane of motion, Eqs. 16-17 must be carefully considered and reduced as appropriate for the selection of the xyz-coordinate system attached to the body. For example, consider a solid circular disk (see Fig. 16-19) mounted on a shaft with its axis inclined at an angle θ with respect to the axis of the shaft. For an xyz-coordinate system with its origin at the mass center G of the disk, $\bar{x} = \bar{y} = 0$, $I_{Gyz} = 0$, and $\mathbf{a}_G = \mathbf{0}$. Thus Eqs. 16-17 reduce to

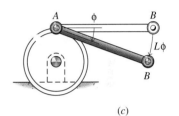

(c)

$$\Sigma F_x = ma_{Gx} = 0 \qquad \Sigma M_{Gx} = -\alpha I_{Gzx}$$
$$\Sigma F_y = ma_{Gy} = 0 \qquad \Sigma M_{Gy} = -\omega^2 I_{Gzx} \qquad (16\text{-}24)$$
$$\Sigma F_z = 0 \qquad \Sigma M_{Gz} = \alpha I_{Gz}$$

A second example of a body that is not symmetric with respect to the plane of motion is illustrated in Fig. 16-20. In this case, a triangular plate of uniform thickness is attached to and rotates with a circular shaft. For an xyz-coordinate system with its origin A on the axis of the shaft, $\bar{y} = 0$, $I_{Ayz} = 0$, and $\mathbf{a}_A = \mathbf{0}$. Thus Eqs. 16-17 reduce to

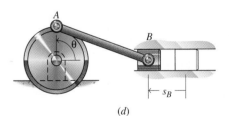

(d)

Fig. 16-18

$$\Sigma F_x = ma_{Gx} = -m\bar{x}\omega^2 \qquad \Sigma M_{Ax} = -\alpha I_{Azx}$$
$$\Sigma F_y = ma_{Gy} = m\bar{x}\alpha \qquad \Sigma M_{Ay} = -\omega^2 I_{Azx} \qquad (16\text{-}25)$$
$$\Sigma F_z = 0 \qquad \Sigma M_{Az} = \alpha I_{Az}$$

These five equations provide sufficient information for the determination of five unknowns, which may include the bearing components B_x, B_y, C_x, and C_y needed at any instant to produce the moments M_x and M_y required to maintain the body in a state of plane motion.

The procedure for solving motion problems involving general plane motion is illustrated in the following examples.

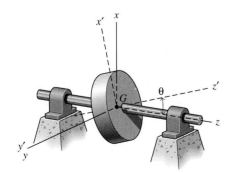

Fig. 16-19

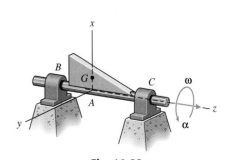

Fig. 16-20

A block and a spool are supported by cables wound around the spool as shown in Fig. 16-21a. The block weighs 95 lb; the spool weighs 50 lb and has a radius of gyration of 4 in. with respect to the mass center. If the system is released from rest in the position shown, determine the acceleration of the mass center G of the spool and the tensions in the three cables.

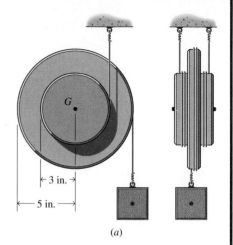

(a)

SOLUTION

Free-body diagrams for the block and spool are shown in Fig. 16-21b. The motion of the block is translation ($\omega = \alpha = \mathbf{a}_{Gx} = 0$); the motion of the spool is general plane motion with $\mathbf{a}_{Gx} = 0$. The origin of an xyz-coordinate system has been placed at the mass center G of each body.

The equations of motion for the block are

$$\Sigma F_x = 0 \qquad \Sigma F_y = ma_{Gy} \qquad \Sigma M_{Gz} = 0$$

$$+\uparrow\Sigma F_y = T_2 - 95 = \frac{95}{32.2}a_{Gyb}$$

$$T_2 - 2.950\, a_{Gyb} = 95 \qquad\qquad (a)$$

The equations of motion for the spool are

$$\Sigma F_x = 0 \qquad \Sigma F_y = ma_{Gy} \qquad \Sigma M_{Gz} = I_{Gz}\alpha$$

$$I_{Gz} = mk^2 = \frac{50}{32.2}\left(\frac{4}{12}\right)^2 = 0.17253 \text{ slug} \cdot \text{ft}^2$$

$$+\uparrow\Sigma F_y = 2T_1 - T_2 - 50 = \frac{50}{32.2}a_{Gys}$$

$$2T_1 - T_2 - 1.5528\, a_{Gys} = 50 \qquad\qquad (b)$$

$$+\downarrow\Sigma M_{Gz} = 2T_1(3/12) - T_2(5/12) = 0.17253\alpha$$

$$6T_1 - 5T_2 = 2.070\alpha \qquad\qquad (c)$$

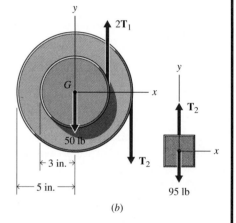

(b)

Since Eqs. a, b, and c contain five unknowns, two additional kinematic equations are required to complete the solution of the problem. From the sketch shown in Fig. 16-21c

$$y_{Gs} = -(3/12)\theta = -(1/4)\theta$$
$$\ddot{y}_{Gs} = -(1/4)\ddot{\theta} \quad \text{or} \quad a_{Gys} = -(1/4)\alpha \qquad\qquad (d)$$
$$y_{Gb} = y_{Gs} + (5/12)\theta = -(1/4)\theta + (5/12)\theta = (1/6)\theta$$
$$\ddot{y}_{Gb} = (1/6)\ddot{\theta} \quad \text{or} \quad a_{Gyb} = (1/6)\alpha \qquad\qquad (e)$$

Solving Eqs. a, b, c, d, and e, simultaneously yields

$$T_1 = 72.0 \text{ lb} \qquad\qquad \text{Ans.}$$
$$T_2 = 90.3 \text{ lb} \qquad\qquad \text{Ans.}$$
$$a_{Gys} = 2.37 \text{ ft/s}^2 \uparrow \qquad\qquad \text{Ans.}$$
$$a_{Gyb} = -1.581 \text{ ft/s}^2 = 1.581 \text{ ft/s}^2 \downarrow$$
$$\alpha = -9.48 \text{ rad/s}^2 = 9.48 \text{ rad/s}^2 \downarrow$$

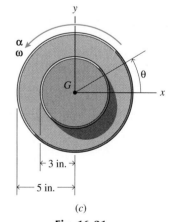

(c)

Fig. 16-21

EXAMPLE PROBLEM 16-8

The slender bar AB shown in Fig. 16-22a has a uniform cross section and weighs 50 lb. It is fastened to collars at ends A and B that slide on smooth horizontal and vertical rods. When the bar is in the position shown, the collar at A has a velocity of 5 ft/s to the right and is accelerating at a rate of 4 ft/s². Determine the force $\mathbf{F}$, the angular velocity $\boldsymbol{\omega}$ and the angular acceleration $\boldsymbol{\alpha}$ of the bar, and the forces exerted on the bar by the pins at A and B.

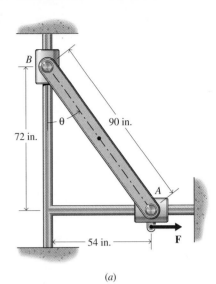

(a)

SOLUTION

A free-body diagram for bar AB is shown in Fig. 16-22b. The positions of ends A and B of the bar for the coordinate system shown are

$$x_A = L \sin \theta \qquad y_B = L \cos \theta$$

Thus

$$v_A = \dot{x}_A = L\dot{\theta} \cos \theta = L\omega \cos \theta$$
$$a_A = \ddot{x}_A = L\ddot{\theta} \cos \theta - L\dot{\theta}^2 \sin \theta = L\alpha \cos \theta - L\omega^2 \sin \theta$$

From the given data

$$v_A = 5 \text{ ft/s} \qquad a_A = 4 \text{ ft/s}^2 \qquad L = 90/12 = 7.5 \text{ ft}$$

$$\omega = \frac{v_A}{L \cos \theta} = \frac{5}{(7.5)(0.8)} = 0.8333 \text{ rad/s} \qquad\qquad \text{Ans.}$$

$$\alpha = \frac{a_A + L\omega^2 \sin \theta}{L \cos \theta} = \frac{4 + 7.5(0.8333)^2(0.6)}{7.5(0.8)} = 1.1875 \text{ rad/s}^2 \qquad \text{Ans.}$$

$$a_G = a_A + a_{G/A}$$

$$a_{Gx} = a_{Ax} + \frac{L}{2}\omega^2 \sin \theta - \frac{L}{2}\alpha \cos \theta$$
$$= 4 + 3.75(0.8333)^2(0.6) - 3.75(1.1875)(0.8) = 2.000 \text{ ft/s}^2$$

$$a_{Gy} = a_{Ay} - \frac{L}{2}\omega^2 \cos \theta - \frac{L}{2}\alpha \sin \theta$$
$$= 0 - 3.75(0.8333)^2(0.8) - 3.75(1.1875)(0.6) = -4.755 \text{ ft/s}^2$$

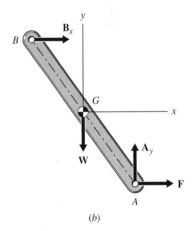

(b)

The equations of motion for the bar are

$$\Sigma F_x = ma_{Gx} \qquad \Sigma F_y = ma_{Gy} \qquad \Sigma M_{Gz} = I_{Gz}\alpha$$

$$I_{Gz} = \frac{1}{12}m L^2 = \frac{1}{12}\frac{50}{32.2}(7.5)^2 = 7.279 \text{ slug} \cdot \text{ft}^2$$

$$+\rightarrow \Sigma F_x = F + B_x = ma_{Gx}$$

$$F + B_x = \frac{50}{32.2}(2.000) = 3.106 \qquad\qquad (a)$$

$$+\uparrow \Sigma F_y = A_y - W = ma_{Gy}$$

$$A_y - 50 = \frac{50}{32.2}(-4.755)$$

$$A_y = +42.62 = 42.6 \text{ lb}\uparrow \qquad\qquad \text{Ans.}$$

$$+\downdownarrows \Sigma M_{Gz} = F(3) + A_y(2.25) - B_x(3) = I_{Gz}\alpha$$

$$F(3) + 42.62(2.25) - B_x(3) = 7.279(1.1875)$$

$$F - B_x = -29.08 \qquad\qquad (b)$$

Solving Eqs. a and b yields

$$F = -12.99 = 12.99 \text{ lb}\leftarrow \qquad\qquad \text{Ans.}$$
$$B_x = 16.093 = 16.09 \text{ lb}\rightarrow \qquad\qquad \text{Ans.}$$

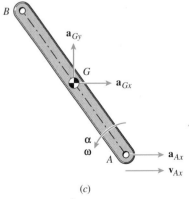

(c)

Fig. 16-22

221

A solid homogeneous cyclinder with a mass of 100 kg rests on an inclined surface as shown in Fig. 16-23a. The coefficient of friction between the cylinder and the inclined surface is 0.40. A cable wrapped around a shallow groove in the cylinder connects it to block, which has a mass of 75 kg. The pulley that the cable passes over has a mass of 10 kg. If the system is released from rest in the position shown, determine the acceleration of the mass center G of the block, the acceleration of the mass center G of the cylinder, and the tensions in the two parts of the cable.

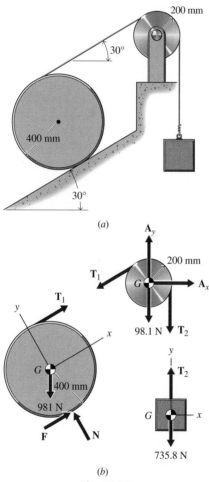

Fig. 16-23

SOLUTION

Free-body diagrams for the cylinder, pulley, and block are shown in Fig. 16-23b. The motion of the block is translation ($\boldsymbol{\omega} = \boldsymbol{\alpha} = \mathbf{a}_{Gx} = \mathbf{0}$); the motion of the pulley is rotation about a fixed axis through the mass center ($\mathbf{a}_{Gx} = \mathbf{a}_{Gy} = \mathbf{0}$); and the motion of the cylinder is general plane motion with $\mathbf{a}_{Gy} = \mathbf{0}$.

The equations of motion for the block are

$$\Sigma F_x = 0 \qquad \Sigma F_y = ma_{Gy} \qquad \Sigma M_{Gz} = 0$$
$$+\uparrow \Sigma F_y = T_2 - 735.8 = 75a_{Gyb}$$
$$T_2 - 75a_{Gyb} = 735.8 \tag{a}$$

The equations of motion for the pulley are

$$\Sigma F_x = 0 \qquad \Sigma F_y = 0 \qquad \Sigma M_{Gz} = I_{Gz}\alpha$$
$$I_{Gzp} = \frac{1}{2}m_p R_p^2 = \frac{1}{2}(10)(0.200)^2 = 0.200 \text{ kg} \cdot \text{m}^2$$
$$+\downarrow \Sigma M_{Gzp} = T_1(0.200) - T_2(0.200) = 0.200\alpha_p$$
$$T_1 - T_2 = \alpha_p \tag{b}$$

The equations of motion for the cylinder are

$$\Sigma F_x = ma_{Gx} \qquad \Sigma F_y = 0 \qquad \Sigma M_{Gz} = I_{Gz}\alpha$$
$$I_{Gzc} = \frac{1}{2}m_c R_c^2 = \frac{1}{2}(100)(0.400)^2 = 8.00 \text{ kg} \cdot \text{m}^2$$
$$+\nwarrow \Sigma F_y = N - m_c g \cos 30°$$
$$= N - 100(9.81)\cos 30° = 0 \qquad N = 849.6 \text{ N}$$
$$+\nearrow \Sigma F_x = T_1 + F - m_c g \sin 30° = ma_{Gxc}$$
$$= T_1 + F - 100(9.81)\sin 30° = 100a_{Gxc}$$
$$T_1 + F - 100a_{Gxc} = 490.5 \tag{c}$$
$$+\downarrow \Sigma M_{Gzc} = F(0.400) - T_1(0.400) = 8.00\alpha_c$$
$$F - T_1 = 20.0\alpha_c \tag{d}$$

Since Eqs. *a*, *b*, *c*, and *d* contain seven unknowns, three additional kinematic equations are required to complete the solution of the problem. If the cable does not slip on the pulley

$$y_{Gb} = 0.200 \; \theta_p$$
$$\ddot{y}_{Gb} = 0.200 \; \ddot{\theta}_p \qquad \text{or} \qquad \alpha_p = 5.000a_{Gyb} \tag{e}$$

If the cylinder does not slip on the inclined surface

$$y_{Gb} = 2(0.400) \; \theta_c$$
$$\ddot{y}_{Gb} = 0.800 \; \ddot{\theta}_c \qquad \text{or} \qquad \alpha_c = 1.250a_{Gyb} \tag{f}$$
$$x_{Gx} = -0.400 \; \theta_c$$
$$\ddot{x}_{Gc} = -0.400 \; \ddot{\theta}_c \qquad \text{or} \qquad a_{Gxc} = -0.500a_{Gyb} \tag{g}$$

Solving Eqs. *a*, *b*, *c*, *d*, *e*, *f*, and *g* simultaneously yields

$$T_1 = 401.8 = 402 \text{ N} \qquad\qquad \text{Ans.}$$
$$T_2 = 422.7 = 423 \text{ N} \qquad\qquad \text{Ans.}$$
$$a_{Gyb} = -4.175 = 4.18 \text{ m/s}^2 \downarrow \qquad\qquad \text{Ans.}$$
$$a_{Gxc} = 2.087 = 2.09 \text{ m/s}^2 \nearrow \qquad\qquad \text{Ans.}$$
$$\alpha_p = -20.87 = 20.9 \text{ rad/s}^2 \downarrow$$
$$\alpha_c = -5.219 = 5.22 \text{ rad/s}^2 \downarrow$$
$$F = 297.4 = 297 \text{ N} \nearrow$$

A check is required to see if the coefficient of friction is sufficient to develop the frictional force **F** required to prevent slipping between the cylinder and the inclined surface. Thus

$$F_{\max} = \mu N = 0.40(849.6) = 339.8 = 340 \text{ N}$$

Since $F_{\max}$ is greater than F, slipping will not occur. If $F_{\max}$ had been less than F, slipping would occur and the problem would have to be re-solved with $F = F_{\max} = \mu N$ in place of Eq. *g* and $0.4\alpha_c - a_{Gxc} = a_{Gyb}$ in place of Eq. *f*.

EXAMPLE PROBLEM 16-10

A system consisting of a flywheel, a connecting rod, and a piston are shown in Fig. 16-24a. The mass of the flywheel is 50 kg and its radius of gyration with respect to its axis of rotation of 155 mm. The connecting rod AB has a uniform cross section and a mass of 10 kg. The mass of the piston is 15 kg. A couple $\mathbf{T}$ rotates the flywheel counterclockwise at a constant angular velocity of 500 rpm. Determine the magnitude of the couple $\mathbf{T}$, the angular velocity $\boldsymbol{\omega}_{AB}$ and the angular acceleration $\boldsymbol{\alpha}_{AB}$ of rod AB, and the vertical and horizontal components of the forces exerted on rod AB by the pins at A and B when angle $\theta = 60°$. Neglect friction between the cylinder wall and the piston.

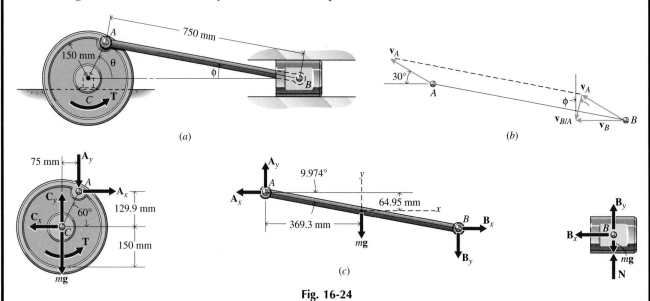

Fig. 16-24

SOLUTION

From the geometry of the system at $\theta = 60°$

$$\frac{\sin 60°}{750} = \frac{\sin \phi}{150} \qquad \phi = \sin^{-1}\left(\frac{150}{750}\sin 60°\right) = 9.974°$$

Since the flywheel is rotating with a constant angular velocity ($N = 500$ rpm), the velocity and acceleration of pin A are easily obtained. Thus, for the flywheel

$$\omega_f = \frac{2\pi N}{60} = \frac{2\pi(500)}{60} = 52.36 \text{ rad/s}\!\downarrow$$

$$\alpha_f = 0$$

For pin A

$$\mathbf{v}_A = r_f\omega_f\, \mathbf{e}_\theta = 0.150(52.36)\, \mathbf{e}_\theta = 7.854\, \mathbf{e}_\theta \text{ m/s}$$
$$\mathbf{a}_A = -r_f\omega_f^2\, \mathbf{e}_r = -0.150(52.36)^2\, \mathbf{e}_r = -411.2\, \mathbf{e}_r \text{ m/s}^2$$

Once $\mathbf{v}_A$ and $\mathbf{a}_A$ are known, the velocity $\mathbf{v}_B$ and acceleration $\mathbf{a}_B$ of pin B and the angular velocity $\boldsymbol{\omega}_{AB}$ and angular acceleration $\boldsymbol{\alpha}_{AB}$ of rod AB can be determined. For pin B

$$\mathbf{v}_B = \mathbf{v}_A + \mathbf{v}_{B/A}$$

The velocities $\mathbf{v}_A$, $\mathbf{v}_B$, and $\mathbf{v}_{B/A}$ are illustrated in Fig. 16-24b. Thus,

$$v_B\mathbf{i} = 7.854(-\cos 30°\, \mathbf{i} + \sin 30°\, \mathbf{j}) + 0.750\, \omega_{AB}(\sin \phi\, \mathbf{i} + \cos \phi\, \mathbf{j})$$
$$= (-6.802 + 0.1299\, \omega_{AB})\, \mathbf{i} + (3.927 + 0.7387\, \omega_{AB})\, \mathbf{j}$$

From which

$$3.927 + 0.7387\,\omega_{AB} = 0 \qquad \omega_{AB} = -5.316 \text{ rad/s}$$
$$\omega_{AB} = 5.32 \text{ rad/s} \downarrow \quad \text{Ans.}$$
$$v_B = -6.802 + 0.1299\,\omega_{AB} = -6.802 + 0.1299(-5.316) = -7.493 \text{ m/s}$$
$$\mathbf{v}_B = 7.493 \text{ m/s} \leftarrow$$

Similarly for the acceleration

$$\mathbf{a}_B = \mathbf{a}_A + \mathbf{a}_{B/A}$$

The acceleration $\mathbf{a}_{B/A}$ has two components that result from the angular velocity $\boldsymbol{\omega}_{AB}$ and the angular acceleration $\boldsymbol{\alpha}_{AB}$ of rod AB. The component along the axis of the rod (directed toward A) is

$$(a_{B/A}) = L_{AB}\omega_{AB}^2 = 0.750(5.316)^2 = 21.19 \text{ m/s}^2$$

The component perpendicular to the axis of the rod is

$$(a_{B/A})_t = L_{AB}\alpha_{AB} = 0.750\alpha_{AB}$$

Cartesian components of the accelerations $\mathbf{a}_A$, $\mathbf{a}_B$, and $\mathbf{a}_{B/A}$ can be combined to yield

$$a_B\mathbf{i} = 411.2(-\cos 60° \,\mathbf{i} - \sin 60° \,\mathbf{j}) + 21.19(-\cos\phi\,\mathbf{i} + \sin\phi\,\mathbf{j})$$
$$+ 0.750\alpha_{AB}(\sin\phi\,\mathbf{i} + \cos\phi\,\mathbf{j})$$
$$= (-226.5 + 0.1299\alpha_{AB})\,\mathbf{i} + (-352.4 + 0.7387\alpha_{AB})\,\mathbf{j}$$

From which

$$-352.4 + 0.7387\alpha_{AB} = 0 \qquad \alpha_{AB} = 477.1 \text{ rad/s}^2$$
$$\boldsymbol{\alpha}_{AB} = 477 \text{ rad/s}^2 \downarrow \qquad \text{Ans.}$$

and

$$a_B = -226.5 + 0.1299\alpha_{AB} = -226.5 + 0.1299(477.1) = -164.5 \text{ m/s}^2$$
$$\mathbf{a}_B = 164.5 \text{ m/s}^2 \leftarrow$$

Finally, the horizontal and vertical components of the acceleration of the mass center of the rod are obtained from

$$\mathbf{a}_G = \mathbf{a}_B + \mathbf{a}_{G/B}$$
$$a_{Gx} = a_{Bx} + \frac{L}{2}\omega_{AB}^2\cos\phi - \frac{L}{2}\alpha_{AB}\sin\phi$$
$$= -164.47 + 0.375(5.316)^2(0.9849) - 0.375(477.1)(0.1732) = -185.02 \text{ m/s}^2$$
$$a_{Gy} = a_{By} - \frac{L}{2}\omega_{AB}^2\sin\phi - \frac{L}{2}\alpha_{AB}\cos\phi$$
$$= 0 - 0.375(5.316)^2(0.1732) - 0.375(477.1)(0.9849) = -178.05 \text{ m/s}^2$$

Free-body diagrams for the flywheel, rod, and piston are shown in Fig. 16-24c. The motion of the piston is pure translation. Therefore,

$$+\rightarrow \Sigma F_x = ma_{Gx} \qquad -B_x = 15(-164.47)$$
$$\mathbf{B}_x = 2467 \text{ N} \leftarrow$$
$$+\uparrow \Sigma F_y = ma_{Gy} \qquad N + B_y - mg = 0$$
$$+\downarrow \Sigma M_{Gz} = I_{Gz}\alpha \qquad 0 = 0$$

The bar is in general plane motion; therefore,

$$+\rightarrow \Sigma F_x = ma_{Gx} \qquad 2467 - A_x = 10(-185.02)$$
$$A_x = 4317 \text{ N} \leftarrow$$
$$+\uparrow \Sigma F_y = ma_{Gy} \qquad A_y - 10(9.81) - B_y = 10(-178.05)$$
$$A_y - B_y = -1682.4 \qquad (a)$$
$$I_{Gz} = \frac{1}{12}m\,L^2 = \frac{1}{12}(10)(0.75))^2 = 0.4688 \text{ kg}\cdot\text{m}^2$$
$$+\downarrow \Sigma M_{Gz} = I_{Gz}\alpha$$
$$4317(0.06495) - A_y(0.3693) + 2467(0.06495) - B_y(0.3693) = 0.4688(477.1)$$
$$A_y + B_y = 587.48 \qquad (b)$$

Solving Eqs. a and b yields

$$A_y = -547.5 = 547.5 \text{ N} \downarrow \qquad \text{Ans.}$$
$$B_y = 1134.9 = 1134.9 \text{ N} \uparrow \qquad \text{Ans.}$$

Also

$$A_x = 4317 \text{ N} \leftarrow \qquad \text{Ans.}$$
$$B_x = 2467 \text{ N} \rightarrow \qquad \text{Ans.}$$

The motion of the flywheel is rotation about a fixed axis; therefore,

$$+\rightarrow \Sigma F_x = ma_{Gx} \qquad A_x - C_x = 0$$
$$+\uparrow \Sigma F_y = ma_{Gy} \qquad C_y - A_y - mg = 0$$
$$+\downarrow \Sigma M_{Gz} = I_{Gz}\alpha$$
$$T - A_y \, r_f \cos\theta - A_x \, r_f \sin\theta = 0 \quad (\text{since } \alpha = 0)$$
$$T = -547.5(0.150)\cos 60° + 4317\,(0.150)\sin 60°$$
$$= 519.7 \text{ N·m} \qquad\qquad\qquad \text{Ans.}$$

EXAMPLE PROBLEM 16-11

Two 120-mm diameter spheres are attached to a shaft and rotated as shown in Fig. 16-25a. Each sphere has a mass of 7.50 kg. The bars connecting the spheres to the shaft have 30 mm diameters, are 220 mm long, and have masses of 1.20 kg. The 40-mm diameter shaft has a mass of 8.50 kg. Determine the components of the bearing reactions at the supports and the applied torque **T** when the shaft is rotating counterclockwise at 600 rpm and increasing in speed at the rate of 60 rpm per second. Assume that the bearing at A resists any motion of the shaft in the z-direction.

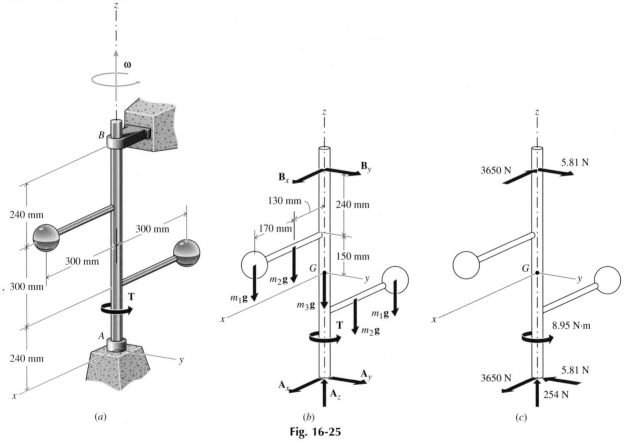

(a)　　　　　　　(b)　　　　　　　(c)

Fig. 16-25

SOLUTION

A free-body diagram of the system is shown in Fig. 16-25b. Since the origin of the xyz-coordinate system coincides with the mass center G of the system, $\bar{x} = \bar{y} = 0$. Also, at the instant shown, $I_{Gyz} = 0$ (symmetry about xz-plane), and $\mathbf{a}_G = \mathbf{0}$ (fixed axis of rotation passes through G). Thus, Eqs. 16-17 reduce to

$$\Sigma F_x = ma_{Gx} = 0 \qquad \Sigma F_y = ma_{Gy} = 0 \qquad \Sigma F_z = 0$$
$$\Sigma M_{Gx} = -\alpha I_{Gzx} \qquad \Sigma M_{Gy} = -\omega^2 I_{Gzx} \qquad \Sigma M_{Gz} = \alpha I_{Gz}$$

The moment of inertia I_{Gz} and the product of inertia I_{Gzx} can be determined by using Table B-5 and the parallel-axis equations (Eqs. 16-11 and 16-16). Thus for each sphere

$$I_{Gz1} = \frac{2}{5} m_1 R_1^2 + m_1 d_1^2$$

$$= \frac{2}{5}(7.50)(0.060)^2 + 7.50(0.300)^2 = 0.6858 \text{ kg} \cdot \text{m}^2$$

$$I_{Gzx1} = 0 + m_1 \bar{z}_1 \bar{x}_1$$
$$= 0 + 7.50(0.150)(0.300) = 0.3375 \text{ kg} \cdot \text{m}^2$$

For each bar

$$I_{Gz2} = \frac{1}{4} m_2 R_2^2 + \frac{1}{12} m_2 L_2^2 + m_2 d_2^2$$

$$= \frac{1}{4}(1.20)(0.015)^2 + \frac{1}{12}(1.20)(0.220)^2 + 1.20(0.130)^2 = 0.02519 \text{ kg} \cdot \text{m}^2$$

$$I_{Gzx2} = 0 + m_2 \bar{z}_2 \bar{x}_2 = 0 + 1.20(0.150)(0.130) = 0.0234 \text{ kg} \cdot \text{m}^2$$

For the shaft

$$I_{Gz3} = \frac{1}{2} m_3 R_3^2$$

$$= \frac{1}{2}(8.50)(0.020)^2 = 0.0017 \text{ kg} \cdot \text{m}^2$$

$$I_{Gzx3} = 0$$

For the system

$$I_{Gz} = 2I_{Gz1} + 2I_{Gz2} + I_{Gz3}$$
$$= 2(0.6858) + 2(0.02519) + 0.0017 = 1.4237 \text{ kg} \cdot \text{m}^2$$
$$I_{Gzx} = 2I_{Gzx1} + 2I_{Gzx2} + I_{Gzx3}$$
$$= 2(0.3375) + 2(0.0234) + 0 = 0.7218 \text{ kg} \cdot \text{m}^2$$
$$\omega = \frac{2\pi(600)}{60} = 62.83 \text{ rad/s} \qquad \alpha = \frac{2\pi(60)}{60} = 6.283 \text{ rad/s}^2$$

$$+\swarrow \Sigma F_x = A_x + B_x = 0 \qquad\qquad (a)$$
$$+\searrow \Sigma F_y = A_y + B_y = 0 \qquad\qquad (b)$$
$$+\uparrow \Sigma F_z = A_z - 2m_1 g - 2m_2 g - m_3 g = 0$$
$$A_z = 2(7.50)(9.81) + 2(1.20)(9.81) + 8.50(9.81)$$
$$= 254.1 = 254 \text{ N}\uparrow \qquad\qquad \text{Ans.}$$
$$+\downarrow \Sigma M_{Gx} = A_y(0.390) - B_y(0.390) = -\alpha I_{Gzx} = -6.283(0.7218)$$
$$A_y - B_y = -11.628 \qquad\qquad (c)$$
$$+\downarrow \Sigma M_{Gy} = B_x(0.390) - A_x(0.390) = -\omega^2 I_{Gzx} = -(62.83)^2(0.7218)$$
$$B_x - A_x = -7306 \qquad\qquad (d)$$
$$+\downarrow \Sigma M_{Gz} = T = \alpha I_{Gz} = 6.283(1.4237) = 8.945 = 8.95 \text{ N} \cdot \text{m} \qquad\qquad \text{Ans.}$$

From Eqs. a and d

$$A_x = +3653 = 3650 \text{ N}\swarrow \qquad\qquad \text{Ans.}$$
$$B_x = -3653 = 3650 \text{ N}\nearrow \qquad\qquad \text{Ans.}$$

From Eqs. b and c

$$A_y = -5.814 = 5.81 \text{ N}\nwarrow \qquad\qquad \text{Ans.}$$
$$B_y = +5.814 = 5.81 \text{ N}\searrow \qquad\qquad \text{Ans.}$$

The results are shown in Fig. 16-25c.

A solid circular steel ($\gamma = 0.284$ lb/in.^{3}) cylinder is mounted on a shaft as shown in Fig. 16-26a. Determine the components of the bearing reactions and the applied torque **T** when the cylinder is in the position shown (the x axis is vertical) and the shaft is rotating counterclockwise at 500 rpm and increasing in speed at the rate of 50 rpm per second. Assume that the bearing at B resists any motion of the shaft in the z-direction and that the mass of the shaft is negligible.

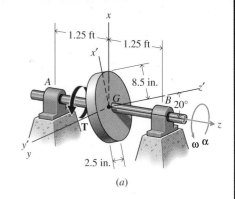

(a)

SOLUTION

A free-body diagram of the cylinder and shaft is shown in Fig. 16-26b. Since the origin of the xyz-coordinate system coincides with the mass center G of the cylinder, $\bar{x} = \bar{y} = 0$. Also, $I_{Gyz} = 0$ (symmetry about xz-plane), and $\mathbf{a}_G = \mathbf{0}$ (fixed axis of rotation passes through G). Thus Eqs. 16-17 reduce to

$$\Sigma F_x = ma_{Gx} = 0 \qquad \Sigma F_y = ma_{Gy} = 0 \qquad \Sigma F_z = 0$$
$$\Sigma M_{Gx} = -\alpha I_{Gzx} \qquad \Sigma M_{Gy} = -\omega^2 I_{Gzx} \qquad \Sigma M_{Gz} = \alpha I_{Gz}$$

The moment of inertia I_{Gz} and the product of inertia I_{Gzx} can be determined by using the principal moments of inertia $I_{x'}$, $I_{y'}$, and $I_{z'}$ from Table B-5 and the transformation equations A-13a and A-13b. Thus

$$W = \gamma V = \gamma \pi R^2 L = 0.284(\pi)(8.5)^2(2.5) = 161.16 = 161.2 \text{ lb}$$

$$I_{x'} = I_{y'} = \frac{1}{4}\left(\frac{161.16}{32.2}\right)\left(\frac{8.5}{12}\right)^2 + \frac{1}{12}\left(\frac{161.16}{32.2}\right)\left(\frac{2.5}{12}\right)^2 = 0.6459 \text{ slug} \cdot \text{ft}^2$$

$$I_{z'} = \frac{1}{2}\left(\frac{161.16}{32.2}\right)\left(\frac{8.5}{12}\right)^2 = 1.2556 \text{ slug} \cdot \text{ft}^2$$

$$I_z = I_{x'} \cos^2 \theta_{x'z} + I_{y'} \cos^2 \theta_{y'z} + I_{z'} \cos^2 \theta_{z'z}$$
$$= 0.6459 \cos^2 110° + 0.6459 \cos^2 90° + 1.2556 \cos^2 20° = 1.1843 \text{ slug} \cdot \text{ft}^2$$

$$I_{zx} = -I_{x'} \cos \theta_{x'z} \cos \theta_{x'x} - I_{y'} \cos \theta_{y'z} \cos \theta_{y'x} - I_{z'} \cos \theta_{z'z} \cos \theta_{z'x}$$
$$= -0.6459 \cos 110° \cos 20° - 0.6459 \cos 90° \cos 90°$$
$$\qquad - 1.2556 \cos 20° \cos 70° = -0.19595 \text{ slug} \cdot \text{ft}^2$$

$$\omega = \frac{2\pi(500)}{60} = 52.36 \text{ rad/s} \qquad \alpha = \frac{2\pi(50)}{60} = 5.236 \text{ rad/s}^2$$

$$+\uparrow \Sigma F_x = A_x + B_x - W = 0$$
$$= A_x + B_x - 161.16 = 0 \qquad (a)$$
$$+\swarrow \Sigma F_y = A_y + B_y = 0 \qquad (b)$$
$$+\rightarrow \Sigma F_z = B_z = 0 \qquad B_z = 0 \qquad \text{Ans.}$$
$$+\downdownarrows \Sigma M_{Gx} = A_y(1.25) - B_y(1.25) = -\alpha I_{Gzx} = -5.236(-0.19595)$$
$$A_y - B_y = 0.8208 \qquad (c)$$
$$+\downdownarrows \Sigma M_{Gy} = B_x(1.25) - A_x(1.25) = -\omega^2 I_{Gzx} = -(52.36)^2(-0.19595)$$
$$B_x - A_x = 429.8 \qquad (d)$$
$$+\downdownarrows \Sigma M_{Gz} = T = \alpha I_{Gz} = 5.236(1.1843) = 6.20 \text{ ft} \cdot \text{lb} \downdownarrows \qquad \text{Ans.}$$

From Eqs. a and d

$$A_x = -134.32 = 134.3 \text{ lb}\downarrow \qquad B_x = +295.4 = 295 \text{ lb}\uparrow \qquad \text{Ans.}$$

From Eqs. b and c

$$A_y = +0.4104 = 0.410 \text{ lb}\swarrow \qquad B_y = -0.4104 = 0.410 \text{ lb}\nearrow \qquad \text{Ans.}$$

Problems of this type can also be solved as three-dimensional problems by resolving the angular velocity $\boldsymbol{\omega}$ and the angular acceleration $\boldsymbol{\alpha}$ into x'- and y'-components. As shown in Section 16-5, this approach simplifies the solution by eliminating the requirement for determining the nonprincipal moments of inertia and the products of inertia.

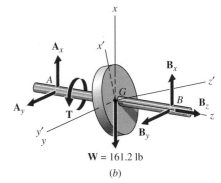

W = 161.2 lb

(b)

Fig. 16-26

PROBLEMS

In the following problems, all ropes, cords, and cables are assumed to be flexible, inextensible, and of negligible mass. All pins and pulleys have negligible mass and are frictionless unless specified otherwise.

16-43* A solid homogeneous sphere that weighs 10 lb rolls without slipping down a plane that is inclined at an angle of 28° with the horizontal as shown in Fig. P16-43. Determine the acceleration of the mass center of the sphere and the minimum coefficient of friction required to prevent slipping.

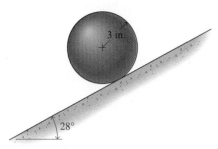

Fig. P16-43

16-44* A solid circular disk with a mass of 15 kg rolls without slipping on the inclined surface shown in Fig. P16-44. In the position shown, the angular velocity of the disk is 10 rad/s clockwise. Determine the angular acceleration of the disk and the minimum coefficient of friction required to prevent slipping.

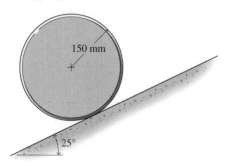

Fig. P16-44

16-45 Two 16-in. diameter disks and an 8-in. diameter cylinder are fastened together to form a spool weighing 50 lb which has a radius of gyration of 5 in. with respect to the axis of the spool. A force **P** of 50 lb is applied to the spool through a cord that is wrapped around the cylinder as shown in Fig. P16-45. If the spool rolls without slipping on the horizontal surface, determine the acceleration of the mass center of the spool and the minimum coefficient of friction required to prevent slipping.

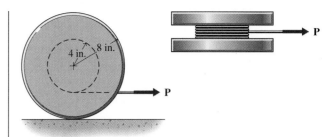

Fig. P16-45

16-46 A solid circular cylinder A with a radius R of 200 mm and a mass m_A of 75 kg is connected with a flexible cable to a body B with a mass m_B of 50 kg as shown in Fig. P16-46. If the cylinder rolls without slipping on the inclined plane, determine the acceleration of body B and the tension in the cable.

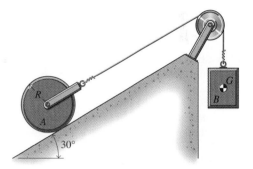

Fig. P16-46

16-47* The wheel shown in Fig. P16-47 is being pulled forward by a constant force **P** of 52 lb. The weight of the wheel is 75 lb, and its radius of gyration with respect to the axis of the wheel is 9.25 in. The wheel rolls without slipping on the horizontal surface and in the position shown has an angular velocity of 15 rad/s clockwise. Determine the angular acceleration of the wheel and the horizontal and vertical components of the force exerted by the surface on the wheel.

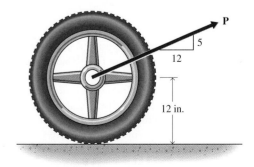

Fig. P16-47

16-48* Two 400-mm diameter disks and a 240-mm diameter disk are joined to form a spool that has a mass of 125 kg and a radius of gyration of 125 mm with respect to an axis through the mass center of the spool. A force **F** of 500 N is applied to the spool through a cable wrapped around the 240-mm disk as shown in Fig. P16-48. Determine the acceleration of the mass center and the angular acceleration of the spool if

a. The horizontal surface is smooth ($\mu = 0$).
b. The horizontal surface is rough ($\mu = 0.25$).

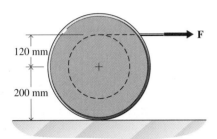

Fig. P16-48

16-49 The 100-lb spool A, shown in Fig. P16-49, has a radius of gyration of 4.75 in. with respect to its mass center G. A cord connects the spool to a 25-lb block B, which rests on an inclined surface. The kinetic coefficient of friction between the block and the inclined surface is 0.10. If the spool rolls without slipping on the horizontal surface, determine the acceleration of the block, the tension in the cable, and the minimum static coefficient of friction required to ensure rolling without slipping of the spool.

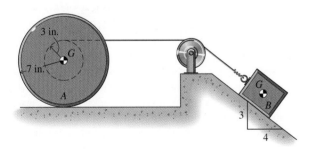

Fig. P16-49

16-50 A 200-mm diameter cylinder A is mounted on a 50-mm diameter axle as shown in Fig. P16-50. The mass of the cylinder and axle is 50 kg and the radius of gyration with respect to the axis of the axle is 70 mm. Flexible cords wrapped around the axle on both sides of the cylinder are connected to a 100-kg block, which rests on a horizontal surface. The kinetic coefficient of friction between the horizontal surface and the block is 0.25. If the cylinder rolls without slipping, determine the acceleration of the block and the tensions in the two cords.

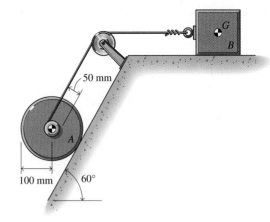

Fig. P16-50

16-51* A spool is supported by a cord wrapped around the inner core of the spool as shown in Fig. P16-51. The spool weighs 10 lb and has a radius of gyration of 4 in. with respect to its mass center G. When the spool is released to move in a vertical plane, determine the angular acceleration of the spool and the tension in the cord.

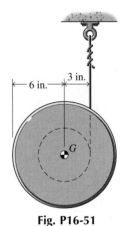

Fig. P16-51

16-52* A 100-kg cylinder A and a 50-kg body B are attached with cables to the compound cable sheave shown in Fig. P16-52. The sheave weighs 20 kg and has a radius of gyration with respect to its axis of rotation of 110 mm. If

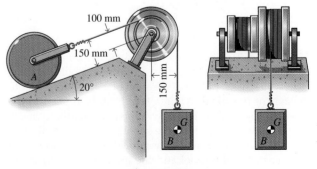

Fig. P16-52

230

the cylinder rolls without slipping on the inclined surface, determine the acceleration of body B and the tensions in the two cables.

16-53 A block A, which weighs 500 lb, is supported on an inclined surface by a platform B, which weighs 100 lb, and a pair of wheels C, which together weigh 200 lb as shown in Fig. P16-53. The wheels have a radius of gyration about their axis of rotation of 8.50 in. The static and kinetic coefficients of friction between the inclined surface and the contacting bodies are 0.20 and 0.15, respectively. During motion of the system, determine the acceleration of the mass center of the wheels and the normal and frictional forces exerted on the platform and wheels by the inclined surface.

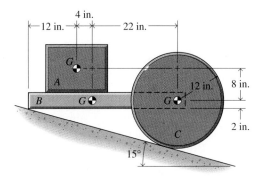

Fig. P16-53

16-54 A continuous cable supports a solid circular disk A and a block B as shown in Fig. P16-54. The masses of disk A and block B are 40 kg and 25 kg, respectively. There is no slipping between the disk and the cable. During motion of the system, determine the acceleration of the block, the angular acceleration of the disk, and the tension in the cable at support C.

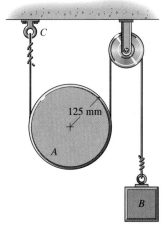

Fig. P16-54

16-55* A 4-ft diameter homogeneous cylinder that weighs 300 lb rests on the bed of a flatbed truck as shown in Fig. P16-55. The truck accelerates from rest at a rate of

3 ft/s² for 20 s and then moves with a constant velocity. If the cylinder rolls without slipping, determine the distance traveled by the truck before the cylinder rolls off the truck.

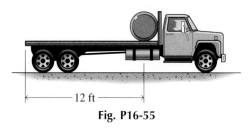

Fig. P16-55

16-56* The 220-mm diameter bowling ball shown in Fig. P16-56 has a mass of 7.25 kg. The instant the ball comes in contact with the alley, it has a forward velocity **v** of 7 m/s and a back spin **ω** of 6 rad/s. If the kinetic coefficient of friction between the ball and the alley is 0.15, determine the elapsed time and the distance traveled before the ball begins to roll without slipping.

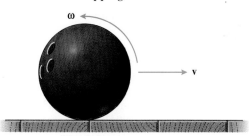

Fig. P16-56

16-57 A 12-in. diameter disk mounted on a 4-in. diameter shaft rolls without slipping on a pair of inclined rails. A cord wrapped around a shallow groove on the outer surface of the disk supports a 20-lb body B as shown in Fig. P16-57. The weight of the disk and shaft is 75 lb and the radius of gyration with respect to an axis through the mass center is 4.25 in. If the system is released from rest, determine the initial tension in the cable, acceleration of body B, and angular acceleration of the disk.

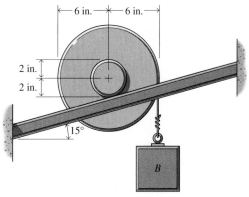

Fig. P16-57

16-58 The solid circular cylinder shown in Fig. P16-58 has a mass of 50 kg. The cylinder rests on a block B, which has a mass of 35 kg. When the force F is applied to the block, the block accelerates at a rate of 5 m/s². If the coefficients of friction between both the block and the cylinder and the block and the horizontal surface are 0.25, determine the magnitude of the force F and the frictional force exerted by the block on the cylinder.

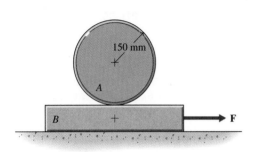

Fig. P16-58

16-59* The slender bar shown in Fig. P16-59 has a uniform cross section and weighs 20 lb. The bar is released at rest when vertical and rotates in a vertical plane under the action of gravity. The coefficient of friction between the bar and the horizontal surface is 0.50. Determine:

a. The angular acceleration of the bar and the reaction at end A when $\theta = 40°$.
b. The angle θ_s when the bar begins to slip.

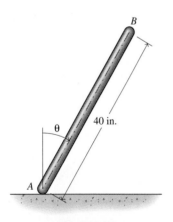

Fig. P16-59

16-60* A solid circular disk A with a mass of 40 kg is connected with a link to a block B, which has a mass of 50 kg. Both rest on a rough ($\mu = 0.25$) inclined plane as shown in Fig. P16-60. During motion of the system, determine the force in the link, the normal components of the

reactions at supports C and D of the block, and the angular acceleration of the disk.

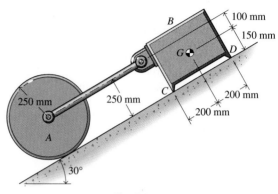

Fig. P16-60

16-61 A spool A and a block B are supported by cables wrapped around the circular surfaces of the spool as shown in Fig. P16-61. The block weighs 75 lb; the spool weighs 104 lb and has a radius of gyration with respect to its axis of rotation of 4.12 in. Immediately after the system is released at rest, determine the acceleration of the mass center of the spool and the tensions in the cables.

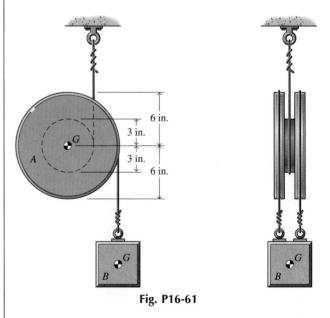

Fig. P16-61

16-62 The half disk shown in Fig. P16-62 has a mass of 25 kg. If the disk rolls without slipping, determine the acceleration of the mass center, the angular acceleration of the disk, and the normal and frictional forces exerted on the disk by the horizontal surface immediately after the cord supporting the disk at point A is cut.

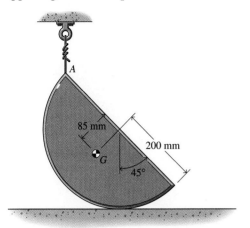

Fig. P16-62

16-63* The slender bar AB shown in Fig. P16-63 has a uniform cross section and weighs 40 lb. The bar is supported by two flexible cables and is held in position by the horizontal cord at support B. Determine the acceleration of the mass center of the bar, the angular acceleration of the bar, and the tensions in the two cables immediately after the horizontal cord at support B is cut.

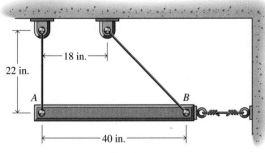

Fig. P16-63

16-64* A slender bar AB with a uniform cross section and a mass of 10 kg is initially held fixed by two cords as shown in Fig. P16-64. Determine the tension in the cord at

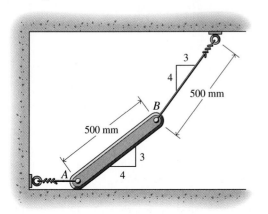

Fig. P16-64

B and the angular acceleration of the bar immediately after the cord at A is cut. Assume that the horizontal surface in contact with end A of the bar is smooth.

16-65 A 4-in. diameter disk shown in Fig. P16-65 weighs 10 lb. The disk is released at rest when $\theta = 30°$ and rolls down the curved surface without slipping. Determine the acceleration of the mass center of the disk and the components of the force exerted on the disk by the curved surface when $\theta = 60°$.

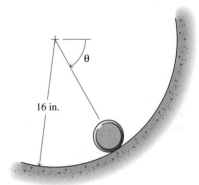

Fig. P16-65

16-66 A 300-mm diameter disk rolls without slipping on a curved surface as shown in Fig. P16-66. The center of the 40-kg disk follows the curve $y = 1 - x^2/4$. If the coefficient of friction between the disk and the surface is 0.5, determine the maximum speed the disk can have at $x = 1$ m without slipping.

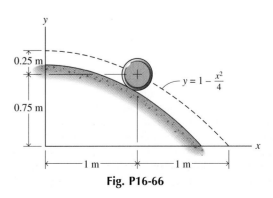

Fig. P16-66

16-67* The unbalanced wheel shown in Fig. P16-67 rolls without slipping down the inclined surface. The wheel weighs 32 lb and has a radius of gyration of 4.25 in. with respect to its axis of rotation. When the wheel is in the position shown, it has an angular velocity of 5 rad/s counterclockwise. At this instant, determine the angular acceleration of the wheel and the normal and frictional forces exerted on the wheel by the inclined surface.

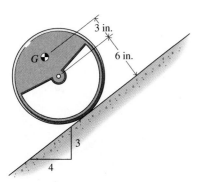

Fig. P16-67

16-68* The unbalanced wheel shown in Fig. P16-68 has a mass of 50 kg and rolls without slipping on the horizontal surface. The radius of gyration of the wheel with respect to an axis through the mass center is 160 mm. In the position shown, the angular velocity of the wheel is 6 rad/s. For this instant, determine the angular acceleration of the wheel and the force acting on the wheel at its point of contact with the horizontal surface.

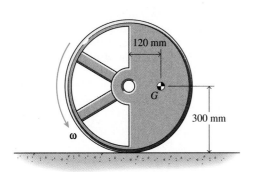

Fig. P16-68

16-69 A 64-lb block is supported by a 15-lb bar with a uniform cross section and a 25-lb solid circular disk of uniform thickness as shown in Fig. P16-69. The vertical surfaces in contact with the block are smooth and the disk rolls without slipping. In the position shown, with **F** equal to 75 lb, the velocity of pin A is 2.5 ft/s to the right. Deter-

mine the angular velocity and angular acceleration of bar AB and the force exerted by the pin at B on the block.

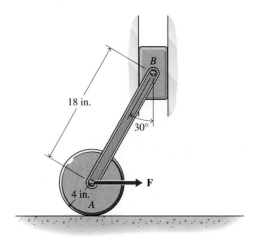

Fig. P16-69

16-70 The slender bar AB shown in Fig. P16-70 rests on a smooth surface at B and is attached to a collar at A that slides freely on the smooth vertical rod. The bar has a uniform cross section and a mass of 20 kg. The bar is initially at rest when vertical ($\theta = 0°$), and when disturbed rotates in a vertical plane under the action of gravity. Determine the angular acceleration of the bar and the reactions at A and B when $\theta = 60°$.

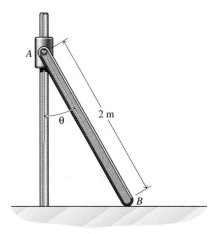

Fig. P16-70

16-71* Bar AB of Fig. P16-71 has a uniform cross section and weighs 25 lb. The collar at end A and the slider at end B have negligible mass and move on smooth surfaces. When in the position shown, bar AB has an angular velocity of 1 rad/s clockwise and an angular acceleration of 2 rad/s² counterclockwise. Determine the force **F** at collar

A and the force exerted by the slot on the slider at end *B* of the bar.

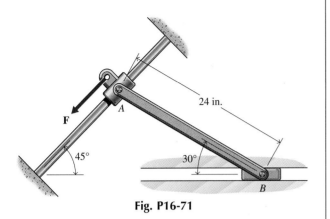

Fig. P16-71

16-72* The slender bar *AB* shown in Fig. P16-72 has a uniform cross section and a mass of 15 kg. Determine the angular acceleration of the bar and the reactions at *A* and *B* in the position shown if *A* is moving to the right with a speed of 2 m/s.

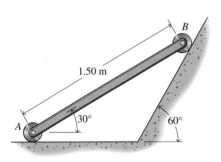

Fig. P16-72

16-73 Bar *AB* of Fig. P16-73 rotates counterclockwise with a constant angular velocity of 10 rad/s. Bar *BC* has a uniform cross section and weighs 10 lb. The solid circular disk *D* weighs 30 lb and rolls without slipping on the horizontal surface. Determine the horizontal and vertical components of the forces exerted by the pins at *B* and *C* on bar *BC* when $\theta = 60°$.

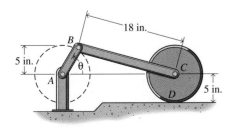

Fig. P16-73

16-74 Bars *AB* and *BC* of Fig. P16-74 have uniform cross sections and masses of 20 and 15 kg, respectively. When the bars are in the position shown, collar *C* is moving to the left with a velocity of 2 m/s and its speed is decreasing at a rate of 4 m/s². Determine the force **F** and the horizontal and vertical components of the force exerted on bar *AB* by the pins at *A* and *B*.

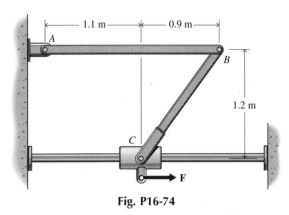

Fig. P16-74

16-75 A circular disk is supported in a vertical plane by the slender bar *AB* shown in Fig. P16-75. The combined bar and disk weigh 161 lb and have a radius of gyration of 6.50 in. with respect to the mass center *G* of the disk. Ends *A* and *B* of the bar roll freely in smooth horizontal and vertical guides. When the bar is in the position shown, end *B* is moving to the left with a velocity of 2.0 ft/s. The force **F** is 50 lb. Determine the angular velocity and angular acceleration of the bar, and the forces exerted on the bar by the pins at *A* and *B*.

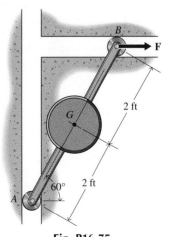

Fig. P16-75

16-76 Bars AB and BC of Fig. P16-76 have uniform cross sections and masses of 10 and 15 kg, respectively. When the bars are in the position shown, collar C is moving to the left with a velocity of 1 m/s and its speed is increasing at a rate of 2 m/s². Determine the force $\mathbf{F}$ and the forces exerted on bar AB by the pins at A and B, respectively.

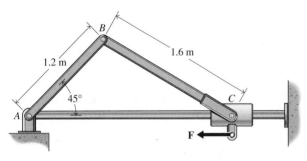

Fig. P16-76

16-77* A thin rectangular plate ($W = 60$ lb) is mounted on a shaft as shown in Fig. P16-77. Determine the bearing reactions when the plate lies in a vertical plane (as shown) and the shaft is rotating at a constant angular velocity of 90 rad/s. Assume that the bearing at B resists any motion of the shaft in the axial direction and that the mass of the shaft is negligible.

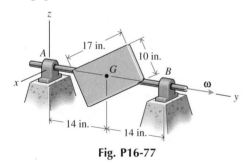

Fig. P16-77

16-78* A 130-mm diameter solid circular cylinder ($m = 50$ kg) is mounted on a shaft as shown in Fig. P16-78. Determine the bearing reactions when the axis of the cylinder lies in a vertical plane (as shown) and the shaft is rotating at a constant angular velocity of 60 rad/s. Assume that the

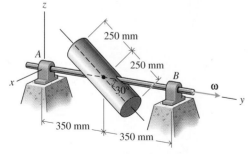

Fig. P16-78

bearing at B resists any motion of the shaft in the axial direction and that the mass of the shaft is negligible.

16-79 Two thin rectangular plates (each weighs 15 lb) are mounted on a shaft as shown in Fig. P16-79. Determine the bearing reactions when the plates lie in a vertical plane (as shown) and the shaft is rotating at a constant angular velocity of 75 rad/s. Assume that the bearing at B resists any motion of the shaft in the axial direction and that the mass of the shaft is negligible.

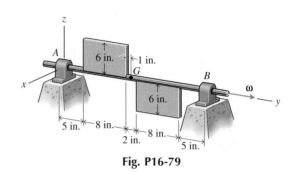

Fig. P16-79

16-80 A thin triangular plate ($m = 10$ kg) is mounted on a shaft as shown in Fig. P16-80. Determine the bearing reactions when the plate lies in a vertical plane (as shown) and the shaft is rotating at a constant angular velocity of 75 rad/s. Assume that the bearing at B resists any motion of the shaft in the axial direction and that the mass of the shaft is negligible.

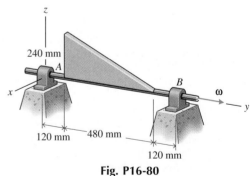

Fig. P16-80

16-81* Solve Problem 16-79 after the plates have rotated 90° and lie in a horizontal plane.

16-82* Solve Problem 16-80 after the plate has rotated 90° and lies in a horizontal plane.

16-83 Solve Problem 16-79 if the shaft is accelerating at a rate of 20 rad/s².

16-84 Solve Problem 16-80 if the shaft is accelerating at a rate of 20 rad/s².

16-85* Two 4-in. diameter spheres (each weighs 10 lb) are attached to a shaft and rotated as shown in Fig. P16-85. The bars connecting the spheres to the shaft have 1 in. diameters, are 7 in. long, and weigh 1.50 lb. The 2-in. diameter shaft weighs 20 lb. Determine the components of the bearing reactions at the supports and the applied torque $\mathbf{T}$ when the angular velocity ω of the shaft is 100 rad/s and increasing at the rate of 20 rad/s^2. Assume that the bearing at A resists any motion of the shaft in the z-direction.

16-86* Two rectangular bars (the mass of each is 5 kg) are attached to a shaft and rotated as shown in Fig. P16-86. The mass of the 40-mm diameter shaft is 6.5 kg. Determine the components of the bearing reactions at the supports and the applied torque $\mathbf{T}$ when the angular velocity ω of the shaft is 150 rad/s and decreasing at the rate of 25 rad/s^2. Assume that the bearing at A resists any motion of the shaft in the z-direction.

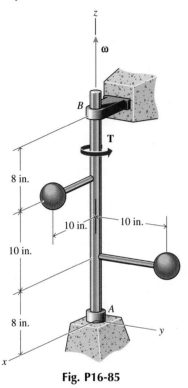

Fig. P16-85

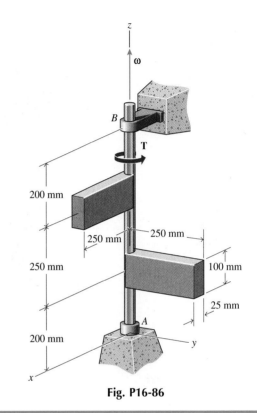

Fig. P16-86

16-5 THREE-DIMENSIONAL MOTION OF A RIGID BODY

The moment $\mathbf{M}_A$ about an arbitrary point A in a body subjected to a system of external forces was developed in Section 16-2 and is given by the equation

$$\mathbf{M}_A = \int_m (\boldsymbol{\rho} \times \mathbf{a}_A)\, dm + \int_m [\boldsymbol{\rho} \times (\dot{\boldsymbol{\omega}} \times \boldsymbol{\rho})]\, dm$$

$$+ \int_m \{\boldsymbol{\rho} \times [\boldsymbol{\omega} \times (\boldsymbol{\omega} \times \boldsymbol{\rho})]\}\, dm \qquad (a)$$

For three-dimensional motion, the different terms appearing in Eq. a are

$$\boldsymbol{\rho} \times \mathbf{a}_A = \begin{vmatrix} \mathbf{i} & \mathbf{j} & \mathbf{k} \\ x & y & z \\ a_{Ax} & a_{Ay} & a_{Az} \end{vmatrix}$$

$$= (ya_{Az} - za_{Ay})\mathbf{i} + (za_{Ax} - xa_{Az})\mathbf{j} + (xa_{Ay} - ya_{Ax})\mathbf{k} \qquad (b)$$

$$\dot{\boldsymbol{\omega}} \times \boldsymbol{\rho} = \begin{vmatrix} \mathbf{i} & \mathbf{j} & \mathbf{k} \\ \dot{\omega}_x & \dot{\omega}_y & \dot{\omega}_z \\ x & y & z \end{vmatrix}$$

$$= (z\dot{\omega}_y - y\dot{\omega}_z)\mathbf{i} + (x\dot{\omega}_z - z\dot{\omega}_x)\mathbf{j} + (y\dot{\omega}_x - x\dot{\omega}_y)\mathbf{k}$$

Similarly

$$\begin{aligned} \boldsymbol{\rho} \times (\dot{\boldsymbol{\omega}} \times \boldsymbol{\rho}) = \ & (y^2\dot{\omega}_x - xy\dot{\omega}_y - xz\dot{\omega}_z + z^2\dot{\omega}_x)\mathbf{i} \\ &+ (z^2\dot{\omega}_y - yz\dot{\omega}_z - xy\dot{\omega}_x + x^2\dot{\omega}_y)\mathbf{j} \\ &+ (x^2\dot{\omega}_z - xz\dot{\omega}_x - yz\dot{\omega}_y + y^2\dot{\omega}_z)\mathbf{k} \end{aligned} \qquad (c)$$

$$\boldsymbol{\omega} \times \boldsymbol{\rho} = (z\omega_y - y\omega_z)\mathbf{i} + (x\omega_z - z\omega_x)\mathbf{j} + (y\omega_x - x\omega_y)\mathbf{k}$$

$$\begin{aligned} \boldsymbol{\omega} \times (\boldsymbol{\omega} \times \boldsymbol{\rho}) = \ & (y\omega_x\omega_y - x\omega_y^2 - x\omega_z^2 + z\omega_x\omega_z)\mathbf{i} \\ &+ (z\omega_y\omega_z - y\omega_z^2 - y\omega_x^2 + x\omega_x\omega_y)\mathbf{j} \\ &+ (x\omega_z\omega_x - z\omega_x^2 - z\omega_y^2 + y\omega_y\omega_z)\mathbf{k} \end{aligned}$$

$$\boldsymbol{\rho} \times [\boldsymbol{\omega} \times (\boldsymbol{\omega} \times \boldsymbol{\rho})]$$

$$\begin{aligned} = \ & (xy\omega_z\omega_x - yz\omega_y^2 + y^2\omega_y\omega_z - z^2\omega_y\omega_z + yz\omega_z^2 - xz\omega_x\omega_y)\mathbf{i} \\ &+ (yz\omega_x\omega_y - zx\omega_z^2 + z^2\omega_z\omega_x - x^2\omega_z\omega_x + zx\omega_x^2 - yx\omega_y\omega_z)\mathbf{j} \\ &+ (zx\omega_y\omega_z - xy\omega_x^2 + x^2\omega_x\omega_y - y^2\omega_x\omega_y + xy\omega_y^2 - zy\omega_z\omega_x)\mathbf{k} \quad (d) \end{aligned}$$

If the moment $\mathbf{M}_A$ is written in Cartesian vector form, the scalar components M_{Ax}, M_{Ay}, and M_{Az} are obtained as follows:

$$\begin{aligned} \mathbf{M}_A &= M_{Ax}\mathbf{i} + M_{Ay}\mathbf{j} + M_{Az}\mathbf{k} \\ &= \int_m (\boldsymbol{\rho} \times \mathbf{a}_A) \, dm + \int_m [\boldsymbol{\rho} \times (\dot{\boldsymbol{\omega}} \times \boldsymbol{\rho})] \, dm + \int_m \{\boldsymbol{\rho} \times [\boldsymbol{\omega} \times (\boldsymbol{\omega} \times \boldsymbol{\rho})]\} \, dm \end{aligned}$$

Substituting Eqs. b, c, and d into Eq. a yields

$$\begin{aligned} M_{Ax} = \ & a_{Az}\int_m y \, dm - a_{Ay}\int_m z \, dm + \dot{\omega}_x\int_m y^2 \, dm \\ &- \dot{\omega}_y\int_m xy \, dm - \dot{\omega}_z\int_m zx \, dm + \dot{\omega}_x\int_m z^2 \, dm \\ &+ \omega_z\omega_x\int_m xy \, dm - \omega_y^2\int_m yz \, dm + \omega_y\omega_z\int_m y^2 \, dm \\ &- \omega_y\omega_z\int_m z^2 \, dm + \omega_z^2\int_m yz \, dm - \omega_x\omega_y\int_m zx \, dm \end{aligned} \qquad (16\text{-}26a)$$

$$\begin{aligned} M_{Ay} = \ & a_{Ax}\int_m z \, dm - a_{Az}\int_m x \, dm + \dot{\omega}_y\int_m z^2 \, dm \\ &- \dot{\omega}_z\int_m yz \, dm - \dot{\omega}_x\int_m xy \, dm + \dot{\omega}_y\int_m x^2 \, dm \\ &+ \omega_x\omega_y\int_m yz \, dm - \omega_z^2\int_m zx \, dm + \omega_z\omega_x\int_m z^2 \, dm \\ &- \omega_z\omega_x\int_m x^2 \, dm + \omega_x^2\int_m zx \, dm - \omega_y\omega_z\int_m xy \, dm \end{aligned} \qquad (16\text{-}26b)$$

$$M_{Az} = a_{Ay} \int_m x \, dm - a_{Ax} \int_m y \, dm + \dot{\omega}_z \int_m x^2 \, dm$$
$$- \dot{\omega}_x \int_m zx \, dm - \dot{\omega}_y \int_m yz \, dm + \dot{\omega}_z \int_m y^2 \, dm$$
$$+ \omega_y\omega_z \int_m zx \, dm - \omega_x^2 \int_m xy \, dm + + \omega_x\omega_y \int_m x^2 \, dm$$
$$- \omega_x\omega_y \int_m y^2 \, dm + \omega_y^2 \int_m xy \, dm - \omega_z\omega_x \int_m yz \, dm \quad (16\text{-}26c)$$

When Eqs. 16-26 are written in terms of first moments, moments of inertia, and products of inertia they become

$$M_{Ax} = a_{Az}\,\bar{y}m - a_{Ay}\,\bar{z}m + I_{Ax}\,\dot{\omega}_x$$
$$- (I_{Ay} - I_{Az})\omega_y\omega_z + I_{Axy}(\omega_z\omega_x - \dot{\omega}_y)$$
$$- I_{Ayz}(\omega_y^2 - \omega_z^2) - I_{Azx}(\omega_x\omega_y + \dot{\omega}_z) \quad (16\text{-}27a)$$
$$M_{Ay} = a_{Ax}\,\bar{z}m - a_{Az}\,\bar{x}m + I_{Ay}\,\dot{\omega}_y$$
$$- (I_{Az} - I_{Ax})\omega_z\omega_x + I_{Ayz}(\omega_x\omega_y - \dot{\omega}_z)$$
$$- I_{Azx}(\omega_z^2 - \omega_x^2) - I_{Axy}(\omega_y\omega_z + \dot{\omega}_x) \quad (16\text{-}27b)$$
$$M_{Az} = a_{Ay}\,\bar{x}m - a_{Ax}\,\bar{y}m + I_{Az}\,\dot{\omega}_z$$
$$- (I_{Ax} - I_{Ay})\omega_x\omega_y + I_{Azx}(\omega_y\omega_z - \dot{\omega}_x)$$
$$- I_{Axy}(\omega_x^2 - \omega_y^2) - I_{Ayz}(\omega_z\omega_x + \dot{\omega}_y) \quad (16\text{-}27c)$$

For most dynamics problems, an instantaneous relationship between the moments and accelerations is desired. A considerable simplification then results by choosing the xyz-coordinate system to coincide with the principal axes through the mass center G of the body at the desired instant. With the origin at the center of mass

$$\bar{x} = \bar{y} = \bar{z} = 0$$

and for principal axes

$$I_{xy} = I_{yz} = I_{zx} = 0$$

Thus

$$M_{Gx} = I_{Gx}\dot{\omega}_x - (I_{Gy} - I_{Gz})\omega_y\omega_z$$
$$M_{Gy} = I_{Gy}\dot{\omega}_y - (I_{Gz} - I_{Gx})\omega_z\omega_x \quad (16\text{-}28)$$
$$M_{Gz} = I_{Gz}\dot{\omega}_z - (I_{Gx} - I_{Gy})\omega_x\omega_y$$

Equations 16-28 are known as Euler's[1] equations.

Euler's equations are valid only instantaneously. If it is necessary to integrate the accelerations to find the velocities, general expressions must be established for the moments of the forces and moments of inertia. Only for highly symmetric bodies will these moments of inertia be constant.

Equations 15-17 together with Eqs. 16-28 provide the relationships required to solve a variety of three-dimensional motion problems. Thus,

$$\Sigma F_x = ma_{Gx} \quad \Sigma M_{Gx} = I_{Gx}\dot{\omega}_x - (I_{Gy} - I_{Gz})\omega_y\omega_z$$
$$\Sigma F_y = ma_{Gy} \quad \Sigma M_{Gy} = I_{Gy}\dot{\omega}_y - (I_{Gz} - I_{Gx})\omega_z\omega_x \quad (16\text{-}29)$$
$$\Sigma F_z = ma_{Gz} \quad \Sigma M_{Gz} = I_{Gz}\dot{\omega}_z - (I_{Gx} - I_{Gy})\omega_x\omega_y$$

The procedure for solving three-dimensional motion problems is illustrated in the following examples.

[1] Leonhard Euler (1707–1783), a Swiss mathematician.

EXAMPLE PROBLEM 16-13

Solve Example 16-11 by using Euler's equations.

SOLUTION

Figure 16-25 from Example 16-11, which shows the geometry of the system and a free-body diagram for the cylinder and shaft, is repeated here for convenience. From Example 16-11

$$W = 161.2 \text{ lb} \qquad I_{Gy'} = 0.6459 \text{ slug} \cdot \text{ft}^2$$
$$I_{Gx'} = 0.6459 \text{ slug} \cdot \text{ft}^2 \qquad I_{Gz'} = 1.2556 \text{ slug} \cdot \text{ft}^2$$
$$\omega_z = 52.36 \text{ rad/s} \qquad \alpha_z = 5.236 \text{ rad/s}^2$$

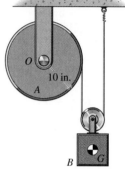

Fig. 16-25

The components of ω_z and α_z about the principal axes are

$$\omega_{x'} = \omega_z \cos \theta_{x'z} = 52.36 \cos 110° = -17.91 \text{ rad/s}$$
$$\omega_{y'} = \omega_z \cos \theta_{y'z} = 52.36 \cos 90° = 0$$
$$\omega_{z'} = \omega_z \cos \theta_{z'z} = 52.36 \cos 20° = 49.20 \text{ rad/s}$$
$$\alpha_{x'} = \alpha_z \cos \theta_{x'z} = 5.236 \cos 110° = -1.791 \text{ rad/s}^2$$
$$\alpha_{y'} = \alpha_z \cos \theta_{y'z} = 5.236 \cos 90° = 0$$
$$\alpha_{z'} = \alpha_z \cos \theta_{z'z} = 5.236 \cos 20° = 4.920 \text{ rad/s}^2$$

Then, from Eqs. 16-28

$$\Sigma M_{Gx'} = I_{Gx'}\alpha_{x'} = 0.6459(-1.791) = -1.1568 \text{ ft} \cdot \text{lb}$$
$$\Sigma M_{Gy'} = -(I_{Gz'} - I_{Gx'})\omega_{z'}\omega_{x'}$$
$$= -(1.2556 - 0.6459)(49.20)(-17.91) = 537.3 \text{ ft} \cdot \text{lb}$$
$$\Sigma M_{Gz'} = I_{Gz'}\alpha_{z'} = 1.2556(4.920) = 6.178 \text{ ft} \cdot \text{lb}$$

Moments about the x-, y-, and z-axes are then obtained as

$$\Sigma M_{Gx} = \Sigma M_{Gx'} \cos \theta_{x'x} + \Sigma M_{Gz'} \cos \theta_{z'x}$$
$$= -1.1568 \cos 20° + 6.178 \cos 70° = 1.026 \text{ ft} \cdot \text{lb}$$
$$\Sigma M_{Gy} = \Sigma M_{Gy'} = 537.3 \text{ ft} \cdot \text{lb}$$
$$\Sigma M_{Gz} = \Sigma M_{Gx'} \cos \theta_{x'z} + \Sigma M_{Gz'} \cos \theta_{z'z}$$
$$= -1.1568 \cos 110° + 6.178 \cos 20° = 6.201 \text{ ft} \cdot \text{lb}$$

Thus, since $\mathbf{a}_G = \mathbf{0}$, Eqs. 16-29 reduce to

$$\Sigma F_x = 0 \qquad\qquad \Sigma F_y = 0 \qquad\qquad \Sigma F_z = 0$$
$$\Sigma M_{Gx} = 1.026 \text{ ft} \cdot \text{lb} \qquad \Sigma M_{Gy} = 537.3 \text{ ft} \cdot \text{lb} \qquad \Sigma M_{Gz} = 6.201 \text{ ft} \cdot \text{lb}$$

$$+\uparrow \Sigma F_x = A_x + B_x - W = A_x + B_x - 161.16 = 0 \qquad (a)$$
$$+\swarrow \Sigma F_y = A_y + B_y = 0 \qquad (b)$$
$$+\rightarrow \Sigma F_z = B_z = 0 \qquad B_z = 0 \qquad \text{Ans.}$$
$$+\downarrow\!\!\!\downarrow \Sigma M_{Gx} = A_y(1.25) - B_y(1.25) = 1.026$$
$$A_y - B_y = 0.8208 \qquad (c)$$
$$+\downarrow\!\!\!\downarrow \Sigma M_{Gy} = B_x(1.25) - A_x(1.25) = 537.3$$
$$B_x - A_x = 429.8 \qquad (d)$$
$$+\downarrow\!\!\!\downarrow \Sigma M_{Gz} = T = 6.201 \qquad T = 6.20 \text{ ft} \cdot \text{lb}\downarrow\!\!\!\downarrow \qquad \text{Ans.}$$

From Eqs. a and d

$$A_x = -134.32 = 134.3 \text{ lb}\downarrow \qquad \text{Ans.}$$
$$B_x = +295.4 = 295 \text{ lb}\uparrow \qquad \text{Ans.}$$

From Eqs. b and c

$$A_y = +0.4104 = 0.410 \text{ lb}\swarrow \qquad \text{Ans.}$$
$$B_y = -0.4104 = 0.410 \text{ lb}\nearrow \qquad \text{Ans.}$$

EXAMPLE PROBLEM 16-14

The combined masses of the shaft and cylinder shown in Fig. 16-27a is 20 kg. The moments of inertia of the combined shaft and cylinder about the x-, y-, and z-axes through the center of mass G are $I_{Gx} = I_{Gz} = 0.1595$ kg·m^2 and $I_{Gy} = 0.0625$ kg·m^2. If the cylinder is rotating at a constant angular velocity of 75 rad/s and the frame is rotating at a constant angular velocity of 25 rad/s, determine the reactions at supports A and B of the shaft. Assume that the bearing at B resists any force along the axis of the shaft.

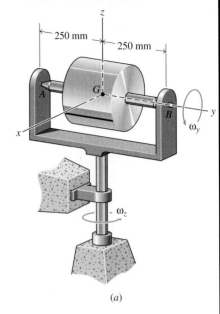

SOLUTION

For the motion illustrated in Fig. 16-27a, $a_{Gx} = a_{Gy} = a_{Gz} = \omega_x = 0$. Although the magnitude of the angular velocity vector is constant ($\dot{\omega}_x = \dot{\omega}_y = \dot{\omega}_z = 0$), its direction is changing. Therefore, the angular acceleration $\boldsymbol{\alpha} \neq \mathbf{0}$ and Eqs. 16-29 reduce to

$$\Sigma F_x = 0 \qquad \Sigma M_{Gx} = I_{Gx}\alpha_x - (I_{Gy} - I_{Gz})\omega_y\omega_z$$

$$\Sigma F_y = 0 \qquad \Sigma M_{Gy} = I_{Gy}\alpha_y$$

$$\Sigma F_z = 0 \qquad \Sigma M_{Gz} = I_{Gz}\alpha_{xz}$$

where the angular acceleration is calculated (see Example Problem 14-12)

$$\begin{aligned}
\boldsymbol{\alpha} &= \dot{\boldsymbol{\omega}} = \dot{\omega}_y\mathbf{e}_{AB} + \omega_y\dot{\mathbf{e}}_{AB} + \dot{\omega}_z\mathbf{k} \\
&= \omega_y(\omega_z\mathbf{k} \times \mathbf{e}_{AB}) = (75)(25)(-\mathbf{e}_x) \\
&= -1875\mathbf{i} \text{ rad/s}^2
\end{aligned}$$

A free-body diagram for the shaft and cylinder is shown in Fig. 16-27b. From the equations of motion

$$+\swarrow\Sigma F_x = A_x + B_x = 0 \tag{a}$$

$$+\rightarrow\Sigma F_y = B_y = 0 \tag{b}$$

$$+\uparrow\Sigma F_z = A_z + B_z - mg = 0 \tag{c}$$

$$\begin{aligned}
+\downarrow\Sigma M_{Gx} &= B_z(0.250) - A_z(0.250) \\
&= I_{Gx}\alpha_x - (I_{Gy} - I_{Gz})\omega_y\omega_z
\end{aligned} \tag{d}$$

$$+\downarrow\Sigma M_{Gy} = 0 \tag{e}$$

$$+\downarrow\Sigma M_{Gz} = A_x(0.250) - B_x(0.250) = 0 \tag{f}$$

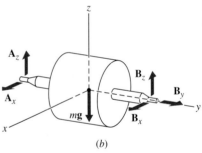

Fig. 16-27

From Eqs. a, b, and f

$$A_x = B_x = B_y = 0 \qquad \text{Ans.}$$

From Eqs. c and d

$$A_z + B_z = 20(9.81) = 196.2$$

$$\begin{aligned}
A_z - B_z &= \frac{1}{0.250}\Big((0.0625 - 0.1595)(75)(25) - (0.1595)(-1875)\Big) \\
&= 468.75
\end{aligned}$$

Therefore:

$$A_z = 332.48 = 332 \text{ N}\uparrow \qquad \text{Ans.}$$

$$B_z = -136.28 = 136.3 \text{ N}\downarrow \qquad \text{Ans.}$$

PROBLEMS

16-87* The two thin quarter-circular plates shown in Fig. P16-87 each weigh 20 lb. Determine the bearing reactions when the plates lie in a vertical plane (as shown) and the angular velocity and angular acceleration of the shaft are 100 rad/s and 25 rad/s², respectively. Assume the bearing at B resists any motion of the shaft in the axial direction and that the mass of the shaft is negligible.

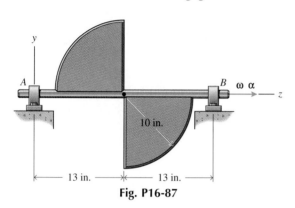

Fig. P16-87

16-88* A 75-mm diameter solid circular cylinder with hemispherical ends is mounted on a shaft as shown in Fig. P16-88. The mass of the cylinder is 6 kg and the mass of each hemisphere is 1 kg. Determine the bearing reactions when the axis of the cylinder lies in a vertical plane (as shown) and the angular velocity and angular acceleration of the shaft are 50 rad/s and 15 rad/s², respectively. Assume the bearing at B resists any motion of the shaft in the axial direction and that the mass of the shaft is negligible.

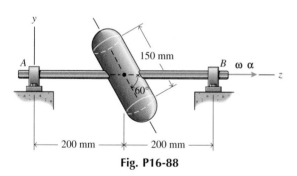

Fig. P16-88

16-89 Repeat Problem 16-87 for the case in which the plates lie in a horizontal plane (rotated 90° about the z-axis). Use the same angular velocity and angular acceleration as before.

16-90 Repeat Problem 16-88 for the case in which the cylinder lies in a horizontal plane (rotated 270° about the z-axis). Use the same angular velocity and angular acceleration as before.

16-91* The crank shown in Fig. P16-91 is rotating counterclockwise with a constant angular velocity of 20 rad/s. Bar AB weighs 20 lb and is connected to the crank at A and to the slider at B with ball and socket joints. Determine the reactions at ends A and B of the bar when the crank is in the position shown in the figure.

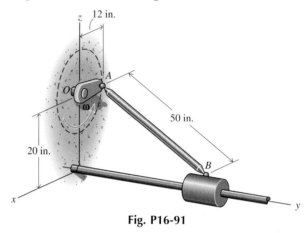

Fig. P16-91

16-92* A slender rod AB with a mass of 5 kg and a sphere with a mass of 6 kg are supported on a vertical shaft as shown in Fig. P16-92. Determine the reaction at A and the tension in cable CD when the shaft is rotating counterclockwise with a constant angular velocity of 30 rad/s.

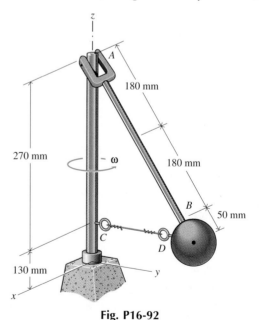

Fig. P16-92

16-93 Solve Problem 16-91 if the crank has a clockwise angular velocity of 25 rad/s and a counterclockwise angular acceleration of 5 rad/s².

16-94 Solve Problem 16-92 if the shaft has a counterclockwise angular velocity of 30 rad/s and a clockwise angular acceleration of 10 rad/s².

16-95* The solid circular disk shown in Fig. P16-95 weighs 25 lb. The disk is rotating on shaft AB with a constant angular velocity of 500 rpm. At the same time, the shaft AB is rotating in a vertical plane about a pin at support A. Determine the reaction at support A when the system is in the position shown and the angular velocity and angular acceleration of the shaft are $\omega_z = 20$ rad/s and $\alpha_z = 5$ rad/s². Assume the mass of the shaft is negligible.

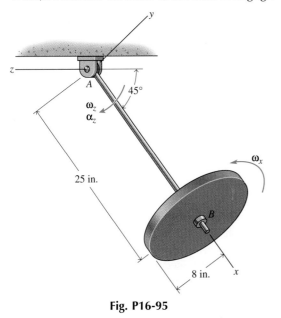

Fig. P16-95

16-96* The solid circular disk shown in Fig. P16-96 rolls without slipping along a circular path on the horizontal surface as the shaft on which it is mounted rotates about the vertical post. The mass of the disk is 50 kg. Assume that the shaft supporting the disk has a negligible mass and the bearings at A and B both slide freely on the vertical post. Determine the bearing reactions and the force between the disk and the horizontal surface when the shaft is rotating with a constant angular velocity $\omega_y = 25$ rad/s.

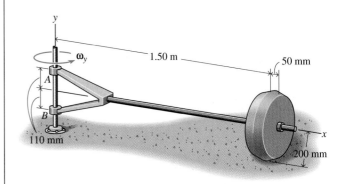

Fig. P16-96

16-6 D'ALEMBERT'S PRINCIPLE—REVERSED EFFECTIVE FORCES

Newton's second law of motion as applied to a particle or to the mass center of a rigid body is given by Eq. 15-16 as

$$\mathbf{R} = \Sigma\mathbf{F} = m\mathbf{a}_G \qquad (15\text{-}16)$$

Jean Le Rond d'Alembert (1717–1783) was the first to suggest that a system of inertia forces ($m\mathbf{a}_G$) can be added to a system of actual forces in a dynamics problem to obtain a force system that is in equilibrium.[2] The process, known as d'Alembert's principle, can be expressed mathematically as

$$\mathbf{R} + (-m\mathbf{a}_G) = \mathbf{R} + \mathbf{F}_{re} = 0 \qquad (16\text{-}30)$$

The term $\mathbf{F}_{re} = (-m\mathbf{a}_G)$ in Eq. 16-30 is known as a reversed effective force. Reversed effective forces are not true forces since they do not represent the action of a different body on the body of interest.

[2] Dr. Ernst Mach, *The Science of Mechanics*, 9th ed., The Open Court Publishing Company, LaSalle, Ill., 1942. Originally published in German in 1893 and translated from German to English by Thomas H. McCormack in 1902.

For problems involving translation of a rigid body, solution by use of d'Alembert's principle is accomplished by placing the reversed effective force $\mathbf{F}_{re} = (-m\mathbf{a}_G)$ at the mass center of the body when the free-body diagram is drawn. The equilibrium equations $\Sigma\mathbf{F} = \mathbf{0}$ and $\Sigma\mathbf{M} = \mathbf{0}$ are then applied using all forces on the free-body diagram (including the reversed effective force). Moment equations used for solution of the problem can be written with respect to points on or off the body. By selecting a moment center to eliminate a number of unknowns from the moment equation, the need to solve simultaneous equations is frequently avoided.

Application of d'Alembert's principle becomes involved when the body has angular motion. For a rigid body experiencing plane motion, reversed effective couples in addition to reversed effective forces must be applied to the free-body diagram. With the xy-plane as the plane of motion and the mass center G as the origin of the xyz-coordinate system, the reversed effective forces and couples that must be placed on the free-body diagram are

$$
\begin{aligned}
\mathbf{F}_{rex} &= -ma_{Gx}\mathbf{i} & \mathbf{C}_{rex} &= -(-\alpha I_{Gzx} + \omega^2 I_{Gyz})\mathbf{i} \\
\mathbf{F}_{rey} &= -ma_{Gy}\mathbf{j} & \mathbf{C}_{rey} &= -(-\alpha I_{Gyz} - \omega^2 I_{Gzx})\mathbf{j} \quad (16\text{-}31) \\
\mathbf{F}_{rez} &= \mathbf{0} & \mathbf{C}_{rez} &= -(\alpha I_{Gz})\mathbf{k}
\end{aligned}
$$

The reversed effective forces must be placed at the mass center of the body. The reversed effective couples can be placed anywhere on the body. Note that the moments and products of inertia in Eqs. 16-31 are with respect to axes through the mass center of the body. Again, the equilibrium equations $\Sigma\mathbf{F} = \mathbf{0}$ and $\Sigma\mathbf{M} = \mathbf{0}$ can be used to solve the motion problem by using the reversed effective forces, the reversed effective couples, and the applied forces and couples shown on the free-body diagram. Proper selection of moment centers for moment equations can simplify the solution.

For general three-dimensional motion problems with principal axes and the origin of the xyz-coordinate system at the mass center G of the body, the reversed effective forces and couples required to solve a dynamics problem by using d'Alembert's principle are

$$
\begin{aligned}
\mathbf{F}_{rex} &= -ma_{Gx}\mathbf{i} & \mathbf{C}_{rex} &= -[I_{Gx}\dot{\omega}_x - (I_{Gy} - I_{Gz})\omega_y\omega_z]\mathbf{i} \\
\mathbf{F}_{rey} &= -ma_{Gy}\mathbf{j} & \mathbf{C}_{rey} &= -[I_{Gy}\dot{\omega}_y - (I_{Gz} - I_{Gx})\omega_z\omega_x]\mathbf{j} \quad (16\text{-}32) \\
\mathbf{F}_{rez} &= -ma_{Gz}\mathbf{k} & \mathbf{C}_{rez} &= -[I_{Gz}\dot{\omega}_z - (I_{Gx} - I_{Gy})\omega_x\omega_y]\mathbf{k}
\end{aligned}
$$

D'Alembert's principle provides an alternative method for solving dynamics problems, which some instructors find appealing. Since the method does not provide additional information, it will not be pursued further in this text. The following two example problems are provided to illustrate the procedure for those with an interest in the method.

An automobile with a wheelbase of 114 in. weighs 3500 lb. The center of mass of the car is 48 in. behind the front axle and 22 in. above the surface of the road. Determine the normal forces exerted by the road on the front and rear wheels as the speed of the car is reduced uniformly from 60 mi/h to 30 mi/h in a distance of 150 ft on a level stretch of road.

SOLUTION

The initial and final velocities v_i and v_f of the car are

$$v_i = \frac{60(5280)}{3600} = 88.00 \text{ ft/s}$$
$$v_f = \frac{30(5280)}{3600} = 44.0 \text{ ft/s}$$

The acceleration of the car is

$$a = \frac{v_f^2 - v_i^2}{2\,s}$$
$$= \frac{44^2 - 88^2}{2(150)} = -19.36 \text{ ft/s}^2$$

The reversed effective force is

$$F_{re} = -ma$$
$$= -\frac{3500}{32.2}(-19.36) = 2104 \text{ lb}$$

A free-body diagram of the car is shown in Fig. 16-28. The reversed effective force F_{re} is directed forward since the acceleration of the car is toward the rear. The front wheel reaction N_B is determined by using a moment equation with the moment center at A to eliminate the frictional forces F_A and F_B. Similarly, the rear wheel reaction N_A is calculated by using a moment equation with the moment center at B. Thus,

$$+\curvearrowleft \Sigma M_A = N_B(114) - 3500(66) - 2104(22) = 0 \qquad N_B = 2432 \text{ lb} \qquad \text{Ans.}$$
$$+\curvearrowright \Sigma M_B = N_A(114) - 3500(48) + 2104(22) = 0 \qquad N_A = 1068 \text{ lb} \qquad \text{Ans.}$$

As a check

$$+\uparrow \Sigma F = N_A + N_B - W = 2432 + 1068 - 3500 = 0$$

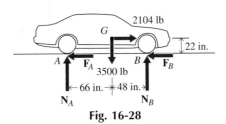

Fig. 16-28

EXAMPLE PROBLEM 16-16

A thin 600-mm diameter circular disk with a mass of 60 kg is supported on an inclined surface by a block and a cable wrapped around a shallow groove in the curved surface of the disk as shown in Fig. 16-29a. Determine the tension T in the cable and the acceleration a_G of the mass center of the disk once the block is removed and the disk is free to slide on the inclined surface. The kinetic coefficient of friction between the disk and the inclined surface is 0.20.

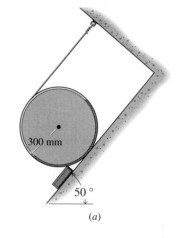

(a)

SOLUTION

A free-body diagram for the disk is shown in Fig. 16-29b. Motion of the disk is possible only if slipping occurs; therefore, $F = \mu N$. Also, the point where the cable leaves the surface of the disk is an instantaneous center of rotation; therefore, $\mathbf{a}_G = R\boldsymbol{\alpha} = 0.300\boldsymbol{\alpha}$ or $\alpha = 3.333 a_G$

From Appendix B

$$I_G = \frac{1}{2}mR^2$$

$$= \frac{1}{2}(60)(0.300)^2 = 2.70 \text{ kg} \cdot \text{m}^2$$

By definition

$$F_{re} = -ma_g = -60a_G$$

$$C_{re} = -I_G\alpha = -2.70(3.333a_G) = -9.00a_G$$

Application of the three equilibrium equations $\Sigma F_x = 0$, $\Sigma F_y = 0$, and $\Sigma M = 0$ about any point on or off the disk yields the remaining unknowns for the problem. Thus,

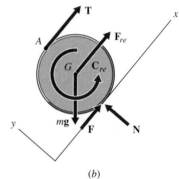

(b)

Fig. 16-29

$$+\nwarrow\Sigma F_y = N - mg \cos 50° = 0$$

$$N = mg \cos 50°$$

$$= 60(9.81) \cos 50° = 378.3 = 378 \text{ N}\nwarrow$$

$$+\curvearrowleft\Sigma M_A = C_{re} + F_{re}R - mg \sin 50°(R) + \mu N(2R) = 0$$

$$-9.00a_G - 60a_G(0.300) = 60(9.81)(\sin 50°)(0.300)$$

$$- 0.20(378.3)(0.600)$$

From which

$$a_G = -3.328 = 3.33 \text{ m/s}^2\swarrow \qquad\qquad \text{Ans.}$$

$$+\nearrow\Sigma F_x = F + T + F_{re} - mg \sin 50° = 0$$

$$T = mg \sin 50° - \mu N + 60a_G$$

$$= 60(9.81) \sin 50° - 0.20(378.3) + 60(-3.328)$$

$$= 175.55 = 175.6 \text{ N}$$

SUMMARY

Any system of forces acting on a rigid body can be replaced by an equivalent system that consists of a resultant force $\mathbf{R}$ whose line of action passes through the mass center G of the body and a resultant couple $\mathbf{C}$. The motion of the mass center G of the body is governed by Newton's second law, which can be expressed in equation form as

$$\mathbf{R} = m\mathbf{a}_G \tag{15-16}$$

Equation 15-16 expresses the fact that the magnitudes of $\mathbf{R}$ and $\mathbf{a}_G$ are proportional and that the vectors $\mathbf{R}$ and $\mathbf{a}_G$ have the same direction. Equation 15-16 is valid both for constant forces and for forces that vary with time. The system of axes used for the measurement of the acceleration $\mathbf{a}_G$ must be a *primary inertial system* (have a constant orientation with respect to the stars). Any *nonrotating system of axes,* however, which *translates* with a constant velocity with respect to the primary system, is equally satisfactory. Equation 15-16 is not valid if $\mathbf{a}_G$ represents a relative acceleration measured with respect to a system of rotating axes.

The actual motion of most rigid bodies consists of the superposition of the translation produced by the resultant force $\mathbf{R}$ and the rotation produced by the couple $\mathbf{C}$. For an *xyz*-coordinate system with its origin at the center of mass G of the body and coordinate directions selected to coincide with the principal axes through the mass center, the instantaneous relationships between the moment components of the couple $\mathbf{C}$ and the angular velocities, angular accelerations, and inertia properties of the rigid body are given by Euler's equations as

$$\begin{aligned} M_{Gx} &= I_{Gx}\dot{\omega}_x - (I_{Gy} - I_{Gz})\omega_y\omega_z \\ M_{Gy} &= I_{Gy}\dot{\omega}_y - (I_{Gz} - I_{Gx})\omega_z\omega_x \\ M_{Gz} &= I_{Gz}\dot{\omega}_z - (I_{Gx} - I_{Gy})\omega_x\omega_y \end{aligned} \tag{16-28}$$

A very large number of dynamics problems involve plane motion. Plane motion of a rigid body is defined as motion in which all elements of the body move in parallel planes. The parallel plane that passes through the mass center G of the body is known as the "plane of motion." When a rigid body is experiencing plane motion, the angular velocity and angular acceleration vectors $\boldsymbol{\omega}$ and $\boldsymbol{\alpha}$, respectively, are parallel to each other and perpendicular to the plane of motion. If the *xyz*-coordinate system is chosen so that the motion is parallel to the *xy*-plane, Eqs. 16-28 reduce to

$$\begin{aligned} M_{Gx} &= 0 \\ M_{Gy} &= 0 \\ M_{Gz} &= I_{Gz}\alpha \end{aligned} \tag{16-5}$$

Plane motion problems are classified in three distinct categories, which depend on the nature of the motion (1) translation, (2) fixed-axis rotation, and (3) general plane motion, which is a combination of translation and fixed-axis rotation. The motion is defined as translation when every line in the body remains parallel to its initial position during the motion. The resultant of the external forces acting on a body in translation is a force $\mathbf{R}$ whose line of action passes through the mass center G of the body. The equations of motion for translation with the origin O of the xyz-coordinate system at the mass center G of the body are

$$\Sigma F_x = ma_{Gx}$$
$$\Sigma F_y = ma_{Gy} \qquad (16\text{-}18)$$
$$\Sigma M_{Gz} = 0$$

When all elements of a body travel in circular paths about a fixed axis, the motion is defined as fixed-axis rotation. The equations of motion for a rigid body symmetric with respect to the plane of motion and rotating about a fixed axis through the mass center G ($\mathbf{a}_G = \mathbf{0}$) are

$$\Sigma F_x = ma_{Gz} = 0$$
$$\Sigma F_y = ma_{Gy} = 0 \qquad (16\text{-}20)$$
$$\Sigma M_{Gz} = I_{Gz}\alpha$$

Rotation about fixed axes other than those through the mass center G of the body are also encountered. The equations of motion for a rigid body symmetric with respect to the plane of motion and rotating about a fixed axis through an arbitrary point A ($\mathbf{a}_A = \mathbf{0}$) on the x-axis are

$$\Sigma F_x = ma_{Gx} = -m\bar{x}\omega^2$$
$$\Sigma F_y = ma_{Gy} = m\bar{x}\alpha \qquad (16\text{-}21)$$
$$\Sigma M_{Az} = I_{Az}\alpha$$

REVIEW PROBLEMS

16-97* The speed of the 3100-lb automobile shown in Fig. P16-97 is reduced uniformly from 60 mi/h to 12 mi/h in a distance of 150 ft. Determine the normal forces exerted by the pavement on the front and rear wheels during the braking action and the minimum coefficient of friction required between the tires and the pavement. The car has four-wheel brakes.

16-98* Two crates A and B are connected by a cable as shown in Fig. P16-98. The mass of crate A is 50 kg; the mass of crate B is 40 kg. Both crates are symmetric with respect to the plane of motion, and the cable AB and the force $\mathbf{P}$ are in the plane of motion. The surface supporting the crates is smooth. Determine the maximum force $\mathbf{P}$ that can be applied to crate B before either crate is on the verge of tipping.

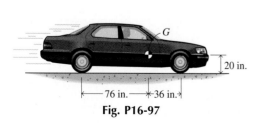

Fig. P16-97

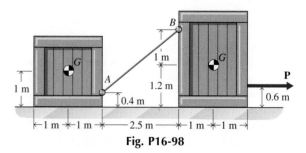

Fig. P16-98

16-99 A 50-lb block rests on a 120-lb platform that is supported by four weightless rods as shown in Fig. P16-99. The coefficient of friction between the platform and the block is 0.10. The platform is released at rest in the position shown by cutting the cable at A. Determine the acceleration of the mass center of the block and the forces exerted on the platform by the rods at A and C at the instant motion begins.

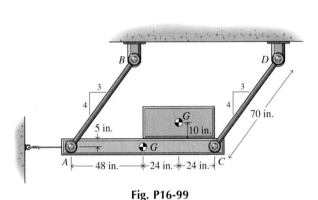

Fig. P16-99

16-100 The platform and lever system shown in Fig. P16-100 is used to transfer boxes between floors in a factory. In the position shown, lever AB is rotating clockwise with an angular velocity of 0.5 rad/s and the velocity is decreasing at a rate of 1.5 rad/s^2. The mass of the box is 500 kg. At this instant, determine the vertical components of the forces exerted on the box at supports C and D and the minimum coefficient of friction required to prevent slipping of the box.

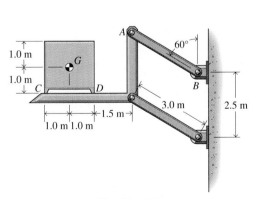

Fig. P16-100

16-101* The slender bar shown in Fig. P16-101 is rotating in a vertical plane about the smooth pin at support B. The bar has a uniform cross section and weighs 42 lb. When the bar is in the position shown, its angular velocity is 20 rad/s counterclockwise and its angular acceleration is 5 rad/s^2 clockwise. Determine the magnitude of the couple C being applied to the bar and the force exerted on the bar by the pin at support B.

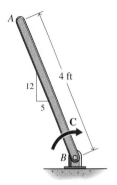

Fig. P16-101

16-102* The mass of the turntable system shown in Fig. P16-102 is 10 kg and the radius of gyration with respect to the axis of rotation is 350 mm. A 1-kg block B rests on the turntable in the position shown. The coefficient of friction between the block and the table is 0.55. If a constant 5 N·m couple C is applied when the system is at rest, determine the number of rotations before the block begins to slip and the angular velocity of the table when slipping begins.

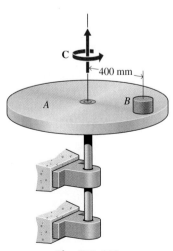

Fig. P16-102

16-103 A semicircular plate of uniform thickness is supported by two cables as shown in Fig. P16-103. The plate weighs 100 lb. If the cable at B breaks, determine the acceleration of the mass center of the plate and the reaction at support A the instant motion begins.

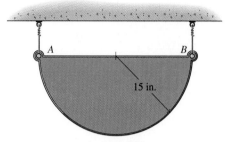

Fig. P16-103

16-104 The slender bar AB shown in Fig. P16-104 has a uniform cross section and a mass of 15 kg. Determine the acceleration of the mass center of the bar and the reaction at support A immediately after the cord at support B is cut if

a. The horizontal surface at A is smooth ($\mu = 0$).
b. The horizontal surface at A is rough ($\mu = 0.25$).

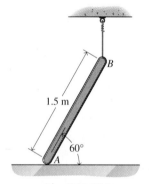

Fig. P16-104

16-105* Bar AB of Fig. P16-105 weighs 7 lb and bar BC weighs 10 lb. The surface at support C is smooth. In the position shown, the angular velocity of bar AB is 5 rad/s and the angular acceleration is 2 rad/s², both counterclockwise. Determine the forces exerted on bar BC by the pin at B and the surface at C.

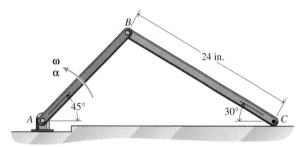

Fig. P16-105

16-106* A bar ($m_B = 20$ kg) with a uniform cross section is attached to a solid circular disk ($m_D = 5$ kg) at end A and to a slider block of negligible mass in a smooth vertical slot at end B, as shown in Fig. P16-106. The 200-mm diameter disk rolls without slipping on the horizontal surface and at the instant shown has an angular velocity of 15 rad/s counterclockwise and an angular acceleration of 25 rad/s² clockwise. Determine the force **F** being applied to pin A and the forces exerted on the bar by the pins at A and B.

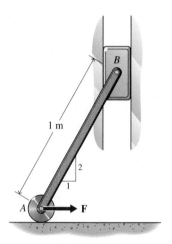

Fig. P16-106

16-107 Bar BC of Fig. P16-107 has a uniform cross section and weighs 25 lb. If disk A is rotating clockwise with a constant angular velocity of 100 rpm, determine the acceleration of the mass center of bar BC and the forces exerted on the bar by the pins at B and D. Assume that the collar at D slides freely on bar BC.

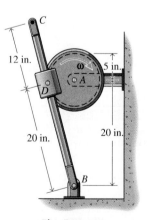

Fig. P16-107

16-108 A bar (m_B = 15 kg) with a uniform cross section is attached to a solid circular disk (m_D = 30 kg) at end A and to a slider block of negligible mass in a smooth horizontal slot at end B, as shown in Fig. P16-108. The 500-mm diameter disk rolls without slipping on the horizontal surface and at the instant shown has an angular velocity of 25 rad/s clockwise and an angular acceleration of 50 rad/s² counterclockwise. Determine the magnitude of the couple **C** being applied to the disk and the forces exerted on the bar by the pins at A and B.

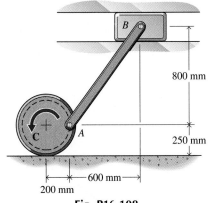

Fig. P16-108

Computer Problems

C16-109 A 16-lb rectangular plate swings at the end of two identical slender links as shown in Fig. P16-109. If the mass of the links can be neglected, and the system starts from rest when $\theta = \theta_0 = 20°$, calculate and plot:

a. The angular velocity of the links $\dot{\theta}$ as a function of their angular position θ ($20° \le \theta \le 160°$).
b. The tensions T_{AB} and T_{CD} in the two links as a function of their angular position θ ($20° \le \theta \le 160°$).
c. What is the smallest initial angle θ_0 for which both tensions T_{AB} and T_{CD} will always be positive?

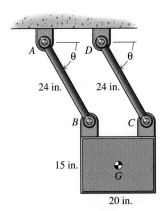

Fig. P16-109

C16-110 The platform and lever system shown in Fig. P16-110 is used to transfer boxes between floors in a factory. The mass of the crate is 120 kg, the mass of the platform is 30 kg, and their combined mass center is at the point G. If the platform is slowly lowered at a constant rate of $\dot{\theta} = 0.5$ rad/s by means of a torque T applied to the lever arm AB, calculate and plot:

a. The required torque T as a function of the angular position θ ($5° \le \theta \le 175°$).
b. The forces exerted on the platform by the lever arms at A and C as a function of θ ($5° \le \theta \le 175°$).

c. The normal and friction forces exerted on the bottom of the crate by the platform as a function of θ ($5° \le \theta \le 175°$).

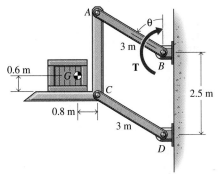

Fig. P16-110

C16-111 The flywheel shown in Fig. P16-111 is rotating at a constant angular velocity of 5 rad/s in a counterclockwise direction. The bar AB is 20 in. long, weighs 5 lb, and has a mass moment of inertia of $I_A = 0.12$ slug · ft² relative to point A. Calculate and plot:

a. The force exerted on the pin P by the bar AB as a function of the angular position θ ($0° \le \theta \le 360°$).
b. The force exerted on the bar AB by the support at A as a function of θ ($0° \le \theta \le 360°$).

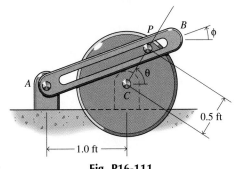

Fig. P16-111

C16-112 The 5-kg uniform bar AB shown in Fig. P16-112 rotates in a vertical plane. If the bar is released from rest when $\theta \cong 0°$, calculate and plot:

a. The acceleration of the mass center of the bar a_{Gx} and a_{Gy} as a function of its angular position θ ($0° \leq \theta \leq 90°$). (Note that $a_{Gx} \neq 0$ and $a_{Gy} \neq -g$.)
b. The force exerted on the bar at A by the support as a function of θ ($0° \leq \theta \leq 90°$).

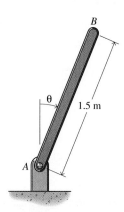

Fig. P16-112

C16-113 The 10-lb uniform bar AB shown in Fig. P16-113 rotates in a vertical plane. The coefficient of friction between the bar and the surface at A is 0.6. If the bar is released from rest when $\theta \cong 0°$, calculate and plot:

a. The normal and friction forces exerted on the bar at A as a function of its angular position θ ($0° \leq \theta \leq 90°$).
b. The location of the mass center of the bar y_G versus x_G as the bar falls.
c. The motion of end A of the bar x_A and y_A as a function of t.
d. At what angle does the bar begin to slip on the surface? When it begins to slip, does A slip to the left or to the right? Does A ever lift away from the surface?

(Note that as the bar falls, the normal force decreases and the friction eventually is not sufficient to keep the bar from sliding. Therefore, it will probably be necessary to use the Euler method described in Appendix C to solve the differential equations of motion.)

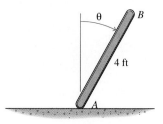

Fig. P16-113

C16-114 The spoked wheel shown in Fig. P16-114 rolls without slipping on a horizontal surface. Because it has lost a couple of its spokes, the center of mass of the 12-kg wheel is 50 mm away from its middle and the radius of gyration relative to the mass center is 0.6 m. If the center of the wheel has a speed of 3 m/s when $\theta = 0$, calculate and plot:

a. The normal and friction forces exerted on the wheel as a function of θ ($0° \leq \theta \leq 360°$).
b. The angular velocity $\dot{\theta}$ of the wheel as a function of θ ($0° \leq \theta \leq 360°$).

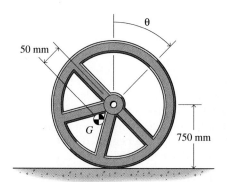

Fig. P16-114

C16-115 The unbalanced wheel A shown in Fig. P16-115 weighs 20 lb, and its radius of gyration with respect to the axis of rotation is 5 in. A cable attached to the wheel supports a 10-lb block B. If the system is released from rest when $\theta = 0$, calculate and plot:

a. The force exerted on the wheel by the axle at A as a function of its angular position θ ($0° \leq \theta \leq 600°$).
b. The tension in the cable as a function of θ ($0° \leq \theta \leq 600°$).

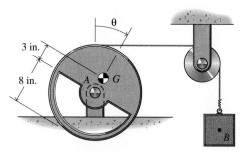

Fig. P16-115

C16-116 The slider-crank mechanism of Fig. P16-116 is an idealization of an automobile crank shaft, connecting rod, and piston. Treat the connecting rod as a uniform rod $\ell_{BC} = 175$ mm long with a mass of 0.12 kg. The crank throw is $\ell_{AB} = 75$ mm long, and the mass of the piston is 0.17 kg. If the crank shaft is rotating at a constant rate $\dot{\theta} = 4800$ rpm, calculate and plot:

a. The force exerted on the connecting rod at B by the crank shaft as a function of θ ($0° \le \theta \le 360°$).

b. The force exerted on the piston by the connecting rod as a function of θ ($0° \le \theta \le 360°$).

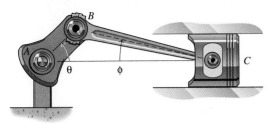

Fig. P16-116

C16-117 The mechanism shown in Fig. P16-117 is a simplification of a printing press. The print drum is a solid circular cylinder weighing 16 lb, bar AB rotates counterclockwise with a constant angular velocity of $\dot{\theta} = 15$ rpm, and the weight of bar BC may be neglected. If the lengths $L_{AB} = 2.5$ ft, $L_{BC} = 4$ ft, and the radius of the drum is 1 ft, calculate and plot:

a. The normal and friction forces exerted on the print drum as a function of θ ($0° \le \theta \le 360°$).

b. The force exerted on the print drum by the arm BC as a function of θ ($0° \le \theta \le 360°$).

c. The torque required to drive arm AB at the constant angular speed as a function of θ ($0° \le \theta \le 360°$).

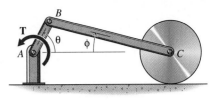

Fig. P16-117

C16-118 A 7-kg bowling ball has a diameter of 220 mm. The ball is released down the alley with an initial velocity of 6 m/s and no angular velocity. If the coefficient of friction between the ball and the alley is 0.1, calculate and plot:

a. The velocity v_G of the mass center of the ball, the velocity v_C of the point on the bottom of the ball in contact with the alley, and the angular velocity ω of the ball, all as functions of its position x_G from the moment the ball is released until it strikes the head pin 18.25 m away.

b. The position x_G and velocity v_G of the mass center of the ball as functions of time t from the moment the ball is released until it strikes the head pin 18.25 m away.

17

KINETICS OF PARTICLES: WORK AND ENERGY METHODS

As a box moves down a chute it exchanges its gravitational potential energy for kinetic energy.

17-1   INTRODUCTION

The preceding two chapters solved kinetics problems using Newton's second law. This chapter and the next will present an alternate method—the method of work and kinetic energy—which is useful for solving certain types of kinetics problems.

In the Newton's second law approach, the instantaneous equations of motion were used to relate the forces acting on a body to the acceleration of the body. If the equations were applied to a specific position of the body, only instantaneous relationships were obtained. If the equations were applied to an arbitrary position of the body, then the resulting acceleration could be integrated using the principles of kinematics discussed in Chapters 13 and 14.

The work–energy method combines the principles of kinematics with Newton's second law to directly relate the position and speed of a body. In this approach, Newton's second law is integrated in a general sense with respect to position. Clearly, for this method to be useful, the forces acting on a body must be a function of position only. For some types of these forces, however, the resulting integrals can be evaluated explicitly. The result is a simple algebraic equation that relates the speed of the body at two different positions of its motion.

Since the work–energy method combines the principles of kinematics with Newton's second law, it is not a new or independent principle. There are no problems that can be solved with the work–energy method that cannot be solved using Newton's second law. However, when the work–energy method does apply, it is usually the easiest method of solving a problem.

17-2   WORK OF A FORCE

In mechanics, a force does work only when the particle to which the force is applied moves. For example, when a constant force $\mathbf{P}$ is applied to a particle, which moves a distance d in a straight line as shown in Fig. 17-1, the work done on the particle by the force $\mathbf{P}$ is defined by the scalar product

$$U = \mathbf{P} \cdot \mathbf{d} = Pd \cos \phi$$
$$= P_x d_x + P_y d_y + P_z d_z \tag{17-1}$$

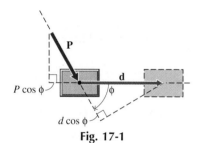

Fig. 17-1

where ϕ is the angle between the vectors $\mathbf{P}$ and $\mathbf{d}$. Equation 17-1 is usually interpreted as: *The work done by the force is the magnitude of the force $\mathbf{P}$ and the rectangular component of the displacement in the direction of the force $d \cos \phi$* (Fig. 17-1). However, the $\cos \phi$ can be associated with the force P instead of with the displacement d. Then Eq. 17-1 would be interpreted as: *The work done by the force is the product of the magnitude of the displacement d and the rectangular component of the force in the direction of the displacement $P \cos \phi$* (Fig. 17-1).

When $0 \le \phi < 90°$, the force and displacement are in the same direction and the work done by the force is positive. When $90° < \phi \le 180°$, the force and displacement are in opposite directions and the work done by the force is negative. When $\phi = 90°$ the force and displacement are perpendicular and the work done by the force is zero. Of course, the work done by the force is also zero if the displacement is zero $d = 0$.

Work has dimensions of force times length. In the SI system of units, this combination of dimensions is called a *joule* (1 J = 1 N·m).[1] In the U.S. Customary system of units, there is no special unit for work. It is expressed simply as ft·lb.

If the force is not constant or if the displacement is not in a straight line, Eq. 17-1 gives the work done by the force only during an infinitesimal part of the displacement, $d\mathbf{r}$:

$$dU = \mathbf{P} \cdot d\mathbf{r} = P \, ds \cos \phi = P_t \, ds$$
$$= P_x \, dx + P_y \, dy + P_z \, dz \qquad (17\text{-}2)$$

where $d\mathbf{r} = ds \, \mathbf{e}_t = dx\mathbf{i} + dy\mathbf{j} + dz\mathbf{k}$. Integrating Eq. 17-2 along the particle path from position 1 to position 2 gives the work done by the force $U_{1 \to 2}$

$$U_{1 \to 2} = \int_1^2 dU = \int_{s_1}^{s_2} P_t \, ds$$
$$= \int_{x_1}^{x_2} P_x \, dx + \int_{y_1}^{y_2} P_y \, dy + \int_{z_1}^{z_2} P_z \, dz \qquad (17\text{-}3)$$

In some cases, the functional relationship between force and displacement may not be known. Instead, the force components (P_t or P_x, P_y, and P_z) may be given in the form of graphs (Fig. 17-2). The integrals in Eq. 17-3 then represent the area under the curve and must be computed numerically.

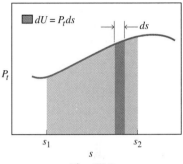

Fig. 17-2

17-2-1 Work Done by a Constant Force

When a constant force $\mathbf{P} = P_x\mathbf{i} + P_y\mathbf{j} + P_z\mathbf{k}$ is applied to a particle, Eq. 17-3 gives the work done by the force on the particle as

$$U_{1 \to 2} = P_x \int_{x_1}^{x_2} dx + P_y \int_{y_1}^{y_2} dy + P_z \int_{z_1}^{z_2} dz$$
$$= P_x(x_2 - x_1) + P_y(y_2 - y_1) + P_z(z_2 - z_1)$$

Note that evaluation of the work done by a constant force depends on the coordinates at the end points of the particle's path but not on the actual path traveled by the particle. For the constant force $\mathbf{P}$ shown in Fig. 17-3, it doesn't matter if the particle moves along path a from position 1 to position 2 or along path b or along some other path. The work done by the force $\mathbf{P}$ is always the same. Forces for which the work done is independent of the path are called conservative forces. They will be studied in more detail in Section 17-5.

The weight of a particle W is a particular example of a constant force. When bodies move near the surface of the earth, the force of the earth's gravity is essentially constant ($P_x = 0$, $P_y = 0$, and $P_z = -W$). Therefore, the work done on a particle by its weight is $-W(z_2 - z_1)$. When $z_2 > z_1$, the particle moves upward (opposite the gravitational force), and the work done by gravity is negative. When $z_2 < z_1$, the particle moves downward (in the direction of the gravitational force), and the work done by gravity is positive.

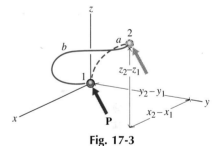

Fig. 17-3

---

[1] It may be noted that work and the moment of a force have the same dimensions: they are both force times length. However, work and moment are two totally different concepts and the special unit *joule* should be used only to describe work and energy. The moment of a force must always be expressed as N·m.

17-2-2 Work Done by a Massless Linear Spring Force

The force required to stretch a massless, linear spring is directly proportional to the stretch in the spring

$$P = k(\ell - \ell_0) = k\delta \tag{17-4}$$

where k is a constant known as the *modulus* of the spring and ℓ, ℓ_0, and δ are the present length, unstretched length, and deformation of the spring from its unloaded position, respectively. When the present length is greater than the unstretched length ($\delta > 0$), the spring is stretched and the force P is positive, as shown in Fig. 17-4. When the present length is less than the unstretched length ($\delta < 0$), the spring is compressed and the force P is negative (the force actually pushes on the spring).

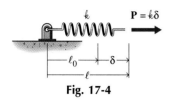

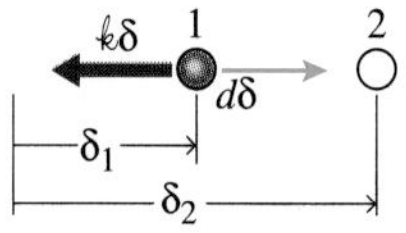

Fig. 17-4

When the spring of Fig. 17-4 is connected to a particle, the force exerted on the particle is equal in magnitude and opposite in direction to that on the spring (Fig. 17-5). If the particle moves from position 1 (where the deformation of the spring is δ_1), to position 2 (where the deformation of the spring is δ_2), the work done on the particle by the spring will be

$$ \tag{17-5}$$

$$U_{1 \to 2} = \int_{\delta_1}^{\delta_2} - k\delta \, d\delta = -\frac{1}{2} k (\delta_2^2 - \delta_1^2)$$

Fig. 17-5

where the minus sign arises because the force $k\delta$ acts to the left on the particle and the displacement $d\delta$ is to the right. If $0 < \delta_1 < \delta_2$, then the net motion of the particle is to the right (opposite the spring force), and the work done on the particle is negative. If $0 < \delta_2 < \delta_1$, then the spring force and the motion are both to the left and the work done on the particle is positive.

17-3 PRINCIPLE OF WORK AND ENERGY

The principle of work and energy is obtained by integrating Newton's second law with respect to position. Consider the particle of mass m whose free-body diagram is shown in Fig. 17-6. The force $\mathbf{R}$ represents the resultant of all external forces acting on the particle. Writing the tangential component of Newton's second law for the particle gives

$$R_t = ma_t = m \frac{dv}{dt} \tag{17-6}$$

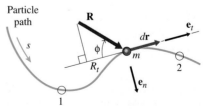

Fig. 17-6

where $\mathbf{v} = v\mathbf{e}_t = (ds/dt)\mathbf{e}_t$ and $d\mathbf{r} = ds\ \mathbf{e}_t$. Using the chain rule of differentiation, Eq. 17-6 can be written

$$R_t = m\frac{dv}{ds}\frac{ds}{dt} = mv\frac{dv}{ds} \qquad (17\text{-}7)$$

Finally, rearranging Eq. 17-7 and integrating along the particle's path from point 1 to point 2 gives

$$\int_{s_1}^{s_2} R_t\ ds = m\int_{v_1}^{v_2} v\ dv = \frac{1}{2}mv_2^2 - \frac{1}{2}mv_1^2 \qquad (17\text{-}8)$$

But the integral on the left-hand side of Eq. 17-8 is just the total work $U_{1\to2} = \int_{s_1}^{s_2} R_t\ ds$ exerted by the resultant force $\mathbf{R}$ on the particle during the motion. The terms on the right-hand side of Eq. 17-8 are called the kinetic energy of the particle $T = \frac{1}{2}mv^2$. Clearly, the basic unit for kinetic energy is the same as for work: joules (J) or ft·lb.

Equation 17-8 expresses the *principle of work and energy:*

$$T_2 - T_1 = U_{1\to2} \qquad (17\text{-}9a)$$

which states that *the net increase in the kinetic energy of a particle in a displacement from position 1 to position 2 is equal to the work done on the particle by external forces during the displacement.* Alternatively, the principle of work and energy is often written in the form

$$T_i + U_{i\to f} = T_f \qquad (17\text{-}9b)$$

which states that *the final kinetic energy of a particle is equal to the sum of its initial kinetic energy and the work done on the particle by external forces.*

Since the mass m and the square of the speed v^2 are both positive quantities, the kinetic energy of a particle will always be positive. If the work done on the particle is positive, the final kinetic energy will be larger than the initial kinetic energy ($0 < T_1 < T_2$). If the work done is negative, the final kinetic energy will be smaller than the initial ($0 < T_2 < T_1$).

The convenience of the method of work and energy is that it directly relates the speed of a particle at two different positions of its motion to the forces that do work on the particle during the motion. If Newton's second law were applied directly, the acceleration would have to be obtained for an arbitrary position of the particle. Then the acceleration would have to be integrated using the principles of kinematics. The work–energy method combines these two steps into one.

The limitations of the method of work–energy are that Eq. 17-9 is a scalar equation and it can be solved for no more than one unknown; the acceleration cannot be calculated directly; and only forces that do work are involved. These are usually not serious limitations, however. The normal component of acceleration is a function of velocity $a_n = mv^2/\rho$, and the velocity is easily found using the work–energy method. Then the normal component of Newton's second law can be used to determine forces that act normal to the path of motion and that do no work.

Finally, a complete free-body diagram must be drawn to ensure that all forces are identified and considered. Although forces that do no work are not needed in the work–energy equation, they are needed for everything else. Always drawing complete free-body diagrams is just a good habit to get into.

17-4   SYSTEMS OF PARTICLES

Applying the principle of work and energy to a particle yields a single scalar equation. Therefore, when a group of particles interact with one another, the principle of work and energy may be applied to each individual particle and the equations added together. The result will be a single scalar equation that can be solved for at most one unknown. Unless the motions of most of the particles are known, this equation will not be very useful. One exception for which the equation is useful is when the particles are rigidly connected so that kinematics can be used to relate the motions of the particles.

17-4-1   Two Particles Connected by a Massless Rigid Link

Consider the two particles of Fig. 17-7a, which are connected by a massless, rigid link. The free-body diagrams of the particles and the link are shown in Fig. 17-7b in which $\mathbf{R}_1$ and $\mathbf{R}_2$ represent the resultants

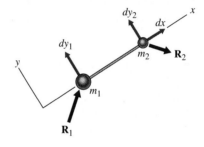

(a)

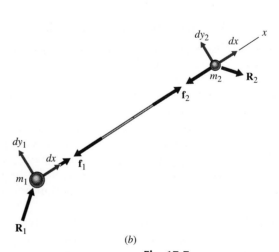

(b)

Fig. 17-7

of all external forces acting on the particles; $\mathbf{f}_1$ and $\mathbf{f}_2$ represent the internal forces between the link and the particles. Since the mass of the link is assumed negligible, application of Newton's second law to the free-body diagram of the link gives $\Sigma\mathbf{F} = m\mathbf{a} = \mathbf{0}$ and $\Sigma\mathbf{M}_G = I_G\boldsymbol{\alpha} = \mathbf{0}$. Therefore, assuming that no external forces act on the rigid link, the forces $\mathbf{f}_1$ and $\mathbf{f}_2$ must be directed along the link, equal in magnitude, and opposite in direction ($\mathbf{f}_2 = -\mathbf{f}_1$) as shown in Fig. 17-7$b$.

During some infinitesimal motion of the pair of particles, particle 1 will move a distance dx along the link and dy_1 perpendicular to the link. Since the link is rigid, particle 2 will move the same distance dx along the link but a different distance dy_2 perpendicular to the link. Then the work done on particle 2 by the link [$dU_2 = \mathbf{f}_2 \cdot (dx\mathbf{i}) = -\mathbf{f}_1 \cdot (dx\mathbf{i})$] will be the negative of the work done on particle 1 by the link [$dU_1 = \mathbf{f}_1 \cdot (dx\mathbf{i})$] during the infinitesimal motion. When the equations expressing the work–energy principle for the two particles are added together, the work done by the internal forces $\mathbf{f}_1$ and $\mathbf{f}_2$ will cancel at every instant of the motion leaving

$$\left(\frac{1}{2}m_1v_{1i}^2 + \int_i^f \mathbf{R}_1 \cdot d\mathbf{r}_1\right) + \left(\frac{1}{2}m_2v_{2i}^2 + \int_i^f \mathbf{R}_2 \cdot d\mathbf{r}_2\right)$$
$$= \frac{1}{2}m_1v_{1f}^2 + \frac{1}{2}m_2v_{2f}^2$$

That is, Eq. 17-9 also applies to the pair of connected particles if T_i and T_f are interpreted as the sum of the initial and final kinetic energies of both particles and $U_{i\rightarrow f}$ is the work done on both particles by all external forces acting on the pair of particles. So long as the particles are rigidly connected, the internal forces between the link and the particles do no work.

Similarly, when two bodies are connected by a flexible inextensible cable, the resultant work done on the bodies by the force in the cable is zero. Assuming that the mass of the cable is negligible and that any pulleys are small, massless, and/or frictionless, the two forces at the ends of the cable will have the same magnitude. Then since the cable is inextensible, the components of the displacements of the two ends in the direction of the forces must have the same magnitude and the resultant work done by the cable is zero.

17-4-2 General System of Interacting Particles

For a general system of interacting particles, the equations of work and kinetic energy for each of the particles can also be added together. The result will again be the same as Eq. 17-9 if T_i and T_f are interpreted as the sum of the initial and final kinetic energies of all the particles making up the system and $U_{i\rightarrow f}$ is the work done on the particles by *all forces*, both *internal and external*. Even though the internal forces always occur in equal-magnitude but opposite-direction pairs, the displacements will usually be different unless the particles are rigidly connected. Therefore, the work done by all the internal forces will usually not be zero.

A 16-lb crate slides down a ramp as shown in Fig. 17-8a. If the crate is released from rest 10 ft above the bottom of the ramp and the coefficient of friction between the crate and the ramp is $\mu_k = 0.20$, determine the speed of the crate when it reaches the bottom of the ramp.

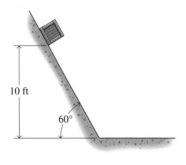

(a)

Fig. 17-8a

SOLUTION

The free-body diagram of the crate is shown in Fig. 17-8b for a general position along the ramp. The normal force N does no work since it is perpendicular to the motion. However, even though it does not affect the work–energy equation, the normal force must still be found to determine the work done by friction. Since the crate does not move in the y-direction, the y-component of Newton's second law $\Sigma F_y = ma_y = 0$ is easily solved giving $N = 16 \cos 60° = 8.00$ lb. Therefore, the friction force is $F = (0.20)(8.00) = 1.600$ lb.

Since the crate starts from rest, its initial kinetic energy is zero $T_i = 0$. The final kinetic energy of the crate is

$$T_f = \frac{1}{2} \frac{16}{32.2} v_f^2 = 0.2484 v_f^2$$

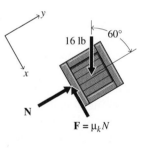

(b)
Fig. 17-8b

and the work done on the crate by the normal, gravitational, and frictional forces as the crate slides the $10/\sin 60° = 11.547$ ft down the ramp are

$$(U_{i \to f})_N = 0$$
$$(U_{i \to f})_g = \int_0^{11.547} (16 \sin 60°)\, dx = 160.00 \text{ ft·lb}$$
$$(U_{i \to f})_F = \int_0^{11.547} -1.600\, dx = -18.48 \text{ ft·lb}$$

respectively. Substituting these values into the work–energy equation gives

$$0 + 160.00 - 18.48 = 0.2484 v_f^2$$

or

$$v_f = 23.9 \text{ ft/s} \qquad\qquad \text{Ans.}$$

EXAMPLE PROBLEM 17-2

The 5-kg block shown in Fig. 17-9a slides along a horizontal floor and strikes the bumper B. The coefficient of friction between the block and the floor is $\mu_k = 0.25$ and the mass of the bumper may be neglected. If the speed of the block is 10 m/s when it is 15 m from the bumper, determine

a. The speed v_C of the block at the instant it strikes the bumper.
b. The maximum deflection δ_{max} of the spring due to the motion of the block.

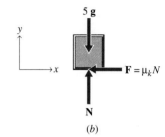

10 m/s

B $k = 2$ kN/m

5 kg

$\leftarrow$ 15 m $\rightarrow$

(a)

SOLUTION

a. The free-body diagram of the block before it contacts the bumper is shown in Fig. 17-9b. Using the y-component of Newton's second law $\Sigma F_y = ma_y = 0$ gives $N = (5)(9.81) = 49.05$ N. Then the friction force is $F = (0.25)(49.05) = 12.263$ N.

The initial kinetic energy of the block is

$$T_i = \frac{1}{2}(5)(10)^2 = 250.0 \text{ J}$$

and the kinetic energy when it is about to strike the bumper is

$$T_f = \frac{1}{2}(5)v_C^2 = 2.500v_C^2$$

The work done on the block by friction as the block slides along the floor is

$$(U_{i \to f})_F = \int_0^{15} -12.263 \, dx = -183.95 \text{ J}$$

and since the normal force is perpendicular to the motion, it does no work. Therefore, the work–energy principle gives

$$250.0 - 183.95 = 2.500v_C^2 \quad \text{or} \quad v_C = 5.14 \text{ m/s} \qquad \text{Ans.}$$

b. After the block contacts the bumper, the block and the bumper will move together. The free-body diagram of the system is shown in Fig. 17-9c. The normal and friction forces are still $N = 49.05$ N and $F = \mu_S N = 12.263$ N, respectively. The initial velocity of the block for this phase of the motion is $v_C = 5.14$ m/s and the initial kinetic energy is

$$T_i = \frac{1}{2}(5)(5.14)^2 = 66.05 \text{ J}$$

At the point of maximum spring deflection, the velocity of the block is zero, and so the final kinetic energy is zero $T_f = 0$. The work done on the block by friction and by the spring force as the block slides along the floor an additional distance δ_{max} is

$$(U_{i \to f})_F = \int_0^{\delta_{max}} -12.263 \, dx = -12.263 \, \delta_{max}$$

$$(U_{i \to f})_s = \int_0^{\delta_{max}} -2000 \, \delta \, d\delta = -1000 \, \delta_{max}^2$$

The work–energy equation then gives

$$66.05 - 12.263 \, \delta_{max} - 1000 \, \delta_{max}^2 = 0 \quad \text{or} \quad \delta_{max} = 0.251 \text{ m} \quad \text{Ans.}$$

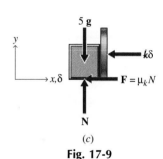

5 g

y

$\rightarrow x$ $F = \mu_k N$

N

(b)

5 g

y

$\rightarrow x, \delta$ $k\delta$ $F = \mu_k N$

N

(c)

Fig. 17-9

Two blocks are connected by a light rope, which passes over a small friction-less pulley of negligible mass (Fig. 17-10a). The coefficient of friction between the block A and the horizontal surface is $\mu_k = 0.4$. If the system is released from rest, determine the speed of the two blocks when block A has moved 6 ft to the right.

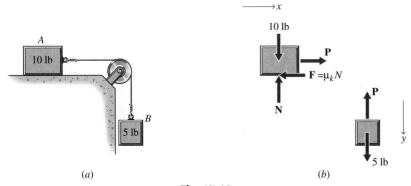

Fig. 17-10

SOLUTION

The free-body diagrams of the blocks are shown in Fig. 17-10b. Applying the y-component of Newton's second law to block A gives $N = 10$ lb. Then the friction force acting on block A is $F = (0.4)(10) = 4$ lb.

The work done on block A by friction during the motion is

$$(U_{i \to f})_F = \int_0^6 -4 \, dx = -24 \text{ ft} \cdot \text{lb}$$

When block A moves 6 ft to the right, block B moves down 6 ft, so that the work done by gravity on block B is

$$(U_{i \to f})_g = \int_0^6 5 \, dy = 30.00 \text{ ft} \cdot \text{lb}$$

Neither the normal force N nor the weight of block A does work since they both act perpendicular to the motion. Finally, considering the two blocks as a system, the work done by the internal force P will cancel out since the cord is inextensible.

Both blocks start from rest so the initial kinetic energy of the system is zero $T_i = 0$. At the final instant, the speeds of both blocks are the same, so that the final kinetic energy of the pair of blocks is

$$T_f = \frac{1}{2} \frac{10}{32.2} v_f^2 + \frac{1}{2} \frac{5}{32.2} v_f^2 = 0.2329 v_f^2$$

Substituting all these values into the work–energy equation gives

$$0 - 24.00 + 30.00 = 0.2329 v_f^2$$

or

$$v_f = 5.08 \text{ ft/s} \qquad \text{Ans.}$$

PROBLEMS

17-1* A 7500-lb truck is traveling on the freeway at 65 mi/h when the driver suddenly notices a moose standing on the road 200 ft straight ahead (Fig. P17-1). If it takes the driver 0.4 s to apply the brakes and the coefficient of friction between the tires and the road is 0.5,

a. Can the driver avoid hitting the moose without steering to one side?
b. Where will the truck come to rest relative to the moose?
c. If the driver must steer to one side, determine the speed of the truck as it passes the moose.

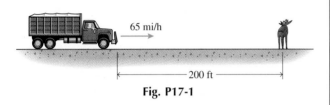

Fig. P17-1

17-2 A 1200-kg car is traveling along a mountain road at 90 km/h when a boulder rolls onto the highway 60 m in front of it (Fig. P17-2). The road is level and the coefficient of friction between the tires and the road is 0.5. If the driver of the car takes 0.4 s to apply the brakes,

a. Can the driver of the car avoid hitting the boulder without steering to one side?
b. If the driver of the car must steer to one side, determine the speed of the car as it passes the boulder.

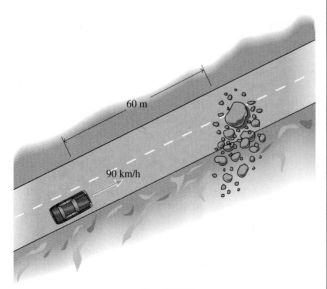

Fig. P17-2

17-3* A fully loaded Boeing 747 has a take-off weight of 660,000 lb, and its engines develop a combined thrust of 200,000 lb. If air resistance and friction between the tires and runway are neglected, determine the required length of the runway for a take-off speed of 140 mi/h (Fig. P17-3).

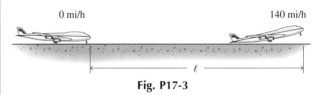

Fig. P17-3

17-4* A train is moving at 30 km/h when a coupling breaks and the last car separates from the train. As soon as the car separates, the brakes are automatically applied, locking all wheels of the runaway car. If the coefficient of friction between the wheels and the rails is 0.2, determine the distance the 180,000 kg car will travel before coming to a stop

a. If the tracks are level.
b. If the tracks slope downward at 5°.

17-5 A 25,000-lb F15 jet is catapulted from the deck of an aircraft carrier by a hydraulic ram (Fig. P17-5). Determine the average force exerted on the jet if it accelerates from rest to 160 mi/h in 300 ft.

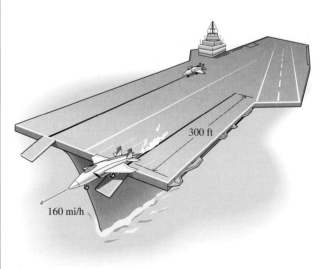

Fig. P17-5

17-6* A 10-g bullet has a horizontal velocity of 400 m/s when it strikes a 25-mm thick wooden target. Although slowed by the target, the bullet passes through the target and lands in a pond 50 m away (Fig. P17-6). Determine the average force exerted on the bullet by the target.

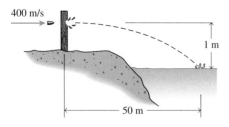

Fig. P17-6

17-7 When the 25,000-lb F15 jet of Problem 17-5 (p. 265) returns to the aircraft carrier, it is stopped by a combination of friction and a cable which exerts a force on the jet like a linear spring. If the landing speed of the jet is 140 mi/h and the coefficient of friction between its tires and the deck of the carrier is 0.6, determine the spring constant k required to stop the jet in a distance of 400 ft.

17-8 A 10-g bullet has a horizontal velocity of 400 m/s when it strikes and lodges in a 2.5-kg wooden block (Fig. P17-8). The block is initially at rest, the mass of the bumper B can be neglected, and the floor is smooth. In the motion after the impact, the maximum compression in the spring is observed to be 73 mm. Determine

a. The percent of the bullet's initial kinetic energy that was lost as a result of the collision.
b. The velocity of the block and the bullet when the block initially impacts the bumper.

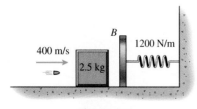

Fig. P17-8

17-9* In a shipping warehouse, packages are moved between levels by sliding them down a chute as shown in Fig. P17-9. The coefficient of friction between the package and the ramp is $\mu_k = 0.25$, the corner at the bottom of the ramp is abrupt but smooth, and $\theta = 30°$. If a 20 lb package starts from rest at $\ell = 10$ ft, determine

a. The speed of the package when it reaches the bottom of the ramp.
b. The distance d the package will slide along the horizontal surface before coming to rest.

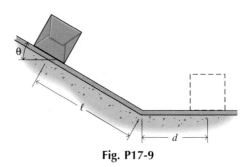

Fig. P17-9

17-10 In a shipping warehouse, packages are moved between levels by sliding them down a chute as shown in Fig. P17-9. The coefficient of friction between the package and the ramp is $\mu_k = 0.20$, the corner at the bottom of the ramp is abrupt but smooth, and $\theta = 30°$. If a 10-kg package starts at $\ell = 3$ m with an initial speed of 5 m/s down the ramp, determine

a. The speed of the package when it reaches the bottom of the ramp.
b. The distance d the package will slide along the horizontal surface before coming to rest.

17-11* In a shipping warehouse, packages are moved between levels by sliding them down a chute as shown in Fig. P17-9. If a 30-lb package starts at $\ell = 25$ ft with an initial speed of 15 ft/s down a $\theta = 10°$ chute, determine the coefficient of friction μ_k for which the package will reach the corner at the bottom of the ramp with zero speed.

17-12* A 10-kg crate slides on a 30° ramp as shown in Fig. P17-9. The coefficient of friction between the crate and the ramp is $\mu_k = 0.25$, and the corner at the bottom of the ramp is abrupt but smooth. If the crate starts at $\ell = 3$ m with an initial speed of 5 m/s up the ramp, determine

a. The speed of the crate when it returns to its starting position.
b. The speed of the crate when it reaches the bottom of the ramp.
c. The distance d the crate will slide along the horizontal surface before coming to rest.

17-13 A 20-lb crate slides on an inclined ramp as shown in Fig. P17-9. The crate starts at $\ell = 10$ ft with a speed of 15 ft/s up the ramp, and the corner at the bottom of the ramp is abrupt but smooth. If the static and kinetic coefficients of friction are 0.40 and 0.30, respectively, determine

a. The minimum angle θ for which the crate will return to its initial position.
b. The speed of the crate when it reaches the bottom of the ramp.
c. The distance d the crate will slide along the horizontal surface before coming to rest.

17-14* If the packages of Problem 17-10 come off the chute with too great of a velocity, a bumper as shown in Fig. P17-14 may be required to "catch" them. The coefficient of friction between the package and the floor is $\mu_k = 0.25$, the spring modulus is $k = 1750$ N/m, and the mass of the bumper B may be neglected. If the speed of a 2.5 kg package is $v_0 = 8$ m/s when it is $\ell = 3$ m from the bumper, determine

a. The maximum deflection δ of the spring.
b. The final resting position of the package.

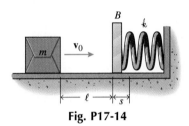

Fig. P17-14

17-15 If the packages of Problem 17-9 come off the chute with too great of a velocity, a bumper as shown in Fig. P17-14 may be required to "catch" them. The spring modulus is $k = 6.0$ lb/ft and the mass of the bumper B may be neglected. If the static and kinetic coefficients of friction between a 15-lb package and the floor are 0.6 and 0.4, respectively, determine the maximum initial speed v_0 of the package when $\ell = 5$ ft such that the package will not rebound off the spring.

17-16 If the packages of Problem 17-10 come off the chute with too great of a velocity, a bumper as shown in Fig. P17-14 may be required to "catch" them. The kinetic coefficient of friction between the package and the floor is $\mu_k = 0.2$, the spring modulus is $k = 250$ N/m, and the mass of the bumper B may be neglected. If the speed of a 5-kg package is $v_0 = 3.5$ m/s when it is $\ell = 3$ m from the bumper, determine the minimum static coefficient of friction such that the package will not rebound off the spring.

17-17* A particle is attached to a nonlinear (softening) spring for which the force–deformation relation is

$$F = 90\,\delta - 120\,\delta^3$$

where F is in pounds and δ is in feet. Determine the work done on the particle by the spring as the stretch increases from $\delta = 1$ in. to $\delta = 2.5$ in.

17-18 A particle is attached to a nonlinear (hardening) spring for which the force–deformation relation is

$$F = 1200\,(\delta + 10\,\delta^2)$$

where F is in Newtons and δ is in meters. Determine the work done on the particle by the spring as the stretch decreases from $\delta = 150$ mm to $\delta = 50$ mm.

17-19* The pressure in the cylindrical, gas-filled chamber of Fig. P17-19 varies inversely with the volume of the gas ($p = $ constant/volume). Initially, the piston is at rest, $x = 6$ in., and $p = 2\,p_{\text{atm}}$ where $p_{\text{atm}} = 14.7$ lb/in.² If air pressure on the outside surface of the piston is constant ($= p_{\text{atm}}$), determine for the subsequent motion

a. The maximum speed $v_{\max}$ of the piston.
b. The maximum displacement $x_{\max}$ of the piston.
c. The minimum constant force F that must be applied to the piston to limit its motion so that $x_{\max} < 18$ in.

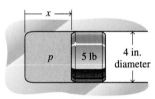

Fig. P17-19

17-20* A 2-kg block slides on a frictionless floor and strikes the bumpers shown in Fig. P17-20. The two linear springs are identical with spring constants of $k = 1.5$ kN/m and the mass of the bumpers may be neglected. If the initial speed of the block is 4 m/s, determine the maximum deformation of the springs.

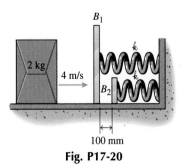

Fig. P17-20

17-21 A 5-lb crate slides up a 25° inclined ramp with an initial speed of 45 ft/s as shown in Fig. P17-21. If the coefficient of friction between the ramp and the crate is $\mu_k = 0.3$ and $\ell = 10$ ft, determine

a. The speed of the crate when it reaches the top of the ramp.
b. The maximum height h attained by the crate.
c. The distance d at which the crate will strike the horizontal surface.

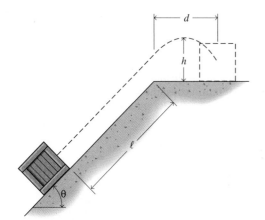

Fig. P17-21

17-22* A 5-kg crate slides up a 25° inclined ramp with $\ell = 5$ m as shown in Fig. P17-21. If the coefficient of friction between the ramp and the crate is $\mu_k = 0.5$, determine the initial speed v_0 for which the crate will reach the top corner with zero speed.

17-23 A 10-lb crate slides up a 30° inclined ramp with $\ell = 13$ ft as shown in Fig. P17-21. If the coefficient of friction between the ramp and the crate is $\mu_k = 0.3$ and the crate strikes the horizontal surface at $d = 5$ ft, determine

a. The initial speed of the crate v_0.
b. The maximum height h attained by the crate.

17-24 A 7-kg crate slides up a 60° inclined ramp with $\ell = 3$ m as shown in Fig. P17-21. If the coefficient of friction between the ramp and the crate is $\mu_k = 0.3$ and the maximum height attained by the crate is $h = 360$ mm, determine

a. The initial speed of the crate v_0.
b. The distance d at which the crate will strike the horizontal surface.

17-25* When a bullet is fired from a rifle, the gas pressure in the barrel varies as shown in Fig. P17-25. The rifle barrel is 2 ft long and has a 0.25 in. diameter. If the bullet weighs 0.025 lb and friction forces may be neglected compared to the gas force, estimate the muzzle velocity (velocity of the bullet as it leaves the barrel) for this rifle.

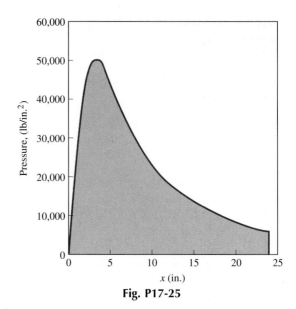

Fig. P17-25

17-26 A 5-kg block slides on the inside of a cylindrical shell as shown in Fig. P17-26. The radius of the cylinder is 3 m. If the block starts from rest when $\theta = 30°$, determine the speed of the block when $\theta = 90°$.

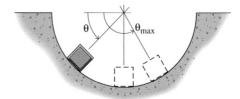

Fig. P17-26

17-27* The pair of blocks shown in Fig. P17-27 are connected by a light inextensible cord and are released from rest with $d = 18$ in. The weights of the blocks are $W_A = 5$ lb

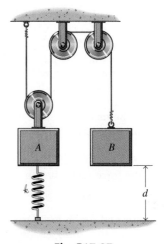

Fig. P17-27

and W_B = 10 lb, and the spring (k = 20 lb/ft) is unstretched in the initial position. Determine the speed of block B as it hits the floor.

17-28 The pair of blocks shown in Fig. P17-27 are connected by a light inextensible cord and are released from rest with d = 500 mm. The masses of the blocks are m_A = 6 kg and m_B = 4 kg and the spring is unstretched in the initial position. Determine the minimum spring modulus such that block B does not hit the floor in the ensuing motion.

17-29* The pair of blocks shown in Fig. P17-29 are connected by a light inextensible cord and are released from rest when the spring is unstretched. The static and kinetic coefficients of friction are 0.2 and 0.1, respectively. For the ensuing motion, determine

a. The maximum velocity of the blocks and the stretch in the spring at that position.
b. The maximum amount that the 5-lb block will drop.
c. If the blocks will rebound from the position of part *b*.

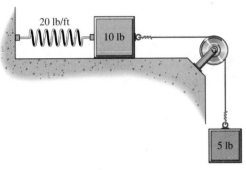

Fig. P17-29

17-30* The pair of blocks shown in Fig. P17-30 are connected by a light inextensible cord and are released from rest when the spring is unstretched. The static and kinetic coefficients of friction are 0.3 and 0.2, respectively. For the ensuing motion, determine

a. The maximum velocity of the blocks and the stretch in the spring at that position.
b. The maximum distance that the 10-kg block will slide down the inclined surface.
c. If the blocks will rebound from the position of part *b*.

17-31 The pair of blocks shown in Fig. P17-31 are connected by a light inextensible cord. The coefficient of friction between the 10-lb block and the floor is μ_k = 0.6. If the blocks are released from rest when the spring is stretched 15 in., determine for the ensuing motion the maximum velocity of the blocks and the stretch in the spring at that position.

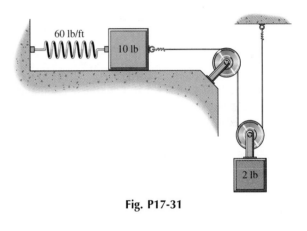

Fig. P17-31

17-32* The pair of blocks shown in Fig. P17-32 are connected by a light inextensible cord. The blocks are released from rest from the position shown with the spring stretched 150 mm, and friction may be neglected. For the ensuing motion, determine the maximum distance that the 2-kg block will rise above the floor.

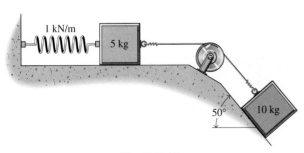

Fig. P17-30

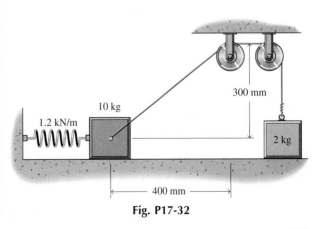

Fig. P17-32

17-33 The pair of blocks shown in Fig. P17-33 are connected by a light inextensible cord. Both the horizontal surface and the vertical pole are frictionless. In the position shown, the 2-lb block has a speed of 5 ft/s to the right. For the ensuing motion, determine the maximum distance that the 5-lb block will rise above its initial position.

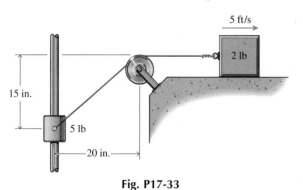

Fig. P17-33

17-34 The pair of blocks shown in Fig. P17-34 are connected by a light inextensible bar. Both the horizontal and vertical guide slots are frictionless. If the blocks are released from rest from the position shown, determine the velocity of the 3-kg block when:

a. It is at the same level as the 2-kg block.
b. It is 150 mm below the 2-kg block.

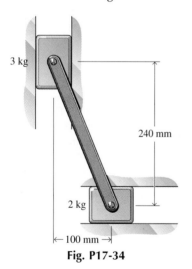

Fig. P17-34

17-35* A 3-lb block slides on a frictionless vertical guide as shown in Fig. P17-35. A constant horizontal force of 12 lb is applied to the end of the light inextensible cord that is attached to the block. If the block is released from rest when $d = 32$ in., determine the velocity of the block when $d = 18$ in.

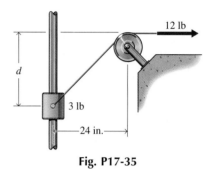

Fig. P17-35

17-36 A 10-kg block slides on a frictionless horizontal surface as shown in Fig. P17-36. A constant horizontal force of 50 N is applied to the end of the light inextensible cord that is attached to the block. If the block is released from rest from the position shown in which the spring is unstretched, determine for the ensuing motion

a. The maximum velocity of the block and the stretch in the spring at that position.
b. The maximum stretch in the spring.

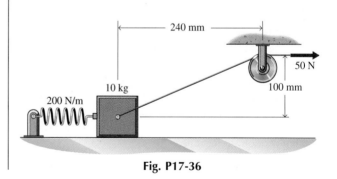

Fig. P17-36

17-5 CONSERVATIVE FORCES AND POTENTIAL ENERGY

Although the kinetic energy of a particle is always calculated as $T = \frac{1}{2}mv^2$, the work done on a particle must be calculated for each force acting on the particle. For a special category of forces called conservative forces, however, the work can also be calculated in a general sense, thus simplifying considerably the computations associated with the work–energy principle. For these conservative forces, the work done on a particle is calculated from a potential energy function, which depends only on the position of the particle.

17-5-1 Potential Energy of a Constant Force

A constant force is a trivial example of a force for which the work done can be replaced with a potential energy function. Consider a constant force **P** applied to the particle of Fig. 17-11. The coordinate system will be chosen with the positive x-axis in the direction of the force. Then, the work done on the particle by the force **P** as the particle moves from point 1 to point 2 is

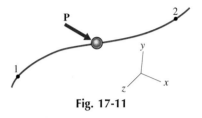

Fig. 17-11

$$U_{1 \to 2} = \int_1^2 \mathbf{P} \cdot d\mathbf{r} = \int_{x_1}^{x_2} P \, dx = P \int_{x_1}^{x_2} dx$$
$$= Px_2 - Px_1 \tag{17-10}$$

The integral always comes out the same independently of the path that the particle follows. Therefore, the work done by the constant force **P** is evaluated by subtracting the value of Px at the initial position (position 1) from the value of Px at the final position (position 2).

For the constant force **P** the scalar function[2]

$$V_P = -Px \tag{17-11}$$

is called the *potential energy* of the force. The value of the potential energy depends on the location of the origin from which x is measured. For a given particle position, the potential energy may be positive, negative, or zero depending on the location of the origin. However, the work done on a particle by a constant force **P** is given by the difference in the potential energy

$$U_{1 \to 2} = (V_P)_1 - (V_P)_2$$

and the difference is the same regardless of the location from which x is measured. The location for which the potential energy is zero is called the datum location. The datum is often chosen to make either the initial or the final potential energy zero.

The basic unit for potential energy is the same as for work and kinetic energy: joules (J) in the SI system of units or ft·lb in the U. S. Customary system of units.

In terms of the potential energy, then, the work–energy principle can be written

$$T_1 + (V_P)_1 - (V_P)_2 + \tilde{U}_{1 \to 2} = T_2$$

or

$$T_1 + (V_P)_1 + \tilde{U}_{1 \to 2} = T_2 + (V_P)_2 \tag{17-12}$$

where $\tilde{U}_{1 \to 2}$ is the work done on the particle by all forces other than the constant force **P**.

---

[2] The choice $V_P = -Px$ rather than $V_P = Px$ is arbitrary. The minus sign is included so that T_1 and $(V_P)_1$ will have the same sign in Eq. 17-12.

17-5-2   Gravitational Potential Energy (Constant g)

The force of gravity acting on bodies near the surface of the Earth may be considered constant. Therefore, the force of gravity acting on a particle is just a particular example of a constant force, and the work done on a particle by the force of gravity may be replaced by a potential energy function

$$U_{1 \to 2} = (V_g)_1 - (V_g)_2$$

where

$$V_g = mgh = Wh \tag{17-13}$$

and h is the height of the particle.[3] Therefore, the work done by gravitational forces can be included in the work–energy principle simply as

$$T_1 + (V_g)_1 + \tilde{U}_{1 \to 2} = T_2 + (V_g)_2 \tag{17-14}$$

where $\tilde{U}_{1 \to 2}$ is the work done on the particle by all forces other than the constant gravitational force.

The choice of the datum level (the height at which the gravitational potential energy is zero) is arbitrary. Although it is often taken to be the surface of the Earth, it is just as often taken to be either the initial or the final height of the particle.

17-5-3   Gravitational Potential Energy (Inverse Square Law)

When particles move such that their height varies a great deal, the gravitational force $\mathbf{W} = -(GMm/r^2)\mathbf{e}_r$ can no longer be approximated as a constant. The value of GM is often replaced by noting that the weight of a body is $W = mg$ at the Earth's surface. Comparing these two expressions for weight gives $GM = gR^2$, where R is the radius of the Earth. Then if the displacement is expressed in cylindrical coordinates

$$d\mathbf{r} = dr\,\mathbf{e}_r + r\,d\theta\,\mathbf{e}_\theta + dz\,\mathbf{e}_z$$

the work done by gravity becomes

$$U_{1 \to 2} = \int_{r_1}^{r_2} \mathbf{W} \cdot d\mathbf{r} = -\int_{r_1}^{r_2} \frac{mgR^2}{r^2}\,dr = \left[\frac{mgR^2}{r}\right]_{r_1}^{r_2}$$
$$= \frac{mgR^2}{r_2} - \frac{mgR^2}{r_1} \tag{17-15}$$

The work done by the gravitational force is again independent of the path followed by the particle and depends only on the position of the particle at the beginning and end of the motion. The gravitational potential energy is then defined as

$$V_g = -\frac{mgR^2}{r} = -\frac{GMm}{r} \tag{17-16}$$

---

[3] The gravitational potential energy is *positive mgh* since the height of the particle increases in the direction opposite that of the direction of the gravitational force. In the previous section, the distance x increased in the same direction as the force $\mathbf{P}$ and so the potential energy was *negative Px*.

and the work done by the gravitational force is

$$U_{1\rightarrow2} = (V_g)_1 - (V_g)_2$$

Except for the definition of the potential energy function, the work–energy equation for this case is the same as Eq. 17-14.

Note that the datum for the gravitational potential energy defined by Eq. 17-16 is at $r = \infty$ and that V_g is negative for $r < \infty$. Of course, a constant can always be added to the potential energy,[4] thus giving a different datum level, if desired.

17-5-4 Potential Energy of a Linear Elastic Spring Force

Next, consider a particle attached to a linear spring as shown in Fig. 17-12a. The force exerted on the particle by the spring is $\mathbf{F}_s = -k\delta\,\mathbf{e}_r$ as shown in Fig. 17-12b. Therefore, when the spring is stretched, δ is positive and the spring pulls on the particle in the negative $\mathbf{e}_r$ direction as drawn. On the other hand, when the spring is compressed, δ will be negative and the spring will actually push in the positive $\mathbf{e}_r$ direction.

Again expressing the displacement in cylindrical coordinates, the work done on the particle by the spring force $\mathbf{F}_s$ is

$$U_{1\rightarrow2} = \int_{r_1}^{r_2} \mathbf{F}_s \cdot d\mathbf{r} = -\int_{r_1}^{r_2} k\delta\,dr$$

But the deformation of the spring is its present length minus its undeformed length $\delta = \ell - \ell_0 = r - \ell_0$ so that $d\delta = dr$. Then the work becomes

$$U_{1\rightarrow2} = -\int_{\delta_1}^{\delta_2} k\delta\,d\delta = -\left[\frac{1}{2}k\delta^2\right]_{\delta_1}^{\delta_2}$$

$$= -\frac{1}{2}k\delta_2^2 + \frac{1}{2}k\delta_1^2 \qquad (17\text{-}17)$$

which again is independent of the path of the particle and depends only on the stretch of the spring at the initial and final positions. The potential energy for the spring force is then defined as[5]

$$V_s = \frac{1}{2}k\delta^2 \qquad (17\text{-}18)$$

and the work done by the spring force can be included in the work–energy principle simply as

$$T_1 + (V_s)_1 + \tilde{U}_{1\rightarrow2} = T_2 + (V_s)_2 \qquad (17\text{-}19)$$

where $\tilde{U}_{1\rightarrow2}$ is the work done on the particle by all forces other than the spring force.

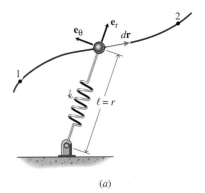

(a)

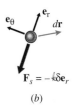

(b)

Fig. 17-12

---

[4]It is important to distinguish between adding a constant to the potential energy function and adding a constant to the radius r. The correct way to set the datum (zero level of potential energy) at the earth's surface is $V_g = mgR - mgR^2/r$ (not $-mgR^2/h$) where $h = r - R$ is the height above the surface of the earth.

[5]Note that it is the *deformation* of the spring and *not* the *length* of the spring that is squared. That is,

$$V_s = \tfrac{1}{2}k\delta^2 = \tfrac{1}{2}k(\ell - \ell_0)^2 \neq \tfrac{1}{2}k(\ell^2 - \ell_0^2)$$

Note that the potential energy of the spring force defined by Eq. 17-18 is zero when the spring is undeformed. The datum can be set to any level by setting the potential energy $V_s = \frac{1}{2}k\delta^2 - \frac{1}{2}k\delta_d^2$ where δ_d is the deformation of the spring at the new datum.[6]

17-5-5   Friction

As an example of a nonconservative force, the work done by friction forces depends on the path. Frictional forces always oppose motion so the work done by friction forces is always negative. Then, the longer the path traveled by the particle, the greater the work done by friction. Since the work done by frictional forces is not independent of the path, frictional forces are not conservative forces and the work done by frictional forces cannot be represented by a potential function.

17-5-6   Conservative Forces

The concept of potential energy may be used whenever the work of the force considered is independent of the path followed by its point of application as this point moves from an initial position to a final position. Such forces are said to be *conservative forces*.

Consider a general force **F** for which the work done as the point of application moves from point 1 to point 2 is

$$U_{1\to2} = \int_1^2 \mathbf{F}\cdot d\mathbf{r} = \int_1^2 (F_x\,dx + F_y\,dy + F_z\,dz) \qquad (17\text{-}20)$$

If the integral is to be independent of the path and depends only on the end points of the path, then the integrand of Eq. 17-20 must be an exact differential

$$U_{1\to2} = \int_1^2 (-dV) = V_1 - V_2 \qquad (17\text{-}21)$$

That is,

$$-dV = F_x\,dx + F_y\,dy + F_z\,dz \qquad (17\text{-}22)$$

where again the minus sign is included so that T_1 and V_1 will have the same sign in Eq. 17-12. Since the potential energy is a function of all three position variables $V = V(x,y,z)$, its differential is given by

$$dV = \frac{\partial V}{\partial x}\,dx + \frac{\partial V}{\partial y}\,dy + \frac{\partial V}{\partial z}\,dz \qquad (17\text{-}23)$$

where $\partial V/\partial x$, $\partial V/\partial y$, and $\partial V/\partial z$ are the partial derivatives of the potential energy function $V(x,y,z)$. Therefore, comparing Eqs. 17-22 and 17-23 gives

$$\frac{\partial V}{\partial x}\,dx + \frac{\partial V}{\partial y}\,dy + \frac{\partial V}{\partial z}\,dz = -(F_x\,dx + F_y\,dy + F_z\,dz) \quad (17\text{-}24)$$

or

$$\left(\frac{\partial V}{\partial x} + F_x\right)dx + \left(\frac{\partial V}{\partial y} + F_y\right)dy + \left(\frac{\partial V}{\partial z} + F_z\right)dz = 0 \quad (17\text{-}25)$$

---

[6] Note that $V_s = \frac{1}{2}k\delta^2 - \frac{1}{2}k\delta_d^2$ is *not* the same as $\frac{1}{2}k(\delta - \delta_d)^2$ nor do these expressions differ by a constant.

But if the work is in fact independent of the path, then Eq. 17-25 must be satisfied for arbitrary choices of dx, dy, and dz from which it follows that

$$F_x = -\frac{\partial V}{\partial x} \qquad F_y = -\frac{\partial V}{\partial y} \qquad F_z = -\frac{\partial V}{\partial z} \qquad (17\text{-}26)$$

That is, the components of the conservative force $\mathbf{F}$ are all derivable from the potential energy function $V(x,y,z)$.

Equation 17-26 may be expressed in vector notation

$$\mathbf{F} = -\left(\frac{\partial V}{\partial x} \mathbf{i} + \frac{\partial V}{\partial y} \mathbf{j} + \frac{\partial V}{\partial z} \mathbf{k} \right) = -\nabla V \qquad (17\text{-}27)$$

in which ∇ is the vector operator

$$\nabla = \mathbf{i}\frac{\partial}{\partial x} + \mathbf{j}\frac{\partial}{\partial y} + \mathbf{k}\frac{\partial}{\partial z} \qquad (17\text{-}28)$$

The vector ∇V is called the *gradient of V*.

Since the components of $\mathbf{F}$ are all derivable from a single scalar function, they are related to each other. For example, taking the partial derivatives of F_x and F_y with respect to y and x, respectively, gives

$$\frac{\partial F_x}{\partial y} = -\frac{\partial^2 V}{\partial y \partial x} \qquad \text{and} \qquad \frac{\partial F_y}{\partial x} = -\frac{\partial^2 V}{\partial x \partial y} \qquad (17\text{-}29)$$

The order of the partial derivatives is immaterial, however, so that

$$\frac{\partial F_x}{\partial y} = \frac{\partial F_y}{\partial x} \qquad (17\text{-}30a)$$

Similar relations hold between F_y and F_z and between F_x and F_z

$$\frac{\partial F_y}{\partial z} = \frac{\partial F_z}{\partial y} \qquad \text{and} \qquad \frac{\partial F_z}{\partial x} = \frac{\partial F_x}{\partial z} \qquad (17\text{-}30b,c)$$

Therefore, it is easy to determine whether a force is conservative or not. If its components satisfy Eq. 17-30, the force is conservative and a potential energy function can be found. If its components do not satisfy Eq. 17-30, the force is not conservative and a potential energy function does not exist.

17-6 GENERAL PRINCIPLE OF WORK AND ENERGY

When the work–energy principle (Eq. 17-9) is used to solve a problem, the work done on a particle must be computed for each force that acts on the particle. Computation of the work term $U_{1\rightarrow 2}$ is simplified considerably by using the concept of potential energy. The work term in the work–energy principle can be divided into two parts

$$U_{1\rightarrow 2} = U_{1\rightarrow 2}^{(c)} + U_{1\rightarrow 2}^{(o)} \qquad (17\text{-}31)$$

The term $U_{1\rightarrow 2}^{(c)}$ represents the work done by all the conservative forces acting on the particle; that is, by all the forces whose potential energy function is known. Similarly, the term $U_{1\rightarrow 2}^{(o)}$ represents the work done by all the other forces acting on the particle; that is, by all of the forces that either have no potential or whose potential energy function is not known. Then the work–energy principle becomes

$$T_1 + U^{(c)}_{1 \to 2} + U^{(o)}_{1 \to 2} = T_2 \tag{17-32}$$

But the conservative part of the work term can be replaced using with the potential energy $U^{(c)}_{1 \to 2} = V_1 - V_2$ where $V = V_g + V_e + \cdots$ is the sum of the potential energies of the conservative forces. This gives

$$T_1 + V_1 + U^{(o)}_{1 \to 2} = T_2 + V_2 \tag{17-33}$$

The combination of terms $T + V$ is called the *total mechanical energy*[7] E. Equation 17-33 expresses that the net increase in the total mechanical energy between the initial and the final position of the particle is equal to the work done on the particle by nonconservative forces.

Equation 17-33 is commonly used to determine the speed or position of a particle before or after a particular displacement has occurred. Since the reference (or datum) level from which to determine the potential energy is arbitrary, the calculations are simplified if the datum location is chosen so as to make one or more of the terms in Eq. 17-33 zero. In fact, different datum locations can be chosen for each of the potential energy terms.

17-7   CONSERVATION OF ENERGY

In many problems of interest, friction forces are often negligible and the only forces acting on a particle are elastic springs and gravity. In these cases, $U^{(o)}_{1 \to 2} = 0$ and Eq. 17-33 becomes simply

$$T_1 + V_1 = T_2 + V_2 \tag{17-34}$$

That is, when the only forces acting on a particle are conservative forces, the total mechanical energy remains constant

$$E_1 = E_2 = \text{constant} \tag{17-35}$$

Equation 17-34 is often referred to as the *Principle of Conservation of Energy*. The total mechanical energy of the particle is conserved in the sense that whatever value $E = T + V$ has at position 1, it has the same value at position 2. Any decrease in the potential energy is accompanied by an equivalent increase in the kinetic energy and vice versa. The constant in Eq. 17-35 is determined from the known position and velocity of the particle at some instant. Since the location of the datum for the potential energy functions is arbitrary, the constant can be positive, zero, or negative.

Conservation of energy is not really a new principle. It is simply a special case of the general principle of work and energy, Eq. 17-33.

17-8   POWER AND EFFICIENCY

Power and efficiency are concepts that are closely related to the concepts of mechanical work and mechanical energy. Any measure of the

---

[7] The total mechanical energy is a relative term. Since the datum or zero level of potential energy is arbitrary, a moving particle may have a zero or even negative value of total mechanical energy.

mechanical output of a machine must take into account the rate at which the work is done (power) as well as the total amount of work done by the machine. Also, the rating of the machine should account for the amount of energy the machines requires to do the work.

17-8-1 Power

The output power of a force $\mathbf{F}$ is given by the rate at which it does work

$$\text{Power} = \frac{dU}{dt} = \frac{\mathbf{F} \cdot d\mathbf{r}}{dt} = \mathbf{F} \cdot \mathbf{v} \qquad (17\text{-}36)$$

Since $d\mathbf{r}$ is the displacement of the particle (the point to which the force is applied), $d\mathbf{r}/dt = \mathbf{v}$ is the velocity of the particle.

Power has dimensions of *work* (*force* times *length*) divided by *time*. In the SI system of units, this combination of units is called a *watt* (1 W = 1 J/s = 1 N·m/s). In the U. S. Customary system of units it is called *horsepower* (1 hp = 550 ft·lb/s = 33,000 ft·lb/min).

17-8-2 Mechanical Efficiency

All real mechanical systems lose energy to friction so that the amount of work done by a machine (output work) is always less than the amount of work done on a machine (input work). The *mechanical efficiency* η_{mech} of the machine is defined by the ratio of these two quantities

$$\eta_{\text{mech}} = \frac{\text{output work}}{\text{input work}} \qquad (17\text{-}37)$$

Dividing the numerator and denominator of Eq. 17-37 by dt gives

$$\eta_{\text{mech}} = \frac{\text{power output}}{\text{power input}} \qquad (17\text{-}38)$$

Since all real systems lose energy to friction, the output work is always less than the input work and η_{mech} is always less than one.

Of course, the energy is not really lost in a thermodynamic sense. In a typical machine, the mechanical energy lost due to the negative work of friction forces between moving parts is converted to heat energy, which in turn is dissipated to the surroundings of the machine. When all forms of energy (mechanical, electrical, thermal, chemical, etc.) are accounted for, however, energy is conserved.

A 5-kg crate slides down a ramp as shown in Fig. 17-13*a* and strikes the bumper *B*. The coefficient of friction between the crate and the floor is $\mu_k = 0.25$ and the mass of the bumper may be neglected. If the crate is released from rest when it is 15 m from the bumper, determine:

a. The speed v_C of the block at the instant it strikes the bumper.
b. The maximum deflection δ_{max} of the spring due to the motion of the block.

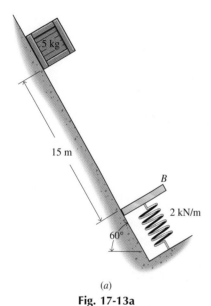

(*a*)
Fig. 17-13a

SOLUTION

a. The free-body diagram of the crate is shown in Fig. 17-13*b* for a general position along the ramp before the crate contacts the bumper. Since the crate does not move in the *y*-direction (the direction perpendicular to the surface), the *y*-component of Newton's second law $\Sigma F_y = ma_y = 0$ gives $N = (5)(9.81) \cos 60° = 24.53$ N. Then the friction force is $F = (0.25)(24.53) = 6.13$ N.

The gravitational force $W = mg$ is conservative, and its work can be determined using its potential energy. Using the initial height of the crate as the datum, the initial gravitational potential energy is zero $V_1 = 0$, and the gravitational potential energy when the crate strikes the bumper is

$$V_2 = (5)(9.81)(-15 \sin 60°) = -637.2 \text{ J}$$

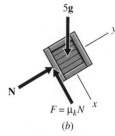

(*b*)
Fig. 17-13b

The friction force is nonconservative, and its work must be computed directly

$$(U_{1\to2})_F = \int_0^{15} -6.13\, dx = -91.95\,\text{J}$$

The remaining force N acts normal to the motion and so does no work on the crate $(U_{i\to f})_N = 0$.

Since the crate starts from rest, its initial kinetic energy is zero $T_1 = 0$ and the kinetic energy of the crate when it is about to strike the bumper is

$$T_2 = \frac{1}{2}(5)v_C^2 = 2.50\, v_C^2$$

Substituting these values into the work–energy equation gives

$$0 + 0 - 91.95 = 2.50\, v_C^2 - 637.2$$

or

$$v_C = 14.77\,\text{m/s} \qquad\qquad \text{Ans.}$$

b. After the block contacts the bumper, the block and the bumper will move together. The free-body diagram of the system is shown in Fig. 17-13c. The normal and friction forces are still $N = 24.53$ N and $F = 6.13$ N, respectively. The initial velocity of the block for this phase of the motion is $v_C = 14.77$ m/s and the initial kinetic energy is

$$T_2 = \frac{1}{2}(5)(14.77)^2 = 545.4\,\text{J}$$

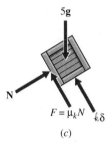

(c)

Fig. 17-13c

At the point of maximum spring deflection, the velocity of the block is zero and so the final kinetic energy is zero $T_3 = 0$.

The work done on the block by friction as the block slides along the floor an additional distance $\delta_{\max}$ is

$$(U_{i\to f})_F = \int_0^{\delta_{\max}} -6.13\, dx = -6.13\, \delta_{\max}$$

Both the spring force and the gravitational force are conservative, and their work can be determined using their potential energies. Using the position when the crate first contacts the crate as the datum, the initial gravitational potential energy is zero $(V_g)_2 = 0$, and the final gravitational potential energy is

$$(V_g)_3 = 5g(-\delta_{\max} \sin 60°) = -42.48\, \delta_{\max}$$

Since the spring is undeformed when the crate first contacts the block, the same position will be used as the datum for the potential energy of the spring force. Then the initial potential energy of the spring force is zero $(V_s)_2 = 0$, and the final potential energy of the spring force is

$$(V_s)_3 = \frac{1}{2}(2000)\delta_{\max}^2 = 1000\, \delta_{\max}^2$$

The work–energy equation then becomes

$$545.4 + 0 + 0 - 6.13\, \delta_{\max} = 0 - 42.48\, \delta_{\max} + 1000\, \delta_{\max}^2$$

which gives

$$\delta_{\max} = 0.757\,\text{m} \qquad\qquad \text{Ans.}$$

A 0.5-lb slider moves along a semicircular wire in a horizontal plane as shown in Fig. 17-14a. The undeformed length of the spring is 8 in. and friction may be neglected. If the slider is released from rest at position A, determine

a. The velocity of the slider at position B.
b. The force exerted on the slider by the wire at position B.

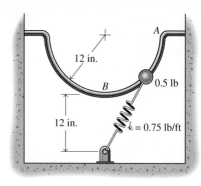

(a)

SOLUTION

a. The free-body diagram of the slider is shown in Fig. 17-14b for a general position along the wire. The weight force acts perpendicular to the motion (perpendicular to the figure) and does no work. The normal force N also acts perpendicular to the motion and does no work. The spring force is a conservative force, and its work can be calculated using its potential energy. The length of the spring at position A is $\ell_A = \sqrt{24^2 + 12^2} = 26.83$ in.; the deformation of the spring at position A is $\delta_A = (26.83 - 8)/12 = 1.5692$ ft. Therefore the potential energy of the spring force at position A is

(b)

$$(V_s)_A = \frac{1}{2}(0.75)(1.5692)^2 = 0.9234 \text{ ft} \cdot \text{lb}$$

At position B the deformation of the spring is $\delta_B = (12 - 8)/12 = 0.3333$ ft, and the potential energy of the spring force at B is

$$(V_s)_B = \frac{1}{2}(0.75)(0.3333)^2 = 0.04166 \text{ ft} \cdot \text{lb}$$

There are no nonconservative forces so that $U_{1 \to 2}^{(o)} = 0$. Since the slider starts from rest its initial kinetic energy is zero ($T_A = 0$); at position B the kinetic energy of the slider is

$$T_B = \frac{1}{2}\left(\frac{0.5}{32.2}\right)v_B^2 = 0.007764v_B^2$$

Substituting these values in the work–energy equation gives

$$0 + 0.9234 + 0 = 0.007764v_B^2 + 0.04166$$

or

$$v_B = 10.66 \text{ ft/s} \qquad \text{Ans.}$$

b. The free-body diagram of the slider is shown in Fig. 17-14c for position B. The component of the slider's acceleration, which is normal to the wire, is $a_n = v^2/r = 10.66^2/1 = 113.64$ ft/s². Then the component of Newton's second law in the direction perpendicular to the wire $\Sigma F_n = ma_n$ gives

$$N - (0.75)(0.3333) = \frac{0.5}{32.2}(113.64)$$

or

$$N = 2.015 \text{ lb} \qquad \text{Ans.}$$

(c)

Fig. 17-14

Two blocks are connected by a light inextensible cord, which passes around small massless pulleys as shown in Fig. 17-15a. If block B is pulled down 500 mm from the equilibrium position and released from rest, determine its speed when it returns to the equilibrium position.

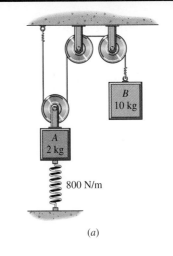

(a)

SOLUTION

The two blocks A and B are a system of interacting particles. The work–energy equations for each particle individually can be added together to get a similar equation for the system

$$T_i + V_i + U^{(o)}_{i \to f} = T_f + V_f \qquad (a)$$

In Eq. a, T represents the sum of the kinetic energies of both particles, V represents the sum of the potential energies of all conservative forces acting on both particles, and $U^{(o)}_{i \to f}$ represents the sum of the work done by all other forces acting on both particles.

The free-body diagrams of the two blocks are shown in Figs. 17-15b and 17-15c. In the equilibrium position the sum of forces is zero for both blocks:

$$\uparrow \Sigma F = 2\,T_{st} - (2)(9.81) - 800\,\delta_{st} = 2a_A = 0$$
$$\uparrow \Sigma F = T_{st} - (10)(9.81) = 10a_B = 0$$

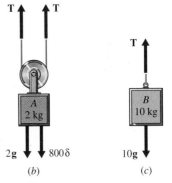

(b)   (c)

Therefore the static tension in the rope is $T_{st} = 98.1$ N and the static deformation of the spring is $\delta_{st} = 0.2207$ m.

Since the length of the cord is a constant, block A must rise whenever block B falls and vice versa. Referring to Fig. 17-15d, the length of the cord in the equilibrium position ($y_A = y_B = 0$) is given by

$$\ell = 2d + b + c \qquad (b)$$

When block A has moved upward a distance y_A and block B has moved downward a distance y_B, the length of the cord is given by

$$\ell = 2(d - y_A) + b + (c + y_B) \qquad (c)$$

Subtracting Eq. b from Eq. c gives the position relationship $y_B = 2y_A$; differentiating this equation gives the velocity relationship $v_B = \dot{y}_B = 2\dot{y}_A = 2v_A$.

Since the system is released from rest, the initial kinetic energies of both blocks are zero $(T_A)_i = (T_B)_i = 0$. When the blocks return to the equilibrium position, the sum of their kinetic energies will be

$$(T_A)_f + (T_B)_f = \tfrac{1}{2}(2)v_A^2 + \tfrac{1}{2}(10)v_B^2 = 5.25v_B^2$$

Since the cord is inextensible, the work done by the tension force cancels out when the work is added up for both particles. Both the gravitational and the spring forces are conservative so that the work done by nonconservative forces is zero $U^{(o)}_{i \to f} = 0$. Measuring the potential energies of the forces acting on each block from its equilibrium position gives

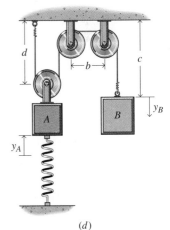

(d)

Fig. 17-15

$$(V_A)_i = (V_{Ag})_i + (V_{As})_i = (2)(9.81)\left(\frac{0.5}{2}\right) + \frac{1}{2}(800)(0.2207 + 0.500)^2 = 212.67 \text{ J}$$

$$(V_A)_f = (V_{Ag})_f + (V_{As})_f = \frac{1}{2}(800)(0.2207)^2 = 19.48 \text{ J}$$

$$(V_B)_i = (V_{Bg})_i = (10)(9.81)(-0.5) = -49.05 \text{ J} \quad \text{and} \quad (V_B)_f = (V_{Bg})_f = 0 \text{ J}$$

Substituting these values into Eq. a gives

$$0 + 212.67 - 49.05 + 0 = 5.25v_B^2 + 19.48 \quad \text{or} \quad v_B = 5.24 \text{ m/s} \quad \text{Ans.}$$

PROBLEMS

17-37* Solve Problem 17-11 using potential energies (see p. 266).

17-38 Solve Problem 17-12 using potential energies (see p. 266).

17-39* Solve Problem 17-13 using potential energies (see p. 266).

17-40* Solve Problem 17-14 using potential energies (see p. 267).

17-41 Solve Problem 17-15 using potential energies (see p. 267).

17-42 Solve Problem 17-22 using potential energies (see p. 268).

17-43* Solve Problem 17-23 using potential energies (see p. 268).

17-44* A 0.5-kg mass slides on a frictionless vertical rod as shown in Fig. P17-44. The undeformed length of the spring is $\ell_0 = 200$ mm and the distance $d = 300$ mm. If the slider is released from rest when $b = 0$, determine the spring modulus k such that $b_{max} = 400$ mm.

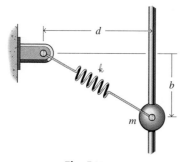

Fig. P17-44

17-45 A 1.5-lb weight slides on a frictionless vertical rod as shown in Fig. P17-44. The undeformed length of the spring is $\ell_0 = 8$ in., the spring modulus is $k = 80$ lb/ft, and the distance $d = 12$ in. If the slider is released from rest when $b = 9$ in., determine the speed of the slider when $b = 0$.

17-46* A 0.5-kg mass slides on a frictionless rod in a vertical plane as shown in Fig. P17-46. The undeformed length of the spring is $\ell_0 = 250$ mm, the spring modulus is $k = 600$ N/m, and the distance $d = 800$ mm. If the slider is re-

leased from rest when $b = 300$ mm, determine the speed of the slider at positions A and B.

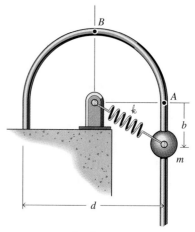

Fig. P17-46

17-47 A 0.5-lb weight slides on a frictionless rod in a vertical plane as shown in Fig. P17-46. The undeformed length of the spring is $\ell_0 = 6$ in., the spring modulus is $k = 5$ lb/ft, and the distance $d = 18$ in. If the slider has a velocity of 2 ft/s to the right at position B, determine the speed of the slider at position A and at the position where $b = 9$ in.

17-48 A singer swings a 0.35-kg microphone in a vertical plane at the end of a 750-mm long cord (Fig. P17-48). Determine

a. The minimum speed that the microphone must have at

Fig. P17-48

282

the position *A* to travel completely around the circular path.

b. The maximum tension in the cord.

17-49* A singer swings a 0.75-lb microphone in a vertical plane at the end of an 18-in. long cord (Fig. P17-48). If the speed of the microphone is 12 ft/s at position *A*, determine

a. The angle θ at which the cord becomes slack (tension = 0).
b. The maximum tension in the cord.

17-50 A singer swings a 0.35-kg microphone in a vertical plane at the end of a 600-mm long cord (Fig. P17-48). If the cord goes slack (tension = 0) when $\theta = 120°$, determine

a. The speed of the microphone at *A* ($\theta = 0°$).
b. The maximum tension in the cord.

17-51* A singer swings a 0.75-lb microphone in a vertical plane at the end of a 24-in. long cord as shown in Fig. P17-48. If the tension in the cord when the microphone is at *A* is twice the tension in the cord when the microphone is at *B*, determine the velocity of the microphone and the tension in the cord when the microphone is at *A*.

17-52* A small sack containing 1.5 kg of marbles is tied to the end of an 800-mm long cord as shown in Fig. P17-52. The maximum tension that the cord can withstand is $P_{max} = 30$ N. If the boy slowly pulls the sack off of the shelf, determine the angle θ through which the sack will swing before the cord breaks.

Fig. P17-52

17-53 A small sack of marbles is tied to the end of an 18-in. long cord as shown in Fig. P17-52. The boy slowly pulls the sack off of the shelf, and the cord is observed to break when $\theta = 70°$. If the maximum tension that the cord can withstand is $P_{max} = 4$ lb, determine the weight of the marbles in the sack.

17-54* A small sack of marbles is tied to the end of a 500-mm long cord as shown in Fig. P17-52. If the maxi-

mum tension that the cord can withstand is $P_{max} = 50$ N, determine the maximum weight of marbles that would not break the cord when the boy slowly pulls the sack off of the shelf.

17-55* A small box is sliding along a frictionless horizontal surface when it comes upon a circular ramp as shown in Fig. P17-55. If the initial speed of the box is $v_0 = 5$ ft/s and $r = 15$ in., determine the angle θ at which the box will lose contact with the circular ramp.

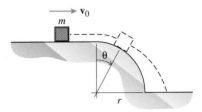

Fig. P17-55

17-56 A small box is sliding along a frictionless horizontal surface when it comes upon a circular ramp as shown in Fig. P17-55. If the radius of the ramp is $r = 750$ mm and the box loses contact with the ramp when $\theta = 25°$, determine the initial speed v_0 of the box.

17-57 A small box is sliding along a frictionless horizontal surface when it comes upon a circular ramp as shown in Fig. P17-55. If the radius of the ramp is $r = 30$ in., determine the maximum angle θ_{max} at which the box can maintain contact with the circular ramp.

17-58* A small toy car rolls down a ramp and through a vertical loop as shown in Fig. P17-58. The mass of the car is $m = 50$ g and the diameter of the vertical loop is $d = 300$ mm. If the car is released from rest, determine:

a. The minimum release height *h* such that the car will travel all the way around the loop.
b. The force exerted on the track by the car when it is at point *B* (one-quarter of the way around the loop).

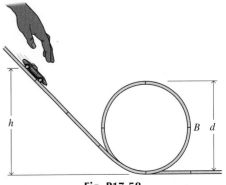

Fig. P17-58

17-59 Solve Problem 17-27 using potential energies (see p. 268).

17-60* Solve Problem 17-28 using potential energies (see p. 269).

17-61 Solve Problem 17-29 using potential energies (see p. 269).

17-62 Solve Problem 17-30 using potential energies (see p. 269).

17-63* Solve Problem 17-31 using potential energies (see p. 269).

17-64 Solve Problem 17-32 using potential energies (see p. 269).

17-65* Solve Problem 17-33 using potential energies (see p. 270).

17-66* Solve Problem 17-34 using potential energies (see p. 270).

17-67 The pair of blocks shown in Fig. P17-67 are connected by a light inextensible cord. The spring has a modulus of $k = 72$ lb/ft and an unstretched length of $\ell_0 = 12$ in. Friction may be neglected. If the system is released from rest when $x = 0$, determine

a. The speed of the blocks when $x = 4$ in.
b. The maximum displacement x for the ensuing motion.

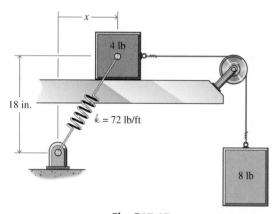

Fig. P17-67

17-68* The pair of blocks shown in Fig. P17-68 are connected by a light inextensible cord. The spring has a modulus of $k = 500$ N/m and an unstretched length of $\ell_0 = 400$ mm. Friction may be neglected. If the system is released from rest when $x = -800$ mm, determine

a. The speed of the blocks when $x = 0$ mm.
b. The maximum displacement x for the ensuing motion.

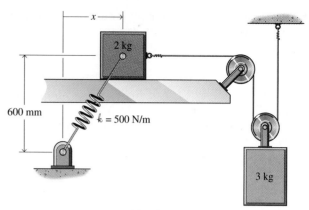

Fig. P17-68

17-69 The drag force due to air resistance on a cyclist moving with speed v is given by

$$F_D = 0.006v^2$$

where F_D is in pounds and v is in feet per second. If the combined weight of the cyclist and bicycle is 180 lb, determine

a. The power required to maintain a steady speed of 20 mi/h on level ground.
b. The maximum speed the cyclist could maintain up a 5° incline for the same amount of power.

17-70 An elevator E is attached by means of an inextensible cable to a 900-kg counterweight C (Fig. P17-70). The combined mass of the man and the elevator is 1000 kg. The elevator is raised and lowered using the motor M, which is

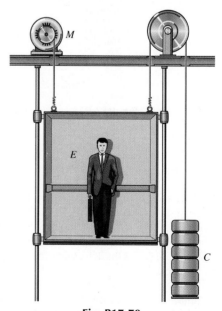

Fig. P17-70

attached to the elevator by means of a second cable. Determine the power the motor must supply if the elevator is

a. Raised at a constant rate of 0.5 m/s.
b. Lowered at a constant rate of 0.5 m/s.

17-71* A 400-lb crate C is attached to a power winch W as shown in Fig. P17-71. If the coefficient of friction between the crate and the 25° incline is 0.2 and the maximum power of the winch is 0.5 hp, determine the maximum constant speed at which the winch can raise the crate.

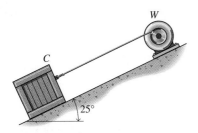

Fig. P17-71

17-72 A cyclist can maintain a speed of 30 km/h on level ground with a power output of 275 watts. The combined weight of the cyclist and bicycle is 800 N. Assuming that the retarding forces remain constant, determine

a. The maximum steady speed that the cyclist could maintain up a 5° incline for the same power.
b. The power that the cyclist would have to deliver to climb the 5° incline at 30 km/h.

17-73* A 2500-lb car requires 20 hp delivered to its wheels to maintain a steady 50 mi/h on a level road. Assuming that the retarding forces remain constant, determine

a. The maximum speed that the car could maintain up a 5° incline with the same 20 hp delivered to its wheels.
b. The horsepower that must be delivered to the wheels for the car to climb the 5° incline at a steady 50 mi/h.

17-74* The sum of all drag forces acting on a 1200-kg car moving with a speed v is given by

$$F_D = 200 + 0.8v^2$$

where F_D is in newtons and v is in meters per second (Fig. P17-74). Determine the power that must be delivered to the wheels to travel

a. At 40 km/h on a horizontal road.
b. At 80 km/h on a horizontal road.
c. At 40 km/h up a 5° incline.

Fig. P17-74

17-75 The sum of all drag forces acting on a 3200-lb car moving with speed v is given by

$$F_D = a + bv^2$$

where a represents the rolling resistance of the tires and bv^2 represents air resistance. If 8 hp must be delivered to the wheels of the car to maintain a constant speed of 30 mi/h and 14 hp must be delivered to the wheels to maintain a constant speed of 40 mi/h (both on a horizontal road), determine the horsepower required to travel

a. At 55 mi/h on a horizontal road.
b. At 40 mi/h up a 5° incline.

SUMMARY

The work–energy method combines the principles of kinematics with Newton's second law to directly relate the position and speed of a body. For this method to be useful, the forces acting on a body must be a function of position only. For some types of these forces, however, the resulting integrals can be evaluated explicitly. The result is a simple algebraic equation that relates the speed of the body at two different positions of its motion.

The work done by a force on a particle is the product of the particle's displacement and the component of force in the direction of the displacement. If there is no displacement or no component of force in the direction of the displacement, then the force does no work on the particle.

For conservative forces, the amount of work done depends only on the position of the particle at the beginning and end of the motion. Examples of conservative forces are: constant forces, gravitational forces, and linearly elastic spring forces.

Conservative forces can always be written as the gradient of a potential energy function. The work done by the force during some motion is the difference in the value of the potential energy function at the beginning and end of the motion.

The kinetic energy of a particle depends only on its speed $T = \frac{1}{2}mv^2$. Since the mass m and the square of the speed v^2 are both positive quantities, the kinetic energy of a particle will always be positive.

The principle of work and energy

$$T_i + U_{i \to f} = T_f \qquad (17\text{-}9b)$$

states that the final kinetic energy of a particle is equal to the sum of its initial kinetic energy and the work done on the particle by external forces. If the work term is divided up into a part due to conservative forces $U^{(c)}_{i \to f} = V_i - V_f$ and a part due to all other forces $U^{(o)}_{i \to f}$, the work–energy principle can be written

$$T_i + V_i + U^{(o)}_{i \to f} = T_f + V_f \qquad (17\text{-}33)$$

The convenience of the method of work and energy is that it directly relates the speed of a particle at two different positions of its motion to the forces that do work on the particle during the motion. If Newton's second law were applied directly, the acceleration would have to be obtained for an arbitrary position of the particle. Then the acceleration would have to be integrated using the principle of kinematics. The work–energy method combines these two steps into one.

The limitations of the method of work and energy are that Eq. 17-9 is a scalar equation and it can be solved for no more than one unknown; the acceleration cannot be calculated directly; and only forces that do work are involved. However, the normal component of acceleration is a function of velocity and the velocity is easily found using the work–energy method. Then the normal component of Newton's second law can be used to determine forces that act normal to the path of motion and those that do no work.

Since the work–energy method combines the principles of kinematics with Newton's second law, it is not a new or independent principle. There are no problems that can be solved with the work–energy method that cannot be solved using Newton's second law. However, when the work–energy method does apply, it is usually the easiest method of solving a problem.

REVIEW PROBLEMS

17-76* A 1.5-kg package is dropped from a height of 600 mm onto the lightweight platform of a scale as shown in Fig. P17-76. The platform is supported by two identical springs ($k = 150$ N/m), which are compressed 50 mm when no object is on the scale (the position shown). If no energy is lost when the package hits the scale, determine the maximum compression in the springs as a result of the package landing on the scale. Compare this value with the static compression of the scale (the package is slowly lowered onto the scale).

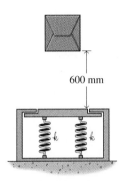

Fig. P17-76

17-77* A 4-lb pendulum bob has its motion interrupted by a small peg located directly under the support (Fig. P17-77). If the pendulum has an angular speed of 3 rad/s when $\theta = 75°$, determine the tension in the cord

a. At position A.
b. At position B.

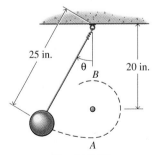

Fig. P17-77

17-78 A rope tow is used to pull skiers up a ski slope as shown in Fig. P17-78. If the coefficient of friction between the skis and the snow is 0.1 and the average weight of the skiers is 650 N, determine

a. The power required to operate the rope tow at 2 m/s if 50 skiers are holding onto the rope.
b. The speed of the rope tow if the power remains constant but an additional 25 skiers are holding onto the rope.

Fig. P17-78

17-79* In a shipping warehouse, packages are slid down a chute and allowed to fall on the floor as shown in Fig. P17-79. The coefficient of friction between the package and the ramp is $\mu_k = 0.40$ and $\theta = 20°$. If a 5-lb package has speed of 8 ft/s at A, determine

a. The speed with which the package hits the floor.
b. The distance d from the end of the chute to the point where the package hits the floor.

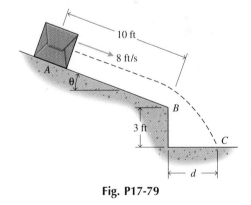

Fig. P17-79

17-80 In a carnival game of skill, players slide a 60-g hockey puck across a horizontal wood floor. The object is to have the puck stop as close to the wall as possible without hitting it. The coefficient of friction between the puck and the floor is 0.4 and the "foul line" (the point at which the player must release the puck) is 2 m from the wall. If a winning play must stop within 100 mm of the wall, determine the range of initial velocities that will win a prize.

17-81* A 10-lb block is attached to a spring having $k =$ 48 lb/ft and an unstretched length $\ell_0 = 18$ in. as shown in Fig. P17-81. The static and dynamic coefficients of friction between the block and the horizontal surface are 0.5 and 0.4, respectively. If the block has an initial speed of 7 ft/s when the spring is unstretched, determine the position x where the block will come to rest and the force F_s in the spring when the block stops for

a. Initial motion to the left.
b. Initial motion to the right.

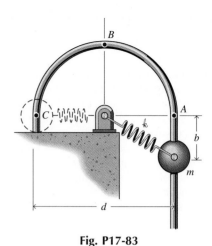

Fig. P17-83

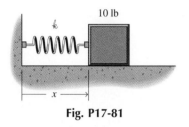

Fig. P17-81

17-82* In a carnival game of skill, players slide nickels along a wooden playing surface as shown in Fig. P17-82. To win a prize, a nickel must stop between the lines C and D on the lower surface. The coefficient of friction between the 5-g nickels and the floor is 0.2, the corners are abrupt but smooth, and the "foul line" (the point at which the player must release the nickels) is 1 m from the corner B. Determine the range of initial velocities that will win a prize in this game.

17-84* The retarding force due to air resistance on a 70-kg cyclist is given by

$$F_D = 0.8v^2$$

where v is in meters per second. If the cyclist can maintain a power output of 200 watts, determine the speed that the cyclist can maintain:

a. On level ground.
b. Up a 5° incline.
c. Down a 5° incline.

17-85 The pressure in the cylindrical, gas-filled chamber of Fig. P17-85 varies inversely with volume ($p =$ constant/volume). The pressure inside the chamber is the same as atmospheric ($p_{atm} = 14.7$ lb/in.^{2}) when the 5-lb piston is at $x = 10$ in. If $\dot{x} = -4$ ft/s when $x = 10$ in., determine the range of travel (x_{min} and x_{max}) of the piston.

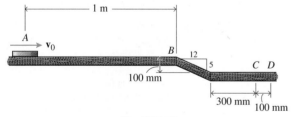

Fig. P17-82

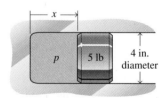

Fig. P17-85

17-83 A 0.5-lb weight slides on a frictionless rod in a vertical plane as shown in Fig. P17-83. The undeformed length of the spring is $\ell_0 = 6$ in., the spring modulus is $k = 5$ lb/ft, and the distance $d = 18$ in. If the weight is pulled down a distance b and released from rest, determine

a. The minimum distance b for which the weight will travel all the way around the rod to C.
b. The speed of the weight when it reaches C.

17-86 In a carnival game of skill, players shoot nickels up an inclined playing surface and into a 50-mm wide slot using a spring operated plunger as shown in Fig. P17-86. The coefficient of friction between the 5-g nickels and the wood floor is 0.2 and the spring modulus is $k = 75$ N/m. If the plunger is pulled back a distance δ and released from rest, determine the range of δs that will win a prize in this game. (Neglect the small size of the nickel when deciding if the nickel will fall through the slot.)

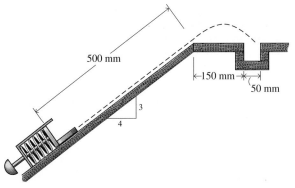

Fig. P17-86

17-87* The drag force on a vehicle moving with speed v is given by

$$F_D = 0.017W + 0.0012Cv^2 \text{ lb}$$

where W is the weight of the vehicle in pounds, C is a constant, which depends on the size and shape of the vehicle, and v is in feet per second. If a 3600-lb pickup carrying a 1500 lb load ($C = 18$) to market maintains a constant speed of 45 mi/h, determine

a. The power that must be delivered to the wheels.
b. The speed the truck can maintain when returning from the market empty ($C = 14$) for the same amount of power.

Computer Problems

C17-88 An ice cube slides down the inside of a frictionless cylinder as shown in Fig. P17-88. The mass of the ice cube is 20 g and the radius of the cylinder is $r = 300$ mm. If the ice cube starts from rest when $\theta = 75°$, calculate and plot the kinetic energy T, the potential energy V, the total energy $E = T + V$, and the normal force N exerted on the ice cube by the cylinder, all as functions of the angle θ ($-60° \leq \theta \leq 75°$). (Let the zero of potential energy be at $\theta = 90°$.)

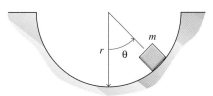

Fig. P17-88

C17-89 The pendulum shown in Fig. P17-89 consists of a 4-lb bob on an 18-in. long string. The pendulum is released from rest when $\theta = \theta_0 = 60°$. Let the zero of potential energy be at $\theta = 90°$.

a. Calculate and plot the kinetic energy T, the potential energy V, the total energy $E = T + V$, and the tension P in the string, all as functions of the angle θ ($-45° \leq \theta \leq 60°$).
b. If the string breaks when the tension equals 7 lb, determine the angle θ_b when the string breaks. Calculate and plot the motion (y versus x) of the pendulum bob from the time when it is released until it strikes the floor.

C17-90 An ice cube is balanced on a beach ball when it is disturbed and starts to slide (Fig. P17-90). The mass of the ice cube is 20 g and the radius of the ball is 300 mm. Let the zero of potential energy be at the center of the ball $\theta = 90°$.

a. Calculate and plot the kinetic energy T, the potential energy V, the total energy $E = T + V$, and the normal force N between the ball and the ice cube, all as functions of the angle θ ($0° \leq \theta \leq \theta_{max}$). For what angle θ_{max} does the ice cube slip off of the ball?
b. Calculate and plot the motion (y versus x) of the ice cube from the time when it is released until it lands in the sand.

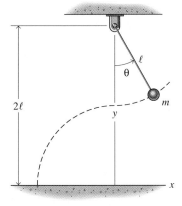

Fig. P17-89

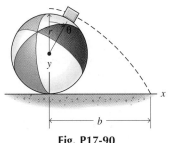

Fig. P17-90

C17-91 A 4-lb block A and an 8-lb block B are connected by an inextensible cord, which passes over a small frictionless pulley as shown in Fig. P17-91. Block A is also attached to a spring, which has a modulus of $k = 48$ lb/ft and an unstretched length of 12 in. When $x = 0$, block A is moving to the right with a speed of 10 ft/s. If the surface at A is smooth and $h = 18$ in.,

a. Determine the range of travel x_{min} and x_{max} of block A. Is $|x_{min}| = |x_{max}|$?
b. Calculate and plot the kinetic energy T, the potential energy V, and the total energy $E = T + V$ of the system, all as functions of x ($x_{min} \leq x \leq x_{max}$).

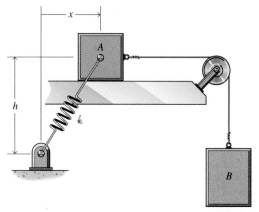

Fig. P17-91

C17-92 A 2-kg block A and a 5-kg block B are connected by an inextensible cord, which passes over a small frictionless pulley as shown in Fig. P17-91. Block A is also attached to a spring, which has a modulus of $k = 750$ N/m and an unstretched length of 300 mm. When $x = 0$, block A is moving to the right with a speed of 3 m/s. If the surface at A is rough ($\mu = 0.35$) and $h = 400$ mm, calculate and plot

a. The normal force N and friction force F exerted on block A as a function of x as it first moves to the right ($0 \leq x \leq x_{max}$) and then moves back to the left ($x_{max} \geq x \geq x_{min}$).
b. The work done by friction $U^{(o)}$ on block A as it moves from its initial position $x = 0$ to position x_{max} and back to position x_{min}.
c. The kinetic energy T, the potential energy V, and the total energy $E = T + V$ of the system, all as functions of x ($0 \leq x \leq x_{max}$ and $x_{max} \geq x \geq x_{min}$).

C17-93 A 5-lb block A and a 1-lb block B are connected by an inextensible cord, which passes over two small frictionless pulleys as shown in Fig. P17-93. Block A is also

attached to a spring, which has a modulus of $k = 85$ lb/ft. The surface at A is smooth and $h = 18$ in. If the system is released from rest with $x = 16$ in. and the spring stretched 6 in.,

a. Determine the range of travel x_{max} of block A.
b. Calculate and plot the kinetic energy T, the potential enervy V, and the total energy $E = T + V$ of the system, all as functions of x (16 in. $\leq x \leq x_{max}$).

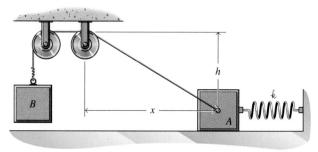

Fig. P17-93

C17-94 A 10-kg block A and a 2-kg block B are connected by an inextensible cord, which passes over two small frictionless pulleys as shown in Fig. P17-94. Block A is also attached to a spring, which has a modulus of $k = 1000$ N/m. The surface at A is rough ($\mu = 0.35$). If the system is released from rest with $x = 400$ mm and the spring stretched 150 mm, calculate and plot

a. The normal force N and friction force F exerted on block A as a function of x as it first moves to the right ($0 \leq x \leq x_{max}$) and then moves back to the left ($x_{max} \geq x \geq x_{min}$).
b. The work done by friction $U^{(o)}$ on block A as it moves from its initial position $x = 0$ to position x_{max} and back to position x_{min}.
c. The kinetic energy T, the potential energy V, and the total energy $E = T + V$ of the system, all as functions of x ($0 \leq x \leq x_{max}$ and $x_{max} \geq x \geq x_{min}$).

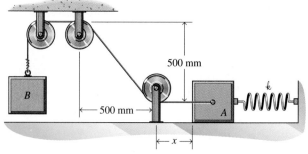

Fig. P17-94

KINETICS OF RIGID BODIES: WORK AND ENERGY METHODS

The gymnast uses her arm strength and body weight to create the forces and moments required to execute gymnastic movements on the uneven parallel bars.

18-1   INTRODUCTION

The method of work and energy combines the principles of kinematics with Newton's second law to directly relate the position and speed of a body. Therefore, the method of work and energy is not a new principle but is merely a special solution of the differential equations that arise when using Newton's second law. Nevertheless, the work–energy method greatly facilitates the solution of a certain class of problems.

In the work and energy approach, Newton's second law is integrated in a general sense with respect to position. For this method to be useful, the forces acting on a body must be a function of position only. For conservative forces, the resulting integrals can be evaluated explicitly in a general sense. The result is a simple algebraic equation that relates the speed of the body at two different positions of its motion.

For a general system of particles, the work–energy method was not found to be particularly useful. The principal reasons for this lack of usefulness were: (1) the motions of the particles are unrelated and must be specified independently; and (2) the work done by internal as well as external forces needed to be considered. However, it was shown in Section 17-4 that when two particles are connected by a rigid link, the work done by the internal forces cancel each other out. Therefore, the work–energy method is quite useful when the system of particles forms a rigid body as in this chapter.

The primary advantages of the work–energy method (the acceleration of a body need not be determined and integrated to determine the change in velocity of the body; forces that do no work have no effect on the work–energy equations and need not be included) are also the primary limitations of the work–energy method (the method of work and energy cannot determine accelerations or forces that do no work on a body).

Since Newton's second law is often used in conjunction with the principle of work and energy, a complete free-body diagram should be drawn. That is, all forces should be shown and not just those forces that do work during a particular motion of the body. Separate free-body diagrams showing the body in its initial and final positions may also be useful.

18-2   WORK OF FORCES AND COUPLES ACTING ON RIGID BODIES

Rigid bodies may be acted on by both forces and couples or pure moments. In addition, the body can rotate as well as translate. The work done by a force depends only on the motion of the point of application of the force. It does not depend on whether the motion is caused by the translation or the rotation of a rigid body. It will be shown, however, that a moment does no work due to a translation of the body on which it acts. Moments do work on a body only when the body rotates.

18-2-1   Work of Forces

The work done by a force **P** during a motion from point 1 to point 2 was defined in Chapter 17 as

$$U_{1 \to 2} = \int_1^2 \mathbf{P} \cdot d\mathbf{r} \qquad (18\text{-}1)$$

The calculation of work using Eq. 18-1 is independent of whether the force is applied to a particle, a translating rigid body, a rotating rigid body, or a translating and rotating rigid body. The work done by various types of forces was treated in detail in Section 17-2 and will not be repeated here.

For conservative forces, the potential energy V is also defined and determined in the same manner as for particles. The work done by conservative forces may be computed by direct integration using Eq. 18-1 or by using potential energy functions as discussed in Section 17-5.

18-2-2   Work of Internal Forces

Work done by internal forces in a rigid body does not have to be considered. Forces of interaction between two particles in a rigid body always occur in equal-magnitude but opposite-direction collinear pairs. Because the body is rigid, however, the two particles always undergo the same displacement in the direction of the forces. Therefore, the work done on one particle by one force always cancels with the work done on the second particle by the second force and the resultant work done by the pair of internal forces is zero.[1] That is, the work done on a *rigid* body by a system of external forces is the algebraic sum of works done by the individual forces.

Work done on a pair of rigid bodies by smooth connecting pins or by flexible, inextensible cables also need not be considered. Again, the forces occur in equal-magnitude but opposite-direction pairs and the points to which the forces are applied undergo equal displacements in the direction of the forces. Therefore, the resultant work done on the bodies by the connecting members of the system is zero.

For example, if the mass of the cable shown in Fig. 18-1 is negligible, then the tensions at the two ends of the cable will be the same. Since the cable is inextensible, however, the displacement in the direction of the cable at B and the displacement in the direction of the cable at C must also be equal. One of the forces will be in the direction of the displacement and will do positive work; the other force will be opposite the direction of the displacement and will do negative work. Therefore, the resultant work done on the pair of bodies by the cable must be zero.

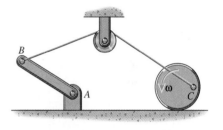

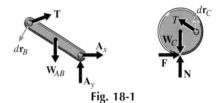

Fig. 18-1

18-2-3   Work of Couples and Moments

The work done by a couple is obtained by calculating the work done by each force of the couple separately and adding their works together. For example, consider a couple $\mathbf{C}$ acting on a rigid body as shown in Fig. 18-2a. During some small time dt the body translates and rotates. If

[1] If the body were not rigid, the internal forces would still occur in equal-magnitude and opposite-direction collinear pairs. However, the components of the displacements in the direction of the forces would, in general, not be the same. Therefore, the work done by the internal forces would not cancel and there would be a net work done by the internal forces.

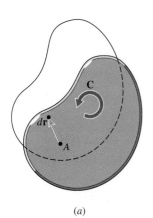

(a)

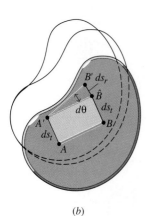

(b)

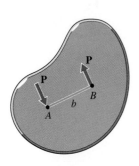

(c)

Fig. 18-2

the displacement of point A is $d\mathbf{r} = ds_t\,\mathbf{e}_t$, choose a second point B such that the line AB is perpendicular to $d\mathbf{r}$. Then the motion that takes point A to A' will take point B to B'. This motion may be considered in two parts: first a translation that takes the line AB to $A'\hat{B}$, followed by a rotation $d\theta$ about A' that takes $\hat{B}$ to B' (Fig. 18-2b).

Now represent the couple by the pair of forces of magnitude $P = C/b$ in the direction perpendicular to the line AB (Fig. 18-2c). During the translational part of the motion, one force will do positive work $P\,ds_t$ and the other will do negative work $-P\,ds_t$; therefore, the sum of the work done on the body by the pair of forces during the translational part of the motion is zero.

During the rotational part of the motion, A' is a fixed point and the force applied at A' does no work. The work done by the force at B is $dU = P\,ds_r \cong Pb\,d\theta$, where $d\theta$ is in radians and $C = Pb$ is the magnitude of the moment of the applied couple. The work is positive if the couple is in the same sense as $d\theta$ and negative if the couple is in the opposite sense. Therefore, the total work done by the couple during the differential rotational motion is

$$dU = C\,d\theta = \mathbf{C} \cdot d\boldsymbol{\theta} \tag{18-2}$$

where $\mathbf{C} = C\,\mathbf{k}$ and $d\boldsymbol{\theta} = d\theta\,\mathbf{k}$.

The work done on the body by the couple as the body rotates through an angle $\Delta\theta = \theta_2 - \theta_1$ is obtained by integrating Eq. 18-2:

$$U_{1\to 2} = \int_1^2 dU = \int_{\theta_1}^{\theta_2} C\,d\theta \tag{18-3}$$

If the couple is constant, then C can be taken outside the integral and Eq. 18-3 becomes

$$U_{1\to 2} = C\int_1^2 d\theta = C(\theta_2 - \theta_1) = C\,\Delta\theta \tag{18-4}$$

18-2-4   Forces That Do No Work

One of the principle advantages of the method of work and energy is that forces that do no work do not enter the equation. Some of the more obvious forces that do no work are:

1.   Forces applied to points that do not move. For example, when a wheel rotates on a fixed, frictionless axle, the forces exerted on the wheel by the axle do no work.
2.   Forces that act perpendicular to the motion. For example, the normal force acting on a body that slides or rolls along a surface does no work.

Not so obvious is the fact that the frictional force acting on a body that rolls without slipping does no work. The reason that the friction force does no work is because the contact point is an instantaneous center of zero velocity; it is instantaneously at rest. Since the point of application of the force is not moving (at the instant considered)

$$dU = \mathbf{F} \cdot d\mathbf{r}_{IC} = \mathbf{F} \cdot (\mathbf{v}_{IC}\,dt) = 0 \tag{18-5}$$

and therefore the friction force does no work.

18-3 KINETIC ENERGY OF RIGID BODIES IN PLANE MOTION

The kinetic energy of a body is obtained by adding together the kinetic energies of the particles that make up the body. For a general body that is not rigid, there is no simple equation relating the motion of the various particles; there is no general expression for the kinetic energy of the body. When the body is rigid, however, the velocities of the various particles are related by the relative velocity equation. This relationship allows for a particularly simple formula expressing the kinetic energy of a rigid body in plane motion.

For example, consider the body shown in Fig. 18-3. Point A is any point in the body and $\mathbf{r} = \mathbf{r}_{P/A} = x\mathbf{i} + y\mathbf{j} + z\mathbf{k}$ is the position vector from A to an arbitrary particle P of mass dm in the body. Then the velocity of dm is related to the velocity of point A by the relative velocity equation

$$\mathbf{v} = \mathbf{v}_A + \mathbf{v}_{P/A} = \mathbf{v}_A + \boldsymbol{\omega} \times \mathbf{r} \qquad (18\text{-}6)$$

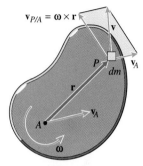

Fig. 18-3

and the kinetic energy of the particle can be written

$$dT = \frac{1}{2}\, dm\, v^2 = \frac{1}{2}\, dm\, \mathbf{v} \cdot \mathbf{v}$$

$$= \frac{1}{2}\, dm\, (\mathbf{v}_A + \boldsymbol{\omega} \times \mathbf{r}) \cdot (\mathbf{v}_A + \boldsymbol{\omega} \times \mathbf{r}) \qquad (18\text{-}7)$$

$$= \frac{1}{2}\, dm\, v_A^2 + dm\, \mathbf{v}_A \cdot (\boldsymbol{\omega} \times \mathbf{r}) + \frac{1}{2}\, dm\, (\boldsymbol{\omega} \times \mathbf{r}) \cdot (\boldsymbol{\omega} \times \mathbf{r})$$

Since the body has plane motion, the angular velocity of the body has only a z-component $\boldsymbol{\omega} = \omega\mathbf{k}$ so that

$$\boldsymbol{\omega} \times \mathbf{r} = \omega(x\mathbf{j} - y\mathbf{i}) \qquad (18\text{-}8)$$

and

$$(\boldsymbol{\omega} \times \mathbf{r}) \cdot (\boldsymbol{\omega} \times \mathbf{r}) = \omega^2(x^2 + y^2) \qquad (18\text{-}9)$$

Therefore, the kinetic energy of the piece of mass dm is

$$dT = \frac{1}{2}\, dm\, v_A^2 + dm\, \mathbf{v}_A \cdot (\boldsymbol{\omega} \times \mathbf{r}) + \frac{1}{2}\, dm\, \omega^2(x^2 + y^2) \quad (18\text{-}10)$$

Integrating Eq. 18-10 over all the mass in the body and recognizing that $\mathbf{v}_A$ and $\boldsymbol{\omega}$ are independent of dm gives

$$T = \frac{1}{2} v_A^2 \int dm + \mathbf{v}_A \cdot \left(\boldsymbol{\omega} \times \int \mathbf{r}\, dm \right) + \frac{1}{2}\omega^2 \int (x^2 + y^2)\, dm \quad (18\text{-}11)$$

But the integral in the first term of Eq. 18-11 is just the total mass of the body ($\int dm = m$); the integral in the second term is the location of the mass center measured from point A ($\int \mathbf{r}\, dm = m\mathbf{r}_{G/A}$); and the integral in the last term is the mass moment of inertia about an axis through point A parallel to the z-axis $\left(\int (x^2 + y^2)\, dm = I_{Az} \right)$; so that[2]

$$T = \frac{1}{2} mv_A^2 + m\mathbf{v}_A \cdot (\boldsymbol{\omega} \times \mathbf{r}_{G/A}) + \frac{1}{2} I_{Az}\omega^2 \qquad (18\text{-}12)$$

Although Eq. 18-12 is the general expression for the kinetic energy of a rigid body in plane motion, it is unnecessarily complex and is seldom used. Equation 18-12 simplifies significantly for the common types of plane motion to be discussed below. Even for general plane motion, Eq. 18-12 can be simplified considerably by an appropriate choice of the arbitrary point A.

18-3-1   Translation of a Rigid Body

When a rigid body moves without rotating, its angular velocity is zero $\boldsymbol{\omega} = \mathbf{0}$ and the velocity of every point in the body is the same. In this case, Eq. 18-12 reduces to

$$T = \frac{1}{2} mv^2 \qquad (18\text{-}13)$$

where v is the speed of any point in the body. Clearly, the idealization of particle motion is just the pure translation of a rigid body.

18-3-2   Rotation of a Rigid Body About a Fixed Axis

If point A is a point on the axis of rotation for a rigid body that is rotating about a fixed axis, then $\mathbf{v}_A = \mathbf{0}$ and Eq. 18-12 simplifies to

$$T = \frac{1}{2} I_{Az}\omega^2 \qquad (18\text{-}14)$$

Note, however, that Eq. 18-14 is also true if point A is an instantaneous center of zero velocity. The reduction of Eq. 18-12 to 18-14 requires merely that point A have zero velocity at the instant of the calculation.

18-3-3   General Plane Motion of a Rigid Body

The general motion of a rigid body consists of a combination of translation and rotation. Even for the case of general plane motion of a rigid body, however, Eq. 18-12 can be simplified substantially by an appropriate choice of point A. For example, when point A is chosen to be the mass center of the body G, then the position vector $\mathbf{r}_{G/A} = \mathbf{0}$, and

$$T = \frac{1}{2} mv_G^2 + \frac{1}{2} I_{Gz}\omega^2 \qquad (18\text{-}15)$$

where v_G is the speed of the mass center of the body and I_{Gz} is the moment of inertia about an axis through the mass center G parallel to the z-axis (perpendicular to the plane of motion). The first term of Eq. 18-15 is just the kinetic energy associated with the translation of the body's mass center, and the second term is the kinetic energy associated with the rotation of the body about an axis through its center of mass.

---

[2] Although it is not obvious from the derivation at this point, the results obtained are not limited to the motion of plane slabs or to the motion of bodies that are symmetrical with respect to the reference plane. Equation 18-12 and the other equations derived from it (Eqs. 18-13, 18-14, and 18-15) may be applied to the study of the plane motion of any rigid body, regardless of its shape. (See Section 18-6.)

The special case of rotation about a fixed axis through an arbitrary point A, Eq. 18-14, is also contained in the general expression for kinetic energy, Eq. 18-15. When the body rotates about a fixed axis through point A, the velocity of the mass center is given by

$$\mathbf{v}_G = \mathbf{v}_A + \mathbf{v}_{G/A} = \mathbf{0} + \boldsymbol{\omega} \times \mathbf{r}_{G/A} \tag{18-16}$$

so that

$$
\begin{aligned}
v_G^2 = \mathbf{v}_G \cdot \mathbf{v}_G &= (\omega \mathbf{k} \times \mathbf{r}_{G/A}) \cdot (\omega \mathbf{k} \times \mathbf{r}_{G/A}) \\
&= (\omega x_G \mathbf{j} - \omega y_G \mathbf{i}) \cdot (\omega x_G \mathbf{j} - \omega y_G \mathbf{i}) \\
&= \omega^2 (x_G^2 + y_G^2) = \omega^2 d^2
\end{aligned}
\tag{18-17}
$$

where $d^2 = x_G^2 + y_G^2$ is the square of the distance between the axis of rotation and the mass center G. Then the kinetic energy is

$$
\begin{aligned}
T &= \frac{1}{2} m v_G^2 + \frac{1}{2} I_{Gz} \omega^2 = \frac{1}{2} (md^2 + I_{Gz}) \omega^2 \\
&= \frac{1}{2} I_{Az} \omega^2
\end{aligned}
\tag{18-18}
$$

where $I_{Az} = I_{Gz} + md^2$ by the parallel-axis theorem for mass moments of inertia.

18-4 PRINCIPLE OF WORK AND ENERGY FOR THE PLANE MOTION OF RIGID BODIES

The principle of work and energy for a rigid body is obtained by adding together the equations of work and energy for each of the particles that make up the rigid body. This gives

$$U_{1 \to 2} = T_2 - T_1$$

in which T_1 and T_2 are the total kinetic energies of all the particles that make up the body (given by Eq. 18-15); $U_{1 \to 2}$ is the total work done by all external forces and couples acting on all the particles; and the work done by the internal forces need not be considered. Rearranging terms gives the principle of work and energy for a rigid body as

$$T_1 + U_{1 \to 2} = T_2 \tag{18-19}$$

which looks exactly the same as Eq. 17-9b for the work and energy of a particle. The difference between these equations is that the kinetic energy terms in Eq. 18-19 include the rotational kinetic energy of the rigid body as well as the translational kinetic energy and that the work term includes the work done by all external moments as well as the work done by all external forces that act on the rigid body.

Just as with a particle, the work term can be divided up into a part done by conservative forces (forces whose potential is known) $U_{1 \to 2}^{(c)}$ and a part done by all other forces (either nonconservative forces that have no potential or conservative forces whose potential is unknown) $U_{1 \to 2}^{(o)}$. The work done by the conservative forces can be expressed in terms of potential functions so that Eq. 18-19 can be written

$$T_1 + V_1 + U_{1 \to 2}^{(o)} = T_2 + V_2 \tag{18-20}$$

When two or more rigid bodies are connected by inextensible cords or cables or by frictionless pins, Eq. 18-19 (or Eq. 18-20) can be written for each of the bodies. When the resulting equations are added together, the work done by the connection forces will cancel in pairs. Therefore, Eqs. 18-19 and 18-20 also express the work–energy principle for a system of connected rigid bodies. For such a system of rigid bodies, T is the kinetic energy of the entire system and $U_{1\to2}$ ($= V_1 + U^{(o)}_{1\to2} - V_2$) includes the work done on the entire system by all external forces and moments.

For either a single rigid body or an interconnected system of rigid bodies, a complete free-body diagram must be drawn to ensure that all forces and moments are identified and considered. While including forces and moments that do no work on the free-body diagram may seem unnecessary, the work–energy principle is often used in conjunction with Newton's second law. Therefore, *all external forces and moments must be shown* on the free-body diagram and not just those forces and moments that do work on the body or bodies.

In addition to a complete free-body diagram of the body or bodies, it may also be helpful to draw diagrams that show the initial and final positions of the system for the given interval of motion.

18-5   POWER

Power, which is the time rate of doing work, was defined and discussed with respect to particle motion in Section 17-8. For a rigid body in plane motion, the work done must include the work done by couples as well as the work done by forces. If a rigid body is acted on simultaneously by a force $\mathbf{P}$ and a couple $\mathbf{C} = C\mathbf{k}$, the work done on the body is

$$dU = \mathbf{P} \cdot d\mathbf{r} + \mathbf{C} \cdot d\boldsymbol{\theta} \qquad (18\text{-}21)$$

in which $d\mathbf{r}$ is the displacement of the point of application of the force $\mathbf{P}$ and $d\boldsymbol{\theta} = d\theta\mathbf{k}$ is the rotation of the body. Then dividing through by dt gives the total power supplied to the rigid body at some instant

$$\text{power} = \mathbf{P} \cdot \mathbf{v} + \mathbf{C} \cdot \boldsymbol{\omega} \qquad (18\text{-}22)$$

where $\mathbf{v} = d\mathbf{r}/dt$ is the velocity of the point of application of the force $\mathbf{P}$ and $\boldsymbol{\omega} = \omega\mathbf{k} = \dfrac{d\theta}{dt}\mathbf{k}$ is the angular velocity of the body.

The turntable of a record player consists of a solid disk 12 in. in diameter and weighing 5 lb. If a constant torque motor accelerates the turntable from rest to its operating speed of $33\frac{1}{3}$ rpm in just one revolution, determine the torque **C** and the maximum power expended by the motor.

SOLUTION

The free-body diagram of the turntable is shown in Fig. 18-4. Only the torque **C** does work $U_{1\to2} = C\,\Delta\theta$, where $\Delta\theta = 1$ rev $= 2\pi$ radians. Since the motion is fixed-axis rotation about an axis through the mass center of the turntable, the kinetic energy is given by $T = \dfrac{1}{2}\,I_G\omega^2$, where the moment of inertia is

$$I_G = \frac{1}{2}\left(\frac{5}{32.2}\right)\left(\frac{6}{12}\right)^2 = 0.01941 \text{ slug}\cdot\text{ft}^2$$

the final angular velocity is $\omega_f = \dfrac{33\frac{1}{3} \text{ rev/min}}{60 \text{ s/min}}\,(2\pi\,\text{rad/rev}) = 3.491$ rad/s, and the initial angular velocity is zero. Therefore, the work–energy equation gives

$$0 + C(2\pi) = \frac{1}{2}\,(0.01941)(3.491)^2$$

or

$$\mathbf{C} = 0.01882 \text{ lb}\cdot\text{ft}\downarrow \qquad\qquad \text{Ans.}$$

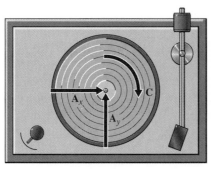

Fig. 18-4

The power of a couple is given by power $= C\omega$. Since the torque **C** is constant, the maximum power occurs where the angular velocity is maximum; that is

$$\text{power}_{\text{max}} = (0.01882)(3.491)$$
$$= 0.0657 \text{ lb}\cdot\text{ft/s} \qquad\qquad \text{Ans.}$$

A 4-kg concentrated mass is attached to the end of a 9-kg uniform slender rod that is rotating in a vertical plane as shown in Fig. 18-5a. The rod AB is 2 m long and has an angular velocity of 3 rad/s clockwise when it is vertical. If the undeformed length of the spring is $\ell_0 = 0.25$ m, determine the spring modulus k such that the angular velocity of AB will be zero when the rod is horizontal.

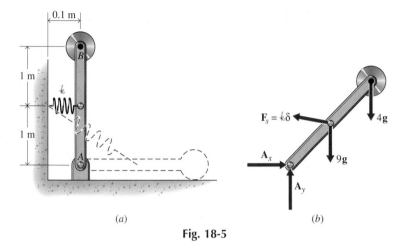

(a)   (b)

Fig. 18-5

SOLUTION

The free-body diagram of the system is shown in Fig. 18-5b. The system is rotating about a fixed axis so the kinetic energy is given by

$$T = \frac{1}{2} I_A \omega^2$$

where the moment of inertia is

$$I_A = \frac{1}{3} 9(2)^2 + 4(2)^2 = 28.00 \text{ kg} \cdot \text{m}^2$$

Then the initial kinetic energy is $T_i = \frac{1}{2}(28.00)(3)^2 = 126.00$ J and the final kinetic energy is $T_f = 0$.

The two gravitational forces and the spring force each have a potential. Using the level of point A as the zero of gravitational potential energy gives

$$V_i = (9)(9.81)(1) + (4)(9.81)(2) + \frac{1}{2} k(0.1 - 0.25)^2$$
$$= 166.77 + 0.01125 \, k \text{ J}$$

and

$$V_f = 0 + 0 + \frac{1}{2} k(1.4866 - 0.25)^2 = 0.7646 \, k \text{ J}$$

The pin forces at A do no work and there are no nonconservative forces acting on the system $U_{1 \rightarrow 2}^{(o)} = 0$. Substituting all these values in the work–energy equation gives

$$126.00 + (166.77 + 0.01125 \, k) + 0 = 0 + 0.7646 \, k$$

or

$$k = 389 \text{ N/m} \qquad\qquad\qquad \text{Ans.}$$

The wheel shown in Fig. 18-6a consists of a uniform half circle of wood weighing 20 lb encircled by an 18-in. diameter circular steel band of negligible thickness and weight. If the wheel rolls without slipping on a horizontal floor and has an angular velocity of 15 rad/s clockwise when the mass center G is directly below the center of the wheel C, determine:

a. The angular velocity of the wheel when G is directly to the left of C.
b. The normal and frictional components of the force exerted on the wheel by the floor when G is directly to the left of C.

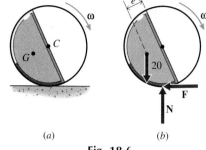

(a)   (b)

Fig. 18-6

SOLUTION

a. The free-body diagram of the wheel is shown in Fig. 18-6b. The only force that does work on the wheel is the gravitational force. Using the level of the center of the wheel as the zero of potential energy, the initial and final gravitational potentials are

$$V_{gi} = (20)(-0.3183) = -6.366 \text{ lb} \cdot \text{ft} \qquad V_{gf} = 0$$

where $e = 4r/3\pi = 0.3183$ ft is the distance between the center of the wheel C and the center of gravity G.

Since the wheel is moving in general plane motion, the kinetic energy of the wheel is given by $T = \frac{1}{2}mv_G^2 + \frac{1}{2}I_G\omega^2$, where the velocity of the mass center G is $\mathbf{v}_G = \mathbf{v}_C + \mathbf{v}_{G/C} = r\omega\mathbf{i} + \mathbf{v}_{G/C}$, ω is the clockwise angular velocity, and the moment of inertia is

$$I_G = \left(\frac{1}{2} - \frac{16}{9\pi^2}\right)mr^2 = \left(\frac{1}{2} - \frac{16}{9\pi^2}\right)\left(\frac{20}{32.2}\right)\left(\frac{9}{12}\right)^2 = 0.11176 \text{ slug} \cdot \text{ft}^2$$

When G is directly below C, $\mathbf{v}_{G/C} = -e\omega_i\mathbf{i}$ and the initial kinetic energy of the wheel is

$$T_i = \frac{1}{2}\left(\frac{20}{32.2}\right)\left[\frac{9}{12}(15) - (0.3183)(15)\right]^2 + \frac{1}{2}(0.11176)(15)^2$$
$$= 25.60 \text{ lb} \cdot \text{ft}$$

When G is directly to the left of C, $\mathbf{v}_{G/C} = e\omega_f\mathbf{j}$ and the final kinetic energy of the wheel is

$$T_f = \frac{1}{2}\left(\frac{20}{32.2}\right)\left[\left(\frac{9}{12}\omega_f\right)^2 + \left(0.3183\,\omega_f\right)^2\right] + \frac{1}{2}0.11176\,\omega_f^2$$
$$= 0.2620\,\omega_f^2 \text{ lb} \cdot \text{ft}$$

Substituting all these quantities in the work–energy equation then gives

$$25.60 - 6.366 + 0 = 0.2620\,\omega_f^2 + 0 \qquad \omega_f = 8.57 \text{ rad/s} \qquad \text{Ans.}$$

b. When G is directly to the left of C, the acceleration of the mass center is $\mathbf{a}_G = \mathbf{a}_C + \mathbf{a}_{G/C} = (r\alpha\mathbf{i}) + (e\alpha\mathbf{j} + e\omega_f^2\mathbf{i})$. Then the equations of motion

$$\rightarrow \Sigma F_x = ma_x: \qquad -F = \frac{20}{32.2}\left(\frac{9}{12}\alpha + 23.37\right)$$

$$\uparrow \Sigma F_y = ma_y: \qquad N - 20 = \frac{20}{32.2}(0.3183\alpha)$$

$$\curvearrowleft \Sigma M_G = I_G\alpha: \qquad \frac{9}{12}F - 0.3183N = 0.11176\alpha$$

give

$$N = 13.49 \text{ lb} \uparrow \qquad F = 0.820 \text{ lb} \leftarrow \qquad \text{Ans.}$$

EXAMPLE PROBLEM 18-4

A 15-kg crate is attached to the end of an inextensible cord wrapped around a 40-kg, 600-mm diameter, uniform drum as shown in Fig. 18-7a. At the instant shown, the crate is falling at a rate of 9 m/s. Determine the constant braking couple **C** that must be applied to the drum to bring the crate to rest after it descends 3 m.

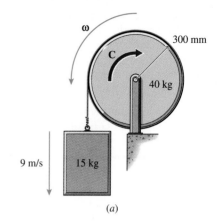

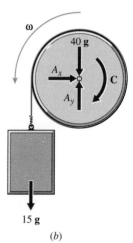

(a)

SOLUTION

KINEMATICS

Since the cord does not slip on the drum, when the crate drops $\Delta y = 3$ m the drum must rotate $\Delta \theta = \Delta y/r = 3/0.3 = 10$ rad. Also, if the crate is descending at a rate of $v = 9$ m/s, then the drum must be rotating at a rate of $\omega = v/r = 9/0.3 = 30$ rad/s.

WORK–ENERGY

Since the crate is attached to the drum by an inextensible cord, the work done by the cord on the drum and on the crate will cancel and need not be considered. Therefore, the crate and the drum will be treated as a single system; the free-body diagram of the entire system is shown in Fig. 18-7b. Neither the weight of the drum nor the forces exerted on the drum by the axle do work. The only nonconservative force acting on the system is the braking moment

$$U^{(o)}_{1 \to 2} = -C \, \Delta \theta = -10C$$

Using the initial height of the crate as the zero level of gravitational potential energy, the initial potential energy is zero $V_i = 0$ and the final potential energy is

$$V_f = (15)(9.81)(-3) = -441.5 \text{ J}$$

The drum is moving in fixed axis rotation about an axis through its mass center so that its initial kinetic energy is $T_d = \frac{1}{2} I_G \omega^2$ where $I_G = \frac{1}{2}(40)(0.3)^2 = 1.800 \text{ kg} \cdot \text{m}^2$. The crate is simply translating so that its initial kinetic energy is $T_C = \frac{1}{2} m v^2$. Therefore, the initial kinetic energy of the system is

$$T_i = \frac{1}{2}(15)(9)^2 + \frac{1}{2}(1.800)(30)^2 = 1417.5 \text{ J}$$

and the final kinetic energy is zero $T_f = 0$. Substituting all these values in the work–energy equation gives

$$1417.5 + 0 - 10C = 0 - 441.5$$

or

$$\mathbf{C} = 185.9 \text{ N} \cdot \text{m} \downarrow \qquad \qquad \text{Ans.}$$

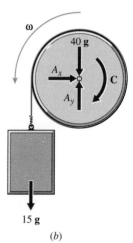

(b)

Fig. 18-7

PROBLEMS

18-1* A 10-lb uniform wheel 16 in. in diameter is at rest when it is placed in contact with a moving belt as shown in Fig. P18-1. The kinetic coefficient of friction between the belt and the wheel is $\mu_k = 0.1$ and the belt moves with a constant speed of 30 ft/s. Determine the number of revolutions the wheel turns before it rolls without slipping on the moving belt.

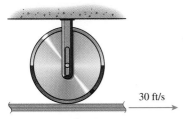

30 ft/s

Fig. P18-1

18-2* A 10-kg uniform flywheel 400 mm in diameter is connected to a constant torque motor with a flexible belt as shown in Fig. P18-2. If the flywheel starts from rest, determine the torque necessary to rotate the flywheel at 4200 rpm after 5 revolutions.

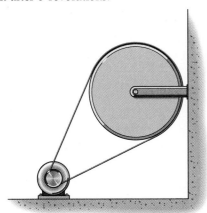

Fig. P18-2

18-3 The 25-lb uniform drum shown in Fig. P18-3 is 20 in. in diameter and is initially at rest. A force of 10 lb is applied to the end of the flexible rope wrapped around the

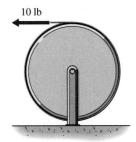

10 lb

Fig. P18-3

drum. If the rope releases from the drum after it has rotated 3 revolutions, determine the final angular velocity of the drum.

18-4* The 5-kg flywheel of Fig. P18-4 has a diameter of 200 mm and a radius of gyration of 90 mm. A flexible rope is wrapped around the flywheel and attached to a spring that has a modulus of $k = 120$ N/m. Initially, the flywheel is rotating clockwise at 20 rad/s and the spring is stretched 800 mm. Determine:

a. The maximum stretch in the spring.
b. The angular velocity of the flywheel when the rope becomes slack.

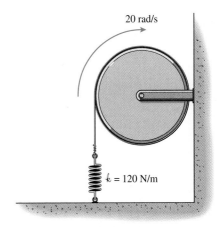

20 rad/s

$k = 120$ N/m

Fig. P18-4

18-5 A 16-lb crate hangs from the end of a cord wrapped around a 36-in. diameter uniform drum as shown in Fig. P18-5. The system starts from rest when the crate is 48 in. above the floor. Determine the weight of the drum that will cause the crate to hit the floor at half the speed it would have if it were simply dropped from the same height.

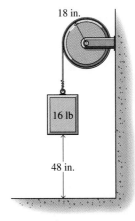

18 in.

16 lb

48 in.

Fig. P18-5

18-6* A 5-kg uniform rod 800 mm long rotates in a vertical plane as shown in Fig. P18-6. If the rod is released from rest when it is horizontal, determine:

a. The angular velocity of the rod when it is vertical.
b. The magnitude and direction of the pin reaction on the rod at B when the rod makes an angle of 75° with the horizontal.

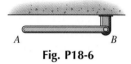

Fig. P18-6

18-7 A 5-lb uniform rod 36 in. long rotates in a vertical plane under the influence of a 2.5 lb·ft couple as shown in Fig. P18-7. If the rod is released from rest when it is horizontal, determine:

a. The angular velocity of the rod when it is vertical.
b. The magnitude and direction of the pin reaction on the rod at B when the rod makes an angle of 60° with the horizontal.

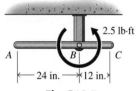

Fig. P18-7

18-8* The pendulum of Fig. P18-8 consists of a 30-kg concentrated mass on the end of a 45-kg uniform bar 2 m long. The pendulum swings in a vertical plane under the influence of a 500 N · m clockwise couple. If the pendulum has an angular velocity of 4 rad/s clockwise when $\theta = 90°$, determine:

a. The angular velocity of the pendulum when $\theta = 180°$, 330°, and 450°.
b. The magnitude and direction of the pin reaction on the pendulum at A when $\theta = 180°$, 330°, and 450°.

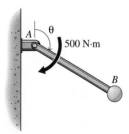

Fig. P18-8

18-9* An 8-lb uniform rectangular door opens upward and is counterbalanced with a spring as shown in Fig. P18-9. The modulus and undeformed length of the spring are $k = 24$ lb/ft and $\ell_0 = 23$ in., respectively. If the door has an

angular velocity of 3 rad/s counterclockwise when it is vertical ($\theta = 0°$), determine the angular velocity of the door and the magnitude and direction of the hinge reaction on the door when it is horizontal ($\theta = 90°$).

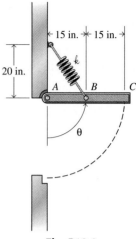

Fig. P18-9

18-10 A 45-kg post rotates in a vertical plane as shown in Fig. P18-10. The modulus and undeformed length of the spring are $k = 140$ N/m and $\ell_0 = 2$ m, respectively. If the angular velocity of the post is 3 rad/s clockwise when it is vertical, determine the angular velocity of the post and the magnitude and direction of the pin reaction on the post when the post is horizontal.

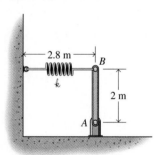

Fig. P18-10

18-11* The 20-lb uniform rod of Fig. P18-11 rotates about a frictionless pin at B. The modulus and undeformed length of the spring are $k = 48$ lb/ft and $\ell_0 = 6$ in., respectively. If the rod has an angular velocity of 3 rad/s counter-

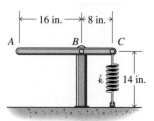

Fig. P18-11

clockwise when it is horizontal, determine the angular velocity of the rod and the magnitude and direction of the pin reaction on the rod when it is vertical with A directly above B.

18-12 A uniform rod is balanced on one end on a flat horizontal surface as shown in Fig. P18-12. The surface is very rough so that the rod will not slip. If the rod is disturbed and begins to tip over, determine the angle θ that the rod makes with the vertical when:

a. The frictional component of the force exerted on the rod by the floor changes direction.
b. The normal component of the force exerted on the rod by the floor becomes zero.

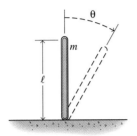

Fig. P18-12

18-13 A uniform rod is balanced on one end on a thin wire as shown in Fig. P18-13. A slight notch in the end of the rod prevents it from slipping off the wire. If the rod is disturbed and begins to tip over, determine the angle θ that the rod makes with the vertical when the rod loses contact with the wire.

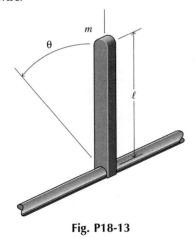

Fig. P18-13

18-14* A uniform cylinder is balanced on the edge of a ledge as shown in Fig. P18-14. The corner is rough so that the cylinder will not slip. If the cylinder is disturbed and begins to roll off the edge, determine the maximum angle θ through which the cylinder will rotate without losing contact with the corner.

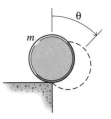

Fig. P18-14

18-15 Repeat Problem 18-14 for a thin hollow cylinder.

18-16* Repeat Problem 18-14 for a sphere.

18-17* A 16-lb bowling ball is placed on a 30° inclined surface and released from rest. Assume that the ball is a uniform sphere 12 in. in diameter. If the coefficient of friction between the ball and the surface is 0.25:

a. Verify that the ball will begin to roll without slipping.
b. Determine the speed v and angular velocity ω of the ball after it has rolled 20 ft down the incline.
c. Compare the speed of part b with the speed of a 16-lb particle that slides (without friction) the same distance down the incline.

18-18 A 12-kg uniform cylinder 600 mm in diameter is rolled up a 20° inclined surface with an initial speed of 10 m/s. If the cylinder rolls without slipping:

a. Determine the maximum distance that the cylinder will roll up the incline.
b. Compare the result of part a with the maximum distance that a 12-kg particle would slide (without friction) up the same incline.

18-19* A 25-lb uniform cylinder 16 in. in diameter rolls without slipping on an inclined surface as shown in Fig. P18-19. A spring having a modulus k = 10 lb/ft is attached to small frictionless pegs at the ends of the cylinder. If the cylinder has a speed of 4 ft/s down the incline when the spring is unstretched, determine:

a. The speed v and angular velocity ω of the cylinder when the spring is stretched 1 ft.
b. The maximum stretch in the spring.

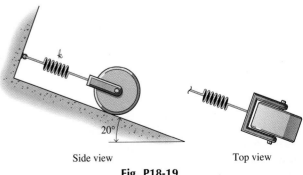

Side view   Top view

Fig. P18-19

18-20 A 15-kg uniform cylinder 800 mm in diameter rolls without slipping on an inclined surface as shown in Fig. P18-20. A light cord wrapped around the cylinder is attached to a spring having $k = 150$ N/m. If the cylinder is released from rest when the spring is stretched 1 m, determine:

a. The speed v and angular velocity ω of the cylinder when the stretch in the spring is 0.5 m.
b. The stretch in the spring when the cylinder is again at rest.

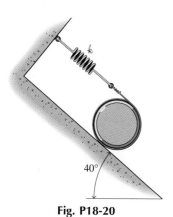

Fig. P18-20

18-21 A 30-lb, 12-in. diameter, uniform cylinder C rolls without slipping on a horizontal surface as shown in Fig. P18-21. A light cord wrapped around the cylinder passes over a small frictionless pulley and is attached to a 30-lb crate A. If the system is released from rest, determine the speed v_C and angular velocity ω_C of the cylinder and the speed v_A of the crate after the crate has dropped 5 ft.

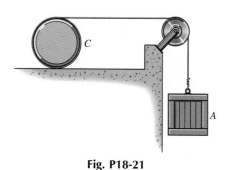

Fig. P18-21

18-22* The 10-kg spool C of Fig. P18-22 has a centroidal radius of gyration of 75 mm. A light cord is attached to the center of the spool, passes over a small frictionless pulley, and is attached to a 25-kg crate A. If the system is released from rest and the spool rolls without slipping, determine

the speed v_C and angular velocity ω_C of the spool and the speed v_A of the crate after the crate has dropped 2 m.

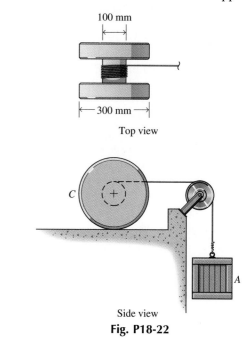

Top view

Side view
Fig. P18-22

18-23 The spool C of Fig. P18-23 consists of a uniform cylinder (8 lb, 6 in. diameter) between two cylindrical disks (each 4 lb, 18 in. diameter). A spring having $k = 24$ lb/ft is attached to cords wrapped around the disks. A second cord is wrapped around the center of the spool, passes over a small frictionless pulley, and is attached to a 30-lb crate A. If the crate has a speed of 4 ft/s downward when the spring is stretched 4 in. and the spool rolls without slipping, determine:

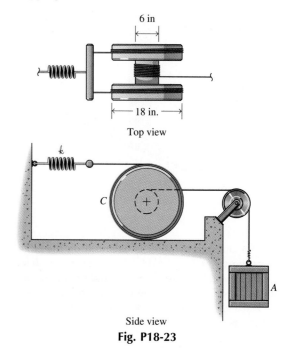

Top view

Side view
Fig. P18-23

a. The maximum distance that the crate will drop.
b. The speed v_C and angular velocity ω_C of the spool and the speed v_A of the crate when the stretch in the spring is zero.
c. The maximum distance that the crate will rise above its initial position.

18-24* The stepped cylinder C of Fig. P18-24 consists of a 5-kg cylindrical annulus (300 mm outer diameter and 100 mm inner diameter) and a 7-kg cylindrical axle (100 mm diameter). A spring having $k = 2$ kN/m is attached to cords wrapped around the axle. A second cord is wrapped around the center of the cylinder, passes over a small frictionless pulley, and is attached to a 15-kg crate A. If the crate has a speed of 1.5 m/s downward when the spring is stretched 100 mm and the cylinder rolls without slipping, determine:

a. The maximum distance that the crate will drop.
b. The speed v_C and angular velocity ω_C of the cylinder and the speed v_A of the crate when the stretch in the spring is zero.
c. The maximum distance that the crate will rise above its initial position.

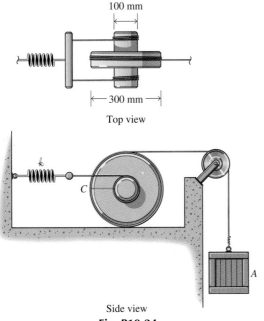

Side view
Fig. P18-24

18-25* The stepped cylinder of Fig. P18-25 consists of a 10-lb cylindrical annulus (16 in. outer diameter and 6 in. inner diameter) and a 5-lb cylindrical axle (6 in. diameter). A spring having $k_1 = 10$ lb/ft is attached to a cord wrapped around the center of the cylinder. A second spring having $k_2 = 10$ lb/ft is attached to cords wrapped around the axle of the cylinder. The system is released from rest with a stretch of 18 in. in spring 1 and spring 2 unstretched. If the coefficient of friction between the axle and the rails is $\mu_s = 0.8$,

a. Verify that the cylinder does not slip when motion starts.
b. Determine the speed v and angular velocity ω of the cylinder when the stretch of spring 1 becomes zero.

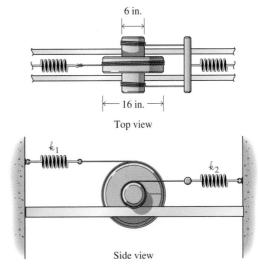

Top view

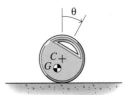

Side view
Fig. P18-25

18-26 A 10-kg nonhomogeneous wheel rolls without slipping on a horizontal surface as shown in Fig. P18-26. The center of gravity of the 500-mm diameter wheel is 50 mm from the center of the wheel. The radius of gyration about the center of mass is 165 mm. If the wheel is rotating clockwise at 9 rad/s when $\theta = 0°$, determine the force (magnitude and direction) exerted on the wheel by the surface when $\theta = 90°$ and $180°$.

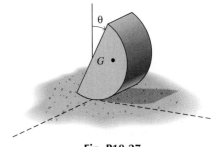

Fig. P18-26

18-27* The 16-lb homogeneous half-cylinder of Fig. P18-27 has a radius of 9 in. If the cylinder is released from rest when $\theta = 30°$ and rolls without slipping, determine the force (magnitude and direction) exerted on the cylinder by the surface when $\theta = 60°$ and $90°$.

Fig. P18-27

18-28 The compound wheel of Fig. P18-28 consists of a 5-kg half circular disk sandwiched between two full circular disks (each 2 kg, 600 mm diameter). If the wheel rolls without slipping and has an angular velocity of 10 rad/s clockwise when $\theta = 0°$, determine the force (magnitude and direction) exerted on the cylinder by the surface when $\theta = 60°$ and $120°$.

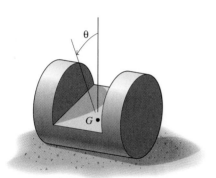

Fig. P18-28

18-29 The 3-lb, 16-in. diameter uniform circular disk of Fig. P18-29 rolls without slipping on a horizontal surface. A 4-lb slender rod penetrates the disk 7 in. from its center. If this system has an angular velocity of 5 rad/s clockwise when $\theta = 0°$, determine the force (magnitude and direction) exerted on the disk by the surface when $\theta = 60°$ and $120°$.

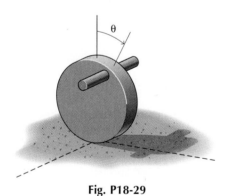

Fig. P18-29

18-30* A solid, homogeneous sphere (2 kg, 100 mm diameter) rolls without slipping on the outside of a fixed cylinder 400 mm in diameter (Fig. P18-30). The coefficient of static friction between the sphere and the cylinder is $\mu_s = 0.7$. If the sphere is released from rest when $\theta = 0°$, determine the angle θ at which the sphere will begin to

slip. (Does the sphere lose contact with the cylinder before or after this angle?)

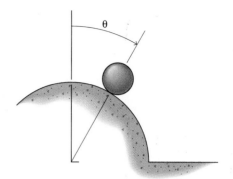

Fig. P18-30

18-31 A horizontal force of 25 lb is applied to one end of a 5-lb uniform rod 10 ft long (Fig. P18-31). A light slider attached to the middle of the rod moves in a frictionless vertical guide, and the surface at C is smooth. If the system is released from rest in the position shown, determine the angular velocity ω of the rod and the velocity $\mathbf{v}_A$ of end A of the rod when the rod is vertical.

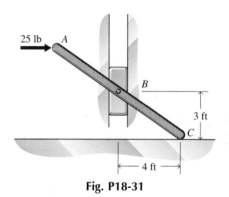

Fig. P18-31

18-32* A horizontal force of 150 N is applied to one end of a 3-kg uniform rod 4 m long (Fig. P18-32). A light slider attached to the middle of the rod moves in a frictionless vertical guide, and the surface at C is smooth. If the system is released from rest when $\theta = 20°$, determine the angular velocity ω of the rod and the velocity $\mathbf{v}_A$ of end A of the rod when $\theta = 80°$.

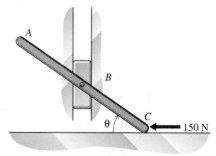

Fig. P18-32

18-33* The 5-lb uniform rod of Fig. P18-33 is 3 ft long. The light sliders at the ends of the rod move in frictionless guides. A spring attached to the slider at A has a modulus of $\ell = 15$ lb/ft and is stretched 2 ft when the rod is vertical. If the system is released from rest when the rod is vertical, determine the angular velocity ω of the rod and the velocity $\mathbf{v}_B$ of end B when:

a. The stretch in the spring is zero.
b. The bar is horizontal.

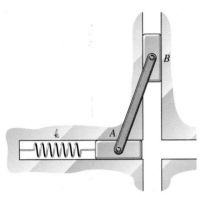

Fig. P18-33

18-34 The 5-kg uniform rod of Fig. P18-34 is 3 m long. The light sliders at the ends of the rod move in frictionless guides. A spring attached to the slider at B has a modulus of $\ell = 600$ N/m and is compressed 500 mm when the rod is horizontal. If the system is released from rest when the rod is horizontal, determine the angular velocity ω of the rod and the velocity $\mathbf{v}_A$ of end A when the stretch in the spring is zero.

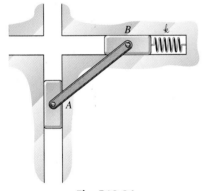

Fig. P18-34

18-35* A horizontal force of 25 lb is applied to one end of a 30-lb uniform rod 10 ft long (Fig. P18-31). A 20-lb slider attached to the middle of the rod moves in a frictionless vertical guide, and the surface at C is smooth. If end C of the rod has a velocity of 3 ft/s to the left in the position shown, determine:

a. The maximum height attained by the slider B.

b. The angular velocity ω of the rod and the velocity $\mathbf{v}_C$ of end C of the rod when the slider height $h_B = 2$ ft.
c. The normal force $\mathbf{N}_C$ exerted on the rod by the surface at C when $h_B = 2$ ft.

18-36 A horizontal force of 150 N is applied to one end of a 20-kg uniform rod 4 m long (Fig. P18-32). A 15-kg slider attached to the middle of the rod moves in a frictionless vertical guide, and the surface at C is smooth. If end C of the rod has a velocity of 1.35 m/s to the left when $\theta = 30°$, determine:

a. The maximum height attained by the slider B.
b. The angular velocity ω of the rod and the velocity $\mathbf{v}_C$ of end C of the rod when $\theta = 20°$.
c. The normal force $\mathbf{N}_C$ exerted on the rod by the surface at C when $\theta = 20°$.

18-37 The 5-lb uniform rod of Fig. P18-33 is 3 ft long. The sliders at the ends of the rod move in frictionless guides and weigh 2 lb each. A spring attached to the slider at A has a modulus of $\ell = 15$ lb/ft and is stretched 2 ft when the rod is vertical. If the system is released from rest when the rod is vertical, determine the angular velocity ω of the rod and the velocity $\mathbf{v}_B$ of end B when:

a. The stretch in the spring is zero.
b. The bar is horizontal.

18-38* The 5-kg uniform rod of Fig. P18-34 is 3 m long. The sliders at the ends of the rod move in frictionless guides and have a mass of 2 kg each. A spring attached to the slider at B has a modulus of $\ell = 600$ N/m and is compressed 500 mm when the rod is horizontal. If the system is released from rest when the rod is horizontal, determine the angular velocity ω of the rod and the velocity $\mathbf{v}_A$ of end A when the stretch in the spring is zero.

18-39 A 16-lb ladder 12 ft long slides in a smooth corner as shown in Fig. P18-39. The ladder is at rest with $\theta = 0°$ when the lower end is disturbed slightly. Determine the angle θ, the angular velocity ω, and the velocity $\mathbf{v}_B$ of end B when end A loses contact with the vertical wall.

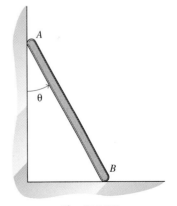

Fig. P18-39

18-40* A 3-kg uniform rod 2 m long is attached to a light slider that moves in a frictionless vertical guide (Fig. P18-40). The other end of the rod slides on a frictionless surface. The system is at rest with $\theta = 0°$ when the lower end is disturbed slightly. Determine the angle θ, the angular velocity ω, and the velocity $\mathbf{v}_B$ of end B when the normal force $\mathbf{N}$ between the horizontal surface and end B becomes zero.

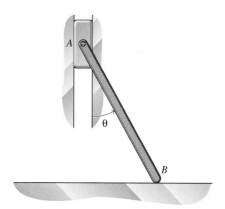

Fig. P18-40

18-41* A 20-lb uniform cylinder 24 in. in diameter rolls without slipping on a horizontal surface as shown in Fig. P18-41. The cylinder is joined to a 30-lb slider by a uniform slender rod (10 lb, 36 in. long). Friction between the slider and the vertical shaft may be neglected. If the system is released from rest when $h_A = 40$ in., determine the velocity $\mathbf{v}_A$ of the slider and the angular velocity ω_C of the cylinder when $h_A = 24$ in.

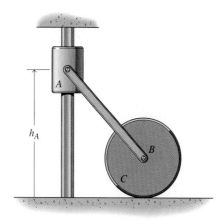

Fig. P18-41

18-42 A 7.5-kg mass A hangs from the frictionless axle of a pulley as shown in Fig. P18-42. The 10-kg pulley may be considered a uniform cylinder 400 mm in diameter. If a constant force $P = 350$ N is applied to the end of the rope and the system starts from rest, determine the velocity $\mathbf{v}_A$ of the mass A after it has risen 3 m.

Fig. P18-42

18-43* A 15-lb weight A hangs from the frictionless axle of a pulley as shown in Fig. P18-42. The 25-lb pulley may be considered a uniform cylinder 16 in. in diameter. If a constant force $P = 10$ lb is applied to the end of the rope and the system starts from rest, determine the velocity $\mathbf{v}_A$ of the weight after it has fallen 10 ft.

18-44 A 5-kg mass A hangs from the end of a rope that passes over a pulley as shown in Fig. P18-44. The 10-kg pulley may be considered a uniform cylinder 400 mm in diameter. If a constant force $P = 65$ N is applied to the axle of the pulley and the system starts from rest, determine the velocity $\mathbf{v}_A$ of the mass after it has fallen 2 m.

Fig. P18-44

18-45 A 10-lb weight A hangs from the end of a rope that passes over a pulley as shown in Fig. P18-44. The 20-lb pulley may be considered a uniform cylinder 16 in. in diameter. If a constant force $P = 150$ lb is applied to the axle of the pulley and the system starts from rest, determine the velocity $\mathbf{v}_A$ of the weight A after it has risen 3 ft.

18-46* A 5-kg uniform half cylinder 300 mm in diameter rotates about a frictionless pin as shown in Fig. P18-46. The half cylinder is joined to a 3-kg slider by a uniform slender rod (2 kg, 500 mm long). Friction between the slider and the horizontal shaft may be neglected. If the system is released from rest when $\theta = 0°$, determine the velocity of the slider $\mathbf{v}_A$ and the angular velocity of the half cylinder ω_C when $\theta = 120°$.

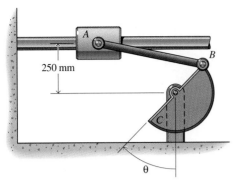

Fig. P18-46

18-47   A 5-lb uniform half cylinder 18 in. in diameter rolls without slipping on a horizontal surface as shown in Fig. P18-47. Friction between the uniform slender rod (8 lb, 30 in. long) and the horizontal surface at A may be neglected. If the system is released from rest when $\theta = 0°$, determine the velocity $\mathbf{v}_B$ and the angular velocity of the cylinder ω_C when $\theta = 90°$.

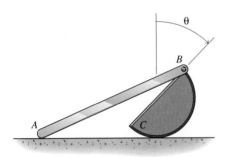

Fig. P18-47

18-48*   The uniform half cylinder shown in Fig. P18-48 has a mass of 8 kg and a radius of 200 mm, and rolls without slipping on the horizontal surface. The uniform bar AB has a mass of 5 kg and is 600 mm long. The system starts from rest when $\theta = 0°$ and the angular velocity of the cylinder is 2 rad/s clockwise when $\theta = 90°$. Neglect friction between the bar and the surface at A and determine the magnitude of the constant horizontal force $\mathbf{P}$.

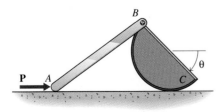

Fig. P18-48

18-49*   A cart of mass m_1 rolls down a 30° incline as shown in Fig. P18-49. The two wheels are uniform cylinders with a mass of m_2 each, are attached to the cart with frictionless axles, and roll without slipping. If the system is released from rest, determine the speed of the cart after it has traveled a distance d down the incline for the case:

a. $m_2 = \dfrac{1}{2} m_1$.

b. $m_2 = 2 m_1$.

(Compare these answers to the speed of a single particle of mass $m_1 + 2m_2$ sliding the same distance down the incline in the absence of friction.)

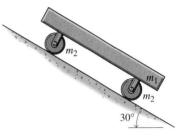

Fig. P18-49

18-50   Re-solve Problem 18-49 for the case where the wheels are not attached to the cart (Fig. P18-50). Friction is sufficient to prevent slipping between the wheels and the cart as well as between the wheels and the surface. Compare the answer with the results of Problem 18-49.

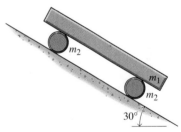

Fig. P18-50

18-51   Re-solve Problem 18-49 for the case where the wheels roll on frictionless axles attached to the surface (Fig. P18-51). Friction is sufficient to prevent slipping between the wheels and the cart. Compare the answer with the results of Problem 18-49.

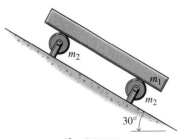

Fig. P18-51

18-52* Bars *AB* and *BC* of Fig. P18-52 each have a mass of 2 kg and a length of 400 mm. The 3-kg slider *C* moves in a frictionless vertical guide. If the system is released from rest in the position shown, determine the velocity $\mathbf{v}_C$ of the slider when it is at the level of *A*.

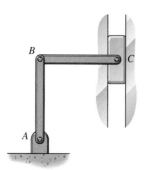

Fig. P18-52

18-53* A 12-lb uniform cylinder 16 in. in diameter rolls without slipping on a horizontal surface as shown in Fig. P18-53. The lightweight slender bars *AB* and *BC* each have a length of 16 in. The system is at rest in the position shown when *C* is displaced slightly to the right. Determine the velocity $\mathbf{v}_C$ of the center of the wheel and the angular velocity ω_{AB} of the crank *AB* when:

a. *AB* is horizontal.
b. *AB* is vertical.

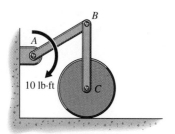

Fig. P18-53

18-54 The 2-kg slider shown in Fig. P18-54 moves in a frictionless guide. Crank arm *AB* has mass 1 kg and length 150 mm; *BC* has mass 3 kg and length 360 mm; and the spring has a modulus $k = 1800$ N/m and an unstretched length of 150 mm. If the system is released from rest when $\theta = 0°$, determine the velocity $\mathbf{v}_C$ of the slider and the angular velocity ω_{AB} of the crank:

a. When $\theta = 90°$.
b. When $\theta = 150°$.

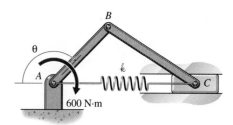

Fig. P18-54

18-55* A torsional spring attached to bar *AB* of Fig. P18-55 exerts a torque $M = k\theta$ where $k = 5$ ft · lb/rad. Bar *AB* weighs 10 lb and is 18 in. long; *BC* weighs 15 lb and is 30 in. long; and the surface at *C* is smooth. If the system is released from rest when $\theta = 0°$, determine the velocity $\mathbf{v}_C$ and the angular velocity ω_{AB}:

a. When $\theta = 60°$.
b. When $\theta = 90°$.

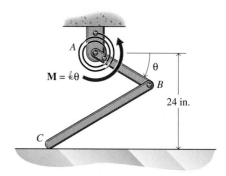

Fig. P18-55

18-56 A torsional spring attached to bar *AB* of Fig. P18-56 exerts a torque $M = k\theta$ where $k = 150$ N · m/rad. Bar *AB* has mass 25 kg and is 3 m long; *BC* has mass 50 kg and is 6 m long; and the surface at *C* is smooth. The system is initially at rest with $\theta = 60°$ and *BC* vertical when *C* is disturbed slightly to the right. Determine the velocity $\mathbf{v}_C$ and the angular velocity ω_{AB}:

a. When $\theta = 120°$.
b. When $\theta = 180°$.

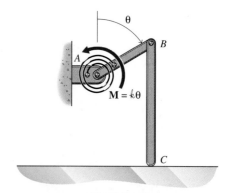

Fig. P18-56

18-6 KINETIC ENERGY OF A RIGID BODY IN THREE DIMENSIONS

In Section 18-3, the kinetic energy was calculated for a group of particles forming a rigid body and moving in a planar motion. In the present section, the restriction of planar motion will be removed.

As in Section 18-3, let point A be any point in the body and $\mathbf{r} = \mathbf{r}_{P/A} = x\mathbf{i} + y\mathbf{j} + z\mathbf{k}$ is the position vector from A to an arbitrary particle of mass dm in the body. Then the velocity of dm is related to the velocity of point A by the relative velocity equation

$$\mathbf{v} = \mathbf{v}_A + \mathbf{v}_{P/A} = \mathbf{v}_A + \boldsymbol{\omega} \times \mathbf{r} \tag{18-23}$$

where $\boldsymbol{\omega} = \omega_x\mathbf{i} + \omega_y\mathbf{j} + \omega_z\mathbf{k}$ is the angular velocity of the body. The kinetic energy of the particle is then

$$\begin{aligned}
dT &= \frac{1}{2}\, dm\, v^2 = \frac{1}{2}\, dm\, \mathbf{v} \cdot \mathbf{v} \\
&= \frac{1}{2}\, dm\, (\mathbf{v}_A + \boldsymbol{\omega} \times \mathbf{r}) \cdot (\mathbf{v}_A + \boldsymbol{\omega} \times \mathbf{r}) \\
&= \frac{1}{2}\, dm\, v_A^2 + dm\, \mathbf{v}_A \cdot (\boldsymbol{\omega} \times \mathbf{r}) + \frac{1}{2}\, dm\, (\boldsymbol{\omega} \times \mathbf{r}) \cdot (\boldsymbol{\omega} \times \mathbf{r}) \tag{18-24}
\end{aligned}$$

where

$$\begin{aligned}
\boldsymbol{\omega} \times \mathbf{r} &= (\omega_y r_z - \omega_z r_y)\mathbf{i} + (\omega_z r_x - \omega_x r_z)\mathbf{j} \\
&\quad + (\omega_x r_y - \omega_y r_x)\mathbf{k} \tag{18-25} \\
\mathbf{v}_A \cdot (\boldsymbol{\omega} \times \mathbf{r}) &= v_{Ax}(\omega_y r_z - \omega_z r_y) + v_{Ay}(\omega_z r_x - \omega_x r_z) \\
&\quad + v_{Az}(\omega_x r_y - \omega_y r_x) \tag{18-26}
\end{aligned}$$

and

$$\begin{aligned}
(\boldsymbol{\omega} \times \mathbf{r}) \cdot (\boldsymbol{\omega} \times \mathbf{r}) &= (\omega_y r_z - \omega_z r_y)^2 + (\omega_z r_x - \omega_x r_z)^2 \\
&\quad + (\omega_x r_y - \omega_y r_x)^2 \\
&= \omega_x^2(r_y^2 + r_z^2) + \omega_y^2(r_x^2 + r_z^2) + \omega_z^2(r_x^2 + r_y^2) \\
&\quad - 2(\omega_x \omega_y r_x r_y + \omega_x \omega_z r_x r_z + \omega_y \omega_z r_y r_z) \tag{18-27}
\end{aligned}$$

Substituting Eqs. 18-25 through 18-27 into Eq. 18-24 gives the kinetic energy of the piece of mass dm

$$\begin{aligned}
dT &= \frac{1}{2}\, dm\, v_A^2 + dm\, [(v_{Ay}\omega_z - v_{Az}\omega_y)\, r_x \\
&\quad + (v_{Az}\omega_x - v_{Ax}\omega_z)\, r_y + (v_{Ax}\omega_y - v_{Ay}\omega_x)\, r_z] \\
&\quad + \frac{1}{2}\, dm\, [\omega_x^2(r_y^2 + r_z^2) + \omega_y^2(r_x^2 + r_z^2) + \omega_z^2(r_x^2 + r_y^2) \\
&\quad - 2\omega_x \omega_y r_x r_y - 2\omega_x \omega_z r_x r_z - 2\omega_y \omega_z r_y r_z] \tag{18-28}
\end{aligned}$$

Integrating Eq. 18-28 over all the mass in the body and recognizing that $\mathbf{v}_A$ and $\boldsymbol{\omega}$ are independent of dm gives the total kinetic energy of the body

$$\begin{aligned}
T = \int dT &= \frac{1}{2}\, v_A^2 \int dm + (v_{Ay}\omega_z - v_{Az}\omega_y) \int r_x\, dm \\
&\quad + (v_{Az}\omega_x - v_{Ax}\omega_z) \int r_y\, dm + (v_{Ax}\omega_y - v_{Ay}\omega_x) \int r_z\, dm \\
&\quad + \frac{1}{2}\, \omega_x^2 \int (r_y^2 + r_z^2)\, dm + \frac{1}{2}\, \omega_y^2 \int (r_x^2 + r_z^2)\, dm \\
&\quad + \frac{1}{2}\, \omega_z^2 \int (r_x^2 + r_y^2)\, dm - \omega_x \omega_y \int r_x r_y\, dm \\
&\quad - \omega_x \omega_z \int r_x r_z\, dm - \omega_y \omega_z \int r_y r_z\, dm \tag{18-29}
\end{aligned}$$

The first integral in Eq. 18-29 is just the mass of the body; the next three integrals give the location of the mass center of the body relative to point A; and the last six integrals are the moments and products of inertia relative to axes through point A. Therefore,

$$
\begin{aligned}
T = \frac{1}{2}\, m v_A^2 &+ (v_{Ay}\omega_z - v_{Az}\omega_y)m r_{Gx} \\
&+ (v_{Az}\omega_x - v_{Ax}\omega_z)\, m r_{Gy} + (v_{Ax}\omega_y - v_{Ay}\omega_x)\, m r_{Gz} \\
&+ \frac{1}{2}\,\omega_x^2 I_{Ax} + \frac{1}{2}\,\omega_y^2 I_{Ay} + \frac{1}{2}\,\omega_z^2 I_{Az} \\
&- \omega_x\omega_y\, I_{Axy} - \omega_x\omega_z\, I_{Axz} - \omega_y\omega_z\, I_{Ayz} \qquad \text{(18-30)}
\end{aligned}
$$

Equation 18-30 can be simplified if point A coincides with the mass center G. Then $r_{Gx} = r_{Gy} = r_{Gz} = 0$ and the second, third, and fourth terms all vanish, leaving

$$
\begin{aligned}
T = \frac{1}{2}\, m v_G^2 &+ \frac{1}{2}\,\omega_x^2 I_{Gx} + \frac{1}{2}\,\omega_y^2 I_{Gy} + \frac{1}{2}\,\omega_z^2 I_{Gz} \\
&- \omega_x\omega_y\, I_{Gxy} - \omega_x\omega_z\, I_{Gxz} - \omega_y\omega_z\, I_{Gyz} \qquad \text{(18-31)}
\end{aligned}
$$

It can now be noted that Eq. 18-31 reduces to Eq. 18-15 for the special case of planar motion with $\omega_x = \omega_y = 0$. No assumptions about symmetry (that is, about the moments or products of inertia) are needed. Equation 18-30 also simplifies for the special case of rotation about a fixed point O. When A coincides with a fixed point O, $\mathbf{v}_O = \mathbf{0}$ and

$$
\begin{aligned}
T = \frac{1}{2}\,\omega_x^2 I_{Ox} &+ \frac{1}{2}\,\omega_y^2 I_{Oy} + \frac{1}{2}\,\omega_z^2 I_{Oz} \\
&- \omega_x\omega_y\, I_{Oxy} - \omega_x\omega_z\, I_{Oxz} - \omega_y\omega_z\, I_{Oyz} \qquad \text{(18-32)}
\end{aligned}
$$

Finally, if the directions of the principal axes are chosen for the xyz-axes, then the products of inertia are all zero and Eqs. 18-31 and 18-32 become

$$
T = \frac{1}{2}\, m v_G^2 + \frac{1}{2}\,\omega_x^2 I_{Gx} + \frac{1}{2}\,\omega_y^2 I_{Gy} + \frac{1}{2}\,\omega_z^2 I_{Gz} \qquad \text{(18-33)}
$$

and

$$
T = \frac{1}{2}\,\omega_x^2 I_{Ox} + \frac{1}{2}\,\omega_y^2 I_{Oy} + \frac{1}{2}\,\omega_z^2 I_{Oz} \qquad \text{(18-34)}
$$

respectively.

Equation 18-31 can also be written in vector form

$$
T = \frac{1}{2}\, m \mathbf{v}_G \cdot \mathbf{v}_G + \frac{1}{2}\,\boldsymbol{\omega} \cdot \mathbf{H}_G \qquad \text{(18-35)}
$$

in which $\mathbf{H}_G$ is called the *angular momentum vector* and has the components

$$
\begin{aligned}
H_{Gx} &= \omega_x I_{Gx} - \omega_y I_{Gxy} - \omega_z I_{Gxz} & \text{(18-36a)} \\
H_{Gy} &= -\omega_x I_{Gyx} + \omega_y I_{Gy} - \omega_z I_{Gyz} & \text{(18-36b)} \\
H_{Gz} &= -\omega_x I_{Gzx} - \omega_y I_{Gzy} + \omega_z I_{Gz} & \text{(18-36c)}
\end{aligned}
$$

The vector expression for the kinetic energy of a rigid body emphasizes that the kinetic energy is the scalar sum of the kinetic energy of translation $\frac{1}{2}\, m \mathbf{v}_G \cdot \mathbf{v}_G$ associated with the center of mass G and the kinetic energy of rotation about the mass center $\frac{1}{2}\,\boldsymbol{\omega} \cdot \mathbf{H}_G$. Equally important, it emphasizes that the reference system used for the computation of the inertia properties has its origin at the center of mass G.

EXAMPLE PROBLEM 18-5

The 16-lb, thin, homogeneous disk of Fig. 18-8a has a diameter of 20 in. and rotates freely on the 24 in. long axle OG. As the disk rolls without slipping on the horizontal surface, the axle rotates freely about point O. Determine the kinetic energy of the disk when its angular speed ω_{GC} is 13 rad/s.

SOLUTION

Coordinate axes are chosen with the x-axis along the axle OG (at the instant shown in Fig. 18-8a) and the z-axis in the horizontal surface.[3] The y-axis then is inclined at an angle of $\theta = \tan^{-1}(10/24) = 22.62°$ to the vertical (Fig. 18-8b). These are principal axes for the disk so that the moments of inertia relative to the mass center G are

$$I_{Gx} = \frac{1}{2}\left(\frac{16}{32.2}\right)\left(\frac{10}{12}\right)^2 = 0.17253 \text{ slug} \cdot \text{ft}^2$$

$$I_{Gy} = I_{Gz} = \frac{1}{4}\left(\frac{16}{32.2}\right)\left(\frac{10}{12}\right)^2 = 0.08627 \text{ slug} \cdot \text{ft}^2$$

$$I_{Gxy} = I_{Gxz} = I_{Gyz} = 0 \text{ slug} \cdot \text{ft}^2$$

As the disk rotates about the axle with an angular speed ω_{GC}, the axle rotates about a vertical axis with an angular speed of ω_{OG} (Fig. 18-8b). Since the disk rolls without slipping, these angular speeds are related by

$$10\omega_{GC} = 26\omega_{OG}$$

which gives $\omega_{OG} = 5$ rad/s. Then, in terms of the xyz-coordinate axes, the angular velocity of the disk is

$$\boldsymbol{\omega} = -\omega_{GC}\mathbf{i} + \omega_{OG}(\sin\theta\,\mathbf{i} + \cos\theta\,\mathbf{j})$$
$$= -11.077\mathbf{i} + 4.615\mathbf{j} \text{ rad/s}$$

Finally, the kinetic energy of the disk is (Eq. 18-33)

$$T = \frac{1}{2}mv_G^2 + \frac{1}{2}\omega_x^2 I_{Gx} + \frac{1}{2}\omega_y^2 I_{Gy} + \frac{1}{2}\omega_z^2 I_{Gz}$$

where the speed of the mass center is $v_G = [(24/12)\cos\theta]\,\omega_{OG} = 9.231$ ft/s. Therefore,

$$T = \frac{1}{2}\left(\frac{16}{32.2}\right)(9.231)^2 + \frac{1}{2}(11.077)^2\,(0.17253)$$

$$+ \frac{1}{2}(4.615)^2\,(0.08627) + \frac{1}{2}\,(0)^2\,(0.08627)$$

$$= 32.67 \text{ lb} \cdot \text{ft} \qquad \text{Ans.}$$

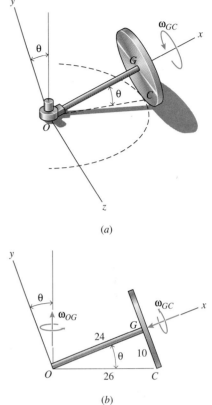

(a)

(b)

Fig. 18-8

[3] It must be noted that the orientation of the coordinate system is fixed; the coordinate system does not rotate with the body. The coordinate system is chosen so that it coincides with the principal axes of the body at the instant shown simply for ease of computing the kinetic energy. However, since the kinetic energy is a scalar quantity, the number obtained is independent of the coordinate system used to compute it. Therefore, different fixed coordinate systems may be used to compute the kinetic energy at the initial and final instants when using the principle of work and energy.

The 5-kg, thin, homogeneous disk of Fig. 18-9a has a diameter of 200 mm and rotates freely on the 300 mm long axle OG. As the disk rolls without slipping on the inclined surface, the axle rotates freely about point O. If the system is released from rest in the position shown (with the disk at its highest position on the inclined surface), determine the angular speed of the disk ω_{GC} when the disk is at its lowest position along the inclined surface.

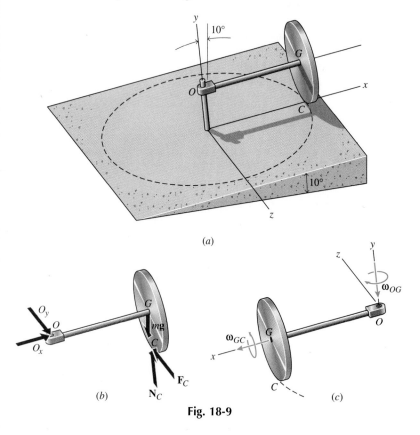

(a)

(b)   (c)

Fig. 18-9

SOLUTION

A free-body diagram of the disk and axle is shown in Fig. 18-9b. The normal force on the disk, the friction force on the disk, and the force at the fixed point

O do no work since they all act at points that are instantaneously at rest. Also, the internal force between the disk and the axle does no work since friction there is negligible. Therefore, the only force that does work is gravity, and it is conservative $U^{(o)}_{1\to2} = 0$. The initial and final gravitational potential energies are

$$V_i = (5)(9.81)(0.3 \sin 10°) = 2.555 \text{ N} \cdot \text{m}$$
$$V_f = (5)(9.81)(-0.3 \sin 10°) = -2.555 \text{ N} \cdot \text{m}$$

respectively.

Since the system starts from rest, the initial kinetic energy is zero $T_i = 0$. To compute the final kinetic energy, a system of axes will be chosen (Fig. 18-9c) in which the x-axis is oriented along the axle OG, the y-axis is perpendicular to the surface, and the z-axis is parallel to the surface. These are principal axes for the disk so that the moments of inertia relative to the fixed point O are

$$I_{Ox} = \frac{1}{2}(5)(0.1)^2 = 0.02500 \text{ kg} \cdot \text{m}^2$$

$$I_{Oy} = I_{Oz} = \frac{1}{4}(5)(0.1)^2 + (5)(0.3)^2 = 0.46250 \text{ kg} \cdot \text{m}^2$$

$$I_{Oxy} = I_{Oxz} = I_{Oyz} = 0 \text{ kg} \cdot \text{m}^2$$

As the disk rotates about the axle with an angular speed ω_{GC}, the axle rotates about a vertical axis with an angular speed of ω_{OG} (Fig. 18-9c). Since the disk rolls without slipping, these angular speeds are related by

$$100\omega_{GC} = 300\omega_{OG}$$

Then, in terms of the xyz-coordinate axes, the angular velocity of the disk is

$$\boldsymbol{\omega} = \omega_{GC}\mathbf{i} - \omega_{OG}\mathbf{j} = \omega_{GC}\mathbf{i} - \frac{\omega_{GC}}{3}\mathbf{j}$$

Then, the final kinetic energy of the disk is (Eq. 18-34)

$$T_f = \frac{1}{2}\omega_x^2 I_{Ox} + \frac{1}{2}\omega_y^2 I_{Oy} + \frac{1}{2}\omega_z^2 I_{Oz}$$
$$= \frac{1}{2}\omega_{GC}^2 (0.02500) + \frac{1}{2}\left(\frac{\omega_{GC}}{3}\right)^2 (0.46250) + 0$$
$$= 0.03819 \, \omega_{GC}^2$$

Finally, substituting all these values in the equation of work and energy $T_i + V_i + U^{(o)}_{i\to f} = T_f + V_f$ gives

$$0 + 2.555 + 0 = 0.03819 \, \omega_{GC}^2 - 2.555$$

or

$$\omega_{GC} = 11.57 \text{ rad/s} \qquad\qquad \text{Ans.}$$

PROBLEMS

18-57* The 5-lb wheel of Fig. P18-57 is rotating with an angular speed of 20 rad/s on a smooth axle. Simultaneously, the arm holding the wheel is rotating about the horizontal shaft with an angular speed of 8 rad/s. If the wheel can be modeled as a uniform thin disk 12 in. in diameter, determine the kinetic energy of the wheel.

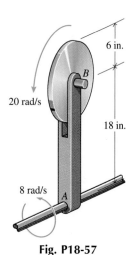

Fig. P18-57

18-58* The 5-kg wheel of Fig. P18-58 is rotating with an angular speed of 25 rad/s on a smooth axle. Simultaneously, the arm holding the wheel is rotating about the horizontal shaft with an angular speed of 10 rad/s. If the wheel can be modeled as a uniform thin disk 400 mm in diameter, determine the kinetic energy of the wheel.

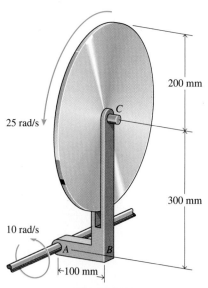

Fig. P18-58

18-59 The 4-lb disk of Fig. P18-59 is mounted at an angle of 40° to the horizontal shaft, which is rotating with an angular speed of 240 rpm. If the disk can be modeled as a uniform thin disk 16 in. in diameter, determine the kinetic energy of the disk.

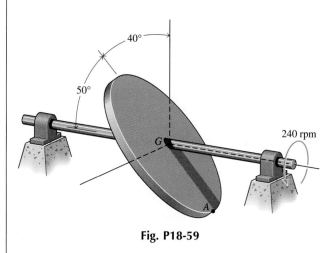

Fig. P18-59

18-60* The 4-kg plate of Fig. P18-60 is mounted at an angle of 40° to the horizontal shaft, which is rotating with an angular speed of 300 rpm. If the plate can be modeled as a uniform thin rectangular plate, determine the kinetic energy of the plate.

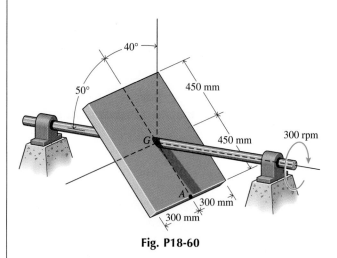

Fig. P18-60

18-61 Repeat Problem 18-59 for the case where the circular disk is mounted to the shaft at point A (on the edge of the disk) rather than at the mass center G.

18-62 Repeat Problem 18-60 for the case where the rectangular plate is mounted to the shaft at point A (on the edge of the plate) rather than at the mass center G.

18-63* The 8-lb disk of Fig. P18-63 rotates freely about the shaft OG as it rolls without slipping on a horizontal surface. If the disk can be modeled as a uniform thin disk 24 in. in diameter, determine the kinetic energy of the disk when ω = 240 rpm.

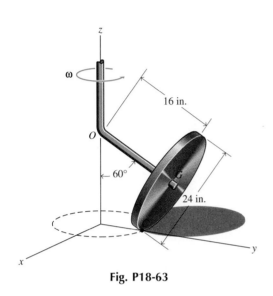

Fig. P18-63

18-64 A 2-kg, thin circular disk 200 mm in diameter (A) and an 8-kg, thin circular disk 400 mm in diameter (B) are rigidly joined by a light shaft AB, which is 120 mm long as shown in Fig. P18-64. Determine the kinetic energy of the pair of disks when they are rotating with an angular speed of 5 rad/s.

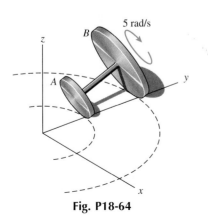

Fig. P18-64

18-65* The 16-lb, thin, homogeneous disk of Example Problem 18-5 (Fig. P18-65) has a diameter of 20 in. and is rigidly attached to the 4-lb axle OG, which is 24 in. long and 1 in. in diameter. If the disk rolls without slipping on a horizontal surface and the axle rotates freely about the fixed point O, determine the kinetic energy of the system when the angular speed of the disk ω_{GC} is 13 rad/s.

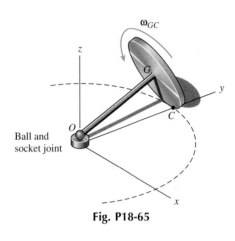

Fig. P18-65

18-66* The 5-kg, thin, homogeneous disk of Example Problem 18-6 (Fig. P18-66) has a diameter of 200 mm and is rigidly attached to the 2-kg axle OG, which is 300 mm long and 25 mm in diameter. If the disk rolls without slipping on a horizontal surface and the axle rotates freely about the vertical shaft at O, determine the kinetic energy of the system when the angular speed of the disk ω_{GC} is 15 rad/s.

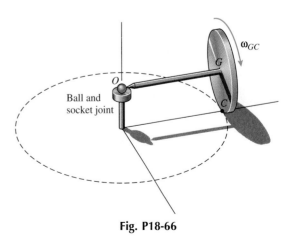

Fig. P18-66

18-67 A 3-lb uniform slender rod AB is attached by means of ball and socket joints to a rotating wheel and a slider as shown in Fig. P18-67. Determine the kinetic energy of the rod in the position shown if the angular speed of the wheel is $\omega = 20$ rad/s.

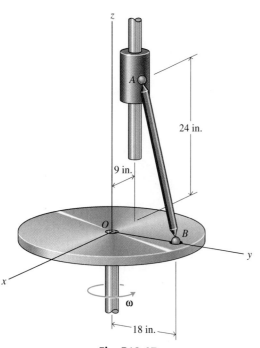

Fig. P18-67

18-68* A 2-kg uniform slender rod AB is attached by means of ball and socket joints to a rotating arm and a slider as shown in Fig. P18-68. Determine the kinetic energy of the rod in the position shown if the angular speed of the arm is $\omega = 15$ rad/s.

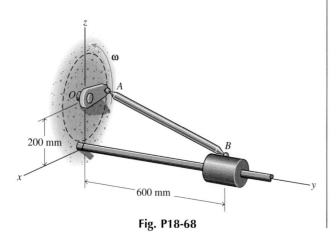

Fig. P18-68

18-69 A 5-lb uniform slender rod AB is attached by means of ball and socket joints to two sliders as shown in Fig. P18-69 ($a = 16$ in., $b = 32$ in., $c = 8$ in.). If slider A is moving upward with a speed of 15 ft/s at the instant shown, determine the kinetic energy of the rod.

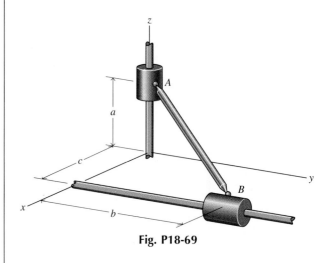

Fig. P18-69

18-70 A 4-kg uniform slender rod AB is attached by means of ball and socket joints to two sliders as shown in Fig. P18-69 ($a = 400$ mm, $b = 800$ mm, $c = 200$ mm). If slider B is moving to the right with a speed of 8 m/s at the instant shown, determine the kinetic energy of the rod.

18-71* The 4-lb disk of Problem 18-59 is initially at rest when a constant moment M_0 is applied to the shaft, causing it to rotate. Determine the magnitude of the moment that will cause the disk to rotate at 240 rpm after 4 revolutions.

18-72 The 4-kg plate of Problem 18-60 is initially at rest when a constant moment M_0 is applied to the shaft, causing it to rotate. Determine the magnitude of the moment that will cause the plate to rotate at 300 rpm after 5 revolutions.

18-73* The 4-lb disk of Problem 18-61 is initially at rest when a constant moment $M_0 = 5$ lb·ft is applied to the shaft, causing it to rotate. Determine the angular speed of the shaft after 2 revolutions.

18-74* The 4-kg plate of Problem 18-62 is initially at rest when a constant moment $M_0 = 5$ N·m is applied to the shaft, causing it to rotate. Determine the angular speed of the shaft after 3 revolutions.

18-75 The 8-lb disk of Problem 18-63 is initially at rest when a constant moment $M_0 = 10$ lb·ft is applied to the vertical shaft, causing it to rotate. Determine the angular speed ω of the vertical shaft after 3 revolutions.

18-76* The pair of disks of Problem 18-64 is placed on an inclined surface and released from rest. If AB initially points directly up the 15° incline, determine the angular speed of the system when AB points directly down the incline.

18-77 The system of Problem 18-65 is initially at rest when a constant moment $M_0 = 15$ lb·ft (about a vertical axis) is applied to the hub O. Determine the angular speed ω_{GC} of the disk after it has rolled 1 revolution about the hub O.

18-78 The system of Problem 18-66 is initially at rest when a constant moment $M_0 = 12$ N·m (about a vertical axis) is applied to the hub O. Determine the angular speed ω_{GC} of the disk after it has rolled 2 revolutions about the vertical shaft.

18-79* The 5-lb slender rod of Problem 18-69 is released from rest when $a = 32$ in., $b = 16$ in., $c = 8$ in. Determine the speed v_B of slider B when $b = 24$ in.

18-80 The 4-kg slender rod of Problem 18-70 is released from rest when $a = 800$ mm, $b = 400$ mm, $c = 200$ mm. Determine the speed v_A of slider A when $a = 200$ mm.

SUMMARY

The work–energy method combines the principles of kinematics with Newton's second law to directly relate the position and speed of a body. For this method to be useful, the forces acting on a body must be a function of position only. For some types of these forces, however, the resulting integrals can be evaluated explicitly. The result is a simple algebraic equation that relates the linear and angular speed of the body at two different positions of its motion.

The work done on a rigid body must include the work done by forces and by couples. The work done by forces acting on a rigid body is calculated in exactly the same way as for forces acting on a particle. Individual forces do work only when the point at which they are applied translates and not when the body rotates about their point of application. Couples do work only when the body rotates.

For conservative forces, the potential energy V is also defined and determined in the same manner as for particles. The work done by conservative forces may be computed by direct integration using Eq. 18-1 or by using potential energy functions as discussed in Section 17-5.

The work done by internal forces in a rigid body occur in equal-magnitude but opposite-direction collinear pairs such that the work of one force cancels that of its counterpart; the resultant work done by the internal forces is zero. By the same token, when two or more rigid bodies are connected by smooth pins or by flexible, inextensible cables, the resultant work done on the bodies by the connecting members of the system is also zero.

The kinetic energy of a body is the sum of the kinetic energies of the particles that make up the body. For the case of a rigid body this gives

$$T = \frac{1}{2}mv_G^2 + \frac{1}{2}I_{Gz}\omega^2 \qquad (18\text{-}15)$$

where v_G is the speed of the mass center of the body and I_{Gz} is the moment of inertia about an axis through the mass center G parallel to the z-axis (perpendicular to the plane of motion). The first term of

Eq. 18-15 is the kinetic energy associated with the translation of the mass center, and the second term is the kinetic energy associated with the rotation of the body about an axis through the center of mass.

The principle of work and energy for a rigid body

$$T_1 + U_{1\to2} = T_2 \tag{18-19}$$

looks exactly the same as Eq. 17-9b for the work and energy of a particle. The difference between these equations is that the kinetic energy terms in Eq. 18-19 include the rotational kinetic energy of the rigid body as well as the translational kinetic energy and that the work term includes the work done by all external moments as well as the work done by all external forces that act on the rigid body.

Just as with a particle, the work term can be divided up into a part done by conservative forces $U^{(c)}_{1\to2}$ and a part done by all other forces $U^{(o)}_{1\to2}$. The work done by the conservative forces can be expressed in terms of potential functions so that Eq. 18-19 can be written

$$T_1 + V_1 + U^{(o)}_{1\to2} = T_2 + V_2 \tag{18-20}$$

The same advantages and limitations will be realized when the method of work and energy is applied to rigid body motion as when it is applied to particle motion. The primary advantage is that the principle of work and energy directly relates the linear and angular speed of the body at two different positions of its motion to the forces and couples that do work on the body during its motion. The primary limitation is that Eq. 18-19 is a scalar equation and it can be solved for no more than a single unknown.

Since the work–energy method combines the principles of kinematics with Newton's second law, it is not a new or independent principle. There are no problems that can be solved with the work–energy method that cannot be solved using Newton's second law. However, when the work–energy method does apply, it is usually the easiest method of solving a problem.

REVIEW PROBLEMS

18-81*   The light tractor shown in Fig. P18-81 is traveling at 30 mi/h up a 10 percent grade at the instant shown. The rear drive wheels are 6 ft in diameter and rotate as a unit having a combined weight of 1000 lb and a centroidal radius of gyration about the axle of $k_G = 2$ ft. The rest of the tractor weighs an additional 2000 lb. If the engine is suddenly disengaged,

a. Determine how much further up the incline the tractor will travel before coming to rest.
b. Determine how far up the incline the tractor would travel if the body weighed 3000 lb and the drive wheels were relatively light.

Fig. P18-81

18-82* At the instant shown in Fig. P18-82 the 70-g yo-yo has a counterclockwise angular velocity of 100 rad/s. If the yo-yo has a centroidal radius of gyration of $k_G = 14$ mm, determine

a. The height h that the yo-yo will climb up the string.
b. The number of revolutions that the yo-yo will rotate before coming to rest.

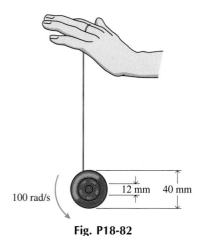

100 rad/s

12 mm   40 mm

Fig. P18-82

18-83 A solid cylinder, a hollow cylinder, and a solid sphere each weigh 16 lb and have outside diameters of 14 in. The inside diameter of the hollow cylinder is 12 in. Determine the speeds that the three objects would attain in rolling 3 revolutions down a 30° inclined plane.

18-84* A brake arm is used to control the motion of a drum and weight as shown in Fig. P18-84. The mass and

centroidal radius of gyration of the drum are 40 kg and 120 mm, respectively; the coefficient of kinetic friction between the brake shoe and the drum is 0.4; and the system is initially at rest with the 50-kg mass 3 m above the floor.

a. If the brake handle is suddenly released, determine the speed of the 50-kg mass after it has fallen 2 m.
b. If the brake is suddenly reapplied after the 50-kg mass has fallen 2 m, determine the minimum force **P** that must be applied to the brake handle to prevent the 50-kg mass from hitting the floor.
c. If the maximum force that can be applied to the brake handle without breaking it is $P = 250$ N, determine the maximum distance that the 50-kg mass can be allowed to fall before the brake is reapplied.

18-85 The 10-lb, 3-ft long bar AB shown in Fig. P18-85a rotates in a vertical plane. A rope attached to the bar at B passes over a small lightweight pulley and supports a 15-lb weight C. If the system is released from rest when $\theta = 0°$, determine the angular velocity ω_{AB}:

a. When $\theta = 30°$.
b. When $\theta = 90°$.
c. Repeat parts a and b when the 15-lb weight is replaced by a constant force as shown in Fig. 18-85b.

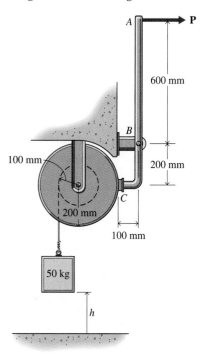

Fig. P18-84

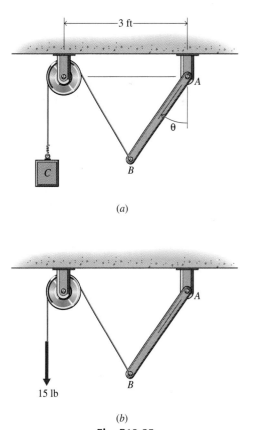

(a)

(b)

Fig. P18-85

18-86 A platform rotates at an angular speed of $\omega_1 = 10$ rad/s while a 5-kg cylinder of radius 50 mm mounted on the platform rotates relative to the platform at an angular speed of $\omega_2 = 25$ rad/s (Fig. P18-86). Determine the kinetic energy of the cylinder for $a = 50$ mm.

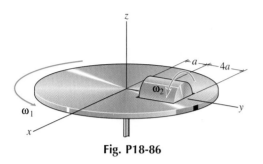

Fig. P18-86

18-87* The system shown in Fig. P18-87 is released from rest. Determine the speed of A after it has fallen 3 ft if:

a. The mass of both pulleys is neglected.
b. Pulley B is a uniform disk weighing 10 lb and pulley C is a uniform disk weighing 20 lb.

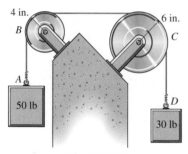

Fig. P18-87

18-88* The spoked wheel shown in Fig. P18-88 rolls without slipping on a horizontal surface. Because it has lost a couple of its spokes, the center of mass of the 12-kg wheel

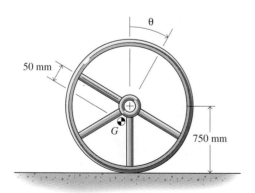

Fig. P18-88

is 50 mm away from its middle and the radius of gyration relative to the mass center is 0.6 m. If the center of the wheel has a speed of 3 m/s when $\theta = 0°$, determine:

a. The angular velocity ω of the wheel when $\theta = 90°$.
b. The normal and friction forces exerted on the wheel when $\theta = 90°$.

18-89 A flywheel is rotated by a flexible rope as shown in Fig. P18-89. The flywheel is a 20-lb solid cylinder 18 in. in diameter and the rope is initially wrapped five full turns around the wheel. If the flywheel starts from rest and the rope disengages when the end of the rope is at the edge of the wheel A, determine the final angular velocity of the flywheel:

a. When a constant force $P = 25$ lb is applied to the end of the flexible rope as shown in Fig. P18-89a.
b. When the force $\mathbf{P}$ is replaced by a load of 25 lb attached to the rope as shown in Fig. P18-89b.

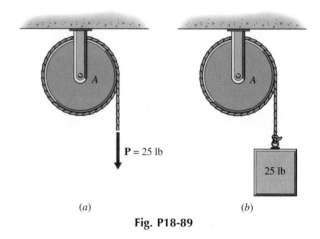

(a)   (b)

Fig. P18-89

18-90* A toy car is driven by a torsional spring as shown in Fig. P18-90. The drive wheel is a 750-g solid disk 150 mm in diameter, the body of the car has a mass of 150 g, and the mass of the front wheels may be neglected. The torsional spring exerts a moment on the drive wheel of $M = k\theta$ where $k = 0.01$ N · m/rad and θ is in radians. Determine the maximum speed of the car if it starts from rest with the spring wound up 8 revolutions. (The spring disengages when $\theta = 0$.)

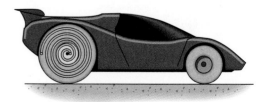

Fig. P18-90

18-91 The uniform cylinder of Fig. P18-91 has a radius of 2 in, a length of 8 in, and weighs 5 lb. The cylinder rotates about its own axis at a constant rate of $\omega_1 = 20$ rad/s. Simultaneously, the yoke holding the cylinder is rotating about a vertical axis at a rate of $\omega_2 = 8$ rad/s. Determine the kinetic energy of the cylinder at this instant.

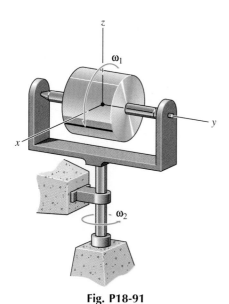

Fig. P18-91

18-92 Figure P18-92 shows a radial-arm wood saw that has an operating speed of 1500 rpm. The blade and motor armature have a combined mass of 1.2 kg and a centroidal radius of gyration of $k_G = 25$ mm. When the saw is turned off, bearing friction and a magnetic brake exert a constant braking torque **T** on the blade and motor.

a. Determine the number of revolutions the blade rotates before coming to rest if $\mathbf{T} = 0.002$ N · m (bearing friction only).

b. Determine the torque **T** necessary to stop the blade in just 1 revolution.

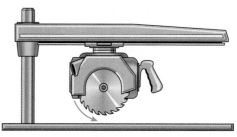

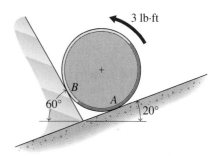

Fig. P18-92

18-93* A 15-lb uniform cylinder rests in a fixed trough as shown in Fig. P18-93. The diameter of the cylinder is 8 in., the coefficient of friction between the cylinder and the trough at A is 0.15, and the surface at B is smooth. If a constant couple of 3 lb · ft is suddenly applied to the cylinder, determine the angular velocity of the cylinder after it rotates 5 revolutions.

Fig. P18-93

Computer Problems

C18-94 A slender uniform rod AB rotates in a vertical plane as shown in Fig. P18-94. The mass of the rod is 5 kg

Fig. P18-94

and its length is 1.2 m. If the rod starts from rest when $\theta = 0°$, calculate and plot the kinetic energy of the mass center $T_v = \frac{1}{2} m v_G^2$, the kinetic energy of rotation about the mass center $T_\omega = \frac{1}{2} I_G \omega^2$, the gravitational potential energy V_g (use the level of A as the zero of potential energy), and the total energy $E = T_v + T_\omega + V_g$, all as functions of the angle θ ($0° \le \theta \le 90°$).

C18-95 A 16-lb bowling ball has a diameter of 8.6 in. The ball is released down the alley with an initial speed of 20 ft/s and no angular velocity. If the coefficient of friction between the ball and the alley is 0.1, calculate and plot the kinetic energy of the mass center $T_v = \frac{1}{2} m v_G^2$, the kinetic energy of rotation about the mass center $T_\omega = \frac{1}{2} I_G \omega^2$, and the total energy $E = T_v + T_\omega$, all as functions of its position x_G from the moment the ball is released until it strikes the head pin 60 ft away.

C18-96 The ends of a slender uniform rod AB are attached to lightweight sliders as shown in Fig. P18-96. A constant force of 10 N is applied to slider B; the mass of the rod is 6 kg and its length is 1.4 m. If the rod starts from rest when $\theta = 0°$, calculate and plot:

a. The velocity $\mathbf{v}_A$ of slider A and the angular velocity ω of the rod as functions of θ ($0° \leq \theta \leq 180°$).
b. The kinetic energy of the mass center $T_v = \frac{1}{2} mv_G^2$, the kinetic energy of rotation about the mass center $T_\omega = \frac{1}{2} I_G \omega^2$, the gravitational potential energy V_g (use the level of A as the zero of potential energy), and the total energy $E = T_v + T_\omega + V_g$, all as functions of the angle θ ($0° \leq \theta \leq 180°$).

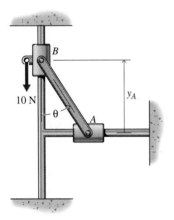

Fig. P18-96

C18-97 A 25-lb uniform cylinder rolls without slipping on an incline (Fig. P18-97). A spring ($k = 12$ lb/ft) is attached by means of a yoke to the axle of the wheel. Friction between the yoke and axle can be neglected. If the cylinder starts from rest when $x = 0$ and the spring is unstretched, calculate and plot the kinetic energy of the mass center $T_v = \frac{1}{2} mv_G^2$, the kinetic energy of rotation about the mass center $T_\omega = \frac{1}{2} I_G \omega^2$, the gravitational potential energy V_g (use the level of the cylinder when $x = 0$ as the zero of potential energy), the elastic potential energy V_s, and the total energy $E = T_v + T_\omega + V_g + V_s$, all as functions of x. ($0 \leq x \leq 30$ in.).

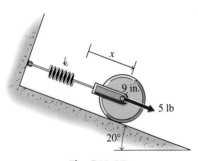

Fig. P18-97

C18-98 A 5-kg uniform half cylinder 800 mm in diameter rolls without slipping on a horizontal surface as shown in

Fig. P18-98. If the cylinder is released from rest when $\theta = 0°$, calculate and plot:

a. The angular velocity ω and the angular acceleration α of the half cylinder as a function of θ ($0° \leq \theta \leq 180°$).
b. The normal force $\mathbf{N}$ and the friction force $\mathbf{F}$ exerted on the half cylinder by the surface as a function of θ ($0° \leq \theta \leq 180°$).
c. The kinetic energy of the mass center $T_v = \frac{1}{2} mv_G^2$, the kinetic energy of rotation about the mass center $T_\omega = \frac{1}{2} I_G \omega^2$, the gravitational potential energy V_g (use the level of the horizontal surface as the zero of potential energy), and the total energy $E = T_v + T_\omega + V_g$, all as functions of the angle θ ($0° \leq \theta \leq 180°$).

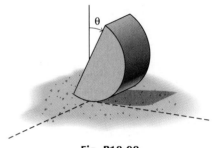

Fig. P18-98

C18-99 The slender uniform rod AB shown in Fig. P18-99 is attached to a lightweight slider at B and slides along a frictionless horizontal surface at A. The rod weighs 15 lb and is 5 ft long. If the rod starts from rest when $\theta = 0°$, calculate and plot:

a. The angular velocity ω and the angular acceleration α of the rod as a function of θ ($0° \leq \theta \leq 90°$).
b. The normal forces $\mathbf{N}_A$ and $\mathbf{N}_B$ exerted on the rod by the surface at A and by the slider at B as a function of θ ($0° \leq \theta \leq 90°$).
c. The kinetic energy of the mass center $T_v = \frac{1}{2} mv_G^2$, the kinetic energy of rotation about the mass center $T_\omega = \frac{1}{2} I_G \omega^2$, the gravitational potential energy V_g (use the level of the horizontal surface as the zero of potential energy), and the total energy $E = T_v + T_\omega + V_g$, all as functions of the angle θ ($0° \leq \theta \leq 90°$).

Fig. P18-99

19

KINETICS OF PARTICLES: IMPULSE AND MOMENTUM

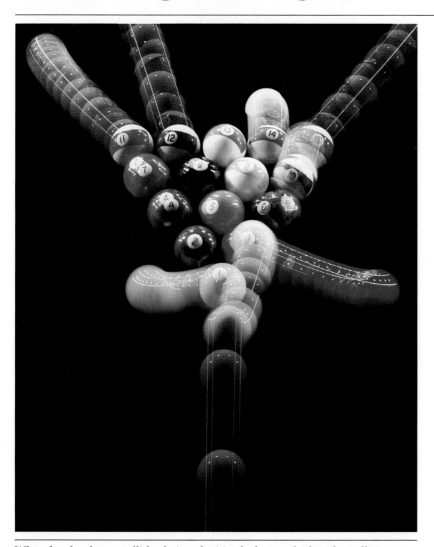

When hard spheres collide their velocities before and after the collision are related by impulse and momentum principles.

19-1 INTRODUCTION

The study of kinetics is based on Newton's second law of motion. In Chapters 15 and 16, Newton's second law was used directly to relate the forces acting on particles and rigid bodies and the resulting acceleration of the particles and rigid bodies. In fact, when information about the acceleration or the value of a force at an instant is desired, Newton's second law is usually the easiest method to use.

In Chapters 17 and 18, Newton's second law was integrated with respect to position to get the principle of work and energy. Since the principle of work and energy is just a combination of Newton's second law and the principles of kinematics, there is no problem that can be solved using the principle of work and energy that cannot also be solved using Newton's second law. However, the principle of work and energy is particularly useful for solving problems in which the speed of a body for two different positions of its motion are to be related and the forces involved can be expressed as functions of the position of the body.

The principles of impulse and momentum developed in this chapter and the next are obtained by integrating Newton's second law with respect to time. The resulting equations are useful for solving problems in which the velocity of a body for two different instants of time are to be related and the forces involved can be expressed as functions of time. Although the principles of impulse and momentum are not required to solve any particular problem, they will be found to be particularly useful for solving problems involving collisions between bodies and variable mass systems.

19-2 LINEAR IMPULSE AND MOMENTUM OF A PARTICLE

Let $\mathbf{R} = \Sigma\mathbf{F}$ be the resultant of all forces acting on a particle of mass m. Then Newton's second law for the particle can be written

$$\mathbf{R} = m\mathbf{a} = m\frac{d\mathbf{v}}{dt}$$

Since the mass of the particle does not depend on time, it can be taken inside the derivative to get

$$\mathbf{R} = \frac{d}{dt}(m\mathbf{v}) \tag{19-1}$$

When the forces are constants or functions of time only, Eq. 19-1 can be integrated to get

$$\int_{t_i}^{t_f} \mathbf{R}\, dt = \int_{mv_i}^{mv_f} d(m\mathbf{v}) = (m\mathbf{v})_f - (m\mathbf{v})_i$$

or

$$(m\mathbf{v})_i + \int_{t_i}^{t_f} \mathbf{R}\, dt = (m\mathbf{v})_f \tag{19-2}$$

where $\mathbf{v}_i$ is the velocity of the particle at some initial time t_i and $\mathbf{v}_f$ is the velocity of the particle at the final time t_f.

19-2-1   Linear Momentum

The vector $m\mathbf{v}$ in Eqs. 19-1 and 19-2 is given the symbol $\mathbf{L}$ and is called the *linear momentum,* or simply the *momentum,* of the particle. Since m is a scalar, the linear momentum vector and the velocity of the particle are in the same direction. The magnitude of the linear momentum is equal to the product of the mass m and the speed v of the particle. In the SI system of measurement, the units of momentum are $kg \cdot m/s$ or equivalently, $N \cdot s$. In the U. S. Customary system of measurement, they are $slug \cdot ft/s$ or $lb \cdot s$.

19-2-2   Linear Impulse

The integral $\int_{t_i}^{t_f} \mathbf{R} \, dt$ is called the *linear impulse* of the force $\mathbf{R}$. The linear impulse is a vector that has units of force-time. In the SI system of measurement the magnitude of the linear impulse of a force is expressed in $N \cdot s = kg \cdot m/s$, which is the same unit obtained for the linear momentum of a particle. Therefore, Eq. 19-2 is dimensionally correct. If U. S. Customary units are used, the linear impulse is expressed in $lb \cdot s = slug \cdot ft/s$, which is also the same unit obtained for the linear momentum of a particle.

In general, the resultant force $\mathbf{R}(t) = R(t)\mathbf{e}_R$ is a vector that changes in both magnitude and direction during the time interval t_i to t_f. However, if the direction $\mathbf{e}_R$ of the force does not change during the time interval, it can be taken outside the integral. Then the value of the integral—which represents the magnitude of the impulse—is just the shaded area under the graph of R versus t (Fig. 19-1). If the magnitude of the force is also constant, then it can also be taken outside the integral, leaving

$$\int_{t_i}^{t_f} \mathbf{R}_c \, dt = \mathbf{R}_c \int_{t_i}^{t_f} dt = \mathbf{R}_c(t_f - t_i) \tag{19-3}$$

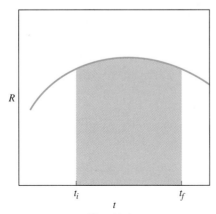

Fig. 19-1

Equation 19-3 is also used to define the time average impulse force $\mathbf{R}_{avg}$—the equivalent constant force that gives the same linear impulse as the original time-varying force $\mathbf{R}(t)$

$$\mathbf{R}_{avg} = \frac{1}{(t_f - t_i)} \int_{t_i}^{t_f} \mathbf{R}\, dt \qquad (19\text{-}4)$$

The average value of the force given by Eq. 19-4 (the time-average value) is usually different from the average value computed from the work done by a force (the distance-average value).

When both the magnitude and direction of the resultant force $\mathbf{R}(t)$ vary during the time interval, evaluation of the impulse integral must be carried out in component form. Rectangular Cartesian components are generally preferred since the unit vectors $\mathbf{i}$, $\mathbf{j}$, and $\mathbf{k}$ do not vary with time. Resolving $\mathbf{R}$ into its rectangular components gives

$$\int_{t_i}^{t_f} \mathbf{R}\, dt = \mathbf{i} \int_{t_i}^{t_f} R_x\, dt + \mathbf{j} \int_{t_i}^{t_f} R_y\, dt + \mathbf{k} \int_{t_i}^{t_f} R_z\, dt$$

Although both the work done by a force (defined in Chapter 17) and the impulse of a force are integrals of a force, they are completely different concepts. Two important differences are as follows:

1. The work done by a force is a scalar quantity; the impulse of a force is a vector.
2. The work done by a force is zero if there is no component of the force in the direction of its displacement. The impulse of a force is never zero even when it is applied at a stationary point.

19-2-3   Principle of Linear Impulse and Momentum

Equation 19-2 expresses the Principle of Linear Impulse and Momentum:

$$\mathbf{L}_i + \int_{t_i}^{t_f} \mathbf{R}\, dt = \mathbf{L}_f$$

The final linear momentum $\mathbf{L}_f$ of a particle is the vector sum of its initial linear momentum $\mathbf{L}_i$ and the impulse $\int \mathbf{R}\, dt$ of the resultant of all forces acting on the particle.

Unlike the work and energy equation, which is a scalar equation, Eq. 19-2 is a vector equation representing three scalar equations. Expressed in rectangular Cartesian coordinates, its three scalar components are

$$mv_{xi} + \int_{t_i}^{t_f} R_x\, dt = mv_{xf}$$

$$mv_{yi} + \int_{t_i}^{t_f} R_y\, dt = mv_{yf}$$

$$mv_{zi} + \int_{t_i}^{t_f} R_z\, dt = mv_{zf}$$

It should be noted at this point that the principle of linear impulse and momentum is not really a new principle. It is merely a combination of Newton's second law and the principles of kinematics for the special case in which force is a function of time. Nevertheless, it is useful for solving for the velocity of a particle when the force is known as a function of time and the acceleration is not of interest.

19-2-4 Conservation of Linear Momentum

It follows from Eq. 19-1 that the rate of change of the linear momentum $m\mathbf{v}$ is zero when $\mathbf{R} = \Sigma\mathbf{F} = 0$. When this happens, the linear momentum is *conserved;* that is, it is constant, in both magnitude and direction:

$$\mathbf{L}_i = \mathbf{L}_f \qquad (19\text{-}5)$$

Linear momentum can be conserved in one direction (if the sum of forces in that direction is zero) independently of any other direction.

Although Eq. 19-5 is often called the *Principle of Conservation of Linear Momentum,* it is only a special case of the general principle of linear impulse and momentum. Conservation of Linear Momentum may be recognized as just an alternative statement of Newton's first law.

Conservation of linear momentum is not related to conservation of kinetic energy. For example, when an elastic sphere bounces off a hard surface, the speed of rebound is nearly the same as the speed at which it struck the surface. Therefore, the kinetic energy of the rebound is essentially the same as the kinetic energy before striking the surface, and kinetic energy is conserved. However, the direction of the velocity after the impact is opposite what it was before the impact. Therefore, the linear momentum after the impact is the negative of what it was before the impact, and linear momentum is not conserved. Similarly, when two particles collide, it is possible to have linear momentum for the pair of particles conserved even though most of their kinetic energy is lost.

Finally, it should be noted that the mass m of the particle is assumed to be constant in Eqs. 19-1 through 19-5. Therefore, these equations should not be used to solve problems involving the motion of bodies, such as rockets, which gain or lose mass. Problems of that type will be considered in Section 19-7.

EXAMPLE PROBLEM 19-1

A Ping-Pong ball weighing 0.2 oz has an initial velocity of $\mathbf{v}_i = 8\,\mathbf{j} + 6\,\mathbf{k}$ ft/s when a gust of wind exerts a force of $\mathbf{F} = 0.5\,t\,\mathbf{i}$ oz on the ball (t is in seconds). Determine the magnitude and direction of the velocity of the ball after 0.5 seconds. (The positive z-direction is up.)

SOLUTION

The free-body diagram of the ball shown in Fig. 19-2 includes the weight of the ball $\mathbf{W} = m\mathbf{g}$ and the force of the wind $\mathbf{F}$. The mass of the ball is

$$\frac{0.2}{(16)(32.2)} = 3.882(10^{-4})\ \text{slug}$$

and the linear impulse on the ball during the 0.5 seconds is

$$\int_0^{0.5} \left[\frac{0.5t\mathbf{i} - 0.2\mathbf{k}}{16} \right] dt = 0.00391\mathbf{i} - 0.00625\mathbf{k}\ \text{lb}$$

Substituting these values into the x-component of the impulse–momentum equation (Eq. 19-2) then gives

$$3.882(10^{-4})(8\mathbf{j} + 6\mathbf{k}) + 39.1(10^{-4})\mathbf{i} - 62.5(10^{-4})\mathbf{k}$$
$$= 3.882(10^{-4})\mathbf{v}_f$$

which gives

$$\mathbf{v}_f = 10.07\mathbf{i} + 8\mathbf{j} - 10.10\mathbf{k}\ \text{ft/s}$$
$$= 16.35(0.616\mathbf{i} + 0.489\mathbf{j} - 0.618\mathbf{k})\ \text{ft/s} \qquad \text{Ans.}$$

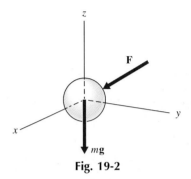

Fig. 19-2

EXAMPLE PROBLEM 19-2

A 10-kg box is resting on a horizontal surface as shown in Fig. 19-3a when a horizontal force $\mathbf{P}$ is applied to it. The magnitude of $\mathbf{P}$ varies with time as shown in Fig. 19-3b. If the static and dynamic coefficients of friction are 0.4 and 0.3, respectively, determine:

a. The velocity of the box at $t = 10$ s.
b. The velocity of the box at $t = 15$ s.
c. The time t_f for which the box stops sliding.

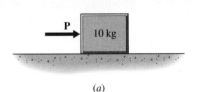

(a)

SOLUTION

a. The free-body diagram of the box is shown in Fig. 19-4. Since there is no motion in the vertical direction, summing forces in the vertical direction gives

$$N = (10)(9.81) = 98.1\ \text{N}$$

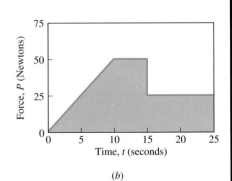

(b)

Fig. 19-3

(Note that the y-component of the linear impulse–momentum equation could have been used to get the same result, although there is no advantage in using it.)

The friction force is then less than or equal to 0.4 N until the box starts moving and is equal to 0.3 N after the box starts moving. When the force **P** is initially applied to the box, the friction available is sufficiently great to keep the box from moving, and the friction force increases at the same rate as the force **P** (Fig. 19-5). When the force **P** reaches 0.4 N = 39.24 N (at t = 7.848 s), however, the friction available is no longer able to prevent the box from moving, the box begins to slide, and the friction force drops to 0.3 N = 29.43 N. The friction force then stays at 29.43 N until t_f, when the box again stops moving and the friction force drops to 25 N, which is the force required to hold the box in equilibrium.

The x-component of the linear impulse due to the force **P** acting on the box from t = 0 to t = 10 s is just the area under the curve

$$\int_0^{10} P\,dt = \frac{1}{2}(50)(10) = 250 \text{ N} \cdot \text{s}$$

while the x-component of the linear impulse due to the friction force acting on the box from t = 0 to t = 10 s is

$$\int_0^{10} F\,dt = \frac{1}{2}(-39.24)(7.848) + (-29.43)(10 - 7.848)$$
$$= -217.31 \text{ N} \cdot \text{s}$$

Then the x-component of the linear impulse–momentum equation gives

$$0 + 250 - 217.31 = 10v_{10}$$

or

$$v_{10} = 3.27 \text{ m/s} \qquad \text{Ans.}$$

b. From t = 10 s to t = 15 s the linear impulses are

$$\int_{10}^{15} P\,dt = (50)(15 - 10) = 250 \text{ N} \cdot \text{s}$$
$$\int_{10}^{15} F\,dt = (-29.43)(15 - 10) = -147.15 \text{ N} \cdot \text{s}$$

and the x-component of the linear impulse–momentum equation gives

$$(10)(3.27) + 250 - 147.15 = 10v_{15}$$

and

$$v_{15} = 13.55 \text{ m/s} \qquad \text{Ans.}$$

c. Between t = 15 s and t = t_f the linear impulses are

$$\int_{15}^{t_f} P\,dt = (25)(t_f - 15)$$
$$\int_{15}^{t_f} F\,dt = (-29.43)(t_f - 15)$$

and the x-component of the linear impulse–momentum equation gives

$$(10)(13.55) + (25 - 29.43)(t_f - 15) = 0$$

or

$$t_f = 45.6 \text{ s} \qquad \text{Ans.}$$

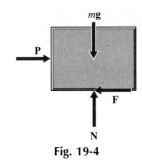

Fig. 19-4

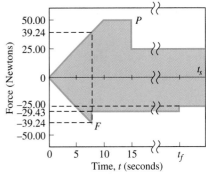

Fig. 19-5

PROBLEMS

19-1* The speed of a 0.4-lb hockey puck sliding across the ice is observed to decrease from 60 ft/s to 40 ft/s in 3 s. Determine the average friction force acting on the puck and the corresponding kinetic coefficient of friction. (Assume that the ice is horizontal.)

19-2* A 1200-kg car is traveling at 75 km/h on an icy road when the driver suddenly applies the brakes. If kinetic coefficient of friction of 0.15 and all four tires slide, determine the time it will take for the car to come to rest.

19-3 A 500-lb boat traveling at 20 mi/h comes to rest 10 s after the motor is shut off. Determine the average drag force of the water on the boat.

19-4* The speed of a toboggan sliding down a hill increases from 0 to 10 m/s in 6 s. The combined mass of the toboggan and riders is 110 kg and the slope of the hill is 20°. Determine the average friction force between the toboggan and the snow and the corresponding coefficient of friction.

19-5 The thrust of a 500-lb rocket sled varies with time as shown in Fig. P19-5. If the sled starts from rest and travels along a straight, horizontal track, determine its velocity when the rocket burns out. (Neglect friction.)

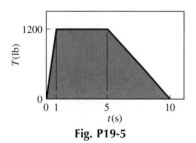

Fig. P19-5

19-6 A 2.0-kg disk is sliding on a smooth, horizontal surface when it is acted on by a cross force (Fig. P19-6a). The force makes an angle of θ with the initial direction of **v**, and its magnitude varies as shown in Fig. P19-6b. If $v = 10$ m/s

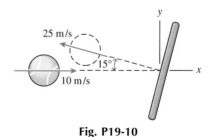

(a)   (b)

Fig. P19-6

and $\theta = 50°$, determine the magnitude and direction of the velocity of the disk when:

a. $t = 5$ s.   b. $t = 10$ s.   c. $t = 15$ s.

19-7* If the disk of Problem 19-6 weighs 5 lb and has an initial speed $v = 25$ ft/s, determine the magnitude and direction of the velocity of the disk when:

a. $t = 5$ s.   b. $t = 15$ s.   c. $t = 20$ s.

The angle $\theta = 105°$, and the magnitude of the force **F** varies as shown in Fig. P19-7.

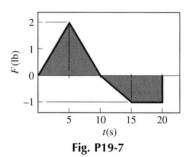

Fig. P19-7

19-8 Determine the maximum velocity of the disk of Problem 19-6.

19-9* If the disk of Problem 19-6 weighs 5 lb and has an initial speed $v = 25$ ft/s, determine the magnitude and direction of the velocity of the disk when:

a. $t = 5$ s.   b. $t = 15$ s.   c. $t = 20$ s.

Assume that the magnitude of the force is constant, $F = 2$ lb, but that the angle θ increases at a constant rate of 0.4 rad/s and that $\theta = 0$ at $t = 0$.

19-10* A 60-g tennis ball has a horizontal speed of 10 m/s when it is struck by a tennis racket (Fig. P19-10). After the impact, the ball's velocity is 25 m/s (still horizontal) and makes an angle of 15° with the initial direction. If the time of contact is 0.05 s, determine the average force (magnitude and direction) of the tennis racket on the ball.

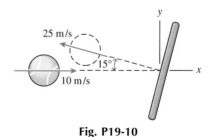

Fig. P19-10

19-11 A 5-oz baseball has a horizontal, initial velocity of 90 ft/s just prior to being hit by a bat. After the impact, the ball's velocity is 110 ft/s at an angle 30° above the horizontal (Fig. P19-11). If the time of impact is 0.01 s, determine the average force (magnitude and direction) of the bat on the ball.

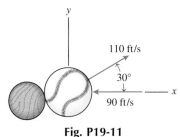

Fig. P19-11

19-12* A 0.2-kg ball has an initial velocity of $\mathbf{v}_0 = 15\mathbf{j} + 20\mathbf{k}$ m/s when a gust of wind exerts a force on it. If the force varies with time according to $\mathbf{F} = 0.5(t^2 - 9)(\cos 30°\mathbf{i} - \sin 30°\mathbf{j})$ where t is in seconds and $\mathbf{F}$ is in Newtons, determine the magnitude and direction of the ball's velocity when:

a. $t = 1$ s.
b. $t = 2$ s.
c. $t = 3$ s.

(Gravity acts in the negative z-direction.)

19-13 A 0.5-lb ball has an initial velocity of $\mathbf{v}_0 = 30\mathbf{i} + 50\mathbf{k}$ ft/s. If a constant force of $\mathbf{F} = (-0.1\mathbf{i} + 0.1\mathbf{j})$ lb acts on the ball, determine the magnitude and direction of the ball's velocity when the velocity is parallel to the y-z plane. (Gravity acts in the negative z-direction.)

19-14 A 10-kg box is resting on a horizontal surface as shown in Fig. P19-14a when a horizontal force $\mathbf{P}$ is applied to it. The magnitude of $\mathbf{P}$ varies with time as shown in Fig. P19-14b. If the static and dynamic coefficients of friction are 0.4 and 0.3, respectively, determine:

a. The time t_1 at which the box starts sliding.
b. The maximum velocity of the box v_{max} and the time t_{max} at which it occurs.
c. The time t_f for which the box stops sliding.

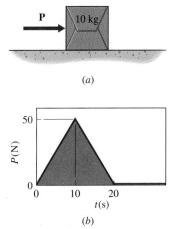

Fig. P19-14

19-15* A 5-lb box is resting on a horizontal surface when a force $\mathbf{P}$ is applied to it (Fig. P19-15a). The magnitude of $\mathbf{P}$ varies with time as shown in Fig. P19-15b. If the static and dynamic coefficients of friction are 0.4 and 0.3, respectively, determine:

a. The time t_1 at which the box starts sliding.
b. The maximum velocity of the box v_{max} and the time t_{max} at which it occurs.
c. The time t_f for which the box stops sliding.

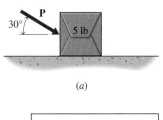

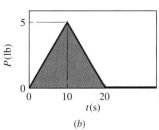

Fig. P19-15

19-16* A 10-kg box is resting on an inclined surface when a force $\mathbf{P}$ is applied to it (Fig. P19-16a). The magnitude of $\mathbf{P}$ varies with time as shown in Fig. P19-16b. If the static and dynamic coefficients of friction are 0.6 and 0.4, respectively, determine:

a. The time t_1 at which the box starts sliding.
b. The maximum velocity of the box v_{max} and the time t_{max} at which it occurs.
c. The time t_f for which the box stops sliding.

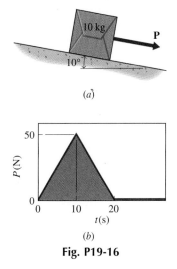

Fig. P19-16

19-17 A 10-lb box is resting on an inclined surface when a force **P** is applied to it (Fig. P19-17a). The magnitude of **P** varies with time as shown in Fig. P19-17b. If the static and dynamic coefficients of friction are 0.6 and 0.4, respectively, determine:

a. The time t_1 at which the box starts sliding.
b. The speed of the box when $t = 5$ s.
c. The speed of the box when $t = 10$ s.
d. The smallest value P_{15} for which the box will be at rest at $t = 20$ s.

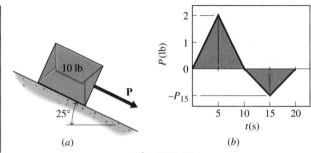

(a)　　　　　(b)

Fig. P19-17

19-3　INTERACTING SYSTEMS OF PARTICLES

When a problem involves two or more interacting particles as in Fig. 19-6, each particle may be considered separately and Eq. 19-2 may be written for each particle.

$$(\mathbf{L}_i)_1 + \int_{t_i}^{t_f} (\mathbf{R}_1 + \mathbf{f}_{12}) \, dt = (\mathbf{L}_f)_1$$

$$(\mathbf{L}_i)_2 + \int_{t_i}^{t_f} (\mathbf{R}_2 + \mathbf{f}_{21}) \, dt = (\mathbf{L}_f)_2$$

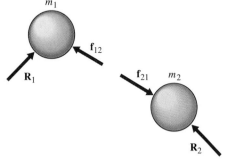

Fig. 19-6

where $\mathbf{R}_1$ is the resultant of all external forces acting on particle 1, and $\mathbf{f}_{12}$ is the force exerted on particle 1 by particle 2, and so on. Since the forces of action and reaction exerted by the particles on each other form pairs of equal and opposite forces ($\mathbf{f}_{21} = -\mathbf{f}_{12}$), and since the time interval from t_1 to t_2 is common to all the forces involved, the impulses of the forces of action and reaction always cancel out when these equations are added together. Therefore, the impulse–momentum equation for a pair of (or for N) interacting particles is

$$\sum_\ell (\mathbf{L}_i)_\ell + \sum_\ell \int_{t_i}^{t_f} \mathbf{R}_\ell \, dt = \sum_\ell (\mathbf{L}_f)_\ell$$

where $(\Sigma \, \mathbf{L})_\ell = \Sigma \, (m\mathbf{v})_\ell$ is the vector sum of the momenta of both (or of all N) particles, $\Sigma \, (\int \mathbf{R}_\ell \, dt)$ is the vector sum of the impulses of all the external forces involved, and the internal forces need not be considered.

Therefore, for a system of N interacting particles: *The final momentum of the system of particles is the vector sum of their initial momenta and the sum of the impulses of the resultants of all external forces acting on the particles:*

$$\sum_\ell (\mathbf{L}_i)_\ell + \sum_\ell \int_{t_i}^{t_f} \mathbf{R}_\ell \, dt = \sum_\ell (\mathbf{L}_f)_\ell \qquad (19\text{-}6)$$

19-3-1　Motion of the Mass Center

The location of the mass center $\mathbf{r}_G$ of a system of N particles is calculated using the first moment of mass

$$m\mathbf{r}_G = \sum_{\ell=1}^{N} m_\ell \mathbf{r}_\ell \qquad (19\text{-}7)$$

where $m = \Sigma\, m_\ell$ is the total mass of all of the particles. Taking the time derivative of Eq. 19-7 and remembering that the mass of each particle is constant gives

$$m\mathbf{v}_G = \sum_{\ell=1}^{N} m_\ell \mathbf{v}_\ell \qquad (19\text{-}8)$$

That is, the total linear momentum of a system of particles is the same as if all the mass were concentrated in a single particle that was moving with the speed of the mass center of the system of particles. Using Eq. 19-8 allows Eq. 19-6 to be rewritten in the form

$$m(\mathbf{v}_G)_i + \sum_{\ell} \int_{t_i}^{t_f} \mathbf{R}_\ell\, dt = m(\mathbf{v}_G)_f \qquad (19\text{-}9)$$

19-3-2   Conservation of Linear Momentum for a System of Particles

If the sum of the impulses of all the external forces acting on the various particles is zero, the linear momentum of the system of particles is conserved

$$\sum_{\ell} (m\mathbf{v}_i)_\ell = \sum_{\ell} (m\mathbf{v}_f)_\ell \qquad (19\text{-}10)$$

or

$$(m\mathbf{v}_G)_i = (m\mathbf{v}_G)_f$$

Dividing through by the total mass of the system of particles (which is constant) gives

$$\mathbf{v}_{Gi} = \mathbf{v}_{Gf}$$

That is, when the impulses of all the external forces acting on a system of particles is zero, the velocity $\mathbf{v}_G$ of the mass center of the particles is constant. This situation occurs, for example, when two particles that are moving freely collide with one another. It should be noted, however, that although the total momentum of the colliding particles is conserved, their total energy is generally not conserved. Problems involving the collision or impact of two particles will be discussed in detail in Section 19-4.

19-3-3   Impulsive and Nonimpulsive Forces

When the linear impulse in a given direction is not zero but is known to be relatively small, it can frequently be neglected in order to obtain an approximate solution sufficiently accurate for many purposes. For example, if the linear momentum of the system composed of blocks A and B in Fig. 19-7 is large compared to the linear impulse of the frictional force, Eq. 19-10 is approximately true during the brief time Δt of impact even though friction exists between the plane and the blocks. Since the frictional forces cannot exceed μN, and the time of impact is small, the impulse of the friction on the blocks during the impact period would not change the large linear momentum of the blocks materially.

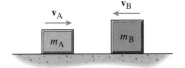

Fig. 19-7

Forces that have very large magnitudes can produce a significant change in momentum even over very short time periods. Such forces are called *impulsive forces.* Motions that result from impulsive forces are called *impulsive motions.* The forces generated when one body strikes another is an example of an impulsive force.

Forces whose magnitudes are small compared to impulsive forces are called *nonimpulsive* forces. Examples are the weight of a body, friction, and spring forces. When the principle of impulse and momentum is applied over a short time interval, the impulse of nonimpulsive forces may be neglected compared to that of impulsive forces.

It is usually not known ahead of time whether unknown reaction forces are impulsive or not. Generally, the reaction force of any support that acts to prevent motion in some direction is just as impulsive as the forces trying to cause motion in that direction.

The final decision of whether or not the linear momentum of a force can be ignored must be based on the required accuracy of the result and on the estimated effect that the term has on the equation. If there is any doubt about whether the impulse of a force is important or not, it should be included in Eqs. 19-6 and 19-9.

19-3-4   Problems Involving Energy and Momentum

It must be remembered that the impulse–momentum principle is not an independent principle. Just like the work–energy principle, it is only a general, first integral of Newton's second law that is applicable for certain special situations. Any problem that can be solved by impulse–momentum methods or by work–energy methods can also be solved directly using Newton's second law. However, when the principle of impulse–momentum or that of work–energy is suitable, these methods will often get the solution more quickly and with less labor.

Not only are the work–energy and the impulse–momentum methods not suitable for solving all problems, few real problems are specifically set up for either one. More generally, both methods will be used in different parts of the same problem. The maximum benefit of these methods is realized by choosing the particular method most suitable for a particular problem or part of a problem. In fact, it is often useful to combine all three methods—impulse–momentum, work–energy, and Newton's second law—to solve particular problems.

Many problems, such as Example Problem 19-5, involve several phases for which different principles are suitable. During the first phase, only conservative forces act and the work–energy equation is the easiest method to use to find the velocity of box *A* just prior to its impact with box *B*. During the impact phase, impulsive forces dominate, and the impulse–momentum principle is the most convenient method to use to find the velocities of the boxes just after the impact. During the final phase, relationships between forces, velocities, and position are again required and work–energy is the most convenient method to use. However, Newton's second law is used to find the normal force so that the work done by friction can be computed.

The vertical component of the linear impulse–momentum equation could also have been used to find the normal force, and it would have given the same result. Since there was no motion in the normal direction, however, the impulse–momentum method offered no advantage over the direct application of Newton's second law.

EXAMPLE PROBLEM 19-3

A 2500-lb car is initially at rest on the deck of a ferry boat, which is tied to a dock as shown in Fig. 19-8. The ferry boat weighs 25,000 lb.

a. If the car accelerates uniformly from rest to 20 mi/h in 4 s, determine the average tension in the cable during this time.

b. At $t = 4$ s the cable between the ferry boat and the dock breaks, the driver of the car applies his brakes, and the car comes to a stop relative to the ferry. Neglect the friction between the ferry boat and the water, and determine the speed at which the boat will hit the dock.

Fig. 19-8

SOLUTION

a. The free-body diagram of the car and boat is shown in Fig. 19-9. The forces between the wheels of the car and the boat are internal forces; they are not needed in the linear impulse–momentum equation and are not shown on the diagram. The only force that has an impulse in the horizontal direction is the tension T, and the x-component of Eq. 19-6 is

$$0 + T_{avg}(4) = \frac{2500}{32.2}(20)\frac{88}{60}$$

which gives

$$T_{avg} = 569 \text{ lb} \qquad\qquad \text{Ans.}$$

b. After the cable to the dock breaks, there are no forces in the x-direction, and so momentum is conserved in the x-direction. Therefore

$$\frac{2500}{32.2}(20)\frac{88}{60} = \frac{27,500}{32.2}v_f\frac{88}{60}$$

and

$$v_f = 1.82 \text{ mi/h} \qquad\qquad \text{Ans.}$$

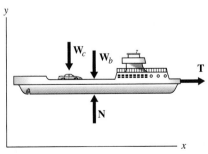

Fig. 19-9

A 3-kg cannonball is fired with an initial velocity of $v_0 = 150$ m/s and $\theta_0 = 60°$ as shown in Fig. 19-10. At the peak of its trajectory, the ball explodes and splits into two pieces. The 1-kg piece hits the ground at $x = 500$ m, and $y = 2500$ m when $t = 35$ s.

a. Determine when and where the 2-kg piece hits the ground.
b. Determine the average magnitude of the explosive force F_{avg} if the duration of the explosion is $\Delta t = 0.005$ s.

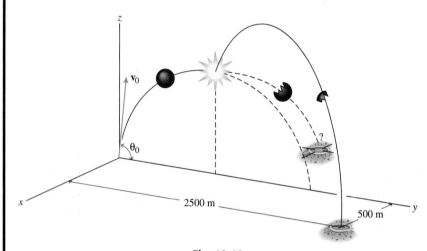

Fig. 19-10

SOLUTION

a. The only force acting on the system of two particles is gravity, and so the motion of the mass center is given by

$$\mathbf{a}_G = -9.81\mathbf{k} \text{ m/s}^2$$
$$\mathbf{v}_G = 75\mathbf{j} + (129.9 - 9.81\,t)\mathbf{k} \text{ m/s}$$
$$\mathbf{r}_G = 75t\mathbf{j} + (129.9t - 4.905t^2)\mathbf{k} \text{ m}$$

The peak of the trajectory corresponds to the time when $dz_G/dt = 129.9 - 9.81t = 0$, which gives $t = 13.24$ s. Then

$$x = 0 \text{ m} \qquad y = 993.2 \text{ m} \qquad z = 860.1 \text{ m}$$

at the instant of the explosion.

After the explosion, the 1-kg mass is acted on only by gravity also, and so its motion is given by

$$\mathbf{a}_1 = -9.81\mathbf{k} \text{ m/s}^2$$
$$\mathbf{v}_1 = v_{10x}\mathbf{i} + v_{10y}\mathbf{j} + [v_{10z} - 9.81(t - 13.24)]\mathbf{k} \text{ m/s}$$
$$x_1 = v_{10x}(t - 13.24) \text{ m}$$
$$y_1 = 993.2 + v_{10y}(t - 13.24) \text{ m}$$
$$z_1 = 860.1 + v_{10z}(t - 13.24) - 4.905(t - 13.24)^2 \text{ m}$$

where the constants of integration ($x_{10} = 0$ m, $y_{10} = 993.2$ m, and $z_{10} = 860.1$ m) were chosen to match the known position immediately after the explosion. The remaining constants of integration are determined by using the known impact time and position:

$$v_{10x} = \frac{500}{(35 - 13.24)} = 22.98 \text{ m/s}$$

$$v_{10y} = \frac{2500 - 993.2}{35 - 13.24} = 69.26 \text{ m/s}$$

$$v_{10z} = \frac{-860.1 + 4.905(35 - 13.24)^2}{35 - 13.24} = 67.21 \text{ m/s}$$

During the explosion, the only external force that has an impulse is gravity so that Eq. 19-6 gives

$$(3)(75\mathbf{j}) + (3)(-9.81\mathbf{k})(0.005) = (1)(22.98\mathbf{i}$$
$$+ 69.26\mathbf{j} + 67.21\mathbf{k}) + (2)\mathbf{v}_{20}$$

and therefore

$$\mathbf{v}_{20} = -11.49\mathbf{i} + 77.87\mathbf{j} - 33.68\mathbf{k} \text{ m/s}$$

Finally, the only force acting on the 2-kg particle after the explosion is gravity so that its motion is given by

$$\mathbf{a}_2 = -9.81\mathbf{k} \text{ m/s}^2$$
$$\mathbf{v}_2 = -11.49\mathbf{i} + 77.87\mathbf{j} - [33.68 + 9.81(t - 13.24)]\mathbf{k} \text{ m/s}$$

and

$$x_2 = -11.49(t - 13.24) \text{ m}$$
$$y_2 = 993.2 + 77.87(t - 13.24) \text{ m}$$
$$z_2 = 860.1 - 33.68(t - 13.24) - 4.905(t - 13.24)^2 \text{ m}$$

This particle hits the ground when $z_2 = 0$, which gives

$t = 23.49$ s	Ans.
$x_2 = -117.7$ m	Ans.
$y_2 = 1791$ m	Ans.

b. Equation 19-6 can also be applied to each of the pieces individually over the duration of the impact. For the 1-kg piece, Eq. 19-6 gives

$$(1)(75\mathbf{j}) + \mathbf{F}(0.005) = (1)[22.98\mathbf{i} + 69.26\mathbf{j} + 67.21\mathbf{k}]$$

Therefore the average force exerted on the 1-kg piece by the explosion is

$$\mathbf{F}_{avg} = 4596\mathbf{i} - 1148\mathbf{j} + 13442\mathbf{k} \text{ N}$$

and a force of equal magnitude but opposite direction is exerted on the 2-kg piece. The average magnitude of the explosive force exerted on each piece is then

$$F_{avg} = 14{,}250 \text{ N} \qquad \text{Ans.}$$

A 20-lb box A slides down a frictionless ramp and strikes a 10-lb box B (Fig. 19-11). As a result of the impact, the two boxes become hooked together and slide as a single unit on the rough surface ($\mu_k = 0.6$). Determine:

a. The velocity of the boxes immediately after impact.
b. The distance d that the boxes will slide before coming to rest.

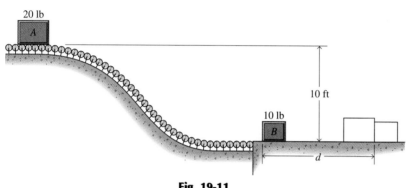

Fig. 19-11

SOLUTION

a. *Work-Energy from 1 → 2:* The free-body diagram of box A as it slides down the ramp is shown in Fig. 19-12a. The normal force N acts perpendicular to the motion and so does no work. The weight force has a potential so that the work–energy equation

$$T_1 + V_1 + U^{(o)}_{1\to2} = T_2 + V_2$$

gives

$$0 + (20)(10) + 0 = \frac{1}{2}\left(\frac{20}{32.2}\right)v_{A2}^2 + 0$$

and

$$v_{A2} = 25.38 \text{ ft/s}$$

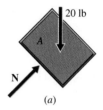

(a)

Linear Impulse–Momentum from 2 → 3: Box B is initially at rest ($v_{B2} = 0$), and after the impact both boxes have the same velocity ($v_{A3} = v_{B3} = v_3$). Over the brief duration of the impact there are no external impulsive forces in the x-direction, and so the x-component of linear momentum is conserved:

$$\left(\frac{20}{g}\right)(25.38) + \left(\frac{10}{g}\right)(0) = \left(\frac{20}{g}\right)v_3 + \left(\frac{10}{g}\right)v_3$$

which gives

$$v_3 = 16.92 \text{ ft/s} \qquad\qquad \text{Ans.}$$

b. *Newton's Second Law:* The free-body diagram of the pair of boxes following the impact is shown in Fig. 19-12b. During this phase of the motion, the boxes slide along a straight horizontal line, and there is no acceleration in the vertical direction. Therefore,

$$\uparrow \Sigma F = N - 30 = 0$$

and $N = 30$ lb. Then the friction force is

$$F = 0.6(30) = 18 \text{ lb}$$

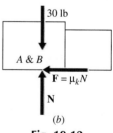

(b)

Fig. 19-12

Work–Energy from 3 → 4: The free-body diagram of Fig. 19-12b still applies. Both the normal force and the weight force act perpendicular to the motion and therefore do no work. The work done by the friction force is computed

$$U_{1 \to 2}^{(o)} = \int_0^d (-18) \, dx = -18d$$

so that the work–energy equation

$$T_1 + V_1 + U_{1 \to 2}^{(o)} = T_2 + V_2$$

gives

$$\frac{1}{2}\left(\frac{30}{32.2}\right)(16.92)^2 - 18d = 0$$

and

$$d = 7.41 \text{ ft} \qquad\qquad \text{Ans.}$$

PROBLEMS

19-18* At some instant of time, the position and velocity of three particles are given by

	Particle		
	1	2	3
m, kg	1	2	3
x, m	3	8	5
y, m	4	3	7
v_x, m/s	10	0	-2
v_y, m/s	-5	5	3

Find the location and velocity of the mass center at this instant.

19-19* At some instant of time, the mass center of three particles weighing 3 lb, 5 lb, and 1 lb, respectively, is located at $\mathbf{r}_G = 8\mathbf{i} + 5\mathbf{j} + 3\mathbf{k}$ ft and has a velocity given by $\mathbf{v}_G = 5\mathbf{i} - 12\mathbf{k}$ ft/s. At this same instant, the 3-lb particle has position $\mathbf{r}_3 = 5\mathbf{i}$ ft and velocity $\mathbf{v}_3 = 3\mathbf{i} + 8\mathbf{k}$ ft/s, and the 1-lb particle has position $\mathbf{r}_1 = 8\mathbf{j} + 3\mathbf{k}$ ft and velocity $\mathbf{v}_1 = 3\mathbf{i} - 3\mathbf{j}$ ft/s. Determine the position and velocity of the 5-lb particle at this instant.

19-20 A 25-kg sled is sliding along a flat, horizontal, frictionless surface with a speed of 5 m/s when a 60-kg man jumps onto it. If the initial speed of the man was 2 m/s at right angles to the motion of the sled, determine the final velocity of the sled and man.

19-21* Two cars collide at an intersection as shown in Fig. P19-21. Car A weighs 2200 lb and has an initial speed $v_A = 15$ mi/h, whereas car B weighs 3500 lb and has an initial speed $v_B = 25$ mi/h. If the cars become entangled and move as a single unit after the impact, determine their speed v_f and direction θ after impact.

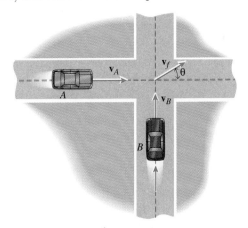

Fig. P19-21

19-22 Two cars collide at an intersection as shown in Fig. P19-21. Car A has a mass of 1000 kg and an initial speed $v_A = 25$ km/h, whereas car B has a mass of 1500 kg. If the cars become entangled and move as a single unit at an angle $\theta = 30°$ after the impact, determine v_B the speed of car B just prior to the impact.

343

19-23 A 2-lb particle is sliding along a flat, horizontal, frictionless surface at $v_i = 10$ ft/s as shown in Fig. P19-23. When the particle is 20 ft from the wall, it explodes and splits into two equal pieces. One piece hits the wall at $y_A = 5$ ft while the other piece hits the wall at $y_B = 10$ ft. Determine:

a. The impulse exerted on particle A by the explosion.
b. The velocity $\mathbf{v}_{A/B}$ of particle A relative to particle B immediately after the explosion.
c. The time difference between when particle A hits the wall and when particle B hits the wall.

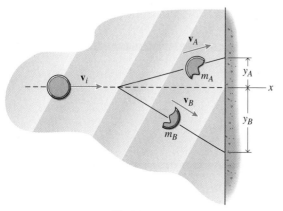

Fig. P19-23

19-24* A 5-kg particle is sliding along a flat, horizontal, frictionless surface at $v_i = 4$ m/s as shown in Fig. P19-23. When the particle is 10 m from the wall, it explodes and splits into two pieces $m_A = 3$ kg and $m_B = 2$ kg. If the 3-kg piece hits the wall 3 s after the explosion at $y_A = 7.5$ m, determine:

a. The impulse exerted on particle A by the explosion.
b. The velocity $\mathbf{v}_{A/B}$ of particle A relative to particle B immediately after the explosion.
c. The position y_B at which particle B hits the wall.
d. The time difference between when particle A hits the wall and when particle B hits the wall.

19-25 A 5-lb particle is sliding along a flat, horizontal, frictionless surface at $v_i = 10$ ft/s as shown in Fig. P19-23. When the particle is 20 ft from the wall, it explodes and splits into two pieces. One piece m_A hits the wall at $y_A = 5$ ft, whereas the other piece m_B hits the wall at $y_B = 10$ ft. If both particles hit the wall at the same time, determine:

a. The size of the two particles m_A and m_B.
b. The impulse exerted on particle A by the explosion.
c. The velocity $\mathbf{v}_{A/B}$ of particle A relative to particle B immediately after the explosion.

19-26* A 5-kg cannonball is fired with an initial velocity of $v_0 = 125$ m/s and $\theta_0 = 75°$ as shown in Fig. P19-26. At the peak of its trajectory, the ball explodes and splits into

two pieces. A 2-kg piece hits the ground at $x = 50$ m, and $y = 350$ m when $t = 25$ s. Determine:

a. When and where the 3-kg piece hits the ground.
b. The impulse exerted on the 2-kg piece by the explosion.
c. The average magnitude of the explosive force F_{avg} if the duration of the explosion is $\Delta t = 0.003$ s.

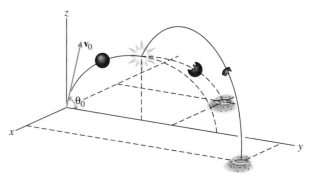

Fig. P19-26

19-27* A 10-lb cannonball is fired with an initial velocity of $v_0 = 450$ ft/s and $\theta_0 = 50°$ as shown in Fig. P19-26. When $t = 5$ s the ball explodes and splits into two pieces. A 6-lb piece hits the ground at $x = 1000$ ft, and $y = 7000$ ft when $t = 25$ s. Determine:

a. When and where the 4-lb piece hits the ground.
b. The impulse exerted on the 6-lb piece by the explosion.
c. The average magnitude of the explosive force F_{avg} if the duration of the explosion is $\Delta t = 0.001$ s.

19-28 A 10-kg box A slides down a frictionless ramp ($\theta = 25°$) and strikes a 5-kg box B, which is attached to a spring of stiffness $k = 8500$ N/m (Fig. P19-28). As a result of the impact, the two boxes become hooked together and slide as a single unit. If box A starts from rest with $d = 5$ m, determine:

a. The velocity of the boxes immediately after impact.
b. The maximum compression in the spring during the resulting motion.
c. The acceleration of the boxes at the instant of maximum compression.

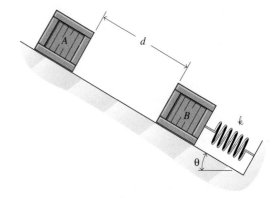

Fig. P19-28

19-29* A 25-lb box A slides down a ramp ($\theta = 25°$) and strikes a 10-lb box B attached to a spring of stiffness $k = 10$ lb/in. (Fig. P19-28). On impact, the two boxes become hooked and slide as a single unit on the rough ($\mu_k = 0.4$) surface. If box A starts from rest with $d = 20$ ft, determine:

a. The velocity of the boxes immediately after impact.
b. The maximum compression in the spring during the resulting motion.
c. The acceleration of the boxes at the instant of maximum compression.

19-30 A 0.40-kg block of wood is at rest on a rough ($\mu_k = 0.3$) horizontal surface when it is struck by a 0.03-kg bullet with an initial speed $v_i = 100$ m/s (Fig. P19-30). On impact, the bullet becomes embedded in the wood. Determine:

Fig. P19-30

a. The speed of the block and the bullet immediately after impact.
b. The distance the block will slide before coming to rest again.

19-31 A 1-lb block of wood is at rest on a rough ($\mu_k = 0.25$) horizontal surface when it is struck by a 0.25-oz bullet (Fig. P19-30). On impact, the bullet becomes embedded in the wood. If the block slides 25 ft before coming to rest again, determine:

a. The speed of the block and the bullet immediately after impact.
b. The initial speed v_i of the bullet.

19-32* A 0.30-kg block of wood is attached to a spring with $k = 7500$ N/m (Fig. P19-32). The block is at rest on a rough ($\mu_k = 0.4$) horizontal surface when it is struck by a 0.030-kg bullet with an initial speed $v_i = 150$ m/s. On impact, the bullet becomes embedded in the wood. Determine:

a. The speed of the block and the bullet immediately after impact.
b. The distance the block will slide before coming to rest again.

Fig. P19-32

19-33 A 12-oz block of wood is attached to a spring with $k = 5$ lb/in. (Fig. P19-32). The block is at rest on a rough ($\mu_k = 0.35$) horizontal surface with the spring undeformed

when it is struck by a 0.25-oz bullet. On impact, the bullet becomes embedded in the wood. If the maximum deflection of the spring after impact is 2.50 in., determine:

a. The speed of the block and the bullet immediately after impact.
b. The initial speed v_i of the bullet.

19-34* A 15-kg block of wood is attached to a spring with $k = 4500$ N/m (Fig. P19-32). The block is at rest on a rough ($\mu_k = 0.3$) horizontal surface when it is struck by a 0.03-kg bullet. On impact, the bullet becomes embedded in the wood. Determine the maximum initial speed v_i of the bullet for which the spring does not rebound.

19-35* A ballistic pendulum consists of a 5-lb box of sand suspended from a light, 5-ft long cord (Fig. P19-35). A 0.5-oz bullet strikes the box and becomes embedded in the sand. If the initial speed of the bullet is 350 ft/s, determine:

a. The speed of the sand and bullet immediately after the impact.
b. The maximum angle through which the pendulum will swing after impact.

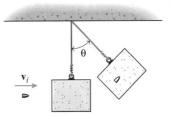

Fig. P19-35

19-36 Two cars collide at an intersection (Fig. P19-36). Car A has a mass of 1200 kg and car B has a mass of 1500 kg. On impact, the wheels of both cars lock and the cars slide ($\mu_k = 0.2$). After impact the cars become entangled and move as a single unit a distance of 10 m at an angle $\theta = 60°$. Determine the speeds of the cars v_A and v_B just prior to the impact.

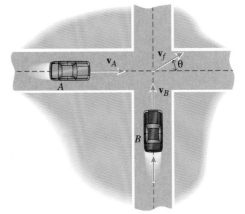

Fig. P19-36

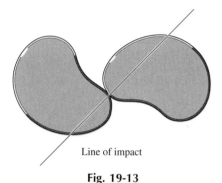

Line of impact

Fig. 19-13

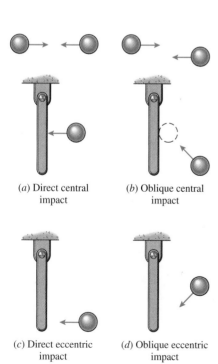

(a) Direct central
impact

(b) Oblique central
impact

(c) Direct eccentric
impact

(d) Oblique eccentric
impact

Fig. 19-14

19-4 COLLISION OF ELASTIC BODIES

An *impact* (collision between two bodies) is an event that usually occurs in a very brief interval of time. It is usually accompanied by relatively large reaction forces between the two bodies and correspondingly large changes of velocity of one or both bodies. The large reaction forces also result in considerable deformation of the impacting bodies and the consequent conversion of mechanical energy into sound and heat.

Impact events are categorized according to the relative location of the mass centers of the bodies, the relative velocity of the mass centers, and the *line of impact*: the straight line normal to the contacting surfaces at the point of impact (Fig. 19-13). If the mass centers of both bodies are on the line of impact, the impact is called a *central impact* (Fig. 19-14a,b). When the mass center of one or both bodies does not lie on the line of impact, the impact is called an *eccentric impact* (Fig. 19-14c,d). Obviously, only central impact can occur between particles since the size and shape of particles is not supposed to affect the calculation of their motion.

Further categorization is based on the orientation of the velocities of the bodies relative to the line of impact. A collision for which the initial velocities of the impacting bodies are along the line of impact is called a *direct impact* (Fig. 19-14a,c). A direct impact is a head-on collision. A collision for which the initial velocities of the impacting bodies are not along the line of impact is called an *oblique impact* (Fig. 19-14b,d).

The impact of two bodies consists of two phases—a deformation or compression phase followed by a restoration or restitution phase—and is accompanied by the generation of heat and sound. During the first phase, which lasts from the instant of contact to the instant of maximum deformation, the two bodies are compressed by the large interaction force. At the end of this phase, the bodies are neither coming closer together nor moving apart: the relative velocity along the line of impact is zero. In the second phase, which lasts from the instant of maximum deformation until the instant of separation, the bodies move apart as the forces within the bodies act to restore the bodies to their original shapes. Generally, not all the deformation is recoverable, however. Because of the permanent deformation of the bodies and because of the sound vibrations that are produced, some of the initial mechanical energy is dissipated in the collision.

There has been a considerable amount of study done on the relationship between the forces of impact and the resulting deformation when two bodies collide. The deformation of the bodies is found to depend on the rate of deformation as well as on the temperature and the material that the bodies are made of. Fortunately, however, the details of the impact can often be avoided; the linear impulse–momentum equation can be used to give a simple relation between the relative velocities of the bodies before and after the impact.

It must be remembered, however, that the impact of two bodies is a complex event. While the simple analysis that follows allows the solution of many impact problems that could not otherwise be solved, always keep in mind that the results of these impact calculations are only approximate.

19-4-1 Direct Central Impact

Consider the motion of two particles A and B along a common line (the line of impact) as in Fig. 19-15a. It is assumed that the speed v_{Ai} is greater than the speed v_{Bi} so that particle A will eventually overtake and collide with particle B. Also, in accordance with general observations, it is commonly assumed that during the brief impact interval, $\Delta t = t_f - t_i$:

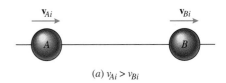

(a) $v_{Ai} > v_{Bi}$

1. The velocity of one or both particles may change greatly.
2. The positions of the particles do not change significantly.
3. Nonimpulsive forces may be neglected.
4. Friction forces between the two bodies are negligible.

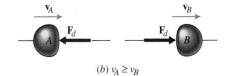

(b) $v_A \geq v_B$

Furthermore, since all motion and forces act along the line of impact, only the component of the impulse–momentum equation along the line of impact need be considered. The positive direction along this line will be taken to the right: forces and velocities to the right will be positive; forces and velocities to the left, negative.

During the collision, there are no impulsive, external forces acting on the pair of particles and thus linear momentum is conserved for the pair of particles

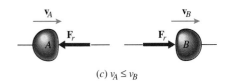

(c) $v_A \leq v_B$

$$m_A v_{Ai} + m_B v_{Bi} = m_A v_{Af} + m_B v_{Bf} \qquad (19\text{-}10)$$

On the other hand, if the particles are examined individually, the internal force is impulsive and must be included in the linear impulse–momentum equation. During the deformation phase of the impact (Fig. 19-15b), the linear impulse–momentum equation gives

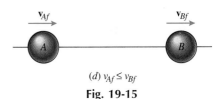

(d) $v_{Af} \leq v_{Bf}$

Fig. 19-15

$$m_A v_{Ai} - \int_{t_i}^{t_c} F_d \, dt = m_A v_c$$

for particle A, where F_d is the interaction force on particle A during the deformation phase, v_c is the common velocity of the two particles at the end of the deformation phase, and t_c is the time at the end of the deformation phase. During the restoration phase (Fig. 19-15c), the linear impulse–momentum equation gives

$$m_A v_c - \int_{t_c}^{t_f} F_r \, dt = m_A v_{Af}$$

for particle A, where F_r is the interaction force on particle A during the restoration phase and v_{Af} is the final velocity of particle A—the velocity of A after the collision is over.

The magnitude of the deformation impulse $\int_{t_i}^{t_c} F_d \, dt$ is generally larger than the magnitude of the restoration impulse $\int_{t_c}^{t_f} F_r \, dt$. The *coefficient of restitution e* is defined as the ratio of these two impulses:

$$e = \frac{\int_{t_c}^{t_f} F_r \, dt}{\int_{t_i}^{t_c} F_d \, dt} = \frac{m_A v_c - m_A v_{Af}}{m_A v_{Ai} - m_A v_c} = \frac{v_c - v_{Af}}{v_{Ai} - v_c}$$

A similar analysis of particle B alone gives a similar equation for the coefficient of restitution:

$$e = \frac{\int_{t_c}^{t_f} F_r \, dt}{\int_{t_i}^{t_c} F_d \, dt} = \frac{m_B v_{Bf} - m_B v_c}{m_B v_c - m_B v_{Bi}} = \frac{v_{Bf} - v_c}{v_c - v_{Bi}}$$

Eliminating the unknown velocity v_c between these two equations gives

$$e = -\frac{v_{Bf} - v_{Af}}{v_{Bi} - v_{Ai}} = -\frac{(v_{B/A})_f}{(v_{B/A})_i} \tag{19-11}$$

That is, the coefficient of restitution e is the negative of the ratio of the relative velocity of the two particles after impact and the relative velocity of the two particles before impact. This ratio is a measure of the elastic properties of the particles and must be measured experimentally.

Equations 19-10 and 19-11 together give two equations for determining the velocities of the two bodies following the impact given the two velocities before the impact.

An impact for which $e = 1$ is called a *perfectly elastic impact*. In this case, Eq. 19-11 gives

$$v_{Ai} + v_{Af} = v_{Bi} + v_{Bf}$$

But Eq. 19-10 can be rewritten

$$m_A(v_{Ai} - v_{Af}) = m_B(v_{Bf} - v_{Bi})$$

and multiplying these two equations together gives

$$m_A(v_{Ai}^2 - v_{Af}^2) = m_B(v_{Bf}^2 - v_{Bi}^2)$$

Finally, rearranging and dividing by 2 gives

$$\frac{1}{2}m_A v_{Ai}^2 + \frac{1}{2}m_B v_{Bi}^2 = \frac{1}{2}m_A v_{Af}^2 + \frac{1}{2}m_B v_{Bf}^2$$

Therefore, for a perfectly elastic impact, $e = 1$, the kinetic energy of the pair of particles is conserved.

On the other hand, an impact for which $e = 0$ is called a *perfectly plastic impact*. In this case, the relative velocity of the two particles after impact is zero and the two particles move together at the same speed. This corresponds to the maximum loss of kinetic energy for the impact. The coefficient of restitution for all real objects always lies between these two limiting cases, $0 \le e < 1$.

The coefficient of restitution is not a property of a single body or a single type of material. Rather, it depends on the types of material of both colliding bodies. Although it is frequently considered a constant, the coefficient of restitution actually varies considerably with the impact velocity as well as with the sizes, shapes, and temperatures of the colliding bodies. Though handbook values for e may be used in the absence of better data, they are generally unreliable.

19-4-2   Oblique Central Impact

The analysis of the oblique central impact of two particles is a simple extension of the analysis of direct central impact. For the case of per-

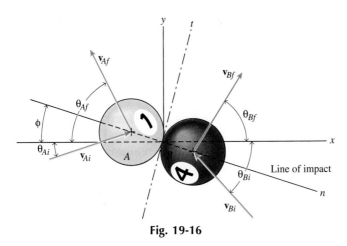

Fig. 19-16

fectly smooth and frictionless particles, oblique central impact will be seen to be merely the superposition of a uniform motion in the direction perpendicular to the line of impact and a direct central impact along the line of impact.

Coordinate axes are chosen along (n-axis) and perpendicular (t-axis) to the line of impact as shown in Fig. 19-16. The impulse–momentum equation applies equally to the combined system of particles (Fig. 19-16) and to each particle individually (Fig. 19-17). During the brief period of the impact, the only impulsive forces that act on either particle are the internal forces that act along the line of impact (n-axis).

Since there is no impulsive force on either particle in the t-direction, the t-component of linear momentum is conserved for each particle individually:

$$m_A(v_{Ai})_t = m_A(v_{Af})_t \tag{19-12a}$$
$$m_B(v_{Bi})_t = m_B(v_{Bf})_t \tag{19-12b}$$

Therefore, the t-components of the particles' velocities are unchanged by the impact:

$$(v_{Ai})_t = (v_{Af})_t \tag{19-13a}$$
$$(v_{Bi})_t = (v_{Bf})_t \tag{19-13b}$$

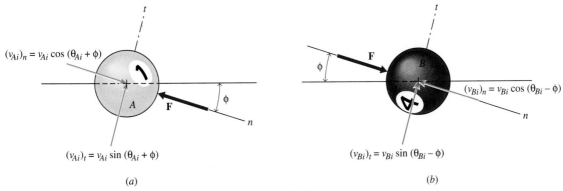

Fig. 19-17

For the system consisting of the pair of particles, there are no impulsive, external forces in any direction, and hence linear momentum is conserved in both the n- and t-directions:

$$m_A(v_{Ai})_t + m_B(v_{Bi})_t = m_A(v_{Af})_t + m_B(v_{Bf})_t \qquad (19\text{-}14a)$$
$$m_A(v_{Ai})_n + m_B(v_{Bi})_n = m_A(v_{Af})_n + m_B(v_{Bf})_n \qquad (19\text{-}14b)$$

Equation 19-14a is the sum of Eqs. 19-12a and 19-12b and contributes no additional information to the problem. Therefore, still another equation is needed to solve for the remaining two unknowns: $(v_{Af})_n$ and $(v_{Bf})_n$. Repeating the analysis of Section 19-4-1 for the n-direction gives

$$e = -\frac{(v_{Bf})_n - (v_{Af})_n}{(v_{Bi})_n - (v_{Ai})_n} = -\frac{(v_{B/A})_{fn}}{(v_{B/A})_{in}} \qquad (19\text{-}14c)$$

where $(v_{B/A})_{fn}$ and $(v_{B/A})_{in}$ are the n-components of the final and initial velocity of B relative to A, respectively. Equations 19-14b and 19-14c give the final two equations necessary to find the normal components of velocity $(v_{Af})_n$ and $(v_{Bf})_n$.

Once the four final velocity components $(v_{Af})_t$, $(v_{Af})_n$, $(v_{Bf})_t$, and $(v_{Bf})_n$ are found, the magnitudes and directions of the final velocities $\mathbf{v}_{Af}$ and $\mathbf{v}_{Bf}$ may be easily determined. Though the n- and t-axes are required to solve for the velocity components, these directions seldom have significance in the overall problem being solved. Therefore the final results should be reported relative to standard horizontal and vertical axes or relative to the initial directions to $\mathbf{v}_A$ and $\mathbf{v}_B$ and not relative to the n- and t-axes.

19-4-3   Constrained Impact

In the foregoing analysis it was assumed that both particles moved freely except for the impact. That is, no external impulsive forces acted on either particle. When one or both of the colliding particles are constrained against motion in some direction as in Example Problem 19-7, the constraint force is likely to be just as impulsive as the internal force. Therefore, linear momentum for the system consisting of the pair of particles may not be conserved in either the n- or t-directions. Also, for the constrained particle, linear momentum in the t-direction will probably not be conserved, and the t-component of its velocity will likely change through the impact.

The equation for the coefficient of restitution, Eq. 19-14c, may still be used to relate the relative velocities along the line of impact. However, the other equations must be replaced with a combination of the following:

1. Conservation of linear momentum for the system of particles in the direction perpendicular to the constraint.
2. Conservation of linear momentum for the unconstrained particle in the t-direction.
3. Kinematic constraints on the direction of the velocity of the constrained particle.

These equations will have to be derived specifically for each particular problem being solved.

EXAMPLE PROBLEM 19-6

Two masses slide on a horizontal frictionless rod as shown in Fig. 19-18. Slider A has a mass of 2 kg and is sliding to the right at 3 m/s, whereas slider B has a mass of 0.75 kg and is sliding to the left at 1 m/s. If the coefficient of restitution for the sliders is 0.6, determine:

a. The velocity of each mass after they collide.
b. The percentage decrease in energy due to the collision.

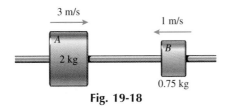

Fig. 19-18

SOLUTION

a. This is a direct central impact problem with the line of impact along the rod. Since there are no impulsive forces in the direction of the rod, linear momentum along the rod is conserved:

$$(2)(3) + (0.75)(-1) = 2v_{Af} + 0.75v_{Bf}$$

Also, Eq. 19-11 gives

$$0.6 = -\frac{v_{Bf} - v_{Af}}{(-1) - (3)}$$

Solving these two equations simultaneously gives

$$v_{Af} = 1.255 \text{ m/s} \qquad \text{Ans.}$$
$$v_{Bf} = 3.65 \text{ m/s} \qquad \text{Ans.}$$

b. The sum of the kinetic energy of the two particles before the collision was

$$T_i = \frac{1}{2}(2)(3)^2 + \frac{1}{2}(0.75)(1)^2 = 9.375 \text{ N} \cdot \text{m}$$

After the collision, the sum of the kinetic energy is

$$T_f = \frac{1}{2}(2)(1.255)^2 + \frac{1}{2}(0.75)(3.65)^2 = 6.571 \text{ N} \cdot \text{m}$$

The percentage decrease in kinetic energy is then

$$\frac{9.375 - 6.571}{9.375}(100) = 29.9\% \qquad \text{Ans.}$$

EXAMPLE PROBLEM 19-7

Two pucks of equal radius sliding on a smooth horizontal surface collide obliquely as shown in Fig. 19-19a. Puck A weighs 5 lb and is traveling to the right at 6 ft/s, whereas puck B weighs 2 lb and is traveling to the left at 3 ft/s. If the coefficient of restitution for the collision is 0.7 and the duration of the contact is 0.001 s, determine:

a. The velocities of the pucks immediately after they collide.
b. The percentage energy loss due to the collision.
c. The average interaction force of puck B on A.

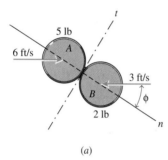

(a)

SOLUTION

a. First, coordinates n and t are drawn along and perpendicular to the line of impact as shown in Fig. 19-19a. As shown in Fig. 19-19b, the vertical distance between the centers is r, the radius of the pucks, and the slant distance between the centers is $2r$. Therefore the angle ϕ between the horizontal and the line of impact is given by

$$\phi = \sin^{-1} \frac{r}{2r} = 30°$$

Next, the initial velocities are resolved into components along and perpendicular to the line of impact:

$$(v_{Ai})_t = 6 \sin 30° = 3.00 \text{ ft/s}$$
$$(v_{Ai})_n = 6 \cos 30° = 5.196 \text{ ft/s}$$
$$(v_{Bi})_t = -3 \sin 30° = -1.500 \text{ ft/s}$$
$$(v_{Bi})_n = -3 \cos 30° = -2.598 \text{ ft/s}$$

Since the only impulsive force acting on the system is the internal reaction force (which acts along the n-direction), linear momentum in the t-direction is conserved for each particle, and their t-component of velocity is unchanged by the impact:

$$(v_{Af})_t = 3.000 \text{ ft/s} \quad \text{and} \quad (v_{Bf})_t = -1.500 \text{ ft/s}$$

Next, considering the pair of particles as a system, there is no external impulsive force in any direction, and therefore linear momentum is conserved in every direction. In particular, for the n-direction, conservation of linear momentum gives

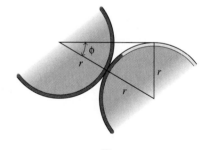

(b)
Fig. 19-19

$$\left(\frac{5}{g}\right)(5.196) + \left(\frac{2}{g}\right)(-2.598) = \left(\frac{5}{g}\right)(v_{Af})_n + \left(\frac{2}{g}\right)(v_{Bf})_n$$

This equation is then combined with the definition of the coefficient of restitution

$$0.7 = -\frac{(v_{Bf})_n - (v_{Af})_n}{(-2.598) - (5.196)}$$

to get

$$(v_{Af})_n = 1.410 \text{ ft/s} \qquad \text{and} \qquad (v_{Bf})_n = 6.866 \text{ ft/s}$$

Finally, the final velocities are expressed relative to their initial horizontal directions. The magnitude of the final velocity of puck A is (Fig. 19-19c)

$$v_{Af} = \sqrt{(1.410)^2 + (3.000)^2} = 3.31 \text{ ft/s}$$

and its direction relative to the horizontal is given by

$$\tan(\theta_{Af} + 30°) = 3.00/1.410$$

or $\theta_{Af} = 34.8°$. Therefore the final velocity of puck A is

$$\mathbf{v}_{Af} = 3.31 \text{ ft/s} \angle 34.8° \qquad \text{Ans.}$$

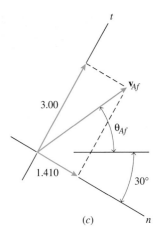

(c)

For puck B (Fig. 19-19d)

$$v_{Bf} = \sqrt{(1.500)^2 + (6.866)^2} = 7.03 \text{ ft/s}$$
$$\tan(\theta_{Bf} - 30°) = 1.500/6.866$$
$$\theta_{Bf} = 42.3°$$

and

$$\mathbf{v}_{Bf} = 7.03 \text{ ft/s} \measuredangle 42.3° \qquad \text{Ans.}$$

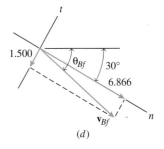

(d)

b. The sum of the kinetic energy of the two particles before the collision was

$$T_i = \frac{1}{2}\left(\frac{5}{32.2}\right)(6)^2 + \frac{1}{2}\left(\frac{2}{32.2}\right)(3)^2 = 3.075 \text{ lb} \cdot \text{ft}$$

After the collision, the sum of the kinetic energy is

$$T_f = \frac{1}{2}\left(\frac{5}{32.2}\right)(3.31)^2 + \frac{1}{2}\left(\frac{2}{32.2}\right)(7.03)^2 = 2.385 \text{ lb} \cdot \text{ft}$$

The percentage decrease in kinetic energy is then

$$\frac{3.075 - 2.385}{3.075}(100) = 22.4\% \qquad \text{Ans.}$$

c. Applying the linear impulse–momentum equation to puck A (Fig. 19-19e) gives

$$\frac{5}{32.2}(6\mathbf{i}) + .001\mathbf{F} = \frac{5}{32.2}(3.31)(\cos 34.8°\mathbf{i} + \sin 34.8°\mathbf{j})$$

which gives

$$\mathbf{F} = -509.6\mathbf{i} + 293.3\mathbf{j} \text{ lb} \qquad \text{Ans.}$$

or

$$= 588 \text{ lb} \measuredangle 30° \qquad \text{Ans.}$$

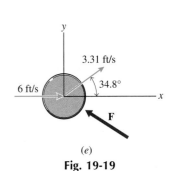

(e)

Fig. 19-19

EXAMPLE PROBLEM 19-8

A 3-kg sphere B is hanging at the end of a 1.5-m long inextensible cord when it is struck by a 2-kg sphere A of the same material (Fig. 19-20a). Sphere A is initially just touching the cord and drops 1 m before striking B. If the coefficient of restitution is 0.8 and the duration of contact is $\Delta t = 0.01$ s, determine:

a. The velocity of each sphere immediately after the collision.
b. The average tensile force in the cord due to the impact.
c. The maximum angle θ that sphere B will swing as a result of the collision.

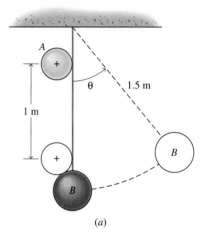

(a)

SOLUTION

a. *Work–Energy:* First, use the work–energy principle to determine the velocity of sphere A just prior to the impact. Taking the initial position of A as the zero for gravitational potential energy,

$$T_1 + V_{g1} + U_{1\rightarrow2}^{(o)} = T_2 + V_{g2}$$

$$0 + 0 + 0 = \frac{1}{2}m_A\,v_{A2}^2 - m_A(9.81)(1)$$

which gives

$$\mathbf{v}_{A2} = 4.429 \text{ m/s} \downarrow$$

where $\mathbf{v}_{A2} = \mathbf{v}_{Ai}$ is the velocity of sphere A just prior to the impact.

Linear Impulse–Momentum: Next, coordinates n and t are drawn along and perpendicular to the line of impact as shown in Fig. 19-20b. The horizontal distance between the centers of the spheres is r_A, and the slant distance between the centers is $r_A + r_B$. Therefore the angle ϕ between the vertical and the line of impact is given by

$$\phi = \sin^{-1}\frac{r_A}{r_A + r_B}$$

But the spheres are made of the same material and thus have the same density $\rho = \text{mass/volume}$

$$\rho = \frac{2 \text{ kg}}{\frac{4}{3}\,\pi\,r_A^3} = \frac{3 \text{ kg}}{\frac{4}{3}\,\pi\,r_B^3}$$

Therefore

$$r_B = \sqrt[3]{3/2}\;r_A = 1.145r_A$$

and

$$\phi = 27.79°$$

Next, the velocities are resolved into components along and perpendicular to the line of impact

$$(v_{Ai})_t = -4.429 \sin 27.79° = -2.065 \text{ m/s}$$
$$(v_{Ai})_n = 4.429 \cos 27.79° = 3.918 \text{ m/s}$$
$$(v_{Bi})_t = (v_{Bi})_n = 0 \text{ m/s}$$

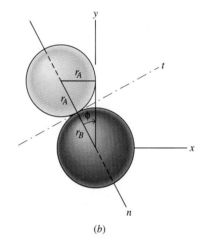

(b)

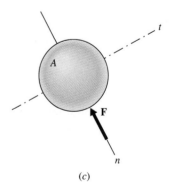

(c)

Fig. 19-20

Furthermore, since the cord is inextensible, sphere B cannot move downward after impact, and so

$$\mathbf{v}_{Bf} = v_{Bf}\mathbf{i}$$

and

$$(v_{Bf})_t = v_{Bf}\cos 27.79° = 0.8847\, v_{Bf}$$
$$(v_{Bf})_n = v_{Bf}\sin 27.79° = 0.4662\, v_{Bf}$$

Because the cord constrains the motion of sphere B, the tension in the cord is just as impulsive as the internal reaction force, and linear momentum for the pair of spheres is not conserved in either the n- or the t-direction. Instead, linear impulse–momentum equations will be written for each of the spheres separately. Referring to the free-body diagrams of Figs. 19-20c and 19-20d, the t- and n-components of these equations are

$$2(-2.065) = 2(v_{Af})_t \qquad (a)$$
$$2(3.918) - F\Delta t = 2(v_{Af})_n \qquad (b)$$
$$0 + T\Delta t \sin 27.79° = 3(0.8847v_{Bf}) \qquad (c)$$
$$0 - T\Delta t \cos 27.79° + F\Delta t = 3(0.4662v_{Bf}) \qquad (d)$$

where F is the average impact force between the spheres, T is the average tensile force in the cord, and $\Delta t = 0.01$ s is the duration of the impact. Equations a through d are combined with the definition of the coefficient of restitution:

$$0.8 = -\frac{0.4662v_{Bf} - (v_{Af})_n}{0 - (3.918)} \qquad (e)$$

and solved to get

$$(v_{Af})_t = -2.065 \text{ m/s} \qquad (v_{Af})_n = -2.242 \text{ m/s}$$
$$\mathbf{v}_{Bf} = 1.915 \text{ m/s} \rightarrow \qquad \text{Ans.}$$

The final velocity of A still needs to be expressed relative to the horizontal direction. The magnitude of the final velocity of sphere A is (Fig. 19-20e)

$$v_{Af} = \sqrt{(-2.065)^2 + (-2.242)^2} = 3.048 \text{ ft/s}$$

and its direction relative to the horizontal is given by

$$\tan (\theta_{Af} + 27.79°) = 2.242/2.065$$

or $\theta_{Af} = 19.56°$. Therefore the final velocity of sphere A is

$$\mathbf{v}_{Af} = 3.05 \text{ ft/s} \,\angle\, 19.56° \qquad \text{Ans.}$$

b. Now that the velocity components have been found, Eq. c gives the average tension in the cord

$$T = 1090 \text{ N} \qquad \text{Ans.}$$

c. *Work–Energy:* Finally, work–energy is used again to find the angle through which sphere B will swing after the collision. Using the upper end of the cord as the zero for gravitational potential energy,

$$T_1 + V_{g1} + U_{1\to2}^{(o)} = T_2 + V_{g2}$$

or

$$\frac{1}{2}m_B(1.915)^2 - m_B(9.81)(1.5) + 0$$

$$= 0 - m_B(9.81)(1.5 \cos \theta)$$

which gives

$$\theta = 28.9° \qquad \text{Ans.}$$

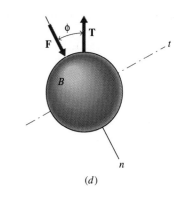

(d)

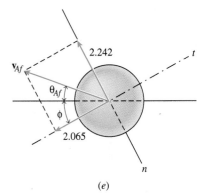

(e)

Fig. 19-20

PROBLEMS

19-37–19-42 Two beads are sliding freely on a horizontal rod as shown in Fig. P19-37. For the conditions specified, determine:

a. The final velocity of both beads.
b. The percentage of the initial kinetic energy lost as a result of the collision.
c. The average interaction force between the beads if the duration of impact is 0.005 s.

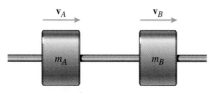

Fig. P19-37

Problem	m_A	v_A	m_B	v_B	e
19-37*	9 lb	3 ft/s	2 lb	0 ft/s	0.3
19-38*	0.5 kg	2 m/s	5 kg	0 m/s	0.7
19-39	7 lb	5 ft/s	3 lb	2 ft/s	0.7
19-40*	3 kg	1 m/s	1 kg	−3 m/s	0.9
19-41	6 lb	3 ft/s	1 lb	−2 ft/s	0.3
19-42	2 kg	3 m/s	3 kg	2 m/s	0.5

19-43* Three beads are sliding freely on a horizontal rod as shown in Fig. P19-43. Initially, beads B and C are at rest and bead A is moving to the right at 5 ft/s. If the coefficient of restitution is 0.8 for all collisions, determine:

a. The final velocity of each of the beads after all collisions have taken place.
b. The percentage of the initial kinetic energy that is lost as a result of the collisions.

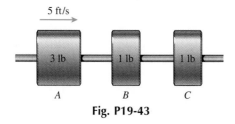

Fig. P19-43

19-44* Three beads are sliding freely on a horizontal rod as shown in Fig. P19-44. Initially, bead B is at rest, bead A is moving to the right at 3 m/s, and bead C is moving to the left at 2 m/s. If the coefficient of restitution is 0.8 for all collisions and the first collision occurs between beads A and B, determine:

a. The final velocity of each of the beads after all collisions have taken place.
b. The percentage of the initial kinetic energy that is lost as a result of the collisions.

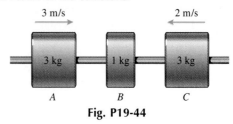

Fig. P19-44

19-45 Repeat Problem 19-43 for the case where bead B weighs 6 lb and all other parameters are unchanged.

19-46 Repeat Problem 19-44 for the case where the first collision occurs between beads B and C and all other conditions are unchanged.

19-47* A 2-lb sphere falls and bounces on a 10-lb plate, which is resting on the ground (Fig. P19-47a). If the sphere starts from rest 6 ft above the plate and rebounds to a height of 5 ft after the impact, determine:

a. The coefficient of restitution for the collision.
b. The height that the sphere would rebound if the 10-lb plate were resting on two springs having $k = 120$ lb/ft each (Fig. P19-47b).

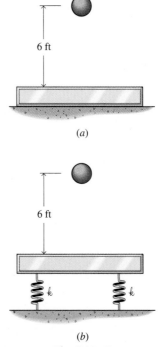

Fig. P19-47

19-48* Two spheres are hanging from cords as shown in Fig. P19-48. The distance from the ceiling to the center of each sphere is 2 m and the coefficient of restitution is 0.75. If sphere A ($m_A = 2$ kg) is drawn back 60° and released from rest, determine:

a. The maximum angle θ_B that sphere B ($m_B = 3$ kg) will swing as a result of the impact.
b. The angle θ_A that sphere A will rebound as a result of the impact.

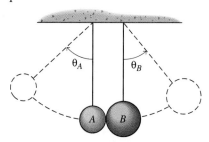

Fig. P19-48

19-49 Two spheres are hanging from cords as shown in Fig. P19-48. The distance from the ceiling to the center of each sphere is 4 ft. When sphere A (5 lb) is drawn back 60° and released from rest, sphere B is observed to swing through a maximum angle $\theta_B = 30°$, and sphere A is observed to rebound through an angle of $\theta_A = 15°$. Determine:

a. The weight of sphere B.
b. The coefficient of restitution of the impact.

19-50* For the two spheres of Problem 19-48, determine the minimum angle θ_A from which sphere A must be released to cause sphere B to swing through an angle of $\theta_B = 50°$ as a result of the impact.

19-51 The 5-lb sphere of Fig. P19-51 is released from rest when $\theta_A = 60°$ and swings down, striking the 10-lb box B. If the distance from the ceiling to the center of the sphere is 3 ft, the coefficient of restitution for the collision is 0.8, and the coefficient of friction between the box and the floor is 0.3, determine:

a. The velocity of the box immediately after the impact.
b. The distance that the box will slide before coming to rest again.

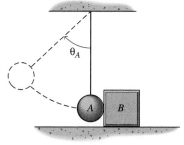

Fig. P19-51

19-52 The 2-kg sphere of Fig. P19-51 is released from rest and swings down, striking the 5-kg box B. The distance from the ceiling to the center of the sphere is 1 m, the coefficient of restitution for the collision is 0.7, and the coefficient of friction between the box and the floor is 0.1. If the box slides 750 mm after the impact before it comes to rest again, determine:

a. The velocity of the box immediately after the impact.
b. The angle θ_A from which the sphere A was released.

19-53* In the system of Problem 19-51, it is required that the average interaction force between the sphere and the box not exceed 1000 lb. If the duration of impact is 0.001 s, determine:

a. The maximum angle θ_A from which the sphere A can be released.
b. The distance that the box will slide as a result of the impact.

19-54 Two identical pucks are sliding on an air hockey table as shown in Fig. P19-54. If puck A has an initial velocity of 5 m/s to the right, puck B is initially at rest, and the coefficient of restitution is 0.9, determine the final velocities (magnitudes and directions) of both pucks.

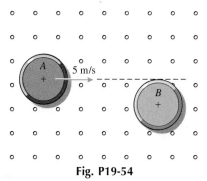

Fig. P19-54

19-55* Two identical pucks are sliding on an air hockey table as shown in Fig. P19-55. If the coefficient of restitution is 0.9, determine the final velocities (magnitudes and directions) of both pucks.

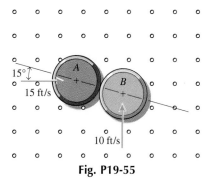

Fig. P19-55

19-56* In a pool shot, the cue ball *A* knocks the 1-ball into the corner pocket as shown in Fig. P19-56. If the coefficient of restitution is 0.95, determine the velocity of the cue ball after the collision.

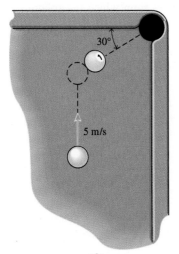

Fig. P19-56

19-57 Two identical spheres collide as shown in Fig. P19-57. If the coefficient of restitution is 0.7 and the velocity of *A* after the collision is in the vertical direction as shown, determine the velocity of *B* after the collision.

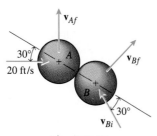

Fig. P19-57

19-58* Two pucks (of different size) collide on an air table as shown in Fig. P19-58. If the coefficient of restitution is 0.7 and the final velocity of each puck is 90° from its

initial direction, determine the magnitudes of the final velocities and the mass of puck *B*.

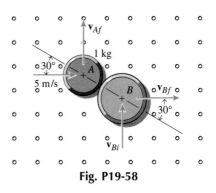

Fig. P19-58

19-59 A ball falls on a hard surface and bounces as shown in Fig. P19-59. If the coefficient of restitution is 0.9 and the ball starts from rest with $h = 4$ ft, determine:

a. The distance c to where the ball will bounce on the horizontal surface.
b. The maximum height d of the bounce measured from the horizontal surface.
c. The distance b along the horizontal surface to the point of the peak of the bounce.

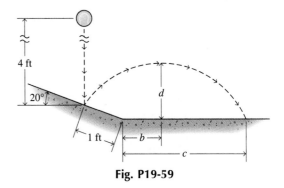

Fig. P19-59

19-60 A ball falls on a hard surface and bounces over a vertical wall as shown in Fig. P19-60. If the coefficient of restitution is 0.8, the ball starts from rest with $h = 1$ m, and the ball just clears the wall at the peak of its bounce, determine the distances b, c, and d in the figure.

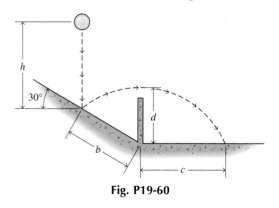

Fig. P19-60

19-61* A ball falls on a hard surface and bounces over a vertical wall as shown in Fig. P19-60. If the coefficient of restitution is 0.8 and the height of the wall is $d = 3$ ft, determine the height h from which the ball must be dropped such that it just clears the wall at the peak of its bounce.

19-62* An attention-getting device in a department store window consists of an air gun that repeatedly bounces a ball off a wall as shown in Fig. P19-62. If the coefficient of restitution is 0.9 and the velocity of the ball as it leaves the air gun is 3 m/s, determine:

a. The distance x that the device must be placed away from the wall.
b. The distance y to where the ball will strike the wall.

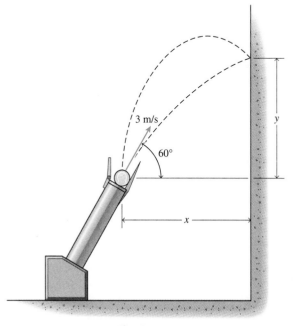

Fig. P19-62

19-63* A 2-lb sphere B is at rest on a ledge when it is struck by an identical sphere (Fig. P19-63). The distance from the ceiling to the center of sphere A is 3 ft and the coefficient of restitution is 0.7. At the moment of impact the cord is vertical and the center of sphere A is level with the bottom of sphere B. If sphere A is released from rest with $\theta_A = 60°$, determine the distance x traveled by sphere B before it bounces on the horizontal surface.

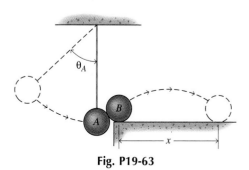

Fig. P19-63

19-64 A 2-kg sphere B is at rest on a ledge when it is struck by an identical sphere (Fig. P19-63). The distance from the ceiling to the center of sphere A is 2 m and the coefficient of restitution is 0.7. At the moment of impact, the cord is vertical and the center of sphere A is level with the bottom of sphere B. Determine the angle θ_A from which sphere A must be released such that sphere B travels a distance $x = 1.5$ m before bouncing on the horizontal surface.

19-65 Suppose that the rigid surface of Problem 19-61 is replaced with a 1.5-lb cart that is free to roll in the horizontal direction as shown in Fig. P19-65. If the coefficient of restitution is 0.8 and the height of the wall is $d = 1$ ft, determine the height h from which the 0.5-lb ball must be dropped such that it just clears the wall at the peak of its bounce.

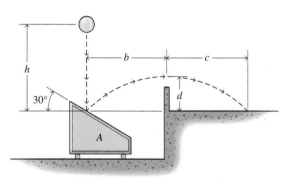

Fig. P19-65

19-5   ANGULAR IMPULSE AND ANGULAR MOMENTUM OF A PARTICLE

The angular impulse–momentum principle, which is developed in the next several sections, relates the moments of the forces acting on a particle and the velocity of the particle when the moments are known as functions of time. Again combining Newton's second law with the principles of kinematics, the angular impulse–momentum principle is particularly useful for solving problems in which several or all of the forces always act through a fixed point.

19-5-1   Angular Momentum

The *angular momentum* $\mathbf{H}_O$ of a particle P about a fixed point O is defined as the moment of the linear momentum $\mathbf{L}$ about the point O. If $\mathbf{r}_{P/O}$ is the position vector from point O to the particle P of mass m and velocity $\mathbf{v}$, then

$$\mathbf{H}_O = \mathbf{r}_{P/O} \times (m\mathbf{v}) \qquad (19\text{-}15)$$

The magnitude of the angular momentum is equal to $r_{P/O}\, mv \sin \theta$, where v is the speed of the particle and θ is the angle between the position vector $\mathbf{r}_{P/O}$ and the velocity $\mathbf{v}$. The direction of the angular momentum vector will be perpendicular to the plane formed by the vectors $\mathbf{r}_{P/O}$ and $\mathbf{v}$ (Fig. 19-21).

Expressing the velocity in terms of polar coordinates in the plane of $\mathbf{r}_{P/O}$ and $\mathbf{v}$ (Fig. 19-21), the angular momentum becomes

$$\mathbf{H}_O = \mathbf{r}_{P/O} \times m(v_r\mathbf{e}_r + v_\theta\mathbf{e}_\theta)$$

But the first term on the right-hand side is zero since the cross-product of two parallel vectors is zero. Therefore only the component of the velocity perpendicular to $\mathbf{r}_{P/O}$ contributes to the angular momentum, and the angular momentum represents a rotation of the particle about an axis through O along the angular momentum vector.

In the SI system of measurement, the units of angular momentum are $\text{kg} \cdot \text{m}^2/\text{s}$ or equivalently $\text{N} \cdot \text{m} \cdot \text{s}$. In the U. S. Customary system of measurement, they are $\text{slug} \cdot \text{ft}^2/\text{s} = \text{lb} \cdot \text{ft} \cdot \text{s}$.

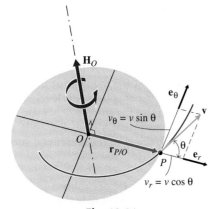

Fig. 19-21

19-5-2   Angular Impulse

The *angular impulse* of the resultant force about a fixed point O is defined as the impulse of the moment

$$\int_{t_i}^{t_f} \mathbf{r}_{P/O} \times \mathbf{R} \, dt = \int_{t_i}^{t_f} \mathbf{M}_O \, dt$$

where $\mathbf{r}_{P/O}$ is again the position of the particle P relative to the fixed point O and $\mathbf{R}$ is the resultant of all forces acting on the particle. In the SI system of measurement the magnitude of the angular impulse is expressed in $\text{N} \cdot \text{m} \cdot \text{s} = \text{kg} \cdot \text{m}^2/\text{s}$, which is the same as the units of angular momentum. If U. S. Customary units are used, the linear impulse is expressed in $\text{lb} \cdot \text{ft} \cdot \text{s} = \text{slug} \cdot \text{ft}^2/\text{s}$.

19-5-3 Principle of Angular Impulse and Momentum

Differentiating the angular momentum with respect to time gives

$$\frac{d\,\mathbf{H}_O}{dt} = \frac{d\,\mathbf{r}_{P/O}}{dt} \times (m\mathbf{v}) + \mathbf{r}_{P/O} \times \left(m\frac{d\,\mathbf{v}}{dt}\right)$$
$$= \mathbf{v} \times (m\mathbf{v}) + \mathbf{r}_{P/O} \times (m\mathbf{a})$$

But the first term $\mathbf{v} \times (m\mathbf{v}) = \mathbf{0}$, since the cross-product of any vector with itself is zero. In the second term, Newton's second law can be used to replace the $m\mathbf{a}$ with the resultant force $\mathbf{R}$. Then the second term is just $\mathbf{r}_{P/O} \times \mathbf{R} = \mathbf{M}_O$, the moment of the resultant force about point O. Therefore

$$\frac{d}{dt}\mathbf{H}_O = \mathbf{M}_O \qquad (19\text{-}16)$$

that is: *The time rate of change of the angular momentum of a particle about a fixed point O is equal to the resultant moment about O of all forces acting on the particle.*

If the moments of the forces are known as functions of time, Eq. 19-16 can be integrated from some initial time t_i to some final time t_f to get the Principle of Angular Impulse and Momentum:

$$\mathbf{H}_{Oi} + \int_{t_i}^{t_f} \mathbf{M}_O\,dt = \mathbf{H}_{Of} \qquad (19\text{-}17)$$

that is: *The final angular momentum $\mathbf{H}_{Of}$ of a particle about a fixed point O is the vector sum of its initial angular momentum $\mathbf{H}_{Oi}$ about O and the angular impulse $\int \mathbf{M}_O\,dt$ about O of the resultant of all forces acting on the particle during the time interval.*

Like the linear impulse–momentum principle, Eqs. 19-16 and 19-17 are vector equations representing three scalar components. Since $\mathbf{M}_O$ may vary in both magnitude and direction, rectangular Cartesian components are usually the most convenient to use. The three components can be applied independently of one another.

19-5-4 Conservation of Angular Momentum

It is unusual to know how the resultant moment about O, $\mathbf{M}_O = \mathbf{r}_{P/O} \times \mathbf{R}$, varies with time except in a few special cases. For example, if all the forces acting on a particle pass through a single point O, then the sum of moments about that point will be zero and the initial and final angular momentum about O will be the same:

$$\mathbf{H}_{Oi} = \mathbf{H}_{Of} \qquad (19\text{-}18)$$

This behavior is called the Principle of Conservation of Angular Momentum.

19-5-5 Systems of Particles

For a system of interacting particles, the angular impulse–momentum equations can be written for each particle separately and the equations added together. For example, for the set of particles in Fig. 19-6, the angular impulse–momentum equations (Eq. 19-16) are

$$\frac{d}{dt}(\mathbf{r}_{1/O} \times m_1\mathbf{v}_1) = \mathbf{r}_{1/O} \times (\mathbf{R}_1 + \mathbf{f}_{12} + \mathbf{f}_{13} + \cdots + \mathbf{f}_{1i} + \cdots)$$

$$\frac{d}{dt}(\mathbf{r}_{2/O} \times m_2\mathbf{v}_2) = \mathbf{r}_{2/O} \times (\mathbf{R}_2 + \mathbf{f}_{21} + \mathbf{f}_{23} + \cdots + \mathbf{f}_{2i} + \cdots)$$

$$\vdots$$

$$\frac{d}{dt}(\mathbf{r}_{\ell/O} \times m_\ell\mathbf{v}_\ell) = \mathbf{r}_{\ell/O} \times (\mathbf{R}_\ell + \mathbf{f}_{\ell 1} + \mathbf{f}_{\ell 2} + \cdots + \mathbf{f}_{\ell i} + \cdots)$$

and so on. Adding these equations together gives

$$\sum_\ell \left(\frac{d}{dt}\mathbf{H}_{\ell/O}\right) = \sum_\ell \left(\mathbf{r}_{\ell/O} \times \mathbf{R}_\ell\right) + (\mathbf{r}_{1/O} \times \mathbf{f}_{12} + \mathbf{r}_{2/O} \times \mathbf{f}_{21})$$
$$+ (\mathbf{r}_{1/O} \times \mathbf{f}_{13} + \mathbf{r}_{3/O} \times \mathbf{f}_{31}) + \cdots$$

But the internal forces always occur in pairs having equal magnitude and the same line of action but opposite direction along the line of action (Fig. 19-22). Therefore, the sum of moments about O for each pair of forces is zero, and for the system of particles

$$\frac{d}{dt}\mathbf{H}_O = \sum_{\ell=1}^{N} \mathbf{M}_{\ell/O} \qquad (19\text{-}19)$$

where $\mathbf{H}_O = \Sigma\, \mathbf{H}_{\ell/O}$ is the total angular momentum about O of the system of particles, $\Sigma\, \mathbf{M}_{\ell/O}$ is the sum of moments about O of all the external forces acting on the system of particles, and the moments of the internal forces need not be considered.

Of course, Eq. 19-19 can be integrated with respect to time from t_i to t_f to get the *Principle of Angular Impulse and Momentum for a System of Interacting Particles:*

$$(\mathbf{H}_O)_i + \int_{t_i}^{t_f} \sum_{\ell=1}^{N} \mathbf{M}_{\ell/O}\, dt = (\mathbf{H}_O)_f \qquad (19\text{-}20)$$

that is: *The final angular momentum $(\mathbf{H}_O)_f$ of a system of particles about a fixed point O is the vector sum of their initial angular momentum $(\mathbf{H}_O)_i$ about the point O and the angular impulse $\int \Sigma\, \mathbf{M}_{\ell/O}\, dt$ about O of all external forces acting on the system of particles.*

A common requirement for a system of particles is to compute the moments and angular momentum about the mass center of the system rather than about a fixed point O. The angular momentum of the system of particles about the mass center G is defined as the moment of the linear momentum

$$\mathbf{H}_G = \Sigma\, \mathbf{H}_{\ell/G} = \Sigma\, \mathbf{r}_{\ell/G} \times (m_\ell\, \mathbf{v}_\ell)$$

where $\mathbf{r}_{\ell/G}$ is the position of the ℓth particle relative to the mass center G (Fig. 19-23) and $\mathbf{v}_\ell = \dot{\mathbf{r}}_{\ell/O}$ is the absolute velocity of the ℓth particle. The absolute velocity can be replaced using the relative velocity equation $\mathbf{v}_\ell = \mathbf{v}_G + \mathbf{v}_{\ell/G}$, where $\mathbf{v}_G$ is the velocity of the mass center of the system of particles, to get

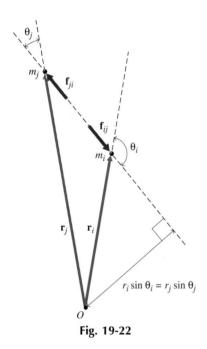

Fig. 19-22

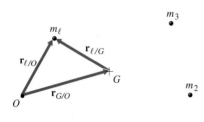

Fig. 19-23

$$\mathbf{H}_G = \Sigma\, \mathbf{r}_{\ell/G} \times m_\ell\, (\mathbf{v}_G + \mathbf{v}_{\ell/G})$$
$$= (\Sigma\, m_\ell \mathbf{r}_{\ell/G}) \times \mathbf{v}_G + \Sigma\, \mathbf{r}_{\ell/G} \times (m_\ell\, \mathbf{v}_{\ell/G})$$

The velocity of the mass center $\mathbf{v}_G$ is the same for every particle and can be taken outside the summation in the first term. But then the quantity inside the parentheses is zero by the definition of the mass center, since the position vectors $\mathbf{r}_{\ell/G}$ are measured from the mass center. Therefore,

$$\mathbf{H}_G = \Sigma\, \mathbf{r}_{\ell/G} \times (m_\ell\, \mathbf{v}_\ell) = \Sigma\, \mathbf{r}_{\ell/G} \times (m_\ell\, \mathbf{v}_{\ell/G}) \qquad (19\text{-}21)$$

That is, the angular momentum of the system of particles about the mass center G can be computed using either the absolute velocity $\mathbf{v}_\ell = \dot{\mathbf{r}}_{\ell/O}$ or the velocity relative to the mass center $\mathbf{v}_{\ell/G}$.

Taking the time derivative of $\mathbf{H}_G$ gives

$$\frac{d\,\mathbf{H}_G}{dt} = \Sigma \left(\frac{d\,\mathbf{r}_{\ell/G}}{dt} \times m_\ell\, \mathbf{v}_\ell + \mathbf{r}_{\ell/G} \times m_\ell\, \frac{d\,\mathbf{v}_\ell}{dt} \right)$$
$$= \Sigma\, \mathbf{v}_{\ell/G} \times m_\ell\, \mathbf{v}_\ell + \Sigma\, \mathbf{r}_{\ell/G} \times m_\ell\, \mathbf{a}_\ell$$
$$= \Sigma\, \mathbf{v}_{\ell/G} \times m_\ell(\mathbf{v}_G + \mathbf{v}_{\ell/G}) + \Sigma\, \mathbf{r}_{\ell/G} \times m_\ell\, \mathbf{a}_\ell$$
$$= (\Sigma\, m_\ell\, \mathbf{v}_{\ell/G}) \times \mathbf{v}_G + \Sigma\, \mathbf{v}_{\ell/G} \times m_\ell\, \mathbf{v}_{\ell/G} + \Sigma\, \mathbf{r}_{\ell/G} \times m_\ell\, \mathbf{a}_\ell$$

But the summation in the first term is again zero by the definition of the mass center since $\Sigma\, m_\ell\, \mathbf{v}_{\ell/G} = \dfrac{d}{dt} \Sigma\, m_\ell\, \mathbf{r}_{\ell/G}$ and the position vectors $\mathbf{r}_{\ell/G}$ are measured from the mass center G. Also, every term in the second summation of this equation is zero since the cross-product of any vector with itself is always zero. Finally, using Newton's second law to replace the factors $m_\ell\, \mathbf{a}_\ell$ as in the derivation of Eq. 19-19 gives

$$\frac{d\,\mathbf{H}_G}{dt} = \Sigma\, \mathbf{M}_{\ell/G} \qquad (19\text{-}22)$$

where $\Sigma\, \mathbf{M}_{\ell/G}$ is the sum of moments of the external forces about the mass center G and the moments of the internal forces need not be considered. Integrating Eq. 19-22 with respect to time gives

$$(\mathbf{H}_G)_i + \int_{t_i}^{t_f} \Sigma\, \mathbf{M}_{\ell/G}\, dt = (\mathbf{H}_G)_f \qquad (19\text{-}23)$$

Note that the form of the angular momentum principle for a system of particles is the same whether moments are summed around a fixed point O or about the moving mass center G. However, the form of the equations will not be the same if moments are summed about an arbitrary moving point P. In this case, the terms involving $\Sigma\, m_\ell\, \mathbf{r}_{\ell/P}$ will not drop out and there will be extra terms in Eqs. 19-22 and 19-23.

Finally, it should be noted that Eqs. 19-19 through 19-23 have been derived for a general system of particles. They apply equally to a system of independently moving particles and a system of particles that make up a rigid body.

A 500-lb satellite is in a circular orbit 100 mi above the earth and is to be moved to another circular orbit 1000 mi above the earth using a rocket engine having thrust of 750 lb (Fig. 19-24). The orbital transfer is effected using an elliptic transfer orbit by firing the maneuvering engine first at A and then at B. If the required velocity at A in the transfer orbit is 26,874 ft/s, determine the length of the engine burns at A and B necessary to effect the orbit change.

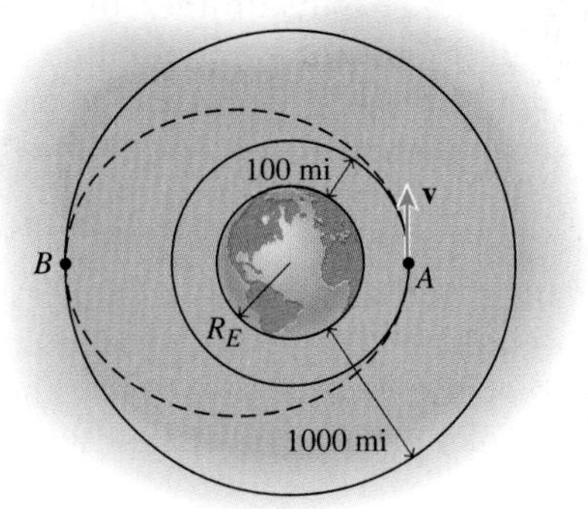

Fig. 19-24

SOLUTION

In the circular orbits, the acceleration of the satellite is

$$a = \frac{v^2}{r}$$

directed toward the center of the earth, and the only force acting on the satellite is

$$F = \frac{GMm}{r^2} = \frac{gmR_E^2}{r^2}$$

also directed toward the center of the earth. Then Newton's law gives

$$\frac{gmR_E^2}{r^2} = \frac{mv^2}{r}$$

which gives the velocity in the circular orbits as

$$v_c = R_E \sqrt{\frac{g}{r}}$$

Therefore, using $R_E = 3960$ mi,

$$v_{100c} = (3960)(5280) \sqrt{\frac{32.2}{(4060)(5280)}} = 25626 \text{ ft/s}$$

for the lower circular orbit, and

$$v_{1000c} = (3960)(5280) \sqrt{\frac{32.2}{(4960)(5280)}} = 23185 \text{ ft/s}$$

for the higher circular orbit.

Since the velocity in a circular orbit is always perpendicular to the radius, the angular momentum about the axis perpendicular to the plane of the circular orbit is just $H = rmv$;

$$H_{100c} = (4060)(5280)\left(\frac{500}{32.2}\right)(25626) = 8.530(10^{12}) \text{ lb} \cdot \text{ft} \cdot \text{s}$$

$$H_{1000c} = (4960)(5280)\left(\frac{500}{32.2}\right)(23185) = 9.428(10^{12}) \text{ lb} \cdot \text{ft} \cdot \text{s}$$

At A in the elliptic orbit, the velocity is also perpendicular to the radius so that the angular momentum of the satellite in the elliptical orbit at A is also $H = rmv$;

$$H_{100e} = (4060)(5280)\left(\frac{500}{32.2}\right)(26874) = 8.946(10^{12}) \text{ lb} \cdot \text{ft} \cdot \text{s}$$

Then applying the angular momentum principle (Eq. 19-17) across the duration of the burn at A gives (gravity acts through the axis and thus has no moment about the axis)

$$H_{100c} + r_A T \Delta t_A = H_{100e}$$
$$8.530(10^{12}) + (4060)(5280)(750)\Delta t_A = 8.946(10^{12})$$

or

$$\Delta t_A = 25.8 \text{ s} \qquad \text{Ans.}$$

Applying the angular momentum principle across the duration of the burn at B gives

$$H_{100e} + r_B T \Delta t_B = H_{1000c}$$
$$8.946(10^{12}) + (4960)(5280)(750)\Delta t_B = 9.428(10^{12})$$

or

$$\Delta t_B = 24.5 \text{ s} \qquad \text{Ans.}$$

EXAMPLE PROBLEM 19-10

A 0.6-kg mass slides on a smooth horizontal surface at the end of an inextensible string (Fig. 19-25a). The other end of the string passes through a hole in the surface and is attached to a spring having $k = 100$ N/m. The spring is unstretched when $\ell = 0$. If $v = 10$ m/s and $\ell = 0.5$ m at the instant shown, determine the minimum and maximum values of ℓ in the resulting motion.

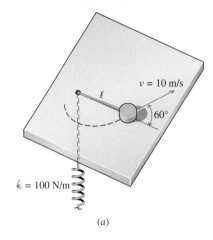

$v = 10$ m/s

ℓ

$60°$

$k = 100$ N/m

(a)

SOLUTION

Angular momentum about a vertical axis through the hole is conserved since none of the three forces acting on the mass has a moment about the axis (Fig. 19-25b). Both the weight W and the normal force N act parallel to the axis and thus have no moment about the axis, while the string tension acts through the axis and has no moment about the axis.

At the instant shown, the angular momentum is

$$H_{Oi} = (0.5)(0.6)(10 \sin 60) \qquad \text{N·m·s}$$

When the string is at its minimum or maximum length, the velocity of the mass is perpendicular to the string and the angular momentum is

$$H_{Of} = \ell(0.6)v$$

Therefore, conservation of angular momentum about a vertical axis through the hole gives

$$(0.5)(0.6)(10 \sin 60) = \ell(0.6)v \qquad (a)$$

A second equation relating the length of the string and the velocity is obtained from the work–energy equation. Neither the weight W nor the normal force N do work, and the work done by the spring has a potential. Therefore,

$$\frac{1}{2}(0.6)(10)^2 + \frac{1}{2}(100)(0.5)^2 = \frac{1}{2}(0.6)v^2 + \frac{1}{2}(100)\ell^2 \qquad (b)$$

Solving equations a and b simultaneously gives

$$\ell_{\max} = 0.828 \text{ m} \qquad \text{Ans.}$$
$$\ell_{\min} = 0.405 \text{ m} \qquad \text{Ans.}$$

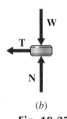

W

T

N

(b)

Fig. 19-25

PROBLEMS

19-66* A 250-g particle slides in a circular path on a smooth horizontal surface at the end of an inextensible string (Fig. P19-66). The other end of the cord is drawn very slowly through the central hole, reducing the radius of the circular path from 500 mm to 200 mm. If the initial velocity of the particle is 5 m/s, determine its velocity when the radius is 200 mm.

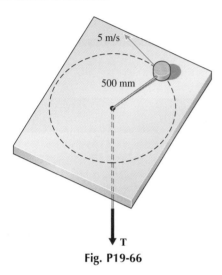

Fig. P19-66

19-67* The vertical shaft of Fig. P19-67 is rotating with an initial angular velocity of 20 rad/s when the 0.5-lb cylinder A starts to slide slowly outward along the lightweight horizontal arm. Determine the decrease in the angular velocity of the shaft as the cylinder A slides from 3 in. out to 24 in. from the axis of the shaft.

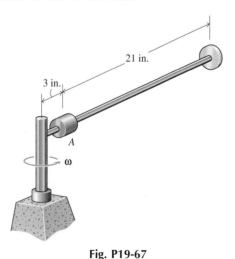

Fig. P19-67

19-68 Show that in central force motion the quantity $r^2\dot{\theta}$ is constant.

19-69 Prove Kepler's second law: "The radius vector from the sun to a planet sweeps equal areas in equal times." That is, prove that $dA/dt = $ constant, where dA is the shaded area in Fig. P19-69.

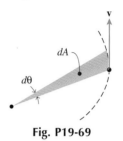

Fig. P19-69

19-70* A satellite is in an elliptical Earth orbit with semimajor axis $a = 17000$ km and semiminor axis $b = 13725$ km (Fig. P19-70). If the velocity of the satellite is 9500 m/s at A, determine the velocity of the satellite at B and at C.

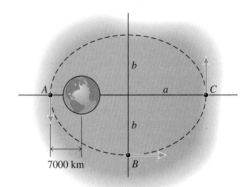

Fig. P19-70

19-71* A 5-lb ball is swinging at the end of a 2-ft long inextensible cord (Fig. P19-71). At the instant shown the velocity is in a horizontal plane with $v = 6$ ft/s and $\theta = 60°$. Determine the minimum angle θ in the resulting motion of the ball.

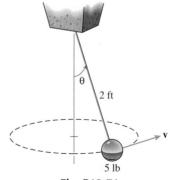

Fig. P19-71

19-72* A marble rolls freely on the inside of a 30° cone as shown in Fig. P19-72. At the instant shown the velocity of the marble is horizontal with $z = 500$ mm. Determine the maximum height to which the marble will rise if the initial velocity $v_i = 4$ m/s. Repeat for $v_i = 1$ m/s.

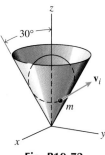

Fig. P19-72

19-73 A marble rolls freely on the inside of a cone as shown in Fig. P19-73. At the instant shown the velocity of the marble is horizontal with $z = 24$ in. If the minimum height in the resulting motion is 12 in., determine the initial velocity of the marble v_i and the velocity of the marble at the lowest point.

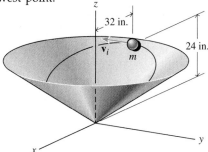

Fig. P19-73

19-74 A 2-kg particle sliding on a smooth horizontal surface is attached to the end of an elastic cord (Fig. P19-74). The other end of the cord, which has an unstretched length of 400 mm and an elastic constant $k = 250$ N/m, is attached at A. At its closest approach to A ($d = 200$ mm), the particle has a speed of 5 m/s. Determine:

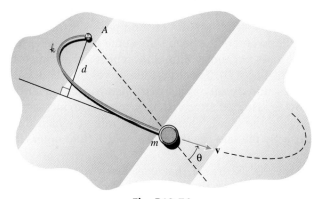

Fig. P19-74

a. The velocity of the particle (speed v and direction θ) when the length of the cord is 750 mm and the particle is moving away from A.
b. The length of the elastic cord and the velocity of the particle when the particle is at its farthest from A.
c. The velocity of the particle when the length of the cord is 600 mm and the particle is moving toward A.

19-75* A 2-lb particle slides on a smooth horizontal surface at the end of an elastic cord (Fig. P19-75). The other end of the cord, which has an unstretched length of 18 in. and an elastic constant $k = 8$ lb/ft, is attached at A. If $v = 10$ ft/s, $\theta = 40°$, and $\ell = 27$ in. at the instant shown, determine:

a. The velocity of the particle (speed v and direction θ) when the tension in the cord is zero.
b. The distance d of closest approach to point A.
c. The length of the elastic cord and the velocity of the particle when the particle is at its farthest from A.

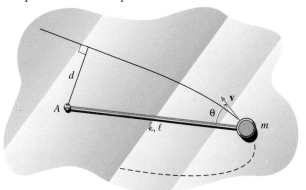

Fig. P19-75

19-76 For a system of n particles, show that the total angular momentum about a fixed point O can be written

$$\mathbf{H}_O = \mathbf{H}_G + \mathbf{r}_G \times m\mathbf{v}_G$$

where $\mathbf{H}_O = \Sigma\, \mathbf{r}_{i/O} \times m_i\, \mathbf{v}_i$, $\mathbf{H}_G = \Sigma\, \mathbf{r}_{i/G} \times m_i\, \mathbf{v}_i$, $m = \Sigma\, m_i$, and $\mathbf{r}_G$ and $\mathbf{v}_G$ are the position and velocity, respectively, of the mass center of the system of particles relative to the fixed point O.

19-77 For a system of n particles, show that the total angular momentum about a moving point P can be written

$$\mathbf{H}_P = \mathbf{H}_G + \mathbf{r}_{G/P} \times m\mathbf{v}_G$$

where $\mathbf{H}_P = \Sigma\, \mathbf{r}_{i/P} \times m_i\, \mathbf{v}_i$, $\mathbf{H}_G = \Sigma\, \mathbf{r}_{i/G} \times m_i\, \mathbf{v}_i$, $m = \Sigma\, m_i$, and $\mathbf{v}_G$ is the absolute velocity of the mass center of the system of particles.

19-78 For a system of n particles, show that the total external moment about an arbitrary (moving) point P can be written

$$\Sigma\, \mathbf{M}_P = \dot{\mathbf{H}}_G + \mathbf{r}_{G/P} \times m\mathbf{a}_G$$

where $\Sigma\, \mathbf{M}_P = \Sigma\, \mathbf{r}_{i/P} \times \mathbf{F}_i$, $\mathbf{H}_G = \Sigma\, \mathbf{r}_{i/G} \times m_i\, \mathbf{v}_i$, $m = \Sigma\, m_i$, and $\mathbf{a}_G$ is the absolute acceleration of the mass center of the system of particles.

19-6   SYSTEMS WITH VARIABLE MASS

The kinetics principles developed in the past several chapters apply only to constant systems of particles—systems that neither gain nor lose particles. Many important dynamics problems, however, consist of large systems of particles in which individual particles are not easily identified (such as fluid flow). For these types of problems, it is often more convenient to study the particles in a fixed region of space—a *control volume*—than to study a fixed system of particles. Two of the more common types of variable-mass problems are:

1.   **Steady Flow of Mass.**   In many fluid flow problems, fluid particles enter and leave a control volume at the same rate. Although the mass (total number of particles) of fluid in the control volume at any time is constant, the particles that make up this mass are constantly changing. Since the particles entering have different momenta than the particles exiting the control volume, external forces must be exerted on the control volume even though the total momentum of the particles within the control volume does not change with time. This type of problem is considered in detail in Section 19-6-1.

2.   **Systems Gaining or Losing Mass.**   In many problems encountered in dynamics, particles leave (or enter) a control volume at a constant rate over some interval of time. For example, in rocket propulsion, the control volume may consist of the rocket shell and the unburned fuel. As the engine burns fuel, the fuel is expelled from the control volume. Not only does the system decrease in mass as particles are taken away from the system, but the particles are expelled at some velocity relative to the rest of the system. Therefore, the total momentum of the system may vary even in the absence of any external applied force. This type of problem is considered in detail in Section 19-6-3.

19-6-1   Steady Flow of Mass

A knowledge of the forces exerted on fan and turbine blades by a steadily moving fluid stream is important in the analysis of many machines. A complete analysis of such problems will be presented in a course in fluid mechanics. The presentation here is just to illustrate how the momentum principles developed in this chapter apply to such steady flow problems.

Consider the problem of finding the force exerted on a fixed reducing bend in a pipe as a steady stream of fluid passes through it as shown in Fig. 19-26. Fluid enters the bend with some velocity $\mathbf{v}_1$, pressure p_1, and density (mass per unit volume) ρ_1, which are assumed to be constant over the inlet area A_1. The fluid then leaves the bend, with velocity $\mathbf{v}_2$, pressure p_2, and density ρ_2 also assumed constant over the exit area A_2. The flow is assumed steady; that is, there is no increase or decrease of fluid inside the bend. Therefore, the rate at which fluid leaves the bend is exactly the same as the rate at which fluid enters the bend.

A control volume $\mathcal{CV}$ is drawn that encloses a region of fluid bounded by the surface on which the force is desired and surfaces on which the forces are known or can be determined. Furthermore, the

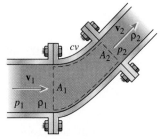

Fig. 19-26

surfaces bounding the control volume are chosen so that the rate of fluid flow across the surface is either zero or is known or can be easily determined. The system of particles enclosed by the control volume is a variable mass system, since it continually gains particles flowing in and loses an equal number of particles flowing out. Therefore, the momentum principles developed earlier in this chapter for fixed systems of particles do not apply directly to the mass that makes up the control volume.

In order to get a fixed system of particles to which the momentum principles do apply, consider the larger group of particles shown in Fig. 19-27a. This system consists of the particles in the original control volume at time t (having total mass m_t) plus the particles that will enter the control volume in the time interval Δt (having total mass Δm_1). Since all particles within a distance $\Delta s_1 = v_1 \Delta t$ will enter the bend in the time Δt, the volume of the additional region is $V_1 = A_1 \Delta s_1$.[1] Then the total mass $\mathcal{M}$ of this larger group of particles is

$$\mathcal{M} = m_t + \Delta m_1 = m_t + \rho_1 V_1 = m_t + \rho_1 A_1 v_1 \Delta t \qquad (19\text{-}24)$$

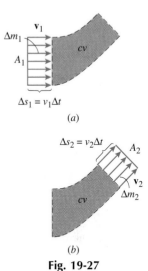

(a)

(b)

Fig. 19-27

At time $t + \Delta t$ this same system of particles will occupy the region shown in Fig. 19-27b. This region consists of the particles in the original control volume at time $t + \Delta t$ (having total mass $m_{t+\Delta t}$) plus those particles that have left the control volume during the time Δt (having total mass Δm_2). The mass of the fixed system of particles is now given by

$$\mathcal{M} = m_{t+\Delta t} + \Delta m_2 = m_{t+\Delta t} + \rho_2 V_2 = m_{t+\Delta t} + \rho_2 A_2 v_2 \Delta t \quad (19\text{-}25)$$

But by the assumption of steady flow, the mass of the particles inside the original control volume is the same at all times; that is, $m_t = m_{t+\Delta t}$. Therefore, combining Eqs. 19-24 and 19-25 gives in the limit as $\Delta t \to 0$

$$\rho_1 A_1 v_1 = \rho_2 A_2 v_2 \qquad (19\text{-}26)$$

which verifies the earlier statement that the rate at which fluid leaves the bend is exactly the same as the rate at which fluid enters the bend.

Equation 19-26 expresses the principle of *conservation of mass*. The terms in the equation are called the *mass flow rate* $\dot{m} = \rho v A$ and represent the rate at which mass is entering or leaving the control volume. In SI units, the mass flow rate has dimensions of kg/s. In U. S. customary units, the mass flow rate has dimensions of slug/s = lb·s/ft. For the flow of incompressible fluids (fluids for which the density is constant) as well as other constant density flows, the *volume flow rate*

$$Q = \frac{\dot{m}}{\rho} = A_1 v_1 = A_2 v_2$$

is often used instead of the mass flow rate. The volume flow rate has dimensions of m³/s in the SI system of units and ft³/s in the U. S. customary system of units.

[1] It is assumed here that the area A_1 was chosen perpendicular to the velocity so that the region is a right cylinder. If the velocity is not perpendicular to the area, then the formula for the volume will include the *sine* of the angle between the velocity and the plane of the area.

At time t the fixed system of particles identified above has linear momentum

$$\mathbf{L}_t = \Delta m_1 \mathbf{v}_1 + (\mathbf{L}_{\mathscr{CV}})_t = (\dot{m}\,\Delta t)\mathbf{v}_1 + (\mathbf{L}_{\mathscr{CV}})_t$$

where $(\mathbf{L}_{\mathscr{CV}})_t$ is the linear momentum of all particles in the control volume at time t and $\Delta m_1 \mathbf{v}_1$ is the linear momentum of the particles that are about to enter the control volume at time t. At time $t + \Delta t$ the same system of particles will have linear momentum

$$\mathbf{L}_{t+\Delta t} = (\mathbf{L}_{\mathscr{CV}})_{t+\Delta t} + \Delta m_2\,\mathbf{v}_2 = (\mathbf{L}_{\mathscr{CV}})_{t+\Delta t} + (\dot{m}\,\Delta t)\mathbf{v}_2$$

But because the flow is steady, $(\mathbf{L}_{\mathscr{CV}})_t = (\mathbf{L}_{\mathscr{CV}})_{t+\Delta t}$ and the linear momentum equation (Eq. 19-6)

$$\mathbf{L}_t + \Sigma\,(\!\int \mathbf{F}\,dt) = \mathbf{L}_{t+\Delta t}$$

gives in the limit as $\Delta t \rightarrow 0$

$$(\dot{m}\,\Delta t)\mathbf{v}_1 + \Sigma\,\mathbf{F}\,\Delta t = (\dot{m}\,\Delta t)\mathbf{v}_2$$

or

$$\Sigma\,\mathbf{F} = \dot{m}(\mathbf{v}_2 - \mathbf{v}_1) \qquad (19\text{-}27)$$

where $\Sigma\,\mathbf{F}$ is the sum of all external forces acting on the system of particles inside the control volume.

In Eq. 19-27 it is important to include *all* external forces acting on the system of particles inside the control volume. Therefore, a correct free-body diagram is just as important for these fluid flow problems as it is for any other particle or rigid body problem. Referring to the free-body diagram of the control volume (Fig. 19-28),

$$\Sigma\,\mathbf{F} = \mathbf{F}_p + \mathbf{W} - p_1 A_1 \mathbf{n}_1 - p_2 A_2 \mathbf{n}_2$$

where $\mathbf{F}_p$ is the force exerted by the pipe bend on the fluid in the control volume (the fluid exerts an equal but opposite force back on the pipe bend); $\mathbf{W}$ is the weight of the fluid in the control volume; $\mathbf{n}_1$ and $\mathbf{n}_2$ are the outward-pointing unit normals to the surfaces A_1 and A_2, respectively, and $-p_1 A_1 \mathbf{n}_1$ and $-p_2 A_2 \mathbf{n}_2$ are the forces exerted on the fluid in the control volume by the adjoining portions of the fluid.

An analogous result can be obtained using the angular momentum principle Eq. 19-20 (or Eq. 19-23). Taking the angular momenta and the moments of all external forces with respect to some arbitrary fixed point O (or about the mass center G), this gives

$$\mathbf{r}_1 \times [(\dot{m}\,\Delta t)\mathbf{v}_1] + \Sigma\,\mathbf{M}_O\,\Delta t = \mathbf{r}_2 \times [(\dot{m}\,\Delta t)\mathbf{v}_2]$$

or

$$\Sigma\,\mathbf{M}_O = \dot{m}(\mathbf{r}_2 \times \mathbf{v}_2 - \mathbf{r}_1 \times \mathbf{v}_1) \qquad (19\text{-}28)$$

where $\Sigma\,\mathbf{M}_O = \Sigma\,(\mathbf{r} \times \mathbf{F})$ is the sum of the moments of all external forces acting on the fluid inside the control volume, and $\mathbf{r}_1$ and $\mathbf{r}_2$ are the position vectors of the centers of the areas A_1 and A_2, respectively. The angular momenta and moments of all forces are to be computed relative to the same fixed point O (or relative to the mass center G).

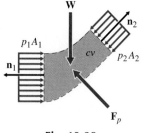

Fig. 19-28

19-6-2   Common Applications of Steady Flow

Equations 19-27 and 19-28 can be used to solve a wide variety of fluid flow problems, including the reducing elbow of Fig. 19-26 and the situations depicted in Fig. 19-29. Some considerations for the application of these principles to some of the more common classes of flows are as follows:

1. **Enclosed Flows.** In flows through pipes (Fig. 19-26), pipe bends (Fig. 19-29a), and nozzles (Fig. 19-29b), the cross-sectional area of the flow is presumed known at any section needed, and the fluid velocities can be obtained from the flow rate $\dot{m}$ or Q. The fluid pressure in the pipe will, in general, not be negligible or constant and must either be given or be determined from other fluid mechanics principles. The weight of the fluid is usually negligible compared to other forces in the problem unless the control volume is very large. If the pipe length is very long and/or if the pipe diameter is very small, fluid friction on the side walls of the pipe (not shown in the free-body diagram of Fig. 19-28) may also have to be included.

2. **Channel Flows.** In flows of water under a sluice gate (Fig. 19-29c) or over a weir (Fig. 19-29d), the fluid pressure is very important

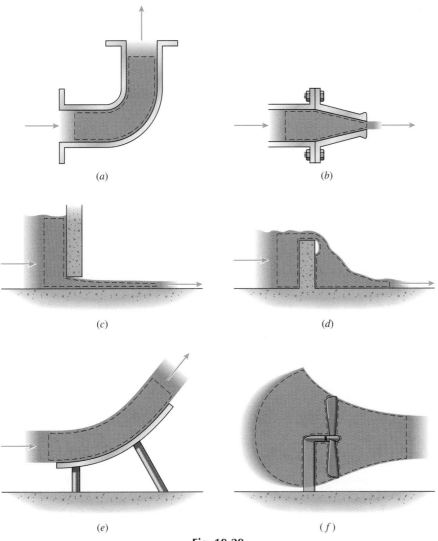

(a) (b)

(c) (d)

(e) (f)

Fig. 19-29

and cannot be neglected. It is shown in fluid mechanics that in a region of the flow where the streamlines are straight and parallel that the fluid pressure increases linearly with depth $p = \rho g h$, where h is depth of a point below the surface of the fluid. Although the weight would be needed to find the force on the bottom of the channel, it is not needed to find the force on vertical surfaces such as the sluice gate or the weir shown in Figs. 19-29c and 19-29d. Fluid friction on the bottom or sides or both of the channel are seldom significant compared to the other forces in the problem.

3. **Free Jet Flows.**   It is shown in fluid mechanics that the pressure in free jets (fluid flow that is not contained by pipe or channel walls) is the same as that of the surrounding fluid. For the fluid being deflected by the fixed vane of Fig. 19-29e, this means that the pressure in the jet of water approaching the vane and the pressure in the jet of water leaving the vane are both zero. It is further shown for flows such as this that the speed of the jet leaving the vane is the same as that of the jet entering the vane. The fluid weight and frictional forces on the vane are seldom significant.

4. **Stationary Fans.**   When air or water passes through a fan, the velocity increases from one side of the blades to the other. Except for a region near the fan, the fluid approaching the fan and the jet of fluid leaving the fan of Fig. 19-29f can be treated as free jets: the pressure can be neglected. While the jet of fluid expelled by the fan (called the slipstream) is usually concentrated and of uniform velocity, the air approaching the fan is usually more dispersed: the intake area is very large and the intake velocity can be neglected.

5. **Moving Vanes and Propellers.**   Flows around moving vanes and moving propellers are not steady flows and the foregoing equations do not directly apply. However, if the vanes or propellers are moving in a straight line with a constant speed, the flow will appear steady to an observer moving with the vane or propeller. Therefore, these problems can be converted to a steady flow situation and the foregoing equations applied by choosing a coordinate system moving with the vane or propeller. The velocities (and therefore the flowrates!) must be expressed relative to the moving coordinate system. If the vane or propeller are not moving in a straight line with a constant speed, then alternative equations must be developed that properly take into account the acceleration of the flow/coordinate system.

19-6-3   Systems Gaining or Losing Mass

The other type of variable mass system that will be analyzed here is the system that gains mass by collecting particles (such as a moving container being filled with water or grain) or that loses mass by expelling particles (such as a rocket burning fuel). The general procedure will be developed for a system that is acquiring mass but will apply equally to both situations.

Consider a body that is absorbing a stream of particles as shown in Fig. 19-30. The body is obviously a variable mass system and the momentum principles developed earlier in this chapter do not directly apply. Instead, define a system of particles consisting of the body (which at time t has mass m and velocity $\mathbf{v}$) and the particle(s)

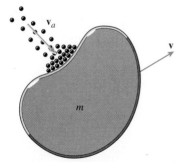

Fig. 19-30

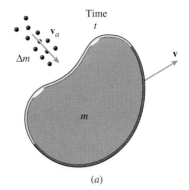

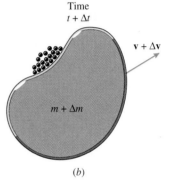

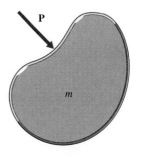

(having total mass Δm and velocity $\mathbf{v}_a$) that will be absorbed in the time Δt (Fig. 19-31). This larger system is a fixed system of particles over the time interval Δt, and the momentum principles can be used to relate the forces acting on the system and the change in momentum of the system.

At time t the momentum of the system is just the sum of the momenta of the parts (Fig. 19-31a)

$$L_t = m\mathbf{v} + \Delta m\, \mathbf{v}_a$$

while the momentum after the particle(s) has (have) been absorbed and all the mass is moving as a single object of mass $m + \Delta m$ and velocity $\mathbf{v} + \Delta \mathbf{v}$ (Fig. 19-31b) is

$$L_{t+\Delta t} = (m + \Delta m)(\mathbf{v} + \Delta \mathbf{v}) = m\mathbf{v} + \Delta m\, \mathbf{v} + m\, \Delta \mathbf{v} + \Delta m\, \Delta \mathbf{v}$$

Then if $\mathbf{R}$ is the resultant of all *external* forces acting on the system, the linear momentum principle (Eq. 19-6) gives

$$m\mathbf{v} + \Delta m\, \mathbf{v}_a + \int \mathbf{R}\, dt = m\mathbf{v} + \Delta m\, \mathbf{v} + m\, \Delta \mathbf{v} + \Delta m\, \Delta \mathbf{v}$$

Dividing through by Δt and rearranging this expression gives

$$\frac{1}{\Delta t}\int_t^{t+\Delta t} \mathbf{R}\, dt = m\frac{\Delta \mathbf{v}}{\Delta t} - \frac{\Delta m}{\Delta t}(\mathbf{v}_a - \mathbf{v}) + \frac{\Delta m\, \Delta \mathbf{v}}{\Delta t}$$

which in the limit as $\Delta t \to 0$ becomes

$$\mathbf{R} = m\mathbf{a} - \dot{m}(\mathbf{v}_a - \mathbf{v}) = m\mathbf{a} - \dot{m}\, \mathbf{v}_{a/m} \qquad (19\text{-}29)$$

where $\mathbf{a} = \dot{\mathbf{v}} = \lim_{\Delta t \to 0} \Delta \mathbf{v}/\Delta t$ is the acceleration of the body due to the action of the external forces and the absorbed particles; $\dot{m} = \lim_{\Delta t \to 0} \Delta m/\Delta t$ is the rate at which the body is absorbing mass from the particle stream; $\mathbf{v}_{a/m} = \mathbf{v}_a - \mathbf{v}$ is the relative velocity of the absorbed particles with respect to the body; and $\lim_{\Delta t \to 0} \Delta m\, \Delta \mathbf{v}/\Delta t = \lim_{\Delta t \to 0} \dot{m}\, \Delta \mathbf{v} = \mathbf{0}$.

The force $\mathbf{R}$ represents the resultant of *all external forces* that act on the system—the body and the piece of mass being absorbed. However, this force resultant *does not include* the forces of action and reaction $\mathbf{P}$ (Fig. 19-32) between the body and the mass being absorbed since these are internal to the system.

It is interesting to compare Eq. 19-29 with Newton's second law of motion. For example, Newton's second law of motion might be written in the form (the resultant force equals the change in linear momentum)

$$\mathbf{R} = \frac{d\,(m\mathbf{v})}{dt} = \dot{m}\mathbf{v} + m\dot{\mathbf{v}}$$

Equation 19-29 agrees with Newton's law in this form only if $\mathbf{v}_a = \mathbf{0}$; that is, the acquired mass is at rest before it is picked up.

If Eq. 19-29 is compared with Newton's second law of motion written in its usual form, $\Sigma \mathbf{F} = m\mathbf{a}$, it is convenient to rearrange the terms of Eq. 19-29

$$\mathbf{R} + \dot{m}\, \mathbf{v}_{a/m} = m\mathbf{a} \qquad (19\text{-}30)$$

That is, the effect on the body of the particles being absorbed is the

Fig. 19-31

Fig. 19-32

same as that of a force in the direction of the relative velocity of magnitude $\dot{m}v_{a/m}$. In fact, applying the linear momentum principle to the single particle Δm (Fig. 19-32a) gives

$$\Delta m\, \mathbf{v}_a + \int (-\mathbf{P})\, dt = \Delta m\, (\mathbf{v} + \Delta \mathbf{v})$$

which after rearranging, dividing through by Δt, and letting $\Delta t \to 0$ gives

$$\mathbf{P} = \dot{m}\,(\mathbf{v}_a - \mathbf{v}) = \dot{m}\mathbf{v}_{a/m} \qquad (19\text{-}31)$$

This "effective force" will tend to accelerate the body if the particles are added "from behind" or will tend to decelerate the body if the particles are added "from in front" of the body.

Equation 19-29 may also be used for a body expelling mass such as a rocket burning fuel. In this case, the mass flow rate $\dot{m}$ is negative.[2] Then, by Eqs. 19-30 and 19-31, the effect on the body of the particles being expelled will be the same as that of a force in the direction opposite that of the relative velocity and of magnitude $P = |\dot{m}\,\mathbf{v}_{a/m}|$. That is, particles expelled "from the rear" will tend to accelerate the body whereas particles expelled "from the front" will tend to decelerate the body. This is the mechanism of propulsion by rockets.

19-6-4   Special Cases of Systems Gaining or Losing Mass

Equations 19-29 through 19-31 may be applied to a wide variety of systems gaining or losing mass from rockets to hoisting cables. The equations simplify for some special cases such as:

1. **A Rocket Sled.**   When a rocket sled accelerates horizontally along a straight track, the weight and track reaction forces will be normal to the velocity and relative velocity. The drag force due to aerodynamic forces is generally proportional to the square of the speed of the rocket kv^2. Therefore, the component of Eq. 19-30 in the direction of the sled's motion is

$$(m_0 - bt)\,\dot{v} = bu - kv^2 \qquad (19\text{-}32)$$

    where b is the constant rate at which the rocket is burning fuel and u is the velocity of the burned gases relative to the sled. If the rocket thrust P is known rather than the mass flow rate and relative velocity, Eq. 19-31 can be combined with Eq. 19-30 to give

$$(m_0 - bt)\,\dot{v} = P - kv^2 \qquad (19\text{-}33)$$

    Equation 19-32 or 19-33 is then solved to find the speed of the sled as a function of time.

2. **All External Forces Are Zero.**   When a space ship travels in outer space, it encounters no air resistance. If the spaceship is also far from any planets or stars, then any gravitational force acting on the space ship is also negligible so that $\mathbf{R} = \mathbf{0}$ and Eq. 19-29 becomes

$$m\mathbf{a} = \dot{m}\mathbf{v}_{a/m}$$

---

[2] That is, for a rocket burning fuel at a constant rate b the rate of change of mass $\dot{m} = -b$. Then, if the initial mass of the rocket is m_0, the mass of the rocket at time t will be $m = m_0 + \dot{m}t = m_0 - bt$. Furthermore, if the velocity of the burned gases with respect to the rocket is $\mathbf{u} = \mathbf{v}_{a/m}$, then the thrust on the rocket, $\mathbf{P} = \dot{m}\mathbf{v}_{a/m} = -b\mathbf{u}$ will be in the direction opposite the relative velocity $\mathbf{u}$.

EXAMPLE PROBLEM 19-11

Water is being expelled from a nozzle at a constant rate of 500 gal/min as shown in Fig. 19-33a. The nozzle is attached to a 4-in. diameter pipe with six bolts and has an exit diameter of 2 in. If the pressure measured in the pipe is 16.45 lb/in.², determine the force in each bolt. (The specific weight of water is 62.4 lb/ft³ and 7.481 gal = 1 ft³.)

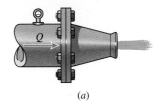

(a)

SOLUTION

The flow rate is given as

$$Q = \frac{500 \text{ gal/min}}{(7.481 \text{ gal/ft}^3)(60 \text{ s/min})}$$

$$= 1.114 \text{ ft}^3/\text{s}$$

$$= v_1 A_1 = v_2 A_2$$

from which the velocities are determined to be

$$v_1 = 12.76 \text{ ft/s}$$

and

$$v_2 = 51.06 \text{ ft/s}$$

Applying the x-component of Eq. 19-27 to the free-body diagram of the water contained in the nozzle (Fig. 19-33b) gives

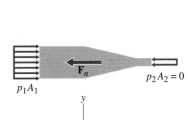

(b)

$$p_1 A_1 - F_n = \rho Q(v_{2x} - v_{1x})$$

or

$$(16.45)\frac{\pi}{4}(4)^2 - F_n = \frac{62.4}{32.2}(1.114)(51.06 - 12.76)$$

which gives

$$F_n = 124.03 \text{ lb}$$

as the force exerted on the water by the nozzle. The water exerts an equal but opposite force back on the nozzle. Then from the free-body diagram of the nozzle (Fig. 19-33c), equilibrium gives the tension in the bolts as

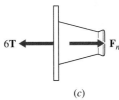

(c)

Fig. 19-33

$$T = \frac{F_n}{6} = 20.67 \text{ lb} \qquad \text{Ans.}$$

Water having a density $\rho = 1000$ kg/m³ flows under a sluice gate as shown in Fig. 19-34a. The width of the channel is 2 m and the flowrate is $Q = 3.10$ m³/s. Compare the force exerted on the gate by the flowing water with the force the water would exert on the gate if it were not moving.

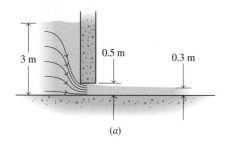

(a)

SOLUTION

The free-body diagram of the water pressing on the gate (Fig. 19-34b) includes the pressure forces of the adjacent water in the channel $\mathbf{F}_1$ and $\mathbf{F}_2$. The magnitudes of these forces are equal to the areas under the pressure loading diagrams (Fig. 19-34c, 19-34d)

$$F_1 = 0.5[(1000)(9.81)(3)](3)(2) = 88290 \text{ N}$$
$$F_2 = 0.5[(1000)(9.81)(0.3)](0.3)(2) = 882.9 \text{ N}$$

The mass flowrate and velocities of the water are obtained from the volume flowrate $\dot{m} = \rho Q = \rho v_1 A_1 = \rho v_2 A_2$ giving

$$\dot{m} = (1000)(3.10) = 3100 \text{ kg/s}$$
$$v_1 = \frac{3.10}{(3)(2)} = 0.5167 \text{ m/s}$$

and

$$v_2 = \frac{3.10}{(0.3)(2)} = 5.167 \text{ m/s}.$$

Then applying the x-component of Eq. 19-27

$$88290 - 882.9 - F_g = (3100)(5.167 - 0.5167)$$

gives the force exerted on the water by the sluice gate

$$\mathbf{F}_g = 73000 \text{ N} \leftarrow \quad \text{on the water}$$

The water exerts an equal and opposite force back on the sluice gate:

$$\mathbf{F}_g = 73000 \text{ N} \rightarrow \quad \text{on the gate} \qquad \text{Ans.}$$

If the fluid were not moving, it would exert a pressure force on the gate that increases linearly with depth as shown in Fig. 19-34e. The magnitude of this force is equal to the area under the pressure-loading diagram

$$F_{gs} = 0.5[(1000)(9.81)(2.5)](2.5)(2)$$
$$= 61,300 \text{ N} \rightarrow \quad \text{on the gate} \qquad \text{Ans.}$$

The difference between these two answers is that the streamlines of the flow near the gate are not straight and parallel. Therefore, the pressure does not vary linearly with depth in this region.

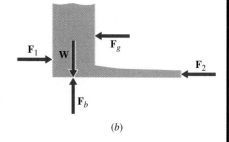

(b)

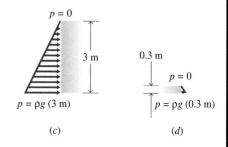

(c) $\qquad$ (d)

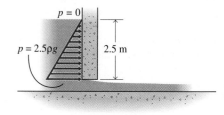

(e)

Fig. 19-34

EXAMPLE PROBLEM 19-13

A jet of water ($\gamma = \rho g = 62.4$ lb/ft^3) is deflected by a turning vane as shown in Fig. 19-35a. The water jet has an absolute velocity of 30 ft/s and a diameter of 1 in. If the turning angle of the vane is 50°, determine the horizontal force **P** required to move the vane to the left at a steady rate of 10 ft/s.

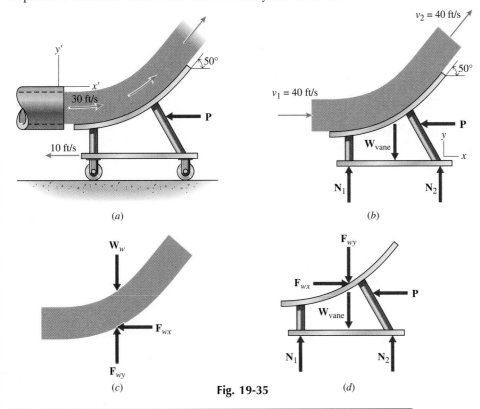

Fig. 19-35

SOLUTION

In a coordinate system moving to the left with the vane (Fig. 19-35b), the flow is steady and Eq. 19-27 applies. In this coordinate system, the water appears to be approaching the vane at a speed of 40 ft/s and the mass flow rate is

$$\rho Q = \rho v_1 A_1 = \left(\frac{62.4}{32.2}\right)(40)\left(\frac{\pi}{4}\right)\left(\frac{1}{12}\right)^2 = 0.4228 \text{ slug/s}$$

In free-jet problems such as this, the speed of the water leaving the vane is the same as the speed of the water approaching the vane, 40 ft/s. Therefore, applying the x-component of Eq. 19-27 to the free-body diagram of Fig. 19-35c gives the force exerted on the water by the vane

$$-F_{wx} = \rho Q(v_{2x} - v_{1x}) = (0.4228)(40 \cos 50 - 40)$$
$$\mathbf{F}_{wx} = 6.04 \text{ lb} \leftarrow \quad \text{on the water}$$

The vane exerts an equal but opposite force back on the vane

$$\mathbf{F}_{wx} = 6.04 \text{ lb} \rightarrow \quad \text{on the vane}$$

Finally, applying the equilibrium equation to the vane gives the force required to keep the vane moving to the left at a constant speed

$$\mathbf{P} = 6.04 \text{ lb} \leftarrow \quad \text{on the vane} \qquad\qquad \text{Ans.}$$

A 1000-kg rocket sled accelerates from rest along a horizontal track. The rocket motor burns fuel at the rate of 15 kg/s, and the velocity of the exhaust gas relative to the sled is 3500 m/s. If 200 kg of the rocket's initial weight is fuel, ignore aerodynamic drag and determine:

a. The initial acceleration of the sled.
b. The thrust exerted on the sled by the rocket motor.
c. The velocity and acceleration of the sled an instant before the rocket motor burns out.

SOLUTION

a. Since there are no external forces acting on the rocket sled in the horizontal direction, the horizontal component of Eq. 19-29 gives

$$R_x = 0 = ma_x - \dot{m}\, v_{a/m}$$

where $m = 1000 + \dot{m}t$, $\dot{m} = -15$ kg/s, and $v_{a/m} = -3500$ m/s. Solving for the initial acceleration ($t = 0$) gives

$$a_x = \frac{-15}{1000}(-3500) = 52.5 \text{ m/s}^2$$
$$= 5.35\, g \qquad\qquad \text{Ans.}$$

b. The thrust on the rocket sled is given by Eq. 19-31

$$P = \dot{m}\, v_{a/m} = (-15)(-3500)$$
$$= 52500 \text{ N} \qquad\qquad \text{Ans.}$$

c. The mass of the rocket sled at the instant that the fuel is burnt up will be 800 kg. The acceleration will then be

$$a_x = \frac{-15}{800}(-3500) = 65.6 \text{ m/s}^2$$
$$= 6.69\, g \qquad\qquad \text{Ans.}$$

The velocity of the rocket sled is obtained by integrating the acceleration

$$\frac{dv}{dt} = a_x = \frac{-15}{1000 - 15t}(-3500)$$

giving

$$v(t) = 3500 \ln\left(\frac{1000}{1000 - 15t}\right)$$

The rocket will burn up the 200 kg of fuel in 200/15 = 13.33 s. Therefore, the speed of the sled at burnout will be

$$v(13.33) = 781 \text{ m/s} \qquad\qquad \text{Ans.}$$

PROBLEMS

19-79* Water ($\gamma = \rho g = 62.4 \text{ lb/ft}^3$) flows through a 90° bend in a 1-in. diameter pipe at a constant rate of $Q = 4.5 \text{ ft}^3/\text{min}$ (Fig. P19-79). If the pressure in the water is constant at 2 lb/in.^{2}, determine the force (magnitude and direction) exerted on the pipe bend by the water flowing through it. (Assume that the bend is in a horizontal plane and neglect the weight of water in the pipe.)

Fig. P19-79

19-80* Water ($\rho = 1000 \text{ kg/m}^3$) flows through a 40-mm diameter pipe at a constant rate of $Q = 0.15 \text{ m}^3/\text{min}$ (Fig. P19-80). A nozzle attached to the pipe has an exit diameter of 20 mm and bends the flow through a 30° angle. If the nozzle is attached to the pipe using four bolts and the pressure in the pipe just before the nozzle is 29,680 N/m^2, determine the average tension in the four bolts. (Assume that the bend is in a horizontal plane and neglect the weight of water in the nozzle.)

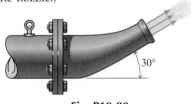

Fig. P19-80

19-81 If the nozzle of Example Problem 19-11 makes a 180° bend (in a horizontal plane) as shown in Fig. P19-81, determine the average tension in each of the six bolts holding the nozzle to the pipe.

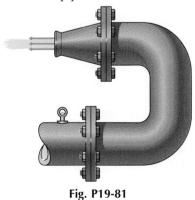

Fig. P19-81

19-82* Water ($\rho = 1000 \text{ kg/m}^3$) flows over a sharp crested weir at a constant rate of $Q = 3.33 \text{ m}^3/\text{s}$ as shown in Fig. P19-82. If the channel is 5 m wide, determine the horizontal component of force exerted on the weir by the water.

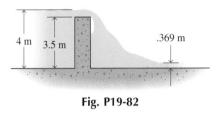

Fig. P19-82

19-83 Water ($\gamma = \rho g = 62.4 \text{ lb/ft}^3$) flows under an inclined sluice gate at a constant rate of $Q = 125 \text{ ft}^3/\text{s}$ as shown in Fig. P19-83. The width of the channel is 10 ft. Given that the force of the water on the gate is perpendicular to the gate, determine the magnitude of the force exerted on the gate by the water.

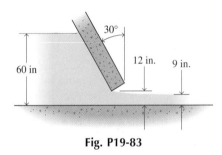

Fig. P19-83

19-84 A 25-mm diameter jet of water ($\rho = 1000 \text{ kg/m}^3$) is deflected through an angle of 50° by a turning vane as shown in Fig. P19-84. The combined mass of the vane and its base is 10 kg. If the coefficient of static friction between the base and the floor is $\mu_s = 0.25$, determine the maximum velocity of the jet of water for which the vane will not move.

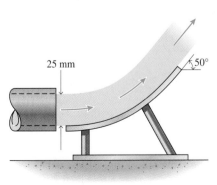

Fig. P19-84

19-85* A child's toy consists of a clown head attached to a garden hose that shoots a jet of water ($\gamma = \rho g = 62.4$ lb/ft^3) vertically upward, suspending a cone-shaped hat as shown in Fig. P19-85. The weight of the hat is 0.5 lb and the water coming out of the hat makes an angle of 30° with the vertical. If the diameter of the jet entering the hat is 0.5 in., determine the flowrate required to support the hat.

Fig. P19-85

19-86 A 40-mm diameter jet of water ($\rho = 1000$ kg/m^3) strikes the middle of a plate that is mounted on four identical springs ($\ell = 1500$ N/m each) as shown in Fig. P19-86. If the incoming jet has a speed of 5 m/s and the water leaving the plate flows radially outward (perpendicular to the direction of the incoming jet of water), determine the deflection of the springs due to the force of the water.

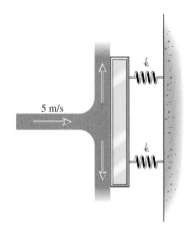

Fig. P19-86

19-87* The table fan of Fig. P19-87 exhausts a 10-in. diameter jet of air ($\gamma = \rho g = 0.0749$ lb/ft^3) having a speed of 20 ft/s. If the fan weighs 5 lb, determine the minimum coef-

ficient of friction between the fan and the table for which the fan will not slide.

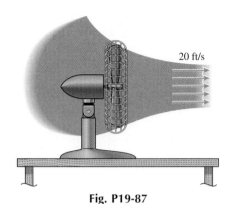

Fig. P19-87

19-88* A 2300-kg helicopter has a slipstream diameter of 10 m. Determine the speed of the air ($\rho = 1.225$ kg/m^3) in the slipstream when the helicopter is hovering.

19-89 A 20,000-lb spacecraft increases its orbital speed using an orbit maneuvering engine. The engine burns fuel at the rate of 50 lb/s, and the velocity of the exhaust gas relative to the spacecraft is 10,000 ft/s. If the initial speed of the spacecraft is 26,000 ft/s, determine:

a. The initial thrust exerted on the spacecraft by the engine.
b. The initial acceleration of the spacecraft.
c. The velocity of the spacecraft after the engine has burned for 10 s.

19-90* A 2200-kg spacecraft fires its retrorocket to decrease its orbital speed. The engine burns fuel at the rate of 10 kg/s, and the velocity of the exhaust gas relative to the spacecraft is 2700 m/s. If the initial speed of the spacecraft is 8000 m/s, determine:

a. The initial thrust exerted on the spacecraft by the engine.
b. The initial deceleration of the spacecraft.
c. The velocity of the spacecraft after 800 kg of fuel have been burned.

19-91 The first stage of a two-stage rocket weighs 100 lb when empty, carries 750 lb of fuel, burns fuel at the rate of 20 lb/s, and has an exhaust velocity of 7000 ft/s. When the first stage's fuel is used up, its shell is discarded and the second stage is ignited. The second stage weighs 75 lb when empty, carries 550 lb of fuel, burns fuel at the rate of 15 lb/s, and has an exhaust velocity of 7000 ft/s. If this rocket is used to launch a 50 lb payload, determine:

a. The initial thrust exerted on the payload.
b. The velocity of the rocket when the first stage burns out and is discarded.
c. The maximum velocity attained by the payload.

19-92   A small model of a rocket sled has a mass of 5 kg including 4 kg of fuel. The sled starts from rest, and its fuel is exhausted at a rate of 0.5 kg/s and a velocity of 150 m/s. If air resistance on the model is proportional to its speed, $F_D = 0.3v$ where v is in meters per second and F_D is in Newtons, determine the maximum speed attained by the sled when it is fired along a horizontal, frictionless track.

19-93*   Suppose that the two stage rocket of Problem 19-91 is replaced with a single-stage rocket that weighs 175 lb when empty, carries 1300 lb of fuel, burns fuel at the rate of 20 lb/s, and has an exhaust velocity of 7000 ft/s. Determine the initial thrust on the payload and the maximum velocity attained by the payload.

19-94   An 18,000-kg railcar is being filled with grain at the rate of 2000 kg/s (Fig. P19-94). If the grain enters the car at an angle of 60° and a speed of 6 m/s, determine the initial acceleration of the railcar.

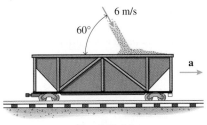

Fig. P19-94

19-95*   A 40,000-lb railcar is being filled with grain at the rate of 5000 lb/s (Fig. P19-95). If the grain is falling straight downward, determine the horizontal force **P** necessary to keep the car moving horizontally at a speed of 1 ft/s.

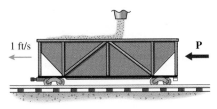

Fig. P19-95

19-96*   Sand ($\rho = 1860$ kg/m³) is flowing out of a dump truck at a rate of 0.7 m³/s (Fig. P19-96). The initial mass of the truck and the sand is 20,000 kg. If the sand is discharged at an angle of 40° to the horizontal and a speed of 5 m/s relative to the truck, determine the force (magnitude and direction) necessary to keep the truck moving forward at a constant speed of 0.6 m/s.

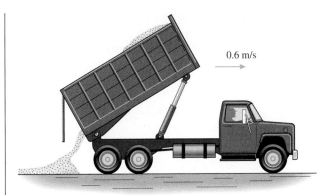

Fig. P19-96

19-97   A chain weighing 0.5 lb/ft is being raised at a constant rate of 8 ft/s by a force **F** (Fig. P19-97). Determine the magnitude of the force **F** when:

a. $y = 1$ ft.
b. $y = 4$ ft.
c. $y = 8$ ft.

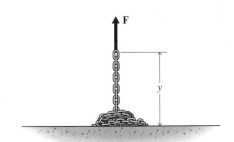

Fig. P19-97

19-98*   An 8-m long chain having a total mass of 4 kg is being pulled along a frictionless horizontal surface at a constant rate of 2 m/s by a force **F** (Fig. P19-98). Determine the magnitude of the force **F** when:

a. $y = 1$ m.
b. $y = 3$ m.
c. $y = 6$ m.

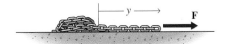

Fig. P19-98

19-99   A 14-ft long chain having a total weight of 21 lb is being raised by a constant force $F = 9$ lb (Fig. P19-97). If the chain starts from rest when $y = 1$ ft, determine:

a.   The speed of the chain when $y = 5$ ft.
b.   The maximum speed attained by the chain.
c.   The maximum height attained by the chain.

19-100 A 6-m long chain having a density of 0.5 kg/m is loosely coiled on the floor (Fig. P19-100). The upper end of the chain is attached to a light cord, which passes over a small, frictionless pulley; the other end of the cord is tied to a 1.5-kg block. If the system is released from rest with $y = 1$ m, determine:

a. The maximum upward velocity $\dot{y}$ attained by the upper end of the chain.

b. The maximum height y_{max} reached by the upper end of the chain.

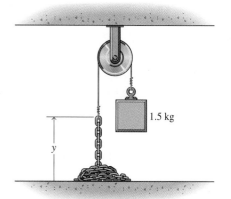

Fig. P19-100

19-101* A 24-ft long chain weighing 0.5 lb/ft is loosely coiled on the floor (Fig. P19-101). The upper end of the chain is attached to a light cord, which passes over a small, frictionless pulley 20 ft above the floor; the other end of the cord is tied to an 8-lb block. If the system is released from rest with $h = 15$ ft and $y = 1$ ft, determine the speed of the block just before it hits the floor.

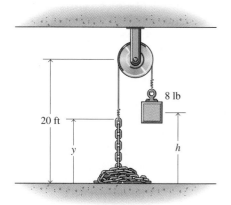

Fig. P19-101

19-102 A long, uniform chain having a density of 0.25 kg/m is piled on a horizontal surface as shown in Fig. P19-102. If one end of the chain falls through a hole, determine the speed $\dot{y}$ of the end of the chain when $y = 3$ m. (Assume that all links are at rest until the moment that they fall through the hole.)

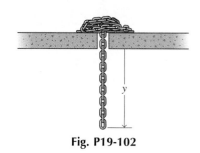

Fig. P19-102

SUMMARY

All of the study of kinetics is based on Newton's second law of motion. In Chapters 15 and 16, Newton's second law was used directly to relate the forces acting on particles and rigid bodies and the resulting acceleration of the particles and the rigid bodies. When information about the acceleration or when the value of a force at an instant is desired, Newton's second law is usually the easiest method to use.

In Chapters 17 and 18, Newton's second law was integrated with respect to position to get the principle of work and energy. Since the principle of work and energy is just a combination of Newton's second law and the principles of kinematics, there is no problem that can be solved using the principle of work and energy that cannot also be solved using Newton's second law. However, the principle of work and energy is particularly useful for solving problems in which the speed of a body for two different positions of its motion are to be related and the forces involved can be expressed as functions of the position of the body.

In this chapter, Newton's second law was integrated with respect to time to get the principle of impulse and momentum. The resulting equations are useful for solving problems in which the velocity of a body for two different instants of time are to be related and the forces involved can be expressed as functions of time. The principles of impulse and momentum were found to be particularly useful for solving problems involving collisions between bodies and variable-mass systems.

The principle of linear impulse and momentum is expressed by

$$\mathbf{L}_i + \int_{t_i}^{t_f} \mathbf{R}\, dt = \mathbf{L}_f$$

The final linear momentum $\mathbf{L}_f = (m\mathbf{v})_f$ of a particle is the vector sum of its initial linear momentum $\mathbf{L}_i = (m\mathbf{v})_i$ and the impulse $\int \mathbf{R}\, dt$ of the resultant of all forces acting on the particle. This equation is a vector equation representing three scalar equations. The three scalar equations are completely independent of each other.

The principle of angular impulse and momentum is expressed by

$$\mathbf{H}_{Oi} + \int_{t_i}^{t_f} \mathbf{M}_O\, dt = \mathbf{H}_{Of} \tag{19-17}$$

that is: *The final angular momentum $\mathbf{H}_{Of} = [\mathbf{r}_{P/O} \times (m\mathbf{v})]_f$ of a particle about a fixed point O is the vector sum of its initial angular momentum $\mathbf{H}_{Oi} = [\mathbf{r}_{P/O} \times (m\mathbf{v})]_i$ about O and the angular impulse $\int \mathbf{M}_O\, dt$ about O of the resultant of all forces acting on the particle during the time interval.* Like the principle of linear impulse and momentum, Eq. 19-17 is a vector equation representing three scalar equations. In planar problems, only the component perpendicular to the plane gives useful information.

An impact or collision between two bodies is an event that occurs in a very brief interval of time. It is usually accompanied by relatively large reaction forces between the two bodies and correspondingly large changes of velocity of one or both bodies.

When two particles collide head-on, the relative velocity of the particles before the impact and after the impact are related by the coefficient of restitution

$$e = -\frac{v_{Bf} - v_{Af}}{v_{Bi} - v_{Ai}} = -\frac{(v_{B/A})_f}{(v_{B/A})_i} \tag{19-11}$$

When two particles collide obliquely, Eq. 19-11 relates the normal components of the relative velocities.

Neither the work–energy method nor the impulse–momentum method is suitable for solving all problems. The maximum benefit of these methods is realized by choosing the particular method most suitable for a particular problem or part of a problem. It is often useful to combine all three methods—impulse–momentum, work–energy, and Newton's second law—to solve particular problems.

REVIEW PROBLEMS

19-103* A 3200-lb car moving at 45 mi/h collides head-on with a stationary 2200-lb car. On impact, the wheels of the cars lock and slide ($\mu_k = 0.5$). If the cars become entangled and move as a single unit after the impact:

a. Estimate the distance the cars move after the collision.
b. Determine the amount of kinetic energy lost in the collision.

19-104* Suppose that the rigid surface of Problem 19-60 is replaced with a 2-kg cart that is free to roll in the horizontal direction as shown in Fig. P19-104. If the coefficient of restitution is 0.8, the 0.5-kg ball starts from rest with $h = 1$ m, and the ball just clears the wall at the peak of its bounce, determine the distances b, c, and d in the figure.

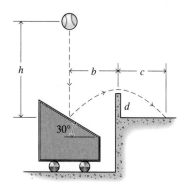

Fig. P19-104

19-105 Sand ($\gamma = \rho g = 120$ lb/ft^3) is flowing out of a dump truck at a rate of 25 ft^3/s (Fig. P19-105). The initial weight of the truck and the sand is 40,000 lb. If the sand is discharged at an angle of 40° to the horizontal and a speed of 15 ft/s relative to the truck, determine the initial acceleration of the truck.

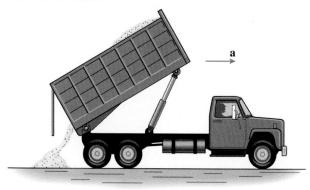

Fig. P19-105

19-106 A ballistic pendulum consists of a 3-kg box of sand suspended from a light 2-m long cord (Fig. P19-106).

A 0.05-kg bullet strikes the box and becomes embedded in the sand. If the maximum angle swing of the pendulum following impact is 25°, determine:

a. The speed of the sand and bullet immediately after the impact.
b. The initial speed v_i of the bullet.

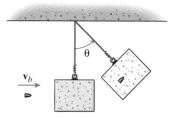

Fig. P19-106

19-107* A 2-lb sphere A is hanging motionless at the end of an inextensible cord as shown in Fig. P19-107 when it is struck by an identical sphere B rolling along the horizontal surface. The distance from the ceiling to the center of the sphere is 3 ft, and the center of the sphere is initially level with the horizontal surface. If the coefficient of restitution is 0.8, determine:

a. The angle that the final velocity of sphere B makes with the horizontal.
b. The maximum angle θ_A through which sphere A will swing as a result of the impact.

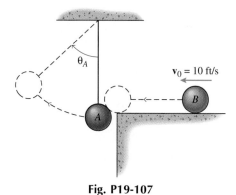

Fig. P19-107

19-108 The orbit maneuvering engine of an 1800-kg spacecraft burns fuel at the rate of 10 kg/s, and the velocity of the exhaust gas relative to the spacecraft is 2700 m/s. Initially, the spacecraft is in a 160-km circular orbit with a velocity of 7810 m/s. If the satellite uses its engine to increase its speed to 8190 m/s, determine:

a. The initial thrust exerted on the spacecraft by the engine.

b. The length of time that the engine must be on if the mass of the spacecraft and the thrust are assumed constant.

c. The length of time that the engine must be on taking into account the decrease in mass of the spacecraft as it burns fuel.

19-109* A small bag of sand weighing 10 lb swings in a vertical x-z plane at the end of a 5-ft long rope as shown in Fig. P19-109. When the bag is at the end of its swing ($\theta = 20°$ and $\dot{\theta} = 0$), it is struck by a 2.1-oz. bullet traveling at 825 ft/s in the y-direction. If the bullet embeds in the sand and the rope remains straight, determine for the ensuing motion:

a. The maximum angle θ_{max} that the rope makes with the vertical.

b. The velocity of the sandbag when $\theta = \theta_{max}$.

c. The tension in the rope when $\theta = \theta_{max}$.

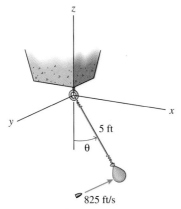

Fig. P19-109

19-110 Two beads are sliding freely on a horizontal rod as shown in Fig. P19-110. Show that the final velocities of the two beads are given by

$$v_{Af} = v_{Ai} - \frac{(1 + e)\, m_B}{m_A + m_B}(v_{Ai} - v_{Bi})$$

$$v_{Bf} = v_{Bi} + \frac{(1 + e)\, m_B}{m_A + m_B}(v_{Ai} - v_{Bi})$$

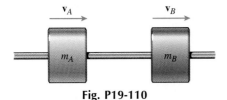

Fig. P19-110

19-111 Two beads are sliding freely on a horizontal rod as shown in Fig. P19-110. Show that the maximum de-

crease in kinetic energy of the system (which corresponds to $e = 0$) is

$$\Delta T = \frac{1}{2}\left(\frac{m_A m_B}{m_A + m_B}\right)(v_{Ai} - v_{Bi})^2$$

19-112* An 8-m long chain having density of 0.5 kg/m is being pulled along a frictionless horizontal surface by a constant force $F = 18$ N (Fig. P19-112). If the chain starts from rest when $y = 0.5$ m, determine the speed of the chain when it becomes fully extended.

Fig. P19-112

19-113* Two cars collide at an intersection as shown in Fig. P19-113. Car A weighs 2500 lb and has an initial speed $v_A = 20$ mi/h whereas car B weighs 3000 lb and has an initial speed $v_B = 15$ mi/h. On impact, the wheels of the cars lock and the cars slide ($\mu_k = 0.2$). If the cars become entangled and move as a single unit after the impact, determine:

a. The speed v_f and direction θ of the cars after the impact.

b. The distance the cars will slide before coming to rest after the impact.

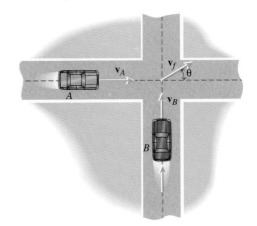

Fig. P19-113

19-114 A 50,000-kg railcar A rolls with an initial speed of 3 m/s along a straight level track. Car A collides with and couples to a second railcar B of mass 80,000 kg, which has an initial speed of 2 m/s. Determine the common final velocity of the two cars if the car B was initially moving:

a. In the same direction as car A.

b. In the opposite direction.

20

KINETICS OF RIGID BODIES: IMPULSE AND MOMENTUM

As the swings of the carnival ride move outward, the moment of inertia of the system increases. Therefore, by the angular momentum principle, the angular velocity must decrease unless a torque is provided to the rotor.

20-1   INTRODUCTION

The linear and angular momentum principles for particle motion were derived in the last chapter. It was seen that these equations are first integrals of the equations of motion with respect to time. The resulting equations relate the forces acting on the particle, the velocity of the particle, and time. Therefore, these principles are particularly useful for solving problems in which the velocity of a body for two different instants of time are to be related and the forces involved can be expressed as functions of time.

The impulse–momentum principles were also applied to an arbitrary system of interacting particles. All of those results apply immediately to a rigid body since a rigid body is just a system of interacting particles. All that remains is to simplify the general results using the relative velocity equation that relates the velocities of particles in a rigid body.

20-2   LINEAR IMPULSE AND MOMENTUM OF A RIGID BODY

The linear momentum of a system of particles—rigid or nonrigid—was defined in Section 19-3 as the sum of the linear momenta of the individual particles. Using the definition of the mass center, this was written

$$\mathbf{L} = \sum_{\ell} \mathbf{L}_{\ell} = \sum_{\ell} (m\mathbf{v})_{\ell} = m\,\mathbf{v}_G$$

For a continuous rigid body, the sum must be replaced with an integral

$$\mathbf{L} = \int d\mathbf{L} = \int \mathbf{v}\,dm = m\,\mathbf{v}_G \tag{20-1}$$

Then the Linear Impulse–Momentum Principle was written (Eq. 19-9)

$$m(\mathbf{v}_G)_i + \sum_{\ell} \int_{t_i}^{t_f} \mathbf{R}_{\ell}\,dt = m(\mathbf{v}_G)_f \tag{20-2}$$

where $\sum_{\ell} \int_{t_i}^{t_f} \mathbf{R}_{\ell}\,dt$ is the impulse of all external forces acting on the system of particles and the impulses due to internal forces have no effect and can be ignored. But the system of particles is arbitrary, and Eq. 20-2 applies equally to a system of independent, interacting particles and to rigid bodies.

20-3   ANGULAR IMPULSE AND MOMENTUM OF A RIGID BODY IN PLANE MOTION

The angular momentum of a system of particles—rigid or nonrigid—was also defined in Section 19-5 as the sum of the angular momenta of the individual particles. For the general system of interacting particles of Section 19-5, this led to the differential statements of the Angular Impulse–Momentum Principle (Eqs. 19-19 and 19-22):

$$\sum_\ell \mathbf{M}_{O\ell} = \frac{d}{dt} \mathbf{H}_O \quad \text{and} \quad \sum_\ell \mathbf{M}_{G\ell} = \frac{d}{dt} \mathbf{H}_G \qquad \text{(20-3)}$$

and the integral statements of the Angular Impulse–Momentum Principle (Eqs. 19-20 and 19-23):

$$(\mathbf{H}_O)_i + \int_{t_i}^{t_f} \sum_\ell \mathbf{M}_{O\ell}\, dt = (\mathbf{H}_O)_f \qquad \text{(20-4a)}$$

$$(\mathbf{H}_G)_i + \int_{t_i}^{t_f} \sum_\ell \mathbf{M}_{G\ell}\, dt = (\mathbf{H}_G)_f \qquad \text{(20-4b)}$$

where $\mathbf{H} = (\Sigma \mathbf{r}_\ell \times m\mathbf{v}_\ell)$, O is a fixed point, and G is the mass center of the system of particles. Again, the system of particles is arbitrary and Eqs. 20-3 and 20-4 apply equally to an arbitrary system of particles and to the system of particles that make up a rigid body.

The angular momentum of a particle can be calculated relative to any point, fixed or moving. For an arbitrary system of interacting particles, the particles move independently and the expression of the Angular Impulse–Momentum Principle relative to a fixed point O is usually the most useful. For rigid bodies, however, velocities of points in the body are related by the angular velocity and the expression of the Angular Impulse–Momentum Principle relative to the mass center is usually the most useful.

Let $\mathbf{r}_{\ell/G}$ be the position of the element of mass dm relative to the mass center G of the rigid body shown in Fig. 20-1. If the absolute velocity of dm is denoted by $\mathbf{v}_\ell = \dot{\mathbf{r}}_{\ell/O}$, the angular momentum of dm about G is the moment of the linear momentum

$$d\mathbf{H}_G = (\mathbf{r}_{\ell/G} \times \mathbf{v}_\ell)dm$$

Then the angular momentum of the entire rigid body is

$$\mathbf{H}_G = \int d\mathbf{H}_G = \int (\mathbf{r}_{\ell/G} \times \mathbf{v}_\ell)\, dm$$

$$= \int \mathbf{r}_{\ell/G} \times (\mathbf{v}_G + \mathbf{v}_{\ell/G})\, dm$$

where the absolute velocity has been replaced using the relative velocity equation $\mathbf{v}_\ell = \mathbf{v}_G + \mathbf{v}_{\ell/G} = \mathbf{v}_G + \boldsymbol{\omega} \times \mathbf{r}_{\ell/G}$ and $\mathbf{v}_G$ is the velocity of the mass center of the rigid body.

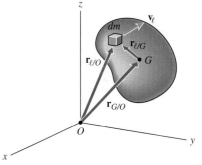

Fig. 20-1

20-3-1 Planar Motion of a Rigid Body

For the planar motion of a rigid body, the angular velocity is perpendicular to the plane of the motion $\boldsymbol{\omega} = \omega\mathbf{k}$. Therefore, the angular momentum of a rigid body about an axis through its mass center G can be written

$$\mathbf{H}_G = \int \mathbf{r}_{\ell/G} \times (\mathbf{v}_G + \omega\mathbf{k} \times \mathbf{r}_{\ell/G})\, dm$$

$$= \left(\int \mathbf{r}_{\ell/G}\, dm \right) \times \mathbf{v}_G + \int \mathbf{r}_{\ell/G} \times (\omega\mathbf{k} \times \mathbf{r}_{\ell/G})\, dm$$

In the first term, the velocity of the mass center $\mathbf{v}_G$ has been taken outside the integral since it is the same for every element of mass dm. But the integral $\int \mathbf{r}_{\ell/G}\, dm$ is zero by the definition of the mass center since the position vector $\mathbf{r}_{\ell/G}$ is measured from the mass center.

In the remaining term let $\mathbf{r}_{\ell/G} = x\mathbf{i} + y\mathbf{j} + z\mathbf{k}$. Then, taking the constant ω outside the integral and expanding the triple cross-product gives

$$\mathbf{H}_G = -\omega \int xz\, dm\, \mathbf{i} - \omega \int yz\, dm\, \mathbf{j} + \omega \int (x^2 + y^2)\, dm\, \mathbf{k}$$

$$= -\omega\, I_{Gxz}\, \mathbf{i} - \omega\, I_{Gyz}\, \mathbf{j} + \omega\, I_{Gz}\, \mathbf{k} \tag{20-5}$$

where $I_{Gz} = \int (x^2 + y^2)\, dm$ is the moment of inertia with respect to the z-axis and $I_{Gxz} = \int xz\, dm$ and $I_{Gyz} = \int yz\, dm$ are the products of inertia of the rigid body with respect to planes through the mass center.[1] If the rigid body is symmetric about the plane of motion (e.g., a slab of uniform thickness in the z-direction or a cylinder with axis parallel to the z-axis) or if the z-axis through G is an axis of symmetry,[2] then the products of inertia I_{Gxz} and I_{Gyz} will be zero and

$$\mathbf{H}_G = \omega I_G \mathbf{k} \tag{20-6}$$

where $I_G = I_{Gz}$ is the moment of inertia of the rigid body with respect to an axis through the mass center G and perpendicular to the plane of motion. Then, substitution of Eq. 20-6 in Eq. 20-3b gives the differential form of the Angular Impulse–Momentum equation for a rigid body in plane motion

$$\Sigma \mathbf{M}_G = \frac{d\, I_G \omega}{dt}\, \mathbf{k} = I_G \alpha \mathbf{k} \tag{20-7}$$

where $\Sigma \mathbf{M}_G$ is the sum of moments of the external forces about the mass center G. For planar motion in which the forces lie in the plane of motion, the moment $\mathbf{M}_G = M_G \mathbf{k}$ so that the x- and y-components of Eq. 20-7 are satisfied identically.[3] Then the z-component of Eq. 20-7 can be written

$$\Sigma M_G = \frac{d\, I_G \omega}{dt} = I_G \alpha \tag{20-8}$$

---

[1] Note that if the angular momentum is calculated relative to a fixed point O rather than the mass center G, then $\int \mathbf{r}_{\ell/O}\, dm = m\mathbf{r}_{G/O} \neq \mathbf{0}$ and the first term will reduce to $m\mathbf{r}_{G/O} \times \mathbf{v}_G$ rather than zero. Also, the position vectors in the second term will be measured from different points and

$$\int \mathbf{r}_{G/O} \times (\omega \mathbf{k} \times \mathbf{r}_{\ell/G})\, dm$$

will not give the moments of inertia in Eq. 20-5.

[2] These two situations are special cases of the more general situation of the z-axis through G being a principal axis of inertia.

[3] Note that if the products of inertia I_{Gxz} and I_{Gyz} are not zero, then the angular momentum will also have x- and y-components even for planar motion. This means that moment components in the x- and/or y-direction will be required to keep the motion in the x-y plane if the angular velocity is changing.

which is just one of the general equations of motion for rigid bodies (Eq. 16-23c). Finally, integration of Eq. 20-8 with respect to time gives the integral form of the Angular Impulse–Momentum equation for a rigid body

$$(I_G\omega)_i + \int_{t_i}^{t_f} \Sigma M_G \, dt = (I_G\omega)_f \tag{20-9}$$

Equation 19-23 (which applies to any system of interacting particles—rigid or nonrigid) and Eq. 20-9 (which applies to a rigid body) both state that the angular impulse $\int_{t_i}^{t_f} \Sigma M_G \, dt$ acting on a system of particles is equal to the change in the angular momentum $(H_G)_f - (H_G)_i$ of the system of particles. The only difference between Eqs. 19-23 and 20-9 is in the manner in which the initial and final angular momentum is calculated. Therefore, the use of Eq. 20-9 requires only that the body behave rigidly at the initial time t_i and at the final time t_f so that H_G can be computed using $I_G\omega$ at these times. Between the initial and final instants of time, parts of the body may move relative to each other and $(I_G)_i$ may not be the same as $(I_G)_f$.

20-3-2 Rotation About a Fixed Axis

A common type of problem encountered in dynamics is the rotation of a rigid body about a fixed axis (Fig. 20-2). In this case, all particles in the body travel in circular paths in planes perpendicular to the axis. Therefore, the motion is a plane motion and Eqs. 20-1 through 20-9 apply. Although the foregoing equations apply directly to this type of problem, the Angular Impulse–Momentum Principle can be simplified by combining it with the Linear Impulse–Momentum Principle.

Select a coordinate system with the z-axis along the axis of rotation and origin O where the axis of rotation penetrates the plane of motion (the plane containing the mass center). Then the mass center G will travel in a circular path about the axis of rotation with a velocity $\mathbf{v}_G = \omega\mathbf{k} \times \mathbf{r}_{G/O}$ where $\mathbf{r}_{G/O} = x_G\mathbf{i} + y_G\mathbf{j}$. Therefore, the sum of the angular momentum about G and the moment of the linear momentum about O gives

$$\mathbf{H}_G + \mathbf{r}_{G/O} \times m\mathbf{v}_G = I_G\omega\mathbf{k} + \mathbf{r}_{G/O} \times m(\omega\mathbf{k} \times \mathbf{r}_{G/O})$$
$$= I_G\omega\mathbf{k} + m(x_G^2 + y_G^2)\omega\mathbf{k} = I_O\omega\mathbf{k} = \mathbf{H}_O$$

Similarly, the sum of the angular impulse about G and the moment of the linear impulse about the fixed point O gives

$$\int_{t_i}^{t_f} \sum_{\ell} M_{G\ell} \, \mathbf{k} \, dt + \mathbf{r}_{G/O} \times \int_{t_i}^{t_f} \Sigma \mathbf{R}_\ell \, dt$$

$$= \int_{t_i}^{t_f} \left(\sum_{\ell} (\mathbf{r}_{\ell/G} \times \mathbf{R}_\ell) + \left(\mathbf{r}_{G/O} \times \Sigma \mathbf{R}_\ell \right) \right) dt$$

$$= \int_{t_i}^{t_f} \sum_{\ell} \left((\mathbf{r}_{\ell/G} + \mathbf{r}_{G/O}) \times \mathbf{R}_\ell \right) dt$$

$$= \int_{t_i}^{t_f} \sum_{\ell} M_{O\ell} \, \mathbf{k} \, dt$$

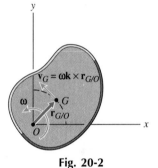

Fig. 20-2

Therefore, adding the moment of Eq. 20-2 about the fixed point O to Eq. 20-9 gives

$$(I_O\omega)_i + \int_{t_i}^{t_f} \sum_\ell M_{O\ell}\, dt = (I_O\omega)_f \qquad (20\text{-}10)$$

That is, for a rigid body rotating about a fixed axis, the change in angular momentum about the axis of rotation is equal to the angular impulse about the axis.

20-3-3 Graphical Representation of the Linear and Angular Momentum Principles

Together, Eqs. 20-2 and 20-9 say that the momentum of the particles making up the rigid body can be replaced with an equivalent "force–couple" system at the mass center G. The equivalent "force" is equal to the linear momentum vector $\mathbf{L} = m\mathbf{v}_G$ and the equivalent "couple" is equal to the angular momentum vector $\mathbf{H}_G = I_G\omega\mathbf{k}$. Then the results of Eqs. 20-2 and 20-9 can be summarized graphically as shown in the kinetic diagrams of Fig. 20-3. That is, the sum of the equivalent "force–couple" of the momentum at time t_i (Fig. 20-3a) and the equivalent force–couple of the impulses (Fig. 20-3b) is equal to the equivalent "force–couple" of the momentum at time t_f (Fig. 20-3c).

The graphical representation of Fig. 20-3 also includes the special case of rotation about a fixed axis. If O is a point on the axis of rotation as described in Section 20-3-2, then computing the moment of the equivalent force–couple systems in each part of the figure gives Eq. 20-10. For example, in Fig. 20-3a the velocity of the mass center $\mathbf{v}_G = \boldsymbol{\omega} \times \mathbf{r}_{G/O}$ is perpendicular to $\mathbf{r}_{G/O}$, and the moment of the force–couple system about O is

$$H_G + r_{G/O} m v_G = I_G\omega + r_{G/O} m r_{G/O}\omega$$
$$= (I_G + m r_{G/O}^2)\omega = I_O\omega$$

as was obtained in Section 20-3-2.

The graphical representation of Fig. 20-3 can also be used to write the Angular Impulse–Momentum Principle about an arbitrary fixed point P. However, if the body is not rotating about an axis through P, then $\mathbf{v}_G \neq \boldsymbol{\omega} \times \mathbf{r}_{G/P}$ and the sum of H_G and the moment of the linear momentum will *not* reduce to $I_P\omega$.

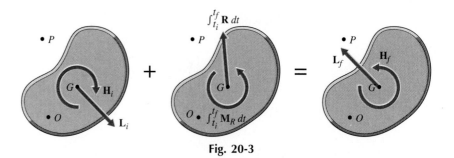

Fig. 20-3

20-3-4 Center of Percussion

Just as a force and couple can be reduced to its simplest form (its resultant), it is possible to reduce the linear and angular momentum vectors of Fig. 20-3 to a single, equivalent linear momentum vector. The "resultant" will be equal to the linear momentum of the mass center $m\mathbf{v}_G$ and will act along a line in the direction of the linear momentum $m\mathbf{v}_G$ and located a distance

$$d = \frac{I_G\omega}{mv_G}$$

away from the mass center (Fig. 20-4).

In particular, for a body rotating about a fixed axis through O, the system linear and angular momenta at the mass center G (Fig. 20-5a) are equivalent to the system linear momentum $m\mathbf{v}_G$ at point P (Fig. 20-5b). The linear momentum is clearly the same for both kinetic diagrams. The angular momentum will also be the same if the position of P is chosen such that

$$r_P(mv_G) = I_G\omega + r_G(mv_G)$$

The point P located in this manner is called the *center of percussion*.

Note that the location of the center of percussion depends on the motion of the body as well as on the size, shape, and mass distribution of the body. Since the body of Fig. 20-5 is rotating about a fixed axis, $v_G = r_G\omega$, and therefore

$$r_P(mr_G\omega) = mk_G^2\omega + r_G(mr_G\omega)$$

where k_G is the radius of gyration of the body relative to an axis

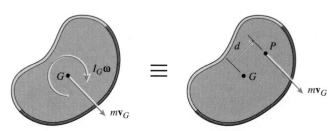

Fig. 20-4

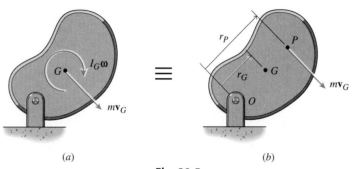

(a)   *(b)*

Fig. 20-5

through its mass center parallel to the axis of rotation. Dividing through by the common factor $m\omega$ gives

$$r_P r_G = k_G^2 + r_G^2$$

or

$$(r_P - r_G)r_G = k_G^2$$

That is, the distance between the center of percussion and the mass center $d = r_P - r_G$ is equal to the ratio of k_G^2, which is a constant for a given body, and r_G, which depends on the location of the axis of rotation.

20-4   SYSTEMS OF RIGID BODIES

As has already been pointed out, the use of Eq. 20-9 requires only that the body behave rigidly at the initial time t_i and at the final time t_f so that the angular momentum H_G can be computed using $I_G\omega$ at these times. Between the initial and final instants of time, parts of the body may move relative to each other and $(I_G)_i$ may not be the same as $(I_G)_f$. If parts of the body are moving relative to each other at the initial and/or the final instants of time, then a separate angular impulse momentum equation should be written for each part that does behave in a rigid manner and the equations added together. If the moments of momentum and moments of forces are all written relative to the same point in each equation, then the moments of the joint forces holding the various parts together will cancel in pairs and need not be computed. This is most easily accomplished using the graphical representation of Fig. 20-6. In the first and last parts of the figure the momentum of each rigid body has been replaced with an equivalent "force–couple" system at its own mass center. In the middle part of the figure, the joint forces are internal forces and need not be shown.

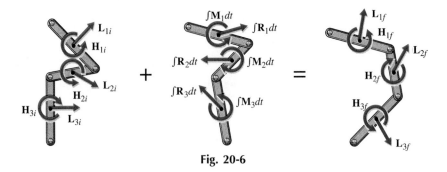

Fig. 20-6

A 20-lb uniform disk rotates about an axle through its center (Fig. 20-7a). The radius of the disk is 9 in. and the initial angular velocity of the disk is 600 rpm clockwise. A counterclockwise torque $M = 10 \sin nt$ acts on the disk where M is in pounds·foot, t is in seconds, and $n = 1$ rad/s. Neglect any friction between the bearings and the axle of the disk and determine the angular velocity of the disk after 1 s; 3 s; 5 s.

SOLUTION

The Impulse–Momentum Principle is shown in graphical form in Fig. 20-7b, in which the initial and final linear momentum of the mass center are both zero; the initial angular velocity of the disk is

$$\omega_0 = 600 \text{ rpm} \left(\frac{2\pi \text{ rad/rev}}{60 \text{ s/min}} \right) = 62.83 \text{ rad/s}$$

and the moment of inertia about the axle (the mass center) is

$$I_G = \frac{1}{2} mR^2 = \frac{1}{2} \frac{20}{32.2} \left(\frac{9}{12} \right)^2$$
$$= 0.17469 \text{ lb} \cdot \text{ft} \cdot \text{s}^2$$

Then the sum of moments about the axle (which is also the mass center) gives

$$\curvearrowright +H_G: \qquad -(0.17469)(62.83) + \int_0^\tau 10 \sin t \, dt = 0.17469 \, \omega_f$$

or

$$\omega_f = 57.24 \, (1 - \cos \tau) - 62.83 \text{ rad/s} \quad \text{(CCW)}$$

Therefore

$$\omega_f(1 \text{ s}) = -36.52 \text{ rad/s}$$
$$= 36.52 \text{ rad/s} \quad \text{(CW)} \qquad \qquad \text{Ans.}$$
$$\omega_f(3 \text{ s}) = 51.08 \text{ rad/s} \quad \text{(CCW)} \qquad \qquad \text{Ans.}$$
$$\omega_f(5 \text{ s}) = -21.83 \text{ rad/s}$$
$$= 21.83 \text{ rad/s} \quad \text{(CW)} \qquad \qquad \text{Ans.}$$

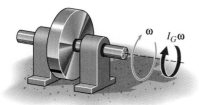

(a)

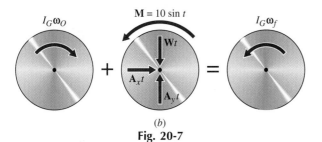

(b)

Fig. 20-7

EXAMPLE PROBLEM 20-2

A bowling ball is modeled as a 7-kg uniform sphere 300 mm in diameter (Fig. 20-8a). The ball is released on a horizontal wood floor with an initial velocity of $v_0 = 6$ m/s and an initial angular velocity of $\omega_0 = 0$. If the coefficient of kinetic friction between the ball and the floor is $\mu_k = 0.1$, determine:

a. The time t_f at which the ball begins to roll without slipping.
b. The velocity v_f and angular velocity ω_f of the ball at time t_f.

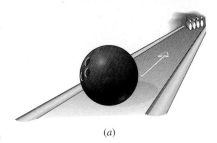

(a)

SOLUTION

The Impulse–Momentum Principle is shown in graphical form in Fig. 20-8b, in which the moment of inertia of the sphere about an axis through its mass center is

$$I_G = \frac{2}{5}mR^2 = \frac{2}{5}(7)(0.15)^2 = 0.0630 \text{ kg} \cdot \text{m}^2$$

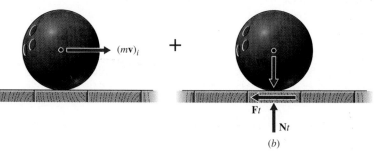

(b)

Fig. 20-8

The initial momentum of the sphere is replaced with an equivalent "force–couple" at the mass center consisting of a "force" in the x-direction of magnitude $L_i = (7 \text{ kg})(6 \text{ m/s}) = 42 \text{ kg} \cdot \text{m/s}$ and a "moment" about the mass center of $(I_G\omega)_i = 0$. Similarly, the final momentum of the sphere is replaced with a "force" in the x-direction of magnitude $L_f = mv_f$ acting through the mass center and a "moment" about the mass center of $I_G\omega_f$. Then, with reference to the figure, the linear and angular momentum principles give

$$+\rightarrow L_x: \qquad 42 - Ft = 7\,v_f \qquad (a)$$
$$+\uparrow L_y: \qquad 0 + Nt - (7)(9.81)t = 0 \qquad (b)$$
$$\curvearrowright + H_G: \qquad 0 + (0.150)Ft = (0.0630)\omega_f \qquad (c)$$

Equation b gives $N = 68.67$ N for any $0 < t < t_f$ (or even for any $t > t_f$).[4] Since the ball slides between $t = 0$ and $t = t_f$, the friction force on the ball is

$$F = \mu_k N = 0.1(68.67) = 6.867 \text{ N}$$

Finally, kinematics is used to relate the linear and angular velocity at time t_f (Fig. 20-9)

$$v_f = 0.150\,\omega_f \qquad (d)$$

Then Eqs. a, c, and d give

$$t_f = 1.747 \text{ s} \qquad v_f = 4.286 \text{ m/s} \rightarrow \qquad \omega_f = 28.57 \text{ rad/s} \quad \text{(CW)} \qquad \text{Ans.}$$

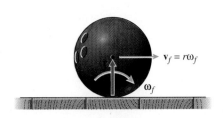

Fig. 20-9

---

[4]This same result is trivially obtained using Newton's second law. Since there is no motion in the y-direction, $a_y = 0$ and $\Sigma F_y = 0$ or $N - (7)(9.81) = 0$.

A 2-ft long uniform rod weighing 3 lb hangs from a frictionless pin at A (Fig. 20-10). A bullet weighing 0.05 lb and traveling with an initial speed of 1800 ft/s strikes the rod and becomes embedded in it. Determine the angular velocity of the rod immediately after the bullet becomes imbedded in it.

Fig. 20-10

SOLUTION

Since the rod is in fixed axis rotation about an axis through A, Eq. 20-10 will be used. The appropriate momentum and free-body diagrams are shown in Fig. 20-11. The initial linear momentum of the bullet is

$$L_{ib} = \frac{0.05}{32.2} 1800 = 2.795 \text{ lb} \cdot \text{s}$$

and its moment about point A is $1.5 L_{ib} = 4.193 \text{ ft} \cdot \text{lb} \cdot \text{s}$. The initial linear and angular momentum of the rod are both zero. Therefore the total angular momentum about A of the system just before the bullet strikes the rod is $H_{Ai} = 4.193 \text{ ft} \cdot \text{lb} \cdot \text{s}$.

When the momenta and impulses are added together, the interaction force of the bullet and the rod is an internal force and need not be shown on the free-body diagram. None of the other three forces have a moment about A, and so the total angular impulse about A is zero.

Immediately after the bullet lodges in the rod, the rod and bullet rotate as a single unit about the fixed pin A. The final angular momemtum of the system is $H_{Af} = (I_A \omega)_f$ where

$$I_A = \frac{1}{3} \frac{3}{32.2} (2)^2 + \frac{0.05}{32.2} (1.5)^2$$
$$= 0.12772 \text{ lb} \cdot \text{ft} \cdot \text{s}^2$$

is the moment of inertia of both the rod and the imbedded bullet about the axis through A.

Finally then, the angular impulse-momentum principle (Eq. 20-10) gives

$$\text{↻} + H_A: 4.193 + 0 = 0.12772 \, \omega_f$$

or

$$\omega_f = 32.8 \text{ rad/s (CCW)} \qquad \text{Ans.}$$

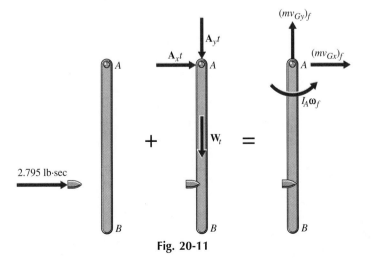

Fig. 20-11

PROBLEMS

20-1* A flywheel consists of a 10-lb uniform disk 15 in. in diameter and 1 in. thick. If bearing friction reduces the angular velocity of the flywheel from 3600 rpm to zero in 3 min, determine the average frictional moment exerted on the flywheel by the bearing.

20-2* The armature of an electric motor coasts to rest from an angular velocity of 2400 rpm in 150 s. If the 3-kg armature has a radius of gyration of 100 mm, determine the average frictional moment exerted on the flywheel by the bearings of the motor.

20-3 The starting moment of an electric motor is given by $M_0 e^{-t}$, where M_0 is a constant and the bearings exert a constant resisting moment of 0.006 ft·lb. If the 5-lb armature has a radius of gyration of 2.3 in. and the motor reaches its operating speed of 3000 rpm in 3 s, determine the value of M_0.

20-4* A force of $P = 50$ N is applied to the end of a rope wrapped around the outside of a hollow drum (Fig. P20-4). The radius of gyration of the 20-kg drum is 175 mm and axle friction may be neglected. If the drum is released from rest, determine the downward velocity of point A on the rope after 10 s.

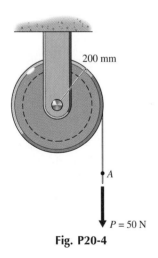

Fig. P20-4

20-5 A rope is wrapped around the outside of a 25-lb uniform drum as shown in Fig. P20-5a. At $t = 0$ the drum is at rest when the force shown in Fig. P20-5b is suddenly applied to the end of the rope. If axle friction may be neglected, determine the downward velocity of point A on the rope after 7 s.

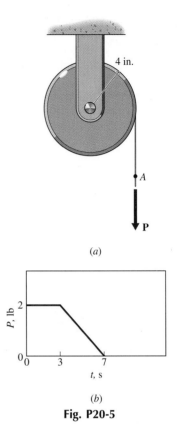

(a)

(b)

Fig. P20-5

20-6 A 50-N weight is tied to the end of a rope wrapped around the outside of a hollow drum (Fig. P20-6). The radius of gyration of the 20-kg drum is 175 mm and axle friction may be neglected. If the drum is released from rest, determine the downward velocity of point A on the rope after 10 s.

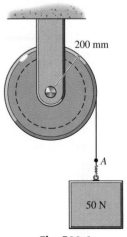

Fig. P20-6

20-7* The 20-lb uniform wheel of Fig. P20-7 is rotating at 3000 rpm when a force of $P = 40(1 - e^{-0.05t})$ lb is applied to the handle of the brake arm. If the coefficient of kinetic friction between the brake arm and the wheel is 0.1, determine the length of time until the wheel stops rotating if:

a. It is rotating clockwise.
b. It is rotating counterclockwise.

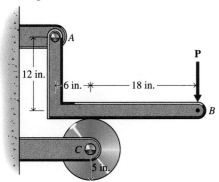

Fig. P20-7

20-8 The 20-kg stepped wheel of Fig. P20-8 has a radius of gyration of 150 mm and an initial angular rotation rate of 3000 rpm counterclockwise. If the coefficient of kinetic friction between the brake arm and the wheel is 0.2, determine the length of time:

a. Until the wheel stops rotating.
b. Until the rotation rate of the wheel is 3000 rpm clockwise.

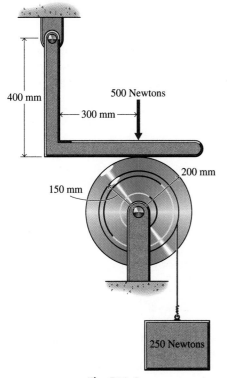

Fig. P20-8

20-9* The uniform wheel A (8-in. diameter, 20 lb) is initially raised and given a counterclockwise angular velocity of 4500 rpm while the uniform wheel B (8-in. diameter, 20 lb) remains at rest (Fig. P20-9). Wheel A is then released and allowed to spin against wheel B. If the coefficient of kinetic friction between the two wheels is 0.1, determine:

a. The length of time before the wheels begin to rotate together without slipping.
b. The final angular velocities of both wheels.

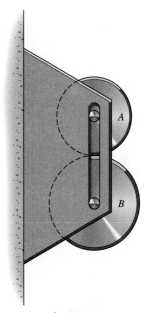

Fig. P20-9

20-10* The uniform wheel A (200-mm diameter, 10 kg) is initially raised and given a counterclockwise angular velocity of 4500 rpm while the uniform wheel B (400-mm diameter, 20 kg) remains at rest (Fig. P20-9). Wheel A is then released and allowed to spin against wheel B. If the coefficient of kinetic friction between the two wheels is 0.1, determine:

a. The length of time before the wheels begin to rotate together without slipping.
b. The final angular velocities of both wheels.

20-11 The two uniform wheels A and B of Fig. P20-9 are initially rotating together without slipping when wheel B is suddenly stopped. Wheel A has an 8 in. diameter, weighs 20 lb, and has a counterclockwise angular velocity of 4500 rpm; wheel B has a 12 in. diameter and weighs 45 lb. If the coefficient of kinetic friction between the two wheels is 0.2, determine the length of time before wheel A stops rotating.

20-12* The two uniform wheels A and B of Fig. P20-9 (p. 399) are initially at rest when a constant counterclockwise moment $M = 2.5$ N·m is suddenly applied to wheel A. Wheel A has a 200 mm diameter and a 10 kg mass whereas wheel B has a 300 mm diameter and a 25 kg mass. If the coefficients of static and kinetic friction between the two wheels are 0.2 and 0.1, respectively, determine the angular velocities of both wheels at $t = 5$ s, 15 s, and 25 s.

20-13 The two uniform wheels A and B of Fig. P20-9 (p. 399) are initially at rest when a constant counterclockwise moment $M = 3$ lb·ft is suddenly applied to wheel A. Wheel A has a 10 in. diameter and weighs 20 lb whereas wheel B has a 16 in. diameter and weighs 30 lb. If the coefficients of static and kinetic friction between the two wheels are 0.2 and 0.1, respectively, determine the angular velocities of both wheels at $t = 5$ s, 15 s, and 25 s.

20-14 A 300-mm diameter 5-kg homogeneous sphere is lowered onto a horizontal surface with an initial angular velocity $\omega_0 = 3000$ rpm and zero linear velocity $v_0 = 0$ (Fig. P20-14). If the coefficient of kinetic friction is 0.15, determine:

a. The time t_f at which the sphere will begin to roll without slipping.
b. The velocity of the mass center at t_f.
c. The angular velocity of the sphere at t_f.

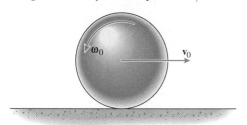

Fig. P20-14

20-15* A 14-in. diameter 16-lb homogeneous sphere is lowered onto a horizontal surface with an initial angular velocity $\omega_0 = 3000$ rpm and a linear velocity $v_0 = 20$ ft/s (Fig. P20-14). If the coefficient of kinetic friction is 0.15, determine:

a. The time t_f at which the sphere will begin to roll without slipping.
b. The velocity of the mass center at t_f.
c. The angular velocity of the sphere at t_f.

20-16 A 200-mm diameter 5-kg homogeneous sphere is lowered onto a horizontal surface with an initial angular velocity $\omega_0 = 3000$ rpm (Fig. P20-14). If the coefficient of kinetic friction is 0.15, determine the initial velocity v_0 for which the sphere will end up at rest (the angular velocity and the linear velocity will both be zero when the sphere stops sliding).

20-17* A 12-in. diameter 16-lb homogeneous sphere rolls without slipping on an inclined surface (Fig. P20-17). If the initial speed of the sphere is 20 ft/s up the incline and $\theta = 10°$, determine:

a. The time t_1 at which the sphere stops rolling up the incline.
b. The time t_2 at which the sphere is rolling down the incline at 30 ft/s.

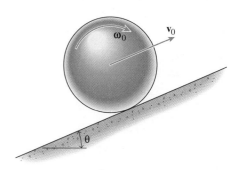

Fig. P20-17

20-18 A 200-mm diameter 5-kg homogeneous sphere is lowered onto an inclined surface with an initial angular velocity $\omega_0 = 3000$ rpm and zero linear velocity $v_0 = 0$ (Fig. P20-17). If the coefficient of kinetic friction is 0.25 and $\theta = 10°$, determine:

a. The time t_1 at which the sphere will begin to roll without slipping.
b. The velocity of the mass center $\mathbf{v}$ and the angular velocity $\boldsymbol{\omega}$ at t_1.
c. The time t_2 at which the sphere stops rolling up the incline.

20-19 A 14-in. diameter 16-lb homogeneous sphere is lowered onto an inclined surface with an initial angular velocity $\omega_0 = 3000$ rpm and zero linear velocity $v_0 = 0$ (Fig. P20-17). If the coefficient of kinetic friction is 0.25 and $\theta = 20°$, determine:

a. The time t_f at which the sphere will begin to roll without slipping.
b. The velocity of the mass center at $\mathbf{v}$ and the angular velocity $\boldsymbol{\omega}$ at t_f.

20-20* A 200-mm diameter 5-kg homogeneous sphere is lowered onto an inclined surface with an initial angular velocity $\omega_0 = 3000$ rpm (Fig. P20-17). If the coefficient of kinetic friction is 0.20 and $\theta = 15°$, determine the smallest initial velocity v_0 for which the sphere will stop sliding and stop rotating at the same instant.

20-21* The uniform slender rod AB ($W = 3$ lb, $\ell = 2$ ft) is resting on a frictionless horizontal surface when it is struck with an impulse of 2 lb·s as shown in Fig. P20-21. If $b = 1.5$ ft, determine:

a. The angular velocity of the rod immediately after the impact.
b. The velocity of end A immediately after the impact.

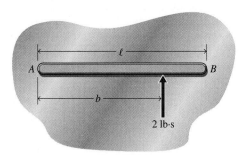

Fig. P20-21

20-22 The uniform slender rod AB ($m = 3$ kg, $\ell = 800$ mm) is resting on a frictionless horizontal surface when it is struck with an impulse of 5 N·s as shown in Fig. P20-22. If $b = 300$ mm and the duration of the impact $\Delta t = 0.002$ s, determine:

a. The angular velocity of the rod immediately after the impact.
b. The average magnitude of the force exerted on the rod by the frictionless pin at A.

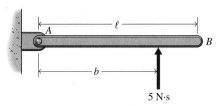

Fig. P20-22

20-23 For the slender rod of Problem 20-21 determine the distance b for which end A is an instantaneous center of zero velocity (the velocity of A is zero immediately after the impact).

20-24* For the slender rod of Problem 20-22 determine the distance b for which the average force exerted on the rod by the frictionless pin at A is zero.

20-25* A 3-lb uniform slender rod AB 4 ft long is rotating in a vertical plane about a frictionless pin through its center as shown in Fig. P20-25. A small piece of putty ($W_p = 0.4$ lb) falls and strikes the rod when it is horizontal. If the initial rotation of the rod is counterclockwise at 120 rpm and the putty starts from rest at $h = 5$ ft, determine:

a. The rotation rate of the rod and putty immediately after the impact.
b. The average force of contact between the rod and the putty for an impact duration $\Delta t = 0.005$ s.
c. The average magnitude of the force exerted on the rod by the frictionless pin for an impact duration $\Delta t = 0.005$ s.
d. The total system energy lost in the collision.

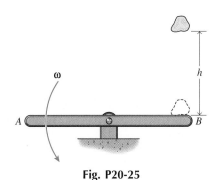

Fig. P20-25

20-26* A 3-kg uniform slender rod AB 800 mm long hangs in a vertical plane from a frictionless pivot when a 0.03 kg bullet strikes the rod and becomes embedded in it (Fig. P20-26). If the initial velocity of the bullet is $v_0 = 350$ m/s, determine:

a. The rotation rate of the rod and bullet immediately after the impact.
b. The average force of contact between the rod and the bullet for an impact duration $\Delta t = 0.001$ s.
c. The average magnitude of the force exerted on the rod by the frictionless pin at A for an impact duration $\Delta t = 0.001$ s.
d. The total system energy lost in the collision.
e. The maximum angle through which the rod will swing after the collision.

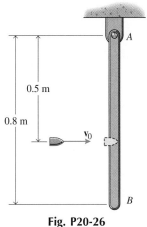

Fig. P20-26

20-27 For the rod and putty system of Problem 20-25 (see p. 401), the height h from which the putty is dropped is adjusted so that the angular velocity of the rod is zero immediately after the impact. Determine:

a. The adjusted height h.
b. The average force of contact between the rod and the putty for an impact duration $\Delta t = 0.005$ s.
c. The average magnitude of the force exerted on the rod by the frictionless pin for an impact duration $\Delta t = 0.005$ s.
d. The angular velocity of the rod and putty when the rod is vertical (the putty end has swung under the pivot).

20-28* The rod of Problem 20-26 is released from rest when it is horizontal as shown in Fig. P20-28. If the bullet strikes the rod when it is vertical, determine:

a. The angular velocity of the rod and bullet immediately after the impact.
b. The average magnitude of the force exerted on the rod by the frictionless pin at A for an impact duration $\Delta t = 0.001$ s.
c. The total system energy lost in the impact.
d. The maximum angle through which the rod will swing after the collision.

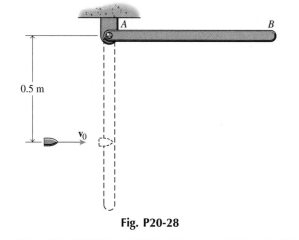

Fig. P20-28

20-29 For the rod and putty system of Problem 20-25 (p. 401), the height h from which the putty is dropped is adjusted so that the angular velocity of the rod is zero when the rod is vertical (the putty is directly over the pivot). Determine:

a. The adjusted height h.
b. The average force of contact between the rod and the putty for an impact duration $\Delta t = 0.005$ s.
c. The average magnitude of the force exerted on the rod by the frictionless pin for an impact duration $\Delta t = 0.005$ s.
d. The angular velocity of the rod and putty immediately after the collision.

20-30 The rod of Problem 20-26 is released from rest when it is horizontal as shown in Fig. P20-28. If the bullet strikes the rod when it is vertical, determine:

a. The initial velocity of the bullet for which the angular velocity of the rod will be zero immediately after the impact.
b. The average magnitude of the force exerted on the rod by the frictionless pin at A for an impact duration $\Delta t = 0.001$ s.

20-31* Determine the location of the center of percussion for a slender rod rotating about a frictionless pin if the pin is:

a. Located at one end of the rod.
b. Located $\ell/4$ from one end of the rod.
c. Located at the middle of the rod.

20-32 The rod of Problem 20-26 is released from rest when it is horizontal as shown in Fig. P20-28. The bullet strikes the rod when it is vertical such that the angular velocity of the rod is zero immediately after the impact and no impulsive force acts on the pivot at A. Determine:

a. The distance below the pivot at which the bullet must strike the rod.
b. The required initial velocity of the bullet v_0.

20-5   IMPACT OF RIGID BODIES

The impact of colliding bodies was discussed in Section 19-4. For particle collisions, only the cases of direct central and oblique central impact were applicable, and these cases were developed in detail there. In the present section, the additional cases of eccentric impact will be developed.

Even for the relatively simple case of particle collisions, the impact phenomena were seen to be complex events. Fortunately, however, the details of the impact can often be avoided; the linear impulse–momentum equation can be used to give a simple relation between the relative velocities of the bodies before and after the impact. Although the impulse–momentum method of approach is only an approximation of a very complex event and should be applied with care, the method allows the solution of some impact problems that would not otherwise be simply solvable.

20-5-1   Impulsive Forces and Impulsive Motion

Although impact events occur in a relatively short time interval, it is observed that the velocities and angular velocities of bodies can change significantly. The changes in momentum and angular momentum then require an impulse that does not go to zero for the brief impact times. Forces characterized by very large magnitudes such that they produce a significant change in momentum (a large impulse) even for very short time periods are called *impulsive forces*. Motions that result from impulsive forces are called *impulsive motions*. The forces generated when one body strikes another is an example of an impulsive force.

Forces that produce a negligible change in momentum (a small impulse) for small time periods are called *nonimpulsive* forces. Examples of nonimpulsive forces are the weight of a body, friction forces, and spring forces. The magnitudes of nonimpulsive forces are always small compared to the magnitudes of impulsive forces. When the principle of impulse and momentum is applied over a short time interval, the impulse of nonimpulsive forces may often be neglected compared to the impulse of the impulsive forces.

It is usually not known ahead of time whether unknown reaction forces are impulsive or not. Generally, the reaction force of any support that acts to prevent motion in some direction is just as impulsive as the forces trying to cause motion in that direction.

The final decision of whether or not the impulse of a force can be ignored must be based on the required accuracy of the result and on the estimated effect that the term has on the equation.

20-5-2   Assumptions for Impact Problems

By their very nature, impact events occur in very brief intervals of time. Based on observations of many impact events, it is assumed that during the brief impact interval $\Delta t = t_f - t_i$:

1.   The positions of the impacting bodies do not change appreciably.
2.   The velocities and/or angular velocities of one or both of the impacting bodies may change greatly.

3. Nonimpulsive forces and moments may be neglected.
4. Friction forces (forces tangent to the plane of impact) may be neglected.[5]

20-5-3   Eccentric Impact of Rigid Bodies

The analysis of particle collision problems in Section 19-4 illustrated the case of central impact. In central impact the line of impact coincides with the line connecting the mass centers. Therefore the contact forces of impact pass through the mass centers of the bodies (Fig. 20-12). These problems were solved using the concept of conservation of linear momentum in conjunction with the coefficient of restitution, e, which compares the relative velocity of separation of the points of contact (after the collision) to their relative velocity of approach (prior to the collision).

The impact problem for rigid bodies is quite similar to that of particles, but it is complicated slightly by the fact that the line of impact usually does not pass through the mass centers of the two bodies (Fig. 20-13). As noted in Section 19-4, such a collision is called an *eccentric impact*.

An additional complication arises if the coefficient of restitution is defined as the ratio of the restitution impulse and the deformation impulse, as in Section 19-4. An analysis similar to that in Section 19-4 again gives the coefficient of restitution in terms of the relative velocity of separation of the points of contact (after the collision) to their relative velocity of approach (prior to the collision). However, the velocity of the body at the point of impact is usually different from the velocity of its mass center. Therefore, when eccentric impact is involved, the relative velocity equations must be used to relate the velocities of the contact points in the equation for the coefficient of restitution and the velocities of the mass centers in the linear and angular momentum principles.

Consider the impact of the rigid bodies shown in Fig. 20-13. Points A and B are the mass centers of the bodies and points C and D are the actual points of contact. Coordinates t and n are chosen in and perpendicular to the plane of contact as shown. The coefficient of restitution is again defined as the ratio of the restitution impulse and the deformation impulse. Then, an analysis similar to that of Section 19-4 gives the coefficient of restitution

$$e = -\frac{(v_{Df})_n - (v_{Cf})_n}{(v_{Di})_n - (v_{Ci})_n} \tag{20-11}$$

where $(v_{Ci})_n$ and $(v_{Di})_n$ are the initial components of the velocities of points C and D (before impact) and $(v_{Cf})_n$ and $(v_{Df})_n$ are the final velocities of the points (after impact). The components of the velocities of points C and D are related to the velocities of the mass centers A and B by the relative velocity equations

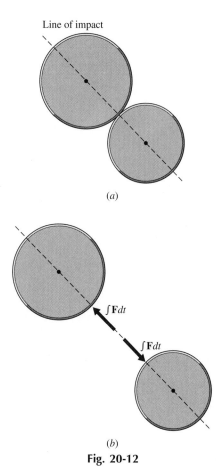

Line of impact

(a)

$\int \mathbf{F} dt$

$\int \mathbf{F} dt$

(b)

Fig. 20-12

---

[5]This last is often *not* a good approximation. For such problems, the reader is referred to Raymond M. Brach, *Mechanical Impact Dynamics: Rigid Body Collisions* (New York: Wiley, 1991).

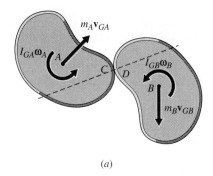

(a)

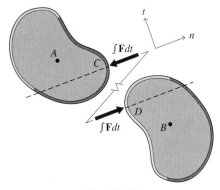

Fig. 20-13

$$\mathbf{v}_C = \mathbf{v}_A + \boldsymbol{\omega}_A \times \mathbf{r}_{C/A} \qquad (20\text{-}12)$$

and

$$\mathbf{v}_D = \mathbf{v}_B + \boldsymbol{\omega}_B \times \mathbf{r}_{D/B} \qquad (20\text{-}13)$$

The result of combining Eqs. 20-11, 20-12, and 20-13 gives one scalar equation relating the velocities $\mathbf{v}_A$ and $\mathbf{v}_B$ and the angular velocities ω_A and ω_B after the impact. Four additional scalar equations (two vector equations) can be obtained from applying the Linear Impulse–Momentum Principle to each body separately. Finally, the Principle of Angular Impulse–Momentum can be applied about the center of mass of each body, giving two more scalar equations for a total of seven equations. These equations are used to solve for the seven unknowns $(v_{Cf})_n$, $(v_{Cf})_t$, $(v_{Df})_n$, $(v_{Df})_t$, ω_{Af}, ω_{Bf}, and the magnitude of the impulsive contact force F exerted between the two bodies. As shown in Fig. 20-13, the force $\mathbf{F}$ is directed along the normal to the plane of contact.

If one or both of the colliding bodies is constrained to rotate about a fixed point or points, an impulsive reaction will be exerted at the fixed point(s). The impulse of these reactions must also be included in the equations of linear and angular impulse–momentum.

A 1.5-kg uniform rod 800 mm long is at rest on a frictionless horizontal surface when it is struck by a 0.5-kg disk as shown in Fig. 20-14. If the disk strikes the rod 200 mm from the end and the coefficient of restitution of the collision is 0.4, determine:

a. The velocity of the disk after the collision.
b. The velocity of the mass center of the rod after the collision.
c. The angular velocity of the rod after the collision.
d. The location of a point on the rod that is instantaneously at rest during the collision.

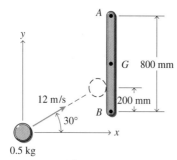

Fig. 20-14

SOLUTION

a. The kinetic diagrams of the rod and the disk are shown in Fig. 20-15, in which the moment of inertia of the rod about an axis through its mass center is

$$I_G = \frac{1}{12}(1.5)(0.8)^2 = 0.0800 \text{ kg} \cdot \text{m}^2$$

Since friction between the disk and rod is neglected, there is no component of linear impulse in the y-direction on either the rod or the disk. Therefore,

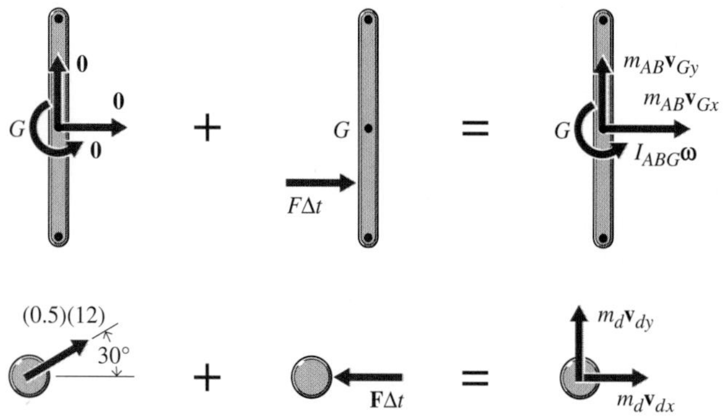

Fig. 20-15

the y-component of the linear momentum principle applied to the rod and disk separately gives:

$$\text{rod:} \qquad [0] + 0 = [1.5v_{Gy}]$$
$$\text{disk:} \qquad [(0.5)(12 \sin 30°)] + 0 = [0.5v_{dy}]$$

or

$$v_{Gy} = 0 \quad \text{and} \quad v_{dy} = 6 \text{ m/s}.$$

For the combined system of the rod and the disk, there is no linear impulse in the x-direction or moment of impulse about the mass center of the rod. Therefore, the x-component of the linear momentum principle applied to the combined system gives

$$[(0.5)(12 \cos 30°)] + 0 = [0.5v_{dx} + 1.5v_{Gx}] \qquad (a)$$

and the angular momentum principle applied to the combined system gives

$$[(0.2)(0.5)(12 \cos 30°) + 0] + 0 = [(0.2)(0.5)v_{dx} + 0.08\omega] \qquad (b)$$

Finally, the coefficient of restitution relates the x-components of velocities of the contact points before and after the collision:

$$e = -\frac{(v_{Gx} + 0.2\omega) - v_{dx}}{0 - 12 \cos 30°} = 0.4 \qquad (c)$$

Solving Eqs. a, b, and c simultaneously gives $v_{dx} = 1.203$ m/s, $v_{Gx} = 3.06$ m/s, and $\omega = 11.49$ rad/s. Then combining the x- and y-components of the velocity of the disk after the collision gives

$$\mathbf{v}_d = 6.12 \text{ m/s} \measuredangle 78.7° \qquad \text{Ans.}$$

b. Combining the x- and y-components of the velocity of the mass center of the rod after the collision gives

$$\mathbf{v}_G = 3.06 \text{ m/s} \rightarrow \qquad \text{Ans.}$$

c. The angular velocity of the rod after the collision is

$$\omega = 11.49 \text{ rad/s} \quad \text{(CCW)} \qquad \text{Ans.}$$

d. If C is an instantaneous center of zero velocity, then (Fig. 20-16)

$$3.06 = 11.49d$$

and the instantaneous center is

$$d = 0.266 \text{ m} \qquad \text{Ans.}$$

away from the mass center G on the side opposite the point of contact. (Note that if the rod were fixed at C with a frictionless pin, then the point of contact would be the center of percussion for the rod. That is, the radius of gyration of the rod is $k = \sqrt{0.08/1.5}$ and $0.2 \, d = k^2$.)

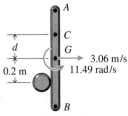

Fig. 20-16

EXAMPLE PROBLEM 20-5

A 25-lb uniform rod 3-ft long is attached to a frictionless hinge at A (Fig. 20-17). The rod starts from rest in the vertical position shown, falls against the bumper C, and rebounds upward. If the coefficient of restitution of the collision is 0.6, determine:

a. The maximum angle θ_{max} that the bar will make with the horizontal after the collision.
b. The average magnitude of the support reaction at A for an impact duration of 0.01 s.
c. The amount of total system energy lost during the collision.

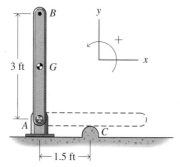

Fig. 20-17

SOLUTION

a. **Work–energy.** From the time t_0 (when the bar is vertical) until the time t_1 (just before the bar collides with the bumper C), the only force that does work on the bar is gravity. Therefore, the Work–Energy Principle can be used to determine the motion of the bar at time t_1. The initial kinetic energy of the bar is zero, and since point A is a fixed axis of rotation, the kinetic energy of the bar at t_1 can be written $T = \dfrac{1}{2}I_A\omega_1^2$ where $I_A = \dfrac{1}{3}m\ell^2 = 2.329$ lb·ft·s². Then the Work–Energy Principle

$$0 + (25)(1.5) = \frac{1}{2}(2.329)\,\omega_1^2 + 0$$

gives $\omega_1 = 5.675$ rad/s (CW) and the velocity of the mass center at t_1 is $v_{G1} = 1.5\omega_1 = 8.512$ ft/s↓.

Impact. The coefficient of restitution is used to determine the change in motion across the impact

$$e = 0.6 = -\frac{v_{G2} - 0}{(-8.512) - 0}$$

where the velocity of the bumper is zero both before and after the impact. Therefore, $v_{G2} = 5.107$ ft/s↑ and the angular velocity of the rod after the impact is $\omega_2 = \dfrac{v_{G2}}{1.5} = 3.405$ rad/s (CCW).

Work–energy. From the time t_2 (just after the collision) until the time t_3 (when the bar is at its maximum angle θ_{max}), the only force that does work on the bar is again gravity. At the maximum angle, the kinetic energy of the bar is zero and the Work–Energy Principle

$$\frac{1}{2}(2.329)(3.405)^2 + 0 = 0 + (25)(1.5 \sin \theta_{max})$$

gives

$$\theta_{max} = 21.1° \qquad \text{Ans.}$$

b. **Impulse–momentum.** Referring to the kinetic diagram of Fig. 20-18, the x-component of the Linear Impulse–Momentum Principle

$$0 + A_x(0.01) = 0$$

gives $A_x = 0$. Using the same kinetic diagram (in which $I_G = \dfrac{1}{12}m\ell^2 = 0.5823$ lb·ft·s², the Angular Impulse–Momentum Principle

$$(0.5823)(-5.675) + (1.5)A_y(0.01) = (0.5823)(3.405)$$

gives $A_y = 352$ lb. Therefore the average magnitude of the support reaction at A is

$$A = 352 \text{ lb} \qquad \text{Ans.}$$

c. Since the kinetic energy at times t_0 and t_3 are both zero, the total system energy at these times is just the potential energies

$$E_0 = (25)(1.5) = 37.5 \text{ lb·ft}$$
$$E_3 = (25)(1.5 \sin 21.1°) = 13.50 \text{ lb·ft}$$

The energy lost is then

$$\frac{37.50 - 13.50}{37.50}(100) = 64.0\% \qquad \text{Ans.}$$

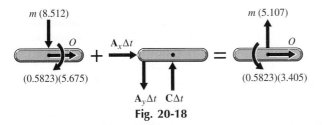

Fig. 20-18

PROBLEMS

20-33* A slender uniform rod (ℓ = 21 in., W_{AB} = 10 lb) is at rest on a frictionless horizontal surface when it is struck by a small disk (W_d = 2 lb) as shown in Fig. P20-33. If b = 3 in., e = 0.6, and the initial velocity of the disk is v_0 = 15 ft/s at an angle of θ = 60°, determine:

a. The velocity of the disk after the collision.
b. The velocity of the mass center of the rod after the collision.
c. The angular velocity of the rod after the collision.
d. The location of a point on the rod that is instantaneously at rest during the collision.

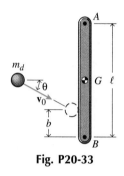

Fig. P20-33

20-34* A slender uniform rod (ℓ = 900 mm, m_{AB} = 5 kg) is at rest on a frictionless horizontal surface when it is struck by a small disk (m_d = 0.5 kg) as shown in Fig. P20-33. If b = 250 mm, e = 0.5, and the initial velocity of the disk is v_0 = 10 m/s at an angle of θ = 40°, determine:

a. The velocity of the disk after the collision.
b. The velocity of the mass center of the rod after the collision.
c. The angular velocity of the rod after the collision.
d. The location of a point on the rod that is instantaneously at rest during the collision.

20-35 For the disk and rod of Problem 20-33, determine the coefficient of restitution e for which the disk has no component of velocity along the line of impact after the collision. (Assume that all other parameters are unchanged.)

20-36 For the disk and rod of Problem 20-34, determine the mass of the disk m_d for which the disk has no component of velocity along the line of impact after the collision. (Assume that all other parameters are unchanged.)

20-37* A slender uniform rod (ℓ = 36 in., W_{AB} = 12 lb) hangs motionless from a frictionless hinge at A as shown in Fig. P20-37. A small ball (W_b = 2 lb) strikes the rod d = 4 in. from the bottom end. If e = 0.5, and the initial velocity of the ball is v_0 = 30 ft/s at an angle of θ = 40° to the horizontal, determine:

a. The velocity of the ball after the collision.
b. The angular velocity of the rod after the collision.
c. The average magnitude of the support reaction at A for an impact duration of 0.005 s.
d. The maximum angle of swing of rod AB after the collision.

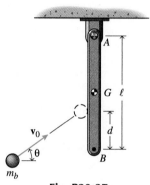

Fig. P20-37

20-38 A slender uniform rod (ℓ = 750 mm, m_{AB} = 10 kg) hangs motionless from a frictionless hinge at A as shown in Fig. P20-37. A small ball (m_b = 2 kg) strikes the rod d = 400 mm from the bottom end. The coefficient of restitution e = 0.8, and the initial velocity of the ball makes an angle of θ = 60° with the horizontal. If the angular velocity of the rod after the impact is 2.5 rad/s (CCW), determine:

a. The initial velocity of the ball v_0.
b. The velocity of the ball after the collision.
c. The average magnitude of the support reaction at A for an impact duration of 0.001 s.
d. The maximum angle of swing of rod AB after the collision.

20-39* A slender uniform rod (ℓ = 30 in., W_{AB} = 10 lb) is attached to a frictionless hinge at A as shown in Fig. P20-39. The rod is released from rest in a horizontal position ϕ_0 = 90° and a small ball (W_b = 3 lb) strikes the rod d = 8 in. from the bottom end when it is vertical ϕ = 0°. If the initial velocity of the disk is v_0 = 20 ft/s at an angle of θ = 30° to the horizontal and the angular velocity of the rod after the collision is zero, determine:

a. The coefficient of restitution of the collision, e.
b. The velocity of the disk after the collision.
c. The average magnitude of the support reaction at A for an impact duration of 0.003 s.

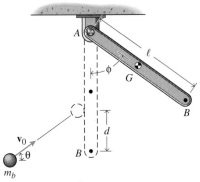

Fig. P20-39

20-40 A slender uniform rod ($\ell = 600$ mm, $m_{AB} = 5$ kg) is attached to a frictionless hinge at A as shown in Fig. P20-39. The rod is released from rest with $\phi_0 = 60°$ and a small ball ($m_b = 0.8$ kg) strikes the rod $d = 100$ mm from the bottom end when it is vertical $\phi = 0°$. The coefficient of restitution $e = 0.7$, and the initial velocity of the ball makes an angle of $\theta = 50°$ with the horizontal. If the maximum swing of the rod after the impact is 30°, determine:

a. The initial velocity of the ball, v_0.
b. The velocity of the ball after the collision.
c. The average magnitude of the support reaction at A for an impact duration of 0.008 s.

20-41 A slender uniform bar ($\ell = 24$ in., $W_{AB} = 5$ lb) is attached to a frictionless hinge at A as shown in Fig. P20-41. The rod is released from rest with $\phi_0 = 90°$ and strikes the bumper at C ($d = 8$ in.). If the coefficient of restitution $e = 0.7$, determine:

a. The angular velocity of the bar immediately after the collision with the bumper.
b. The maximum angle of rebound of the bar.

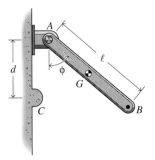

Fig. P20-41

c. The average magnitude of the support reaction at A for an impact duration of 0.005 s.
d. The amount of total system energy lost in the collision.

20-42* A slender uniform bar ($\ell = 750$ mm, $m_{AB} = 8$ kg) is attached to a frictionless hinge at A as shown in Fig. P20-41. The rod is released from rest with $\phi_0 = 60°$ and strikes the bumper at C ($d = 600$ mm). If the bar rebounds 30° after the impact, determine:

a. The coefficient of restitution of the impact.
b. The angular velocity of the bar immediately after the collision with the bumper.
c. The average magnitude of the support reaction at A for an impact duration of 0.008 s.

20-43* A slender uniform rod ($\ell = 30$ in., $W_{AB} = 4$ lb) is released from rest at an angle of $\theta = 70°$ to the horizontal and strikes a hard horizontal surface as shown in Fig. P20-43. If the initial height of the rod is $h = 60$ in. and the coefficient of restitution $e = 0.7$, determine:

a. The angular velocity of the rod after the impact.
b. The velocity of the mass center of the rod immediately after the impact.
c. If end B of the rod will strike the surface as the rod rotates immediately after the impact.

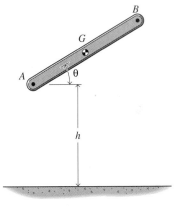

Fig. P20-43

20-44 A slender uniform rod ($\ell = 800$ mm, $m_{AB} = 2$ kg) is released from rest from an initial height of $h = 2$ m and strikes a hard horizontal surface as shown in Fig. P20-43. If the coefficient of restitution $e = 0.7$ and end B of the rod just clears the surface as the rod rotates immediately after the impact, determine:

a. The angle θ at which the bar was released.
b. The angular velocity of the rod after the impact.
c. The velocity of the mass center of the rod immediately after the impact.

20-45* Bar *AB* of Fig. P20-45*a* is attached to a frictionless pin at *A*; bar *CD* is attached to a frictionless pin at *E* and rests on a frictionless support at *F*. Both *AB* and *CD* are uniform slender bars 36 in. long and weighing 5 lb. Both bars are initially at rest when a slight disturbance causes bar *AB* to fall to the right and strike bar *CD* as shown in Fig. P20-45*b*. If the coefficient of restitution is *e* = 0.6, determine:

a. The angular velocities of both bars immediately after the impact.
b. The maximum angle of rebound of bar *AB* after the impact.
c. The average magnitude of the support reaction at *E* for an impact duration of 0.005 s.

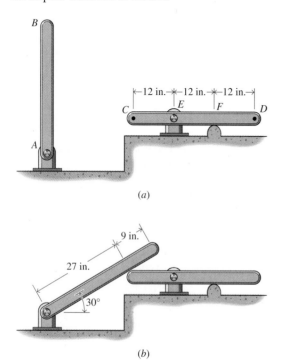

(a)

(b)

Fig. P20-45

20-46* The uniform slender bar *CD* at Fig. P20-46 is attached to a frictionless pin at *E* and rests on a frictionless support at *F*. The bar is 800 mm long with a mass of 4 kg and is initially at rest. The uniform slender bar *AB* is 500 mm long with a mass of 3 kg. Bar *AB* is released from rest with *h* = 2.5 m and strikes bar *CD* as shown. If the coefficient of restitution is *e* = 0.6, determine:

a. The angular velocities of both bars immediately after the impact.
b. The average magnitude of the support reaction at *E* for an impact duration of 0.003 s.

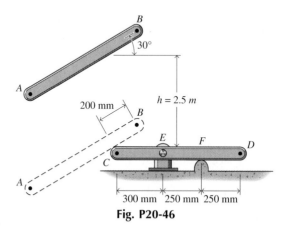

Fig. P20-46

20-47 Repeat Problem 20-45 for the case where bar *CD* is just resting on the pin at *E* instead of being attached to the pin. Determine the velocity of the mass center of bar *CD* immediately after the impact.

20-48 Repeat Problem 20-46 for the case where bar *CD* is just resting on the pin at *E* instead of being attached to the pin. Determine the velocity of the mass center of bar *CD* immediately after the impact.

20-6 ANGULAR IMPULSE AND MOMENTUM OF A RIGID BODY IN THREE-DIMENSIONAL MOTION

The general form of the impulse–momentum equations stated earlier in this chapter (Eqs. 20-2, 20-3, and 20-4) apply equally to an arbitrary system of particles and to the system of particles that make up a rigid body. The general form of the equations also apply equally to two-dimensional or three-dimensional motion. In fact, not only is the form of the linear impulse–momentum equation (Eq. 20-2) exactly the same for an arbitrary system of interacting particles, a rigid body in planar motion, and a rigid body in general three-dimensional motion, but the terms in the equation are also calculated in exactly the same manner for all three cases.

Calculation of the angular momentum terms in Eqs. 20-3 and 20-4, however, depends on whether the particles move independently or form a rigid body. For the planar motion of a symmetrical rigid body, the angular momentum was simply (Eq. 20-6)

$$H_G\mathbf{k} = I_G\omega\mathbf{k}$$

For the three-dimensional motion of a rigid body, however, the angular momentum has a number of additional components that are absent in plane motion.

20-6-1 Angular Momentum

The angular momentum of a particle about a point is just the moment of the linear momentum of the particle about that point. Let A be an arbitrary point in a rigid body (Fig. 20-19). Then the angular momentum of the particle P of mass dm about point A is given by

$$d\mathbf{H}_A = \mathbf{r}_{P/A} \times \mathbf{v}_P \, dm = \mathbf{r}_{P/A} \times (\mathbf{v}_A + \mathbf{v}_{P/A}) \, dm$$
$$= \mathbf{r}_{P/A} \times [\mathbf{v}_A + (\boldsymbol{\omega} \times \mathbf{r}_{P/A})] \, dm \qquad (20\text{-}14)$$

where $\mathbf{v}_A$ and $\mathbf{v}_P = \mathbf{v}_A + \mathbf{v}_{P/A}$ are the absolute velocities of point A and particle P; $\mathbf{r}_{P/A}$ and $\mathbf{v}_{P/A} = \boldsymbol{\omega} \times \mathbf{r}_{P/A}$ are the position and velocity of particle P relative to point A; and $\boldsymbol{\omega}$ is the angular velocity of the rigid body. Integration of Eq. 20-14 over all the particles in the rigid body gives the angular momentum of the entire rigid body about A as

$$\mathbf{H}_A = \int \mathbf{r}_{P/A} \times [\mathbf{v}_A + (\boldsymbol{\omega} \times \mathbf{r}_{P/A})] \, dm$$
$$= \left(\int \mathbf{r}_{P/A} \, dm \right) \times \mathbf{v}_A + \int \mathbf{r}_{P/A} \times (\boldsymbol{\omega} \times \mathbf{r}_{P/A}) \, dm \qquad (20\text{-}15)$$

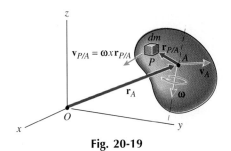

Fig. 20-19

where $\mathbf{v}_A$ is independent of dm and has been taken outside of the integral in the first term. Then by the definition of the mass center, the first integral can be written $\int \mathbf{r}_{P/A} \, dm = m\mathbf{r}_{G/A}$. Therefore,

$$\mathbf{H}_A = \mathbf{r}_{G/A} \times (m\mathbf{v}_A) + \int \mathbf{r}_{P/A} \times (\boldsymbol{\omega} \times \mathbf{r}_{P/A}) \, dm \qquad (20\text{-}16)$$

Equation 20-16 can be simplified even further for special choices of the point A. For example, if point A is a fixed point about which the body is rotating, then $\mathbf{v}_A = \mathbf{0}$ and Eq. 20-16 becomes

$$\mathbf{H}_A = \int \mathbf{r}_{P/A} \times (\boldsymbol{\omega} \times \mathbf{r}_{P/A}) \, dm \qquad (20\text{-}17)$$

Similarly, if point A is the center of mass G, then $\mathbf{r}_{G/A} = \mathbf{r}_{G/G} = \mathbf{0}$, and Eq. 20-16 becomes

$$\mathbf{H}_G = \int \mathbf{r}_{P/G} \times (\boldsymbol{\omega} \times \mathbf{r}_{P/G}) \, dm \qquad (20\text{-}18)$$

Even when point A is an arbitrary point, Eq. 20-16 can be written in a slightly more convenient form. Using the substitution $\mathbf{r}_{P/A} = \mathbf{r}_{P/G} + \mathbf{r}_{G/A}$ in Eq. 20-16 gives

$$\mathbf{H}_A = \mathbf{r}_{G/A} \times (m\mathbf{v}_A) + \int (\mathbf{r}_{P/G} + \mathbf{r}_{G/A}) \times [\boldsymbol{\omega} \times (\mathbf{r}_{P/G} + \mathbf{r}_{G/A})] \, dm$$

But $\mathbf{r}_{G/A}$ and $\boldsymbol{\omega}$ are independent of dm and can be taken outside of the integral. Therefore,

$$\mathbf{H}_A = \mathbf{r}_{G/A} \times (m\mathbf{v}_A) + \int (\mathbf{r}_{P/G} \times (\boldsymbol{\omega} \times \mathbf{r}_{P/G}) \, dm$$
$$+ \left(\int \mathbf{r}_{P/G} \, dm \right) \times (\boldsymbol{\omega} \times \mathbf{r}_{G/A}) + \mathbf{r}_{G/A} \times \left(\boldsymbol{\omega} \times \int \mathbf{r}_{P/G} \, dm \right)$$
$$+ \mathbf{r}_{G/A} \times (\boldsymbol{\omega} \times \mathbf{r}_{G/A}) \int dm \qquad (20\text{-}19)$$

But the first integral in Eq. 20-19 is just $\mathbf{H}_G$ (Eq. 20-18), the second and third integrals are zero since $\int \mathbf{r}_{P/G} \, dm = m\mathbf{r}_{G/G} = \mathbf{0}$ by the definition of the mass center, and the last term is $\mathbf{r}_{G/A} \times m\mathbf{v}_{G/A}$. Finally, combining the first and last terms and using the relative velocity equation $\mathbf{v}_G = \mathbf{v}_A + \mathbf{v}_{G/A}$ gives

$$\mathbf{H}_A = \mathbf{r}_{G/A} \times (m\mathbf{v}_G) + \mathbf{H}_G \qquad (20\text{-}20)$$

That is, the momentum properties of a rigid body may be represented by the equivalent "force–couple" system shown in the kinetic diagram of Fig. 20-20. Although the resultant angular momentum vector $\mathbf{H}_G$ is a free vector, it is represented as acting about the mass center G for convenience. The resultant linear momentum vector $\mathbf{L} = m\mathbf{v}_G$ acts through the mass center G.

Equations 20-17 and 20-18 have similar structure and can be developed simultaneously by writing $\mathbf{r} = x\mathbf{i} + y\mathbf{j} + z\mathbf{k}$ and $\boldsymbol{\omega} = \omega_x\mathbf{i} + \omega_y\mathbf{j} + \omega_z\mathbf{k}$. For Eq. 20-17 $\mathbf{H} = \mathbf{H}_A$, $\mathbf{r} = \mathbf{r}_{P/A}$, and the positions x, y, and z are measured relative to coordinate axes centered at the fixed point A.

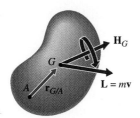

Fig. 20-20

For Eq. 20-18 $\mathbf{H} = \mathbf{H}_G$, $\mathbf{r} = \mathbf{r}_{P/G}$, and the positions x, y, and z are measured relative to coordinate axes centered at the mass center G.

Now, expanding the triple vector product in Eqs. 20-17 and 20-18 gives

$$\mathbf{H} = \left(\omega_x \int (y^2 + z^2)\, dm - \omega_y \int xy\, dm - \omega_z \int xz\, dm \right) \mathbf{i}$$
$$+ \left(-\omega_x \int xy\, dm + \omega_y \int (x^2 + z^2)\, dm - \omega_z \int yz\, dm \right) \mathbf{j}$$
$$+ \left(-\omega_x \int xz\, dm - \omega_y \int yz\, dm + \omega_z \int (x^2 + y^2)\, dm \right) \mathbf{k} \quad (20\text{-}21)$$

where the components of the angular velocity are also independent of dm and have again been taken outside of the integrals. The integrals in Eq. 20-21 represent the mass moments of inertia and products of inertia of the body with respect to the xyz-axes:

$$I_x = \int (y^2 + z^2)\, dm \qquad I_{xy} = \int xy\, dm = I_{yx}$$
$$I_y = \int (x^2 + z^2)\, dm \qquad I_{yz} = \int yz\, dm = I_{zy} \qquad (20\text{-}22)$$
$$I_z = \int (x^2 + y^2)\, dm \qquad I_{xz} = \int xz\, dm = I_{zx}$$

The calculation of the moments and products of inertia was covered in the volume on *Statics*. Much of that material has been repeated in Appendix A of this volume.

Substituting the moments of inertia from Eqs. 20-22 into Eqs. 20-17 and 20-18 gives the angular momentum of the body about a fixed point A or about its mass center G:

$$\mathbf{H} = (I_x \omega_x - I_{xy} \omega_y - I_{xz} \omega_z)\, \mathbf{i}$$
$$+ (-I_{yx} \omega_x + I_y \omega_y - I_{yz} \omega_z)\, \mathbf{j}$$
$$+ (-I_{zx} \omega_x - I_{zy} \omega_y + I_z \omega_z)\, \mathbf{k} \quad (20\text{-}23)$$

where the moments and products of inertia are relative to axes through the fixed point A for $\mathbf{H}_A$ or relative to axes through the mass center G for $\mathbf{H}_G$. Equation 20-23 is valid for a particular position of the body. Since the orientation of the coordinate axes is fixed, the moments of inertia and products of inertia will, in general, change as the body rotates relative to xyz.

Equation 20-23 appears rather complicated, but it can be simplified considerably for special orientations of the coordinate axes. The ideal set of axes is the principal axes of inertia $\hat{x}\hat{y}\hat{z}$. If the coordinate axes coincide with the principal axes of inertia, then all the products of inertia are zero $I_{\hat{x}\hat{y}} = I_{\hat{y}\hat{x}} = I_{\hat{y}\hat{z}} = I_{\hat{z}\hat{y}} = I_{\hat{x}\hat{z}} = I_{\hat{z}\hat{x}} = 0$. Then Eq. 20-23 becomes (for that instant only, in most cases)

$$\mathbf{H} = (I_{\hat{x}}\omega_{\hat{x}})\,\mathbf{i} + (I_{\hat{y}}\omega_{\hat{y}})\,\mathbf{j} + (I_{\hat{z}}\omega_{\hat{z}})\,\mathbf{k} \tag{20-24}$$

where $I_{\hat{x}}$, $I_{\hat{y}}$, and $I_{\hat{z}}$ are the principal moments of inertia. Even though use of the principal axes simplifies the expression for the angular momentum, it is not always convenient for geometric reasons to use these axes to compute $\mathbf{H}$.

Finally, it must be noted that the angular momentum vector $\mathbf{H}$ and the angular velocity $\boldsymbol{\omega}$ will have different directions unless $\boldsymbol{\omega}$ is directed along a principal axis of inertia. For example, for the plane motion of a rigid body that is symmetrical with respect to the xy-plane, the z-axis is a principal direction, I_z is a principal moment of inertia, $\boldsymbol{\omega} = \omega_z\mathbf{k}$, and Eq. 20-24 gives

$$\mathbf{H} = (I_z\omega_z)\,\mathbf{k} = I_z(\omega_z\mathbf{k}) = I_z\boldsymbol{\omega} \tag{20-25}$$

Therefore, the vectors $\mathbf{H}$ and $\boldsymbol{\omega}$ are collinear. Actually, if the three principal moments of inertia are equal, $I_{\hat{x}} = I_{\hat{y}} = I_{\hat{z}} = \hat{I}$, then Eq. 20-24 gives

$$\mathbf{H} = \hat{I}(\omega_{\hat{x}}\mathbf{i} + \omega_{\hat{y}}\mathbf{j} + \omega_{\hat{z}}\mathbf{k}) = \hat{I}\boldsymbol{\omega} \tag{20-26}$$

and the vectors $\mathbf{H}$ and $\boldsymbol{\omega}$ are also collinear. However, if the three principal moments of inertia are equal, then every axis is a principal axis. Therefore, Eq. 20-26 is just a special case of Eq. 20-25 since no matter what direction $\boldsymbol{\omega}$ is in, it will coincide with a principal direction.

20-6-2   Angular Impulse–Momentum Principle

The Angular Impulse–Momentum Principle for a system of particles (Eq. 20-4)

$$(\mathbf{H}_O)_i + \int_{t_i}^{t_f} \Sigma\mathbf{M}_{\ell/O}\,dt = (\mathbf{H}_O)_f \tag{20-27a}$$

$$(\mathbf{H}_G)_i + \int_{t_i}^{t_f} \Sigma\mathbf{M}_{\ell/G}\,dt = (\mathbf{H}_G)_f \tag{20-27b}$$

applies to any system of particles whether it consists of independently moving, interacting particles (where $\mathbf{H}$ is computed by summing $\mathbf{r}_{\ell} \times m\mathbf{v}_{\ell}$ for all the particles) or the particles that make up a rigid body. Restricting the point about which moments and angular momentum is calculated to either the mass center G or a fixed point O about which the rigid body is rotating allows the angular momentum $\mathbf{H}$ to be computed using Eq. 20-23. If the rigid body is rotating about a fixed point O, then Eq. 20-27a can be used, with the moments of inertia in Eq. 20-23 calculated relative to coordinate axes centered at the fixed point O. Otherwise, Eq. 20-27b must be used, with the moments of inertia in Eq. 20-23 calculated relative to coordinate axes centered at the mass center G.

The angular impulse–momentum equations (Eq. 20-27) are particularly useful when the moment of external forces about a specified axis is known. For example, if no external force acting on a body has a moment about a particular axis, then the angular momentum about that axis is constant. Usually, when the body is acted on by impulsive forces, only the moments of the impulsive forces need be considered.

When a rigid body is rotating about a fixed point O that is not the center of mass, the impulse of the reaction must be included in the analysis if the mass center G is used as a reference (Eq. 20-27b). In such cases it is generally more convenient to use the fixed point O for reference since the reaction at O has no moment about O and will not enter into Eq. 20-27a.

As was the case with the application of the angular impulse momentum principle to a rigid body in planar motion, the body need move rigidly only at the initial and final instants of time for Eq. 20-23 to be used to calculate the angular momentum. Between the initial and final instants, parts of the body may move freely relative to each other.

20-6-3 Graphical Representation of the Linear and Angular Momentum Principles

The linear and angular momentum (linear and angular impulse) are analogous to a force and a couple. The impulse momentum principles expressed by

$$m(\mathbf{v}_G)_i + \int_{t_i}^{t_f} \mathbf{R}\, dt = m(\mathbf{v}_G)_f$$

$$(\mathbf{H}_G)_i + \int_{t_i}^{t_f} \Sigma \boldsymbol{M}_{\ell/G}\, dt = (\mathbf{H}_G)_f$$

can be represented by the kinetic diagrams of Fig. 20-21, in which the linear and angular momenta have been replaced with an equivalent "force–couple" $\mathbf{L} = m\mathbf{v}_G$ and $\mathbf{H}_G$ applied at the mass center. That is, the sum of the equivalent "force–couple" of the momenta at time t_1 (Fig. 20-21a) and the equivalent force–couple of the impulses (Fig. 20-21b) is equal to the equivalent "force–couple" of the momenta at time t_f (Fig. 20-21c).

The graphical representation of Fig. 20-21 can also be used to write the Angular Impulse–Momentum Principle relative to an arbitrary fixed point P as well as relative to a fixed point O about which the body is rotating. The former statement is immediately verified by comparing the moments about P of the equivalent "force–couple" in each part of the figure with the sum of $\mathbf{r}_{G/P} \times$ (Eq. 20-2) and Eq. 20-27b. The latter statement is similarly verified by adding $\mathbf{r}_{G/O} \times$ (Eq. 20-2) to Eq. 20-27b and making use of the parallel axis theorems for moments and products of inertia (Problem 20-79).

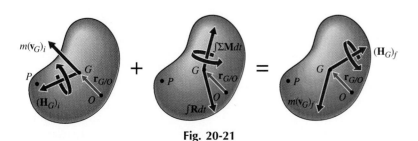

Fig. 20-21

20-6-4   Systems of Rigid Bodies

As has already been pointed out, the use of Eq. 20-27 requires only that the body behave rigidly at the initial time t_i and at the final time t_f so that the angular momentum **H** can be calculated using Eq. 20-23 at these times. Between the initial and final instants of time, parts of the body may move relative to each other. If parts of the body are moving relative to each other at the initial and/or the final instants of time, then a separate angular impulse momentum equation should be written for each part that does behave in a rigid manner and the equations added together. If the moments of momentum and moments of forces are all written relative to the same point in each equation, then the moments of the joint forces holding the various parts together will cancel in pairs and need not be computed. This is most easily accomplished using the graphical representation of Fig. 20-22. In the first and last parts of the figure the momentum of each rigid body has been replaced with an equivalent "force–couple" system at its own mass center. In the middle part of the figure, the joint forces are internal forces and need not be shown.

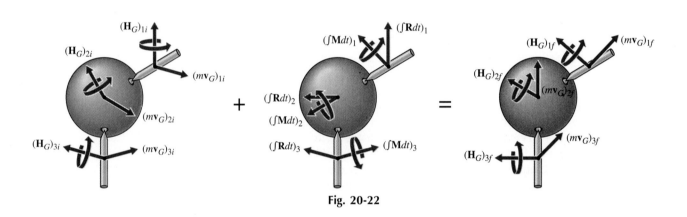

Fig. 20-22

EXAMPLE PROBLEM 20-6

A homogeneous wheel of diameter 800 mm, thickness 50 mm, and mass 40 kg is rigidly attached to an axle OG of length 1200 mm, diameter 50 mm, and mass 10 kg as shown in Fig. 20-23. The axle is pivoted at the fixed point O, and the wheel rolls without slipping on a horizontal floor. If the center of the wheel G has a speed of 2 m/s as the wheel rolls, determine for the instant shown:

a. The angular velocity $\boldsymbol{\omega}$ of the wheel and axle system.
b. The angular momentum $\mathbf{H}_O$ about O of the wheel and axle system.
c. The angle between $\boldsymbol{\omega}$ and $\mathbf{H}_O$.

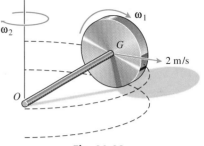

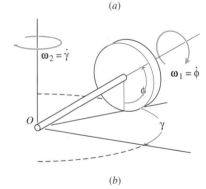

Fig. 20-23

SOLUTION

a. Set up a coordinate system with the x-axis along OG, the y-axis in the vertical plane containing OG, and the z-axis in the horizontal plane (Fig. 20-24a). These axes are principal axes for the wheel and axle system. As the wheel rotates about the axle OG at a rate of ω_1, it also rotates with the axle about a vertical axis at a rate ω_2. The total angular velocity of the system is therefore

$$\boldsymbol{\omega} = \omega_1 \mathbf{i} - \omega_2 (\sin \theta \, \mathbf{i} + \cos \theta \, \mathbf{j})$$

where $\theta = \tan^{-1}(400/1200) = 18.43°$. As the wheel rotates through an angle ϕ, the axle rotates through an angle γ and the arc lengths $(1.2/\cos \theta)\gamma$ and $(0.4)\phi$ are equal (Fig. 20-24b). Therefore, $(1.2/\cos \theta)\omega_2 = (0.4)\omega_1$ or $\omega_1 = 3.162\omega_2$. But since O is fixed,

$$\mathbf{v}_G = 2.0\mathbf{k} = \mathbf{v}_O + \mathbf{v}_{G/O} = \mathbf{0} + \boldsymbol{\omega} \times \mathbf{r}_{G/O}$$
$$= [(3.162 - \sin \theta)\omega_2 \mathbf{i} - (\cos \theta)\omega_2 \mathbf{j}] \times (1.2\mathbf{i})$$

Therefore, $\omega_2 = 1.7568$ rad/s, $\omega_1 = 5.5556$ rad/s, and

$$\boldsymbol{\omega} = 5.00\mathbf{i} - 1.667\mathbf{j} \text{ rad/s} \qquad \text{Ans.}$$

b. Then, since the xyz-axes are principal axes, the angular momentum of the system about O is

$$\mathbf{H}_O = (I_x \omega_x)\mathbf{i} + (I_y \omega_y)\mathbf{j} + (I_z \omega_z)\mathbf{k}$$

where

$$I_x = \frac{1}{2}(10)(0.025)^2 + \frac{1}{2}(40)(0.400)^2$$
$$= 3.203 \text{ kg} \cdot \text{m}^2$$

and

$$I_y = I_z = \frac{1}{4}(10)(0.025)^2 + \frac{1}{3}(10)(1.200)^2$$
$$+ \frac{1}{4}(40)(0.400)^2 + \frac{1}{12}(40)(0.050)^2 + (40)(1.225)^2$$
$$= 66.43 \text{ kg} \cdot \text{m}^2$$

Therefore,

$$\mathbf{H}_O = 16.03\mathbf{i} - 110.7\mathbf{j} \text{ kg} \cdot \text{m}^2/\text{s} \qquad \text{Ans.}$$

c. The angle between $\boldsymbol{\omega}$ and $\mathbf{H}_O$ is given by

$$\cos^{-1}\left(\frac{\boldsymbol{\omega} \cdot \mathbf{H}_O}{\omega H_O}\right) = \cos^{-1}\left(\frac{264.6}{(5.270)(111.88)}\right)$$
$$= 63.33° \qquad \text{Ans.}$$

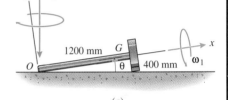

(a)

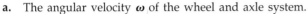

(b)

Fig. 20-24

A 2-lb uniform sign is suspended by two wires as shown in Fig. 20-25. The thickness of the sign is negligible compared to its surface dimensions. If a bullet weighing 0.05 lb and traveling with a speed of 500 ft/s in the negative x-direction strikes the sign at the corner C and becomes embedded in it, determine the velocity $\mathbf{v}_G$ of the mass center of the sign and the angular velocity $\boldsymbol{\omega}$ of the sign immediately after the impact.

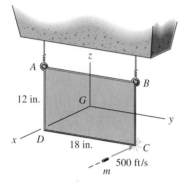

Fig. 20-25

SOLUTION

Separate linear and angular impulse–momentum equations will be written for the bullet and the sign and then added together. When the equations are added together, the linear and angular impulses of the force of interaction between the bullet and the sign will cancel. The combined kinetic diagrams are then shown in Fig. 20-26, in which the impulse of the force of interaction between the bullet and the sign has been omitted.

Linear Momentum: Initially, the sign is at rest and has no linear momentum $(\mathbf{L}_s)_i = \mathbf{0}$. After the impact, the linear momentum of the sign is just $(\mathbf{L}_s)_f = M\mathbf{v}_G$ where $M = 2/32.2$ slugs is the mass of the sign and $\mathbf{v}_G$ is the velocity of the mass center of the sign immediately after the impact. The initial and final linear momentum of the bullet are:

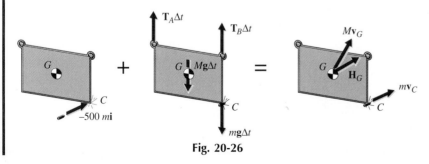

Fig. 20-26

$$(\mathbf{L}_b)_i = -500m\mathbf{i} = -0.7764\mathbf{i} \text{ lb} \cdot \text{s}$$
$$(\mathbf{L}_b)_f = m[\mathbf{v}_G + (\omega_x\mathbf{i} + \omega_y\mathbf{j} + \omega_z\mathbf{k}) \times (0.75\mathbf{j} - 0.50\mathbf{k})]$$
$$= m[\mathbf{v}_G - (0.5\omega_y + 0.75\omega_z)\mathbf{i} + 0.5\omega_x\mathbf{j} + 0.75\omega_x\mathbf{k}]$$

where $m = 0.05/32.2$ slugs is the mass of the bullet, $\boldsymbol{\omega} = \omega_x\mathbf{i} + \omega_y\mathbf{j} + \omega_z\mathbf{k}$ is the angular velocity of the sign immediately after the impact, and the velocity of the bullet has been replaced using the relative velocity equation $\mathbf{v} = \mathbf{v}_G + \boldsymbol{\omega} \times \mathbf{r}_{C/G}$. Then the x-, y-, and z-components of the linear impulse–momentum equation are

$$-500m = (m + M)v_{Gx} - m(0.5\omega_y + 0.75\omega_z) \qquad (a)$$
$$0 = (m + M)v_{Gy} + 0.5\,m\omega_x \qquad (b)$$
$$(T_A + T_B - 2.05)\Delta t = (m + M)v_{Gz} + 0.75\,m\omega_x \qquad (c)$$

Angular Momentum About G: Initially, the sign is at rest and has no angular momentum $(\mathbf{H}_{Gs})_i = \mathbf{0}$. Since the x-, y-, z-axes are principal axes, the angular momentum of the sign after the impact is just $(\mathbf{H}_{Gs})_f = I_x\omega_x\mathbf{i} + I_y\omega_y\mathbf{j} + I_z\omega_z\mathbf{k}$ where

$$I_x = \frac{1}{12}\left(\frac{2}{32.2}\right)(1.5^2 + 1.0^2) = 0.016822 \text{ lb} \cdot \text{ft} \cdot \text{s}^2$$

$$I_y = \frac{1}{12}\left(\frac{2}{32.2}\right)(1.0^2) = 0.005176 \text{ lb} \cdot \text{ft} \cdot \text{s}^2$$

$$I_z = \frac{1}{12}\left(\frac{2}{32.2}\right)(1.5^2) = 0.011646 \text{ lb} \cdot \text{ft} \cdot \text{s}^2$$

The angular momentum of the bullet about G before the impact is the moment of its linear momentum

$$(\mathbf{H}_{Gb})_i = (0.75\mathbf{j} - 0.50\mathbf{k}) \times m(-500\mathbf{i})$$
$$= 0.3882\mathbf{j} + 0.5823\mathbf{k} \text{ lb} \cdot \text{ft} \cdot \text{s}$$

Similarly,

$$(\mathbf{H}_{Gb})_f = (0.75\mathbf{j} - 0.50\mathbf{k}) \times m[\mathbf{v}_G + \boldsymbol{\omega} \times (0.75\mathbf{j} - 0.50\mathbf{k})]$$

Then, the x-, y-, and z-components of the angular impulse–momentum equation are

$$(T_B - T_A - 0.05)(0.75)\Delta t = I_x\omega_x + 0.75\,mv_{Gz}$$
$$+ 0.5\,mv_{Gy} + (0.75^2 + 0.5^2)m\omega_x \qquad (d)$$
$$0.3882 = I_y\omega_y - 0.5\,mv_{Gx} + 0.5\,m(0.5\,\omega_y + 0.75\,\omega_z) \qquad (e)$$
$$0.5823 = I_z\omega_z - 0.75\,mv_{Gx} + 0.75\,m(0.5\,\omega_y + 0.75\,\omega_z) \qquad (f)$$

Equations a through f are to be solved subject to the additional constraints: the wires are inextensible (neither A nor B can have a component of velocity in the negative z-direction immediately after the impact); the tensions T_A and T_B must not be negative (the wires cannot withstand compressive forces); and if either A or B has a component of velocity in the positive z-direction immediately after impact, then the corresponding tension T_A or T_B must be zero. Guessing that the wires both remain taut (T_A and T_B both remain positive) and that $v_{Gz} = \omega_x = 0$ gives the solution

$$v_{Gx} = -10.64 \text{ ft/s} \qquad \text{Ans.}$$
$$v_{Gy} = v_{Gz} = \omega_x = 0 \qquad \text{Ans.}$$
$$\omega_y = 63.8 \text{ rad/s} \qquad \text{Ans.}$$
$$\omega_z = 42.6 \text{ rad/s} \qquad \text{Ans.}$$

Finally, substituting this solution into Eqs. c and d gives $T_A = 1.00$ lb and $T_B = 1.05$ lb, thus verifying that both wires remain taut. Furthermore, it is easily verified that $\mathbf{v}_A = 53.2\mathbf{i}$ ft/s and $\mathbf{v}_B = -10.64\mathbf{i}$ ft/s so that neither A nor B has a z-component of velocity.

PROBLEMS

20-49–20-59 For each of the objects shown, determine the angular momentum $\mathbf{H}_O$ (where O is the origin of the coordinate system) and the angle between the angular velocity and the angular momentum vectors at the instant shown. Neglect the mass of the shafts that the spheres, plates, and so on are mounted to.

20-49* The slender bent rod of Fig. P20-49 weighs 0.2 lb/ft.

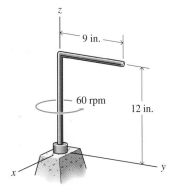

Fig. P20-49

20-50* The slender branched rod of Fig. P20-50 has a mass of 0.25 kg/m.

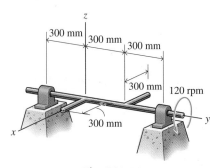

Fig. P20-50

20-51 The thin disk of Fig. P20-51 has a radius of 10 in., weighs 2 lb, and is mounted 7 in. off center.

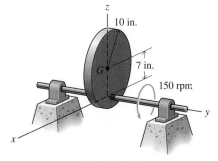

Fig. P20-51

20-52* The thin rectangular plate of Fig. P20-52 is 300 mm high, 800 mm long, and has a mass of 5 kg.

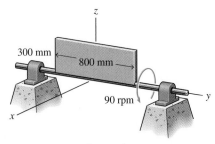

Fig. P20-52

20-53 The three identical spheres of Fig. P20-53 each weigh 2 lb and have a diameter of 4 in. The centers of the spheres are 10 in. from the center of the shaft, and they are symmetrically located around the shaft.

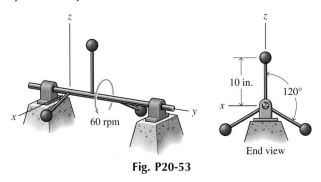

Fig. P20-53

20-54 The 1.5-kg cylinder AB of Fig. P20-54 has a diameter of 50 mm and a length of 200 mm. It is mounted to a thin circular disk having a radius of 400 mm and a mass of 0.5 kg. The distance between the axis of the cylinder and the axis of the shaft is 300 mm.

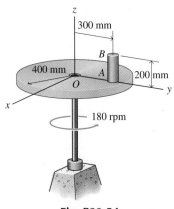

Fig. P20-54

20-55* The bent rod of Fig. P20-55 has a diameter of 0.5 in., weighs 0.2 lb/ft, and is 18 in. long

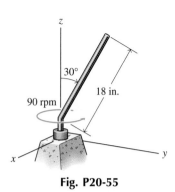

Fig. P20-55

20-56 The 100-mm long rod of Fig. P20-56 has a diameter of 20 mm, a mass of 3 kg, and is mounted at an angle of 30° to the axis of the shaft.

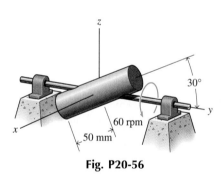

Fig. P20-56

20-57* The thin circular plate of Fig. P20-57 has a radius of 9 in. and weighs 1.5 lb. The plane of the plate makes an angle of 60° with the axis of the shaft.

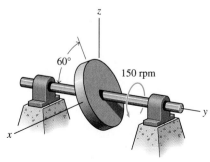

Fig. P20-57

20-58* The thin rectangular plate (300 mm by 800 mm) of Fig. P20-58 has a mass of 5 kg and rotates about a shaft along its diagonal.

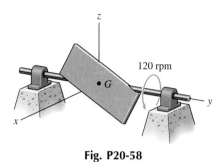

Fig. P20-58

20-59 The 12-in. long circular cylinder of Fig. P20-59 has a diameter of 6 in., weighs 4 lb, and rotates about a shaft along its "diagonal" as shown.

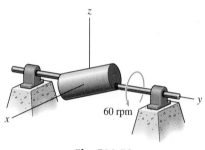

Fig. P20-59

20-60* The thin rectangular plate of Fig. P20-60 is 300 mm tall, 800 mm wide, and has a mass of 5 kg. The plate is balanced on edge when it is struck by an impulse of $\mathbf{F}\Delta t = -10\mathbf{i}\ \text{N}\cdot\text{s}$ at corner C. Assume that Δt is sufficiently small that nonimpulsive forces can be neglected and determine:

a. The velocity of the mass center of the plate immediately after the impact.
b. The angular velocity of the plate immediately after the impact.
c. The angle between the velocity and the angular momentum vectors immediately after the impact.

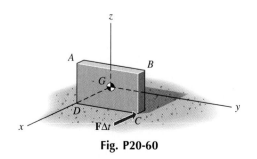

Fig. P20-60

20-61 The thin plate of Fig. P20-61 is an equilateral triangle 18 in. on a side and weighs 5 lb. The plate is balanced on edge when it is struck by an impulse $\mathbf{F}\Delta t = -0.15\mathbf{i}$ lb·s at corner A. Assume that Δt is sufficiently small that nonimpulsive forces can be neglected and determine:

a. The velocity of the mass center of the plate immediately after the impact.
b. The angular velocity of the plate immediately after the impact.
c. The angle between the angular velocity and the angular momentum vectors immediately after the impact.

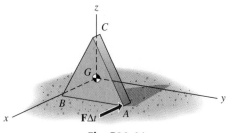

Fig. P20-61

20-62 The thin circular plate of Fig. P20-62 has a radius of 300 mm and a mass of 2 kg. The plate is suspended on a wire when it is struck by an impulse of $\mathbf{F}\Delta t = -1.4\mathbf{i}$ N·s at point A. Assume that Δt is sufficiently small that nonimpulsive forces can be neglected and determine:

a. The angle between the angular velocity and the angular momentum vectors immediately after the impact.
b. The velocity of point A immediately after the impact.

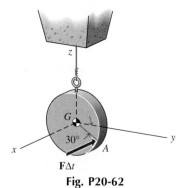

Fig. P20-62

20-63* The thin plate of Fig. P20-63 is a square 6 in. on a side and weighs 2 lb. The plate is suspended on a wire attached to the middle of side AB when it is struck by an impulse of $\mathbf{F}\Delta t = -0.05\mathbf{i}$ lb·s at corner C. Assume that Δt

is sufficiently small that nonimpulsive forces can be neglected and determine:

a. The angle between the angular velocity and the angular momentum vectors immediately after the impact.
b. The velocity of point C immediately after the impact.

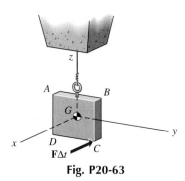

Fig. P20-63

20-64 The thin plate of Fig. P20-64 is a square 200 mm on a side and has a mass of 1.5 kg. The plate is suspended on a wire attached to corner A when it is struck by an impulse of $\mathbf{F}\Delta t = -2.5\mathbf{i}$ N·s at E (the midpoint of side AD). Assume that Δt is sufficiently small that nonimpulsive forces can be neglected and determine:

a. The angle between the angular velocity and the angular momentum vectors immediately after the impact.
b. The velocity of corner C immediately after the impact.

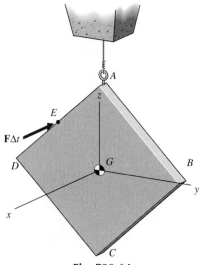

Fig. P20-64

20-65* The thin plate of Fig. P20-65 is a rectangle 9 in. high and 18 in. wide and weighs 2.5 lb. The plate is suspended on a wire attached to corner A when it is struck by an impulse of $\mathbf{F}\Delta t = -0.1\mathbf{i}$ lb·s at corner D. Assume that Δt is sufficiently small that nonimpulsive forces can be neglected and determine:

a. The angle between the angular velocity and the angular momentum vectors immediately after the impact.
b. The velocity of corner D immediately after the impact.

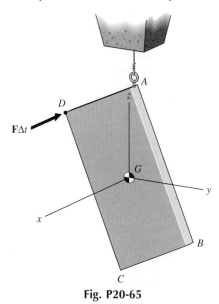

Fig. P20-65

20-66* The thin plate of Fig. P20-66 has a mass of 1.2 kg. The plate is suspended on a wire attached to its edge when it is struck by an impulse $\mathbf{F}\Delta t = 0.5\mathbf{i} + \mathbf{j} - 0.8\mathbf{k}$ N·s at corner C. Assume that Δt is sufficiently small that nonimpulsive forces can be neglected and determine:

a. The angle between the angular velocity and the angular momentum vectors immediately after the impact.
b. The velocity of corner C immediately after the impact.

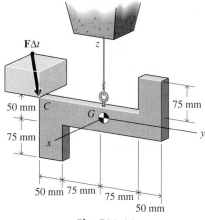

Fig. P20-66

20-67 The assembly of Fig. P20-67 is made by joining five identical slender rods (12 in. long and 0.5 lb each). The assembly is suspended on a wire attached to the midpoint of segment CD when it is struck by an impulse of $\mathbf{F}\Delta t = 0.1\mathbf{j} - 0.05\mathbf{k}$ lb·s at A. Assume that Δt is sufficiently small that nonimpulsive forces can be neglected and determine:

a. The angle between the angular velocity and the angular momentum vectors immediately after the impact.
b. The velocity of end A immediately after the impact.

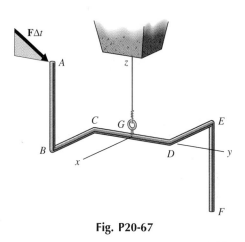

Fig. P20-67

20-68* The assembly of Fig. P20-68 is made by joining five identical slender rods (300 mm long and 0.25 kg each). The assembly is suspended on a wire attached to the midpoint of segment CD when it is struck by an impulse of $\mathbf{F}\Delta t = 0.5\mathbf{j} - 0.25\mathbf{k}$ N·s at A. Assume that Δt is sufficiently small that nonimpulsive forces can be neglected and determine:

a. The angle between the angular velocity and the angular momentum vectors immediately after the impact.
b. The velocity of end A immediately after the impact.

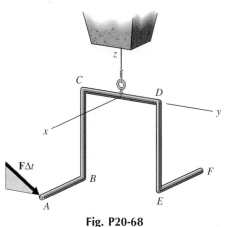

Fig. P20-68

20-69 The assembly of Fig. P20-69 is made by joining six identical slender rods (15 in. long and 0.6 lb each) to another slender rod (*DK*: 30 in. long and 1.2 lb). The assembly is suspended on a wire attached to joint *K* when it is struck by an impulse of $\mathbf{F}\Delta t = \mathbf{0.15j} - 0.1\mathbf{k}$ lb·s at *A*. Assume that Δt is sufficiently small that nonimpulsive forces can be neglected and determine:

a. The angle between the angular velocity and the angular momentum vectors immediately after the impact.
b. The velocity of end *A* immediately after the impact.

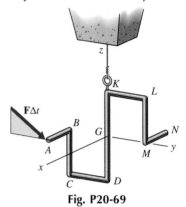

Fig. P20-69

20-70 A 50-g bullet traveling with a speed of 150 m/s in the negative *x*-direction strikes the rectangular plate of Problem 20-60 at the corner *C* (see p. 423). If the bullet becomes embedded in the plate, determine:

a. The velocity of the mass center of the plate immediately after the impact.
b. The angular velocity of the plate immediately after the impact.
c. The velocity of corner *C* immediately after the impact.

20-71* A 0.08-lb bullet traveling with a speed of 600 ft/s in the negative *x*-direction strikes the triangular plate of Problem 20-61 at the corner *A* (see p. 424). If the bullet becomes embedded in the plate, determine:

a. The velocity of the mass center of the plate immediately after the impact.
b. The angular velocity of the plate immediately after the impact.
c. The velocity of corner *A* immediately after the impact.

20-72 A 30-g bullet traveling with a speed of 180 m/s in the negative *x*-direction strikes the circular plate of Problem 20-62 at the point *A* (see p. 424). If the bullet becomes embedded in the plate, determine:

a. The magnitude of the impulse exerted on the plate by the bullet.
b. The angular momentum $\mathbf{H}_G$ of the plate immediately after the impact.
c. The velocity of point *A* immediately after the impact.

20-73* A 0.05-lb bullet traveling with a speed of 800 ft/s in the negative *x*-direction strikes the square plate of Problem 20-63 at the corner *C* (see p. 424). If the bullet becomes embedded in the plate, determine:

a. The magnitude of the impulse exerted on the plate by the bullet.
b. The angular momentum $\mathbf{H}_G$ of the plate immediately after the impact.
c. The velocity of corner *C* immediately after the impact.

20-74* A 125-g arrow traveling with a speed of 100 m/s in the negative *x*-direction strikes the rectangular plate of Problem 20-60 at the corner *A* (see p. 423). (The arrow may be modeled as a uniform slender rod 800 mm long.) If the arrowhead becomes embedded in the plate, determine:

a. The velocity of the mass center of the plate immediately after the impact.
b. The angular velocity of the plate immediately after the impact.
c. The velocity of corner *A* immediately after the impact.

20-75 A 0.2-lb arrow traveling with a speed of 300 ft/s in the negative *x*-direction strikes the triangular plate of Problem 20-61 at the corner *A* (see p. 424). (The arrow may be modeled as a uniform slender rod 32 in. long.) If the arrowhead becomes embedded in the plate, determine:

a. The velocity of the mass center of the plate immediately after the impact.
b. The angular velocity of the plate immediately after the impact.
c. The velocity of corner *A* immediately after the impact.

20-76* A 100-g arrow traveling with a speed of 150 m/s in the negative *x*-direction strikes the circular plate of Problem 20-62 at the point *A* (see p. 424). (The arrow may be modeled as a uniform slender rod 800 mm long.) If the arrowhead becomes embedded in the plate, determine:

a. The magnitude of the impulse exerted on the plate by the arrow.
b. The angular momentum $\mathbf{H}_G$ of the plate immediately after the impact.
c. The velocity of point *A* immediately after the impact.

20-77* A 0.25-lb arrow traveling with a speed of 400 ft/s in the negative *x*-direction strikes the square plate of Problem 20-63 at the corner *C* (see p. 424). (The arrow may be modeled as a uniform slender rod 32 in. long.) If the arrowhead becomes embedded in the plate, determine:

a. The magnitude of the impulse exerted on the plate by the arrow.
b. The angular momentum $\mathbf{H}_G$ of the plate immediately after the impact.
c. The velocity of corner *C* immediately after the impact.

SUMMARY

The linear and angular momentum principles are integrals of the equations of motion with respect to time. They are particularly useful for solving problems in which the velocity of a body for two different instants of time are to be related and the forces involved can be expressed as functions of time.

The linear momentum of a system of particles whether rigid or nonrigid is the product of its mass and the velocity of its mass center $\mathbf{L} = m\mathbf{v}_G$. Therefore, the Linear Impulse–Momentum Principle as expressed by Eq. 20-2

$$m(\mathbf{v}_G)_i + \sum_{\ell} \int_{t_i}^{t_f} \mathbf{R}_{\ell}\, dt = m(\mathbf{v}_G)_f \tag{20-2}$$

applies equally to a system of independent, interacting particles and to rigid bodies.

The angular momentum of a particle can be calculated relative to any point, fixed or moving. For an arbitrary system of interacting particles, the particles move independently, and the expression of the Angular Impulse–Momentum Principle relative to a fixed point O is usually the most useful. For rigid bodies, however, velocities of points in the body are related by the angular velocity and the expression of the Angular Impulse–Momentum Principle relative to the mass center is usually the most useful.

For planar motion, $\boldsymbol{\omega} = \omega\mathbf{k}$, the angular momentum of a rigid body is

$$\mathbf{H}_G = -\omega I_{Gxz}\mathbf{i} - \omega I_{Gyz}\mathbf{j} + \omega I_{Gz}\mathbf{k} \tag{20-5}$$

If the products of inertia I_{Gxz} and I_{Gyz} are not zero, then the angular momentum will also have x- and y-components even for planar motion. This means that moment components in the x- and/or y-direction will be required to keep the motion in the x-y plane if the magnitude of the angular velocity is changing.

When rigid bodies collide, the coefficient of restitution relates the relative velocity of the points of contact before and after the collision. Since the principles of linear and angular impulse–momentum involve the velocities of the mass centers of the rigid bodies, the velocities of the contact points must be related to the velocities of the mass centers of the rigid bodies using the relative velocity equations.

REVIEW PROBLEMS

20-78* Each of the four wheels of the cart shown in Fig. P20-78 is a uniform disk 500 mm in diameter having a mass of 5 kg. If the cart and its load add an additional 30 kg to the total mass, determine the speed the cart will attain in 10 s after starting from rest.

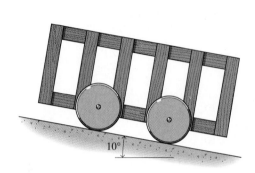

Fig. P20-78

20-79* When an LP record drops onto a spinning turntable (Fig. P20-79), it slips for a short time before attaining its final speed of $33\frac{1}{3}$ rpm. The record has a diameter of 12 in. and weighs 2.0 oz. If the coefficient of friction between the record and the turntable is $\mu_k = 0.2$, determine the length of time that the record will slip. (Assume that the contact force between the turntable and the record is uniformly distributed over the surface of the record.)

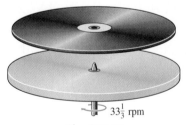

$33\frac{1}{3}$ rpm

Fig. P20-79

20-80 A 1.5-kg uniform slender rod AB hangs in a vertical plane from a friction pivot as shown in Fig. P20-80. A 125-g arrow traveling at 100 m/s strikes the rod at a point 500 mm below the pivot. The arrow may be modeled as a uniform slender rod 800 mm long. If the arrowhead becomes embedded in the rod, determine:

a. The angular velocity ω of the system immediately after the impact.

b. The average moment exerted on the arrow by the bar AB for an impact duration of $\Delta t = 0.01$ s.

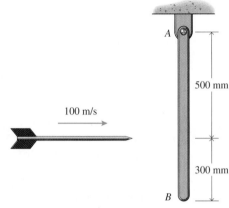

Fig. P20-80

20-81* A square crate sliding across a frictionless floor strikes a small obstacle A as shown in Fig. P20-81. If the crate rotates about A after impact, determine:

a. The minimum speed v_0 for which the crate will tip all the way over.

b. The velocity $\mathbf{v}_G$ and angular velocity ω of the crate immediately after the impact.

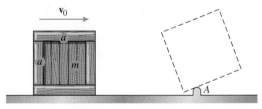

Fig. P20-81

20-82* Determine the height h at which the cue stick shown in Fig. P20-82 must strike the billiard ball if the ball is to roll without slipping and without the aid of friction. (Assume that the cue stick exerts only a horizontal force on the ball.)

Fig. P20-82

20-83 The rear drive wheels of the light tractor sketched in Fig. P20-83 are 5 ft in diameter. The two wheels rotate as a unit having a combined weight of 1000 lb and a centroidal radius of gyration about the axle of $k_G = 1.5$ ft. The rest of the tractor weighs an additional 2000 lb and its center of mass is 1.5 ft above the rear axle. If the tractor accelerates from zero to 20 mi/h in 3 s, determine:

a. The average frictional force exerted by the ground on the drive tires.
b. The average torque applied to the axle of the rear wheels.
c. The minimum distance a between the rear axle and the center of gravity of the tractor body for which the tractor will not overturn.

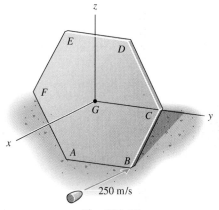

Fig. P20-83

20-84 The thin plate of Fig. P20-84 is a regular hexagon 300 mm on a side and has a mass of 2 kg. The plate is balanced on edge when a 60-g bullet traveling with a speed of 250 m/s in the negative x-direction strikes the plate at the corner B. If the bullet becomes embedded in the plate, determine:

a. The velocity $\mathbf{v}_G$ of the mass center of the plate immediately after the impact.
b. The angular velocity $\boldsymbol{\omega}$ of the plate immediately after the impact.
c. The velocity of corner B immediately after the impact.

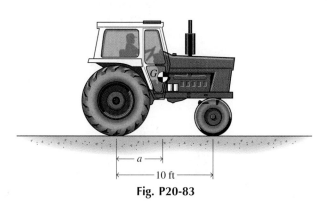

Fig. P20-84

20-85* Sphere A of Fig. P20-85 is rolling without slipping on a horizontal surface when it collides head-on with an identical stationary sphere B. If the coefficient of kinetic friction between the spheres and the horizontal surface is $\mu_k = 0.4$ and the collision is perfectly elastic $e = 1$, determine:

a. The linear and angular velocity of both spheres immediately after the impact.
b. The velocity of both spheres when they both roll without slipping after the impact.

Fig. P20-85

20-86* A 2-m long slender rod of mass 5 kg is initially balanced on end as shown in Fig. P20-86. When the rod is disturbed it first falls against the sharp corner C and then rotates about C. Determine:

a. The angular velocity of the rod $\boldsymbol{\omega}$ and the velocity of the mass center of the rod $\mathbf{v}_G$ immediately after the impact with the corner C.
b. The angular velocity of the rod when B strikes the horizontal surface.

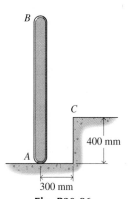

Fig. P20-86

429

20-87 The uniform rod of Fig. P20-87 falls horizontally and strikes the two rigid corners A and B. Corner A is slightly lower than corner B and so the rod strikes corner B first. If the velocity of the rod just before impact is v_0 and the impact is perfectly elastic $e = 1$, determine the angular velocity ω and the velocity of the mass center $\mathbf{v}_G$:

a. Immediately after striking corner B.
b. Immediately after striking corner A.

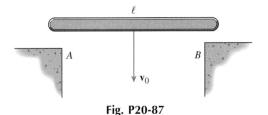

Fig. P20-87

20-88 The thin plate of Fig. P20-88 is an equilateral triangle 300 mm on a side and has a mass of 2 kg. The plate is suspended from a thin wire and is at rest when a 125-g arrow traveling with a speed of 100 m/s in the negative x-direction strikes the plate at the corner B. (The arrow may be modeled as a uniform slender rod 800 mm long.) If the arrowhead becomes embedded in the plate, determine:

a. The velocity of the mass center of the plate immediately after the impact.

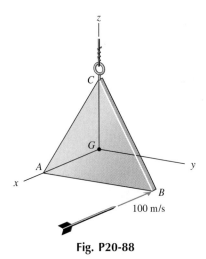

Fig. P20-88

b. The angular velocity of the plate immediately after the impact.
c. The average moment exerted on the arrow by the plate for an impact duration of $\Delta t = 0.02$ s.

20-89* The two uniform, slender bars shown in Fig. P20-89 rotate in a vertical plane. Bar AB (5 lb, 2 ft long) is released from rest when it is horizontal and impacts bar CD (8 lb, 2 ft long) when it is vertical. If the coefficient of restitution $e = 0.8$, determine:

a. The angular velocity of both bars immediately after the impact.
b. The maximum angle through which CD swings after the impact.
c. The rebound angle of bar AB.

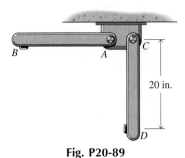

Fig. P20-89

20-90* An 8-kg uniform sphere 400 mm in diameter rolls without slipping on a horizontal surface and strikes a 100-mm tall step as shown in Fig. P20-90. The collision with the step is perfectly plastic and the sphere rotates about the corner of the step after impact. Determine:

a. The angular velocity ω and the velocity of the mass center $\mathbf{v}_G$ of the sphere immediately after the collision if the initial speed of the sphere is $v_0 = 2.5$ m/s.
b. The kinetic energy lost in the collision.
c. The minimum initial speed v_0 for which the sphere will rotate all the way over to the upper level.

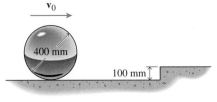

Fig. P20-90

MECHANICAL VIBRATIONS

The vibratory motion of a pendulum is used to regulate a clock.

21-1   INTRODUCTION

A mechanical vibration is the repeated oscillation of a particle or rigid body about an equilibrium position. In many devices vibratory motions are desirable and are deliberately generated; for example, a pendulum used to control a clock, a plucked string on a guitar or piano, the vibrator used to compact concrete in a form, and so on. The task of the engineer in such problems is to create and to control the vibrations. However, most vibrations in rotating machinery and in structures are undesirable. If rotating machine parts are not carefully balanced, they will vibrate. The vibrations can cause discomfort to the machine operator as well as damage to the machine or its support. Vibrations in structures due to earthquakes or to traffic of vehicles nearby may cause damage to or even collapse of the structure. In these cases the task of the engineer is to eliminate the vibrations (or at least to reduce the effect of the vibrations as much as possible) by appropriate design.

When a particle or rigid body in stable equilibrium is displaced by the application of an additional force, a mechanical vibration will result. Some common examples are the following:

1. The horizontal oscillation of a body attached to a spring (Fig. 21-1a) when it is displaced from its equilibrium position and then released.
2. The vertical oscillation of a flexible board or rod (Fig. 21-1b) when it is displaced from its equilibrium position and then released.
3. The rotational oscillation of a pendulum bob supported by an inextensible cord of negligible weight (Fig. 21-1c) when it is displaced from its equilibrium position and then released.

The common characteristic in each of these examples is that restoring forces act on the body, causing it to return to its equilibrium position (Fig. 21-2a). However, when the body reaches its equilibrium position, it has a non-zero velocity and passes through it (Fig. 21-2b). The process is then repeated as the restoring forces again act to cause the body to return to its equilibrium position (Fig. 21-2c). The motion is repeated over and over again as the body moves back and forth through its equilibrium position.

In many cases, the position or motion of a body can be completely specified by one coordinate (e.g., x in Fig. 21-1a; θ in Fig. 21-1b; etc.).

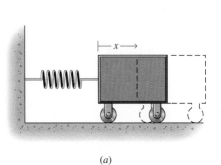

(a)

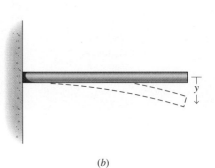

(b)

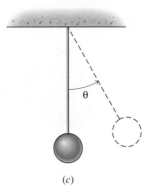

(c)

Fig. 21-1

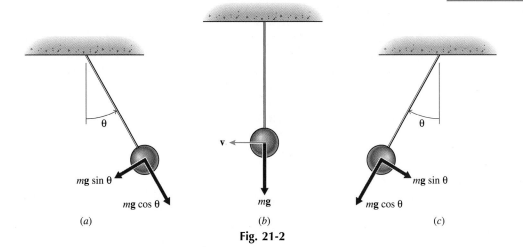

Fig. 21-2

(a)   (b)   (c)

Such bodies have one *degree of freedom.* In other cases, a body can vibrate independently in two directions (Fig. 21-3a), or two bodies can be connected together but each can vibrate independently in a single direction (Fig. 21-3b). Since two coordinates are required to completely specify the position or motion of these systems, they have two degrees of freedom. Only single-degree-of-freedom systems are covered in this first course in dynamics.

Figure 21-4 shows typical graphs of the displacement (x or y or θ) from the equilibrium position versus time. Oscillations that repeat uniformly as in Figs. 21-4a and 21-4b are called *periodic;* oscillations that do not repeat uniformly (Fig. 21-4c) are called *aperiodic* or *random vibrations.* Random vibrations are not covered in this first course in Dynamics.

One of the more important characteristics of a periodic oscillation is the *period τ,* which is the minimum amount of time before the motion repeats itself. The motion completed in one period is called a *cycle.* The period is expressed in *seconds per cycle* or just seconds. The *frequency f* of an oscillation is the reciprocal of the period

$$f = \frac{1}{\tau} \qquad (21\text{-}1)$$

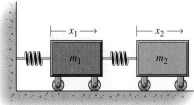

(a)

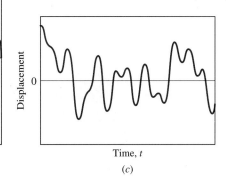

(b)

Fig. 21-3

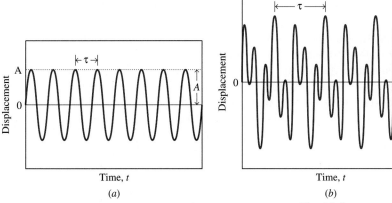

(a)   (b)   (c)

Fig. 21-4

or the number of cycles that occur per unit of time. The customary unit of frequency, *cycles per second (cps)*, is also called *hertz (Hz)*. The *amplitude A* of an oscillation is the maximum displacement of the body from its equilibrium position.

Finally, it must be noted that the study of vibrations is merely an application of the principles developed earlier. In the earlier chapters the acceleration was usually obtained only for a particular position of the body and at a particular instant of time. Here, the acceleration will be obtained for an arbitrary position of the body and then integrated to get the velocity and position for all future times.

21-2   UNDAMPED FREE VIBRATIONS

Mechanical vibrations are generally categorized as either *free vibrations* (also called *natural vibrations*) or *forced vibrations*. A free vibration is produced and maintained by forces such as elastic and gravitational forces, which depend only on the position and motion of the body. A forced vibration is produced and maintained by an externally applied periodic force, which does not depend on the position or motion of the body.

Free vibrations and forced vibrations may be further categorized as either damped or undamped. When forces that oppose the restoring force (friction, air resistance, viscous damping, etc.) are negligible, the vibration is called *undamped*. When resisting forces are not negligible, the vibration is called *damped*. An undamped free vibration will repeat itself indefinitely; a damped free vibration will eventually die out.

Of course, all real systems contain frictional forces, which will eventually stop a free vibration. In many systems, however, the energy loss due to air resistance, the internal friction of springs, or other friction forces are small enough that an analysis based on negligible damping often gives quite satisfactory engineering results. In particular, the frequency and period of vibration obtained for a freely vibrating system are very close to the values obtained for a system that has a small amount of damping.

21-2-1   Undamped Free Vibration of Particles

Consider a block of mass m sliding on a frictionless horizontal surface as shown in Fig. 21-5a. Vibration is induced by displacing the block a distance x_0 and then releasing it with an initial velocity of $\dot{x}_0 = v_0$.

The free-body diagram of the block is shown in Fig. 21-5b, in which the block has been displaced an arbitrary amount in the positive coordinate direction. The elastic restoring force of the spring $F_s = kx$ is always directed toward the equilibrium position, whereas the acceleration $a_x = d^2x/dt^2 = \ddot{x}$ acts in the direction of *positive displacement*. It is important to remember that since acceleration is the second time derivative of displacement, both the displacement x and the acceleration $\ddot{x}$ must be measured positive in the same direction. Applying Newton's second law $\Sigma F_x = ma_x = m\ddot{x}$ to the block gives the differential equation of motion for the block

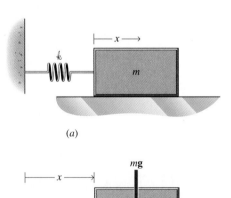

(a)

(b)

Fig. 21-5

$$-kx = m\ddot{x} \qquad \text{or} \qquad \ddot{x} = -\frac{k}{m}\,x \qquad (21\text{-}2)$$

Therefore, when the block is to the right of the equilibrium position (x is positive), its acceleration is to the left ($\ddot{x}$ is negative) or toward the equilibrium position. Similarly, when the block is to the left of the equilibrium position (x is negative), its acceleration is to the right ($\ddot{x}$ is positive), also toward the equilibrium position. That is, the acceleration of the block is proportional to its displacement from the equilibrium position and is directed toward that position.

21-2-2   Simple Harmonic Motion

Equation 21-2 describes *simple harmonic motion:* a motion for which the acceleration is proportional to the displacement from a fixed point and is directed toward the fixed point. Most of the vibrations encountered in engineering applications may be represented by a simple harmonic motion. Many other vibrations may be closely approximated by simple harmonic motion. A thorough knowledge of this concept is most helpful when analyzing such systems.

Equation 21-2 is a standard type of differential equation (a homogeneous, second-order, linear differential equation with constant coefficients) and is usually written in the form

$$\ddot{x} + \omega_n^2 x = 0 \tag{21-3}$$

The coefficient $\omega_n = \sqrt{k/m}$, which has units of rad/s, is related to the frequency of the oscillation and is called the *natural circular frequency*.[1] The general solution of Eq. 21-3 is[2]

$$x(t) = B \cos \omega_n t + C \sin \omega_n t \tag{21-4}$$

in which B and C are constants of integration to be determined from the initial conditions of the problem ($x = x_0$ and $\dot{x} = v_0$ when $t = 0$).

The solution (Eq. 21-4) can also be written as either

$$x(t) = A \cos (\omega_n t - \phi_c) \tag{21-5a}$$

or

$$x(t) = A \sin (\omega_n t - \phi_s) \tag{21-5b}$$

To verify that Eq. 21-5a is equal to Eq. 21-4, first expand Eq. 21-5a to get

$$x(t) = A (\cos \omega_n t \cos \phi_c + \sin \omega_n t \sin \phi_c) \tag{21-6}$$

Then, setting Eq. 21-4 equal to Eq. 21-6 gives

$$(B - A \cos \phi_c) \cos \omega_n t + (C - A \sin \phi_c) \sin \omega_n t = 0 \tag{21-7}$$

But if Eq. 21-4 is truly equal to Eq. 21-5a, then Eq. 21-7 must hold for any and all values of t. In particular, when $t = 0$, $\cos \omega_n t = 1$ and $\sin \omega_n t = 0$ so that

$$B = A \cos \phi_c \tag{21-8a}$$

---

[1]While the natural circular frequency ω_n is often equal to $\sqrt{k/m}$ as in the present example, it is not always so. More generally, ω_n^2 is the ratio of the *effective spring constant* (the coefficient of the x term) and the *effective mass* (the coefficient of the $\ddot{x}$ term) in the differential equation of motion.

[2]It is easily verified by direct substitution that the solution (Eq. 21-4) satisfies the differential equation (Eq. 21-3) for any values of the constants B and C.

Similarly, when $t = \pi/2\omega_n$, $\cos \omega_n t = 0$ and $\sin \omega_n t = 1$ so that

$$C = A \sin \phi_c \qquad (21\text{-}8b)$$

Therefore, Eqs. 21-4 and 21-5a will be equal if

$$A = \sqrt{B^2 + C^2} \qquad \text{and} \qquad \tan \phi_c = \frac{C}{B} \qquad (21\text{-}9)$$

(The equality of Eqs. 21-4 and 21-5b is verified in a similar fashion.) Since $\cos(\omega_n t - \phi_c)$ oscillates between -1 and $+1$, the amplitude of the oscillation is $A = \sqrt{B^2 + C^2}$. The *phase angle* ϕ_c (or ϕ_s) is the amount by which the solution must be shifted to make it a simple cosine (or sine) curve.

The velocity and acceleration of the block are obtained by differentiating Eq. 21-4 or 21-5 with respect to time. For example, the velocity of the block is

$$\begin{aligned}
v(t) = \dot{x}(t) &= -\omega_n B \sin \omega_n t + \omega_n C \cos \omega_n t & (21\text{-}10a)\\
&= -\omega_n A \sin (\omega_n t - \phi_c) & (21\text{-}10b)\\
&= \omega_n A \cos (\omega_n t - \phi_s) & (21\text{-}10c)
\end{aligned}$$

and the acceleration of the block is

$$\begin{aligned}
a(t) = \ddot{x}(t) &= -\omega_n^2 B \cos \omega_n t - \omega_n^2 C \sin \omega_n t & (21\text{-}11a)\\
&= -\omega_n^2 A \cos (\omega_n t - \phi_c) & (21\text{-}11b)\\
&= -\omega_n^2 A \sin (\omega_n t - \phi_s) & (21\text{-}11c)
\end{aligned}$$

Since the cosine curve (Eq. 21-5a) and the sine curve (Eq. 21-5b) repeat whenever their argument increases by an angle of 2π rad, the period of the oscillation is given by $\omega_n \tau_n = 2\pi$ or

$$\tau_n = \frac{2\pi}{\omega_n} \qquad (21\text{-}12)$$

where the natural circular frequency ω_n is obtained from the differential equation of motion. The natural frequency of the oscillation in *hertz* (cycles per second) is then

$$f_n = \frac{1}{\tau_n} = \frac{\omega_n}{2\pi} \qquad (21\text{-}13)$$

and the natural frequency f_n and the natural circular frequency ω_n are related by $\omega_n = 2\pi f_n$. That is, a natural frequency of $f_n = 1$ Hz is equivalent to a natural circular frequency of $\omega_n = 2\pi$ rad/s.

It must be pointed out that the results obtained in this section are not limited to the vibration of a particle on a horizontal surface. They may be used to analyze the vibrational motion of a particle whenever the equations of motion reduce to the form (Eq. 21-3)

$$\ddot{x} + \omega_n^2 x = 0$$

which characterizes simple harmonic motion.

On the other hand, if the equations of motion do not reduce to the form of Eq. 21-3, the motion may still be an oscillatory motion but it will not be a simple harmonic motion. In this case new expressions for the period, frequency, and so on must be obtained from solving the differential equation of motion.

21-2-3 Displaced Equilibrium Position

Simple harmonic motion also occurs if the block is suspended from the spring (Fig. 21-6a) rather than sliding on a frictionless surface so long as the y-coordinate is measured from the equilibrium position of the system. To see that this is so, draw the free-body diagrams of the block in its equilibrium position (Fig. 21-6b) and in an arbitrary displaced position (Fig. 21-6c). In the equilibrium position (before the block has been displaced and released), the sum of forces acting on the block must add to zero

$$mg - k\delta_{eq} = 0$$

where δ_{eq} is the static deformation of the spring (the elongation of the spring in the static equilibrium position $y = 0$). Therefore, the static deformation of the spring is $\delta_{eq} = mg/k$.

When the block has been displaced downward (in the positive y-direction) some amount y, the spring will be stretched a total amount of $y + \delta_{eq}$ and the force exerted on the block will be $k(y + \delta_{eq}) = k(y + mg/k)$ upward. Then writing Newton's second law gives

$$mg - k(y + mg/k) = m\ddot{y}$$

or

$$m\ddot{y} + ky = 0 \qquad (21\text{-}14)$$

which is again the equation for simple harmonic motion and has the solution

$$y = B \cos \omega_n t + C \sin \omega_n t \qquad (21\text{-}15)$$

The circular frequency ω_n, the natural frequency f_n, the period of vibration τ_n, and other vibrational characteristics of the block are then obtained as in Section 21-2-2 above.

If the position of the block were measured from the position where the spring is unstretched ($\hat{y} = 0$ when the spring is unstretched) rather than from the equilibrium position, then the force in the spring would be $k\hat{y}$ and Eq. 21-14 would become

$$m\ddot{\hat{y}} + k\hat{y} = mg \qquad (21\text{-}16)$$

But the solution to Eq. 21-16 is just a constant plus Eq. 21-4

$$\hat{y}(t) = \frac{mg}{k} + B \cos \omega_n t + C \sin \omega_n t$$
$$= \delta_{eq} + B \cos \omega_n t + C \sin \omega_n t$$

where $\hat{y}(t) = y(t) + \delta_{eq}$. That is, the oscillation consists of simple harmonic motion about the equilibrium position $\hat{y} = \delta_{eq}$.

21-2-4 Approximately Simple Harmonic Motion

If the equations of motion do not reduce to the form of Eq. 21-3,

$$\ddot{x} + \omega_n^2 x = 0$$

then the motion is not simple harmonic motion. Many motions, however, are well approximated by Eq. 21-3 as long as the amplitude of the motion is small. Such motions can be approximated as simple harmonic motions and all the results of Section 21-2-2 directly apply.

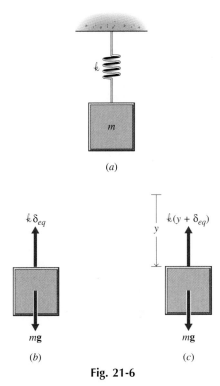

(a)

(b)   (c)

Fig. 21-6

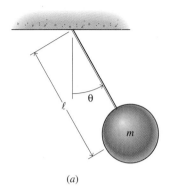

(a)

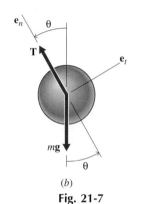

(b)

Fig. 21-7

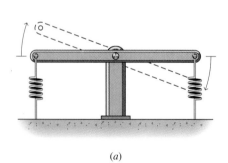

(a)

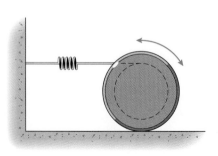

(b)

Fig. 21-8

For example, consider the oscillation of the simple pendulum shown in Fig. 21-7a. The pendulum consists of a particle of mass m swinging at the end of an inextensible, lightweight cord of length ℓ. The pendulum is released with an initial angle θ_0 and an initial speed $\dot{\theta}_0 = \omega_0$. Since the cord is inextensible, the particle will travel along a circular path with an acceleration

$$\mathbf{a} = \ell\ddot{\theta}\mathbf{e}_t + \ell\dot{\theta}^2\,\mathbf{e}_n$$

where the normal direction is toward the suspension point and the tangential direction is in the direction of increasing θ (Fig. 21-7b). The tangential component of Newton's second law, $\Sigma F_t = ma_t$ then gives the differential equation of motion

$$-mg\,\sin\,\theta = m\ell\ddot{\theta}$$

or

$$\ddot{\theta} = -\frac{g}{\ell}\,\sin\,\theta \qquad (21\text{-}17)$$

As long as the angle θ is small, $\sin\,\theta \cong \theta$ (where θ is in radians) and Eq. 21-17 becomes

$$\ddot{\theta} = -\frac{g}{\ell}\theta \qquad (21\text{-}18)$$

Therefore, the pendulum has simple harmonic motion with a natural circular frequency $\omega_n = \sqrt{g/\ell}$ and period $\tau_n = 2\pi/\omega_n = 2\pi\sqrt{\ell/g}$. If the angle θ does not remain small, the resulting motion will still be an oscillatory motion, but it will not be a simple harmonic motion. The solution in this case must be obtained from solving the differential equation of motion Eq. 21-17.[3]

2-2-5 Undamped Free Vibration of Rigid Bodies

A rigid body oscillating about a fixed axis (Fig. 21-8a) and a wheel oscillating on a flat surface (Fig. 21-8b) are also one-degree-of-freedom vibrating systems. The analysis of such rigid body systems is essentially the same as for a particle. First, the free-body diagram is drawn for an arbitrary position of the rigid body. Next, the equations of motion are written. Finally, the principles of kinematics are used to reduce the equations of motion to a single differential equation involving a single variable that describes the position and motion of the rigid body. If the resulting differential equation can be written in the form of Eq. 21-3

$$\ddot{x} + \omega_n^2 x = 0$$

then the motion of the rigid body is a simple harmonic motion and all the results of Section 21-2-2 apply. If the equation of motion cannot be written in the form of Eq. 21-3, the resulting motion may still be an oscillatory motion, but it will not be a simple harmonic motion. The solution in this case must be obtained from solving the differential equation of motion.

---

[3]If the angle θ does not remain small, the solution to Eq. 21-17 can still be found, although it cannot be written in terms of simple functions such as polynomials or trigonometric functions. For $\theta_{max} = 5^0$ the difference between the exact and approximate solutions is only about 0.05% in the value of the period; $\theta_{max} = 10^0$, 0.19%; $\theta_{max} = 20^0$, 0.76%; and $\theta_{max} = 40^0$, 3.15%.

The pendulum of Fig. 21-9a consists of a 2-kg uniform bar 0.8 m long suspended from a frictionless pin at one end. Determine the natural frequency and the period of the resulting oscillation. (Assume small angle of oscillation.)

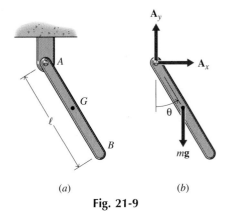

(a)   (b)

Fig. 21-9

SOLUTION

The free-body diagram of the pendulum is drawn in Fig. 21-9b. Since the motion is a rotation about a fixed axis, Newton's second law can be written in the form $\Sigma M_A = I_A \alpha = I_A \ddot{\theta}$ which gives

$$-\frac{\ell}{2}(mg \sin \theta) = \left(\frac{1}{3}m\ell^2\right)\ddot{\theta}$$

But if the angle of oscillation is small, $\sin \theta \cong \theta$ (in radians) so that the differential equation of motion of the pendulum is

$$\ddot{\theta} + \frac{3g}{2\ell}\theta = 0$$

Therefore, the natural circular frequency, the natural frequency, and the period of the oscillation are

$$\omega_n = \sqrt{(3)(9.81)/(2)(0.8)} = 4.289 \text{ rad/s} \qquad \text{Ans.}$$

$$f_n = \frac{\omega_n}{2\pi} = 0.683 \text{ Hz} \qquad \text{Ans.}$$

$$\tau_n = \frac{1}{f_n} = 1.465 \text{ s} \qquad \text{Ans.}$$

(These results can be compared with the results for a simple pendulum in which all the mass is concentrated at the end of a massless rod or string. In Section 21-2-4 the natural circular frequency of the simple pendulum was found to be $\omega_n = \sqrt{g/\ell} = \sqrt{(9.81)/(0.8)} = 3.502$ rad/s. Therefore, the natural frequency and period of the simple pendulum would be $f_n = \dfrac{\omega_n}{2\pi} = 0.557$ Hz and $\tau_n = \dfrac{1}{f_n} = 1.794$ s, respectively.)

A 10-lb cart is attached to three springs and rolls on an inclined surface as shown in Fig. 21-10a. The elastic modulus of the springs are $k_1 = k_2 = 5$ lb/ft and $k_3 = 15$ lb/ft. If the cart is moved 3 in. up the incline from its equilibrium position and released with an initial velocity of 15 in./s up the incline when $t = 0$, determine

a. The period τ_n, the frequency f_n, and the circular frequency ω_n of the resulting vibration.
b. The position of the cart as a function of time.
c. The amplitude A of the resulting vibration.

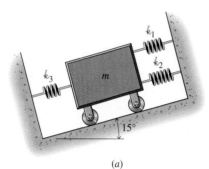

(a)

Fig. 21-10

SOLUTION

a. The free-body diagram of the cart is drawn in Fig. 21-10b in which the x-coordinate measures the position of the cart along the incline with $x = 0$ at the equilibrium position. In the equilibrium position (before the cart has been disturbed), the spring forces are proportional to their deformation $F_1 = k_1\delta_{eq1}$, $F_2 = k_2\delta_{eq2}$, and $F_3 = k_3\delta_{eq3}$ so that equilibrium gives

$$k_1\delta_{eq1} + k_2\delta_{eq2} - k_3\delta_{eq3} - mg\sin 15° = 0 \qquad (a)$$

Since it is not known how much the springs were stretched or compressed before being attached to the cart, it is not possible to determine values for the static deflections δ_{eq1}, δ_{eq2}, and δ_{eq3}, individually. However, Eq. a does give a relationship between the static deflections and the weight of the cart.

When the cart is at an arbitrary (positive) location x, the stretch in springs 1 and 2 will be reduced ($F_1 = k_1[\delta_{eq1} - x]$ and $F_2 = k_2[\delta_{eq2} - x]$)

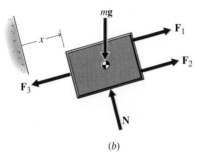

(b)

Fig. 21-10

and the stretch in spring 3 will be increased ($F_3 = k_3[\delta_{eq3} + x]$). Therefore, Newton's second law of motion $\Sigma F_x = m\ddot{x}$ gives

$$k_1(\delta_{eq1} - x) + k_2(\delta_{eq2} - x) - k_3(\delta_{eq3} + x) - mg \sin 15° = m\ddot{x}$$

or

$$(k_1\delta_{eq1} + k_2\delta_{eq2} - k_3\delta_{eq3} - mg \sin 15°) - (k_1 + k_2 + k_3)x = m\ddot{x}$$

The term in the first set of parentheses is zero by Eq. *a*, however, so the differential equation of motion is

$$m\ddot{x} + (k_1 + k_2 + k_3)x = 0$$

or

$$\ddot{x} + 80.50x = 0$$

Then the natural circular frequency, the natural frequency, and the period are

$$\omega_n = \sqrt{80.50} = 8.972 \text{ rad/s} \qquad \text{Ans.}$$

$$f_n = \frac{\omega_n}{2\pi} = 1.428 \text{ Hz} \qquad \text{Ans.}$$

$$\tau_n = \frac{1}{f_n} = 0.700 \text{ s} \qquad \text{Ans.}$$

b. The displacement and velocity of the cart can be written in the form

$$x(t) = B \cos 8.972t + C \sin 8.972t$$
$$\dot{x}(t) = -8.972B \sin 8.972t + 8.972C \cos 8.972t$$

But at $t = 0$, $x = B = 3$ in. and $\dot{x} = 8.972C = 15$ in./s. Therefore, $B = 3$ in. and $C = 1.672$ in., and

$$x(t) = 3 \cos 8.972t + 1.672 \sin 8.972t \text{ in.} \qquad \text{Ans.}$$

This solution is shown in Fig. 21-10c.

Alternatively, the position and velocity of the cart can be written in the form

$$x(t) = A \cos(8.972t - \phi_c)$$
$$\dot{x}(t) = -8.972A \sin(8.972t - \phi_c)$$

Then, applying the initial conditions $x(0) = A \cos \phi_c = 3$ in. and $\dot{x}(0) = -8.972A (-\sin \phi_c) = 15$ in./s gives $A = 3.43$ in. and $\phi_c = 29.13° = 0.508$ rad. Therefore, the equation describing the position of the cart is

$$x(t) = 3.43 \cos(8.972t - 0.508) \text{ in.} \qquad \text{Ans.}$$

The displacement and velocity of the cart can also be written in the form

$$x(t) = A \sin(8.972t - \phi_s)$$
$$\dot{x}(t) = 8.972A \cos(8.972t - \phi_s)$$

Then, applying the initial conditions $x(0) = A (-\sin \phi_s) = 3$ in. and $\dot{x}(0) = 8.972A \cos \phi_s = 15$ in./s gives $A = 3.43$ in. and $\phi_s = -60.87° = -1.062$ rad. Therefore, the equation describing the position of the cart is

$$x(t) = 3.43 \sin(8.972t + 1.062) \text{ in.} \qquad \text{Ans.}$$

(The solution described by these two equations is exactly the same as that shown in Fig. 21-10c. The phase angles $\phi_c = 0.508$ rad and $\phi_s = -1.062$ are also indicated on Fig. 21-10c.)

c. Since the maximum value of the cosine function is 1, the amplitude of the vibration is

$$A = 3.43 \text{ in.} \qquad \text{Ans.}$$

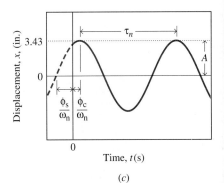

Fig. 21-10

A 5-lb uniform cylinder 12 in. in diameter rolls without slipping on an inclined plane as shown in Fig. 21-11a. A linear spring (k = 24 lb/ft) is attached to the cylinder at point A (which is e = 3 in. from the center of the cylinder), and the spring is unstretched in the position shown. If the cylinder is released from rest in the position shown, determine

a. The period τ_n, the frequency f_n, and the circular frequency ω_n of the resulting vibration.

b. The position of the center of mass of the cylinder as a function of time.

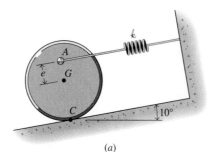

(a)

Fig. 21-11

SOLUTION

a. Figure 21-11b shows the free-body diagram of the cylinder in its equilibrium position. To get from its initial position to the equilibrium position, the cylinder has to roll counterclockwise through an angle θ_{eq}, the center of mass of the cylinder has to move down the incline a distance x_{Geq}, and the spring is stretched by the amount δ_{eq}. If θ_{eq} is small, then $\sin\theta_{eq} \cong \theta_{eq}$ (in radians); $\cos\theta_{eq} \cong 1$; $\theta_{eq} \cong \tan\theta_{eq} \cong \dfrac{\delta_{eq}}{r+e} \cong \dfrac{x_{Geq}}{r}$ (Fig. 21-11c); and the spring force $k\delta_{eq}$ remains parallel to the surface. Then the equilibrium equations

$$\Sigma F_x = 0: \quad mg\sin 10° - F - k\delta_{eq} = 0$$
$$\Sigma M_G = 0: \quad Fr - k\delta_{eq}\, e\cos\theta_{eq} = 0$$

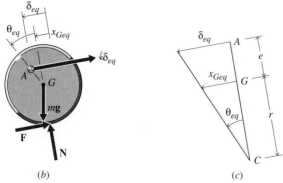

(b)   (c)

Fig. 21-11

can be combined to get

$$mgr \sin 10° - k(r + e)\delta_{eq} = 0 \qquad (a)$$

Equation a gives $\delta_{eq} = 0.02412$ ft = 0.2894 in. from which $x_{Geq} = 0.01608$ ft = 0.1929 in. and $\theta_{eq} = 0.03216$ rad = 1.842° can be found. (As a check of the small angle approximations, note that $\sin \theta_{eq} = 0.03215 \cong \theta_{eq}$ and $\cos \theta_{eq} = 0.9995 \cong 1$.)

Next, the free-body diagram of the cylinder is drawn (Fig. 21-11d) for an arbitrary position in which the center of mass has moved an additional distance x_G down the incline, the cylinder has rotated through an additional angle θ, and (still assuming small angles) the spring has been stretched an additional amount $[(r + e)/r]x_G$. Then the equations of motion for the cylinder are

$$\Sigma F_x = ma_{Gx}: \qquad mg \sin 10° - F - k\left(\delta_{eq} + \frac{9}{6}x_G\right) = m\ddot{x}_G$$

$$\Sigma M_G = I_G\alpha: \qquad Fr - k\left(\delta_{eq} + \frac{9}{6}x_G\right)e \cos(\theta_{eq} + \theta) = \frac{1}{2}mr^2\ddot{\theta}$$

To get the differential equation describing the vibration, replace $\cos(\theta_{eq} + \theta)$ with 1; multiply the first equation by r and add it to the second to get

$$[mgr \sin 10° - k(r + e)\delta_{eq}] - k(r + e)\left(\frac{9}{6}x_G\right) = mr\ddot{x}_G + \frac{1}{2}mr^2\ddot{\theta}$$

But the term in the brackets is zero by Eq. a and the accelerations are related by $\ddot{x}_G = r\ddot{\theta}$. Therefore, the differential equation of motion is

$$mr\ddot{x}_G + \frac{1}{2}mr\ddot{x}_G + k(r + e)\left(\frac{9}{6}x_G\right) = 0$$

or

$$0.11646\ddot{x}_G + 27.00x_G = 0$$

and the natural circular frequency, natural frequency, and period of the vibration are

$$\omega_n = \sqrt{(27.00)/(0.11646)} = 15.23 \text{ rad} \qquad \text{Ans.}$$

$$f_n = \frac{\omega_n}{2\pi} = 2.423 \text{ Hz} \qquad \text{Ans.}$$

$$\tau_n = \frac{1}{f_n} = 0.413 \text{ s} \qquad \text{Ans.}$$

b. The position and velocity of the center of mass of the cylinder can be written in the form

$$x_G(t) = B \cos 15.23t + C \sin 15.23t$$
$$\dot{x}_G(t) = -15.23B \sin 15.23t + 15.23C \cos 15.23t$$

But when $t = 0$, $x_B = B = -x_{Geq} = -0.1929$ in. and $\dot{x}_G = 15.23C = 0$. Therefore, the position of the center of mass of the cylinder is

$$x_G(t) = -0.1929 \cos 15.23t \text{ in.} \qquad \text{Ans.}$$

(As a final check that the small angle of oscillation approximation remains valid, note that the amplitude of the oscillation is 0.1929 in. Therefore, the maximum angle of rotation from the initial position is $\theta_{max} = \theta_{eq} + \dfrac{0.1929}{6} = 0.06431$ rad = 3.685°. But $\sin \theta_{max} = 0.06428 \cong \theta_{max}$ and $\cos \theta_{max} = 0.9979 \cong 1$.

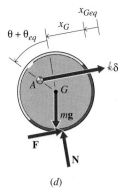

(d)
Fig. 21-11

PROBLEMS

21-1–21-6 The following equations represent the position of a particle in simple harmonic motion. For each equation, plot the position, velocity, and acceleration of the particle versus time for two complete cycles of the oscillation.

21-1* $x(t) = 8 \cos \pi t$ in.

21-2* $x(t) = 5 \sin \pi t/4$ mm

21-3 $x(t) = 3 \cos(\pi t/2 - \pi/4)$ in.

21-4* $x(t) = 10 \sin(3\pi t/4 + \pi/8)$ mm

21-5 $x(t) = 4 \cos 5t - 3 \sin 5t$ in.

21-6 $x(t) = 5 \sin 3t + 12 \cos 3t$ mm

21-7–21-12 The following equations represent the position of a particle in simple harmonic motion. For each equation

a. Write the equation for the particle motion in the form $x(t) = A \cos(\omega_n t - \phi_c)$.

b. Find the maximum velocity and the particle position when it occurs.

c. Find the maximum acceleration and the particle position when it occurs.

21-7* $x(t) = 3 \cos \pi t - 4 \sin \pi t$ in.

21-8 $x(t) = 12 \cos \pi t/2 + 5 \sin \pi t/2$ mm

21-9* $x(t) = 8 \cos 10t + 6 \sin 10t$ in.

21-10* $x(t) = 10 \cos 3\pi t/4 - 24 \sin 3\pi t/4$ mm

21-11 $x(t) = 5 \sin \pi t$ in.

21-12 $x(t) = 4 \sin(3t + \pi/3)$ mm

21-13–21-18 The following equations represent the position of a particle in simple harmonic motion. For each equation

a. Write the equation for the particle motion in the form $x(t) = A \sin(\omega_n t - \phi_s)$.

b. Find the earliest value of t for which the particle's position is zero.

c. Find the earliest value of t for which the particle's velocity is zero.

21-13* $x(t) = 5 \cos \pi t - 12 \sin \pi t$ in.

21-14* $x(t) = 4 \cos \pi t/2 + 3 \sin \pi t/2$ mm

21-15 $x(t) = 8 \cos 3\pi t/4 + 6 \sin 3\pi t/4$ in.

21-16 $x(t) = 5 \cos 10t - 5 \sin 10t$ mm

21-17* $x(t) = 5 \cos \pi t$ in.

21-18 $x(t) = 8 \cos(3\pi t/2 + 2\pi/3)$ mm

21-19* An instrument used to measure the vibration of a particle indicates a simple harmonic motion with a natural frequency of 5 Hz and a maximum acceleration of 160 ft/s². Determine the amplitude and maximum velocity of the vibration.

21-20 An instrument used to measure the vibration of a particle indicates a simple harmonic motion with a period of 0.025 s and a maximum acceleration of 150 m/s². Determine the amplitude and maximum velocity of the vibration.

21-21 A particle vibrates with a simple harmonic motion that has a period of 0.333 s and a maximum velocity of 75 ft/s. Determine the amplitude and maximum acceleration of the vibration.

21-22* A particle vibrates with a simple harmonic motion. When the particle passes through the equilibrium position, its velocity is 2 m/s. When the particle is 20 mm away from its equilibrium position, its acceleration is 50 m/s². Determine the magnitude of the velocity at this position.

21-23 A particle vibrates with a simple harmonic motion. When the particle passes through the equilibrium position, its velocity is 10 ft/s. When the particle is 1.6 in. away from its equilibrium position, its velocity is 6 ft/s. Determine the magnitude of the acceleration at this position.

21-24* A block that has a mass m slides on a frictionless horizontal surface as shown in Fig. P21-24. Determine the modulus k of the single spring that could replace the two springs shown without changing the frequency of vibration of the block.

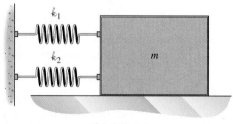

Fig. P21-24

21-25* A block that has a mass m slides on a frictionless horizontal surface as shown in Fig. P21-25. Determine the modulus k of the single spring that could replace the two springs shown without changing the frequency of vibration of the block.

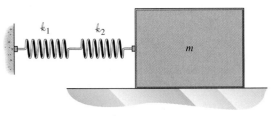

Fig. P21-25

21-26 A 2-kg mass is suspended in a vertical plane by three springs as shown in Fig. P21-26. If the mass is displaced 5 mm below its equilibrium position and released with an upward velocity of 250 mm/s when $t = 0$, determine

a. The differential equation governing the motion.
b. The period and amplitude of the resulting vibration.
c. The position of the mass as a function of time.
d. The earliest time $t_1 > 0$ when the mass passes through its equilibrium position.

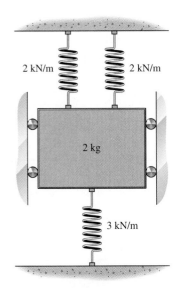

Fig. P21-26

21-27* A 10-lb block is suspended in a vertical plane by three springs as shown in Fig. P21-27. If the block is displaced 7 in. above its equilibrium position and released with an upward velocity of 150 in./s when $t = 0$, determine

a. The differential equation governing the motion.
b. The period and amplitude of the resulting vibration.

c. The position of the block as a function of time.
d. The earliest time $t_1 > 0$ when the block passes through its equilibrium position.

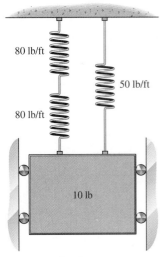

Fig. P21-27

21-28 A 4-kg mass is suspended in a vertical plane as shown in Fig. P21-28. The two springs remain in tension at all times and the two pulleys are both small and frictionless. If the mass is displaced 15 mm above its equilibrium position and released with a downward velocity of 750 mm/s when $t = 0$, determine:

a. The differential equation governing the motion.
b. The period and amplitude of the resulting vibration.
c. The position of the mass as a function of time.
d. The earliest time $t_1 > 0$ when the velocity of the mass is zero.

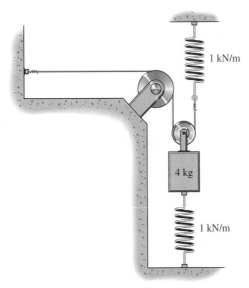

Fig. P21-28

21-29 A 20-lb block slides on a frictionless horizontal surface as shown in Fig. P21-29. The two springs remain in tension at all times and the two pulleys are both small and frictionless. If the block is displaced 3 in. to the left of its equilibrium position and released with a velocity of 50 in./s to the right when $t = 0$, determine:

a. The differential equation governing the motion.
b. The period and amplitude of the resulting vibration.
c. The position of the block as a function of time.
d. The earliest time $t_1 > 0$ when the velocity of the block is zero.

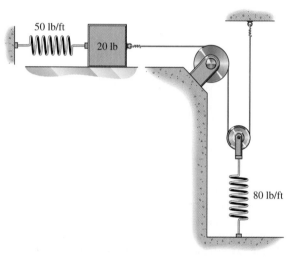

Fig. P21-29

21-30* An 8-kg mass slides on a frictionless horizontal surface as shown in Fig. P21-30. The two springs remain in tension at all times and the pulleys are small and frictionless. If the mass is displaced 25 mm to the right of its equilibrium position and released with a velocity of 800 mm/s to the right when $t = 0$, determine

a. The differential equation governing the motion.
b. The period and amplitude of the resulting vibration.
c. The position of the mass as a function of time.
d. The earliest time $t_1 > 0$ when the acceleration of the mass is zero.

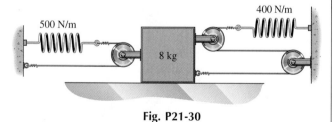

Fig. P21-30

21-31 The 10-lb block of Fig. P21-31 slides on a frictionless horizontal surface while the 5 lb block moves in a vertical plane. The springs remain in tension at all times and

the two pulleys are both small and frictionless. Write the differential equation of motion for $x(t)$, the position of the 10-lb block, and determine the frequency and period of the resulting vibratory motion.

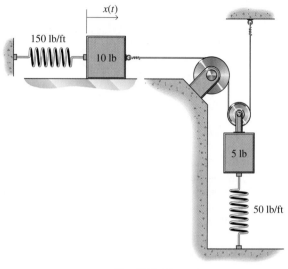

Fig. P21-31

21-32* The two masses of Fig. P21-32 each slide on a frictionless horizontal surface. The springs remain in tension at all times, and the pulleys are small and frictionless. Write the differential equation of motion for $x(t)$, the position of the 10-kg block, and determine the frequency and period of the resulting vibratory motion.

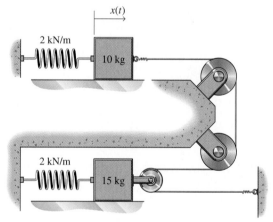

Fig. P21-32

21-33* The two blocks of Fig. P21-33 hang in a vertical plane from a massless bar, which is horizontal in the equilibrium position. If the springs remain in tension at all times, write the differential equation of motion for $y(t)$, the position of the 10-lb block, and determine the frequency and period of the resulting vibratory motion. (Assume small oscillations.)

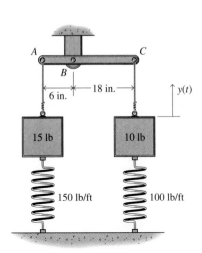

Fig. P21-33

21-34 The two masses of Fig. P21-34 each slide on a frictionless horizontal surface. The bar ABC is vertical in the equilibrium position and has negligible mass. If the springs remain in tension at all times, write the differential equation of motion for $x(t)$, the position of the 10-kg mass, and determine the frequency and period of the resulting vibratory motion. (Assume small oscillations.)

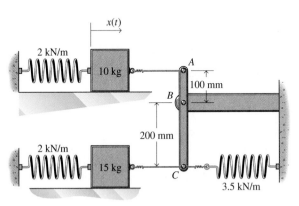

Fig. P21-34

21-35* The 5-lb block of Fig. P21-35 slides on a frictionless horizontal surface while the 3-lb block hangs in a vertical plane. The bar ABC has negligible mass and arm AB is horizontal in the equilibrium position. If the springs remain in tension at all times, write the differential equation of motion for $y(t)$, the position of the 3-lb block, and deter-

mine the frequency and period of the resulting vibratory motion. (Assume small oscillations.)

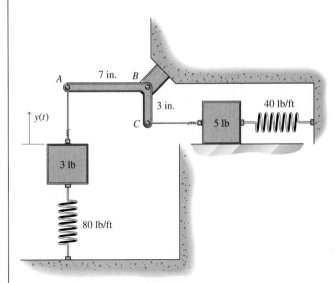

Fig. P21-35

21-36 A 0.5-kg plunger is at rest in a vertical frictionless guide when a 0.3-kg ball rolls off a 4-m high step and bounces off it as shown in Fig. P21-36. If the collision is perfectly elastic ($e = 1$) and the spring modulus is $k = 200$ N/m, determine $y(t)$, the position of the plunger, as a function of the time after the ball strikes it.

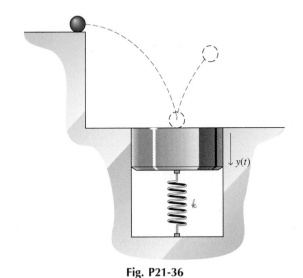

Fig. P21-36

21-37 A 2.5-lb plunger is at rest in a vertical frictionless guide when a 2.0-lb ball rolls off a 15-ft high step and bounces off it as shown in Fig. P21-36. If the coefficient of restitution of the collision is $e = 0.6$ and the spring modulus is $k = 30$ lb/ft, determine $y(t)$, the position of the plunger, as a function of time after the ball strikes it.

447

21-38* A 0.5-kg plunger is at rest in a vertical frictionless guide when a 0.3-kg ball rolls off a 4-m high step and bounces off it as shown in Fig. P21-36. If the collision is perfectly plastic ($e = 0$) and the spring modulus is $k = 200$ N/m, determine $y(t)$, the position of the plunger, as a function of time after the ball strikes it.

21-39 A 7-lb uniform cylinder rolls without slipping on a horizontal surface as shown in Fig. P21-39. The two springs are connected to a small frictionless pin at G the center of the 8-in. diameter cylinder. Write the differential equation of motion for $x_G(t)$, the position of the center of mass of the cylinder, and determine the frequency and period of the resulting vibratory motion.

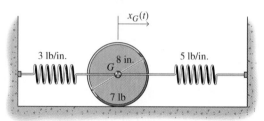

Fig. P21-39

21-40* A 4-kg uniform cylinder is suspended in the vertical plane in the loop of a light cord as shown in Fig. P21-40. If the 500-mm diameter cylinder does not slip on the cord, write the differential equation of motion for $y_G(t)$, the position of the mass center of the cylinder, and determine the frequency and period of the resulting vibratory motion.

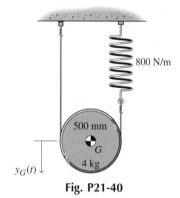

Fig. P21-40

21-41* An 18-lb stepped cylinder rolls without slipping on a horizontal plane as shown in Fig. P21-41. The two

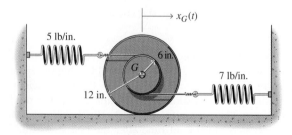

Fig. P21-41

springs are attached to cords, which are wrapped securely around the 12-in. diameter central hub. If the radius of gyration of the stepped cylinder is 9 in., write the differential equation of motion for $x_G(t)$, the position of the mass center of the cylinder, and determine the frequency and period of the resulting vibratory motion.

21-42 A 2-kg thin circular disk ($r = 200$ mm) hangs from a small frictionless pin on its rim as shown in Fig. P21-42. Write the differential equation of motion for $\theta(t)$, the angular position of the disk, and determine the frequency and period of the resulting vibratory motion.

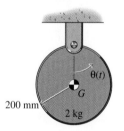

Fig. P21-42

21-43* A 15-lb thin rectangular plate (18 in. by 12 in.) hangs from a small frictionless pin at the middle of its long edge as shown in Fig. P21-43. Write the differential equation of motion for $\theta(t)$, the angular position of the plate, and determine the frequency and period of the resulting vibratory motion.

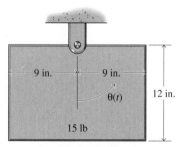

Fig. P21-43

21-44 The circular disk of Problem 21-42 is replaced with a thin circular ring of the same mass and radius as shown in Fig. P21-44. Write the differential equation of motion for $\theta(t)$, the angular position of the ring, and determine the frequency and period of the resulting vibratory motion.

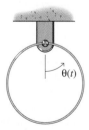

Fig. P21-44

21-45 A 3-lb uniform slender bar 5 ft long is attached to a frictionless pivot at A as shown in Fig. P21-45. The bar is horizontal in the equilibrium position. If end C is pulled down 5 in. and released from rest, determine:

a. The differential equation of motion for $\theta(t)$, the angular position of the bar.
b. The maximum velocity of end C in the resulting vibratory motion.

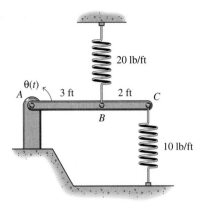

Fig. P21-45

21-46* A 2-kg uniform slender bar 500 mm long is attached to a frictionless pivot at B as shown in Fig. P21-46. The bar is horizontal in the equilibrium position. If end C is pulled down 15 mm and released from rest, determine:

a. The differential equation of motion for $\theta(t)$, the angular position of the bar.
b. The maximum velocity of end C in the resulting vibratory motion.

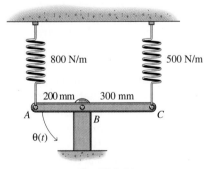

Fig. P21-46

21-47 Two slender uniform bars are welded together as shown in Fig. P21-47. Bar ABC weighs 2 lb and is horizontal in the equilibrium position; bar BD weighs 3 lb and is vertical in the equilibrium position; and the pivot is frictionless. If end D is pulled to the left 3 in. and released from rest, determine:

a. The differential equation of motion for $\theta(t)$, the angular position of the bar.
b. The maximum velocity of end D in the resulting vibratory motion.

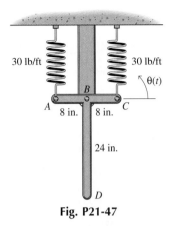

Fig. P21-47

21-48* A brass (8750 kg/m³) paperweight in the form of a half cylinder (75 mm long, 100 mm in diameter) rests on a flat horizontal surface as shown in Fig. P21-48. If the cylinder rolls without slipping, write the differential equation of motion for $\theta(t)$, the angular position of the paperweight, and determine the frequency and period of the resulting vibratory motion.

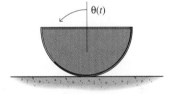

Fig. P21-48

21-49* The light cord attached to the 10-lb block of Fig. P21-49 is wrapped around a 7-lb uniform cylinder. If the cord does not slip on the cylinder, write the differential equation of motion for $y(t)$, the position of the 10-lb block, and determine the frequency and period of the resulting vibratory motion.

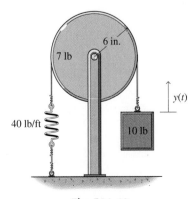

Fig. P21-49

21-50 A 6-kg weight is attached to the cylinder of Problem 21-40 as shown in Fig. P21-50. If the weight is hung from a frictionless pin through the center of the cylinder, write the differential equation of motion for $y_G(t)$, the position of the center of mass of the cylinder, and determine the frequency and period of the resulting vibratory motion.

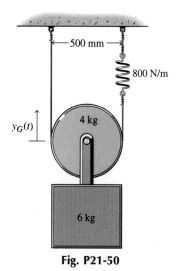

Fig. P21-50

21-51* Repeat Problem 21-33 (p. 446) if *ABC* is a uniform slender bar weighing 12 lb.

21-52 Repeat Problem 21-34 (p. 447) if *ABC* is a uniform slender bar with a mass of 12 kg.

21-53 Repeat Problem 21-35 (p. 447) if *AB* and *BC* are uniform slender bars weighing 2 lb and 1 lb, respectively.

21-54* A 5-kg uniform slender bar 400 mm long is rigidly attached to an 8-kg uniform cylinder 300 mm in diameter as shown in Fig. P21-54. If the cylinder rolls without slipping on the horizontal surface, write the differential equation of motion for $\theta(t)$, the angular position of the cylinder, and determine the frequency and period of the resulting vibratory motion.

Fig. P21-54

21-3 DAMPED FREE VIBRATIONS

The analysis of undamped free vibrations in the previous sections is only an idealization of real systems because it does not account for the energy lost to friction. Once set in motion, such idealized systems vibrate forever with a constant amplitude. All real systems, however, lose energy to friction and will eventually stop unless there is a source of energy to keep them going. When the amount of energy loss in the system is small, the results of the previous sections are often in good agreement with real systems, at least for short intervals of time. For longer intervals of time or for cases in which the energy loss is not small, the effects of friction forces must be included.

There are several types of friction forces that can remove mechanical energy from a vibrating system. Some of the more common friction forces are: *fluid friction* (also called *viscous damping force*), which arises when bodies move through viscous fluids; *dry friction* (also called *coulomb friction*), which arises when a body slides across a dry surface; and *internal friction*, which arises when a solid body is deformed. Damping caused by fluid friction is quite common in engineering work and only linear viscous damping is considered in this first course in Dynamics.

21-3-1 The Linear Viscous Damper

Viscous damping occurs naturally when mechanical systems such as pendulums vibrate in air or water. Viscous damping is also exhibited by devices called *dashpots* (represented symbolically by Fig. 21-12), which are intentionally added to mechanical systems to limit or control vibration. A typical dashpot consists of a piston moving in a cylinder filled with a viscous fluid. Movement of the piston is opposed by the fluid, which must either flow through small holes in the piston or flow through a narrow gap around the piston. Devices such as door closers and shock absorbers are examples of actual dashpots. In addition, symbolic dashpots are sometimes used to represent the frictional loss in systems that do not have distinct damping devices. The mass of a dashpot, like the mass of a spring, is generally neglected.

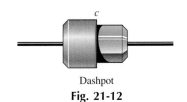

Dashpot
Fig. 21-12

The viscous dampers considered here are linear. That is, the magnitude of the viscous damping force is directly proportional to the speed with which the damper is being extended or compressed

$$F = c\dot{x} \qquad (21\text{-}19)$$

The constant of proportionality c is called the *coefficient of viscous damping*. Typical units for c are $N \cdot s/m$ in the SI system of units or $lb \cdot s/ft$ in the U. S. Customary system of units. The direction of the viscous damping force is always opposite to the direction of the velocity.

21-3-2 Viscous Damped Free Vibration

To illustrate viscous damped free vibration, damping is added to the block-spring system of Fig. 21-5a as shown in Fig. 21-13a. The free-body diagram of the block is now shown in Fig. 21-13b in which the block again has been displaced an arbitrary amount in the positive coordinate direction. The elastic restoring force of the spring $F_s = kx$ is still directed toward the equilibrium position (the negative coordinate direction). Since the positive directions for the velocity $\dot{x}$ and the acceleration $\ddot{x}$ are the same as the positive coordinate direction, the damping force $F_d = c\dot{x}$ also acts in the negative coordinate direction. Applying Newton's second law $\Sigma F = ma_x = m\ddot{x}$ to the block gives the differential equation of motion for the block

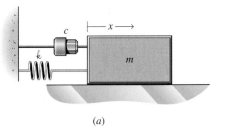

(a)

$$-kx - c\dot{x} = m\ddot{x} \qquad (21\text{-}20a)$$

or

$$m\ddot{x} + c\dot{x} + kx = 0 \qquad (21\text{-}20b)$$

which is again a second-order linear differential equation with constant coefficients.

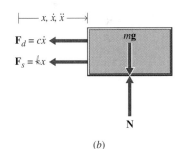

(b)
Fig. 21-13

It is well known from the theory of ordinary differential equations that the solution of any linear, ordinary differential equation with constant coefficients is always of the form

$$x(t) = De^{\lambda t} \qquad (21\text{-}21)$$

where the constants D and λ must be chosen to satisfy the differential equation and the initial conditions. Substituting Eq. 21-21 into Eq. 21-20 gives the characteristic equation[4]

$$m\lambda^2 + c\lambda + k = 0 \qquad (21\text{-}22)$$

[4]If the constant D is zero, then Eq. 21-21 gives the trivial solution $x = 0$, which is of no interest. Clearly, the exponential $e^{\lambda t}$ is never zero. Therefore, the factor $De^{\lambda t}$ is never zero and it is permissible to divide the equation by it.

which has the roots

$$\lambda_{1,2} = \frac{-c \pm \sqrt{c^2 - 4mk}}{2m} \tag{21-23}$$

The displacement of the block is then given by[5]

$$x(t) = D_1 e^{\lambda_1 t} + D_2 e^{\lambda_2 t} \tag{21-24}$$

where the constants D_1 and D_2 are determined from the initial conditions (at $t = 0$; $x = D_1 + D_2 = x_0$ and $\dot{x} = D_1\lambda_1 + D_2\lambda_2 = v_0$) and λ_1 and λ_2 are given by Eq. 21-23.

Before discussing the solution, however, the roots (Eq. 21-23) will be rewritten in terms of more convenient variables. The dimensionless combination of constants

$$\zeta = \frac{c}{2\sqrt{mk}} = \frac{c}{2m\omega_n} \tag{21-25}$$

is called the *damping ratio*.[6] In terms of the damping ratio ζ and the natural circular frequency ω_n Eq. 21-23 becomes

$$\lambda_{1,2} = -\zeta\omega_n \pm \omega_n\sqrt{\zeta^2 - 1} \tag{21-26}$$

The behavior of the system depends on whether the quantity under the radical in Eq. 21-26 is positive, zero, or negative. The value of c that makes the radical zero is called the *critical damping coefficient* c_{cr}. Therefore,

$$c_{cr} = 2m\omega_n = 2\sqrt{mk} \tag{21-27}$$

The solution (Eq. 21-24) will have three totally distinct types of behavior depending on whether the actual system damping c is greater than, equal to, or less than c_{cr}.[7] Each possibility is analyzed separately in the next three sections.

21-3-3   Overdamped Systems

When the damping coefficient c is greater than c_{cr}, then the damping ratio ζ is greater than one, the radical Eq. 21-26 is real, and the two roots λ_1 and λ_2 are both real and are unequal. Furthermore, since

---

[5] When the quantity under the radical in Eq. 21-23 is zero, the two roots are identical $\lambda_1 = \lambda_2 = \lambda = -c/2m = -\sqrt{k/m} = -\omega_n$ (since $c = 2\sqrt{mk}$). In this case the general solution of Eq. 21-20 is

$$x(t) = (B + Ct)e^{\lambda t}$$

as can be easily verified by direct substitution.

[6] In the present example, the constants that appear in the definitions of the damping ratio ζ and the natural circular frequency ω_n are the actual system mass m, damping coefficient c, and spring constant k. In general, however, they should be interpreted as the *effective mass* of the system (the coefficient of the $\ddot{x}$ term in Eq. 21-20), the *effective damping coefficient* (the coefficient of the $\dot{x}$ term in Eq. 21-20), and the *effective spring constant* (the coefficient of the term x term in Eq. 21-20), respectively.

[7] As always, the constants m, c, and k may or may not be the actual system values. Rather, they should be interpreted as the coefficients of the differential equation of motion when written in the standard form—Eq. 21-20b.

$\sqrt{\zeta^2 - 1} < \zeta$ both roots will be negative. Therefore, the displacement (Eq. 21-24) simply decreases to zero as t increases, and the motion is nonvibratory.

The displacement given by Eq. 21-24 is shown in Fig. 21-14 for representative initial conditions. The damping is so severe in this case that an overdamped system returns slowly to its equilibrium position. Since the system does not actually oscillate, there is no period or frequency associated with *overdamped* or *supercritically damped* motions.

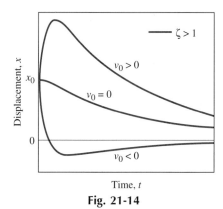

Fig. 21-14

21-3-4 Critically Damped Systems

When the damping coefficient c is equal to c_{cr}, then the damping ratio ζ is equal to one, the radical in Eq. 21-26 is zero, and the two roots $\lambda_1 = \lambda_2 = -\omega_n$ are equal and negative. The solution in this case has the special form

$$x(t) = (B + Ct)e^{-\omega_n t} \qquad (21-28)$$

Again, the displacement (Eq. 21-24) simply decreases to zero as t increases, and the motion is nonvibratory.

Qualitatively, the motion described by Eq. 21-28 for critical damping is the same as the motion for supercritical damping. Critical damping is of special importance only because it is the dividing point between nonvibratory motions and damped oscillatory motions. That is, critical damping is the smallest amount of damping for which a system will not oscillate. In addition, a critically damped system will come to rest in less time than any other system starting from the same initial conditions.[8] Figure 21-15 shows representative displacement curves for both an overdamped and a critically damped system starting from the same initial displacement and the same initial velocity.

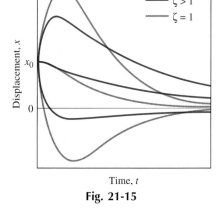

Fig. 21-15

---

[8]Strictly speaking, the system described by Eq. 21-28 does not really come to rest for any finite value of time. Practically speaking, however, the motion will become imperceptible after some finite time and the system may be said to be at rest.

21-3-5   Underdamped Systems

When the damping coefficient c is less than c_{cr}, then the damping ratio ζ is less than one, the radical in Eq. 21-23 is imaginary, and the two roots λ_1 and λ_2 are complex conjugates,

$$\lambda_1 = -\zeta\omega_n + i\omega_d \qquad (21\text{-}29a)$$

$$\lambda_2 = -\zeta\omega_n - i\omega_d \qquad (21\text{-}29b)$$

where

$$i = \sqrt{-1} \qquad \text{and} \qquad \omega_d = \omega_n\sqrt{1 - \zeta^2}$$

When these values are substituted back into Eq. 21-24, the equation for the displacement becomes

$$x(t) = e^{-\zeta\omega_n t}(D_1 e^{i\omega_d t} + D_2 e^{-i\omega_d t}) \qquad (21\text{-}30)$$

By making use of the Euler formula, $e^{ix} = \cos x + i \sin x$, Eq. 21-30 can be rewritten

$$x(t) = e^{-\zeta\omega_n t}[(D_1 + D_2) \cos \omega_d t + i(D_1 - D_2) \sin \omega_d t]$$
$$= e^{-\zeta\omega_n t}(B \cos \omega_d t + C \sin \omega_d t) \qquad (21\text{-}31a)$$
$$= Ae^{-\zeta\omega_n t} \cos (\omega_d t - \phi_c) \qquad (21\text{-}31b)$$

where the constants $B = D_1 + D_2$ and $C = i(D_1 - D_2)$ or $A = \sqrt{B^2 + C^2}$ and $\phi_c = \tan^{-1} B/C$ are to be determined from the initial conditions. A typical displacement curve given by Eq. 21-31 is shown in Fig. 21-16. Just as in the preceding cases, the displacement goes to zero as t goes to infinity. However, here the response oscillates within the bounds of the exponential decay curves $Ae^{-\zeta\omega_n t}$ and $-Ae^{-\zeta\omega_n t}$ as it goes to zero.

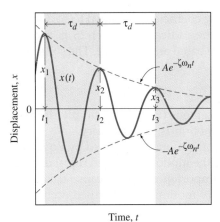

Time, t

Fig. 21-16

The motion described by Eq. 21-31 is called *time-periodic*. The motion oscillates about the equilibrium position, but the amplitude $Ae^{-\zeta\omega_n t}$ decreases since the exponent $-\zeta\omega_n = -c/2m$ is negative. Since the amplitude of a damped oscillation decreases monotonically with time, the oscillation will never repeat itself exactly. Therefore, a damped oscillation does not have a period in the same sense as defined for free undamped vibrations. Because of the similarity of Eqs. 21-31b and 21-5a, however, it is customary to call the constant $\omega_d = \omega_n\sqrt{1 - \zeta^2}$ the *damped natural circular frequency*. Since $0 < \zeta < 1$ for underdamped vibrations, the damped natural circular frequency ω_d will always be less than the undamped natural circular frequency ω_n. Also by analogy with free, undamped vibrations, a *damped natural frequency* f_d and a *damped period* τ_d may be defined as

$$f_d = \frac{\omega_d}{2\pi} = \frac{\omega_n\sqrt{1 - \zeta^2}}{2\pi} \tag{21-32a}$$

$$\tau_d = \frac{2\pi}{\omega_d} = \frac{2\pi}{\omega_n\sqrt{1 - \zeta^2}} \tag{21-32b}$$

The period as defined by Eq. 21-32b is seen to be the time interval between two successive points where the curve of Eq. 21-31 touches one of the limiting curves shown in Fig. 21-16 or twice the time interval between two successive passings through equilibrium. It is interesting to note that the damped period τ_d and the damped natural frequencies f_d and ω_d are constant (independent of time) even though the amplitude is not.

The linear viscous damper in many physical systems is not a real physical element but is merely a mathematical concept used to explain the energy dissipation. For this and other reasons, it is usually necessary to determine the value of the viscous damping ratio ζ experimentally. The determination is easily accomplished by measuring the displacement at two successive "peaks" of the motion; for example, at x_1 and x_2 in Fig. 21-16. Since $\cos(\omega_d t - \phi) = 1$ at both t_1 and t_2, the ratio of these two amplitudes is

$$\frac{x_1}{x_2} = \frac{Ae^{-\zeta\omega_n t_1}}{Ae^{-\zeta\omega_n(t_1+\tau_d)}} = e^{\zeta\omega_n\tau_d}$$

Then taking the natural logarithm of both sides and defining the *logarithmic decrement* $\delta = \ln\left(\frac{x_1}{x_2}\right)$ gives

$$\delta = \zeta\omega_n\tau_d = \zeta\omega_n\frac{2\pi}{\omega_n\sqrt{1 - \zeta^2}} = \frac{2\pi\zeta}{\sqrt{1 - \zeta^2}} \tag{21-33}$$

Note that δ depends only on the damping ratio ζ and not on t_1 or t_2. That is, the logarithmic decrement does not depend on which two successive peaks are used to measure it. Finally, solving for ζ gives

$$\zeta = \frac{\delta}{\sqrt{(2\pi)^2 + \delta^2}} \tag{21-34}$$

When the amount of damping in the system is small, the displacements x_1 and x_2 will be nearly equal $x_1 \cong x_2$ so that $\delta = \ln(x_1/x_2)$ will be very small. Then $\sqrt{(2\pi)^2 + \delta^2} \cong 2\pi$ so that $\zeta \cong \delta/2\pi$ or $\delta \cong 2\pi\zeta$.

A 5-kg block slides on an inclined frictionless surface as shown in Fig. 21-17a. The elastic modulus of the springs are $k_1 = k_2 = 2$ kN/m, and the viscous damping coefficients are $c_1 = c_2 = 25$ N·s/m. If the cart is moved 50 mm up the incline from its equilibrium position and released with an initial velocity of 1.25 m/s down the incline when $t = 0$, determine

a. The damped period τ_d, the damped frequency f_d, and the damped circular frequency ω_d of the resulting vibration.

b. The position of the block as a function of time.

c. The time t_1 at which the amplitude is reduced to 1 percent of its initial value.

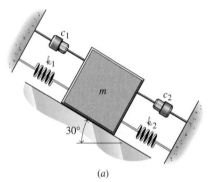

(a)

Fig. 21-17

SOLUTION

a. The free-body diagram of the block is drawn in Fig. 21-17b in which the x-coordinate measures the position of the block along the incline with $x = 0$ at the equilibrium position. In the equilibrium position (before the block

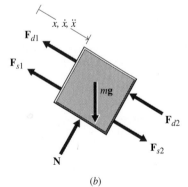

(b)

Fig. 21-17

has been disturbed), the spring forces are proportional to their static deformation $F_{s1} = k_1\delta_{eq1}$, $F_{s2} = k_2\delta_{eq2}$. Since $\dot{x} = 0$ in the static equilibrium position, $F_{d1} = F_{d2} = 0$. Therefore, equilibrium gives

$$mg \sin 30° - k_1\delta_{eq1} + k_2\delta_{eq2} = 0 \qquad (a)$$

Equation a gives a relationship between the static deformations of the springs and the weight of the block.

When the block is at an arbitrary (positive) location x, the stretch in spring 1 will be increased $F_1 = k_1[\delta_{eq1} + x]$ and the stretch in spring 2 will be reduced $F_2 = k_2[\delta_{eq2} - x]$ from their equilibrium values. Both dashpots resist the motion and exert a force $F_{d1} = c_1\dot{x}$ and $F_{d2} = c_2\dot{x}$ in the negative coordinate direction (when $\dot{x}$ is positive). Therefore, Newton's second law of motion $\Sigma F_x = m\ddot{x}$ gives

$$mg \sin 30° - k_1(\delta_{eq1} + x) + k_2(\delta_{eq2} - x) - (c_1 + c_2)\dot{x} = m\ddot{x}$$

or

$$mg \sin 30° - k_1\delta_{eq1} + k_2\delta_{eq2} = m\ddot{x} + (c_1 + c_2)\dot{x} + (k_1 + k_2)x \qquad (b)$$

The left side of Eq. b is zero by Eq. a, however, so the differential equation of motion is

$$m\ddot{x} + (c_1 + c_2)\dot{x} + (k_1 + k_2)x = 0$$
$$5\ddot{x} + 50\dot{x} + 4000x = 0$$

Then the natural circular frequency, the damping ratio, the damped circular frequency, the damped natural frequency, and the damped period are

$$\omega_n = \sqrt{4000/5} = 28.284 \text{ rad/s}$$

$$\zeta = \frac{50}{2(5)(28.284)} = 0.17678$$

$$\omega_d = \omega_n\sqrt{1 - (0.17678)^2} = 27.84 \text{ rad/s} \qquad \text{Ans.}$$

$$f_d = \frac{\omega_d}{2\pi} = 4.431 \text{ Hz} \qquad \text{Ans.}$$

$$\tau_d = \frac{1}{f_d} = 0.2257 \text{ s} \qquad \text{Ans.}$$

b. The displacement and velocity of the block can be written in the form

$$x(t) = e^{-5.000t}(B \cos 27.84t + C \sin 27.84t)$$
$$\dot{x}(t) = -5.000e^{-5.000t}(B \cos 27.84t + C \sin 27.84t)$$
$$+ e^{-5.000t}(-27.84 B \sin 27.84t + 27.84C \cos 27.84t)$$

But at $t = 0$, $x = B = -50$ mm and $\dot{x} = -5B + 27.84C = 1250$ mm/s. Therefore, $B = -50$ mm and $C = 35.92$ mm, and

$$x(t) = e^{-5.000t}(-50 \cos 27.84t + 35.92 \sin 27.84t) \qquad \text{Ans.}$$

This solution is shown in Fig. 21-17c. Also shown, for comparison, are the two parts of the damped solution $x_u(t) = -50 \cos 27.84t + 35.92 \sin 27.84t$ and the amplitude $Ae^{-5.00t}$ where $A = \sqrt{B^2 + C^2} = 61.56$ mm.

c. The initial amplitude of the oscillation is just A. Therefore, at time t_1

$$Ae^{-5.000t_1} = 0.01A$$

which gives

$$t_1 = 0.921 \text{ s} \qquad \text{Ans.}$$

(or a little over 4 cycles of the oscillation).

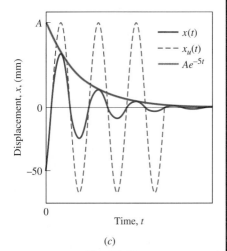

(c)

Fig. 21-17

EXAMPLE PROBLEM 21-5

A 20-lb cart rolls on a flat, horizontal surface as shown in Fig. 21-18a. The cart is pulled 15 in. to the right and released with a velocity of 15 ft/s to the left at $t = 0$. If the spring constant $k = 40$ lb/ft and the damping coefficient c corresponds to critical damping, determine

a. The value of the damping coefficient c.
b. If the cart will overshoot the equilibrium position before coming to rest.

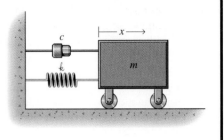

(a)

SOLUTION

a. The free-body diagram of the cart is drawn in Fig. 21-18b for an arbitrary (positive) location x. Applying Newton's second law of motion $\Sigma F = m\ddot{x}$ gives

$$-c\dot{x} - kx = m\ddot{x}$$

or

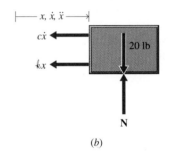

(b)

$$\frac{20}{32.2}\ddot{x} + c\dot{x} + 40x = 0$$

Then, the natural circular frequency

$$\omega_n = \sqrt{\frac{40}{20/32.2}} = 8.025 \text{ rad/s}$$

and the damping ratio is

$$\zeta = \frac{c_{cr}}{2\dfrac{20}{32.2}8.025} = 1$$

Therefore, the damping coefficient is

$$c = c_{cr} = 9.97 \text{ lb} \cdot \text{s/ft} \qquad \text{Ans.}$$

b. For the case of critical damping, the displacement and velocity of the cart are given by

$$x(t) = (B + Ct)e^{-\omega_n t} = (B + Ct)e^{-8.025t} \text{ in.}$$
$$\dot{x}(t) = [C - 8.025(B + Ct)]e^{-8.025t} \text{ in./s}$$

But at $t = 0$; $x = B = 15$ in. and $\dot{x} = C - 8.025B = -180$ in./s. Therefore, $B = 15$ in., $C = -59.63$ in./s, and

$$x(t) = (15 - 59.63 \, t)e^{-8.025t} \text{ in.}$$

Since B and C have opposite signs, there will exist a time $t_1 = 15/59.63 = 0.252$ s at which the position of the cart will be zero. Before t_1 the cart will be on one side of the equilibrium position, and after t_1 it will be on the other side. Therefore, the cart will pass through the equilibrium position before coming to rest. Ans.

The position and velocity of the cart are shown in Fig. 21-18c for $0 \leq t \leq 1.5$ s.

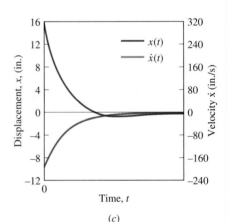

(c)

Fig. 21-18

A 3-kg uniform slender bar 100 mm long is in equilibrium in the horizontal position shown in Fig. 21-19a. When E is displaced down a small amount and released, the amplitude of each peak of the oscillation is observed to be 0.9 of the amplitude of the previous peak. If the spring constant is $k = 400$ N/m, determine

a. The value of the damping coefficient c.
b. The damped period τ_d, the damped frequency f_d, and the damped circular frequency ω_d of the resulting vibration.

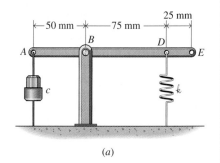

(a)

Fig. 21-19

SOLUTION

a. The logarithmic decrement is determined from the ratio of successive amplitudes $\delta = \ln(x_1/x_2) = \ln(1/0.9) = 0.10536$. Then the damping ratio is

$$\zeta = \frac{\delta}{\sqrt{(2\pi)^2 + \delta^2}} = 0.01677 = \frac{c_{eff}}{2\sqrt{m_{eff}k_{eff}}} \qquad (a)$$

where m_{eff}, c_{eff}, and k_{eff} are the coefficients in the differential equation of motion.

The free-body diagram of the bar is drawn in Fig. 21-19b, in which θ measures the angular position of the bar, with θ positive counterclockwise and $\theta = 0$ at the equilibrium position. In the equilibrium position (before the bar has been disturbed), the dashpot force is $F_d = 0$ and the spring force is $F_s = k\delta_{eq}$ where δ_{eq} is the extension of the spring in the equilibrium position. Therefore, the moment equilibrium equation

$$\curvearrowleft + \Sigma M_B = 0: \qquad -0.075 k\delta_{eq} - 0.025mg = 0 \qquad (b)$$

gives $\delta_{eq} = -24.53$ mm.

When the bar is rotated counterclockwise (in the positive θ direction), the extension in the spring is $\delta_{eq} + \delta_D$ where for small angles of rotation $\delta_D \cong 0.075\,\theta$. Similarly, the dashpot is compressed at the rate of $\dot{\delta}_A \cong 0.050\,\dot{\theta}$. Therefore, the differential equation of motion is

$$\curvearrowleft + \Sigma M_B = 0: \qquad -0.025mg - 0.075 k(\delta_{eq} + \delta_D) - 0.050c\dot{\delta}_A = I_B\ddot{\theta}$$

or

$$I_B\ddot{\theta} + (0.050)^2 c\dot{\theta} + (0.075)^2 k\theta = -0.075 k\delta_{eq} - 0.025mg \qquad (c)$$

where $I_B = \dfrac{1}{12}(3)(0.150)^2 + (3)(0.025)^2 = 7.5(10^{-3})$ kg $\cdot$ m². But the right-hand side of Eq. c is zero by Eq. b so that

$$\ddot{\theta} + 0.3333c\dot{\theta} + 300\theta = 0$$

Substituting the coefficients $m_{eff} = 1$, $c_{eff} = 0.3333c$, and $k_{eff} = 300$ into Eq. a gives the viscous damping coefficient

$$c = \frac{0.01677}{0.3333}(2)\sqrt{300} = 1.743 \text{ N}\cdot\text{s/m} \qquad \text{Ans.}$$

b. Then the natural circular frequency is $\omega_n = \sqrt{300} = 17.321$ rad/s and

$$\omega_d = \omega_n\sqrt{1 - \zeta^2} = 17.318 \text{ rad/s} \qquad \text{Ans.}$$

$$f_d = \frac{\omega_d}{2\pi} = 2.756 \text{ Hz} \qquad \tau_d = \frac{1}{f_d} = 0.363 \text{ s} \qquad \text{Ans.}$$

PROBLEMS

21-55–21-68 The following equations represent the position of a particle in damped vibratory motion. For each equation

a. Classify the motion as underdamped, overdamped, or critically damped.
b. Plot the position, velocity, and acceleration of the particle versus time from $t = 0$ until the amplitude has decreased to 5 percent of its initial value or for three cycles of the oscillation, whichever comes first.

21-55* $x(t) = 10e^{-0.1t} \cos(5t - 1.2)$ in.

21-56* $x(t) = (5 + 3t)e^{-2t}$ mm

21-57 $x(t) = 10e^{-0.5t} - 8e^{-1.5t}$ rad

21-58* $x(t) = e^{-0.05t}(8 \cos 3t - 6 \sin 3t)$ mm

21-59 $x(t) = 8e^{-0.5t} - 8e^{-2t}$ in.

21-60* $x(t) = (-2 + 5t)e^{-1.5t}$ rad

21-61 $x(t) = e^{-0.02t}(12 \sin 12t - 5 \cos 12t)$ in.

21-62 $x(t) = -8e^{-0.02t} \sin(15t + 2.5)$ mm

21-63* $x(t) = -(5 + 10t)e^{-0.2t}$ in.

21-64 $x(t) = 7e^{-2t} + 5e^{-3t}$ rad

21-65* $x(t) = (4 - t)e^{-1.2t}$ rad

21-66* $x(t) = 6e^{-0.15t} \sin(10t - 2.5)$ mm

21-67 $x(t) = 3e^{-0.06t} \cos(8t + 1.8)$ in.

21-68 $x(t) = 5e^{-0.5t} - 8e^{-1.5t}$ mm

21-69–21-76 The following differential equations and initial conditions represent the motion of a particle in damped vibratory motion. For each equation

a. Classify the motion as underdamped, overdamped, or critically damped.
b. Plot the position, velocity, and acceleration of the particle versus time from $t = 0$ until the amplitude has decreased to 5 percent of its initial value or for three cycles of the oscillation, whichever comes first.

21-69* $0.5\ddot{x} + 5\dot{x} + 40x = 0;$ x, in.
$x(0) = 3$ in.; $\dot{x}(0) = 15$ in./s

21-70* $3\ddot{x} + 60\dot{x} + 240x = 0;$ x, mm
$x(0) = -30$ mm; $\dot{x}(0) = 150$ mm/s

21-71 $0.25\ddot{x} + 5\dot{x} + 25x = 0;$ x, in.
$x(0) = -5$ in.; $\dot{x}(0) = 50$ in./s

21-72* $2\ddot{x} + 4\dot{x} + 40x = 0;$ x, mm
$x(0) = 100$ mm; $\dot{x}(0) = 150$ mm/s

21-73 $0.1\ddot{x} + 5\dot{x} + 5x = 0;$ x, in.
$x(0) = 8$ in.; $\dot{x}(0) = 25$ in./s

21-74 $4\ddot{x} + 100\dot{x} + 200x = 0;$ x, mm
$x(0) = -100$ mm; $\dot{x}(0) = -250$ mm/s

21-75* $0.2\ddot{x} + 2\dot{x} + 5x = 0;$ x, in.
$x(0) = -15$ in; $\dot{x}(0) = 0$ in./s

21-76 $5\ddot{x} + 10\dot{x} + 50x = 0;$ x, mm
$x(0) = 0$ mm; $\dot{x}(0) = 500$ mm/s

21-77* A block that has a mass m slides on a frictionless horizontal surface as shown in Fig. P21-77. Determine the damping coefficient c of the single dashpot that could replace the two dashpots shown without changing the frequency of vibration of the block.

Fig. P21-77

21-78 A block that has a mass m slides on a frictionless horizontal surface as shown in Fig. P21-78. Determine the damping coefficient c of the single dashpot that could replace the two dashpots shown without changing the frequency of vibration of the block.

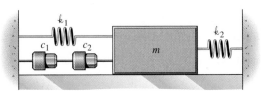

Fig. P21-78

21-79 A 10-lb block is suspended in a vertical plane by two springs and a dashpot as shown in Fig. P21-79. If the block is displaced 7 in. above its equilibrium position and released with an upward velocity of 150 in./s when $t = 0$, determine

a. The differential equation governing the motion.
b. The period of the resulting vibration.

460

c. The position of the block as a function of time.
d. The earliest time $t_1 > 0$ when the block passes through its equilibrium position.

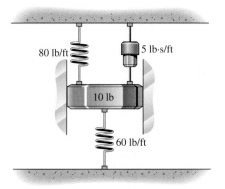

Fig. P21-79

21-80* A 2-kg mass is suspended in a vertical plane by two springs and a dashpot as shown in Fig. P21-80. If the mass is displaced 5 mm below its equilibrium position and released with an upward velocity of 250 mm/s when $t = 0$, determine

a. The differential equation governing the motion.
b. The period of the resulting vibration.
c. The position of the mass as a function of time.
d. The earliest time $t_1 > 0$ when the mass passes through its equilibrium position.

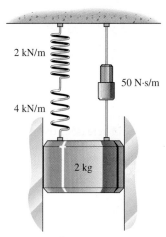

Fig. P21-80

21-81 A 20-lb block slides on a frictionless horizontal surface as shown in Fig. P21-81. The two springs remain in tension at all times and the two pulleys are both small and frictionless. If the block is displaced 3 in. to the left of

its equilibrium position and released with a velocity of 50 in./s to the right when $t = 0$, determine

a. The differential equation governing the motion.
b. The period of the resulting vibration.
c. The position of the block as a function of time.
d. The earliest time $t_1 > 0$ when the velocity of the block is zero.

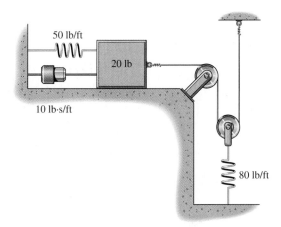

Fig. P21-81

21-82* A 4-kg mass is suspended in a vertical plane as shown in Fig. P21-82. The spring remains in tension at all times and the two pulleys are both small and frictionless. If the mass is displaced 15 mm above its equilibrium position and released with a downward velocity of 750 mm/s when $t = 0$, determine

a. The differential equation governing the motion.
b. The period of the resulting vibration.
c. The position of the mass as a function of time.
d. The earliest time $t_1 > 0$ when the velocity of the mass is zero.

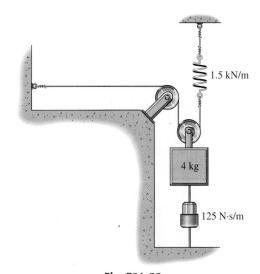

Fig. P21-82

21-83* The two blocks of Fig. P21-83 hang in a vertical plane from a massless bar, which is horizontal in the equilibrium position. If a = 6 in., assume small oscillations and determine

a. The damping ratio ζ.
b. The type of motion (underdamped, critically damped, or overdamped).
c. The frequency and period of the motion (if any exists).
d. The value of a that gives critical damping.

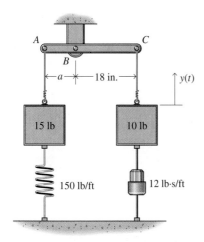

Fig. P21-83

21-84 The two masses of Fig. P21-84 each slide on a frictionless horizontal surface. The bar ABC is vertical in the equilibrium position and has negligible mass. If a = 100 mm, assume small oscillations and determine

a. The damping ratio ζ.
b. The type of motion (underdamped, critically damped, or overdamped).
c. The frequency and period of the motion (if any exists).
d. The value of a that gives critical damping.

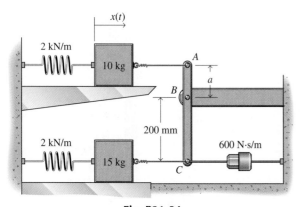

Fig. P21-84

21-85* The 5-lb block of Fig. P21-85 slides on a frictionless horizontal surface while the 3-lb block hangs in a vertical plane. The bar ABC has negligible mass and arm AB is horizontal in the equilibrium position. If c = 15 lb·s/ft, assume small oscillations and determine

a. The damping ratio ζ.
b. The type of motion (underdamped, critically damped, or overdamped).
c. The frequency and period of the motion (if any exists).
d. The value of c that gives critical damping.

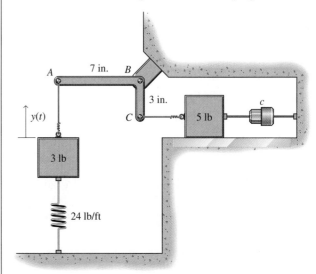

Fig. P21-85

21-86 For the mass–spring–dashpot system shown in Fig. 21-13 (p. 451), determine the ratio of amplitudes of the vibration at

a. The second positive and third positive peaks.
b. The first positive and third positive peaks.
c. The third positive and fifth positive peaks.
d. The first positive and the following negative peak.

21-87 The damping coefficient c of a dashpot is to be determined by observing the oscillation of a 10-lb block suspended from it as shown in Fig. P21-87. When the block is pulled downward and released, the amplitude of the resulting vibration is observed to decrease from 5 in. to 3 in. in 20 cycles of oscillation. Determine the value of c if the 20 cycles are completed in 5 s.

Fig. P21-87

21-88* The damping coefficient c of a dashpot is to be determined by observing the oscillation of a block suspended from it as shown in Fig. P21-87. When the block is pulled downward and released, the amplitude of the resulting vibration is observed to decrease from 75 mm to 20 mm in 10 cycles of oscillation. Determine the value of c if the spring constant is $k = 1.5$ kN/m and the 10 cycles are completed in 8 s.

21-89 At $t = 0$ the 10 lb mass of Problem 21-83 is y_o in. above its equilibrium position. If the system is released with zero initial velocity, determine the amount of time and/or the number of cycles required to reduce the amplitude of the motion to 1 percent of its initial value for:

a. $a = 6$ in.
b. $a = 24$ in.

21-90* At $t = 0$ the 10 kg mass of Problem 21-84 is x_o mm to the left of its equilibrium position. If the system is released with zero initial velocity, determine the amount of time and/or the number of cycles required to reduce the amplitude of the motion to 1 percent of its initial value for:

a. $a = 100$ mm.
b. $a = 500$ mm.

21-91* A 7-lb uniform cylinder rolls without slipping on a horizontal surface as shown in Fig. P21-91. The spring and the dashpot are connected to a small frictionless pin at G the center of the 8-in. diameter cylinder. For this system, determine:

a. The damping ratio ζ.
b. The type of motion (underdamped, critically damped, or overdamped).
c. The frequency and period of the motion (if any exists).

attached to a light inextensible cord, which is wrapped around the cylinder, and the dashpot is connected to a small frictionless pin at G the center of the 400-mm diameter cylinder. For this system, determine:

a. The damping ratio ζ.
b. The type of motion (underdamped, critically damped, or overdamped).
c. The frequency and period of the motion (if any exists).

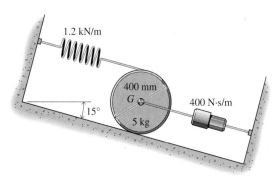

Fig. P21-92

21-93* A 3-lb uniform slender bar 5 ft long is attached to a frictionless pivot at A as shown in Fig. P21-93. The bar is horizontal in the equilibrium position. For this system, determine:

a. The damping ratio ζ.
b. The type of motion (underdamped, critically damped, or overdamped).
c. The frequency and period of the motion (if any exists).

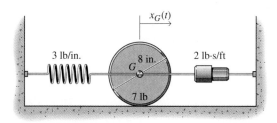

Fig. P21-91

21-92 A 5-kg uniform cylinder rolls without slipping on an inclined surface as shown in Fig. P21-92. The spring is

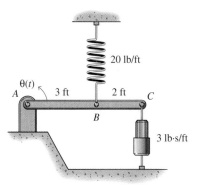

Fig. P21-93

21-94* A 2-kg uniform slender bar 500 mm long is attached to a frictionless pivot at B as shown in Fig. P21-94. The bar is horizontal in the equilibrium position. For this system, determine:

a. The damping ratio ζ.
b. The type of motion (underdamped, critically damped, or overdamped).
c. The frequency and period of the motion (if any exists).

21-95 Two slender uniform bars are welded together as shown in Fig. P21-95. Bar ABC weighs 2 lb and is horizontal in the equilibrium position; bar BD weighs 3 lb and is vertical in the equilibrium position; and the pivot is frictionless. For this system, determine:

a. The damping ratio ζ.
b. The type of motion (underdamped, critically damped, or overdamped).
c. The frequency and period of the motion (if any exists).

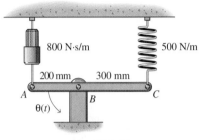

Fig. P21-94

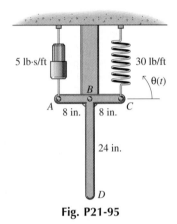

Fig. P21-95

21-4 FORCED VIBRATIONS

A forced vibration is produced and maintained by an externally applied periodic force that does not depend on the position or motion of the body. The force may be applied directly to the body, as in the force that keeps the pendulum of a clock moving. The force may be generated when the support to which the body is attached oscillates, as in the force applied by car springs to a car when the wheels roll over a bumpy road. The force may also be generated internally by the motion of unbalanced rotating parts, as in the force transmitted to an axle when a wheel rotates about an axis that does not pass through its center of mass.

Forced vibrations will occur whenever a periodically varying force is applied to a body. Since any nonharmonic, periodic function of time may be expressed by a Fourier series (a series of simple harmonic functions), a harmonic function of time

$$P = P_0 \sin \Omega t \quad \text{or} \quad P = P_0 \cos \Omega t$$

will be considered here. The constants P_0 and Ω are the amplitude and frequency (rad/s), respectively, of the driving force.

21-4-1 Harmonic Excitation Force

To illustrate viscous damped forced vibration, a harmonic excitation force $P_0 \sin \Omega t$ is added to the block–spring–dashpot system of Fig. 21-13a as shown in Fig. 21-20a. The free-body diagram of the block is

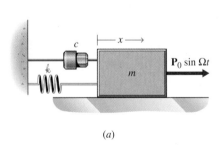

(a)

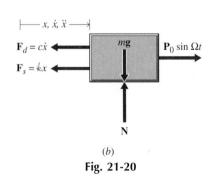

(b)

Fig. 21-20

shown in Fig. 21-20b, in which the block has been displaced an arbitrary amount in the positive coordinate direction. The elastic restoring force of the spring $F_s = kx$ is directed toward the equilibrium position (the negative coordinate direction), and the damping force $F_d = c\dot{x}$ acts opposite the velocity (also in the negative coordinate direction). Then applying Newton's second law of motion $\Sigma F = ma_x = m\ddot{x}$ to the block gives the differential equation of motion for the block

$$-c\dot{x} - kx + P_0 \sin \Omega t = m\ddot{x}$$

or

$$m\ddot{x} + c\dot{x} + kx = P_0 \sin \Omega t \tag{21-35}$$

Equation 21-35 is a nonhomogeneous, linear, second-order differential equation with constant coefficients. Its general solution consists of two parts, a *particular solution* and a *complementary solution*. The particular solution is any function $x_p(t)$ that satisfies the differential equation as written. The complementary solution is the function $x_c(t)$ that satisfies the homogeneous part of the differential equation, which is identical to Eq. 21-20. The complementary solution, therefore, is given by Eqs. 21-24, 21-28, or 21-31, depending on the value of the damping ratio ζ. The general solution to Eq. 21-35 is then

$$x(t) = x_c(t) + x_p(t) \tag{21-36}$$

The complementary part of the solution has already been discussed in considerable detail in Section 21-3. Therefore, the complementary solution will not be considered further except to note that:

1. Whether the system is overdamped, underdamped, or critically damped, $x_c(t)$ contains two constants that must be chosen to satisfy the initial conditions. In evaluating the constants, however, care must be taken to include the particular solution. That is, if the initial position and velocity are $x(0) = x_0$ and $\dot{x}(0) = v_0$, respectively, then $x_c(0) = x_0 - x_p(0)$ and $\dot{x}_c(0) = v_0 - \dot{x}_p(0)$.
2. No real system is completely frictionless. Therefore, the complementary solution $x_c(t)$ will always decay away with time. Since the complementary solution is visible only for some (usually short) period of time after the motion is started, it is called a transient solution.

The particular part of the solution is any function $x_p(t)$ that satisfies Eq. 21-35. Since the periodic exciting force is harmonic, it seems reasonable to guess that $x_p(t)$ is also harmonic

$$\begin{aligned} x_p(t) &= D \sin (\Omega t - \psi_s) \tag{21-37} \\ &= D \sin \Omega t \cos \psi_s - D \sin \psi_s \cos \Omega t \end{aligned}$$

where the constants D and ψ_s are to be chosen to make the solution $x_p(t)$ satisfy the differential equation (Eq. 21-35). Taking the appropriate derivatives and substituting them into the differential equation gives

$$\begin{aligned} D[(k - m\Omega^2) \cos \psi_s + c\Omega \sin \psi_s] \sin \Omega t \\ + D[c\Omega \cos \psi_s - (k - m\Omega^2) \sin \psi_s] \cos \Omega t = P_0 \sin \Omega t \end{aligned} \tag{21-38}$$

But the solution (Eq. 21-37) is supposed to satisfy the differential equation at every instant of time. Therefore, Eq. 21-38 must hold for every instant of time. In particular, when $t = 0$, $\sin \Omega t = 0$ and $\cos \Omega t = 1$ so that[9]

$$\tan \psi_s = \frac{c\Omega}{k - m\Omega^2} = \frac{2\zeta\Omega/\omega_n}{1 - (\Omega/\omega_n)^2} \qquad (21\text{-}39)$$

The *phase angle* ψ_s represents the amount by which the response $D \sin(\Omega t - \psi_s)$ lags the applied force $P_0 \sin \Omega t$. That is, the response hits its peak ψ_s/Ω seconds later than the applied force hits its peak.

When $\Omega t = \pi/2$, $\sin \Omega t = 1$ and $\cos \Omega t = 0$ so that Eq. 21-38 gives

$$D = \frac{P_0}{(k - m\Omega^2) \cos \psi_s + c\Omega \sin \psi_s}$$

where (with reference to Fig. 21-21)

$$\sin \psi_s = \frac{c\Omega}{\sqrt{(k - m\Omega^2)^2 + (c\Omega)^2}}$$

$$\cos \psi_s = \frac{k - m\Omega^2}{\sqrt{(k - m\Omega^2)^2 + (c\Omega)^2}}$$

Therefore, the amplitude of the particular solution is

$$D = \frac{P_0}{\sqrt{(k - m\Omega^2)^2 + (c\Omega)^2}}$$

$$= \frac{P_0/k}{\sqrt{[1 - (\Omega/\omega_n)^2]^2 + (2\zeta\Omega/\omega_n)^2}} \qquad (21\text{-}40)$$

Since the amplitude of the particular solution is constant, the particular solution is called the *steady-state* vibration. That is, after the transient part of the solution x_c has decayed away, the system will oscillate according to $x_p(t) = D \sin(\Omega t - \psi_s)$ as long as the driving force $P_0 \sin \Omega t$ is applied.

Note, however, that $\delta_p = P_0/k$ is the deflection of the spring that would result if the force P_0 were applied statically to the spring.[10] Then the ratio D/δ_p represents the factor by which the magnitude of the dynamic oscillation is greater than the static deflection. This ratio is called the *dynamic magnification factor* and is given by

$$\frac{D}{\delta_p} = \frac{D}{P_0/k} = \frac{1}{\sqrt{[1 - (\Omega/\omega_n)^2]^2 + (2\zeta\Omega/\omega_n)^2}} \qquad (21\text{-}41)$$

Figures 21-22 and 21-23 show the variation of the magnification factor D/δ_p and the phase angle ψ_s with the frequency ratio Ω/ω_n for various values of the damping ratio ζ. When the disturbing force $P_0 \sin \Omega t$ is applied at low frequencies ($\Omega/\omega_n < 1$), the steady-state response is mostly *in phase* with the disturbing force ($0 < \psi_s < 90°$). That is, the disturbing force generally acts to the right ($P_0 \sin \Omega t > 0$) when

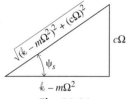

Fig. 21-21

[9] As always, the coefficients m, k, c, and P_0 in the solutions Eqs. 21-38 through 21-42 must be interpreted as the coefficients of the differential equation Eq. 21-35. They may or may not refer to actual system values of mass, spring constant, etc.

[10] The static deflection δ_p should not be confused with the equilibrium deflection δ_{eq} of Sections 21-2 and 21-3. The static deflection δ_p describes the deflection that would occur if the force P_0 were applied statically to the spring and has nothing to do with the equilibrium of the system.

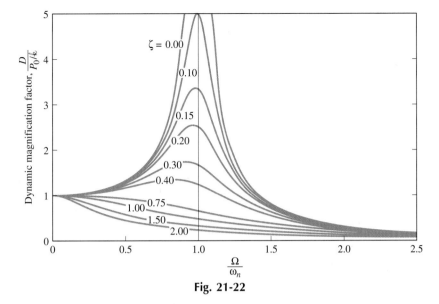

Fig. 21-22

Fig. 21-23

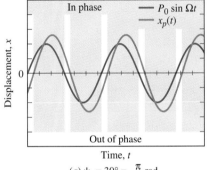

(a) $\psi_s = 30° = \dfrac{\pi}{6}$ rad

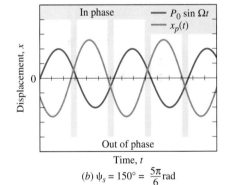

(b) $\psi_s = 150° = \dfrac{5\pi}{6}$ rad

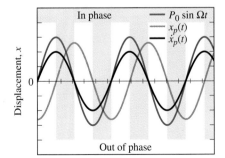

(c) $\psi_s = 90° = \dfrac{\pi}{2}$ rad

Fig. 21-24

the block is to the right of its equilibrium position ($x_p > 0$) and vice versa (Fig. 21-24a). In fact, for very low frequencies ($\Omega/\omega_n \cong 0$), the system is essentially in static equilibrium; the phase angle is nearly zero ($\psi_s \cong 0$), the magnification factor is approximately one ($D/\delta_p \cong 1$), and the steady-state response is $x_p(t) \cong (P_0 \sin \Omega t)/k$.

When the disturbing force is applied at high frequencies ($\Omega/\omega_n > 1$), the steady-state response is mostly *out of phase* with the disturbing force ($90° < \psi_s < 180°$). That is, the disturbing force generally acts to the right ($P_0 \sin \Omega t > 0$) when the block is to the left of its equilibrium position ($x_p < 0$) and vice versa (Fig. 21-24b). For very high frequencies ($\Omega/\omega_n >> 1$), the response is nearly totally out of phase with the disturbing force ($\psi_s \cong 180°$), and the magnification factor is approximately zero independent of the damping ratio. The block is held essentially stationary by the inertial resistance of the block.

When the disturbing force is applied at a frequency close to the natural frequency of the system ($\Omega/\omega_n \cong 1$) and the damping is light ($\zeta \cong 0$), the amplitude of the vibration is magnified substantially. In fact, if the system possesses no damping and is excited by a harmonic force with a frequency very close to the natural frequency $\Omega \cong \omega_n$, the amplitude of the vibration becomes very large according to Eq. 21-40.[11] This condition is called *resonance*. Figure 21-22 suggests that the amplitude of an oscillation may be controlled either by avoiding the condition of resonance or (if resonance cannot be avoided) by increasing the damping ζ.

When the frequency of the disturbing force matches the natural frequency of the system ($\Omega/\omega_n = 1$), as Fig. 21-23 shows, the response lags 90° behind the disturbing force independent of the damping ratio. Therefore the displacement $x_p(t) = D\sin(\Omega t - \pi/2) = -D\cos\Omega t$ is maximum when the disturbing force $P_0 \sin \Omega t$ is zero and vice versa (Fig. 21-24c). However, the velocity $\dot{x}_p(t) = D\Omega\cos(\Omega t - \pi/2) = D\Omega \sin \Omega t$ is in phase with the disturbing force $P_0 \sin \Omega t$. Also when $\Omega/\omega_n = 1$, Eq. 21-41 shows that the magnification factor $D/\delta_p = 1/2\zeta$. These features are often used to determine the natural frequency and the damping ratio of a system experimentally.

It must be noted, however, that, except for $\zeta = 0$, the magnification factor curves (and hence the amplitude of the vibration) do not peak at exactly $\Omega/\omega_n = 1$. Increasing the amount of damping decreases the resonant frequency—the frequency at which the magnification curve peaks. When $\zeta = \sqrt{1/2}$, the maximum amplitude occurs at $\Omega = 0$. When $\zeta \geq \sqrt{1/2}$, the amplitude of vibration D is less than the static displacement δ_p for all frequencies $\Omega > 0$. The exact location of the resonant frequency for any given value of ζ can be calculated by setting the derivative of the magnification factor with respect to Ω/ω_n equal to zero.

In summary then, the complete solution consists of two superimposed vibrations $x(t) = x_c(t) + x_p(t)$. For underdamped systems $\zeta < 1$, the displacement is

$$x(t) = Ae^{-\zeta\omega_n t}\cos(\omega_d t - \phi_c) + D\sin(\Omega t - \psi_s) \qquad (21\text{-}42)$$

The first term in Eq. 21-42 represents a free vibration of the system. The frequency of this vibration depends only on properties of the system (the spring constant k, the damping coefficient c, and the mass m) and is independent of the applied disturbing force. The amplitude of the free vibration (or transient vibration) term decays with time due to friction (damping) forces. The constants A and ϕ_c are chosen to fit the complete solution to the initial conditions.

The last term in Eq. 21-42, which represents the steady-state vibration of the system, is the part of the solution that is usually of primary interest. The frequency of the steady-state vibration is the same as that of the applied disturbing force, and its amplitude depends on the frequency ratio Ω/ω_n.

---

[11] Of course, all real systems possess some damping, and so the amplitude of the vibration cannot become infinite. In addition, physical constraints such as the length of the spring also limit the amplitude of the vibration. Still, resonance is a dangerous condition and should always be avoided.

The cause of forced vibrations need not be a periodic force applied directly to the mass in the system. In many systems such as car suspensions, the forced vibrations are caused by the periodic movement of the system's support rather than by a directly applied force. It will be shown that the periodic movement of the support is directly equivalent to a periodic disturbing force. So long as the coefficients m, k, c, and P_0 in the solutions, Eqs. 21-38 through 21-42, are interpreted as the coefficients of the system's differential equation of motion, those solutions apply to this case as well.

For example, suppose that the support to which the spring of Fig. 21-13a is attached is given a periodically varying displacement $x_b(t) = b \sin \Omega t$ as shown in Fig. 21-25a. The free-body diagram of the block is shown in Fig. 21-25b, in which the block has been displaced an arbitrary amount in the positive coordinate direction. The elongation of the spring is the difference between the displacements of the block and the movable support $x_b(t) - x(t) = b \sin \Omega t - x(t)$. Therefore, the elastic restoring force of the spring is $F_s = k(b \sin \Omega t - x)$ to the right (the spring is stretched and pulls on the block whenever $b \sin \Omega t > x$). Since the dashpot is attached to a fixed support, the rate of extension of the dashpot is just $\dot{x}(t)$ and the damping force $F_d = c\dot{x}$ acts opposite the velocity (in the negative coordinate direction). Then applying Newton's second law of motion $\Sigma F = ma_x = m\ddot{x}$ to the block gives the differential equation

$$-c\dot{x} + k(b \sin \Omega t - x) = m\ddot{x}$$

or

$$m\ddot{x} + c\dot{x} + kx = kb \sin \Omega t \tag{21-43}$$

But by comparison, Eq. 21-43 can be transformed into Eq. 21-35 simply by replacing P_0 with kb. Therefore, Eq. 21-43 has the same solution as Eq. 21-35. That is, the solutions defined by Eqs. 21-38 through 21-42 also describe the motion of the block when subjected to the support displacement $x_b(t) = b \sin \Omega t$ when the constants m, c, k, and P_0 are interpreted as the coefficients of the differential equation of motion when written in the form of Eq. 21-35.

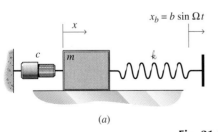

(a)

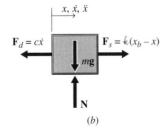

(b)

Fig. 21-25

21-4-3   Rotating Imbalance

Another common source of forced vibrations is an imbalance in a rotating machine piece. For example, the small mass m_s of Fig. 21-26a rotates with angular frequency Ω about an axis fixed in the larger block of mass M. When the block is displaced an arbitrary amount $x(t)$ in the positive coordinate direction, the position of the small mass is $x(t) + e \sin \Omega t$. In the free-body diagram of Fig. 21-26b, the internal forces between the mass and block need not be shown. The elastic restoring force of the spring $F_s = \mathit{k}x$ is directed toward the equilibrium position (the negative coordinate direction). The damping force $F_d = c\dot{x}$ acts opposite the velocity—also in the negative coordinate direction. Then applying Newton's second law of motion $\Sigma F = ma_x = m\ddot{x}$ to the block and mass gives the differential equation

$$-c\dot{x} - \mathit{k}x = M\ddot{x} + m_s\frac{d^2(x + e \sin \Omega t)}{dt^2}$$

or

$$(M + m_s)\ddot{x} + c\dot{x} + \mathit{k}x = em_s\Omega^2 \sin \Omega t \qquad (21\text{-}44)$$

Again by comparison, Eq. 21-44 can be transformed into Eq. 21-35 simply by replacing P_0 with $em_s\Omega^2$ and m with $M + m_s$. That is, the solutions defined by Eqs. 21-38 through 21-42 also describe the motion of the block when subjected to the rotational imbalance of the small mass m_s when the constants m, c, k, and P_0 are interpreted as the coefficients of the differential equation of motion when written in the form of Eq. 21-35.

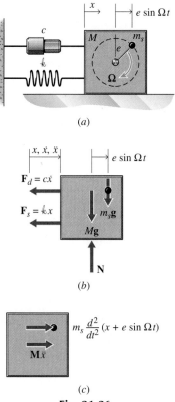

(a)

(b)

(c)

Fig. 21-26

EXAMPLE PROBLEM 21-7

A 3-kg motor sits on a spring (k = 150 kN/m) and a dashpot (c = 120 N·s/m) as shown in Fig. 21-27a. A small mass (m_s = 0.5 kg) is attached to the edge of the motor's pulley (e = 25 mm). Determine the maximum amplitude of the resulting forced vibration of the motor.

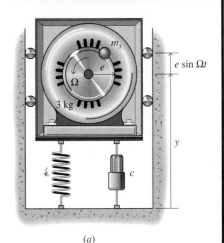

(a)

SOLUTION

The free body diagram of the motor is shown in Fig. 21-27b for an arbitrary (positive) location y. The downward force in the spring is $F_s = k(y + \delta_{eq})$ where δ_{eq} is the stretch of the spring in the equilibrium position where $y = 0$. In the equilibrium position (before the motor is started running), $y = \dot{y} = 0$ and the vertical component of equilibrium ($\uparrow\Sigma F_y = 0$)

$$-(3 + 0.5)(9.81) - 150000\delta_{eq} = 0 \qquad (a)$$

gives the static deflection of the spring $\delta_{eq} = -2.289(10^{-4})$ m = -0.2289 mm. After the motor is started running, Newton's second law of motion $\Sigma F_y = m\ddot{y}$ gives

$$- (3 + 0.5)(9.81) - 150000(y + \delta_{eq}) - 120\dot{y} \qquad (b)$$
$$= 3\frac{d^2y}{dt^2} + 0.5\frac{d^2}{dt^2}(y + 0.025 \sin \Omega t)$$

Then, substituting $\delta_{eq} = -2.289(10^{-4})$ m into Eq. *b* or equivalently subtracting Eq. *a* from Eq. *b* gives the differential equation of motion for the motor

$$3.5\ddot{y} + 120\dot{y} + 150000y = 0.0125\Omega^2 \sin \Omega t$$

Therefore, the natural circular frequency, and damping ratio of the motion are

$$\omega_n = \sqrt{150000/3.5} = 207.0 \text{ rad/s}$$
$$\zeta = \frac{120}{2(3.5)(207.0)} = 0.08282$$

and the amplitude of the steady-state vibration is

$$D = \frac{0.0125\Omega^2/150000}{\sqrt{[1 - (\Omega/207.0)^2]^2 + [2(0.08282)\Omega/207.0]^2}}$$

To find the value of Ω that gives the maximum amplitude, set the derivative $dD/d\Omega = 0$ which gives $\Omega = 208.4$ rad/s. Then

$$D_{\max} = 0.02163 \text{ m} = 21.63 \text{ mm} \qquad \text{Ans.}$$

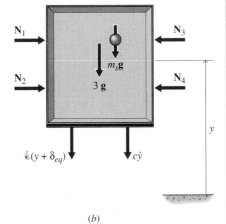

(b)

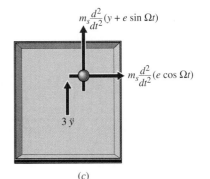

(c)

Fig. 21-27

A 24-lb block slides on a frictionless surface as shown in Fig. 21-28a. The spring is unstretched when bar AB is vertical and bar BC is horizontal. The weights of bars AB and BC may be neglected. Assume that oscillations remain small and determine

a. The range of frequencies Ω for which the angular steady-state motion of bar AB is less than $\pm 5°$.

b. The position of the block as a function of time if the block is pulled 2 in. to the right and released from rest when $t = 0$ and $\Omega = 25$ rad/s.

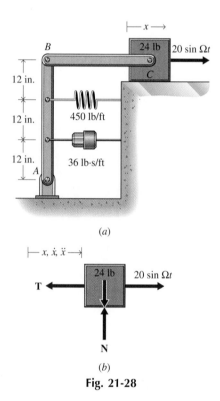

(a)

(b)

Fig. 21-28

SOLUTION

a. Free-body diagrams of the block and of the bar AB are drawn in Figs. 21-28b and 21-28c, in which the block has been displaced an arbitrary distance in the positive coordinate direction (to the right). When the block moves a distance x to the right, the bar AB rotates clockwise through an angle θ. If oscillations remain small, then $\sin \theta \cong \theta$, $\cos \theta \cong 1$, the compression of the spring will be $2x/3$, and the rate of compression of the dashpot will be $\dot{x}/3$. Since the mass of the bar is negligible, the moment of inertia of the bar is negligible and

$$\langle \Sigma M_A = \left(36\frac{\dot{x}}{3}\right) + 2\left(450\frac{2x}{3}\right) - 3T = 0$$

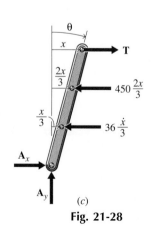

(c)

Fig. 21-28

or

$$T = 4\dot{x} + 200x \qquad (a)$$

Then, applying Newtons' second law of motion $\Sigma F = ma_x = m\ddot{x}$ to the block gives

$$20 \sin \Omega t - T = \frac{24}{32.2}\ddot{x} \qquad (b)$$

Adding Eqs. a and b gives the differential equation of motion of the block

$$0.7453\ddot{x} + 4\dot{x} + 200x = 20 \sin \Omega t$$

Therefore, the natural circular frequency and damping ratio for the system are

$$\omega_n = \sqrt{200/0.7453} = 16.381 \text{ rad/s}$$
$$\zeta = \frac{4}{2(0.7453)(16.381)} = 0.1638$$

Since it is desired to keep the angular motion of bar AB less than $5° = 0.08727$ rad, the maximum amplitude of the steady-state vibration of the block is

$$D \cong (3 \text{ ft})(0.08727 \text{ rad}) = \frac{20}{\sqrt{(200 - 0.7453\Omega^2)^2 + (4\Omega)^2}}$$

which corresponds to the limiting frequencies

$$\Omega = 14.12 \text{ rad/s} \qquad \text{or} \qquad 17.57 \text{ rad/s}$$

Frequencies between these two values give amplitudes that are too great so that the allowable range of frequencies is

$$0 < \Omega < 14.12 \text{ rad/s} \qquad 17.57 \text{ rad/s} < \Omega \qquad \text{Ans.}$$

b. When $\Omega = 25$ rad/s, the equation of motion of the block is

$$x(t) = Ae^{-\zeta\omega_n t}\cos(\omega_d t - \phi_c) + D\sin(\Omega t - \psi_s)$$

where

$$\omega_d = 16.381\sqrt{1 - (0.1638)^2} = 16.16 \text{ rad/s}$$
$$D = \frac{20}{\sqrt{[200 - 0.7453(25)^2]^2 + [4(25)]^2}}$$
$$= 0.07042 \text{ ft} = 0.8451 \text{ in.}$$

and

$$\psi_s = \tan^{-1}\frac{(4)(25)}{200 - (0.7453)(25)^2} = 159.4° = 2.782 \text{ rad}$$

But at $t = 0$

$$x(0) = 2 = A\cos\phi_c - 0.8451\sin 159.4°$$
$$\dot{x}(0) = A[16.16\sin\phi_c - (0.1638)(16.381)\cos\phi_c]$$
$$+ (0.8451)(25)\cos 159.4° = 0$$

Therefore $A = 2.803$ in., $\phi_c = 34.94° = 0.610$ rad, and

$$x(t) = 2.803e^{-2.68t}\cos(16.16t - 0.610) + 0.845\sin(25t - 2.782) \text{ in.} \qquad \text{Ans.}$$

This solution is sketched in Fig. 21-28d. For comparison, a unit loading curve ($\sin 25t$) and the steady-state portion of the response $x_p(t) = 0.845\sin(25t - 2.782)$ are also shown in the figure.

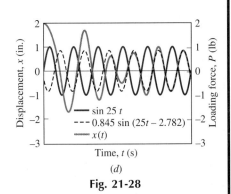

$$\begin{array}{l}\text{---} \quad \sin 25\,t \\ \text{- - -} \quad 0.845\sin(25t - 2.782) \\ \cdots\cdots \quad x(t)\end{array}$$

Time, t (s)

(d)

Fig. 21-28

PROBLEMS

21-96* The particular solution to the differential equation of motion

$$m\ddot{x} + c\dot{x} + kx = P_0 \cos \Omega t$$

can be written in the form

$$x_p(t) = D \cos(\Omega t - \psi_c)$$

Determine expressions for D and ψ_c similar to Eqs. 21-39 and 21-40 for this case.

21-97 Determine the maximum magnification factor given by Eq. 21-41 (p. 466) and the frequency ratio (Ω/ω_n) at which it occurs as a function of the damping ratio ζ.

21-98* A 20-kg block slides on a frictionless surface as shown in Fig. P21-98. The spring (k = 500 N/m) and the dashpot (c = 40 N · s/m) are attached to an oscillating wall. Determine

a. The differential equation governing the motion of the block.
b. The particular solution in the form $x_p(t) = D \sin(\Omega t - \psi_s)$.

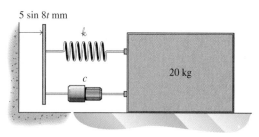

Fig. P21-98

21-99* An upward force $P(t) = 70 \sin 30t$ lb is applied to the 10-lb block of Problem 21-79 (p. 460). For the same initial conditions given in Problem 21-79, determine

a. The differential equation governing the motion.
b. The position of the block as a function of time.

21-100 A downward force $P(t) = 600 \sin 20t$ N is applied to the 2-kg block of Problem 21-80 (p. 461). For the same initial conditions given in Problem 21-80, determine

a. The differential equation governing the motion.
b. The position of the block as a function of time.

21-101 A force $P(t) = 40 \sin 12t$ lb to the right is applied to the 20-lb block of Problem 21-81 (p. 461). For the same initial conditions given in Problem 21-81, determine

a. The differential equation governing the motion.
b. The position of the block as a function of time.

21-102* A downward force $P(t) = 150 \sin 18t$ N is applied to the 4-kg block of Problem 21-82 (p. 461). For the same initial conditions given in Problem 21-82, determine

a. The differential equation governing the motion.
b. The position of the block as a function of time.

21-103 The two blocks of Fig. P21-103 hang in a vertical plane from a massless bar, which is horizontal in the equilibrium position. If an upward force $P(t) = 4 \sin \Omega t$ lb is applied at point D of the bar, determine

a. The maximum amplitude of the steady-state oscillation of the 10-lb block.
b. The range of frequencies Ω that must be avoided if the amplitude of the steady-state oscillation of the 10-lb block is not to exceed 1.5 in.

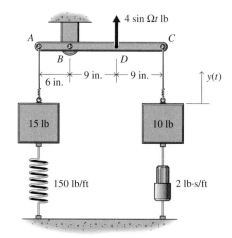

Fig. P21-103

21-104* The two masses of Fig. P21-104 each slide on a frictionless horizontal surface. The bar ABC is vertical in the equilibrium position and has negligible mass. If a force

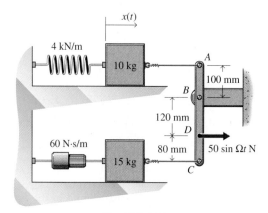

Fig. P21-104

474

$P(t) = 50 \sin \Omega t$ N is applied at point D of the bar, determine

a. The maximum amplitude of the steady-state oscillation of the 10-kg block.
b. The range of frequencies Ω that must be avoided if the amplitude of the steady-state oscillation of the 10-kg block is not to exceed 25 mm.

21-105* For the 10-lb block of Problem 21-79 (p. 460) determine the amplitude of the resulting steady-state oscillation when the lower support oscillates vertically according to $y_\ell = 7 \sin 30t$ in.

21-106 For the 4-kg block of Problem 21-82 (p. 461) determine the amplitude of the resulting steady-state oscillation when the upper support oscillates vertically according to $y_u = 80 \sin 35t$ mm.

21-107* For the 20-lb block of Problem 21-81 (p. 461) determine the amplitude of the resulting steady-state oscillation when the lower support oscillates vertically according to $y_\ell = 15 \sin 12t$ in.

21-108 For the 4-kg block of Problem 21-82 (p. 461) determine the amplitude of the resulting steady-state oscillation when the lower support oscillates vertically according to $y_\ell = 200 \cos 18t$ mm.

21-109 For the system of Problem 21-83 (p. 462) determine the amplitude of the resulting steady-state oscillation of the 10-lb block when $a = 9$ in. and

a. The lower left support oscillates vertically according to $y_\ell = 4 \sin 9t$ in.
b. The lower right support oscillates vertically according to $y_r = 4 \cos 9t$ in.

21-110* For the system of Problem 21-84 (p. 462) determine the amplitude of the resulting steady-state oscillation of the 10-kg block when $a = 150$ mm and

a. The upper left support oscillates horizontally according to $x_u = 5 \sin 8t$ mm.
b. The lower left support oscillates horizontally according to $x_\ell = 5 \sin 8t$ mm.

21-111 A 2-lb weight rotating around a 6-in. radius circle at a rate of $\Omega = 30$ rad/s is added to the 10-lb block of Problem 21-79 (p. 460). Determine the amplitude of the resulting steady-state oscillation.

21-112* A 0.6-kg mass rotating around a 150-mm radius circle at a rate of $\Omega = 20$ rad/s is added to the 2-kg block of Problem 21-80 (p. 461). Determine the amplitude of the resulting steady-state oscillation.

21-113* A 1.5-lb weight rotating around an 18-in. radius circle at a rate of $\Omega = 12$ rad/s is added to the 20-lb block of Problem 21-81 (p. 461). Determine the amplitude of the resulting steady-state oscillation.

21-114 A 0.8-kg mass rotating around a 400-mm radius circle at a rate of $\Omega = 25$ rad/s is added to the 4-kg block of Problem 21-82 (p. 461). Determine the amplitude of the resulting steady-state oscillation.

21-5   ENERGY METHODS

The approach taken in the first few sections of this chapter has been to get the differential equations of motion by applying Newton's second law of motion to the appropriate free-body diagram(s). These differential equations were then solved to get the frequency, period, and amplitude of vibration, as well as the equations of position, velocity, and acceleration of the system. In this direct approach all forces (including internal connection forces and friction or viscous damping forces) must be included in the analysis.

If no friction or viscous damping forces act on the system, however, the work–energy principle as described in Chapters 17 and 18 may be a simpler approach. When the only forces acting on the system are conservative (as in the undamped free vibration of particles and rigid bodies), the work–energy principle reduces to the *conservation of energy: the total mechanical energy of the system is constant*

$$T + V = \text{constant}$$

The conservation of energy principle can be manipulated to give both the differential equation of motion and the natural frequency of the vibration.

Although all real systems lose energy to friction, the damping in many real systems is light. Only very small errors in determining the systems' natural frequency and period result from approximating these systems as undamped. The work–energy approach is particularly suited to problems involving systems of particles connected by rigid links and systems of connected rigid bodies. Using the work–energy method, the system need not be taken apart and the motion of the individual pieces need not be considered separately.

21-5-1   Differential Equation of Motion by Energy Methods

Consider again the block of Fig. 21-5a, which slides on a frictionless, horizontal surface. When the block is displaced a distance x in the positive coordinate direction, the kinetic energy of motion is $T = \frac{1}{2}mv^2 = \frac{1}{2}m\dot{x}^2$ and the potential energy of the elastic spring force will be $V = \frac{1}{2}kx^2$. Then, differentiating the conservation of energy principle with respect to time gives

$$\frac{d}{dt}(T + V) = \frac{d}{dt}\left(\frac{1}{2}kx^2 + \frac{1}{2}m\dot{x}^2\right) = (kx + m\ddot{x})\dot{x} = 0 \quad (21\text{-}45)$$

But since the velocity $\dot{x}$ is not zero at every instant of time, Eq. 21-45 gives the differential equation of motion

$$m\ddot{x} + kx = 0 \qquad (21\text{-}46)$$

which is the same as Eq. 21-2. Then the natural frequency ω_n, period τ_n, and so on all follow from the differential equation as in Section 21-2.

21-5-2   Frequency of Vibration by Energy Methods

The natural frequency and period of vibration can also be determined using the conservation of energy principle without first deriving the differential equation of motion. In Section 21-2 it was seen that for a system that vibrates with simple harmonic motion about its equilibrium position (where $x = 0$), the position and velocity of the system can be written

$$x(t) = A \sin(\omega_n t - \phi_c)$$
$$v(t) = \dot{x}(t) = A\omega_n \cos(\omega_n t - \phi_c)$$

But it is noted from these two expressions that the position is a maximum ($x_{\max} = A$) when the velocity is zero. That is, the potential energy is a maximum when the kinetic energy is zero. Likewise, the maximum velocity ($v_{\max} = A\omega_n = \omega_n x_{\max}$) occurs where the position is zero and hence the kinetic energy is a maximum when the potential energy is zero. Therefore, the total mechanical energy of the system is

$$T + V = T_{\max} + 0 = 0 + V_{\max} = \frac{1}{2}m\dot{x}_{\max}^2 = \frac{1}{2}kx_{\max}^2$$

$$\frac{1}{2}m(\omega_n A)^2 = \frac{1}{2}kA^2$$

Solving for the natural angular frequency ω_n gives

$$\omega_n = \sqrt{k/m}$$

which is again the same as in Section 21-2.

Determine the differential equation of motion for the cart of Example Problem 21-2 using the energy method.

SOLUTION

The free-body diagram of the cart (Fig. 21-10b) shows that four of the five forces acting on the cart are conservative and the fifth does no work. Therefore, the differential equation can be obtained using conservation of energy.

Before the motion is started, the cart is in its static equilibrium position and equilibrium ($\Sigma F_x = 0$) gives that

$$k_1\delta_{eq1} + k_2\delta_{eq2} - k_3\delta_{eq3} - mg \sin 15° = 0 \qquad (a)$$

where δ_{eq1}, δ_{eq2}, and δ_{eq3} are the deformation of the springs in the static equilibrium position ($x = 0$). While the individual values of δ_{eq1}, δ_{eq2}, δ_{eq3} cannot be determined, Eq. a relates them to the weight of the cart.

For the arbitrary position shown in Fig. 21-10b, the kinetic energy of the cart is

$$T = \frac{1}{2}mv^2 = \frac{1}{2}m\dot{x}^2$$

When the cart moves to the right, its center of gravity raises so that the gravitational potential energy of the cart is

$$V_\mathbf{g} = mgx \sin 15°$$

Also when the cart moves to the right, the stretch in springs 1 and 2 are decreased and the stretch in spring 3 is increased. Therefore, the elastic potential energies of the three springs are

$$V_{e1} = \frac{1}{2}k_1(\delta_{eq1} - x)^2 \qquad V_{e2} = \frac{1}{2}k_2(\delta_{eq2} - x)^2$$

$$V_{e3} = \frac{1}{2}k_3(\delta_{eq3} + x)^2$$

and the conservation of energy equation becomes

$$T + V = \frac{1}{2}m\dot{x}^2 + mgx \sin 15° + \left[\frac{1}{2}k_1(\delta_{eq1} - x)^2\right]$$
$$+ \left[\frac{1}{2}k_2(\delta_{eq2} - x)^2\right] + \left[\frac{1}{2}k_3(\delta_{eq3} + x)^2\right]$$
$$= \text{const} \qquad (b)$$

Taking the time derivative of Eq. b gives

$$\frac{d}{dt}(T + V) = [m\ddot{x} + mg \sin 15° - k_1(\delta_{eq1} - x)$$
$$- k_2(\delta_{eq2} - x) + k_3(\delta_{eq3} + x)]\dot{x} = 0 \qquad (c)$$

But since the velocity of the cart $\dot{x}$ is not always zero, the term inside the square brackets must be zero. Finally, adding Eqs. a and c gives the differential equation of motion of the cart

$$m\ddot{x} + (k_1 + k_2 + k_3)x = 0 \qquad \text{Ans.}$$

which is exactly the same as derived in Example Problem 21-2.

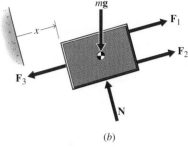

(b)

Fig. 21-10

A 5-kg block slides on a horizontal frictionless surface as shown in Fig. 21-29a. The elastic modulus of the springs are $k_1 = 600$ N/m and $k_2 = 300$ N/m. Assume that the cords remain in tension and determine the natural frequency of the free, undamped vibration of the block using the energy method.

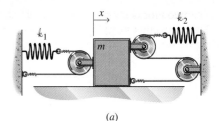

(a)

SOLUTION

The free-body diagram of the block (Fig. 21-29b) shows that two of the forces ($m\mathbf{g}$ and $\mathbf{N}$) do no work and the other forces are due to the springs and are conservative. Therefore, the block oscillates with simple harmonic motion, and the natural frequency of the vibration can be obtained using conservation of energy.

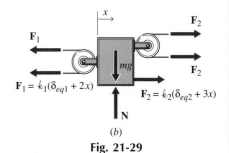

(b)

Fig. 21-29

Before the motion is started, the block is in its static equilibrium position and equilibrium ($\Sigma F_x = 0$) gives that

$$3k_2\delta_{eq2} - 2k_1\delta_{eq1} = 0 \qquad (a)$$

where δ_{eq1} and δ_{eq2} are the deformation of the springs in the static equilibrium position ($x = 0$). When the block has moved to the right an amount x, the stretch in spring 1 is increased by $2x$ and the stretch in spring 2 is decreased by $3x$. Therefore, the elastic potential energy of the springs is

$$V_e = \frac{1}{2}k_1(\delta_{eq1} + 2x)^2 - \frac{1}{2}k_1\delta_{eq1}^2 + \frac{1}{2}k_2(\delta_{eq2} - 3x)^2 - \frac{1}{2}k_2\delta_{eq2}^2 \qquad (b)$$

where the constants have been subtracted so that the zero of potential energy is at the equilibrium position. Expanding Eq. b and simplifying using Eq. a gives

$$V_e = \frac{1}{2}(4k_1x^2 + 4k_1x\delta_{eq1} + 9k_2x^2 - 6k_2x\delta_{eq2}) = \frac{1}{2}(4k_1 + 9k_2)x^2$$

The kinetic energy of the block is just

$$T = \frac{1}{2}mv^2 = \frac{1}{2}m\dot{x}^2$$

For a body oscillating in simple harmonic motion, however, the position and velocity can be written

$$x = A\sin(\omega_n t - \phi_s)$$
$$\dot{x} = A\omega_n \cos(\omega_n t - \phi_s)$$

Therefore, when the position is zero ($x = 0$), the potential energy is zero ($V = 0$), the velocity is a maximum ($\dot{x} = \dot{x}_{\max} = A\omega_n$), and the kinetic energy is also a maximum ($T = T_{\max} = \frac{1}{2}mA^2\omega_n^2$). On the other hand, when the position is a maximum ($x = x_{\max} = A$), the potential energy is also a maximum ($V = V_{\max} = \frac{1}{2}(4k_1 + 9k_2)A^2$), the velocity is zero ($\dot{x} = 0$), and the kinetic energy is also zero ($T = 0$). Writing the conservation of energy equation between these two positions ($T_{\max} + 0 = 0 + V_{\max}$) gives

$$\frac{1}{2}mA^2\omega_n^2 = \frac{1}{2}(4k_1 + 9k_2)A^2 \qquad (c)$$

Finally, solving Eq. c for the natural circular frequency gives

$$\omega_n = \sqrt{\frac{4k_1 + 9k_2}{m}} = 39.5 \text{ rad/s} \qquad \text{Ans.}$$

EXAMPLE PROBLEM 21-11

The two blocks shown in Fig. 21-30a slide on horizontal, frictionless surfaces. The connecting links have negligible weight, and ABC is vertical in the equilibrium position. Assume small oscillations and use the energy method to determine

a. The differential equation of motion of the 15-lb block.
b. The natural frequency of the oscillation.

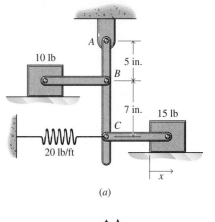

(a)

SOLUTION

a. The free-body diagram of the assembly is shown in Fig. 21-30b. Since the connecting links are all rigid, the work done by the forces at the connections need not be considered. Then the only force that does work is the spring force, and it is conservative. Therefore, the energy method may be used to determine the differential equation of motion and the natural frequency of vibration.

When the 15-lb block moves a distance x to the right, the 10-lb block moves a distance $5x/12$ to the right. Therefore, the kinetic energy of the system is

$$T = \frac{1}{2}\frac{15}{32.2}\dot{x}^2 + \frac{1}{2}\frac{10}{32.2}\left(\frac{5}{12}\dot{x}\right)^2 = 0.2599\dot{x}^2$$

The elastic potential energy of the spring is

$$V = \frac{1}{2}20x^2$$

Then taking the time derivative of the total mechanical energy of the system ($T + V =$ constant) gives

$$\frac{d}{dt}(T + V) = (0.5198\ddot{x} + 20x)\dot{x} = 0$$

Since the velocity $\dot{x}$ is not zero at every instant of time, the quantity inside the parentheses must be zero, which gives the differential equation of motion of the 15-lb block

$$0.5198\ddot{x} + 20x = 0 \qquad\qquad \text{Ans.}$$

b. Once the differential equation of motion is determined, the natural frequency of vibration is just

$$\omega_n = \sqrt{20/0.5198} = 38.5 \text{ rad/s} \qquad\qquad \text{Ans.}$$

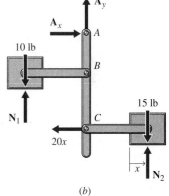

(b)

Fig. 21-30

PROBLEMS

SUMMARY

A mechanical vibration is the repeated oscillation of a particle or rigid body about an equilibrium position. In many devices vibratory motions are desirable and are deliberately generated. The task of the engineer in such problems is to create and to control the vibrations. However, most vibrations in rotating machinery and in structures are undesirable. In these cases the task of the engineer is to eliminate the vibrations (or at least reduce the effect of the vibrations as much as possible) by appropriate design.

The study of vibrations is a direct application of the principles developed earlier. In the earlier chapters, the acceleration was obtained only for a particular position of the body and at a particular instant of time. In this chapter, the acceleration was obtained for an arbitrary position of the body and then integrated to get the velocity and position for all future times.

An undamped free vibration will repeat itself indefinitely. Once set in motion, such idealized systems vibrate forever with a constant amplitude. Of course, all real systems contain frictional forces that will eventually stop a free vibration. In many systems, however, the energy loss due to air resistance, the internal friction of springs, or other friction forces are small enough that an analysis based on negligible damping often gives quite satisfactory engineering results. In particular, the frequency and period of vibration obtained for a freely vibrating system are very close to the values obtained for a system that has a small amount of damping.

A forced vibration is produced and maintained by an externally applied periodic force that does not depend on the position or motion of the body. A damped, forced vibration is maintained as long as the periodic force that produces the vibration is applied.

When a periodic force is applied to a body, the body will begin to oscillate with a combination of free and forced vibrations. Because some friction is present in all real systems, however, the free vibration part of the motion will always decay away. Therefore, the free vibration part of the motion is called the *transient* motion. The frequency of steady-state forced vibrations is the same as that of the applied disturbing force and is independent of the natural frequency and other characteristics of the vibrating body. The amplitude of steady-state forced vibrations, however, does depend on the natural frequency of the system as well as on the frequency of the applied load.

When the disturbing force is applied at a frequency close to the natural frequency of the system and the damping is light, the amplitude of the vibration is magnified substantially. This condition is called *resonance*. The amplitude of an oscillation may be controlled either by avoiding the condition of resonance or (if resonance cannot be avoided) by increasing the damping ζ.

In all cases, the constants m, c, k, and P_0 appearing in the solutions are to be interpreted as the coefficients in the differential equation of motion rather than the actual system mass, the actual damping coefficient, and so on.

REVIEW PROBLEMS

21-139* A 60-lb child bounces up and down on a pair of elastic bands as shown in Fig. P21-139. If the amplitude of the oscillation is observed to decrease by 3 percent every 5 cycles and the 5 cycles take 6.5 s, determine the elastic modulus k and damping coefficient c of the elastic bands.

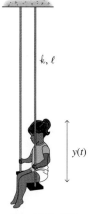

Fig. P21-139

21-140* The pendulum shown in Fig. P21-140 consists of a 5-kg mass on the end of a 0.9-m long lightweight stick. The other end of the stick oscillates along a horizontal guide. Assume small oscillations and determine

a. The differential equation of motion for θ, the angular position of the pendulum.

b. The amplitude of the steady-state oscillation.

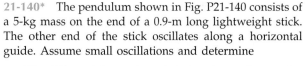

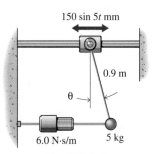

Fig. P21-140

21-141 A small particle slides along the bottom of a 10-in. radius circular bowl. Neglect friction and assume small oscillations. If the particle has a speed of 15 in./s when it is at the bottom of the bowl, determine

a. The differential equation governing the motion.
b. The period and amplitude of the resulting vibration.
c. The position of the particle as a function of time.

21-142* A 10-kg mass slides on a frictionless horizontal surface as shown in Fig. P21-142. At $t = 0$ the mass passes through its equilibrium position with a speed of 2.5 m/s to the right. If $k = 1.2$ kN/m and $c = 180$ N·s/m, determine:

a. The force F_k exerted on the mass by the spring when it reaches its maximum extension.
b. The force F_c exerted on the mass by the dashpot when the mass returns to its equilibrium position.

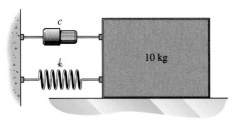

Fig. P21-142

21-143 The particular solution Eq. 21-37 does not satisfy the differential equation of motion Eq. 21-35 (p. 465) when the frequency of the applied force is exactly the same as the system natural frequency.

a. Show that the particular solution has the form

$$x_p(t) = Dt \sin\left(\Omega t - \frac{\pi}{2}\right)$$

when $\Omega = \omega_n$ and $c = 0$.
b. Determine the value of D in terms of the system parameters m, k, ω_n, and P_0.

21-144 A 10-kg mass slides on a frictionless horizontal surface as shown in Fig. P21-144. If $k = 800$ N/m, $c = 30$ N·s/m, $\Omega = 1.5$ Hz; and $P_0 = 80$ N, determine

a. The amplitude of the steady-state oscillation.

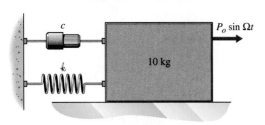

Fig. P21-144

b. The dynamic magnification factor.
c. The magnitude of the total force transmitted to the wall F_w.
d. The transmissibility—the ratio of F_w and P_0 (the amplitude of the applied force).

21-145* A 2-lb piston is initially at rest on two identical springs ($k = 6$ lb/ft each) when a 0.5-lb ball of putty is dropped onto it (Fig. P21-145). If the collision is perfectly plastic ($e = 0$) and $h = 16$ ft, determine

a. The differential equation governing the motion of the piston.
b. The period and amplitude of the resulting vibration.
c. The force exerted on the putty by the piston when the springs are at maximum compression.
d. The force exerted on the putty by the piston when the system passes through equilibrium on the way up.
e. The maximum height h_{max} from which the putty could be dropped and not lose contact with the piston during the subsequent oscillation.

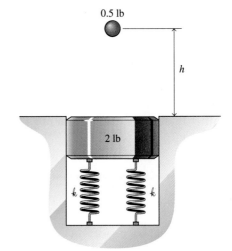

Fig. P21-145

21-146 A 4-kg mass hangs from an elastic band as shown in Fig. P21-146. The unstretched length of the band is 1.5 m

Fig. P21-146

and the equilibrium length is 2.0 m. If the elastic band is to remain taut when the upper support is oscillated according to $\delta = a \sin \Omega t$, determine

a. The maximum amplitude a_{max} when $\Omega = 4$ rad/s.
b. The allowed range of frequencies Ω when $a = 0.7$ m.

21-147* A 1-in. diameter marble ($W = 1$ oz) rolls without slipping in the bottom of a 12-in. radius circular bowl. Assume small oscillations. If the marble has a speed of 15 in./s when it is at the bottom of the bowl, determine

a. The differential equation governing the motion.
b. The frequency and amplitude of the resulting vibration.
c. The position of the marble as a function of time.

21-148* When the system shown in Fig. P21-148 is in equilibrium, spring 1 ($k_1 = 1.2$ kN/m) is stretched 50 mm and spring 2 ($k_2 = 1.8$ kN/m) is stretched 90 mm. If the mass m is pulled down a distance δ and released from rest, determine

a. The differential equation governing the motion.
b. The maximum distance δ_{max} so that all cords remain in tension.
c. The frequency and amplitude of the resulting vibration.
d. The position of the mass as a function of time.

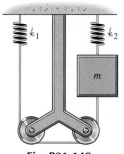

Fig. P21-148

21-149 A 0.1-lb coin sits on top of a 2-lb piston as shown in Fig. P21-149. If the bottom end of the spring is oscillated according to $\delta = a \sin \Omega t$ where $a = 1.5$ in. and $\Omega = 2\pi$ rad/s, determine

a. The differential equation governing the motion of the piston.
b. The amplitude of the resulting vibration.
c. The force exerted on the coin by the piston when the spring is at maximum compression.
d. The force exerted on the coin by the piston when the spring is at maximum extension.

e. The maximum amplitude a_{max} when $\Omega = 10$ rad/s if the coin is to remain in contact with the piston.
f. The allowed range of frequencies Ω when $a = 2$ in. if the coin is to remain in contact with the piston.

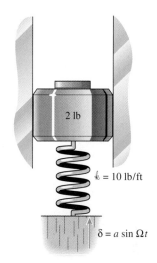

Fig. P21-149

21-150 A 6-kg mass is suspended from a cord, which is wrapped around the outside of a 10-kg cylinder 600 mm in diameter (Fig. P21-150). The system is in equilibrium with point A 200 mm directly above the frictionless axle. If the mass is pulled down 50 mm and the system released from rest, determine

a. The differential equation governing the vertical motion of the mass.
b. The frequency and amplitude of the resulting vibration.
c. The position of the mass as a function of time.

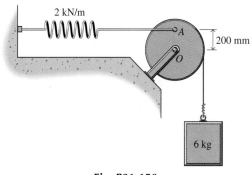

Fig. P21-150

Computer Problems

C21-151 A simple pendulum consists of a concentrated mass m on the end of a lightweight rod AB as shown in Fig. P21-151. If the hinge at A is frictionless, the differential equation of motion for the pendulum is given by

$$\ell\ddot{\theta} + g \sin\theta = 0 \qquad (a)$$

The solution of this equation is only approximated by simple harmonic motion when the angle θ is small so that $\sin\theta \cong \theta$. If $\ell = 4$ ft, $mg = 2$ lb, and the pendulum is released from rest when $\theta = \theta_0$,

a. Use the Euler method of solving differential equations (see Appendix C) to solve Eq. a for the angle θ as a function of time for various initial angles $\theta_0 (10° \leq \theta_0 \leq 120°)$.

b. Then for $\theta_0 = 80°$ plot θ as a function of time t through one complete cycle of the oscillation. On the same graph, plot the solution using the simple harmonic approximation.

c. For each initial angle $\theta_0 = 10°, 20°, 30°, \ldots, 120°$, determine τ, the period of the oscillation. For example, determine the elapsed time for 10 swings through $\theta = 0$ and divide by 5.

d. Plot Err, the percent relative error, in using the small angle approximation as a function of θ_0 $(10° \leq \theta_0 \leq 120°)$, where $Err = [(\tau - \tau_n)/\tau] \times 100$, and $\tau_n = \sqrt{\ell/g}$ is the natural period of the simple harmonic motion solution.

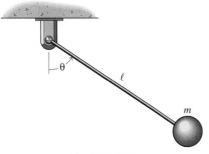

Fig. P21-151

C21-152 A 10-kg block slides on a smooth horizontal surface as shown in Fig. P21-152. At time $t = 0$ the position and velocity of the block are $x_0 = 0.175$ m and $v_0 = 3$ m/s, respectively. If $k = 1000$ N/m and $c = 15$ N$\cdot$s/m, calculate and plot

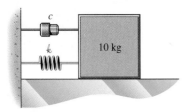

Fig. P21-152

a. The position x of the block as a function of time t $(0 \leq t \leq 5$ s$)$.

b. The velocity v of the block as a function of its position x $(0 \leq t \leq 5$ s$)$.

C21-153 When the forcing frequency Ω of a forced oscillation is close to the natural frequency ω_n of the system, the amplitude of the oscillation varies sinusoidally at a frequency $|\Omega - \omega_n|$. This phenomenon is known as *beating*.

Consider the 25-lb block that slides on a smooth horizontal surface as shown in Fig. P21-153. At time $t = 0$ the position and velocity of the block are $x_0 = 0$ ft and $v_0 = 8$ ft/s, respectively. If $k = 40$ lb/ft, $P_0 = 10$ lb, and $\Omega = 8$ rad/s, calculate and plot the position x of the block as a function of time t $(0 \leq t \leq 25$ s$)$. (Try some other values of k; say, $k = 48$ lb/ft or $k = 50$ lb/ft.)

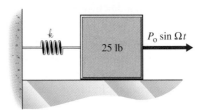

Fig. P21-153

C21-154 A 50-g coin sits on top of a 2-kg piston as shown in Fig. P21-154. The bottom end of the spring is oscillated according to $\delta = \delta_0 \sin\Omega t$. At time $t = 0$ the position and velocity of the block are both zero $x_0 = 0$ m and $v_0 = 0$ m/s. If $k = 205$ N/m, $\delta_0 = 20$ mm, and $\Omega = 8$ rad/s,

a. Calculate and plot the force F_s that must be exerted on the lower end of the spring to produce the motion as a function of time t $(0 \leq t \leq 10$ s$)$.

b. Calculate and plot the force F_c exerted on the coin by the piston as a function of t $(0 \leq t \leq 10$ s$)$.

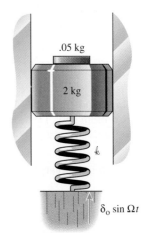

Fig. P21-154

c. Determine the maximum value of δ_0 for which the coin stays in contact with the piston; that is, for which $F_c > 0$ always.

C21-155 A 25-lb block slides on a smooth horizontal surface as shown in Fig. P21-155. At time $t = 0$, the position and velocity of the block are $x_0 = 6$ in. and $v_0 = 0$ in./s, respectively. If $k = 40$ lb/ft, $c = 1$ lb·s/ft, $\Omega = 8$ rad/s, and $P_0 = 10$ lb, calculate and plot

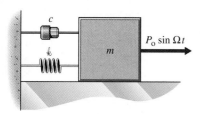

Fig. P21-155

a. The position x of the block as a function of time t ($0 \le t \le 10$ s). On the same graph, plot the steady state part of the solution.

b. The velocity v of the block as a function of x ($0 \le t \le 5$ s). On the same graph, plot the steady-state part of the solution.

C21-156 A 5-kg block slides on a smooth horizontal surface as shown in Fig. P21-155. At time $t = 0$, the position and velocity of the block are $x_0 = 25$ mm and $v_0 = 0$ mm/s, respectively. If $k = 125$ N/m, $c = 5$ N·s/m, $\Omega = 8$ rad/s, and $P_0 = 1000$ N, calculate and plot

a. F/P (the ratio of the force F exerted by the system on the wall) and P_0 (the magnitude of the variable force P) as a function of time t ($0 \le t \le 10$ s).

b. $(F/P)_{max}$ for the steady-state part of the solution as a function of Ω/ω_n ($0.1 \le \Omega/\omega_n \le 3$) for $c = 5$ N·s/m, 10 N·s/m, 15 N·s/m, . . . , 50 N·s/m.

MOMENTS AND PRODUCTS OF INERTIA

A-1 MOMENT OF INERTIA

In analyses of the motion of rigid bodies, expressions are often encountered that involve the product of the mass of a small element of the body and the square of its distance from a line of interest. This product is called the *second moment of the mass* of the element or more frequently the *moment of inertia* of the element. Thus, the moment of inertia dI of an element of mass dm about the axis OO shown in Fig. A-1 is defined as

$$dI = r^2 \, dm$$

The moment of inertia of the entire body about axis OO is defined as

$$I = \int_m r^2 \, dm \tag{A-1}$$

Since both the mass of the element and the distance squared from the axis to the element are always positive, the moment of inertia of a mass is always a positive quantity.

Moments of inertia have the dimensions of mass multiplied by length squared, ML^2. Common units for the measurement of moment of inertia in the SI system are $kg \cdot m^2$. In the U. S. Customary system, force, length, and time are selected as the fundamental quantities, and mass has the dimensions FT^2L^{-1}. Therefore, moment of inertia has the units $lb \cdot s^2 \cdot ft$. If the mass of the body W/g is expressed in slugs ($lb \cdot s^2/ft$), the units for measurement of moment of inertia in the U. S. Customary system are $slug \cdot ft^2$.

The moments of inertia of a body with respect to an xyz-coordinate system can be determined by considering an element of mass as shown in Fig. A-2. From the definition of moment of inertia,

$$dI_x = r_x^2 \, dm = (y^2 + z^2) \, dm$$

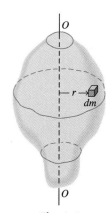

Fig. A-1

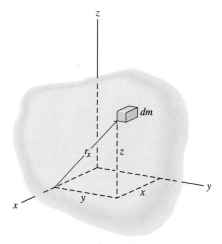

Fig. A-2

Similar expressions can be written for the y- and z-axes. Thus,

$$I_x = \int_m r_x^2 \, dm = \int_m (y^2 + z^2) \, dm$$

$$I_y = \int_m r_y^2 \, dm = \int_m (z^2 + x^2) \, dm \qquad \text{(A-2)}$$

$$I_z = \int_m r_z^2 \, dm = \int_m (x^2 + y^2) \, dm$$

A-1-1   Radius of Gyration

The definition of moment of inertia (Eq. A-1) indicates that the dimensions of moment of inertia are mass multiplied by a length squared. As a result, the moment of inertia of a body can be expressed as the product of the mass m of the body and a length k squared. This length k is defined as the *radius of gyration* of the body. Thus, the moment of inertia I of a body with respect to a given line can be expressed as

$$I = mk^2 \qquad \text{or} \qquad k = \sqrt{\frac{I}{m}} \qquad \text{(A-3)}$$

The radius of gyration of the mass of a body with respect to any axis can be viewed as the distance from the axis to the point where the total mass must be concentrated to produce the same moment of inertia with respect to the axis as does the actual (or distributed) mass.

The radius of gyration for masses is very similar to the radius of gyration for areas discussed in Section 10-2-3. The radius of gyration for masses is not the distance from the given axis to any fixed point in the body such as the mass center. The radius of gyration of the mass of a body with respect to any axis is always greater than the distance from the axis to the mass center of the body. There is no useful physical interpretation for a radius of gyration; it is merely a convenient means of expressing the moment of inertia of the mass of a body in terms of its mass and a length.

A-1-2   Parallel-Axis Theorem for Moments of Inertia

The parallel-axis theorem for moments of inertia is very similar to the parallel-axis theorem for second moments of area discussed in Section 10-2-1. Consider the body shown in Fig. A-3, which has an xyz-coordinate system with its origin at the mass center G of the body and a parallel $x'y'z'$-coordinate system with its origin at point O'. Observe in the figure that

$$x' = \bar{x} + x$$
$$y' = \bar{y} + y$$
$$z' = \bar{z} + z$$

The distance d_x between the x'- and x-axes is

$$d_x = \sqrt{\bar{y}^2 + \bar{z}^2}$$

The moment of inertia of the body about an x'-axis that is parallel to the x-axis through the mass center is by definition

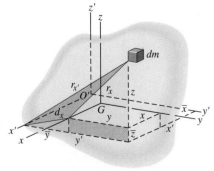

Fig. A-3

$$I_{x'} = \int_m r_{x'}^2 \, dm = \int_m [(\bar{y} + y)^2 + (\bar{z} + z)^2] \, dm$$

$$= \int_m (y^2 + z^2) \, dm + \bar{y}^2 \int_m dm + 2\bar{y} \int_m y \, dm + \bar{z}^2 \int_m dm + 2\bar{z} \int_m z \, dm$$

However,

$$\int_m (y^2 + z^2) \, dm = I_{xG}$$

and, since the x- and y-axes pass through the mass center G of the body,

$$\int_m y \, dm = 0 \qquad \int_m z \, dm = 0$$

Therefore,

$$\begin{aligned} I_{x'} &= I_{xG} + (\bar{y}^2 + \bar{z}^2)m = I_{xG} + d_x^2 m \\ I_{y'} &= I_{yG} + (\bar{z}^2 + \bar{x}^2)m = I_{yG} + d_y^2 m \\ I_{z'} &= I_{zG} + (\bar{x}^2 + \bar{y}^2)m = I_{zG} + d_z^2 m \end{aligned} \qquad \text{(A-4)}$$

Equation A-4 is the parallel-axis theorem for moments of inertia. The subscript G indicates that the x-axis passes through the mass center G of the body. Thus, if the moment of inertia of a body with respect to an axis passing through its mass-center is known, the moment of inertia of the body with respect to any parallel axis can be found, without integrating, by use of Eqs. A-4.

A similar relationship exists between the radii of gyration for the two axes. Thus, if the radii of gyration for the two parallel axes are denoted by k_x and $k_{x'}$, the foregoing equation may be written

$$k_{x'}^2 m = k_{xG}^2 m + d_x^2 m$$

Hence

$$\begin{aligned} k_{x'}^2 &= k_{xG}^2 + d_x^2 \\ k_{y'}^2 &= k_{yG}^2 + d_y^2 \\ k_{z'}^2 &= k_{zG}^2 + d_z^2 \end{aligned} \qquad \text{(A-5)}$$

Note: Equations A-4 and A-5 are valid only for transfers to or from xyz-axes passing through the mass center of the body. *They are not valid for two arbitrary axes.*

A-1-3 Moments of Inertia by Integration

When integration methods are used to determine the moment of inertia of a body with respect to an axis, the mass of the body can be divided into elements in various ways. Depending on the way the element is chosen, single, double, or triple integration may be required. The geometry of the body usually determines whether Cartesian or polar coordinates are used. In any case, the elements of mass should always be selected, so that

1. All parts of the element are the same distance from the axis with respect to which the moment of inertia is to be determined, or
2. If condition 1 is not satisfied, the element should be selected so that the moment of inertia of the element with respect to the axis

about which the moment of inertia of the body is to be found is known. The moment of inertia of the body can then be found by summing the moments of inertia of the elements.

3. If the location of the mass center of the element is known and the moment of inertia of the element with respect to an axis through its mass center and parallel to the given axis is known, the moment of inertia of the element can be determined by using the parallel-axis theorem. The moment of inertia of the body can then be found by summing the moments of inertia of the elements.

When triple integration is used, the element always satisfies the first requirement, but this condition is not necessarily satisfied by elements used for single or double integration.

In some instances, a body can be regarded as a system of particles. The moment of inertia of a system of particles with respect to a line of interest is the sum of the moments of inertia of the particles with respect to the given line. Thus, if the masses of the particles of a system are denoted by m_1, m_2, m_3, $\cdots$, m_n, and the distances of the particles from a given line are denoted by r_1, r_2, r_3, $\cdots$, r_n, the moment of inertia of the system can be expressed as

$$I = \Sigma m r^2 = m_1 r_1^2 + m_2 r_2^2 + m_3 r_3^2 + \cdots + m_n r_n^2$$

Moments of inertia for thin plates are relatively easy to determine. For example, consider the thin plate shown in Fig. A-4. The plate has a uniform density ρ, a uniform thickness t, and a cross-sectional area A. The moments of inertia about x-, y-, and z-axes are by definition

$$I_{xm} = \int_m y^2 \, dm = \int_V y^2 \rho \, dV = \int_A y^2 \rho t \, dA = \rho t \int_A y^2 \, dA = \rho t \, I_{xA}$$

$$I_{ym} = \int_m x^2 \, dm = \int_V x^2 \rho \, dV = \int_A x^2 \rho t \, dA = \rho t \int_A x^2 \, dA = \rho t \, I_{yA} \quad \text{(A-6)}$$

$$I_{zm} = \int_m (x^2 + y^2) \, dm = \rho t \, I_{yA} + \rho t \, I_{xA} = \rho t \, (I_{yA} + I_{xA})$$

where the subscripts m and A denote moments of inertia and second moments of area, respectively. Since the equations for the moments of

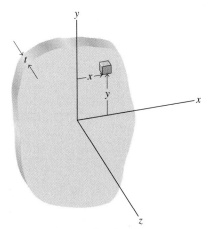

Fig. A-4

inertia of thin plates contain the expressions for the second moments of area, the results listed in Appendix B (Table B-3) for second moments of areas can be used for moments of inertia by multiplying the results listed in the table by ρt.

For the general three-dimensional body, moments of inertia with respect to x-, y-, and z-axes are

$$I_x = \int_m r_x^2 \, dm = \int_m (y^2 + z^2) \, dm$$

$$I_y = \int_m r_y^2 \, dm = \int_m (z^2 + x^2) \, dm \qquad \text{(A-2)}$$

$$I_z = \int_m r_z^2 \, dm = \int_m (x^2 + y^2) \, dm$$

If the density of the body is uniform, the element of mass dm can be expressed in terms of the element of volume dV of the body as $dm = \rho \, dV$. Equations A-2 then become

$$I_x = \rho \int_V (y^2 + z^2) \, dV$$

$$I_y = \rho \int_V (z^2 + x^2) \, dV \qquad \text{(A-7)}$$

$$I_z = \rho \int_V (x^2 + y^2) \, dV$$

If the density of the body is not uniform, it must be expressed as a function of position and retained within the integral sign.

The specific element of volume to be used depends on the geometry of the body. For the general three-dimensional body, the differential element $dV = dx \, dy \, dz$, which requires a triple integration, is usually used. For bodies of revolution, circular plate elements, which require only a single integration, can be used. For some problems, cylinder elements and polar coordinates are useful. Procedures for determining moments of inertia are illustrated in the following examples.

EXAMPLE PROBLEM A-1

Determine the moment of inertia of a homogeneous right circular cylinder with respect to the axis of the cylinder.

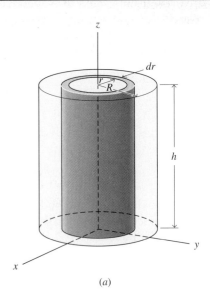

(a)

SOLUTION

The moment of inertia of the cylinder can be determined from the definition of moment of inertia (Eq. A-1) by selecting a cylindrical tube type of element as shown in Fig. A-5a. Thus,

$$dI_{zm} = r^2 \, dm = r^2 \, (\rho \, dV) = r^2 \rho \, (2\pi r h \, dr) = 2\pi \rho h r^3 \, dr$$

Therefore,

$$I_z = \int_m dI_{zm} = \int_0^R 2\pi \rho h r^3 \, dr = \left[\frac{\pi \rho h r^4}{2} \right]_0^R = \frac{1}{2} \pi \rho h R^4$$

Alternatively, a thin circulate plate type of element, such as the one shown in Fig. A-5b, can be used. The moment of inertia for this type of element is given by Eq. A-6 as

$$dI_{zm} = \rho t(I_{yA} + I_{xA})$$

Substituting the second moments for a circular area from Table B-3 yields

$$dI_{zm} = \rho \left(\frac{\pi R^4}{4} + \frac{\pi R^4}{4} \right) dz = \frac{1}{2} \pi \rho R^4 \, dz$$

Therefore,

$$I_z = \int_m dI_{zm} = \int_0^h \frac{1}{2} \pi \rho R^4 \, dz = \left[\frac{1}{2} \pi \rho R^4 z \right]_0^h = \frac{1}{2} \pi \rho h R^4$$

The mass of the cylinder is

$$m = \rho V = \rho(\pi R^2 h) = \rho \pi R^2 h$$

Therefore,

$$I_z = \frac{1}{2} (\rho \pi R^2 h) R^2 = \frac{1}{2} m R^2 \qquad \text{Ans.}$$

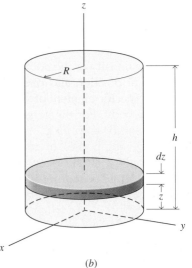

(b)

Fig. A-5

EXAMPLE PROBLEM A-2

Determine the moment of inertia for the homogeneous rectangular prism shown in Fig. A-6a with respect to

a. Axis y through the mass center of the prism.
b. Axis y' along an edge of the prism.
c. Axis x through the centroid of an end of the prism.

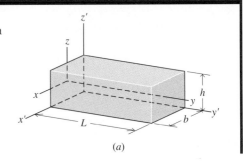

(a)

SOLUTION

a. A thin rectangular plate type of element, such as the one shown in Fig. A-6b, will be used. The moment of inertia for this type of element is given by Eq. A-6 as

$$dI_{ym} = \rho t(I_{zA} + I_{xA})$$

Substituting the second moments for a rectangular area from Table B-3 yields

$$dI_{ym} = \rho\left(\frac{hb^3}{12} + \frac{bh^3}{12}\right) dy = \rho\frac{bh}{12}(b^2 + h^2)\, dy$$

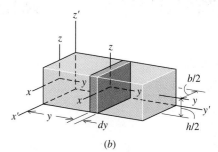

(b)

Fig. A-6

Therefore,

$$I_y = \int_m dI_{ym} = \int_0^L \rho\frac{bh}{12}(b^2 + h^2)\, dy$$

$$= \left[\rho\frac{bh}{12}(b^2 + h^2)\, y\right]_0^L = \frac{\rho bhL}{12}(b^2 + h^2)$$

The mass of the prism is

$$m = \rho V = \rho(bhL) = \rho bhL$$

Therefore,

$$I_y = \frac{\rho bhL}{12}(b^2 + h^2) = \frac{1}{12}m\,(b^2 + h^2) \qquad\qquad \text{Ans.}$$

b. The parallel-axes theorem (Eq. A-4) can be used to determine the moment of inertia about the y'-axis along an edge of the prism. Thus,

$$I_{y'} = I_{yG} + (\overline{x}^2 + \overline{z}^2)\, m$$

$$= \frac{1}{12}m\,(b^2 + h^2) + \left(\frac{b^2}{4} + \frac{h^2}{4}\right)m$$

$$= \frac{1}{3}m\,(b^2 + h^2) \qquad\qquad \text{Ans.}$$

c. The moment of inertia about an x-axis through the mass center of the thin rectangular plate type of element shown in Fig. A-6b is given by Eq. A-6 as

$$dI_{xm} = \rho t I_{xA}$$

Substituting the second moment for a rectangular area from Table B-3 yields

$$dI_{xG} = \rho\frac{bh^3}{12}\, dy = \frac{\rho bh^3}{12}\, dy$$

The parallel-axis theorem (Eq. A-4) with $d_x = y$ then gives the moment of inertia for the thin rectangular plate element about the x-axis shown in Fig. A-6b as

$$dI_x = dI_{xG} + d_x^2 m = \frac{\rho bh^3}{12}\, dy + y^2(\rho bh\, dy) = \frac{\rho bh}{12}(h^2 + 12y^2)\, dy$$

$$I_x = \int_m dI_x = \int_0^L \frac{\rho bh}{12}(h^2 + 12y^2)\, dy$$

$$= \frac{\rho bh}{12}\left[h^2 y + 4y^3\right]_0^L = \frac{\rho bhL}{12}(h^2 + 4L^2)$$

But

$$m = \rho bhL$$

Therefore,

$$I_x = \frac{\rho bhL}{12}(h^2 + 4L^2) = \frac{1}{12}m\,(h^2 + 4L^2) \qquad\qquad \text{Ans.}$$

PROBLEMS

A-1* Determine the moment of inertia of the homogeneous right circular cone shown in Fig. PA-1 with respect to the axis of the cone.

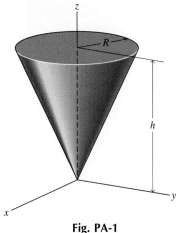

Fig. PA-1

A-2* Determine the moment of inertia of the homogeneous right circular cone shown in Fig. PA-1 with respect to an axis perpendicular to the axis of the cone at the apex of the cone.

A-3 Determine the moment of inertia of the homogeneous right circular cone shown in Fig. PA-1 with respect to an axis perpendicular to the axis of the cone at the base of the cone.

A-4 Determine the moment of inertia of a solid homogeneous sphere of radius R with respect to a diameter of the sphere.

A-5* Determine the moment of inertia of a solid homogeneous cylinder of radius R and length L with respect to a diameter in the base of the cylinder.

A-6* Determine the moment of inertia of the solid homogeneous hemisphere shown in Fig. PA-6 with respect to the x-axis shown on the figure.

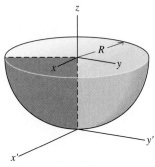

Fig. PA-6

A-7 Determine the moment of inertia of the solid homogeneous hemisphere shown in Fig. PA-6 with respect to the x'-axis shown on the figure.

A-8 Determine the moment of inertia of the solid homogeneous triangular prism shown in Fig. PA-8 with respect to the x-axis shown on the figure.

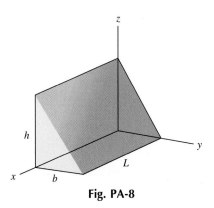

Fig. PA-8

A-9* Determine the moment of inertia of the solid homogeneous triangular prism shown in Fig. PA-8 with respect to a y-axis through the mass center of the prism.

A-10* Determine the moment of inertia of the solid homogeneous tetrahedron shown in Fig. PA-10 with respect to the x-axis shown on the figure.

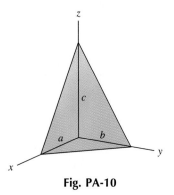

Fig. PA-10

A-11 Determine the moment of inertia of the solid homogeneous tetrahedron shown in Fig. PA-10 with respect to a y-axis through the mass center of the body.

A-12 A homogeneous solid of revolution is formed by revolving the area shown in Fig. PA-12 around the y-axis.

Determine the moment of inertia of the body with respect to the y-axis.

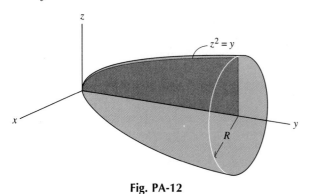

Fig. PA-12

A-13* A homogeneous solid of revolution is formed by revolving the area shown in Fig. PA-12 around the y-axis. Determine the moment of inertia of the body with respect to the x-axis.

A-14* A homogeneous octant of a sphere is formed by rotating the quarter circle shown in Fig. PA-14 for 90°

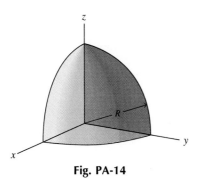

Fig. PA-14

around the z-axis. Determine the moment of inertia of the body with respect to a y-axis through the mass center of the body.

A-15 A homogeneous octant of a cone is formed by rotating the triangle shown in Fig. PA-15 for 90° around the z-axis. Determine the moment of inertia of the body with respect to the y-axis shown on the figure.

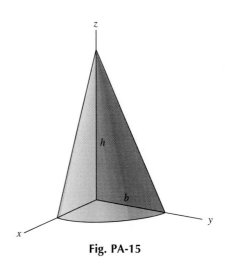

Fig. PA-15

A-16 A homogeneous octant of a cone is formed by rotating the triangle shown in Fig. PA-15 for 90° around the z-axis. Determine the moment of inertia of the body with respect to an x-axis through the mass center of the body.

A-1-4   Moment of Inertia of Composite Bodies

Frequently in engineering practice, a body of interest can be broken up into a number of simple shapes, such as cylinders, spheres, plates, and rods, for which the moments of inertia have been evaluated and tabulated. The moment of inertia of the composite body, with respect to any axis, is equal to the sum of the moments of inertia of the separate parts of the body with respect to the specified axis. For example,

$$
\begin{aligned}
I_x &= \int_m (y^2 + z^2)\, dm \\
&= \int_{m_1} (y^2 + z^2)\, dm_1 + \int_{m_2} (y^2 + z^2)\, dm_2 + \cdots + \int_{m_n} (y^2 + z^2)\, dm_n \\
&= I_{x1} + I_{x2} + I_{x3} + \cdots + I_{xn}
\end{aligned}
$$

When one of the component parts is a hole, its moment of inertia must be subtracted from the moment of inertia of the larger part to obtain the moment of inertia for the composite body. A listing of the moments of inertia for some frequently encountered shapes such as rods, plates, cylinders, spheres, and cones is presented in Appendix B (Table B-5). Procedures for determining moments of inertia for composite bodies by using known values for the parts are illustrated in the following example.

Determine the moment of inertia of the cast-iron flywheel shown in Fig. A-7 with respect to the axis of rotation of the flywheel. The specific weight of the cast iron is 460 lb/ft^3.

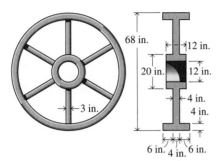

Fig. A-7

SOLUTION

The rim and hub of the flywheel are hollow cylinders, and the spokes are rectangular prisms. The density of the cast iron is

$$\rho = \frac{w}{g} = \frac{460}{32.2} = 14.29 \text{ slugs/ft}^3$$

With all dimensions converted to feet, the moment of inertia of the rim is

$$
\begin{aligned}
I_R &= \frac{1}{2} m_o R_o^2 - \frac{1}{2} m_i R_i^2 \\
&= \frac{1}{2}\left[\pi\left(\frac{34}{12}\right)^2\left(\frac{16}{12}\right)(14.29)\right]\left(\frac{34}{12}\right)^2 - \frac{1}{2}\left[\pi\left(\frac{30}{12}\right)^2\left(\frac{16}{12}\right)[14.29]\right]\left(\frac{30}{12}\right)^2 \\
&= 1929 - 1169 = 760 \text{ slug} \cdot \text{ft}^2
\end{aligned}
$$

The moment of inertia of the hub is

$$
\begin{aligned}
I_H &= \frac{1}{2} m_o R_o^2 - \frac{1}{2} m_i R_i^2 \\
&= \frac{1}{2}\left[\pi\left(\frac{10}{12}\right)^2\left(\frac{12}{12}\right)(14.29)\right]\left(\frac{10}{12}\right)^2 - \frac{1}{2}\left[\pi\left(\frac{6}{12}\right)^2\left(\frac{12}{12}\right)(14.29)\right]\left(\frac{6}{12}\right)^2 \\
&= 10.82 - 1.40 = 9.42 \text{ slug} \cdot \text{ft}^2
\end{aligned}
$$

The moment of inertia of each spoke is

$$
\begin{aligned}
I_S &= I_G + d^2 m \\
&= \frac{1}{12}\left[\frac{3}{12}\left(\frac{4}{12}\right)\left(\frac{20}{12}\right)(14.29)\right]\left[\left(\frac{3}{12}\right)^2 + \left(\frac{20}{12}\right)^2\right] \\
&\quad + \left(\frac{20}{12}\right)^2\left[\frac{3}{12}\left(\frac{4}{12}\right)\left(\frac{20}{12}\right)(14.29)\right] \\
&= 0.4698 + 5.5131 = 5.9829 = 5.98 \text{ slug} \cdot \text{ft}^2
\end{aligned}
$$

The total moment of inertia for the flywheel is

$$
\begin{aligned}
I &= I_R + I_H + 6I_S \\
&= 760 + 9.42 + 6(5.98) = 805 \text{ slug} \cdot \text{ft}^2 \qquad \text{Ans.}
\end{aligned}
$$

PROBLEMS

A-17* A composite body is constructed by attaching a steel ($w = 490$ lb/ft³) hemisphere to an aluminum ($w = 175$ lb/ft³) right circular cone as shown in Fig. PA-17. Determine the moment of inertia of the composite body with respect to the y-axis shown on the figure.

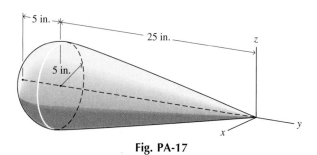

Fig. PA-17

A-18* A composite body consists of a rectangular brass ($\rho = 8.75$ Mg/m³) block attached to a steel ($\rho = 7.87$ Mg/m³) cylinder as shown in Fig. PA-18. Determine the moment of inertia of the composite body with respect to the y-axis shown on the figure.

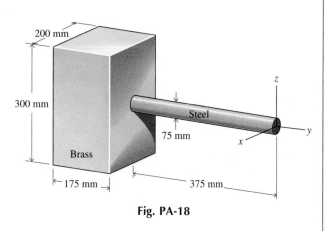

Fig. PA-18

A-19 A composite body is constructed by attaching a steel ($w = 490$ lb/ft³) hemisphere to an aluminum ($w = 175$ lb/ft³) right circular cone as shown in Fig. PA-17. Determine the moment of inertia of the composite body with respect to the x-axis shown on the figure.

A-20 A composite body consists of a cylinder attached to a rectangular block as shown in Fig. PA-18. Determine the moment of inertia of the composite body with respect to the x-axis shown on the figure if the body is made of cast iron ($\rho = 7.37$ Mg/m³).

A-21* Two steel ($w = 490$ lb/ft³) cylinders and a brass ($w = 546$ lb/ft³) sphere form the composite body shown in Fig. PA-21. Determine the moment of inertia of the composite body with respect to the x-axis shown on the figure.

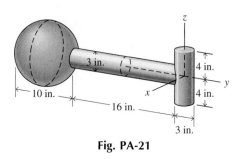

Fig. PA-21

A-22* Determine the moment of inertia of the composite body shown in Fig. PA-22 with respect to the x-axis shown on the figure. The density of the material is 7.87 Mg/m³.

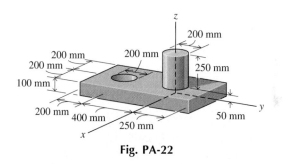

Fig. PA-22

A-23 Two brass ($w = 546$ lb/ft³) cylinders and a bronze ($w = 553$ lb/ft³) sphere form the composite body shown in Fig. PA-21. Determine the moment of inertia of the composite body with respect to the y-axis shown on the figure.

A-24 Determine the moment of inertia of the composite body shown in Fig. PA-22 with respect to the y-axis shown on the figure. The density of the material is 2.80 Mg/m³.

A-25 Two steel ($w = 490$ lb/ft³) cylinders and an aluminum ($w = 173$ lb/ft³) sphere form the composite body shown in Fig. PA-21. Determine the moment of inertia of the composite body with respect to the z-axis shown on the figure.

A-26 Determine the moment of inertia of the composite body shown in Fig. PA-22 with respect to the z-axis shown on the figure. The density of the material is 7.87 Mg/m³.

A-27* Determine the moment of inertia of the composite body shown in Fig. PA-27 with respect to the y-axis shown on the figure. The specific weight of the material is 175 lb/ft^3.

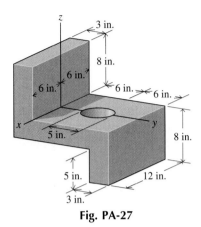

Fig. PA-27

A-28* Determine the moment of inertia of the composite body shown in Fig. PA-28 with respect to the y-axis shown on the figure. The density of the material is 7.37 Mg/m^3.

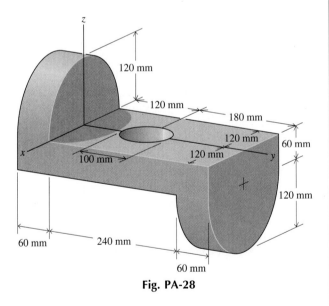

Fig. PA-28

A-29 Determine the moment of inertia of the composite body shown in Fig. PA-27 with respect to the x-axis shown on the figure. The specific weight of the material is 546 lb/ft^3.

A-30 Determine the moment of inertia of the composite body shown in Fig. PA-28 with respect to the x-axis shown on the figure. The density of the material is 2.77 Mg/m^3.

A-31* Determine the moment of inertia of the composite body shown in Fig. PA-31 with respect to the y-axis shown on the figure. The specific weight of the material is 553 lb/ft^3.

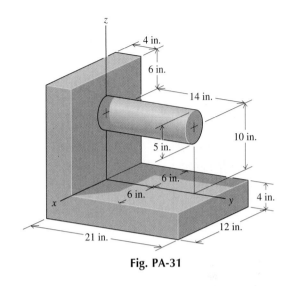

Fig. PA-31

A-32* Determine the moment of inertia of the composite body shown in Fig. PA-32 with respect to the y-axis shown on the figure. The density of the material is 7.87 Mg/m^3.

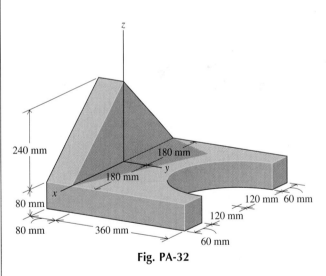

Fig. PA-32

A-33 Determine the moment of inertia of the composite body shown in Fig. PA-31 with respect to the x-axis shown on the figure. The specific weight of the material is 490 lb/ft^3.

A-34 Determine the moment of inertia of the composite body shown in Fig. PA-32 with respect to the x-axis shown on the figure. The density of the material is 8.75 Mg/m^3.

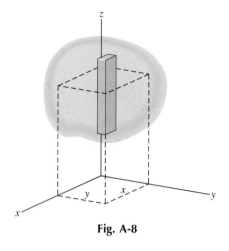

Fig. A-8

A-2   PRODUCT OF INERTIA

In analyses of the motion of rigid bodies, expressions are sometimes encountered that involve the product of the mass of a small element and the coordinate distances from a pair of orthogonal coordinate planes. This product, which is similar to the mixed second moment of an area, is called the *product of inertia* of the element. For example, the product of inertia of the element shown in Fig. A-8 with respect to the *xz*- and *yz*-planes is by definition

$$dI_{xy} = xy \, dm \tag{A-8}$$

The sum of the products of inertia of all elements of mass of the body with respect to the same orthogonal planes is defined as the product of inertia of the body. The three products of inertia for the body shown in Fig. A-8 are

$$I_{xy} = \int_m xy \, dm$$

$$I_{yz} = \int_m yz \, dm \tag{A-9}$$

$$I_{zx} = \int_m zx \, dm$$

Products of inertia, like moments of inertia, have the dimensions of mass multiplied by a length squared, mL^2. Common units for the measurement of product of inertia in the SI system are $kg \cdot m^2$. In the U. S. Customary system, common units are $slug \cdot ft^2$.

The product of inertia of a body can be positive, negative, or zero, since the two coordinate distances have independent signs. The product of inertia will be positive for coordinates with the same sign and negative for coordinates with opposite signs. The product of inertia will be zero if either of the planes is a plane of symmetry, since pairs of elements on opposite sides of the plane of symmetry will have positive and negative products of inertia that will add to zero in the summation process.

The integration methods used to determine moments of inertia apply equally well to products of inertia. Depending on the way the element is chosen, single, double, or triple integration may be required. Moments of inertia for thin plates were related to second moments of area for the same plate. Likewise, products of inertia can be related to the mixed second moments for the plates. If the plate has a uniform density ρ, a uniform thickness t, and a cross-sectional area A, the products of inertia are by definition

$$I_{xym} = \int_m xy \, dm = \int_V xy \, \rho \, dV = \int_A xy \, \rho \, t \, dA = \rho t \int_A xy \, dA = \rho t \, I_{xyA}$$

$$I_{yzm} = \int_m yz \, dm = 0 \tag{A-10}$$

$$I_{zxm} = \int_m zx \, dm = 0$$

where the subscripts m and A denote products of inertia of mass and mixed second moments of area, respectively. The products of inertia

I_{yzm} and I_{zxm} for a thin plate are zero since the x- and y-axes are assumed to lie in the midplane of the plate (plane of symmetry).

A parallel-axis theorem for products of inertia can be developed that is very similar to the parallel-axis theorem for mixed second moments of area discussed in Section 10-2-5. Consider the body shown in Fig. A-9, which has an xyz-coordinate system with its origin at the mass center G of the body and a parallel $x'y'z'$-coordinate system with its origin at point O'. Observe in the figure that

$$x' = \bar{x} + x$$
$$y' = \bar{y} + y$$
$$z' = \bar{z} + z$$

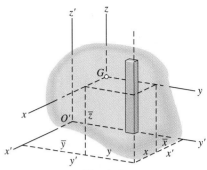

Fig. A-9

The product of inertia $I_{x'y'}$ of the body with respect to the $x'z'$- and $y'z'$-planes is by definition

$$I_{x'y'} = \int_m x'y' \, dm = \int_m (\bar{x} + x)(\bar{y} + y) \, dm$$
$$= \int_m \bar{x}\bar{y} \, dm + \int_m \bar{x}y \, dm + \int_m \bar{y}x \, dm + \int_m xy \, dm$$

Since $\bar{x}$ and $\bar{y}$ are the same for every element of mass dm,

$$I_{x'y'} = \bar{x}\bar{y} \int_m dm + \bar{x} \int_m y \, dm + \bar{y} \int_m x \, dm + \int_m xy \, dm$$

However,

$$\int_m xy \, dm = I_{xy}$$

and, since the x- and y-axes pass through the mass center G of the body,

$$\int_m y \, dm = 0 \qquad \int_m z \, dm = 0$$

Therefore,

$$I_{x'y'} = I_{xyG} + \bar{x}\bar{y} \, m$$
$$I_{y'z'} = I_{yzG} + \bar{y}\bar{z} \, m \qquad\qquad \text{(A-11)}$$
$$I_{z'x'} = I_{zxG} + \bar{z}\bar{x} \, m$$

Equations A-11 are the parallel-axis theorem for products of inertia. The subscript G indicates that the x- and y-axes pass through the mass center G of the body. Thus, if the product of inertia of a body with respect to a pair of orthogonal planes that pass through its mass center is known, the product of inertia of the body with respect to any other pair of parallel planes can be found, without integrating, by use of Eqs. A-11.

Procedures for determining products of inertia are illustrated in the following examples.

EXAMPLE PROBLEM A-4

Determine the product of inertia I_{xy} for the homogeneous quarter cylinder shown in Fig. A-10a.

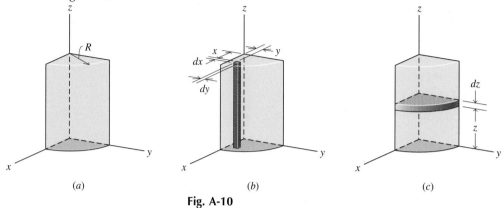

Fig. A-10

SOLUTION

All parts of the element of mass dm, shown in Fig. A-10b, are located at the same distances x and y from the xz- and yz-planes; therefore, the product of inertia dI_{xy} for the element is by definition

$$dI_{xy} = xy \, dm$$

Summing the elements for the entire body yields,

$$I_{xy} = \int_m dI_{xy} = \int_m xy \, dm = \int_V xy \, \rho dV$$

$$= \int_0^R \int_0^{\sqrt{R^2-x^2}} \rho xy \, (h \, dy \, dx)$$

$$= \int_0^R \rho hx \left[\frac{y^2}{2} \right]_0^{\sqrt{R^2-x^2}} dx$$

$$= \int_0^R \frac{1}{2} \rho h (R^2 x - x^3) \, dx$$

$$= \frac{1}{2} \rho h \left[\frac{R^2 x^2}{2} - \frac{x^4}{4} \right]_0^R = \frac{1}{2} \rho h \left(\frac{R^4}{4} \right) = \frac{1}{8} \rho R^4 h$$

Alternatively, the thin-plate element shown in Fig. A-10c could be used to determine I_{xy}. From Eq. A-10 and the data from Table B-5,

$$dI_{xym} = \rho t \, dI_{xyA} = \frac{1}{8} \rho R^4 \, dz$$

Therefore,

$$I_{xym} = \rho t \int_A dI_{xyA} = \int_0^h \frac{1}{8} \rho R^4 \, dz = \frac{1}{8} \rho R^4 h$$

Since the mass of the body is

$$m = \rho V = \rho \left(\frac{1}{4} \pi R^2 h \right) = \frac{1}{4} \rho \pi R^2 h$$

the product of inertia I_{xy} can be written as

$$I_{xy} = \frac{1}{2\pi} \left(\frac{1}{4} \rho \pi R^2 h \right) R^2 = \frac{1}{2\pi} m R^2 \qquad \text{Ans.}$$

EXAMPLE PROBLEM A-5

Determine the products of inertia I_{xy}, I_{yz}, and I_{zx} for the homogeneous flat-plate steel ($\rho = 7870$ kg/m³) washer shown in Fig. A-11. The hole is located at the center of the plate.

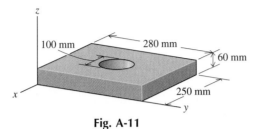

Fig. A-11

SOLUTION

The products of inertia are zero for the planes of symmetry through the mass centers of the plate and hole. Since the xy-, yz-, and zx-planes shown in Fig. A-11 are parallel to these planes of symmetry, the parallel-axis theorem for products of inertia (Eqs. A-11) can be used to determine the required products of inertia. The masses of the plate, hole, and washer are

$$m_P = \rho V = \rho bht = 7870(0.280)(0.250)(0.060) = 33.05 \text{ kg}$$
$$m_H = \rho V = \rho \pi R^2 t = 7870\pi(0.050)^2(0.060) = 3.71 \text{ kg}$$
$$m_W = m_P - m_H = 33.05 - 3.71 = 29.34 \text{ kg}$$

From Eqs. A-11,

$$I_{xy} = I_{xyG} + \bar{x}\bar{y}\, m$$
$$= 0 + (-0.125)(0.140)(29.34) = -0.513 \text{ kg} \cdot \text{m}^2 \qquad \text{Ans.}$$

$$I_{yz} = I_{yzG} + \bar{y}\bar{z}\, m$$
$$= 0 + (0.140)(0.030)(29.34) = 0.1232 \text{ kg} \cdot \text{m}^2 \qquad \text{Ans.}$$

$$I_{zx} = I_{zxG} + \bar{z}\bar{x}\, m$$
$$= 0 + (0.030)(-0.125)(29.34) = -0.1100 \text{ kg} \cdot \text{m}^2 \qquad \text{Ans.}$$

PROBLEMS

A-35* Determine the product of inertia I_{xy} for the homogeneous rectangular block shown in Fig. PA-35.

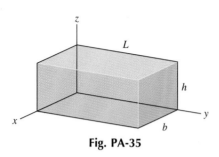

Fig. PA-35

A-36* Determine the product of inertia I_{xy} for the homogeneous octant of a sphere shown in Fig. PA-36.

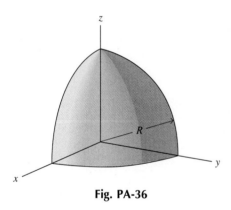

Fig. PA-36

A-37 Determine the products of inertia I_{xy} and I_{yz} for the homogeneous triangular block shown in Fig. PA-37.

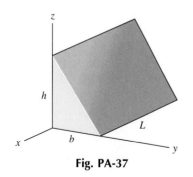

Fig. PA-37

A-38 Determine the products of inertia I_{yz} and I_{zx} for the homogeneous half cylinder shown in Fig. PA-38.

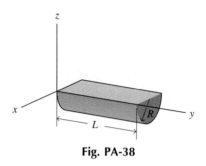

Fig. PA-38

A-39* Determine the products of inertia I_{xy}, I_{yz}, and I_{zx} for the homogeneous steel ($w = 490$ lb/ft^3) bracket shown in Fig. PA-39.

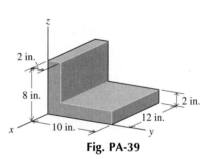

Fig. PA-39

A-40* Determine the products of inertia I_{xy} and I_{zx} for the homogeneous right circular quarter cone shown in Fig. PA-40.

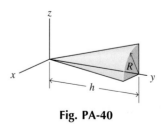

Fig. PA-40

A-41 Determine the products of inertia I_{yz} and I_{zx} for the homogeneous body shown in Fig. PA-41.

A-42 Determine the products of inertia I_{xy}, I_{yz}, and I_{zx} for the homogeneous steel ($\rho = 7870$ kg/m³) block shown in Fig. PA-42.

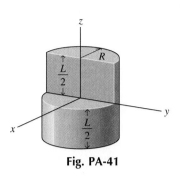

Fig. PA-41

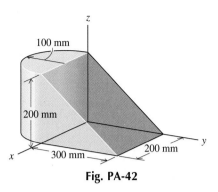

Fig. PA-42

A-3 PRINCIPAL MOMENTS OF INERTIA

In some instances, in the dynamic analysis of bodies, principal axes and maximum and minimum moments of inertia, which are similar to maximum and minimum second moments of an area, must be determined. Again, the problem is one of transforming known or easily calculated moments and products of inertia with respect to one coordinate system (such as an xyz-coordinate system along the edges of a rectangular prism) to a second $x'y'z'$-coordinate system through the same origin O but inclined with respect to the xyz system.

For example, consider the body shown in Fig. A-12 where the x'-axis is oriented at angles $\theta_{x'x}$, $\theta_{x'y}$, and $\theta_{x'z}$ with respect to the x-, y-, and z-axes, respectively. The moment of inertia $I_{x'}$, is by definition

$$I_{x'} = \int_m r^2 \, dm$$

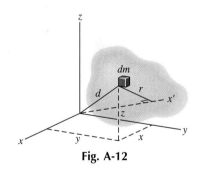

Fig. A-12

The distance d from the origin of coordinates to the element dm is given by the expression

$$d^2 = x^2 + y^2 + z^2 = x'^2 + y'^2 + z'^2 = x'^2 + r^2$$

Therefore,

$$r^2 = x^2 + y^2 + z^2 - x'^2$$

and since

$$x' = x \cos \theta_{x'x} + y \cos \theta_{x'y} + z \cos \theta_{x'z}$$
$$r^2 = x^2 + y^2 + z^2 - (x \cos \theta_{x'x} + y \cos \theta_{x'y} + z \cos \theta_{x'z})^2$$

Recall that

$$\cos^2 \theta_{x'x} + \cos^2 \theta_{x'y} + \cos^2 \theta_{x'z} = 1$$

Therefore,

$$r^2 = (x^2 + y^2 + z^2)(\cos^2 \theta_{x'x} + \cos^2 \theta_{x'y} + \cos^2 \theta_{x'z})$$
$$- (x \cos \theta_{x'x} + y \cos \theta_{x'y} + z \cos \theta_{x'y})^2$$

which reduces to

$$r^2 = (y^2 + z^2) \cos^2 \theta_{x'x} + (z^2 + x^2) \cos^2 \theta_{x'y} + (x^2 + y^2) \cos^2 \theta_{x'z}$$
$$- 2xy \cos \theta_{x'x} \cos \theta_{x'y} - 2yz \cos \theta_{x'y} \cos \theta_{x'z} - 2zx \cos \theta_{x'z} \cos \theta_{x'x}$$

Therefore,

$$I_{x'} = \int_m r^2 \, dm$$

$$= \cos^2 \theta_{x'x} \int_m (y^2 + z^2) \, dm + \cos^2 \theta_{x'y} \int_m (z^2 + x^2) \, dm$$

$$+ \cos^2 \theta_{x'z} \int_m (x^2 + y^2) \, dm - \cos \theta_{x'x} \cos \theta_{x'y} \int_m 2xy \, dm$$

$$- \cos \theta_{x'y} \cos \theta_{x'z} \int_m 2yz \, dm - \cos \theta_{x'z} \cos \theta_{x'x} \int_m 2zx \, dm$$

From Eqs. A-2 and A-9,

$$I_x = \int_m (y^2 + z^2) \, dm \qquad I_{xy} = \int_m xy \, dm$$

$$I_y = \int_m (z^2 + x^2) \, dm \qquad I_{yz} = \int_m yz \, dm$$

$$I_z = \int_m (x^2 + y^2) \, dm \qquad I_{zx} = \int_m zx \, dm$$

Therefore,

$$I_{x'} = I_x \cos^2 \theta_{x'x} + I_y \cos^2 \theta_{x'y} + I_z \cos^2 \theta_{x'z} - 2I_{xy} \cos \theta_{x'x} \cos \theta_{x'y}$$
$$- 2I_{yz} \cos \theta_{x'y} \cos \theta_{x'z} - 2I_{zx} \cos \theta_{x'z} \cos \theta_{x'x} \qquad \text{(A-12a)}$$

In a similar fashion the product of inertia

$$I_{x'y'} = \int_m x'y' \, dm$$

can be expressed in terms of I_x, I_y, I_z, I_{xy}, I_{yz}, and I_{zx} as

$$I_{x'y'} = -I_x \cos \theta_{x'x} \cos \theta_{y'x} - I_y \cos \theta_{x'y} \cos \theta_{y'y} - I_z \cos \theta_{x'z} \cos \theta_{y'z}$$
$$+ I_{xy}(\cos \theta_{x'x} \cos \theta_{y'y} + \cos \theta_{x'y} \cos \theta_{y'x})$$
$$+ I_{yz}(\cos \theta_{x'y} \cos \theta_{y'z} + \cos \theta_{x'z} \cos \theta_{y'y})$$
$$+ I_{zx}(\cos \theta_{x'z} \cos \theta_{y'x} + \cos \theta_{x'x} \cos \theta_{y'z}) \qquad \text{(A-12b)}$$

If the original xyz-axes are principal axes (such as those shown for the figures in Table B-5),

$$I_{xy} = I_{yz} = I_{zx} = 0$$

and Eqs. A-12 reduce to

$$I_{x'} = I_x \cos^2 \theta_{x'x} + I_y \cos^2 \theta_{x'y} + I_z \cos^2 \theta_{x'z} \qquad \text{(A-13a)}$$

and

$$I_{x'y'} = -I_x \cos \theta_{x'x} \cos \theta_{y'x} - I_y \cos \theta_{x'y} \cos \theta_{y'y}$$
$$- I_z \cos \theta_{x'z} \cos \theta_{y'z} \qquad \text{(A-13b)}$$

Equation A-12a for moments of inertia is the three-dimensional equivalent of Eq. 10-14 for second moments of area. By using a similar but much more complicated procedure than the one used with Eq. 10-14 to locate principal axes and determine maximum and minimum second moments of area, principal axes can be located and maximum and minimum moments of inertia can be determined. The procedure yields the following equations:

$$(I_x - I_P) \cos \theta_{Px} - I_{xy} \cos \theta_{Py} - I_{zx} \cos \theta_{Pz} = 0$$
$$(I_y - I_P) \cos \theta_{Py} - I_{yz} \cos \theta_{Pz} - I_{xy} \cos \theta_{Px} = 0 \qquad \text{(A-14)}$$
$$(I_z - I_P) \cos \theta_{Pz} - I_{zx} \cos \theta_{Px} - I_{yz} \cos \theta_{Py} = 0$$

This set of equations has a nontrivial solution only if the determinant of the coefficients of the direction cosines is equal to zero. Expansion of the determinant yields the following cubic equation for determining the principal moments of inertia of the body for the particular origin of coordinates being used:

$$I_P^3 - (I_x + I_y + I_z)I_P^2 + (I_x I_y + I_y I_z + I_z I_x - I_{xy}^2 - I_{yz}^2 - I_{zx}^2)I_P$$
$$- (I_x I_y I_z - I_x I_{yz}^2 - I_y I_{zx}^2 - I_z I_{xy}^2 - 2I_{xy} I_{yz} I_{zx}) = 0 \qquad \text{(A-15)}$$

Equation A-15 yields three values I_1, I_2, and I_3 for the principal moments of inertia. One value is the maximum moment of inertia of the body for the origin of coordinates being used, a second value is the minimum moment of inertia of the body for the origin of coordinates being used, and the third value is an intermediate value of the moment of inertia of the body that has no particular significance.

The direction cosines for the principal inertia axes can be obtained by substituting the three values I_1, I_2, and I_3 obtained from Eq. A-15, in turn, into Eqs. A-14 and using the additional relation

$$\cos^2 \theta_{Px} + \cos^2 \theta_{Py} + \cos^2 \theta_{Pz} = 1$$

Equations A-14 and A-15 are valid for bodies of any shape. The procedure for locating principal axes and determining maximum and minimum moments of inertia is illustrated in the following example.

EXAMPLE PROBLEM A-6

Locate the principal axes and determine the maximum and minimum moments of inertia for the rectangular steel ($w = 490$ lb/ft³) block shown in Fig. A-13.

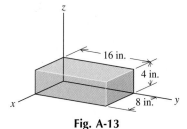

Fig. A-13

SOLUTION

The moments and products of inertia for the block are given by Eqs. A-4 and A-11 as

$$I_x = I_{xG} + (\bar{y}^2 + \bar{z}^2)m \qquad I_{xy} = I_{xyG} + \bar{x}\bar{y}\,m$$
$$I_y = I_{yG} + (\bar{z}^2 + \bar{x}^2)m \qquad I_{yz} = I_{yzG} + \bar{y}\bar{z}\,m$$
$$I_z = I_{zG} + (\bar{x}^2 + \bar{y}^2)m \qquad I_{zx} = I_{zxG} + \bar{z}\bar{x}\,m$$

The mass of the block is

$$m = \rho V = \frac{W}{g}bhL = \frac{490}{32.2}\left(\frac{8}{12}\right)\left(\frac{4}{12}\right)\left(\frac{16}{12}\right) = 4.509 \text{ slugs}$$

Thus, from the results listed in Table B-5,

$$I_x = \frac{1}{12}m(b^2 + h^2) + \left[\left(\frac{h}{2}\right)^2 + \left(\frac{b}{2}\right)^2\right]m$$
$$= \frac{1}{3}m(b^2 + h^2) = \frac{1}{3}(4.509)\left[\left(\frac{8}{12}\right)^2 + \left(\frac{4}{12}\right)^2\right] = 0.835 \text{ slug}\cdot\text{ft}^2$$

$$I_y = \frac{1}{12}m(b^2 + L^2) + \left[\left(\frac{b}{2}\right)^2 + \left(\frac{L}{2}\right)^2\right]m$$
$$= \frac{1}{3}m(b^2 + L^2) = \frac{1}{3}(4.509)\left[\left(\frac{8}{12}\right)^2 + \left(\frac{16}{12}\right)^2\right] = 3.340 \text{ slug}\cdot\text{ft}^2$$

$$I_z = \frac{1}{12}m(h^2 + L^2) + \left[\left(\frac{h}{2}\right)^2 + \left(\frac{L}{2}\right)^2\right]m$$
$$= \frac{1}{3}m(h^2 + L^2) = \frac{1}{3}(4.509)\left[\left(\frac{4}{12}\right)^2 + \left(\frac{16}{12}\right)^2\right] = 2.839 \text{ slug}\cdot\text{ft}^2$$

$$I_{xy} = 0 + \left(\frac{L}{2}\right)\left(\frac{h}{2}\right)m = \frac{1}{4}mLh = \frac{1}{4}(4.509)\left(\frac{16}{12}\right)\left(\frac{4}{12}\right) = 0.501 \text{ slug}\cdot\text{ft}^2$$

$$I_{yz} = 0 + \left(\frac{h}{2}\right)\left(\frac{b}{2}\right)m = \frac{1}{4}mhb = \frac{1}{4}(4.509)\left(\frac{4}{12}\right)\left(\frac{8}{12}\right) = 0.251 \text{ slug}\cdot\text{ft}^2$$

$$I_{zx} = 0 + \left(\frac{b}{2}\right)\left(\frac{L}{2}\right)m = \frac{1}{4}mbL = \frac{1}{4}(4.509)\left(\frac{8}{12}\right)\left(\frac{16}{12}\right) = 1.002 \text{ slug}\cdot\text{ft}^2$$

Once the moments and products of inertia have been determined, the principal moments of inertia can be determined by using Eq. A-15. Thus,

$$I_P^3 - (I_x + I_y + I_z)I_P^2 + (I_xI_y + I_yI_z + I_zI_x - I_{xy}^2 - I_{yz}^2 - I_{zx}^2)I_P$$
$$- (I_xI_yI_z - I_xI_{yz}^2 - I_yI_{zx}^2 - I_zI_{xy}^2 - 2I_{xy}I_{yz}I_{zx}) = 0$$

Substituting values for the moments and products of inertia gives

$$I_P^3 - 7.014I_P^2 + 13.324I_P - 3.548 = 0$$

which has the solution

$$I_1 = I_{max} = 3.451 \text{ slug} \cdot \text{ft}^2 \qquad\qquad \text{Ans.}$$
$$I_2 = I_{int} = 3.246 \text{ slug} \cdot \text{ft}^2$$
$$I_3 = I_{min} = 0.317 \text{ slug} \cdot \text{ft}^2 \qquad\qquad \text{Ans.}$$

The principal directions are obtained by substituting the principal moments of inertia, in turn, into Eqs. A-14. With $I_P = I_1 = I_{max} = 3.451 \text{ slug} \cdot \text{ft}^2$:

$$(I_x - I_P) \cos \theta_{Px} - I_{xy} \cos \theta_{Py} - I_{zx} \cos \theta_{Pz} = 0$$
$$(I_y - I_P) \cos \theta_{Py} - I_{yz} \cos \theta_{Pz} - I_{xy} \cos \theta_{Px} = 0$$
$$(I_z - I_P) \cos \theta_{Pz} - I_{zx} \cos \theta_{Px} - I_{yz} \cos \theta_{Py} = 0$$
$$-2.616 \cos \theta_{1x} - 0.501 \cos \theta_{1y} - 1.002 \cos \theta_{1z} = 0$$
$$-0.501 \cos \theta_{1x} - 0.111 \cos \theta_{1y} - 0.251 \cos \theta_{1z} = 0 \qquad (a)$$
$$-1.002 \cos \theta_{1x} - 0.251 \cos \theta_{1y} - 0.612 \cos \theta_{1z} = 0$$

Equations a together with the required relationship for direction cosines

$$\cos^2 \theta_{1x} + \cos^2 \theta_{1y} + \cos^2 \theta_{1z} = 1$$

have the solution

$$\cos \theta_{1x} = 0.0891 \qquad \text{or} \qquad \theta_{1x} = 84.9°$$
$$\cos \theta_{1y} = -0.9643 \qquad \text{or} \qquad \theta_{1y} = 164.6° \qquad\qquad \text{Ans.}$$
$$\cos \theta_{1z} = 0.2495 \qquad \text{or} \qquad \theta_{1z} = 75.6°$$

With $I_P = I_2 = I_{int} = 3.246 \text{ slug} \cdot \text{ft}^2$:

$$-2.411 \cos \theta_{2x} - 0.501 \cos \theta_{2y} - 1.002 \cos \theta_{2z} = 0$$
$$-0.501 \cos \theta_{2x} + 0.094 \cos \theta_{2y} - 0.251 \cos \theta_{2z} = 0$$
$$-1.002 \cos \theta_{2x} - 0.251 \cos \theta_{2y} - 0.407 \cos \theta_{2z} = 0$$
$$\cos^2 \theta_{2x} + \cos^2 \theta_{2y} + \cos^2 \theta_{2z} = 1$$

the solution becomes

$$\cos \theta_{2x} = -0.4105 \qquad \text{or} \qquad \theta_{2x} = 114.2°$$
$$\cos \theta_{2y} = 0.1925 \qquad \text{or} \qquad \theta_{2y} = 78.9° \qquad\qquad \text{Ans.}$$
$$\cos \theta_{2z} = 0.8913 \qquad \text{or} \qquad \theta_{2z} = 27.0°$$

With $I_P = I_3 = I_{min} = 0.317 \text{ slug} \cdot \text{ft}^2$:

$$0.518 \cos \theta_{3x} - 0.501 \cos \theta_{3y} - 1.002 \cos \theta_{3z} = 0$$
$$-0.501 \cos \theta_{3x} + 3.023 \cos \theta_{3y} - 0.251 \cos \theta_{3z} = 0$$
$$-1.002 \cos \theta_{3x} - 0.251 \cos \theta_{3y} + 2.522 \cos \theta_{3z} = 0$$
$$\cos^2 \theta_{3x} + \cos^2 \theta_{3y} + \cos^2 \theta_{3z} = 1$$

the solution becomes

$$\cos \theta_{3x} = 0.9075 \qquad \text{or} \qquad \theta_{3x} = 24.8°$$
$$\cos \theta_{3y} = 0.1818 \qquad \text{or} \qquad \theta_{3y} = 79.5° \qquad\qquad \text{Ans.}$$
$$\cos \theta_{3z} = 0.3786 \qquad \text{or} \qquad \theta_{3z} = 67.8°$$

Thus, the unit vectors associated with the three principal directions are

$$\mathbf{n}_1 = 0.0891\mathbf{i} - 0.9643\mathbf{j} + 0.2495\mathbf{k} \qquad \mathbf{n}_1 \cdot \mathbf{n}_2 = 0$$
$$\mathbf{n}_2 = -0.4105\mathbf{i} + 0.1925\mathbf{j} + 0.8913\mathbf{k} \qquad \mathbf{n}_2 \cdot \mathbf{n}_3 = 0$$
$$\mathbf{n}_3 = 0.9075\mathbf{i} + 0.1818\mathbf{j} + 0.3786\mathbf{k} \qquad \mathbf{n}_3 \cdot \mathbf{n}_1 = 0$$

which verifies that the three principal axes are orthogonal.

PROBLEMS

A-43* Locate the principal axes and determine the maximum and minimum moments of inertia for the triangular steel ($w = 490$ lb/ft³) block shown in Fig. PA-37 (p. 504) if $b = 8$ in., $h = 6$ in., and $L = 10$ in.

A-44* Locate the principal axes and determine the maximum and minimum moments of inertia for the steel ($\rho = 7870$ kg/m³) half cylinder shown in Fig. PA-38 (p. 504) if $R = 100$ mm and $L = 150$ mm.

A-45 Locate the principal axes and determine the maximum and minimum moments of inertia for the steel ($w = 490$ lb/ft³) angle bracket shown in Fig. PA-39 (p. 504).

A-46 Locate the principal axes and determine the maximum and minimum moments of inertia for the steel ($\rho = 7870$ kg/m³) washer shown in Fig. A-11 (p. 503).

A-47* Locate the principal axes and determine the maximum and minimum moments of inertia for the aluminum ($w = 175$ lb/ft³) cylinder shown in Fig. PA-41 (p. 505) if $R = 4$ in. and $L = 6$ in.

A-48* Locate the principal axes and determine the maxi-

mum and minimum moments of inertia for the brass ($\rho = 8750$ kg/m³) block shown in Fig. PA-42 (p. 505).

A-49 Locate the principal axes and determine the maximum and minimum moments of inertia for the cast iron ($w = 460$ lb/ft³) tetrahedron shown in Fig. PA-10 (p. 494) if $a = 6$ in., $b = 8$ in., and $c = 10$ in.

A-50 Locate the principal axes and determine the maximum and minimum moments of inertia for the brass ($\rho = 8750$ kg/m³) sphere shown in Fig. PA-50. The radius of the sphere is 200 mm. The surface of the sphere is tangent to the three coordinate planes.

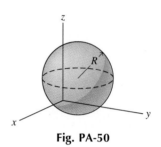

Fig. PA-50

CENTROIDS
CENTERS OF MASS
SECOND MOMENTS
OF PLANE AREAS
MOMENTS OF
INERTIA

Circular arc

$L = 2r\alpha$

$\bar{x} = \dfrac{r \sin \alpha}{\alpha}$

$\bar{y} = 0$

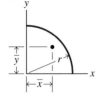

Circular sector

$A = r^2\alpha$

$\bar{x} = \dfrac{2r \sin \alpha}{3\alpha}$

$\bar{y} = 0$

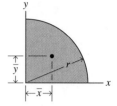

Quarter circular arc

$L = \dfrac{\pi r}{2}$

$\bar{x} = \dfrac{2r}{\pi}$

$\bar{y} = \dfrac{2r}{\pi}$

Quadrant of a circle

$A = \dfrac{\pi r^2}{4}$

$\bar{x} = \dfrac{4r}{3\pi}$

$\bar{y} = \dfrac{4r}{3\pi}$

Semicircular arc

$L = \pi r$

$\bar{x} = r$

$\bar{y} = \dfrac{2r}{\pi}$

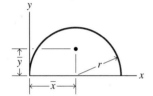

Semicircular area

$A = \dfrac{\pi r^2}{2}$

$\bar{x} = r$

$\bar{y} = \dfrac{4r}{3\pi}$

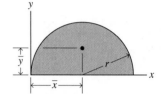

Rectangular area

$A = bh$

$\bar{x} = \dfrac{b}{2}$

$\bar{y} = \dfrac{h}{2}$

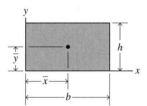

Quadrant of an ellipse

$A = \dfrac{\pi ab}{4}$

$\bar{x} = \dfrac{4a}{3\pi}$

$\bar{y} = \dfrac{4b}{3\pi}$

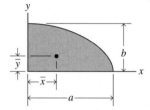

Triangular area

$A = \dfrac{bh}{2}$

$\bar{x} = \dfrac{2b}{3}$

$\bar{y} = \dfrac{h}{3}$

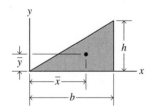

Parabolic spandrel

$A = \dfrac{bh}{3}$

$\bar{x} = \dfrac{3b}{4}$

$\bar{y} = \dfrac{3h}{10}$

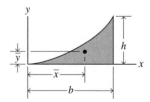

Triangular area

$A = \dfrac{bh}{2}$

$\bar{x} = \dfrac{a+b}{3}$

$\bar{y} = \dfrac{h}{3}$

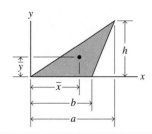

Quadrant of a parabola

$A = \dfrac{2bh}{3}$

$\bar{x} = \dfrac{5b}{8}$

$\bar{y} = \dfrac{2h}{5}$

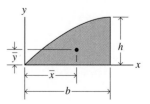

TABLE B-2   CENTROID LOCATIONS FOR A FEW COMMON VOLUMES

Rectangular parallelepiped

$V = abc$

$\bar{x} = \dfrac{a}{2}$

$\bar{y} = \dfrac{b}{2}$

$\bar{z} = \dfrac{c}{2}$

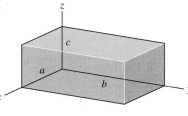

Rectangular tetrahedron

$V = \dfrac{abc}{6}$

$\bar{x} = \dfrac{a}{4}$

$\bar{y} = \dfrac{b}{4}$

$\bar{z} = \dfrac{c}{4}$

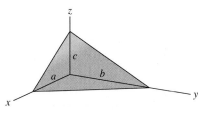

Circular cylinder

$V = \pi r^2 L$

$\bar{x} = 0$

$\bar{y} = \dfrac{L}{2}$

$\bar{z} = 0$

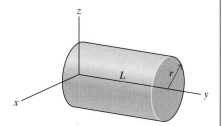

Semicylinder

$V = \dfrac{\pi r^2 L}{2}$

$\bar{x} = 0$

$\bar{y} = \dfrac{L}{2}$

$\bar{z} = \dfrac{4r}{3\pi}$

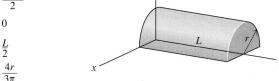

Hemisphere

$V = \dfrac{2\pi r^3}{3}$

$\bar{x} = 0$

$\bar{y} = 0$

$\bar{z} = \dfrac{3r}{8}$

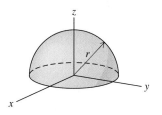

Paraboloid

$V = \dfrac{\pi r^2 h}{2}$

$\bar{x} = 0$

$\bar{y} = \dfrac{2h}{3}$

$\bar{z} = 0$

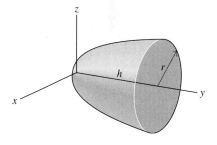

Right circular cone

$V = \dfrac{\pi r^2 h}{3}$

$\bar{x} = 0$

$\bar{y} = \dfrac{3h}{4}$

$\bar{z} = 0$

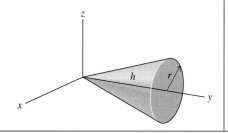

Half cone

$V = \dfrac{\pi r^2 h}{6}$

$\bar{x} = 0$

$\bar{y} = \dfrac{3h}{4}$

$\bar{z} = \dfrac{r}{\pi}$

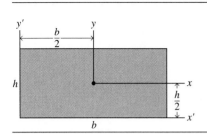

$$I_x = \frac{bh^3}{12}$$

$$I_{x'} = \frac{bh^3}{3}$$

$$A = bh$$

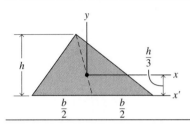

$$I_x = \frac{bh^3}{36}$$

$$I_{x'} = \frac{bh^3}{12}$$

$$A = \frac{1}{2}bh$$

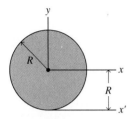

$$I_x = \frac{\pi R^4}{4}$$

$$I_{x'} = \frac{5\pi R^4}{4}$$

$$A = \pi R^2$$

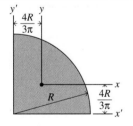

$$I_x = \frac{\pi R^4}{8} - \frac{8R^4}{9\pi}$$

$$I_y = \frac{\pi R^4}{8}$$

$$I_{x'} = \frac{\pi R^4}{8}$$

$$A = \frac{1}{2}\pi R^2$$

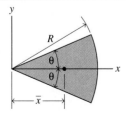

$$I_x = \frac{\pi R^4}{16} - \frac{4R^4}{9\pi}$$

$$I_{x'} = \frac{\pi R^4}{16}$$

$$A = \frac{1}{4}\pi R^2$$

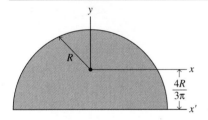

$$I_x = \frac{R^4}{4}\left(\theta - \frac{1}{2}\sin 2\theta\right)$$

$$I_y = \frac{R^4}{4}\left(\theta + \frac{1}{2}\sin 2\theta\right)$$

$$\bar{x} = \frac{2}{3}\frac{R\sin\theta}{\theta}$$

$$A = \theta R^2$$

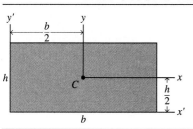

$$I_{xy} = 0 \qquad\qquad I_{x'y'} = \frac{b^2h^2}{4}$$

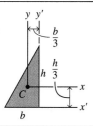

$$I_{xy} = -\frac{b^2h^2}{72} \qquad\qquad I_{x'y'} = \frac{b^2h^2}{24}$$

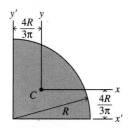

$$I_{xy} = \frac{b^2h^2}{72} \qquad\qquad I_{x'y'} = -\frac{b^2h^2}{24}$$

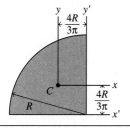

$$I_{xy} = \frac{(9\pi - 32)R^4}{72\pi} \qquad\qquad I_{x'y'} = \frac{R^4}{8}$$

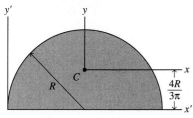

$$I_{xy} = -\frac{(9\pi - 32)R^4}{72\pi} \qquad\qquad I_{x'y'} = -\frac{R^4}{8}$$

$$I_{xy} = 0 \qquad\qquad I_{x'y'} = \frac{2R^4}{3}$$

Slender rod

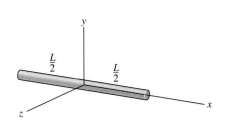

$$I_x = 0$$

$$I_y = I_z = \frac{1}{12}mL^2$$

Thin rectangular plate

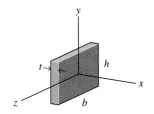

$$I_x = \frac{1}{12}m(b^2 + h^2)$$

$$I_y = \frac{1}{12}mb^2$$

$$I_z = \frac{1}{12}mh^2$$

Thin circular plate

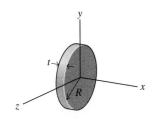

$$I_x = \frac{1}{2}mR^2$$

$$I_y = I_z = \frac{1}{4}mR^2$$

Rectangular prism

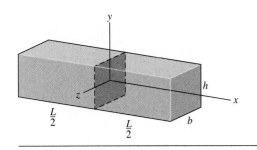

$$V = bhL$$

$$I_x = \frac{1}{12}m(b^2 + h^2)$$

$$I_y = \frac{1}{12}m(b^2 + L^2)$$

$$I_z = \frac{1}{12}m(h^2 + L^2)$$

Right circular cone

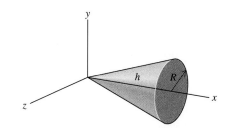

$$V = \frac{1}{3}\pi R^2 h$$

$$\bar{x} = \frac{3}{4}h$$

$$I_x = \frac{3}{10}mR^2$$

$$I_y = I_z = \frac{3}{20}m(R^2 + 4h^2)$$

$$I_{yG} = I_{zG} = \frac{3}{80}m(4R^2 + h^2)$$

Circular cylinder

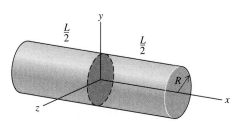

$$V = \pi R^2 L$$

$$I_x = \frac{1}{2}mR^2$$

$$I_y = I_z = \frac{1}{12}m(3R^2 + L^2)$$

Hemisphere

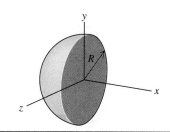

$$V = \frac{2}{3}\pi R^3$$

$$\bar{x} = \frac{3}{8}R$$

$$I_x = I_y = I_z = \frac{2}{5}mR^2$$

$$I_{yG} = I_{zG} = \frac{83}{320}mR^2$$

Sphere

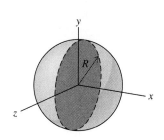

$$V = \frac{4}{3}\pi R^3$$

$$I_x = I_y = I_z = \frac{2}{5}mR^2$$

TABLE B-6   DENSITY ρ OF SELECTED ENGINEERING MATERIALS

	kg/m^3	$slug/ft^3$
Solids		
Aluminum	2,770	5.35
Brass	8,750	16.91
Concrete	2,410	4.66
Cast iron	7,370	14.24
Copper	8,910	17.21
Earth (wet)	1,760	3.40
(dry)	1,280	2.47
Glass	2,590	5.00
Gold	19,300	37.29
Lead	11,370	21.97
Steel	7,870	15.21
Wood (soft pine)	480	0.93
(hard oak)	800	1.55
Liquids		
Ice	900	1.74
Mercury	13,570	26.22
Oil	900	1.74
Water (fresh)	1,000	1.94
(sea)	1,030	1.99
Gasses		
Air	1.225	$2.377(10^{-3})$

TABLE B-7   MISCELLANEOUS CONVERSION FACTORS AND DEFINITIONS

Length
1 ft = 12 in.
1 mi = 5280 ft
1 nautical mile = 6080 ft
1 in. = 25.40 mm
1 mi = 1.609 km

Volume
1 cup = 8 fl oz (fluid ounce)
1 pint = 2 cup = 16 fl oz
1 quart = 2 pint = 32 fl oz
1 gal = 4 quart = 128 fl oz = 231 in.^{3}
1 ft^3 = 7.48 gal
1 barrel (petroleum) = 42 gal
1 liter = 10^{-3} m^3 = (100 mm)^3

Mass
1 metric ton = 1000 kg
1 slug = 14.59 kg

Force
1 lb = 16 oz
1 kip (kilo-pound) = 1000 lb
1 ton = 2000 lb
1 lb = 4.448 Newton

Energy
1 BTU (British Thermal Unit) = 778 ft · lb

Power
1 hp (horsepower) = 550 ft · lb/s
1 ft · lb/s = 1.356 watt

TABLE B-8   ASTRONOMICAL DATA

Universal Gravitational Constant

$G = 6.673(10^{-11})$ m^3/(kg·s^2) $= 3.439(10^{-8})$ ft^4/(lb·s^4)

The Sun

Mass	$1.990(10^{30})$ kg	$1.364(10^{29})$ lb·s^2/ft
Mean radius	696,000 km	432,000 mi

The Earth

Mass	$5.976(10^{24})$ kg	$4.095(10^{23})$ lb·s^2/ft
Mean radius	6370 km	3960 mi
Rotation rate	23.93 hr	

The Moon

Mass	$7.350(10^{22})$ kg	$5.037(10^{21})$ lb·s^2/ft
Mean radius	1740 km	1080 mi
Mean distance to the Earth (center to center)		
	384,000 km	239,000 mi
Eccentricity (e)	0.055	

The Solar System

Planet	Mean Distance to Sun A.U.[a]	e	Mean Diameter (relative to Earth)	Mass (relative to Earth)
Mercury	0.387	0.206	0.380	0.05
Venus	0.723	0.007	0.975	0.81
Earth	1.000	0.017	1.000	1.00
Mars	1.524	0.093	0.532	0.11
Jupiter	5.203	0.048	11.27	317.8
Saturn	9.539	0.056	9.49	95.2

[a]Astronomical Unit (A.U.) is equal to the mean distance from the Earth to the sun = $149.6(10^6)$ km $= 92.96(10^6)$ mi.

COMPUTATIONAL METHODS

C-1 INTRODUCTION

The purpose of this appendix is to provide a few simple numerical methods that can be used to help solve (to take some of the drudgery out of the solution of) mechanics problems. No attempt has been made to provide a collection of computer programs to be used for the solution of the various types of problems encountered in a course in Dynamics. The problems designated as computer problems are basically simple applications of the elementary principles. They simply require solving the same problem over and over again as some parameter in the problem varies. Although they could be solved by hand or with a programmable calculator, these problems are most conveniently solved using a computer—either a microcomputer programmed in BASIC or a mainframe computer programmed in FORTRAN. In any event, the parametric study is intended to display characteristics of the problems that cannot be seen by solving for just one particular value. For example, Problem C13-170 examines the effect of initial angle on shooting baskets in basketball.

This appendix is not designed to teach the student all there is to know about numerical methods. The numerical methods presented here are purposely kept simple—simple for understanding and simple for use. Much more sophisticated methods exist. Students interested in these more sophisticated methods should take a course in numerical methods and/or see some of the references listed at the end of this appendix.

This appendix addresses four types of problems encountered in various mechanics problems:

1. Nonlinear equations (root solving)
2. Systems of linear equations
3. Numerical integration
4. Ordinary differential equations

One or two simple methods are presented for the solution of each of these types of problems. The methods presented are purposely kept simple so that they can be used in hand calculation with a calculator or with a programmable calculator as well as with a computer.

In each case, a simple program is included to demonstrate the use of the numerical method. Versions are supplied in both BASIC and FORTRAN. A companion disk is not included, as the programs are short enough to type in without much effort. The programs are not elegant; like the numerical methods they illustrate, they have been kept as simple as possible so that they will be easy to understand, easy to modify, and easy to customize, and also so that they will run on the widest possible selection of computers. Students will probably want to modify and enhance these programs to improve the transfer of data into the programs and to improve the format of the results output by the programs. Students may also want to modify the programs to take advantage of special features of their individual computers, such as graphical output.

C-2 NONLINEAR EQUATIONS

Problems in mechanics (Dynamics) often require the solution of non-linear equations, such as

$$x^3 - 7.014x^2 + 13.324x - 3.548 = 0 \qquad \text{(C-1a)}$$

These problems are sometimes stated in the form: Find the zeros or roots of the function

$$f(x) = x^3 - 7.014x^2 + 13.324x - 3.548 \qquad \text{(C-1b)}$$

[that is, find the values of x that make $f(x) = 0$]. Therefore, they are sometimes called *root solving* problems. Equation C-1 is a typical equation encountered in the problem of finding the maximum moment of inertia of a body (Eq. C-1a is taken directly from Example Problem A-6).

While such equations can be solved by trial and error (simply guessing values until the left-hand side of the equation is nearly zero), there exist simple, systematic ways to solve such problems. Two such methods—the *Newton–Raphson* method and the *Method of False Position*—will be discussed here.

C-2-1 Newton–Raphson Method

The Newton–Raphson Method of root solving is an iterative method that for most functions converges very quickly. In fact, if the initial guess is reasonably close to the correct value, the Newton–Raphson method often converges within three or four iterations. Because of its quick convergence and ease of use, the Newton–Raphson method is the method of choice in most cases.

At each step n in the iteration process, the Newton–Raphson method uses the tangent line to the curve at the point x_n (Fig. C-1) to estimate the location of the root. The slope of the tangent line at x_n is just the derivative of the function evaluated at x_n

$$\text{slope} = f'(x_n)$$

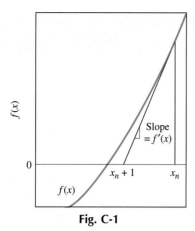

Fig. C-1

But from the geometry of Fig. C-1, the slope is also given by

$$\text{slope} = \frac{f(x_n) - 0}{x_n - x_{n+1}}$$

Setting these two expressions equal and solving for x_{n+1} gives

$$x_{n+1} = x_n - \frac{f(x_n)}{f'(x_n)} \qquad (C-2)$$

Equation C-2 is used iteratively to get improved estimates of the location of the root. A rough graph of the function $f(x)$ should be used to get a first estimate of the root location x_0.

EXAMPLE PROBLEM C-1

Solve Eq. C-1

$$x^3 - 7.014x^2 + 13.324x - 3.548 = 0$$

using the Newton–Raphson Method to a relative accuracy of 0.001 percent.

SOLUTION

The function

$$f(x) = x^3 - 7.014x^2 + 13.324x - 3.548$$

is a cubic polynomial so the equation $f(x) = 0$ should have three roots. A rough sketch of the function (Fig. C-2) indicates that the three roots are near the points $x = 0$, $x = 3$, and $x = 3.5$. Starting from the initial point $x_0 = 3.5$, Eq. C-3 is then used to generate the next several points

$$x_1 = x_0 - \frac{x_0^3 - 7.014x_0^2 + 13.324x_0 - 3.548}{3x_0^2 - 14.028x_0 + 13.324}$$

$$= 3.5 - \frac{0.0395}{0.9760} = 3.4595$$

$$x_2 = 3.4595 - \frac{0.0056}{0.6986} = 3.4514$$

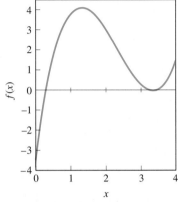

Fig. C-2

and so on. The process is repeated until the relative error

$$\left| \frac{x_{n+1} - x_n}{x_{n+1}} \right|$$

(the ratio of the difference between successive approximations and the current approximation) is less than or equal to 0.00001 (0.001 percent). The result of this process is shown in Fig. C-3a. The column labeled DL is just the second term in the iteration formula

$$DL = -\frac{f(x_n)}{f'(x_n)} = x_{n+1} - x_n$$

After just three iterations, the root is located as $x = 3.4511$. Similar iterations for starting values of 0 and 3.0 give the other two roots $x = 0.31670$ (Fig. C-3b) and $x = 3.24619$ (Fig. C-3c), respectively.

Xo = 3.50000
Error = 0.00001

Xn	DL	ABS(DL/X1)
3.50000	−0.04047	0.01170
3.45953	−0.00808	0.00234
3.45145	−0.00034	0.00010
3.45111	−0.00000	0.00000

The root is 3.45111

(a)

Xo = 0.00000
Error = 0.00001

Xn	DL	ABS(DL/X1)
0.00000	0.26629	1.00000
0.26629	0.04882	0.15492
0.31510	0.00160	0.00504
0.31670	0.00000	0.00001

The root is 0.31670

(b)

Xo = 3.00000
Error = 0.00001

Xn	DL	ABS(DL/X1)
3.00000	0.16932	0.05342
3.16932	0.06170	0.01910
3.23102	0.01426	0.00439
3.24528	0.00090	0.00028
3.24619	0.00000	0.00000

The root is 3.24619

(c)

Fig. C-3

Simple programs in BASIC and FORTRAN to solve nonlinear equations using the Newton–Raphson method are given in Programs C-1a and C-1b, respectively. The function statements defining the function and its derivative in lines 100 and 110 must be changed for the particular problem being solved.

C-2-2   Method of False Position

Although it is not as sophisticated as the Newton–Raphson method, the Method of False Position is a good method to use on functions for which the Newton–Raphson method has difficulty. The Newton–Raphson method has trouble when the derivative $f'(x)$ goes to zero at or near the root being located. This commonly happens when a function has two roots at the same value of x as, for example, $f(x) = (x + 1)(x - 1)^2$. Also, the Newton–Raphson method can be difficult to use for complex functions whose derivative $f'(x)$ may not be easily calculated. Therefore, an alternate method of solution is desirable, and the Method of False Position is a good alternative.

The Method of False Position is a systematic method of narrowing down the region in which a root exists. For example, Fig. C-4 shows a graph of a typical function $f(x)$ as a function of x. Point L lies on the left side of the point where $f(x) = 0$ and point R lies on the right side. Note that for two such points, L and R, which bracket a simple root, $f(x_L)f(x_R) < 0$, always. A rough graph of the function $f(x)$ may be needed to find the initial points L and R.

Construct point C—the point where the straight line joining point

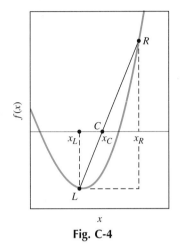

Fig. C-4

```
100 DEF FNY (X) = X*X*X − 7.014*X*X + 13.324*X − 3.548
110 DEF FNYP (X) = 3*X*X − 14.028*X + 13.324
120 PRINT "Enter Xo = ";
130 INPUT X0
140 PRINT "Enter Error = ";
150 INPUT ER
160 CLS
170 PRINT USING "    Xo = ###,###.#####"; X0
180 PRINT USING "  Error =      ##.#####"; ER
190 PRINT "————————————————————————"
200 PRINT "  Xn        DL     ABS(DL/X1) "
210 PRINT "————————————————————————"
220    DL = −FNY(X0) / FNYP(X0)
230    X1 = X0 + DL
240    PE = ABS(DL / X1)
250    PRINT USING "###,###.#####   ###,###.#####   ###.#####"; X0; DL; PE
260    IF ABS(DL / X1) < ER THEN 290
270      X0 = X1
280      GOTO 220
290 PRINT "————————————————————————"
300 PRINT USING "The root is ###,###.#####"; X1
310 END
```

Program C-1a. BASIC program listing for performing the Newton–Raphson method.

```
100 Y (X) = X*X*X − 7.014*X*X + 13.324*X − 3.548
110 YP (X) = 3*X*X − 14.028*X + 13.324
    PRINT *, ' Enter Xo = '
    READ *, X0
    PRINT *, ' Enter Error = '
    READ *, ER
    PRINT
    PRINT 1, X0
  1 FORMAT ('      Xo = ',F13.5)
    PRINT 2, ER
  2 FORMAT ('  Error =        ',F8.5)
    PRINT *, '————————————————————————————'
    PRINT *, '    Xn          DL      ABS(DL/X1) '
    PRINT *, '————————————————————————————'
220    DL = −Y(X0) / YP(X0)
    X1 = X0 + DL
    PE = ABS(DL / X1)
    PRINT 3, X0, DL, PE
  3 FORMAT (1X, F13.5,4X,F13.5,4X,F9.5)
    IF (ABS(DL/X1) .LT. ER) THEN
        GOTO 290
      ELSE
        X0 = X1
        GOTO 220
      END IF
290 PRINT *, '————————————————————————————'
    PRINT 4, X1
  4 FORMAT (' The root is ',F13.5)
    CALL EXIT
    END
```

Program C-1*b***.** FORTRAN program listing for performing the Newton–Raphson method.

L and point R goes through zero—and use it as an estimate of the root. Point C can be found by similar triangles

$$\frac{f(x_R)}{x_R - x_C} = \frac{f(x_R) - f(x_L)}{x_R - x_L} \tag{C-3}$$

Rearranging and solving for x_C gives

$$x_C = x_R - \frac{f(x_R)(x_R - x_L)}{f(x_R) - f(x_L)} \tag{C-4}$$

However, since $f(x)$ is not a straight line, $f(x_C)$ will not be zero. If $f(x_C)f(x_R) < 0$, then x_C is on the left side of the root. In this case, point L is moved to point C and the process is repeated. If $f(x_C)f(x_L) < 0$, then x_C is on the right side of the root. In this case, point R is moved to point C and the process is repeated. The process ends when point C becomes the same as one of the end points (within the limits of numerical round-off error). Then point C is the desired root.

EXAMPLE PROBLEM C-2

Solve

$$200 = \frac{T_0}{5}\left[\cosh\frac{5(800)}{2T_0} - 1\right]$$

using the Method of False Position to a relative accuracy of 0.001 percent.

SOLUTION

First, rewrite the equation in the form of Eq. C-1:

$$f(x) = x\left[\cosh\frac{2000}{x} - 1\right] - 1000 = 0$$

Based on a rough sketch of $f(x)$ (Fig. C-5), the initial points $x_L = 2000$ and $x_R = 2500$ are chosen. Note that $f(x_L) = 86.1614$, $f(x_R) = -156.4125$, and $f(x_L)f(x_R) < 0$. Point C is constructed using Eq. C-4, giving

$$x_C = 2500 - \frac{(-156.4125)(2500 - 2000)}{(-156.4125) - (86.1614)} = 2177.5982$$

Since $f(x_C) = -15.1522$, the product $f(x_C)f(x_R) > 0$ and the product $f(x_C)f(x_L) < 0$. Therefore, point R is moved to point C and the process is repeated. The process ends when the relative error, either

$$\frac{x_R - x_C}{x_C} \qquad \text{or} \qquad \frac{x_C - x_L}{x_C}$$

is less than or equal to 0.00001 (0.001 percent). The result of this process is shown in Fig. C-6. After just five iterations, the root is located as $x = 2148.642$.

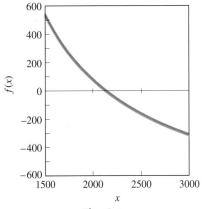

Fig. C-5

Error = 0.00001

XL = 2,000.00000	F(XL) =	86.16138
XR = 2,500.00000	F(XR) =	−156.41248
XC = 2,177.59814	F(XC) =	−15.15216
XL = 2,000.00000	F(XL) =	86.16138
XR = 2,177.59814	F(XR) =	−15.15216
XC = 2,151.03711	F(XC) =	−1.27262
XL = 2,000.00000	F(XL) =	86.16138
XR = 2,151.03711	F(XR) =	−1.27262
XC = 2,148.83862	F(XC) =	−0.10550
XL = 2,000.00000	F(XL) =	86.16138
XR = 2,148.83862	F(XR) =	−0.10550
XC = 2,148.65649	F(XC) =	−0.00865
XL = 2,000.00000	F(XL) =	86.16138
XR = 2,148.65649	F(XR) =	−0.00865
XC = 2,148.64160	F(XC) =	−0.00072

The root is 2148.642

Fig. C-6

527

```
 90 DEF FNCOSH(X) = (EXP(X) + EXP(−X))/2
100 DEF FNY(X) = X*(FNCOSH(2000/X) − 1) − 1000
110 PRINT "Enter XL = ";
120 INPUT XL
130 PRINT "Enter XR = ";
140 INPUT XR
150 PRINT "Enter Error = ";
160 INPUT ER
170 CLS
180 PRINT USING "   Error = ###,###.#####"; ER
190 PRINT "————————————————————————————"
200 XC = XR − FNY(XR)*(XR-XL)/(FNY(XR)-FNY(XL))
210 PRINT USING " XL = ###,###.#####   F(XL) = ###,###.#####"; XL; FNY(XL)
220 PRINT USING " XR = ###,###.#####   F(XR) = ###,###.#####"; XR; FNY(XR)
230 PRINT USING " XC = ###,###.#####   F(XC) = ###,###.#####"; XC; FNY(XC)
240   IF FNY(XC)*FNY(XL) < = 0 THEN 320
250   IF FNY(XC)*FNY(XR) < = 0 THEN 290
260 PRINT
270 PRINT "*** ERROR ***"
280 END
290 IF ABS((XC-XL)/XC) < ER THEN 350
300   XL = XC
310   GOTO 190
320 IF ABS((XC-XR)/XC) < ER THEN 350
330   XR = XC
340   GOTO 190
350 PRINT "————————————————————————————"
360 PRINT USING "     The root is ###,###.#####"; XC
370 END
```

Program C-2a. BASIC program listing for performing the method of false position.

Simple programs in BASIC and FORTRAN to solve nonlinear equations are given in Programs C-2a and C-2b, respectively. The function statement in line 100 must be changed for the particular problem being solved.

C-3   SYSTEMS OF LINEAR EQUATIONS

Many problems in mechanics require the solution of a system of linear equations, such as

$$5x + 3y + 4z = 23 \qquad (C\text{-}5a)$$
$$2x + 1y + 1z = 7 \qquad (C\text{-}5b)$$
$$1x + 3y + 5z = 22 \qquad (C\text{-}5c)$$

One possible scheme for solving such equations would be to:

First, use Eq. C-5a to eliminate x from Eqs. C-5b and C-5c. For example, subtract 2/5 times Eq. C-5a from Eq. C-5b and subtract 1/5 times Eq. C-5a from Eq. C-5c. (If the coefficient of x in Eq. C-5a were zero, the equations would first be reordered so that the coefficient was not zero.)

Next, use the resulting Eq. C-5b to eliminate y from Eqs. C-5a and C-5c.

Finally, use Eq. C-5c to eliminate z from Eqs. C-5a and C-5b.

```
100 Y(X) = X*(COSH(2000/X) − 1) − 1000
      PRINT *, ' Enter XL = '
      READ *, XL
      PRINT *, ' Enter XR = '
      READ *, XR
      PRINT *, ' Enter Error = '
      READ *, ER
      PRINT 1, ER
  1 FORMAT ('   Error = ',F13.5)
190 PRINT *, ' ————————————————————————————'
      XC = XR − Y(XR)*(XR − XL)/(Y(XR) − Y(XL))
      PRINT 2, XL, Y(XL)
  2 FORMAT ('   XL = ',F13.5,'   F(XL) = ',F13.5)
      PRINT 3, XR, Y(XR)
  3 FORMAT ('   XR = ',F13.5,'   F(XR) = ',F13.5)
      PRINT 4, XC, Y(XC)
  4 FORMAT ('   XC = ',F13.5,'   F(XC) = ',F13.5)
      IF (Y(XC)*Y(XL) .LE. 0) GOTO 320
      IF (Y(XC)*Y(XR) .LE. 0) GOTO 290
      PRINT
      PRINT *, " *** ERROR ***"
      CALL EXIT
290 IF (ABS((XC-XL)/XC) .LT. ER) GOTO 350
      XL = XC
      GOTO 190
320 IF (ABS((XC-XR)/XC) .LT. ER) GOTO 350
      XR = XC
      GOTO 190
350 PRINT *, ' ————————————————————————————'
      PRINT 5, XC
  5 FORMAT ('      The root is ',F13.5)
      CALL EXIT
      END
```

Program C-2b. FORTRAN program listing for performing the method of false position.

At this point, Eq. C-5a will give the value of x; Eq. C-5b, the value of y; and Eq. C-5c, the value of z.

The procedure described above is called the Gauss–Jordan method for the solution of systems of linear equations. Since the letters representing the unknowns (x, y, and z) serve no function in the solution scheme except to hold the coefficients in their proper places, the procedure is conveniently carried out on a matrix of the coefficients

$$\begin{bmatrix} 5 & 3 & 4 & 23 \\ 0 & 1 & 1 & 7 \\ 1 & 3 & 5 & 22 \end{bmatrix} \tag{C-6}$$

where the rows of the matrix represent the equations being solved. The first column of each row contains the coefficients of x, the second column contains the coefficients of y, the third column contains the coefficients of z, and the last column contains the right-hand sides of the equations. The rows of the matrix are to be multiplied by constants and added to other rows in the same manner that the equations are in the procedure described above.

EXAMPLE PROBLEM C-3

Solve the system of linear equations (Eqs. C-5a, C-5b, and C-5c) using the Gauss–Jordan method.

SOLUTION

The equations are first written in matrix form as in Eq. C-6. Then 2/5 times the first row (equation) is subtracted from the second row (equation), and 1/5 times the first row (equation) is subtracted from the third row (equation) to give

$$\begin{bmatrix} 5 & 3 & 4 & 23 \\ 0 & -0.2 & -0.6 & -2.2 \\ 0 & 2.4 & 4.2 & 17.4 \end{bmatrix}$$

Then, 15 times the second row is added to the first row and 12 times the second row is added to the third row to give

$$\begin{bmatrix} 5 & 0 & -5.0 & -10.0 \\ 0 & -0.2 & -0.6 & -2.2 \\ 0 & 0 & -3.0 & -9.0 \end{bmatrix}$$

Finally, 5/3 times the third row is subtracted from the first row and 2/10 times the third row is subtracted from the second row to give

$$\begin{bmatrix} 5 & 0 & 0 & 5 \\ 0 & -0.2 & 0 & -0.4 \\ 0 & 0 & -3.0 & -9.0 \end{bmatrix}$$

The answer is more easily interpreted if the first row is divided by 5; the second row by $-2/10$, and the third row by -3, which gives the matrix

$$\begin{bmatrix} 1 & 0 & 0 & 1 \\ 0 & 1 & 0 & 2 \\ 0 & 0 & 1 & 3 \end{bmatrix}$$

The rows of this matrix represent the three equations

$$x = 1 \qquad y = 2 \qquad z = 3$$

which is the solution to the original system of equations.

```
100 DATA 3                                      500 TEMP = ABS(A(I,I))
110 DATA 5, 3, 4, 23                            510 KT = I
120 DATA 2, 1, 1,  7                            520 FOR K = I TO N
130 DATA 1, 3, 5, 22                            530    TT = ABS(A(K,I))
140 DIM A(20,21)                                540    IF TT < = TEMP THEN 570
150 GOSUB 310                                   550       KT = K
160 FOR I = 1 TO N                              560       TEMP = TT
170    GOSUB 500                                570    NEXT K
180    GOSUB 720                                580 IF KT = I THEN 680
190    NEXT I                                   590 REM
200 PRINT                                       600 REM    Interchange rows if necessary to make the
210 PRINT                                       610 REM    'pivot' element as large as possible
220 PRINT "The solution is:"                    620 REM
230 PRINT                                       630 FOR K = 1 TO N+1
240 FOR I = 1 TO N                              640    TEMP = A(I,K)
250    PRINT "x(";I;") = "; A(I,N+1)            650    A(I,K) = A(KT,K)
260    NEXT I                                   660    A(KT,K) = TEMP
270 END                                         670    NEXT K
280 REM                                         680 RETURN
290 REM    Read the input matrix from DATA statements    690 REM
300 REM                                         700 REM    'Normalize' the pivot row
310 READ N                                      710 REM
320 FOR I = 1 TO N                              720 PV = A(I,I)
330    FOR J = 1 TO N+1                         730 FOR K = I TO N+1
340       READ A(I,J)                           740    A(I,K) = A(I,K)/PV
350       NEXT J                                750    NEXT K
360    NEXT I                                   760 REM
370 CLS                                         770 REM    Eliminate all entries from the Ith column
380 PRINT "The input matrix is:"               780 REM    except the pivot element which has been
390 PRINT                                       790 REM    normalized to 1
400 FOR I = 1 TO N                              800 REM
410    FOR J = 1 TO N+1                         810 FOR K = 1 TO N
420       PRINT A(I,J),                         820    IF K = I THEN 870
430       NEXT J                                830    PV = A(K,I)
440    PRINT                                     840    FOR KK = I TO N+1
450    NEXT I                                   850       A(K,KK) = A(K,KK) − PV*A(I,KK)
460 RETURN                                      860       NEXT KK
470 REM                                         870    NEXT K
480 REM    Search for row with largest element in column I    880 RETURN
490 REM
```

Program C-3a. BASIC program listing for solving a system of linear equations using the Gauss–Jordan elimination method.

Simple programs in BASIC and FORTRAN to solve systems of linear equations using the Gauss–Jordan elimination procedure are given in Programs C-3a and C-3b, respectively. The number of equations being solved and the coefficients of the matrix are included in DATA statements (lines 100–130) and must be changed for the specific problem being solved. A desirable modification of these programs would be to have these values entered directly from the keyboard for small numbers of equations or read from a file for large numbers of equations.

```
     REAL A(20,21)                              570  CONTINUE
100 DATA N/3/                                        IF (KT .EQ. I) GOTO 680
110 DATA A(1,1),A(1,2),A(1,3),A(1,4)/5, 3, 4, 23/  C    Interchange rows if necessary to make the 'pivot'
120 DATA A(2,1),A(2,2),A(2,3),A(2,4)/2, 1, 1,  7/  C    element as large as possible
130 DATA A(3,1),A(3,2),A(3,3),A(3,4)/1, 3, 5, 22/     DO 670 K = 1, N+1
     PRINT *, ' The input matrix is:'                    TEMP = A(I,K)
     PRINT *, '  '                                       A(I,K) = A(KT,K)
     DO 450 I = 1, N                                     A(KT,K) = TEMP
       PRINT *, (A(I,J), J = 1, N+1)             670  CONTINUE
450  CONTINUE                                    680 RETURN
     DO 190 I = 1, N                                  END
       CALL PIVOT(N,A,I)                             SUBROUTINE ELIM(N,A,I)
       CALL ELIM(N,A,I)                              REAL A(20,21)
190  CONTINUE                                    C       'Normalize' the pivot row
     PRINT *, '  '                                    PV = A(I,I)
     PRINT *, '  '                                    DO 750 K = I, N+1
     PRINT *, ' The solution is:'                       A(I,K) = A(I,K)/PV
     PRINT *, '  '                               750  CONTINUE
     DO 260 I = 1, N                             C       Eliminate all entries from the Ith column
       PRINT *, 'X(', I, ') = ', A(I,N+1)        C       except the pivot element which has been
260  CONTINUE                                    C       normalized to 1
     CALL EXIT                                        DO 870 K = 1, N
     END                                                IF (K .NE. I) THEN
     SUBROUTINE PIVOT(N,A,I)                              PV = A(K,I)
     REAL A(20,21)                                        DO 860 KK = I, N+1
C       Search for row with largest element in column I     A(K,KK) = A(K,KK) − PV*A(I,KK)
     TEMP = ABS(A(I,I))                          860      CONTINUE
     KT = I                                               END IF
     DO 570 K = I, N                             870  CONTINUE
       TT = ABS(A(K,I))                          880 RETURN
       IF (TT .GT. TEMP) THEN                        END
         KT = K
         TEMP = TT
       END IF
```

Program C-3b. FORTRAN program listing for solving a system of linear equations using the Gauss–Jordan elimination method.

C-4 NUMERICAL INTEGRATION

Most of the functions to be integrated in Dynamics problems are polynomials or other simple functions and are easily evaluated analytically. Sometimes, however, the function to be integrated will be complex enough to require advanced integration techniques. Other times, the function to be integrated will not be given explicitly. Instead, the function may be given by experimentally determined values at a few points. In these last two cases, numerical methods may be useful in evaluating the integrals.

In the simplest form, numerical integration derives from the physical interpretation of an integral as the area under a curve. The area may be approximated using several rectangles, trapezoids, or other simple shapes whose area is easily determined. The approximate value of the integral is obtained by adding together the areas of the several pieces. The method to be described here uses trapezoids to approximate the area under the curve; hence the name: Trapezoidal Rule.

For example, the value of the integral

$$I = \int_a^b f(x)\, dx \qquad \text{(C-7)}$$

is represented by the shaded area in Fig. C-7a. Approximating this area by one large trapezoid of width $h = b - a$, as in Fig. C-7b, gives

$$I \cong T_1 = \frac{h}{2}[f_a + f_b] \qquad \text{(C-8)}$$

where $f_a = f(a)$ and $f_b = f(b)$. This approximation is obviously in error by the amount of the shaded area in Fig. C-7b between the top of the trapezoid and the curve. The amount of error can be reduced, however, by using two trapezoids of width $h = (b - a)/2$, as in Fig. C-7c, and adding their areas together

$$I \cong T_2 = \frac{h}{2}[f_a + f_1] + \frac{h}{2}[f_1 + f_b] = \frac{h}{2}[f_a + f_b + 2f_1] \qquad \text{(C-9)}$$

where, $f_1 = f(x_1)$ and $x_1 = a + h$. The tops of the two trapezoids follow the curve more closely than did the single trapezoid. The error in the approximation T_2 (represented by the shaded area in Fig. C-7c) is less than the error in the approximation T_1.

Continuing with this logic and dividing the interval into N trapezoids of equal width h gives the Trapezoidal Rule approximation

$$I \cong T_n = \frac{h}{2}[f_0 + f_n + 2\Sigma f_i] \qquad \text{(C-10)}$$

where

$$f_0 = f(a) \qquad f_n = f(b) \qquad h = \frac{b - a}{n}$$

$$f_i = f(x_i) = f(a + ih) \qquad i = 1, 2, \ldots, n - 1$$

As the number of panels used is increased, the width of the panels will decrease, the tops of the trapezoids will more closely fit the function being integrated, and the total error of the approximation will be reduced. It can be shown that the error in using the Trapezoidal Rule is approximately proportional to h^2. Therefore, reducing h by a factor of 2 should reduce the error by a factor of 4. Typically, the integral is evaluated several times using smaller and smaller panel widths until the value for two different panel widths is nearly the same. This value then is taken to be the value of the integral.

For experimental data given only at discrete points (possibly not equal-spaced), the Trapezoidal Rule is applied to each pair of points and the sum is then added together. In this case, however, it is usually not possible to vary the number of panels and get an estimate of the error. Instead, a sketch of the function may be used to get a rough estimate of how well the trapezoidal rule approximates the function (see Example Problem C-5). If it appears that the Trapezoidal Rule does not accurately represent the function, then either an interpolation procedure could be used to generate additional points at a large number of equal-spaced values as required for the Trapezoidal Rule or a more accurate integration procedure must be used.

A listing of the BASIC and FORTRAN programs used to generate the output of Fig. C-8 are given in Programs C-4a and C-4b, respectively. The function statement and integration limits in lines 100–120 need to be changed for the function being integrated.

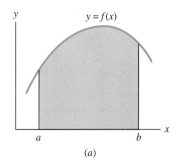

(a)

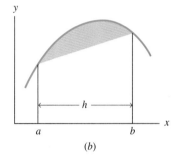

(b)

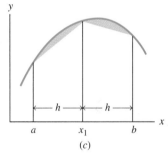

(c)

Fig. C-7

EXAMPLE PROBLEM C-4

Evaluate the integral

$$\int_0^2 x e^{-x^2} dx$$

using the Trapezoidal Rule to an estimated relative error of 0.1 percent.

SOLUTION

Evaluating the integral using Eq. C-6 and a single panel gives

$$T_1 = \frac{2}{2}[0 + 2 e^{-4}] = 0.0366$$

while using two panels gives

$$T_2 = \frac{1}{2}[0 + 2 e^{-4} + 2(e^{-1})] = 0.3862$$

The absolute error (just the difference between the current estimate and the correct value) is *estimated* to be

$$E_{abs} = |T_2 - T_1| = 0.3496$$

and the relative error (the ratio of the absolute error and the correct value) is *estimated* to be

$$E_{rel} = \left| \frac{T_2 - T_1}{T_2} \right| \times 100 = 90.5\%$$

Using four panels gives

$$T_4 = \frac{0.5}{2}\left[0 + 2 e^{-4} + 2(0.5 e^{-0.25} + e^{-1} + 1.5 e^{-2.25})\right] = 0.4688$$

and the relative error is estimated to be

$$E_{rel} = \left| \frac{0.4668 - 0.3862}{0.4668} \right| \times 100 = 17.28\%$$

Continuing using more and more panels until the desired accuracy has been reached gives the results in Fig. C-8. Using 64 panels, the integral is evaluated as 0.4908 with an estimated error of 0.06 percent. (Notice in Fig. C-8 that reducing the width of the panels by a factor of 2 reduced the error by approximately a factor of 4, in accordance with the error estimate mentioned earlier).

No of Panels	Value of Integral	% rel Error
1	0.036631	
2	0.386195	90.514820
4	0.466847	17.275900
8	0.484937	3.730329
16	0.489371	0.906150
32	0.490475	0.224978
64	0.490750	0.056155
128	0.490819	0.014032

Note: The result is 0.4908 with an estimated error of 0.1 percent.

Fig. C-8 Evaluation of the integral $\int_0^2 xe^{-x^2} dx$, using the Trapezoidal Rule (Eq. C-10) and varying numbers of panels.

```
100 DEF FNY(X) = X*EXP(−X*X)
110 A = 0
120 B = 2
130 CLS
140 N = 1
150 H = B − A
160 T = H*(FNY(A) + FNY(B))/2
170 PRINT "——————————————————"
180 PRINT "No of   Value of     % rel"
190 PRINT "Panels   Integral    Error"
200 PRINT "——————————————————"
210 PRINT USING " ####   ###.######"; N, T
220 FOR K = 1 TO 7
230    TOLD = T
240    H = H/2
250    T = FNY(A) + FNY(B)
260    N = (B − A)/H
270    FOR I = 1 TO N-1
280       X1 = A + I*H
290       T = T + 2*FNY(X1)
300    NEXT I
310    T = T*H/2
320    ER = 100*ABS((T − TOLD)/T)
330    PRINT USING " ####   ###.######   ###.######"; N, T, ER
340 NEXT K
350 PRINT "——————————————————"
```

Program C-4a. BASIC program listing for evaluating integrals using the Trapezoidal Rule.

```
100 Y(X) = X*EXP(−X*X)
110 A = 0
120 B = 2
    N = 1
    H = B − A
    T = H*(Y(A) + Y(B))/2
    PRINT *, " ——————————————————"
    PRINT *, " No of   Value of     % rel"
    PRINT *, " Panels   Integral    Error"
    PRINT *, " ——————————————————"
    PRINT 210, N, T
210 FORMAT (3X,I4,2(3X,F10.6))
    DO 340 K = 1, 7
      TOLD = T
      H = H/2
      T = Y(A) + Y(B)
      N = (B − A)/H
      DO 300 I = 1, N-1
        X1 = A + I*H
        T = T + 2*Y(X1)
300     CONTINUE
      T = T*H/2
      ER = 100*ABS((T − TOLD)/T)
      PRINT 210, N, T, ER
340   CONTINUE
    PRINT *, ' ——————————————————'
    CALL EXIT
    END
```

Program C-4b. FORTRAN program listing for evaluating integrals using the Trapezoidal Rule.

Barrels are sometimes placed around bridge abutments to absorb some of the energy and reduce the severity of auto accidents. In a test of one such barrier, a car is driven into the barrels, and instrumentation in the car measures the acceleration of the car as a function of time. These data can be manipulated to give the force on the car as a function of position:

x, ft	F, lb
0.000	0
2.898	4,503
5.237	7,202
6.679	7,016
7.248	4,019
7.316	0

Determine the total energy absorbed by the barrels in this test.

SOLUTION

The energy absorbed is equal to the work done on the car

$$E = \int F \, dx$$

The force is not given at equal intervals of position, however, since the data were collected at equal intervals of time rather than at equal intervals of position. Therefore, the Trapezoidal Rule will have to be applied to each of the five intervals separately, and the sums added together:

$$E \cong \frac{2.898 - 0}{2}(4503 + 0) + \frac{5.237 - 2.898}{2}(7202 + 4503)$$

$$+ \frac{6.679 - 5.237}{2}(7016 + 7202) + \frac{7.248 - 6.679}{2}(4019 + 7016)$$

$$+ \frac{7.316 - 7.248}{2}(0 + 4019)$$

$$= 33{,}740 \text{ lb} \cdot \text{ft}$$

Since the data were obtained experimentally, it is not possible to estimate the error by reevaluating the integral with a smaller step size. Instead, the force F is plotted in Fig. C-9 as a function of x. The solid line indicates the area used by the Trapezoidal Rule, while the dotted line indicates the expected true function shape. The expected error is the shaded region between the two curves. It appears that the Trapezoidal Rule has underestimated the integral by a small amount (perhaps 3 or 4 percent). If a more accurate value of the integral were required, a more accurate integration method would need to be used.

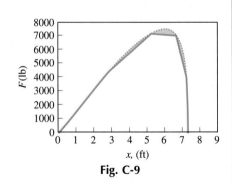

Fig. C-9

C-5 ORDINARY DIFFERENTIAL EQUATIONS

Many phenomena in mechanics require the solution of differential equations. Many of these problems are simple, standard equations that have well-known solutions. Some differential equations, however, are difficult to solve analytically but are easily solved numerically using a simple procedure called *Euler's method*. The procedure will be described in three parts:

1. First-order ordinary differential equations.
2. General *initial-value problems* in which the order of the ordinary differential equation is greater than one, but for which all data are given at a single point.
3. *Boundary-value problems* in which the order of the ordinary differential equation is greater than one and the data are given at two or more points.

C-5-1 First-Order Ordinary Differential Equations

A problem governed by a first-order, ordinary differential equation is of the form

$$\frac{dy}{dt} = f(t, y) \qquad a \leq t \leq b \qquad \text{(C-11a)}$$

$$y(a) = y_0 \qquad \text{(C-11b)}$$

The solution of the differential equation is the function $y(t)$ that satisfies Eq. C-11. But since $f(t, y) = dy/dt$ is the slope of the function $y(t)$, it seems reasonable to expect that $f(a, y_0)$ can be used to predict the value of $y_1 = y(t_1)$ where $t_1 = a + \Delta t$ (see Fig. C-10). Then knowing the value $y_1, f(t_1, y_1)$ can be used to predict $y_2 = y(t_2)$ where $t_2 = a + 2\Delta t$, and so on. The solution $y(t)$ is generated in a marching fashion:

$$y_{n+1} = y_n + f(t_n, y_n)\Delta t \qquad \text{(C-12)}$$

where $t_n = a + n\Delta t$ and $y_n = y(t_n)$. A more theoretical justification of Eq. C-12 could be obtained from the Taylor series for $y(t)$. In fact, rewriting Eq. C-12 in the form

$$y(t_n + \Delta t) = y(t_n) + y'(t_n)\,\Delta t$$

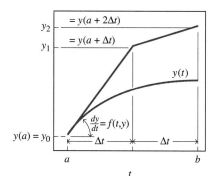

Fig. C-10

shows that it is just the first two terms in the Taylor series expansion of $y(t)$ about the point t_n.

The Euler method for solving differential Eq. C-11 consists of using the recursion relation, Eq. C-12, to generate the points (t_n, y_n) starting from (a, y_0) and marching along the solution curve to $t = b$. The points generated are connected by straight-line segments to define the function $y(t)$. Enough points should be generated so that the resulting curve looks smooth.

Besides generating enough points so that the curve looks smooth, points must be generated close enough together so that the method does not stray too far from the solution curve. If the step size Δt is too large, Euler's method will not follow the solution curve very closely (see Example Problem C-6). The smaller the step size, the better Euler's method works, but more steps and more computation time are required to generate the solution.

It can be shown that the error in using Euler's method is proportional to the step size. Therefore, reducing the step size by a factor of 2 should reduce the error by a factor of 2 also. The normal procedure is to solve the problem with successively smaller step sizes until the solution no longer changes.

EXAMPLE PROBLEM C-6

Solve the differential equation

$$\frac{dy}{dt} = y^2 e^{-t}$$

for $y(t)$, $1 \leq t \leq 5$ and $y(1) = 0.5$.

SOLUTION

Starting from the initial condition, $t_0 = 1$ and $y_0 = 0.5$, use Eq. C-12 and $\Delta t = 1$ to generate the points

$$y_1 = y_0 + (y_0^2 \, e^{-t_0}) \, \Delta t$$
$$= 0.5 + (0.5)^2 \, e^{-1}(1) = 0.5920$$

$$t_1 = t_0 + \Delta t = 1 + 1 = 2$$

$$y_2 = y_1 + (y_1^2 \, e^{-t_1}) \, \Delta t$$
$$= 0.5920 + (0.5920)^2 \, e^{-2}(1) = 0.6394$$

$$t_2 = t_1 + \Delta t = 2 + 1 = 3$$

$$y_3 = y_2 + (y_2^2 \, e^{-t_2}) \, \Delta t$$
$$= 0.6394 + (0.6394)^2 \, e^{-3}(1) = 0.6597$$

$$t_3 = t_2 + \Delta t = 3 + 1 = 4$$

$$y_4 = y_3 + (y_3^2 \, e^{-t_3}) \, \Delta t$$
$$= 0.6597 + (0.6597)^2 \, e^{-4}(1) = 0.6677$$

$$t_4 = t_3 + \Delta t = 4 + 1 = 5$$

These points are plotted in Fig. C-11 and connected by straight-line segments. The resulting curve is very coarse and probably does not follow the true solution very closely. The calculations were repeated using smaller step sizes ($\frac{1}{2}$, $\frac{1}{4}$, $\frac{1}{8}$, and $\frac{1}{16}$) until the curve became smooth and no longer changed. These results are also shown in Fig. C-11.

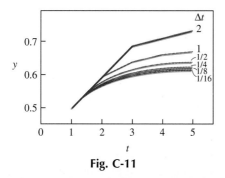

Fig. C-11

```
100 DIM X(100), Y(100)
110 DEF FNF (X, Y) = Y * Y * EXP(−X)
120 XI = 1
130 YI = .5
140 H = 2
150 FOR M = 1 TO 5
160    NS = 4 / H
170    X = XI
180    Y = YI
190    X(0) = X
200    Y(0) = Y
210    FOR N = 1 TO NS
220      Y = Y + H * FNF(X, Y)
230      X = X + H
240      X(N) = X
250      Y(N) = Y
260    NEXT N
270  PRINT
280  PRINT
290  PRINT
300  A$ = " ###.####   ##.#####"
310  FOR N = 0 TO NS
320    PRINT USING A$; X(N); Y(N)
330    NEXT N
340  H = H / 2
350  NEXT M
360 END

(a)
```

```
REAL X(100), Y(100)
FNF (X, Y) = Y * Y * EXP(−X)
XI = 1
YI = .5
H = 2
DO 350 M = 1, 5
   NS = 4 / H
   XC = XI
   YC = YI
   X(0) = XC
   Y(0) = YC
   DO 260 N = 1, NS
     YC = YC + H * FNF(XC, YC)
     XC = XC + H
     X(N) = XC
     Y(N) = YC
260    CONTINUE
   PRINT
   PRINT
   PRINT
300 FORMAT (3X,2F10.4)
   DO 330 N = 0, NS
     PRINT 300, X(N),Y(N)
330    CONTINUE
   H = H / 2
350 CONTINUE
   STOP
   END

(b)
```

Program C-5. Program listings for solving a first-order ordinary differential equation using the Euler method. (*a*) BASIC. (*b*) FORTRAN.

The programs in BASIC and FORTRAN used to generate the data used to draw Fig. C-11 are given as Programs C-5*a* and C-5*b*, respectively. The function $f(t, y)$ that defines the problem and the initial data, t_0 and y_0, in statements 110–130 must be changed for the particular problem being solved.

C-5-2 General Initial-Value Problems

Euler's method can also be applied to higher order differential equations such as

$$\frac{d^4y}{dt^4} + t\left(\frac{d^2y}{dt^2}\right)^2 + y^2\frac{dy}{dt} = t \sin y \tag{C-13}$$

$$y(1) = 1.0 \qquad \frac{dy}{dt}(1) = 0.0$$

$$\frac{d^2y}{dt^2}(1) = -1.0 \qquad \frac{d^3y}{dt^3}(1) = 0.5$$

In order to apply Euler's method to this type of problem, the differential equation (Eq. C-13) must first be reduced to a system of first-order differential equations. One way of doing this is to define $n - 1$ new variables (the dependent variable and its first $n - 1$ derivatives where n is the order of the differential equation)

$$x_1 = y$$

$$x_2 = \frac{dy}{dt} = \frac{dx_1}{dt}$$

$$x_3 = \frac{d^2y}{dt^2} = \frac{dx_2}{dt}$$

$$x_4 = \frac{d^3y}{dt^3} = \frac{dx_3}{dt}$$

The last three equations are three first-order differential equations relating the variables t, x_1, x_2, x_3, and x_4. A fourth equation comes from the original differential equation (Eq. C-13). Thus

$$\frac{dx_1}{dt} = x_2 = f_1(t, x_1, x_2, x_3, x_4) \tag{C-14a}$$

$$\frac{dx_2}{dt} = x_3 = f_2(t, x_1, x_2, x_3, x_4) \tag{C-14b}$$

$$\frac{dx_3}{dt} = x_4 = f_3(t, x_1, x_2, x_3, x_4) \tag{C-14c}$$

$$\frac{dx_4}{dt} = t \sin x_1 - tx_3^2 + x_1^2 x_2 = f_4(t, x_1, x_2, x_3, x_4) \tag{C-14d}$$

In terms of these four new variables, the initial conditions are

$$x_1(1) = 1.0 \qquad x_2(1) = 0.0$$
$$x_3(1) = -1.0 \qquad x_4(1) = 0.5$$

After the nth-order differential equation (Eq. C-13) is reduced to a system of n first-order differential equations (Eqs. C-14), Euler's method (Eq. C-12) is applied to each of the first-order differential equations, in turn, generating the sequence of points

$$x_{11} = x_{10} + f_1(t_0, x_{10}, x_{20}, x_{30}, x_{40}) \, \Delta t$$
$$x_{21} = x_{20} + f_2(t_0, x_{10}, x_{20}, x_{30}, x_{40}) \, \Delta t$$
$$x_{31} = x_{30} + f_3(t_0, x_{10}, x_{20}, x_{30}, x_{40}) \, \Delta t$$
$$x_{41} = x_{40} + f_4(t_0, x_{10}, x_{20}, x_{30}, x_{40}) \, \Delta t$$
$$x_{12} = x_{11} + f_1(t_1, x_{11}, x_{21}, x_{31}, x_{41}) \, \Delta t$$
$$x_{22} = x_{21} + f_2(t_1, x_{11}, x_{21}, x_{31}, x_{41}) \, \Delta t$$

and so on, where $x_{mn} = x_m(t_n)$. Once the solution to the system of first-order equations (Eqs. C-14) is complete, the solution to the original differential equation (Eq. C-13) is obtained simply as

$$y(t) = x_1(t)$$

Suppose that a small ball is thrown upward with an initial speed v_0 and an initial angle of θ_0 to the horizontal. Because of air resistance, the ball will experience a drag force proportional to the square of its speed

$$D = C_D \frac{1}{2}\rho v^2 A$$

where C_D is the drag coefficient (C_D may be taken as approximately one for spheres at low to moderate speeds), ρ is the mass density of the air through which the ball is moving, and $A = \pi r^2$ is the cross-sectional area of the ball. If a Ping-Pong ball ($W = 0.01$ lb, $r = 0.75$ in.) is thrown with an initial speed $v_0 = 20$ ft/s and $\theta_0 = 40°$ through air ($\rho = 0.002377$ slug/ft³):

a. Plot the trajectory of the Ping-Pong ball from $t = 0$ until it returns to its initial level.
b. Determine the range (the total horizontal distance traveled by the ball).

SOLUTION

a. The free-body diagram of the Ping-Pong ball is shown in Fig. C-12 for some time in its motion. The drag force is always in the direction opposite the velocity. Writing Newton's second law for the Ping-Pong ball gives the pair of differential equations

$$-D \cos\theta = m\ddot{x} \tag{a}$$
$$-D \sin\theta - W = m\ddot{y} \tag{b}$$

Equations a and b can be reduced to a system of first-order differential equations by introducing the new pair of variables $v_x = \dot{x}$ and $v_y = \dot{y}$. Then the four first-order differential equations are

$$\dot{x} = v_x$$
$$\dot{v}_x = -\frac{D \cos\theta}{m}$$
$$\dot{y} = v_y$$
$$\dot{v}_y = -\frac{D \sin\theta + W}{m}$$

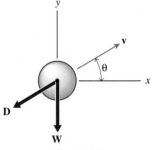

Fig. C-12

where $m = 0.01/32.2 = 3.106(10^{-4})$, $D = (1)(0.5)(0.002377)v^2\pi(0.75/12)^2 = 1.459(10^{-5})v^2$, $v^2 = v_x^2 + v_y^2$, $\sin\theta = v_y/v$, and $\cos\theta = v_x/v$. These equations are to be solved subject to the initial conditions that

$$x = y = 0$$
$$v_x = 20 \cos 40° \qquad v_y = 20 \sin 40°$$

when $t = 0$.

Applying the Euler method with a step size $\Delta t = 0.01$ s gives the initial sequence of points

$$x_0 = y_0 = 0 \qquad \theta_0 = 40°$$
$$v_{x0} = 15.321 \text{ ft/s} \qquad v_{y0} = 12.846 \text{ ft/s}$$
$$v^2 = 15.321^2 + 12.846^2 = 399.753 \text{ ft}^2/\text{s}^2$$

$$x_1 = 0 + (15.321)(0.01) = 0.15321 \text{ ft}$$
$$v_{x1} = 15.321 - \frac{1.459(10^{-5})(399.753)\cos 40°}{3.106(10^{-4})}(0.01)$$
$$= 15.177 \text{ ft/s}$$

$$y_1 = 0 + (12.846)(0.01) = 0.12846 \text{ ft}$$

$$v_{y1} = 12.846 - \frac{1.459(10^{-5})(399.753 \sin 40° + 0.01}{3.106(10^{-4})}(0.01)$$

$$= 12.403 \text{ ft/s}$$

$$v^2 = 15.177^2 + 12.403^2 = 384.176$$

$$\sin \theta = 12.403/(384.176)^{1/2} = 0.6328$$

$$\cos \theta = 15.177/(384.176)^{1/2} = 0.7743$$

$$t_1 = 0 + 0.01 = 0.01 \text{ s}$$

$$x_2 = 0.15321 + (15.177)(0.01) = 0.3050 \text{ ft}$$

$$v_{x2} = 15.177 - \frac{1.459(10^{-5})(384.176)(0.7743)}{3.106(10^{-4})}(0.01)$$

$$= 15.037 \text{ ft/s}$$

$$y_2 = 0.12846 + (12.403)(0.01) = 0.2525 \text{ ft}$$

$$v_{y2} = 12.403 - \frac{1.459(10^{-5})(384.176)(0.6328) + 0.01}{3.106(10^{-4})}(0.01)$$

$$= 11.967 \text{ ft/s}$$

and so on. The iteration is repeated until $y_2 = 0$. Of course, the entire iteration process should be repeated for smaller and smaller step sizes until the solution no longer changes with decreasing step size.

These equations were solved using the programs in BASIC and FOR-TRAN listed as Programs C-6a and C-6b, respectively, and a step size of $\Delta t = 1/256$ s. The solution (y versus x) is shown in Fig. C-13. The trajectory of the Ping-Pong ball for the case where drag has been neglected is included on Fig. C-13 for comparison purposes.

b. The range of the ball is just the x-coordinate when y again becomes zero. From Fig. C-13 the range is

$$R = 8.77 \text{ ft} \quad \text{(when drag is included)}$$
$$R = 12.23 \text{ ft} \quad \text{(when drag is neglected)}$$

(Note that the trajectory is not a parabola when drag is included. If the trajectory were a parabola, its peak would occur at the midpoint $8.77/2 = 4.38$ ft whereas Fig. C-13 indicates that the peak occurs much closer to 5 ft.)

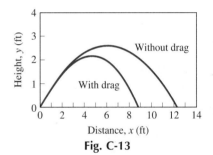

Fig. C-13

```
100 PI = 4 * ATN(1)
110 G = 32.2
120 W = .01
130 M = W / G
140 R = 0.75/12
150 RHO = .002377
160 A = PI * R * R
170 VO = 20
180 T1 = 40
190 DT = 1 / 256
200 T = 0
210 X = 0
220 Y = 0
230 VX = VO * COS(T1 * PI / 180)
240 VY = VO * SIN(T1 * PI / 180)
250 FM$ = "###.###   #,###.###   #,###.###   #,###.###   #,###.###   ###.##"
260 REM
270 REM First calculate the NO DRAG solution
280 REM
290 TI = 0
300 XI = VX * TI
310    YI = VY * TI - G * TI * TI / 2
320    PRINT USING FM$; TI; XI; YI
330    TI = TI + DT
340    IF YI > = 0 THEN 300
350 REM
360 REM Next calculate the DRAG solution
370 REM
380 TH = ATN(VY / VX)
390    VS = VX * VX + VY * VY
400    DR = RHO * VS * A / 2
410    X = X + VX * DT
420    Y = Y + VY * DT
430    VX = VX - (DR * COS(TH) / M) * DT
440    VY = VY - ((DR * SIN(TH) / M) + G) * DT
450    T = T + DT
460    PRINT USING FM$; T; X; Y; VX; VY; TH * 180 / PI
470    IF Y > = 0 THEN 380
```

Program C-6a. BASIC program listing for solving a system of first-order ordinary differential equations using the Euler method.

```
      REAL M
      PI = 4 * ATAN(1)
      G = 32.2
      W = .01
      M = W / G
      R = 0.75/12
      RHO = .00238
      A = PI * R * R
      VO = 20
      T1 = 40
      DT = 1. / 256.
      VX = VO * COS(T1 * PI / 180)
      VY = VO * SIN(T1 * PI / 180)
250   FORMAT (3X,F6.3,5F12.3)
C     First calculate the NO DRAG solution
      T = 0
300      X = VX * T
         Y = VY * T − G * T * T / 2
         T = T + DT
         PRINT 250, T, X, Y
         IF (Y .GE. 0) GOTO 300
C     Next calculate the DRAG solution
         T = 0
         X = 0
         Y = 0
370      TH = ATAN(VY / VX)
         VS = VX * VX + VY * VY
         DR = RHO * VS * A / 2
         X = X + VX * DT
         Y = Y + VY * DT
         VX = VX − (DR * COS(TH) / M) * DT
         VY = VY − ((DR * SIN(TH) / M) + G) * DT
         T = T + DT
         PRINT 250, T, X, Y, VX, VY, TH * 180 / PI
         IF (Y .GE. 0) GOTO 370

      STOP
      END
```

Program C-6*b*. FORTRAN program listing for solving a system of first-order ordinary differential equations using the Euler method.

C-5-3 Boundary-Value Problems

A boundary-value problem consists of a differential equation of order 2 or higher (such as Eq. C-13) that has the data given at two or more points. Euler's method, however, requires all the data be given at the initial point, so that it can march along the solution. Therefore, Euler's method cannot be used directly.

One approach in solving boundary-value problems is to guess values for the derivatives needed by the Euler method, solve the problem using the Euler method, and see how close the solution comes to the second boundary value. If necessary, the guesses for the initial derivatives are adjusted and the problem solved again until the solution meets the second boundary value. This is called a *shooting method* by analogy with the method of aiming an artillery gun: guess a barrel angle, fire the gun, see if the shell goes beyond the target or falls short, adjust the angle, fire again, and so on.

EXAMPLE PROBLEM C-8

An air gun is used to fire a Ping-Pong ball at a target as shown in Fig. C-14. Although the angle of the air gun can be varied, the initial velocity of the ball is fixed at $v_0 = 8$ m/s. Determine the required firing angle θ_0. Include the effects of air resistance as in Example Problem C-7. Use mass $= 5$ g, diameter $= 38$ mm, $C_D = 1$, and air density $= 1.225$ kg/m^3.

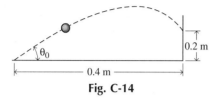

Fig. C-14

SOLUTION

The differential equations are the same as derived for Example Problem C-7:

$$\dot{x} = v_x$$

$$\dot{v}_x = -\frac{D \cos \theta}{m}$$

$$\dot{y} = v_y$$

$$\dot{v}_y = -\frac{D \sin \theta + W}{m}$$

where $m = 0.005$, $D = (1)(0.5)(1.225)v^2\pi(0.019)^2 = 6.946(10^{-4})v^2$, $v^2 = v_x^2 + v_y^2$, $\sin \theta = v_y/v$, and $\cos \theta = v_x/v$. These equations are to be solved subject to the initial conditions that $x = y = 0$ when $t = 0$ and also subject to the second boundary condition that $y = 0.2$ m when $x = 0.4$ m.

Application of the Euler method requires initial values for all four variables. The missing initial values will be obtained by guessing a value for the initial angle θ_0 and adjusting the guess as necessary to attain the second boundary condition. Since the straight line from the initial position to the target makes an angle of about 27° with the horizontal, a first guess of $\theta_0 = 35°$ will be tried. Then

$$v_{x0} = 8 \cos 35° = 6.553 \text{ m/s}$$
$$v_{y0} = 8 \sin 35° = 4.589 \text{ m/s}$$

and applying the Euler method as in Example Problem C-7 gives that $y = 0.253$ m when $x = 0.4$ m. Since this guess was too high, a second guess of $\theta_0 = 30°$ will be tried. Then

$$v_{x0} = 8 \cos 30° = 6.928 \text{ m/s}$$
$$v_{y0} = 8 \sin 30° = 4.000 \text{ m/s}$$

and applying the Euler method gives $y = 0.206$ m when $x = 0.4$ m. This guess is still too high, so a third guess of $\theta_0 = 29.5°$ will be tried, and so on. After a few more guesses, a value of $\theta_0 = 29.28°$ is obtained. The trajectory for this initial angle is shown in Fig. C-15.

The solution just obtained passes through the point $x = 0.4$ m and $y = 0.2$ m on the way up. A second solution is possible in which the ball passes through $x = 0.4$ m and $y = 0.2$ m on the way down. Trying a few more guesses around 70° to 80° yields the solution $\theta_0 = 78.18°$. The trajectory for this initial angle is also shown in Fig. C-15.

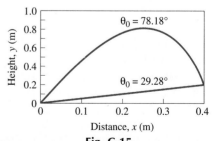

Fig. C-15

C-6 FURTHER READING

The texts listed here are just a few of the elementary numerical methods texts available. All the texts listed include the methods discussed in this appendix as well as more sophisticated methods and other related numerical methods that the student may find useful.

1. Chapra, S. C., and R. P. Canale (1985). *Numerical Methods for Engineers with Personal Computer Applications*, McGraw-Hill, New York.
2. Cheney, W., and D. Kincaid (1980). *Numerical Mathematics and Computing*, Brooks/Cole, Monterey, Calif.
3. James, M. L., G. M. Smith, and J. C. Wolford (1985). *Applied Numerical Methods for Digital Computation*, 3rd ed., Harper & Row, New York.
4. Johnston, R. L. (1982). *Numerical Methods: A Software Approach*, Wiley, New York.
5. Mathews, J. H. (1987). *Numerical Methods for Computer Science, Engineering and Mathematics*, Prentice Hall, Englewood Cliffs, N.J.
6. Shoup, T. E. (1983). *Numerical Methods for the Personal Computer*, Prentice Hall, Englewood Cliffs, N.J.

ANSWERS TO SELECTED PROBLEMS

Chapter 13

13-1 $v(t) = 10t - 8$ ft/s
$a(t) = 10$ ft/s^2
91 ft; 42 ft/s; 10 ft/s^2
91.4 ft

13-2 $v(t) = -4$ m/s
$a(t) = 0$ m/s^2
-5 m; -4 m/s; 0 m/s^2
20 m

13-4 $v(t) = 4 \cos t$ m/s
$a(t) = -4 \sin t$ m/s^2
-3.84 m; 1.135 m/s; 3.84 m/s^2
12.16 m

13-7 $x(t) = 10 + 10t - 8t^2$ ft
$a(t) = -16$ ft/s^2
-422 ft; -118.0 ft/s; -16.0 ft/s^2
282 ft

13-8 $x(t) = -20{,}873 + 8t^3/3 - 20t$ m
$a(t) = 16t$ m/s^2
$-19{,}668$ m; 492 m/s; 128 m/s^2
972 m

13-11 $x(t) = 18.424 - 2t \cos 3t + (2/3)\sin 3t$ ft
$a(t) = 18t \cos 3t + 6 \sin 3t$ ft/s^2
11.03 ft; -43.47 ft/s; 55.65 ft/s^2
74.76 ft

13-13 $x(t) = 5 + 5t^2/2 - t^3/2$ ft
$v(t) = 5t - 3t^2/2$ ft/s
14.00 ft; 1.500 ft/s; -4.00 ft/s^2
105.52 ft

13-14 $x(t) = -37.62 + 31.62t - 4.905t^2$ m
$v(t) = 31.62 - 9.81t$ m/s
13.095 m; 2.190 m/s; -9.81 m/s^2
112.2 m

13-16 $x(t) = -86.25 + 9.081t - 5 \sin 2t$ m
$v(t) = 9.081 - 10 \cos 2t$ m/s
-57.61 m; -0.521 m/s; -5.588 m/s^2
45.93 m

13-19 3.00 ft/s

13-22 $\sqrt{25/2} \sin (\sqrt{2}\, t - 3.229)$ m
$\sqrt{25} \cos (\sqrt{2}\, t - 3.229)$ m/s
$-\sqrt{50} \sin (\sqrt{2}\, t - 3.229)$ m/s^2

13-24 1 m

13-25 1217 mi

13-27 $\sqrt{32{,}200 - 1000\, e^{1734.4 - y/500}}$ ft/s
179.4 ft/s

13-30 $15 - 0.5x$ m/s
5 m/s

13-32 $15e^{-0.50t}$ m/s
$30(1 - e^{-0.50t})$ m
10.02 s; 29.8 m

13-33 3.23 ft/s^2

13-35 72.2 s

13-38 4.818 m

13-44

t	x	v	a
0	0	50	-5
10	250	0	-5
20	0	-50	$-5/0$
30	-500	-50	$0/5$
40	-750	0	5
50	-500	50	$5/0$
60	0	50	0

13-45

t	x	v	a
0	0	0	10
10	500	100	10
20	2000	200	$10/-5$
30	3750	150	$-5/0$
40	5250	150	0
50	6750	150	0
60	8250	150	0

13-49 140 mi/h$\rightarrow$; 140 mi/h$\leftarrow$

13-50 35 m/s$\downarrow$; 35 m/s$\uparrow$

13-52 100 m/s$\leftarrow$; 300 m/s$\leftarrow$

13-55 12.5 mi (from where B starts)
12:30 PM

13-57 100 s; 2500 ft

13-58 4 s; 40 m

13-60 12.54 km (from 1st town)
1:30 PM

13-63 2.217 min; 2.125 mi

13-65 813.2 ft

13-66 2 m/s$\leftarrow$; 1 m/s^2$\rightarrow$
1 m/s$\leftarrow$; 0.5 m/s^2$\rightarrow$

13-68 6 m/s$\downarrow$; 0.6 m/s^2$\uparrow$
8 m/s$\downarrow$; 0.8 m/s^2$\uparrow$

13-71 1.6667 ft/s$\rightarrow$

13-73 1.5 ft/s$\uparrow$; 0.3 ft/s^2$\uparrow$

13-74 3.333 m/s$\uparrow$; 0.1333 m/s^2$\downarrow$

13-76 4 m/s$\leftarrow$; 1.2 m/s^2$\rightarrow$
3 m/s$\leftarrow$; 0.7 m/s^2$\rightarrow$
6 m/s$\leftarrow$; 1.4 m/s^2$\rightarrow$

13-79 2/3 ft/s$\leftarrow$
1/3 ft/s$\rightarrow$; 0 ft/s^2

13-80 1682.7 m

13-81 33.87° or 82.69°

13-84 2.219 m/s; 2.429 m/s

13-87 4.75 in.

13-89 53.89° $\leq \theta \leq$ 58.23°
or 80.76° $\leq \theta \leq$ 81.11°

13-90 39.74 m/s; 3.332 s; 15.38 m

3-92 $(50/3t)\, [\mathbf{e}_r + \theta\mathbf{e}_\theta]$ mm/s
$(50/3t^2)\, [-1(1 + 10\,\theta)\mathbf{e}_r +$
$(20 - \theta)\mathbf{e}_\theta]$ mm/s^2
$8.891\,\mathbf{e}_r + 55.867\,\mathbf{e}_\theta$ mm/s
$-302.8\,\mathbf{e}_r + 65.07\,\mathbf{e}_\theta$ mm/s^2

13-95 2.307 ft
-0.0605 ft/s; 0.1053 ft/s
-0.6883 ft/s^2; 1.1977 ft/s^2

13-97 24 rad/s

13-98 33.33 rad/s

13-100 5.477 rad/s

13-103 40.6 mi/m

13-105 29.7 mi/h

13-106 298.3 m

13-108 0.257 m/s^2; 5 m/s^2
2.95°

13-111 4.745 ft/s^2; 4 ft/s^2
49.87°

13-113 96.42 s; 19.47° ⟋

13-114 6.51 m/s; 50.19° ⟍

13-116 73.69°; 138.8 km/h

13-119 623.4 ft; 20.78 ft/s

13-121 $\mathbf{r}_A = \sqrt{25 - s^2}\,\mathbf{j}$ ft
$\mathbf{v}_A = -s/[25 - s^2]^{1/2}\,\mathbf{j}$ ft/s
$\mathbf{a}_A = -25/[25 - s^2]^{3/2}\,\mathbf{j}$ ft/s^2
$-3\,\mathbf{i} + 4\,\mathbf{j}$ ft
$-\mathbf{i} - 0.75\,\mathbf{j}$ ft/s
$-0.3906\,\mathbf{j}$ ft/s^2

13-122 14.278 m
$5\,\mathbf{i} - 14.01\,\mathbf{j}$ m/s

13-124 6.401 m/s; 9.139 m
$1.401\,\mathbf{i} - 14.01\,\mathbf{j}$ m/s

13-127 31.02 ft/s; 35.82 ft
$18.03\,\mathbf{i} - 26.46\,\mathbf{j}$ ft/s

13-129 0.3141 rad/s
239.1 ft/s ⟋ 15°

13-130 $-216 \sin 6t\,\mathbf{i} - 108\sqrt{3} \cos 6t$
$\mathbf{j} - 108 \cos 6t\,\mathbf{k}$

13-131 $10t\,\mathbf{i} + 3\,\mathbf{j} + 45t^2\,\mathbf{k}$ ft/s
$10\,\mathbf{i} + 90t\,\mathbf{k}$ ft/s^2

13-133 $5.531\,\mathbf{i} + 1.5\,\mathbf{j} + 2.327\,\mathbf{k}$ ft/s
$6.980\,\mathbf{i} - 16.592\,\mathbf{k}$ ft/s^2

13-136 $2\pi\,\mathbf{e}_\theta$ m/s
$-2\pi^2\,\mathbf{e}_r + 32\pi^2\,\mathbf{k}$ m/s^2

13-138 $0.1443\,\mathbf{e}_r + 0.25\,\mathbf{k}$ m/s
$1.8138\,\mathbf{e}_\theta$ m/s^2
$0.1443\,\mathbf{e}_r + 3.628\,\mathbf{e}_\theta + 0.25\,\mathbf{k}$ m/s
$-22.79\,\mathbf{e}_r + 1.814\,\mathbf{e}_\theta$ m/s^2

13-139 $0.1820\,\mathbf{e}_r + 0.5\,\mathbf{k}$ ft/s
$2.287\,\mathbf{e}_\theta$ ft/s^2
$0.1820\,\mathbf{e}_r + 6.861\,\mathbf{e}_\theta + 0.5\,\mathbf{k}$ ft/s
$-43.11\,\mathbf{e}_r + 2.287\,\mathbf{e}_\theta$ ft/s^2

13-141 3.58 ft/s^2

13-144 0.11227 rad/s

13-146 0.0500 rad/s; 0 rad/s^2
0 rad/s; 0.00120 rad/s^2

13-147 44.0 ft/s^2; 23.47 ft/s^2

13-148 2.0 m/s

13-150 205.8 m

13-153 37.82 ft/s; 15.176 ft; 21.87 ft

13-155 843.1 ft; 75.1 mi/h

13-156 $-4\,\mathbf{i} - 2.667\,\mathbf{j}$ m/s
$0.15\,\mathbf{i} + 0.1\,\mathbf{j}$ m/s^2

13-158 115.2 s; 18.58 s

13-161 29.87 ft/s; 45.24°

13-162 460 s; 0.114 m/s^2; 28.75 km

Chapter 14

14-1 -0.3725 rad/s
3.3056 rad/s^2

14-2 $\alpha = -9\theta/4$
$\theta = 5/3 \sin(1.5t + C)$

14-4 2.513 s; 75.4 rev

14-7 151.3 rad/s

14-9 4.944 s; 91.13 rad/s

14-10 21.63 s; 16.00 rad/s

14-12 1.787 s; 1.271 rev
8.936 rad/s

14-15 $a = \sqrt{64 + 256\,\theta^2}$
$\phi = \tan^{-1}\left(\dfrac{1}{2}\theta\right)$

14-17 16 rad/s^2; 80 rad/s
$1600\,\mathbf{i} + r\,\mathbf{j}$ ft/s^2

14-18 1 m/s^2↓
$-26.67\,\mathbf{i} - 1.5\,\mathbf{j}$ m/s^2

14-20 -3.665 rad/s^2; 2.182 rad/s^2
208 rpm

14-23 4.620 s
529 rpm; 353 rpm

14-24 40.4 rad/s↓

14-25 15.71 rad/s↓

14-27 $\omega = \dfrac{v_C\, r}{x\sqrt{x^2 - r^2}}$

14-30 2.078 m/s←; 14.40 m/s^2←

14-32 0.0324 rad/s↓; 0.00028 rad/s^2↓

14-33 0.1768 ft/s↓; 0.09945 ft/s^2↓

14-35 $v_A = \dfrac{-v_B x}{\sqrt{d^2 - x^2}}$
$a_A = \dfrac{-v_B^2 - a_B x - \dfrac{x^2 v_B^2}{d^2 - x^2}}{\sqrt{d^2 - x^2}}$

14-38 0.1607 m/s ⟍ 50°

14-40 257.2 mm/s↑; 805 mm/s^2↓

14-41 $v = \dfrac{yb\omega\cos\theta}{y - b\sin\theta}$

$a = \dfrac{(2v\omega + y\alpha)b\cos\theta - v^2 - yb\omega^2\sin\theta}{y - b\sin\theta}$

14-42 0.03235 rad/s↓; 6.699 mm/s↓

14-43 0.1000 rad/s↓

14-45 0.6944 rad/s↓; 281.7 in./s←

14-48 1.764 rad/s↓; 825.6 mm/s→

14-50 2.500 rad/s↓; 2.083 rad/s↓
250 mm/s→

14-51 1.1452 rad/s↓; 24.57 in./s ↘ 40°

14-53 2.041 rad/s↓
2.356 **i** − 4.081 **j** in/s

14-54 0.592 rad/s↓; 1.130 rad/s↓
32.03 **i** + 77.66 **j** mm/s

14-56 0 rad/s; 523.6 mm/s←

14-57 1.0000 rad/s ↓; 12.00 **j** in./s

14-59 5.6346 rad/s↓; 1.2971 rad/s↓

14-62 3 rad/s

14-64 8.504 rad/s

14-65 6.69 rad/s

14-67 1.2826 rad/s↓

14-68 0.03235 rad/s↓; 6.699 mm/s↓

14-69 0.1000 rad/s↓

14-71 2.041 rad/s↓; 4.712 in./s ↘ 60°

14-74 4.1888 rad/s↓; 360.0 mm/s

14-76 2 rad/s↓; 1120 mm/s←

14-77 3.118 rad/s↓; 1.5 rad/s↓
2.598 ft/s→

14-79 0 rad/s; 10 ft/s→

14-82 30 rad/s↓; 13.5 m/s→
4.5 m/s←; 11.906 m/s ∠ 40.89°

14-84 6 rad/s↓

14-85 45.0 rad/s↓; 15 rad/s↓
71.15 in./s ↘ 71.57°

14-86 0.00028 rad/s²↓; 0.00087 m/s²↓

14-87 0 rad/s²
0.0313 **i** − 0.0626 **j** ft/s²

14-89 6.570 rad/s²↓; 0.6034 rad/s²↓

14-92 18.277 rad/s²↓
−600 **i** − 1827.7 **j** mm/s²

14-94 1210 **i** − 3720 **j** mm/s²
−1230 **i** − 2440 **j** mm/s²

14-95 17.97 rad/s²↓; 35.28 rad/s²↓

14-97 3.6 rad/s²↓; 1.800 rad/s²↓
0.750 ft/s²←

14-100 13.301 rad/s²↓

14-102 2 rad/s²↓; 12.00 m/s²↑

14-103 12.677 rad/s² ↓

14-104 13.678 m/s ∠ 23.97°
4.280 m/s² ↘ 22.25°

14-105 104.35 ft/s ↗ 23.99°
27.565 ft/s² ↗ 78.37°

14-106 11.565 m/s ∠ 46.10°
4.394 m/s² ↗ 89.81°

14-109 98.78 ft/s ↗ 1.95°
29.85 m/s² ∠ 51.54°

14-111 5.692 in./s ↘ ; 0.6500 rad/s↓

14-112 0.962 rad/s↓; 173.08 mm/s ↙

14-114 2.214 rad/s²↓; 657.97 mm/s² ↘

14-117 0.6283 rad/s↓; 5.527 rad/s²↓

14-119 0.8976 rad/s↓; 66.98 rad/s²↓

14-120 2.514 m/s ↘ 88.57°
15.811 m/s² ↗ 2.862°

14-122 2900 mm/s←; 42,000 mm/s²↑
3100 mm/s→; 48,000 mm/s²↑

14-125 120.4 in./s ↘ 85.24°
1825 in./s² ∠ 9.46°
120.4 in./s ↗ 85.24°
1825 in./s² ↙ 9.46°

14-127 10 in./s←; 287.5 in./s²↑
10 in./s←; 312.5 in./s²↓

14-128 2924 mm/s ∠ 1.26°
42,800 mm/s² ↙ 86.08°
3078 mm/s ↘ 1.20°
47,390 mm/s² ↗ 86.46°

14-130 2900 mm/s→
42,060 mm/s² ↙ 88.64°
3100 mm/s←
48,060 mm/s² ↗ 88.81°

14-131 $\Delta\theta y = 90°$

14-132 $\Delta\theta y = 90°$

14-134 No change

14-137 −5 **j** + 3 **k** rad/s
15 **i** − 10 **k** rad/s²
−61 **i** − 45 **j** in./s
255 **i** − 108 **j** − 125 **k** in./s²

14-139 −3 **i** − 5 **j** rad/s
10 **i** + 15 **k** rad/s²
−11 **k** in./s
−125 **i** + 42 **j** + 120 **k** in./s²

14-140 −563.8 **j** − 210.2 **k** rpm
30.91 **i** rad/s²

14-142 88.54 mm/s↓; 21.06 mm/s²↓
0.10494 **i** − 0.02460 **j** + 0.02778 **k** rad/s
0.01097 **i** − 0.00585 **j** rad/s²

14-145 6.067 in./s↓; 8.713 in./s²↓
−0.06320 **i** − 0.39546 **j** + 0.21940 **k** rad/s
−0.09076 **i** + 0.15597 **j** − 0.11986 **k** rad/s²

14-147 18 **j** in./s; 6 **j** in./s²
−0.16667 **i** + 0.16667 **j** − 0.66667 **k** rad/s
0.9444 **i** + 1.5556 **j** − 1.2222 **k** rad/s²

14-148 −6.928 $\mathbf{e}_x$ m/s; −0.2076 $\mathbf{e}_y$ − 14.24 $\mathbf{e}_z$ m/s²

14-150 −519.6 $\mathbf{e}_x$ + 216.5 $\mathbf{e}_y$ − 125.0 $\mathbf{e}_z$ mm/s
−2685 $\mathbf{e}_x$ − 1602 $\mathbf{e}_y$ + 25 $\mathbf{e}_z$ mm/s²

14-151 $40\,\mathbf{e}_x + 50\,\mathbf{e}_y + 86.6\,\mathbf{e}_z$ in./s
$1445.6\,\mathbf{e}_x + 866.0\,\mathbf{e}_y - 820.0\,\mathbf{e}_z$ in./s^2

14-153 0.4744 rad/s$\downarrow$; 0.16422 rad/s$^2\downarrow$
$-2.3720\,\mathbf{i} + 2.3720\,\mathbf{j}$ in./s
$-1.9464\,\mathbf{i} + 0.8211\,\mathbf{j}$ in./s^2

14-158 $40.6991\,\mathbf{i}$ m/s; $-3553\,\mathbf{j}$ m/s^2
$-34.6991\,\mathbf{i}$ m/s; $3553\,\mathbf{j}$ m/s^2
31.8 mm from center

14-160 2.341 rev; 5.423 rad/s; 1.947°
0.2203 rev; 5.261 rad/s; 19.864°
0.0859 rev; 4.627 rad/s; 42.811°

14-161 $-21.321\,\mathbf{i} + 777.62\,\mathbf{j} + 429.33\,\mathbf{k}$ ft/s
$-568.56\,\mathbf{i} + 20{,}736\,\mathbf{j}$ ft; $19{,}432\,\mathbf{j}$ ft

14-163 42.46 ft/s $\searrow$ 0.727°; 4.863 ft/s^2 $\searrow$ 1.370°
298.6 ft below A

Chapter 15

15-1 422 lb; 242 lb

15-2 2000 N

15-3 37.46 lb
3.047 ft/s^2

15-4 84.4 N
0.573

15-7 73.73 ft/s; 184.34 ft
27.88 ft/s; 69.71 ft

15-8 2.406 m/s^2
3.531 s
6.937 m/s

15-11 2.229 ft/s^2
53.85 lb; 94.23 lb

15-12 67.80 N

15-15 21.567 ft/s
18.06 ft

15-16 4.905 m/s^2 $\nearrow$
2.548 m

15-19 4.954 ft/s^2 $\uparrow$; 92.31 lb

15-20 210.2 N; 420.4 N

15-23 2.222 ft/s$^2\downarrow$
44.40 lb
41.34 lb

15-24 1.249 m/s^2 $\nearrow$
256.8 N
6.243 m/s

15-27 9.164 lb
0.3389 s
0.8472 ft$\rightarrow$

15-28 46.25 N
2.803 m/s$\downarrow$
7.007 m$\downarrow$

15-31 5.486 s
63.11 ft$\leftarrow$
0 ft

15-32 41.2 m/s$^2\leftarrow$
2.153 s
60.265 m

15-35 5.675 ft/s$\downarrow$
10.733 ft/s^2

15-36 54.75 m/s
10.222 s

15-39 63.425 ft/s
0.02841 s

15-40 0.9574 m/s$\downarrow$

15-43 28,490 ft; 43.17 s

15-44 1146.8 m
7945 m
30.58 s

15-47 70,618 ft
66.58 s

15-48 4988 m
25.48 s

15-51 8.165 lb $\searrow$ 50.659°

15-52 20.65 N $\nearrow$ 54.462°

15-55 5.216 lb; 18.390 ft/s
30.82 lb; 31.588 ft/s

15-56 3141 N; 24,360 N

15-59 17.41°; 0.314

15-60 0.213

15-63 20.218 lb; 5.749 lb
3.715 rad/s

15-64 65.534°, 11.844 N

15-67 1.8104 lb $\measuredangle$ 26.565°
2.963 lb $\searrow$ 63.435°

15-68 31.215 N
20.00 m

15-69 24,450 ft/s
34,580 ft/s

15-70 7405 m/s
11,107 m/s

15-73 10,018 ft/s

15-74 0.280
5843 km
4848.1 m/s

15-77 3428 ft/s
4323 mi; 15,014 ft/s

15-78 2630 km
4465 m/s; 7442 m/s
218.0 min

15-81 4210 mi; 4791 mi
4080 mi; 4348 mi

15-82 2083 m/s; 5544 m/s

15-85 460.9 ft/s; 44.6 min

15-86 168.3 m/s; 164.7 m/s

15-89 $r = \dfrac{21.8419(10^6)}{1 + 0.02394 \cos \theta}$ ft

$r = \dfrac{21.6760(10^6)}{1 + 0.09037 \cos \theta}$ ft

24,784 ft/s; 278.2 mi
24,400 ft/s; 553.2 mi
−195.1 mi

15-90 0.37214
409 km
8983.9 m/s; 4110.8 m/s
172.2 min

15-93 67,288 lb
274.3 ft

15-94 2.190 m/s²
21.90 m/s
95.6 s

15-97 13.565 ft/s↑
25.71 lb

15-98 75.52 N
4.995 m/s² ↘ 40.892°
1.888 m/s²

15-101 52,407 ft

15-102 10 N←
80 kN/m

15-107 23,189 ft/s
118.3 min

15-108 7118.3 m/s
10,067 m/s
11,028 m/s

Chapter 16

16-1 2.504 ft/s²→
275.3 lb ↘ 78.69°
642.5 lb ↘ 78.69°

16-2 0.372 m/s²→
1659.1 N ↘ 81.47°
2111.6 N ↘ 81.47°

16-5 10.628 s

16-6 5.924 s; 2.998 s

16-9 547.7 lb; 402.3 lb

16-10 7479.5 N; 2330.5 N

16-13 300.42 lb; 60.08 lb
173.93 lb; 34.79 lb

16-14 2.453 m/s²; 0.533

16-17 37.5 − 485.25 sin θ

16-18 458.7 N; 56.35 N
147.15 N ↘ 30°

16-21 2.927 ft/s²↓
454.5 lb

16-22 11.412 rad/s²↓
126.63 N

16-25 12.880 rad/s²↓
8.333 lb
0 lb→; 58.33 lb↑

16-26 781.4 N
292.4 N ↗ 66.732°

16-29 6.192 rad/s²↓
19.00 lb ↗ 60.113°

16-30 441.45 N←; 73.575 N↑

16-33 24.150 ft/s²↓
5 lb↑

16-34 33.333 rad/s²↓
166.67 N→; 1754.8 N↓

16-37 0.59515 ft/s²↑
50.92 lb

16-38 69.06 N
139.54 N←; 696.2 N↑

16-43 10.798 ft/s² ↗ 28°
0.152

16-44 18.426 rad/s²↓
0.155

16-47 12.927 rad/s²↓
17.891 lb←; 55 lb↑

16-48 4 m/s²→; 30.72 rad/s²↓
4.602 m/s²→; 23.011 rad/s²↓

16-51 46.368 rad/s²
6.400 lb

16-52 2.0940 m/s²↓
544.9 N; 385.8 N

16-55 18.00 ft

16-56 1.487 s; 8.784 m

16-59 9.314 rad/s²↓
8.903 lb ∠ 71.16°
51.98°

16-60 12.191 rad/s²↓
13.334 N
255.5 N; 169.24 N

16-63 3.6225 rad/s²↓
13.50 ft/s² ↗ 26.565°
17.500 lb; 21.213 lb

16-64 9.443 rad/s²↓
60.33 N

16-67 23.216 rad/s²↓
19.389 lb ↘ ; 1.896 lb ∠

16-68 19.025 rad/s²↓
382.7 N ∠ 79.556°

16-71 6.324 lb ∠ ; 10.962 lb ↑

16-72 3.398 rad/s²↓
74.266 N ↑; 59.551 N ↘

16-77 339.2 **k** lb
−279.2 **k** lb

16-78 −1956.9 **k** N
2447.4 **k** N

17-74	3320 w	**18-33**	11.70 rad/s, 23.4 ft/s↓
	13,220 w		11.35 rad/s, 34.0 ft/s↓
	14,720 w	**18-35**	3.58 ft
17-76	0.2917 m		0.865 rad/s↙
	0.0098 m		3.46 ft/s→
17-77	45.3 lb		17.98 lb
	21.3 lb	**18-38**	3.00 rad/s↓
17-79	15.34 ft/s		7.49 m/s↓
	2.25 ft	**18-40**	70.53°
17-81	0.014 ft (comp)		3.13 rad/s
	0.014 ft (ext)		2.09 m/s
17-82	2.02 m/s	**18-41**	3.165 rad/s↙
	2.11 m/s		8.95 ft/s↓
17-84	6.30 m/s	**18-43**	15.66 ft/s↓
	2.99 m/s	**18-46**	11.503 rad/s↙
	9.99 m/s		0.415 m/s→
17-87	21.69 hp	**18-48**	22.33 N
	51.6 mi/h	**18-49**	$v = \sqrt{4gd/5}$
			$v = \sqrt{5gd/7}$
Chapter 18			$v = \sqrt{2gd}$
		18-52	2.506 m/s↓
18-1	16.68 rev	**18-53**	5.62 rad/s↓, 4.33 ft/s→
18-2	616 N·m		6.49 rad/s↓, 8.66 ft/s←
18-4	0.8803 m	**18-55**	4.654 rad/s↓, 7.06 ft/s←
	47.9 rad/s		4.432 rad/s↓, 6.65 ft/s←
18-6	6.07 rad/s	**18-57**	85.25 lb·ft
	118.5 N ∡ 76.5°	**18-58**	199.4 J
18-8	6.014 rad/s, 4543 N ⬎ 86.319°	**18-60**	137.41 J
	5.419 rad/s, 2472 N ⬏ 51.648°	**18-63**	171.64 lb·ft
	7.135 rad/s, 5348 N ⬈ 1.680°	**18-65**	34.44 lb·ft
18-9	3.929 rad/s	**18-66**	9.363 J
	2.678 lb ⬎ 26.632°	**18-68**	3.33 J
18-11	4.32 rad/s	**18-71**	0.275 lb·ft
	32.1 lb ⬎ 53.3°	**18-73**	15.99 rad/s
18-14	55.15°	**18-74**	15.81 rad/s
18-16	53.97°	**18-76**	5.96 rad/s
18-17	21.45 ft/s; 42.9 rad/s	**18-79**	4.92 ft/s→
	25.38 ft/s	**18-81**	347 ft
18-19	4.70 ft/s; 7.05 rad/s		302 ft
	2.47 ft	**18-82**	118.3 mm
18-22	4.15 m/s; 27.67 rad/s		3.14 rev
	5.53 m/s	**18-84**	4.27 m/s
18-24	0.2387 m		490.5 N
	0.515 m/s; 3.432 rad/s		1.038 m
	1.029 m/s	**18-87**	6.950 ft/s
	0.2076 m		6.378 ft/s
18-25	11.84 rad/s	**18-88**	2.770 m/s→
	2.96 ft/s		1.565 N←
18-27	17.26 lb ∡ 81.44°		117.1 N↑
	20.43 lb↑	**18-90**	4.45 m/s
18-30	45.00°	**18-93**	75.54 rad/s
	slips before		
18-32	9.38 rad/s		
	18.48 **i** + 6.52 **j** m/s		

19-1   0.08282 lb
           0.207

19-2   14.16 s

19-4   185.7 N
           0.183

19-7   35.3 ft/s $\measuredangle$ 61.82°
           48.3 ft/s $\measuredangle$ 75.00°
           26.0 ft/s $\measuredangle$ 36.74°

19-9   70.89 ft/s $\measuredangle$ 40.034°
           16.054 ft/s $\measuredangle$ 4.582°
           67.77 ft/s $\measuredangle$ 32.973°

19-10 41.7 N $\searrow$ 10.72°

19-12 33.5(-0.5599 **i** $+ 0.7708$ **j** $+ 0.3040$ **k**)m/s
           47.6(-0.6968 **i** $+ 0.7172$ **j** $+ 0.0080$ **k**)m/s
           54.9(-0.7099 **i** $+ 0.6831$ **j** $+ 0.1718$ **k**)m/s

19-15 6.01 s
           15.81 s; 74.0 ft/s
           25.6 s

19-16 8.186 s
           15.68 s; 12.39 m/s
           23.57 s

19-18 5.667 **i** $+ 5.167$ **j** m
           0.6667 **i** $+ 2.333$ **j** m/s

19-19 11.40 **i** $+ 7.40$ **j** $+ 4.80$ **k** ft
           6.60 **i** $+ 0.60$ **j** $- 26.40$ **k** ft/s

19-21 16.41 mi/h $\measuredangle$ 69.34°

19-24 -2.000 **i** $+ 7.50$ **j** N·s
           -1.666 **i** $+ 6.250$ **j** m/s
           7.50 m
           1 s

19-26 $(-31.66, 1078.9)$ m
           7.88 **i** $- 72.30$ **j** $+ 7.43$ **k** N·s
           24,370 N

19-27 -877.3 **i** $+ 5032$ **j** ft
           9.32 **i** $- 2.15$ **j** $+ 13.46$ **k** lb·s
           16,510 lb

19-29 6.284 ft/s
           0.6159 ft
           42.71 ft/s $\searrow$

19-32 13.636 m/s
           0.0903 m

19-34 147.6 m/s

19-35 2.174 ft/s
           9.83°

19-37 2.291 ft/s, 3.191 ft/s
           16.54%
           -39.6 lb

19-38 -1.0909 m/s, 0.3091 m/s
           46.4%
           -309 N

19-40 -0.900 m/s, 2.700 m/s
           19.00%
           -1140 N

19-43 1.816 ft/s, 3.476 ft/s
           6.075 ft/s, 21.5%

19-44 -0.9454 m/s, -0.3086 m/s
           2.0483 m/s, 60.6%

19-47 0.913
           1.765 ft

19-48 40.97°, 2.87°

19-50 74.28°

19-53 31.69°, 0.537 ft

19-55 4.25 ft/s $\searrow$ 80.93°
           16.71 ft/s $\measuredangle$ 20.31°

19-56 4.331 m/s $\searrow$ 60.83°

19-58 2.887 m/s
           2.091 kg; 2.391 m/s

19-61 6.86 ft

19-62 $x = 0.376$ m
           $y = 0.343$ m

19-63 1.839 ft

19-66 12.5 m/s

19-67 0.3125 rad/s

19-70 2463 m/s, 4845 m/s

19-71 37.04°

19-72 1.1654 m
           0.1871 m

19-75 13.132 ft/s, 47.24°
           13.216 in.
           30.5 in.

19-79 5.05 lb $\searrow$ 45°

19-80 6.26 N

19-82 383.6 kN

19-85 0.01373 ft³/s

19-87 0.101

19-88 15.31 m/s

19-90 $-27,000$ N
           -12.272 m/s²
           6780 m/s

19-93 4348 lb
           11,303 ft/s

19-95 155.3 lb

19-96 4987 N←

19-98 2 N, 2 N, 2 N

19-101 12.476 ft/s

19-103 47.51 ft
           40.7%

19-104 $b = 0.397$ m
           $c = 0.397$ m
           $h = 0.07322$ m

19-107 5.006 ft/s $\measuredangle$ 117.17°
           46.21°

19-109 48.84°
           4.855 ft/s
           2.62 lb

19-112 5.988 m/s

19-113 17.938 ft/s $\measuredangle$ 41.99°
24.98 ft

Chapter 20

20-1 0.1270 lb · ft

20-2 0.0503 N · m

20-4 32.65 m/s

20-7 12.160 s
9.758 s

20-9 12.20 s
235.6 rad/s↺
235.6 rad/s↻

20-10 16.01 s
157.1 rad/s↺
78.5 rad/s↻

20-12 47.62 rad/s↻; 31.75 rad/s↺
95.24 rad/s↺; 142.86 rad/s↻
158.73 rad/s↺; 238.10 rad/s↻

20-15 12.023 s
38.1 ft/s←
65.3 rad/s↺

20-17 5.008 s
12.519 s

20-20 4.269 m/s

20-21 32.2 rad/s↺
10.73 ft/s↓

20-24 0.5333 m

20-25 6.412 rad/s↺
76.45 lb
76.45 lb
69.7%

20-26 8.108 rad/s↺
10,378 N
648 N←
98.8%
141.78°

20-28 2.113 rad/s↺
654 N←
99.9%
28.51°

20-31 $2\ell/3$
$5\ell/6$
does not exist

20-33 12.999 ft/s $\searrow$ 87.94°
1.5935 ft/s→
3.902 rad/s↺
4.900 in. above G

20-34 6.811 m/s $\nearrow$ 70.70°
0.991 m/s→
2.937 rad/s↺
338 mm above G

20-37 19.361 ft/s $\searrow$ 84.88°
3.661 rad/s↻
102.3 lb
54.28°

20-39 0.217
11.78 ft/s $\searrow$ 58.10°
73.12 lb

20-42 0.518
2.293 rad/s↻
420 N→

20-43 18.535 rad/s↻
4.637 ft/s
will not hit

20-45 0.4321 rad/s↻; 5.1322 rad/s↻
30.38°
256.2 lb $\searrow$ 68.95°

20-46 12.135 rad/s↻; 15.556 rad/s↻
6930 N $\searrow$ 68.61°

20-49 $-0.01098 \mathbf{j} + 0.00549 \mathbf{k}$ ft · lb · s
63.43°

20-50 $0.04241 \mathbf{i} + 0.05655 \mathbf{j}$ N · m · s
36.87°

20-52 $1.4137 \mathbf{j}$ N · m · s
0°

20-55 $0.9177 \mathbf{i} + 0.5323 \mathbf{k}$ ft · lb · s
59.89°

20-57 $0.18008 \mathbf{j} + 0.04456 \mathbf{k}$ ft · lb · s
13.90°

20-58 $0.8263 \mathbf{j} + 0.9468 \mathbf{k}$ N · m · s
48.89°

20-60 $-2 \mathbf{i}$ m/s
$40 \mathbf{j} + 15 \mathbf{k}$ rad/s
90°

20-63 90°; $-5.635 \mathbf{i}$ ft/s

20-65 0°; $-1.528 \mathbf{i}$ ft/s

20-66 54.09°
$3.179 \mathbf{i} + 1.010 \mathbf{j} - 0.882 \mathbf{k}$ m/s

20-68 7.33°
$0.3816 \mathbf{i} + 1.3249 \mathbf{j} - 0.7470 \mathbf{k}$ m/s

20-71 $-8.392 \mathbf{i}$ ft/s
$38.76 \mathbf{j} + 67.13 \mathbf{k}$ rad/s
$-75.52 \mathbf{i}$ ft/s

20-73 1.0572 lb · s
$0.2643 \mathbf{j} + 0.2643 \mathbf{k}$ lb · ft · s
$-119.15 \mathbf{i}$ ft/s

20-74 $-2.197 \mathbf{i} - 0.1491 \mathbf{j} + 0.3159 \mathbf{k}$ m/s
$0.7897 \mathbf{i} + 26.172 \mathbf{j} + 14.985 \mathbf{k}$ rad/s
$-12.117 \mathbf{i} - 0.0306 \mathbf{j} + 0.6317 \mathbf{k}$ m/s

20-76 12.82 N · s
$1.301 \mathbf{j} + 2.253 \mathbf{k}$ N · m · s
$-23.658 \mathbf{i} - 0.954 \mathbf{j} + 0.551 \mathbf{k}$ m/s

20-77 2.688 lb · s
$0.0986 \mathbf{j} + 0.0986 \mathbf{k}$ lb · ft · s
$-78.307 \mathbf{i} - 11.287 \mathbf{j} + 11.287 \mathbf{k}$ ft/s

20-78	14.20 m/s
20-79	0.203 s
20-81	$v_0 = 1.051\sqrt{ag}$
	$\mathbf{v}_G = 0.5574\sqrt{ag} \angle 45°$
	$\omega = 0.7882\sqrt{g/a}$
20-82	$h = 7r/5$
20-85	$v_A = 0; \; \omega_A = v_0/r$
	$v_B = v_0; \; \omega_B = 0$
	$v_{Af} = 2v_0/7$
	$v_{Bf} = 5v_0/7$
20-86	2.451 rad/s↓; 1.225 m/s ⬊ 36.780°
	4.411 rad/s↓
20-89	1.894 rad/s↓; 3.666 rad/s↓
	43.80°
	22.22°
20-90	8.036 rad/s; 1.6071 m/s
	58.7%
	1.841 m/s

Chapter 21

21-1	$\dot{x}(t) = -8\pi \sin \pi t$ in./s
	$\ddot{x}(t) = -8\pi^2 \cos \pi t$ in./s²
21-2	$\dot{x}(t) = (5\pi/4)\cos \pi t/4$ mm/s
	$\ddot{x}(t) = -(5\pi^2/16)\sin \pi t/4$ mm/s²
21-4	$\dot{x}(t) = (30\pi/4)\cos(3\pi t/4 + \pi/8)$ mm/s
	$\ddot{x}(t) = -(90\pi^2/16)\sin(3\pi t/4 + \pi/8)$ mm/s²
21-7	$x(t) = 5\cos(\pi t + 0.9273)$ in.
	$v_{max} = 5\pi$ in./s at $x = 0$ in.
	$a_{max} = 5\pi^2$ in./s² at $x = -5$ in.
21-9	$x(t) = 10\cos(10t - 0.6435)$ in.
	$v_{max} = 100$ in/s at $x = 0$ in.
	$a_{max} = 1000$ in/s² at $x = -10$ in.
21-10	$x(t) = 26\cos(3\pi t/4 + 1.1760)$ mm
	$v_{max} = 61.26$ mm/s at $x = 0$ mm
	$a_{max} = -144.34$ mm/s² at $x = -26$ mm
21-13	$x(t) = 13\sin(\pi t + 2.7468)$ in.
	0.1257 s, 0.6257 s
21-14	$x(t) = 5\sin(\pi t/2 + 0.9273)$ mm
	1.4097 s; 0.4097 s
21-17	$x(t) = 5\sin(\pi t + \pi/2)$ in.
	0.5 s; 0 s
21-19	1.9454 in.; 5.083 ft/s
21-22	1.732 m/s
21-24	$k = k_1 + k_2$
21-25	$k = \dfrac{k_1 k_2}{k_1 + k_2}$
21-27	$\ddot{y} + 289.80y = 0$
	0.3691 s; 11.253 in.
	$y(t) = -8.811\sin 17.024t - 7\cos 17.024t$ in.
	0.1451 s
21-30	$\ddot{x} + 700x = 0$
	0.2375 s; 39.23 mm
	$x(t) = 30.24\sin 26.458t + 25\cos 26.458t$ mm
	0.09263 s

21-32	$\ddot{x} + 181.818x = 0$
	13.484 rad/s; 0.4660 s
21-33	$\ddot{y} + 322.0y = 0$
	17.944 rad/s; 0.3502 s
21-35	$\ddot{y} + 717.79y = 0$
	26.79 rad/s; 0.2345 s
21-38	$y(t) = 210.11\sin 15.811t - 14.715\cos 15.811t$ mm
21-40	$\ddot{y}_G + 533.33y_G = 0$
	23.094 rad/s; 0.2721 s
21-41	$\ddot{x}_G + 178.60x_G = 0$
	13.364 rad/s; 0.4702 s
21-43	$\ddot{\theta} + 30.912\,\theta = 0$
	5.560 rad/s; 1.1301 s
21-46	$\ddot{\theta} + 1650\,\theta = 0$
	0.6093 m/s
21-48	$\ddot{\theta} + 127.87\,\theta = 0$
	11.31 rad/s; 0.556 s
21-49	$\ddot{y} + 95.407y = 0$
	9.768 rad/s; 0.6433 s
21-51	$\ddot{y} + 254.21y = 0$
	15.944 rad/s; 0.394 s
21-54	$\ddot{\theta} + 28.095\,\theta = 0$
	5.301 rad/s; 1.1854 s
21-55	underdamped
	$\dot{x}(t) = e^{-0.1t}[-\cos(5t - 1.2) -$
	$50\sin(5t - 1.2)]$ in./s
	$\ddot{x}(t) = e^{-0.1t}[-249.9\cos(5t - 1.2) +$
	$10\sin(5t - 1.2)]$ in./s²
21-56	critically damped
	$\dot{x}(t) = (-7 - 6t)e^{-2t}$ mm/s
	$\ddot{x}(t) = (8 + 12t)e^{-2t}$ mm/s²
21-58	underdamped
	$\dot{x}(t) = e^{-0.05t}[-18.4\cos 3t - 23.7\sin 3t]$ mm/s
	$\ddot{x}(t) = e^{-0.05t}[-70.18\cos 3t + 56.39\sin 3t]$ mm/s²
21-60	critically damped
	$\dot{x}(t) = (8 - 7.5t)e^{-1.5t}$ rad/s
	$\ddot{x}(t) = (-19.5 + 11.25t)e^{-1.5t}$ rad/s²
21-63	critically damped
	$\dot{x}(t) = (-9 + 2t)e^{-0.2t}$ in./s
	$\ddot{x}(t) = (3.8 - 0.4t)e^{-0.2t}$ in./s²
21-65	critically damped
	$\dot{x}(t) = (-5.8 + 1.2t)e^{-1.2t}$ rad/s
	$\ddot{x}(t) = (8.16 - 1.44t)e^{-1.2t}$ rad/s²
21-66	underdamped
	$\dot{x}(t) = e^{-0.15t}[-0.9\sin(10t - 2.5) +$
	$60\cos(10t - 2.5)]$ mm/s
	$\ddot{x}(t) = e^{-0.15t}[-600\sin(10t - 2.5) -$
	$18\cos(10t - 2.5)]$ mm/s²
21-69	underdamped
	$x(t) = e^{-5t}[3\cos 7.416t + 4.045\sin 7.416t]$ in.
	$\dot{x}(t) = e^{-5t}[15\cos 7.416t - 42.47\sin 7.416t]$ in./s
	$\ddot{x}(t) = e^{-5t}[-390\cos 7.416t +$
	$101.13\sin 7.416t]$ in./s²

21-70 overdamped
$x(t) = -31.771e^{-5.53t} + 1.771e^{-14.47t}$ mm
$\dot{x}(t) = 175.63e^{-5.53t} - 25.63e^{-14.47t}$ mm/s
$\ddot{x}(t) = -970.9e^{-5.53t} + 370.9e^{-14.47t}$ mm/s²

21-72 underdamped
$x(t) = e^{-t}[100 \cos 4.359t + 57.35 \sin 4.359t]$ mm
$\dot{x}(t) = e^{-t}[150 \cos 4.359t - 493.2 \sin 4.359t]$ mm/s²
$\ddot{x}(t) = e^{-t}[-2300 \cos 4.359t - 160.6 \sin 4.359t]$ mm/s²

21-75 critically damped
$x(t) = (-15 - 75t)e^{-5t}$ in.
$\dot{x}(t) = 375te^{-5t}$ in./s
$\ddot{x}(t) = (375 - 1875t)e^{-5t}$ in./s²

21-77 $c = c_1 + c_2$

21-80 $\ddot{y} + 25\dot{y} + 666.7y = 0$
0.278 s
$y(t) = e^{-12.5t}[-5 \cos 22.592t + 8.299 \sin 22.592t]$ mm
0.0240 s

21-82 $4\ddot{y} + 125\dot{y} + 6000y = 0$
0.1773 s
$y(t) = e^{-15.625t}[15 \cos 35.438t - 14.550 \sin 35.438t]$ mm
0.0552 s

21-83 2.44
overdamped
No frequency or period exists
12.182 in

21-85 0.8061
underdamped
8.311 rad/s; 0.756 s
18.61 lb·s/ft

21-88 8.032 N·s/m

21-90 0.986 s
1.190 s; 2.48 cycles

21-91 0.2919
underdamped
10.050 rad/s; 0.625 s

21-93 3.172
overdamped
No frequency or period exists

21-94 11.041
overdamped
No frequency or period exists

21-96 Same as Eqs. 21-39 and 21-40

21-98 $20\ddot{x} + 40\dot{x} + 500x = 2500 \sin 8t - 1600 \cos 8t$
$x(t) = 3.521 \sin(8t + 2.962)$ mm

21-99 $0.31056\ddot{y} + 5\dot{y} + 140y = 70 \sin 30t$
$y(t) = e^{-8.05t}[10 \cos 19.647t + 16 \sin 19.647t] + 4.101 \sin(30t - 2.320)$ in.

21-102 $4\ddot{y} + 125\dot{y} + 6000y = 150 \sin 18t$
$y(t) = e^{-15.625t}[-2.587 \cos 35.438t + 6.842 \sin 35.438t] + 28.766 \sin(18t - 0.4462)$ mm

21-104 33.95 mm
$5.298 \leq \Omega \leq 8.629$ rad/s

21-105 2.05 in.

21-107 4.936 in.

21-110 1.10 mm
1.46 mm

21-112 34.54 mm

21-113 0.983 in.

21-115 $\ddot{x} + 465.111x = 0$

21-116 $\ddot{x} + 181.818x = 0$

21-118 $\ddot{x} + 342.86x = 0$

21-121 $\ddot{\theta} + 553.84\theta = 0$

21-123 $\ddot{y} + 95.407y = 0$

21-124 $\ddot{y}_G + 266.67y_G = 0$

21-125 10.616 rad/s

21-126 59.161 rad/s

21-128 35.355 rad/s

21-131 17.944 rad/s

21-133 13.364 rad/s

21-134 5.718 rad/s

21-136 4.952 rad/s

21-137 9.768 rad/s

21-139 21.76 lb/ft
0.00873 lb·s/ft

21-140 $4.5\ddot{\theta} + 5.4\dot{\theta} + 49.05\theta = 19.282 \sin(5t - 0.2355)$
0.2798 rad

21-142 $F_k = 105.04$ N
$F_c = 26.248$ N

21-145 $\ddot{y} + 154.5y = 0$
0.5054 s; 6.217 in.
1.743 lb
0.5 lb
2.5 ft

21-147 $\ddot{\theta} + 24\theta = 0$
4.899 rad/s; 0.2663 rad
$\theta(t) = 2.2663 \sin 4.899t$ rad

21-148 $\ddot{y} + 288.53y = 0$
16.986 rad/s; 50 mm
$y(t) = -50 \cos 16.986t$ mm

Appendix A

A-1 $I_z = 3mR^2/10$

A-2 $I_y = 3m(R^2 + 4h^2)/20$

A-5 $I_y = m(3R^2 + 4L^2)/12$

A-6 $I_x = 2mR^2/5$

A-9 $I_{yG} = m(2h^2 + 3L^2)/36$

A-10 $I_x = m(b^2 + c^2 - /10$

A-13 $I_x = mR^2(1 + 3R^2)/6$

A-14 $I_{yG} = 19mR^2/160$

A-17 $I_y = 0.267$ slug·ft²

A-18 $I_y = 1.004$ kg·m²

A-21 $I_x = 19.22 \text{ slug} \cdot \text{ft}^2$

A-22 $I_x = 22.9 \text{ kg} \cdot \text{m}^2$

A-27 $I_y = 0.555 \text{ slug} \cdot \text{ft}^2$

A-28 $I_y = 0.464 \text{ kg} \cdot \text{m}^2$

A-31 $I_y = 8.32 \text{ slug} \cdot \text{ft}^2$

A-32 $I_y = 1.616 \text{ kg} \cdot \text{m}^2$

A-35 $I_{xy} = mbL/4$

A-36 $I_{xy} = 2mR^2/5\pi$

A-39 $I_{xy} = -0.493 \text{ slug} \cdot \text{ft}^2$
$I_{zy} = 0.1174 \text{ slug} \cdot \text{ft}^2$
$I_{zx} = -0.352 \text{ slug} \cdot \text{ft}^2$

A-40 $I_{xy} = 4mRh/5\pi$
$I_{zx} = 3mR^2 \ /10\pi$

A-43 $I_{max} = 0.695 \text{ slug} \cdot \text{ft}^2$
$\theta_x = 68.6°$
$\theta_y = 81.5°$
$\theta_z = 23.1°$
$I_{int} = 0.665 \text{ slug} \cdot \text{ft}^2$
$\theta_x = 71.0°$
$\theta_y = 25.2°$
$\theta_z = 105.9°$
$I_{min} = 0.1078 \text{ slug} \cdot \text{ft}^2$
$\theta_x = 29.2°$
$\theta_y = 113.5°$
$\theta_z = 106.4°$

A-44 $I_{max} = 0.402 \text{ kg} \cdot \text{m}^2$
$\theta_x = 101.3°$
$\theta_y = 76.5°$
$\theta_z = 17.8°$
$I_{int} = 0.378 \text{ kg} \cdot \text{m}^2$

$\theta_x = 52.9°$
$\theta_y = 37.4°$
$\theta_z = 94.1°$
$I_{min} = 0.0546 \text{ kg} \cdot \text{m}^2$
$\theta_x = 39.3°$
$\theta_y = 124.0°$
$\theta_z = 72.7°$

A-47 $I_{max} = 0.0419 \text{ slug} \cdot \text{ft}^2$
$\theta_x = 60.2°$
$\theta_y = 90.0°$
$\theta_z = 29.8°$
$I_{int} = 0.0346 \text{ slug} \cdot \text{ft}^2$
$\theta_x = 90.0°$
$\theta_y = 0°$
$\theta_z = 90.0°$
$I_{min} = 0.0322 \text{ slug} \cdot \text{ft}^2$
$\theta_x = 29.8°$
$\theta_y = 90.0°$
$\theta_z = 119.8°$

A-48 $I_{max} = 2.42 \text{ kg} \cdot \text{m}^2$
$\theta_x = 130.6°$
$\theta_y = 75.8°$
$\theta_z = 44.1°$
$I_{int} = 1.925 \text{ kg} \cdot \text{m}^2$
$\theta_x = 100.6°$
$\theta_y = 29.9°$
$\theta_z = 117.6°$
$I_{min} = 0.892 \text{ kg} \cdot \text{m}^2$
$\theta_x = 42.6°$
$\theta_y = 64.3°$
$\theta_z = 58.7°$

INDEX

PHOTO CREDITS

Chapter 12   John Elk III/Bruce Coleman

Chapter 13   Tim Defrisco/Allsport

Chapter 14   H.P. Merten/Stock Market

Chapter 15   Comstock

Chapter 16   Courtesy NASA

Chapter 17   Courtesy Fleming Companies, Inc.

Chapter 18   Focus on Sports

Chapter 19   Henry Groskinsky/Peter Arnold

Chapter 20   Werner H. Muller/Peter Arnold

Chapter 21   Robert Mathena/Fundamental Photos

Slender rod

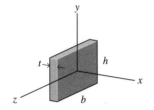

$$I_x = 0$$

$$I_y = I_z = \frac{1}{12}mL^2$$

Thin rectangular plate

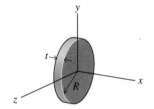

$$I_x = \frac{1}{12}m(b^2 + h^2)$$

$$I_y = \frac{1}{12}mb^2$$

$$I_z = \frac{1}{12}mh^2$$

Thin circular plate

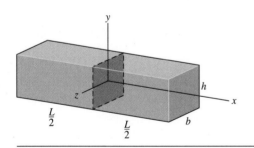

$$I_x = \frac{1}{2}mR^2$$

$$I_y = I_z = \frac{1}{4}mR^2$$

Rectangular prism

$$V = bhL$$

$$I_x = \frac{1}{12}m(b^2 + h^2)$$

$$I_y = \frac{1}{12}m(b^2 + L^2)$$

$$I_z = \frac{1}{12}m(h^2 + L^2)$$

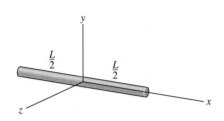